JACARANDA

Geoactive 1

STAGE 4 | FIFTH EDITION

LOUISE SWANSON

NICOLE GRAY

KAREN BOWDEN

ADRIAN HARRISON

KYMBERLY GOVERS

STEVEN NEWMAN

CONTRIBUTORS AND CONSULTANTS

Hayley Saunders | Rob Bell | Laura Sproule | Adeba Qasim | Danielle Creeley

Natasha Craig | Cath McIntyre | Fletcher Williamson

Sandra Duncanson | Alex Kopp | Ben de Vries

Judy Mraz | Jill Price | Cathy Bedson | Jeana Kriewaldt | Denise Miles

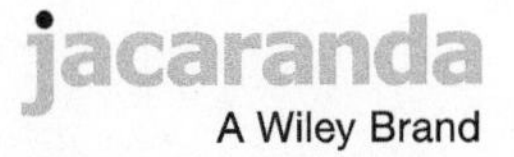

Fifth edition published 2021 by
John Wiley & Sons Australia, Ltd
42 McDougall Street, Milton, Qld 4064

First edition published 1998
Second edition published 2005
Third edition published 2010
Fourth edition published 2017

Typeset in 11/14 pt Times Ltd Std

ISBN: 978-0-7303-9429-7

Cover image: © Andrew James Goh/Shutterstock

Illustrated by various artists, diacriTech and Wiley Composition Services

Typeset in India by diacriTech

A catalogue record for this book is available from the National Library of Australia

Printed in Singapore
M WEP304775 270824

The Publishers of this series acknowledge and pay their respects to Aboriginal Peoples and Torres Strait Islander Peoples as the traditional custodians of the land on which this resource was produced.

This suite of resources may include references to (including names, images, footage or voices of) people of Aboriginal and/or Torres Strait Islander heritage who are deceased. These images and references have been included to help Australian students from all cultural backgrounds develop a better understanding of Aboriginal and Torres Strait Islander Peoples' history, culture and lived experience.

It is strongly recommended that teachers examine resources on topics related to Aboriginal and/or Torres Strait Islander Cultures and Peoples to assess their suitability for their own specific class and school context. Teachers should also know and follow the guidelines laid down by the relevant educational authorities and local Elders or community advisors regarding content about all First Nations Peoples.

All activities in this resource have been written with the safety of both teacher and student in mind. Some, however, involve physical activity or the use of equipment or tools. All due care should be taken when performing such activities. Neither the publisher nor the authors can accept responsibility for any injury or loss that may be sustained when completing activities described in this resource.

The Publisher acknowledges ongoing discussions related to gender-based population data. At the time of publishing, there was insufficient data available to allow for the meaningful analysis of trends and patterns to broaden our discussion of demographics beyond male and female gender identification.

Contents

Chapter and lesson structure

Consistent, inclusive learning structure

Our strong instructional design has been developed and refined by experienced teachers. It provides best-practice pedagogy and easy navigation through each unit – for you and all your students.

Each chapter begins with ...

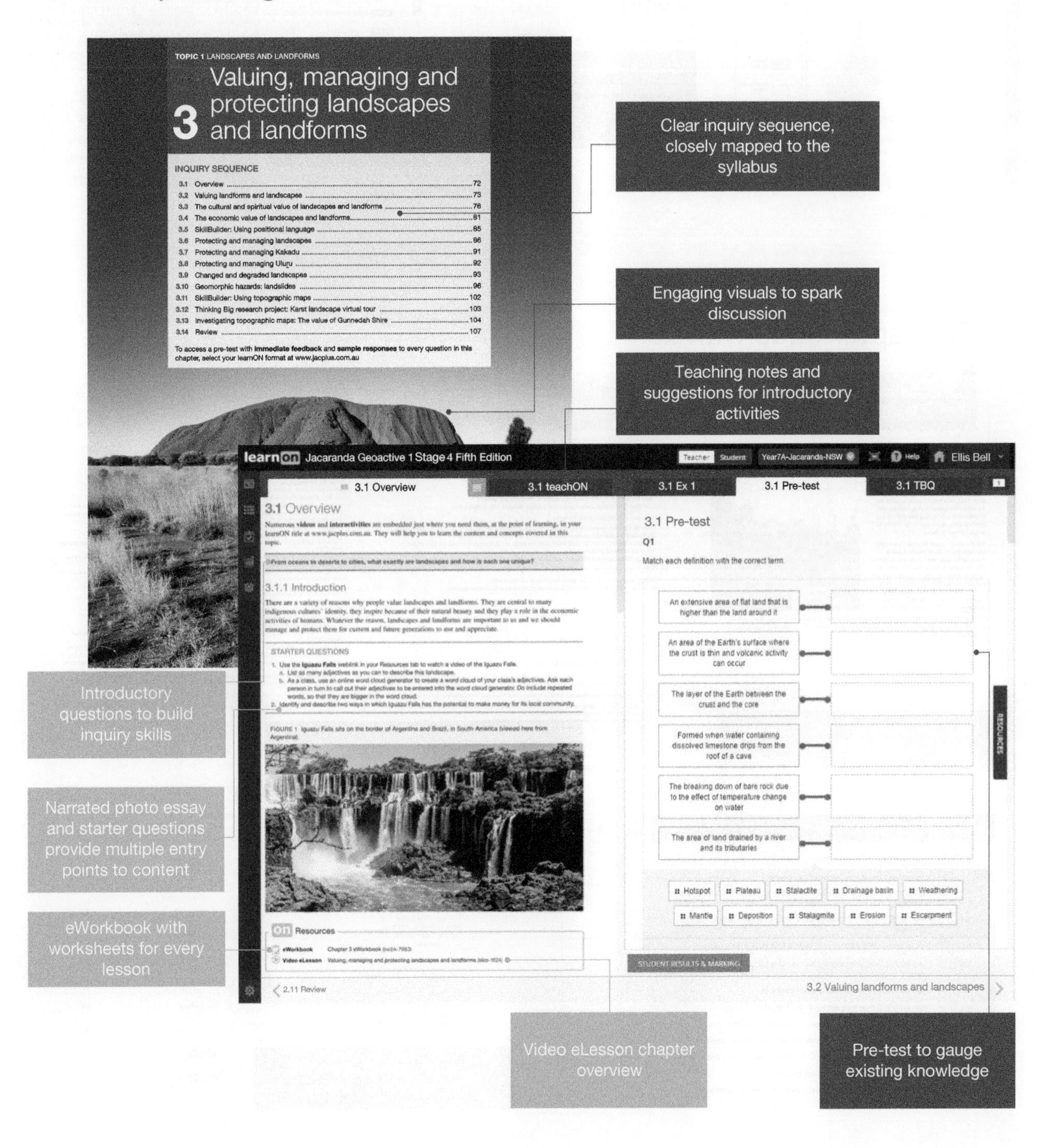

Each chapter includes ...

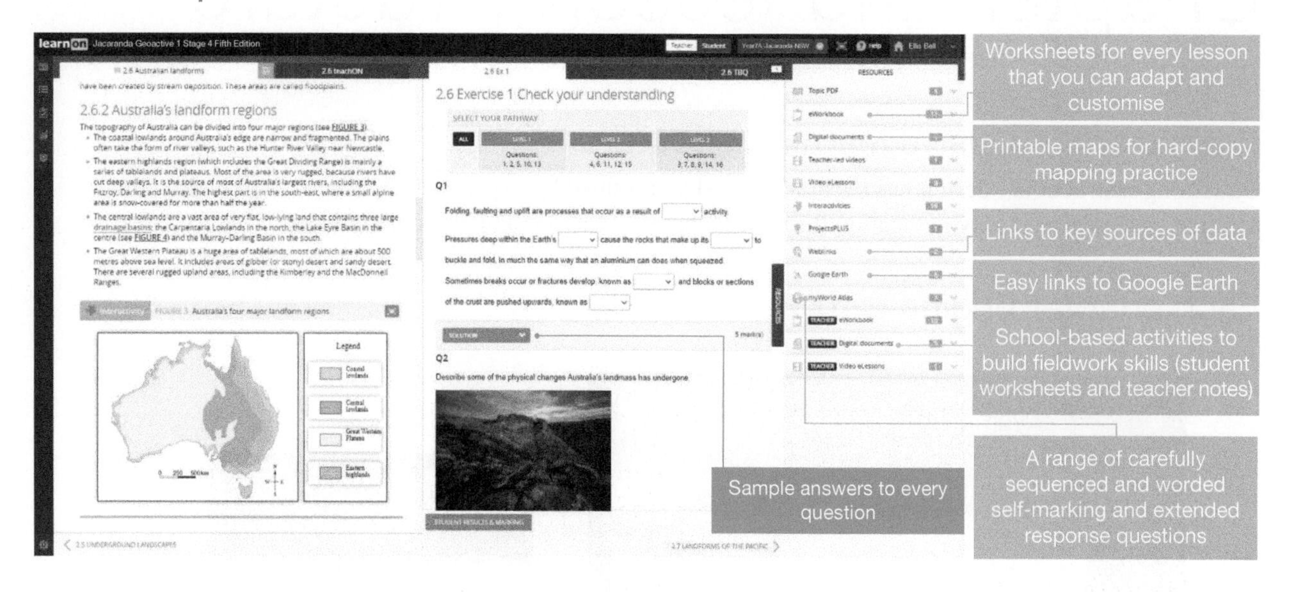

Each chapter ends with ...

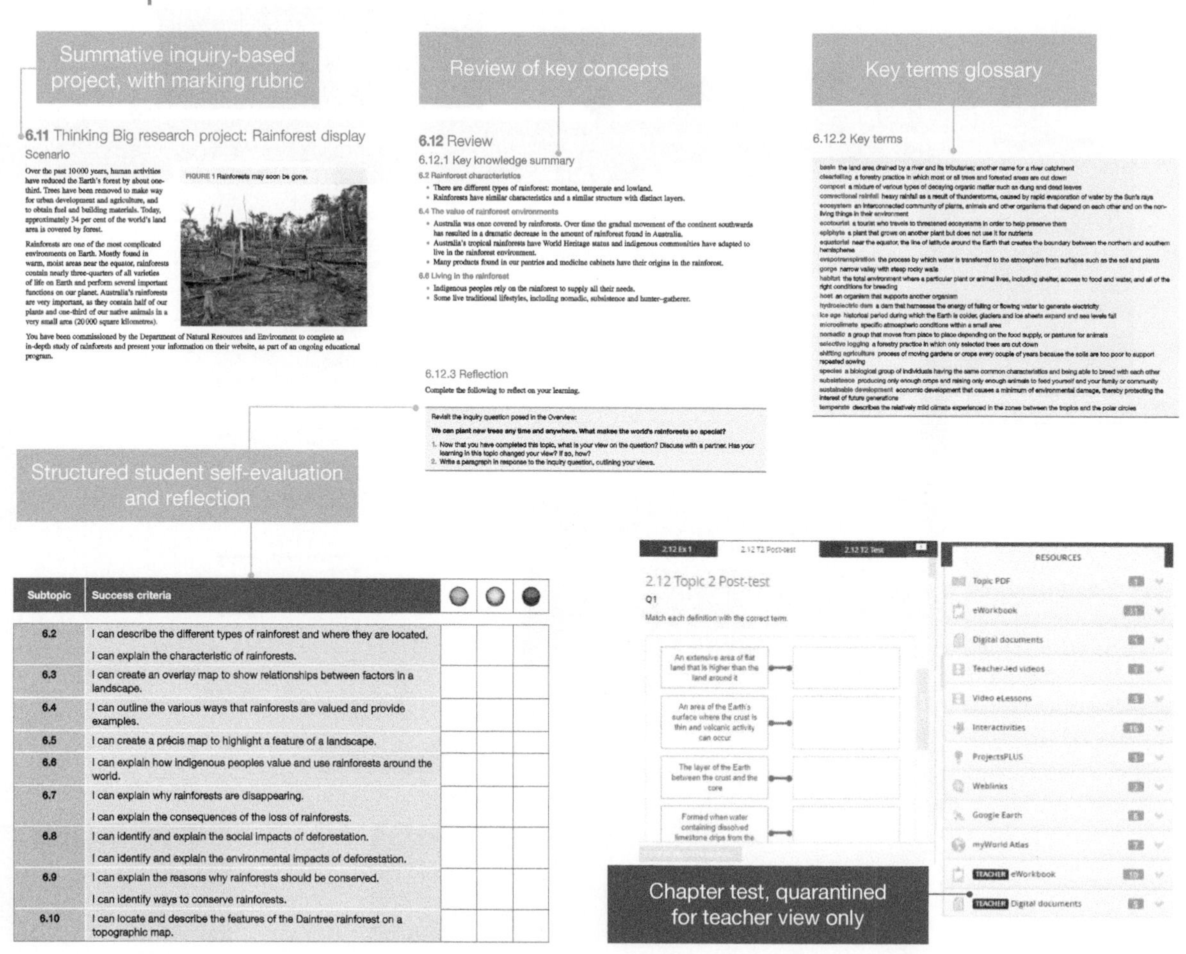

Comprehensive teacher support

Teachers will find the resources and support they need, no matter their level of experience or background in Geography. One simple click on the teachON tab or the Teacher Resources menu provides access to a full suite of differentiated planning, assessment and teaching materials.

Each chapter is supported by ...

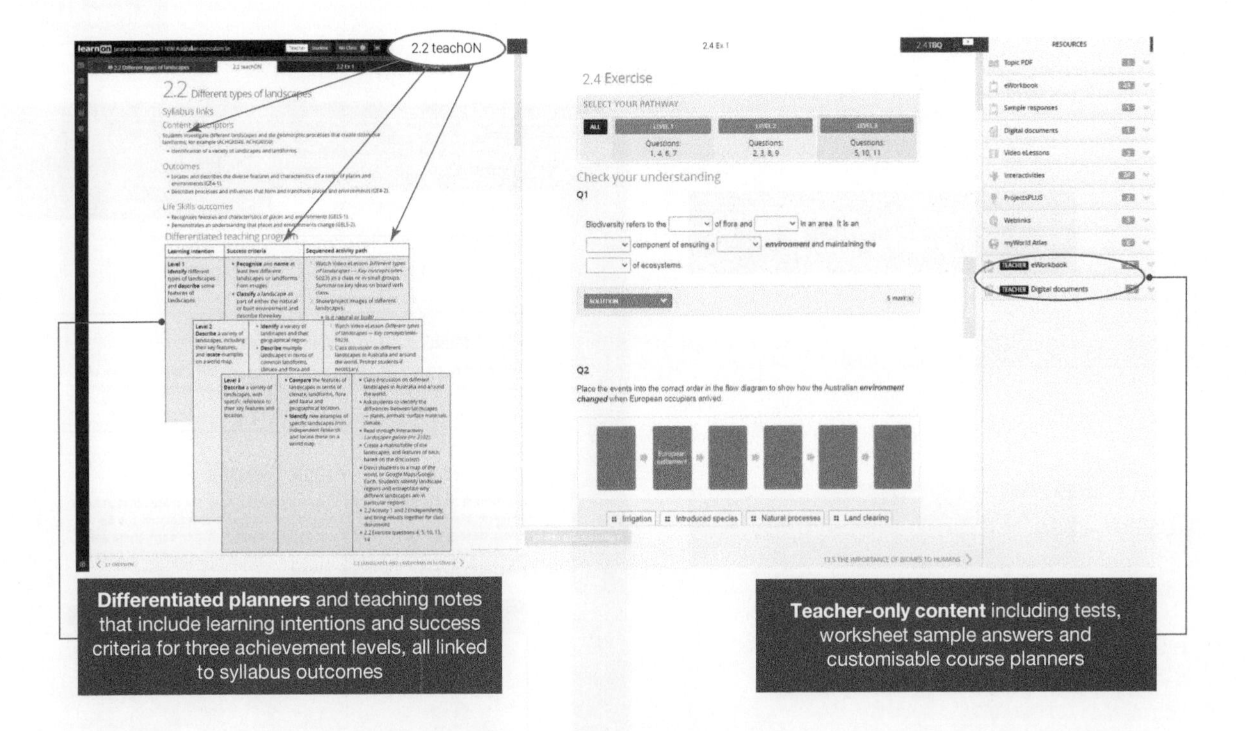

Every task students complete provides ...

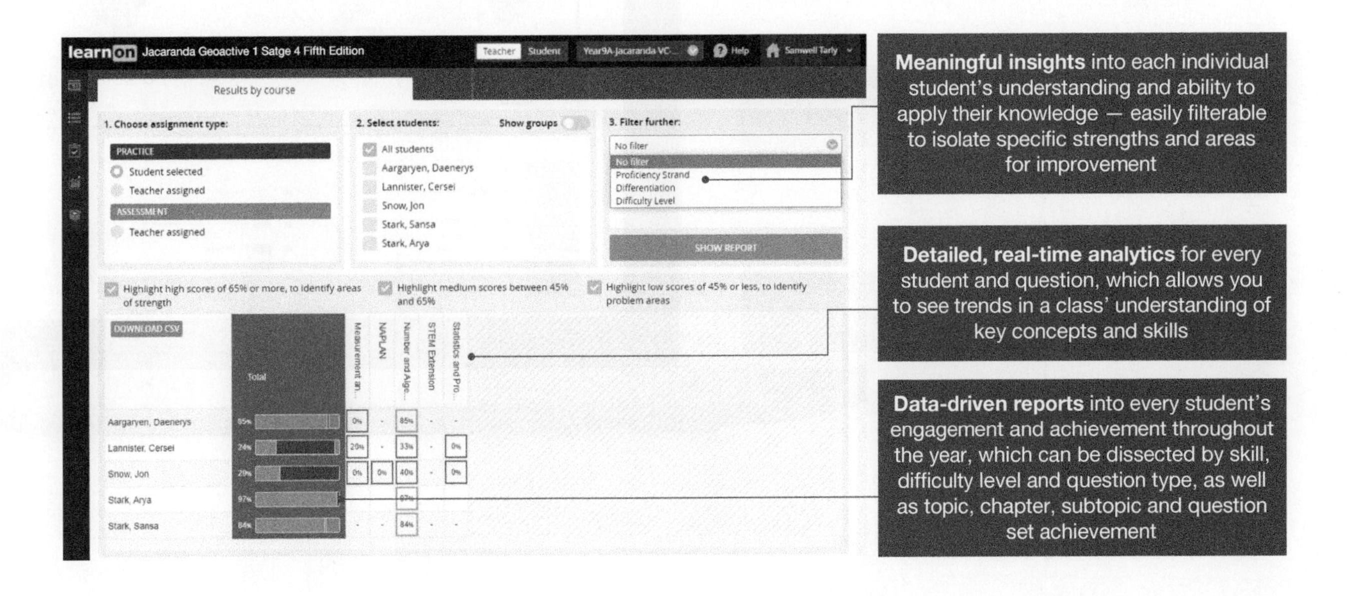

The Geoactive package

Jacaranda Geoactive 1 Stage 4 NSW Ac Fifth Edition has been revised and reimagined to provide students and teachers with the most comprehensive and engaging Geography resource package on the market. This engaging and purposeful suite of resources is fully aligned to the NSW Australian curriculum Geography Stage 4. *Jacaranda Geoactive* is available in digital and print formats.

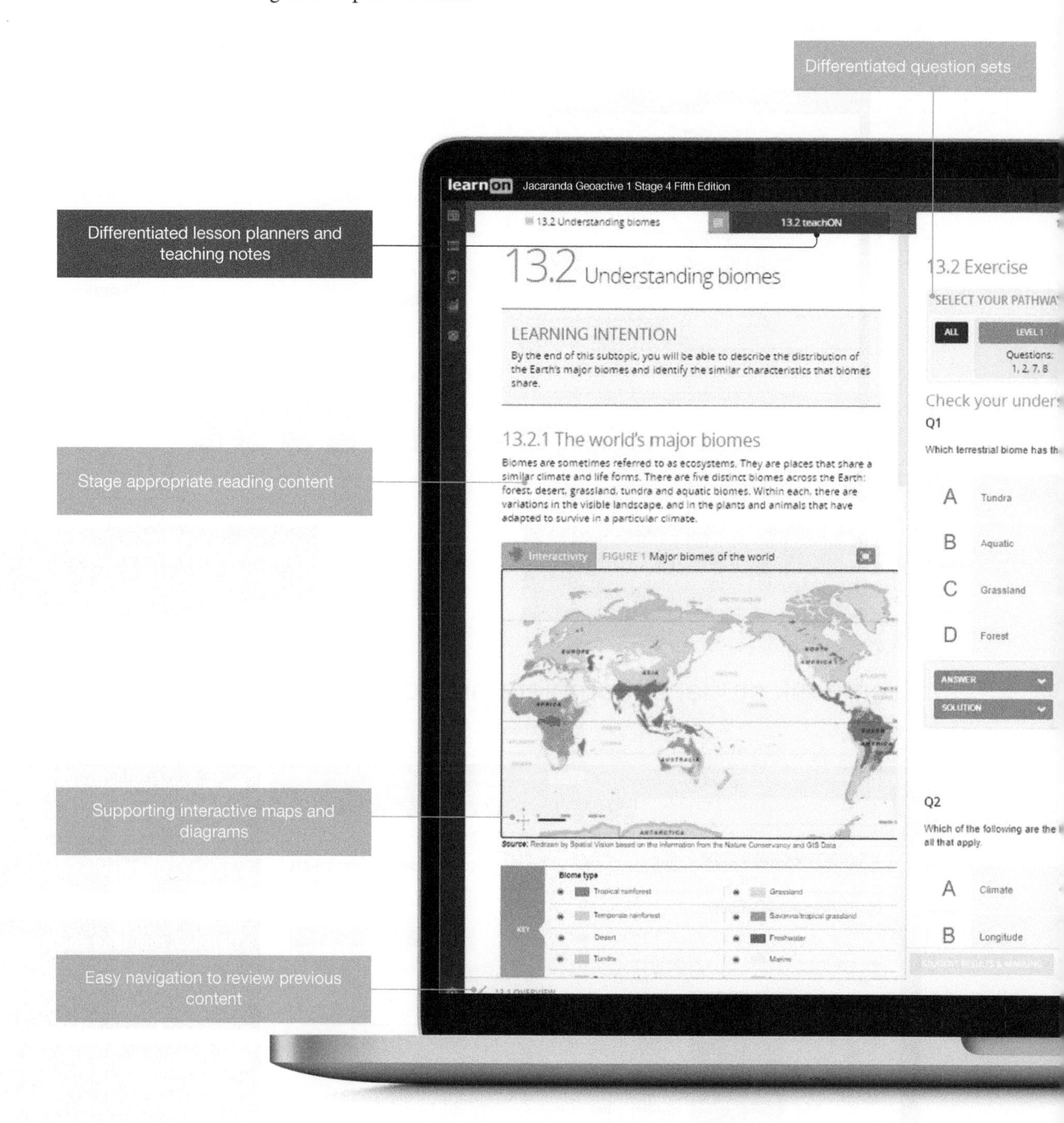

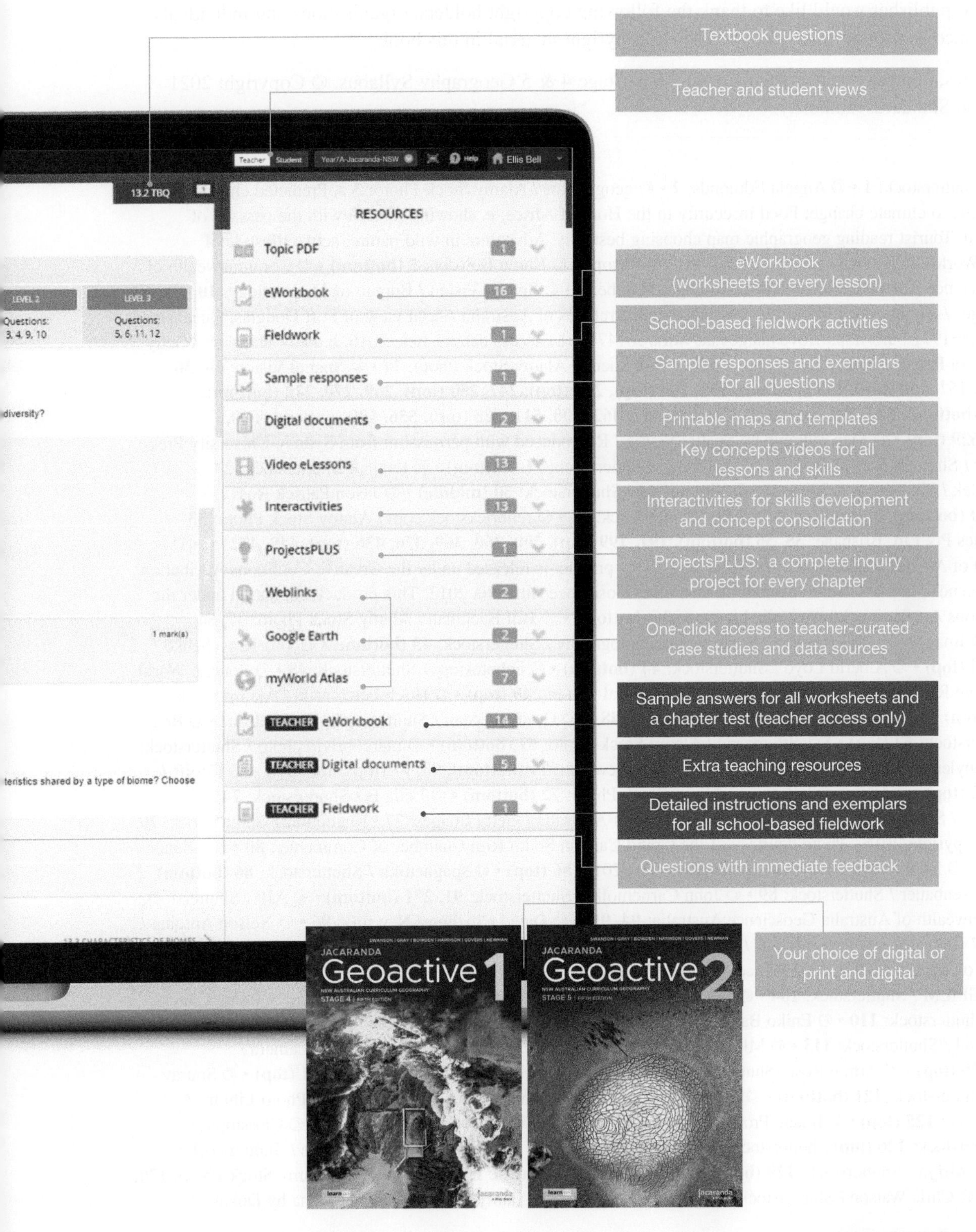
Textbook questions
Teacher and student views
eWorkbook (worksheets for every lesson)
School-based fieldwork activities
Sample responses and exemplars for all questions
Printable maps and templates
Key concepts videos for all lessons and skills
Interactivities for skills development and concept consolidation
ProjectsPLUS: a complete inquiry project for every chapter
One-click access to teacher-curated case studies and data sources
Sample answers for all worksheets and a chapter test (teacher access only)
Extra teaching resources
Detailed instructions and exemplars for all school-based fieldwork
Questions with immediate feedback
Your choice of digital or print and digital
RESOURCES
Topic PDF
eWorkbook
Fieldwork
Sample responses
Digital documents
Video eLessons
Interactivities
ProjectsPLUS
Weblinks
Google Earth
myWorld Atlas
TEACHER eWorkbook
TEACHER Digital documents
TEACHER Fieldwork
13.2 TBQ
Teacher
Student
Ellis Bell
JACARANDA
Geoactive 1
STAGE 4
Geoactive 2
STAGE 5

Acknowledgements

The Geoactive team would like to thank those people who have played a key role in the production of this text. Their families and friends were always patient and supportive, especially when deadlines were imminent. This project was greatly enhanced by the generous cooperation of many academic colleagues and friends. The professionalism and expertise of Wiley staff is also appreciated.

The authors and publisher would like to thank the following copyright holders, organisations and individuals for their assistance and for permission to reproduce copyright material in this book.

Subject outcomes, objectives and contents from NSW Stage 4 & 5 Geography Syllabus. © Copyright 2021 NSW Education Standards Authority

Images

• © withGod / Shutterstock: **1** • © Angela Edmonds: **2** • © geogphotos / Alamy Stock Photo: **3** a. Predicted change in annual run-off due to climate change; Food insecurity in the Horn of Africa; c. showing rainfall with the passage of typhoon Nesat; d. Tourist reading geographic map choosing best way. Adventure in wild nature, active lifestyle: **4** • © Data from World Trade Organization / Spatial Vision: **5 (top)** • © Karen Bowden: **5 (bottom)** • © Commonwealth of Australia. Geoscience Australia 1982: **7** • © 2016 Digital Globe: **8** • © Spatial Vision / Bureau of Meteorology: **10** • © Pulsar Images / Alamy Stock Photo: **11** a. • © National Archives of Australia Aerial view of Port Douglas, Queensland | publication-date=1971 | url=http://trove.nla.gov.au/work/161474801 | access date=9 June 2016; b. Peter Harrison / Getty Images; c. Danitza Pulgar / Alamy Stock Photo: **13** • © ton koene / Alamy Stock Photo: **14** • © Spatial Vision: **16, 36 (top), 131, 147, 157, 168 (top), 168 (bottom), 171, 180 (top), 211 (top), 247, 250 (top), 269, 270, 272 (bottom), 289 (top), 381 (bottom), 382, 391 (bottom), 426, 438, 457, 462, 500, 511, 528 (top), 556, 599 (bottom), 600, 622 (bottom), 629 (top)** • © Map redrawn by Spatial Vision / Reproduced with permission from Sydney University Press: **20** • © Mmaxer / Shutterstock: **21 (top)** • © FotoDuets / Shutterstock: **21 (bottom)** • © totajla / Shutterstock: **27** • © Daniel Prudek / Shutterstock: **30 (top)** • © VittoriaChe / Shutterstock: **30 (middle)** • © Jason Patrick Ross / Shutterstock: **30 (bottom), 43 (top)** • © vichie81 / Shutterstock: **31** • © redbrickstock.com / Alamy Stock Photo: **33** • © MAPgraphics Pty Ltd, Brisbane: **35, 36 (bottom), 191, 199 (top), 200, 260, 369, 376, 436 (top), 449, 472** a. • © Commonwealth of Australia Geoscience Australia 2012. This product is released under the Creative Commons Attribution 3.0 Australia Licence.; b. • © Commonwealth of Australia Geoscience Australia 2012. This product is released under the Creative Commons Attribution 3.0 Australia Licence.: **36 (bottom)** • © Bill Bachman / Alamy Stock Photo: **37, 54** • © AFP / Getty Images: **41, 484 (top)** • © Sarah Fields Photography / Shutterstock: **43 (bottom)** • © Yuri Kravchenko / Shutterstock: **44 (top)** • © Alberto Loyo / Shutterstock: **44 (bottom)** • © aphotostory / Shutterstock: **48** • © Source: World Map of Carbonate Rock Outcrops v3.0. Map created by Spatial Vision.: **49 (top)** • © Hugh Peterswald / Alamy Stock Photo: **49 (bottom)** Alamy Stock Photo: **53, 288, 291, 509, 548** • © Guilbaud Stan / Shutterstock: **57 (bottom)** • © Rex Ellacott / Shutterstock: **57 (top)** • © keith morris / Alamy Stock Photo: **57 (bottom)** • © marcobrivio.photo / Shutterstock: **69** • © Olga Danylenko / Shutterstock: **71 (top)** • © Steven Newman: **71 (bottom), 90, 97 (top), 100, 576** • © CoolR / Shutterstock: **72 (top)** • © Historic Collection / Alamy Stock Photo: **72 (bottom)** • © Lella B / Shutterstock: **74** • © Natalie Maro / Shutterstock: **76** • © Southern Lightscapes-Australia / Getty Images: **77** • © goodluz / Shutterstock: **79, 519, 520** • © Copyright 2016 - Photo courtesy of the Grand Canyon South Rim Chamber of Commerce.: **80** • © c Sandra Duncanson: **81** • © NPWS data 2012 and Forests NSW data 2011: **86 (top)** • © spaghettikk / Shutterstock: **86 (bottom)** • © Stanislav Fosenbauer / Shutterstock: **89** • © John Carnemolla / Shutterstock: **91, 271 (bottom)** • © AJP / Shutterstock: **93** • © Commonwealth of Australia Geoscience Australia: **94, 95** • © David Crosling / Newspix: **96** • © Nelson Antoine / Shutterstock: **97 (bottom)** • © mark higgins / Shutterstock: **101 (top)** • © aeropix / Alamy Stock Photo: **101 (bottom)** • © Data based on Spatial Services 2019. Licensed under CC BY 4.0 https://creativecommons.org/licenses/by/4.0/: **102** • © LIUSHENGFILM / Shutterstock: **104** • © Delphotos / Alamy Stock Photo: **109** a. Joe Cornish / Getty Images, Inc.; b. Irina Mos / Shutterstock: **110** • © Eniko Balogh / Shutterstock: **111** • © MAPgrahic Pty Ltd Brisbane: **112** • © xamnesiacx84 / Shutterstock: **113** • © Mlenny / Getty Images: **118, 125 (middle)** • © AustralianCamera / Shutterstock: **119 (top)** • © Armin Rose /Shutterstock: **119 (bottom)** • © gyn9037 / Shutterstock: **121 (top)** • © Sourav and Joyeeta / Shutterstock: **121 (bottom)** • © Atmotu Images / Alamy Stock Photo: **124** • © Science Photo Library / Alamy Stock Photo: **125 (top)** • © Image Professionals GmbH / Alamy Stock Photo: **125 (bottom)** • © Christopher Halloran / Shutterstock: **126 (top)** Shutterstock / Jana Schoenknecht: **126 (bottom)** • © Janelle Lugge / Shutterstock.: **129 (top)** • © N Mrtgh / Shutterstock: **129 (bottom)** • © National Geographic Image Collection / Alamy Stock Photo: **130, 136 (bottom)** • © Chris Watson / Shutterstock: **132** a. Jorge Fernandez / Alamy Stock Photo; b. Prisma by Dukas

Presseagentur GmbH / Alamy Stock Photo: **135** a. Science Photo Library / Alamy Stock Photo; b. Tim Roberts Photography / Shutterstock: **136 (top)** • © Gaearon Tolon / Shutterstock: **138 (top)** • © Universal Images Group North America LLC / Alamy Stock Photo: **138 (bottom)** • © Michael Wheatley / Alamy Stock Photo: **145** • © Manfred Thürig / Alamy Stock Photo: **146** • © Fineart1 / Shutterstock: **149** • © miramis / Shutterstock: **150** • © Daniel Prudek / Shutterstock: **151 (top)** • © Sean Pavone / Shutterstock: **151 (bottom)** • © Man Down Media / Shutterstock: **152 (top)** • © Nicole Gray: **152 (middle)** • © Asya Karmazin / Shutterstock: **152 (bottom)** • © Shawn Lu / Shutterstock: **153** • © guenterguni / Getty Images: **160 (top)** • © John A Cameron / Shutterstock: **160 (bottom)** • © Denis Mironov / Shutterstock: **161 (top)** • © Josemaria Toscano / Shutterstock: **161 (bottom)** • © US Geological Survey: **162 (top)** • © Thomas Dutour / Alamy Stock Photo: **162 (bottom)** • © Robert Zehetmayer / Alamy Stock Photo: **163** • © mTaira / Shutterstock: **168 (top)** • © MPAK / Alamy Stock Photo: **170** • © The Asahi Shimbun / Getty Images Australia: **173** • © Mark Lincoln: **174 (bottom)** • © Diarmuid / Alamy Stock Photo: **174 (top)** • © Nigel Spiers / Alamy Stock Photo: **176** • © Sergio Thor / Shutterstock: **178** • © BorneoRimbawan / Shutterstock: **179** • © krug_100 / Shutterstock: **180 (bottom)** • © NASA / Image Science and Analysis Laboratory, NASA-Johnson Space Center: **180 (top)** • © Contains data sourced from the LINZ Data Service and licensed for reuse under CC BY 4.0: **181** • © Stephen Sykes / Alamy Stock Photo: **182** • © Amazon-Images / Alamy Stock Photo: **189** • © ronnybas / Shutterstock: **190** • © AustralianCamera / Shutterstock: **192 (top)** • © Paulo Vilela / Shutterstock: **192 (middle)** • © Ashley Whitworth / Shutterstock: **192 (bottom)** a. VectorMine / Shutterstock; b. BorneoRimbawan / Shutterstock; c. Firuz Ashumov / Shutterstock; d. Natali22206 / Shutterstock; e. Hugh Lansdown / Shutterstock; f. Taewphoto / Shutterstock: **193** • © Dirk Ercken / Shutterstock: **197, 198 (top)** a. Ammit Jack / Shutterstock; b. OlegDoroshin / Shutterstock; c. Lukas Gojda / Shutterstock: **198 (bottom)** • © Anna Rogers / Newspix: **199 (bottom)** • © Amy Nichole Harris / Shutterstock: **203** • © YIFANG NIE / Shutterstock: **204** • © Dr. Morley Read / Shutterstock: **205 (top)** • © Eye Ubiquitous / Alamy Stock Photo: **205 (middle)** • © Tarcisio Schnaider / Shutterstock: **205 (bottom)** • © MAPgraphics Pty Ltd Brisbane: **206, 616, 659** • © Photodisc: **207 (bottom), 207 (top)** • © Claudia Weinmann / Alamy Stock Photo: **207 (bottom)** • © DEEPU SG / Alamy Stock Photo: **209** • © Kjersti Joergensen / Shutterstock: **210** • © Hugo Ahlenius, UNEP/GRID-Arendal http://www.grida.no/graphicslib/detail/extent-of-deforestation-in-borneo-1950-2005-and-projection-towards-2020_119c: **211 (bottom)** • © Sue Cunningham Photographic / Alamy Stock Photo: **212 (top)** • © RAUL ARBOLEDA / AFP / Getty Images: **212 (bottom)** • © Pictorial Press Ltd / Alamy Stock Photo: **213 (bottom)** • © dpa picture alliance / Alamy Stock Photo: **213 (top)** • © Jeff Baumgart / Shutterstock: **213 (bottom)** Shutterstock / Txanbelin: **215** a. Hilke Maunder / Alamy Stock Photo; b. Dave and Sigrun Tollerton / Alamy Stock Photo; c. AustralianCamera / Alamy Stock Photo: **216** • © State of Queensland Department of Natural Resources, Mines and Energy, Department of Environment and Science 2020: **217** • © Marco Saracco / Shutterstock: **218** • © guentermanaus / Shutterstock: **207** • © Wirestock Creators / Shutterstock: **225** • © Jonathan Steinbeck / Shutterstock: **227** • © Merrillie Redden / Shutterstock: **228** • © StevenK / Shutterstock: **231** • © Radiokafka / Shutterstock: **232** a. Leah-Anne Thompson / Shutterstock; b. Kamira / Shutterstock; c. annarevoltosphotography / Shutterstock: **233** • © marchello74 / Shutterstock: **234 (left)** • © VIAVAL TOURS / Shutterstock: **234 (top)** • © Adam Kazmierski / iStockphoto: **234 (bottom)** • © Anton_Ivanov / Shutterstock: **234 (top)** • © Eric Nathan / Alamy Stock Photo: **234 (middle)** • © Victor Mafferoyalty free - extended use: **234 (bottom)** • © Guillem Lopez / Alamy Stock Photo: **235 (bottom)** • © Janelle Lugge / Shutterstock: **235 (top)** a. • © Dollar Street 2015; b. • © Dollar Street 2015; c. • © Dollar Street 2015: **236** a. • © Dollar Street 2015; b. • © Dollar Street 2015: **237** • © Nils Versemann / Shutterstock: **239** • © Artokoloro / Alamy Stock Photo: **240** • © Philip Schubert / Shutterstock: **241 (top), 241 (middle)** • © Niranjan Casinader: **241 (bottom), 320 (top), 320 (bottom), 323 (top), 324 (bottom), 324 (top), 324 (bottom), 325 (top)** • © DAE Photo / Shutterstock: **243 (top)** • © sturti / Getty Images: **243 (bottom)** • © f11photo / Shutterstock: **244** • © Michael Fay / Getty Images: **245** • © Martin Cohen Wild About Australia / Getty Images: **250 (bottom)** • © sigurcamp / Shutterstock: **251 (top)** • © Yuen Man Cheung / Alamy Stock Photo: **251 (middle)** • © Dan Breckwoldt / Shutterstock: **251 (bottom)** • © LingHK / Shutterstock: **252** • © Aboriginal Languages / Nations in NSW & ACT • © Reconciliation NSW, www.reconciliationnsw.org.au: **253** • © tunart / Getty Images: **254** • © Michael Evans / Alamy Stock Photo: **257 (top)** • © Olga Kashubin / Shutterstock: **257 (bottom)** a. Andrew Sole / Alamy Stock Photo; b. Andrew Sole / Alamy Stock Photo: **258 (top)** • © Sk Hasan Ali / Shutterstock: **258 (bottom)** • © ollirg / Shutterstock: **259** • © 1989studio / Shutterstock: **262** a. Ian Nellist / Alamy Stock Photo; b. haveseen / Shutterstock; c. Nickolay Vinokurov / Shutterstock; d. Alexander Chaikin / Shutterstock; e. Anki Hoglund / Shutterstock; f. Mohamed Shareef / Shutterstock; g. Andrey Bayda / Shutterstock; h. stocker1970 / Shutterstock; i. Francesco R. Iacomino / Shutterstock; j. Jane Rix / Shutterstock; k. Aleksandar Todorovic / Shutterstock; l. Kenneth Dedeu / Shutterstock; m. chuyu / Shutterstock; n. JeniFoto / Shutterstock; o. Vladimir Melnik / Shutterstock; p. leoks / Shutterstock: **267** a. haveseen / Shutterstock; b. Nickolay Vinokurov / Shutterstock: **268** • © Bloomberg / Getty Images: **271 (top)** • © Bob Barker / Newspix: **271 (top)** • © edella / Shutterstock: **272 (top)** • © Bureau of Meteorology: **273 (top), 334, 479 (left), 479 (right), 480 (top)** • © David Wall / Alamy Stock Photo: **273 (middle)** • © Marina Kryuchina / Shutterstock: **273 (bottom)** • © Commonwealth of Australia. Australian Bureau of Statistics March 2021 Estimated Resident Population - As at 30 June 2014 and 2018. Bega Valley A LGA 10550, accessed 9 March 2021.: **276**

Shutterstock / FiledIMAGE: **277** • © Travelscape Images / Alamy Stock Photo: **278** • © alexnika / Shutterstock: **279** • © a. Niranjan Casinader; b. philipjbigg / Alamy Stock Photo: **281** • © Burleigh, Queensland. 1:50,000 Sheet 9541-I Edition 1 Series R 733 Printed by Royal Australian Survey Corps 1969 Crown Copyright Reserved: **283** • © Gold Coast SUNS / Metricon Stadium: **284 (top)** • © kali9 / E+ / Getty Images: **284 (bottom)** • © Image Courtesy of c 2016 DigitalGlobe: **286** • © Josef Hanus / Shutterstock: **287** • © George Holton /Science Source: **289 (bottom)** • © Liba Taylor / Alamy Stock Photo: **290 (top)** • © Quick Shot / Shutterstock: **290 (bottom)** • © Spatial Vision / Geoscience Australia: **293, 410** • © ArliftAtoz2205 / Shutterstock: **294 (top)** a. Caden Ballantine / Shutterstock; b. ArliftAtoz2205 / Shutterstock: **294 (middle)** • © Oz Aerial Photography: **294 (bottom)** • © National Archives of Australia Aerial view of Port Douglas, Queensland | publication-date=1971 | url=http://trove.nla.gov.au/work/161474801 | access date=9 June 2016: **297** • © a. c Sandra Duncanson; b. c Sandra Duncanson: **300** Shutterstock / Elias Bitar: **303** • © MAGNIFIER / Shutterstock: **307** • © Sadik Gulec / Shutterstock: **308** • © Global Footprint Network: **310** • © Spatial Vision / Food and Agriculture Organisation: **312** a. Ryanzo W. Perez / Shutterstock; b. Friedrich Stark / Alamy Stock Photo; c. Johnny Greig Int / Alamy Stock Photo; d. Piti Anchalee / Alamy Stock Photo; e. USAID / Alamy Stock Photo; f. Irene Abdou / Alamy Stock Photo: **313** • © Department of Foreign Affairs and Trade: **315** a. • © AusAID Department of Foreign Affairs and Trade; b. Kaveh Kazemi / Getty Images Australia; c. Photo as taken by L.Limalevu PACE-SD.: **316** a. ECCA Nepal; b. • © Grant Gibbs www.hipporoller.org: **317** a. • © Cynthia Wardle; b. Niranjan Casinader; c. Niranjan Casinader: **321 (top)** • © a. Niranjan Casinader; b. Niranjan Casinader: **321 (bottom), 323 (middle)** a. Niranjan Casinader; b. Cynthia Wardle; c. Cynthia Wardle: **322 (top)** a. Niranjan Casinader; b. Niranjan Casinader; c. Niranjan Casinader: **322 (bottom)** • © a. Niranjan Casinader; b. Niranjan Casinader; c. Niranjan Casinader: **323 (bottom), 326 (top)** a. Niranjan Casinader; b. Niranjan Casinader; c. Niranjan Casinader; d. Niranjan Casinader; e. Niranjan Casinader: **325 (bottom)** • © Caiaimage / Trevor Adeline / Getty Images: **326 (bottom)** a. Geoff Smith / Alamy Stock Photo; b. Thomas Cockrem / Alamy Stock Photo: **329** • © Glen Berlin / Shutterstock: **330 (top)** • © Nils Versemann / Shutterstock: **330 (bottom)** • © a. Based on data from Roland Berger and Statista; b. Based on data from Roland Berger and Statista: **331 (top)** • © pbk-pg / Shutterstock: **331 (bottom)** • © TSRA: **335** • © Geoscience Australia: **336** • © Gordon Bell / Shutterstock: **338 (top)** • © Travel Stock / Shutterstock: **338 (bottom)** • © Willowtreehouse / Shutterstock: **343** • © Taras Vyshnya / Alamy Stock Photo: **344 (top)** • © Taras Vyshnya / Shutterstock: **344 (bottom), 346** • © Taras Vyshnya / Shutterstock: **345** • © patronestaff / Shutterstock: **349** • © J Marshall - Tribaleye Images / Alamy Stock Photo: **350** • © Monkey Business Images / Shutterstock: **351** • © freya-photographer / Shutterstock: **352** • © WorlcClim, licensed under a Creative Commons Attribution-ShareAlike 4.0 International License.: **358** • © Geophysical Fluid Dynamics Laboratory, National Oceanic and Atmospheric Administration: **359** • © Juan Vilata / Alamy Stock Photo: **360** • © Dept. of Climate Change and Energy Efficiency The Garnaut Climate Change Review 2008, published by the Commonwealth of Australia: **364** • © Taras Vyshnya / Shutterstock Stock Photo: **365** a. Nils Versemann / Shutterstock b. Reeva / Shutterstock; c. Jiratsung / Shutterstock: **368** • © Wisit Tongma / Shutterstock: **374** a. manasesistvan / Shutterstock; b. Lost River Photo / Shutterstock; c. Rainer Fuhrmann / Shutterstock; d. Elizaveta Galitckaia / Shutterstock; e. SherSS / Shutterstock; f. ymphotos / Shutterstock: **377** • © Louise Swanson: **381 (top), 417 (top), 435** • © Martchan / Shutterstock: **384** • © Sandra Duncanson: **389** • © david pearson / Alamy Stock Photo: **390** • © Data source based on WWAP UNESCO World Water Assessment Programme. 2019. The United NationsWorld Water Development Report 2019: Leaving No One Behind. Paris, UNESCO. Available in Open Access under the Attribution-ShareAlike 3.0 IGO CC-BY-SA 3.0 IGO: **391 (top)** • © Taras Vyshnya / Alamy Stock Photo: **393** • © Bastian AS / Shutterstock: **396 (top)** • © matasakti / Alamy Stock Photo: **396 (bottom)** • © Monkey Business Images / Shutterstock: **397** • © ChinaFotoPress / ChinaFotoPress / Getty Images: **400 (top)** • © Stuart Jenner / Getty Images Australia: **400 (bottom)** • © ADALBERTO ROQUE / AFP / Getty Images: **401** Shutterstock: **402, 409, 583 (bottom), 665 (bottom)** • © Amors photos / Shutterstock: **403** • © c UNICEF. State of the World's Sanitation: An urgent call to transform sanitation for better health, environments, economies and societies. New York: United Nations Children's Fund UNICEF and the World Health Organization, 2020.: **404 (top)** • © Spatial Vision / Natural Earth Data: **404 (bottom), 455, 493 (top)** • © Gunditj Mirring Traditional Owners Aboriginal Corporation: **407 (top)** Shutterstock / John Carnemolla: **407 (bottom)** • © c Mirima Dawang Woorlab-gerring Language and Culture Centre: **408** • © Keith Wheatley / Shutterstock: **410** • © Redrawn with permission from the SA Arid Lands Natural Resources Management Board / • © Copyright Commonwealth of Australia Geoscience Australia 2006: **412** Shutterstock / Mandy Creighton: **411** • © ABC Radio Sydney / Matt Bamford: **415** a. BlueOrange Studio / Shutterstock; b. leungchopan / Shutterstock: **417 (bottom)** a. nancymiao.id / Shutterstock; b. Andreas Verginer / Shutterstock; c. CoolR / Shutterstock; Louise Swanson; d. Finpat / Shutterstock; e. Martin Valigursky / Shutterstock f. Katie Purling / Shutterstock: **418** • © DisobeyArt / Alamy Stock Photo: **419 (top)** Shutterstock / Hypervision Creat: **419 (bottom)** • © vallefrias / Shutterstock: **421** • © Data based on Department for Infrastructure and Transport, SA Country Fire Service, Department for Environment and Water and State of Victoria Department of Environment, Land, Water and Planning: **423** • © Istimages / Getty Images: **425 (top)** • © Peter Harrison / Getty Images Australia: **425 (middle)** • © Kym McLeod / Shutterstock: **425 (bottom)** • © Taras Vyshnya / Shutterstock: **428** • © Pulsar Images / Alamy Australia Pty Ltd: **433** • © idiz / Shutterstock: **434** • © Vadym Zaitsev / Shutterstock: **436 (bottom)** • © Spatial Vision / Bureau of

Meterology: **439 (bottom), 439 (top)** • © Sunset And Sea Design / Shutterstock: **440** • © Paul Chesley / Getty Images: **442** • © c Clean Energy Council. Clean Energy Australia Report 2020, page 11.: **443** • © Used under license from IWMI: **444** • © Juan A. Salgado / Shutterstock: **445** • © Reprinted by permission from Macmillan Publishers Ltd: [Nature] 488, 197–200 August 9, 2012, copyright 2012: **446** • © c Frank R. Rijsberman: **447 (top)** • © c Thomas L. Kelly: **447 (bottom)** • © www.CartoonStock.com: **448** • © africa924 / Shutterstock: **452 (left)** • © Artush / Shutterstock: **452 (top)** • © Coffs Harbour City Council 2014: **452 (bottom)** • © graham jepson / Alamy Stock Photo: **454** • © GRAEME MCCRABB / AAP: **456** • © Source: Bureau of Meteorology: **458** • © Bob Hilscher / Shutterstock: **461 (bottom)** • © Dmitry Pichugin / Shutterstock: **461 (top)** • © Ashley Cooper pics / Alamy Stock Photo: **464** • © Stephen Locke: **469** • © Chris Ison / Alamy Stock Photo: **470** • © Phillip Wittke / Shutterstock: **471** • © Stuart Perry / Shutterstock: **473** • © Leah-Anne Thompson / Shutterstock: **477, 521 (top)** • © Newspix / John Grainger: **479 (top)** • © Zoltán Csipke / Alamy Stock Photo: **480 (bottom)** • © Action Plus Sports Images / Alamy Stock Photo: **482 (top)** • © model10 / Alamy Stock Photo: **482 (middle)** • © Dave Hunt / AAP Newswire: **482 (bottom)** • © David Martinelli / Newspix: **484 (bottom)** a. Courtesy NASA / JPL-Caltech; b. Courtesy NASA / JPL-Caltech.: **486 (top)** • © Reproduced by permission of Bureau of Meteorology, • © 2021 Commonwealth of Australia.: **486 (bottom), 492 (top), 654** • © Lauren Dauphin / NASA Earth Observatory: **488** • © paintings / Shutterstock: **492 (bottom)** • © VANDERLEI ALMEIDA / Getty Images: **493 (bottom)** • © David Kapernick / Newspix: **495** • © yampi / Shutterstock: **496 (top)** • © AFP / Stringer / Getty Images: **496 (bottom)** • © KONGKOON / Shutterstock: **497 (top)** • © m.malinika / Shutterstock: **497 (bottom)** • © NASA: **502 (bottom)** • © Rakesh Roul / Shutterstock: **502 (top)** • © Joshua Stevens / NASA Earth Observatory: **502 (middle)** • © dkroy / Shutterstock: **502 (bottom)** • © Kevin Frayer / Getty Images: **504** • © Universal Images Group North America LLC / Alamy Stokc Photo: **505** • © c Fiji Meteorological Service.: **506 (top)** • © Department of Foreign Affairs and Trade website – www.dfat.gov.au: **506 (bottom), 619, 620 (top), 620 (bottom), 622 (top), 625, 671, 672 (bottom), 673** • © Dr Yutaro SETOYA / WHO Tonga Office: **507** • © Steve Marsh / Alamy Stock Photo: **512** • © Harvepino / Shutterstock: **513** • © Elias Bitar / Shutterstock: **521 (bottom)** • © Dale Lorna Jacobsen / Shutterstock: **522** • © GRANT ROONEY PREMIUM / Alamy Stock Photo: **525** • © AVAVA / Shutterstock: **528 (bottom)** • © Andre B. Adur / Shutterstock: **530** • © potowizard / Shutterstock: **531** • © Alex Bascuas / Shutterstock: **532** • © Data source from UN World Tourism Organization: **533 (top)** • © UNWTO, 92844/09/21. World Tourism Organization 2015, UNWTO Tourism Highlights, 2015 Edition, UNWTO, Madrid, pp. 2, 13 and 14, DOI: https://doi.org/10.18111/9789284416899: **533 (left), 533 (right), 553** • © William West / AFP / Getty Images: **538** • © Mistervlad / Shutterstock: **539 (top)** • © Byelikova Oksana / Shutterstock: **539 (middle)** • © BiksuTong / Shutterstock: **539 (bottom)** • © 123Nelson / Shutterstock: **541** • © Gwoeii / Shutterstock: **542** • © Shutterstock / ChameleonsEye: **545** • © mimagephotography / Shutterstock: **549** • © Hananeko_Studio / Shutterstock: **550 (top), 566** • © ssrik / Shutterstock: **550 (bottom)** • © Evgeniyqw / Shutterstock: **551** • © Adapted from Cosmetic Surgeon India and Rowena Ryan / News.com.au: **552** • © e2dan / Shutterstock: **555** • © Vadim Petrakov / Shutterstock: **557** • © Anton Watman / Shutterstock: **558** • © Iakov Kalinin / Shutterstock: **561** • © ChameleonsEye / Shutterstock: **560** • © metamorworks / Shutterstock: **563** • © Andrew Douglas-Clifford: **568 (top)** • © slava296 / Shutterstock: **568 (bottom)** • © ben bryant / Shutterstock: **570** • © Exhibit from “Elements of success: Urban transportation systems of 24 global cities”, June 2018, McKinsey & Company, www.mckinsey.com. Copyright c 2021 McKinsey & Company. All rights reserved. Reprinted by permission.: **571 (top), 571 (bottom)** • © RossHelen editorial / Alamy Stock Photo: **572 (top)** • © Source: Roy Morgan Single Source, April 2019 – March 2020, sample n = 13,208. Base: Australians aged 14+.: **572 (bottom)** • © State of New South Wales Transport for NSW: **573** • © SNAMUTS 2016 www.snamuts.com: **575** • © Nicku / Shutterstock: **579** • © Fekete Tibor / Shutterstock: **580 (top)** • © Data by Greg Mahlknecht, www.cablemap.info. Map drawn by Spatial Vision.: **580 (bottom)** • © VladFree / Shutterstock: **581** Shutterstock / Ververidis Vasili: **583 (top)** • © Average internet performance data from January 2021 based on the Speedtest Global Index™ from Ookla®: https://www.speedtest.net/global-index: **586** • © LUCARELLI TEMISTOCLE / Shutterstock: **587** • © Andy Singer / www.andysinger.com/: **588 (top)** • © Aurora Photos / Alamy Stock Photo: **588 (bottom)** • © Ascannio / Shutterstock: **589** • © Benedicte Desrus / Alamy Stock Photo: **591** a. • © Spatial Vision; b. Based on information from JUMIA 2018: **592** • © Mark52 / Shutterstock: **593 (top)** • © Australia Post. Inside Australian Online Shopping update, December 2020.: **593 (bottom)** • © cigdem / Shutterstock: **595** • © c Picodi: **596** • © Australian Bureau of Statistics, 2011, 2011 Australian Statistical Geography Standard: Remoteness Structure - Remoteness Area Boundaries, Australian Statistical Geography Standard ASGS: Volume 5 - Remoteness Structure, Map, 1270.0.55.005.: **597** • © superjoseph / Shutterstock: **599 (top)** • © EQRoy / Shutterstock: **601** • © Michael H / Getty Images: **602 (top)** Shutterstock / Natalie Board: **602 (middle)** Shutterstock / Dorothy Chiron: **602 (bottom)** • © ABCDstock / Shutterstock: **609** • © cozyta / Shutterstock: **610** • © a. Johan Larson / Shutterstock; b. Jetta Productions Inc / Getty Images; c. I. Pilon / Shutterstock; d. Andor Bujdoso / Shutterstock: **611** • © a. Jimmy Tran / Shutterstock; b. TFoxFoto / Shutterstock; c. monticello / Shutterstock; d. StudioSmart / Shutterstock: **612 (top)** • © Irina Borsuchenko / Shutterstock: **612 (bottom)** • © Nils Versemann / Shutterstock: **613 (bottom)** • © Stanislav Fosenbauer / Shutterstock: **613 (top)** • © Transports for NSW and Australia Bureau of Statistics: **613 (bottom)** • © Skycolors / Shutterstock: **614 (top)** • © Data from Wikimedia Commons, User: Steff. Map drawn by Spatial Vision.: **614 (bottom)** a. Pressmaster /

Shutterstock; b. bikeriderlondon / Shutterstock; c. Roman Kosolapov / Shutterstock; d. Levent Konuk / Shutterstock; e. • © Amanda Sutcliffe / Shutterstock; f. Pressmaster / Shutterstock; g. Goodluz / Shutterstock: **618** • © Commonwealth of Australia, Department of Education, Skills and Employment: **621** • © Source: WTO and UNCTAD: **626** • © Pablo Kaluza, Andrea Ko¨lzsch, Michael T. Gastnerand Bernd Blasius. The complex network of global cargoship movements. J. R. Soc. Interface 2010 7, doi:10.1098/rsif.2009.0495. Figure 1a.: **627 (top)** • © Source: Allianz Global Corporate & Specialty Safety and Shipping Review 2016: **627 (bottom)** • © phadungsak sawasdee / Shutterstock: **628** • © c Karen Bowden: **629 (bottom)** • © George Rudy / Shutterstock: **630** • © Australian Made Campaign Limited: **632 (top)** • © testing / Shutterstock: **632 (bottom)** • © Data from Statista. Map drawn by Spatial Vision: **633 (top)** • © HOANG DINH NAM / AFP / Getty Images: **633 (bottom)** • © a katz / Shutterstock: **634** a. Daniel Kaesler / Alamy Stock Photo; b. Richard Milnes / Alamy Live News: **635** • © TK Kurikawa / Shutterstock: **636 (top)** • © Dmitry Kalinovsky / Shutterstock: **636 (bottom)** • © rSnapshotPhotos / Shutterstock: **637** • © Brisbane / Shutterstock: **638 (top)** • © yankane / Shutterstock: **638 (bottom)** • © Pieter Beens / Shutterstock: **639** • © RosalreneBetancourt 7 / Alamy Stock Photo: **642** • © Land and Property Information NSW: **643** • © Halfpoint / Shutterstock: **645** • © YULIYAPHOTO / Shutterstock: **647** • © Panos Pictures / Andrew McConnell: **651** • © Kev Gregory / Shutterstock: **652** • © Data sourced from Quarterly Update of Australia's National Greenhouse Gas Inventory: December 2019, Commonwealth of Australia 2020: **653** • © Beckman, Eric. August 9, 2018 'The world of plastics, in numbers' https://theconversation.com/the-world-of-plastics-in-numbers-100291 viewed 30/12/2020cites data from: Roland Geyer, Jenna R. Jambeck and Kara Lavender Law, 'Production, use, and fate of all plastics ever made', Science Advances 19 Jul 2017: Vol. 3, no. 7, e1700782, DOI: 10.1126/sciadv.1700782: **655 (top)** • © Pavlovska Yevheniia / Shutterstock: **655 (bottom)** • © c Primary microplastics in the oceans, IUCN 2017: **656 (top)** • © Pro_Vector / Shutterstock: **656 (bottom)** • © Maxim Blinkov / Shutterstock: **656 (top)** • © a. Steve Meese / Shutterstock; b. AP Photo / Maritime New Zealand / AAP Newswire; c. Tappasan Phurisamrit / Shutterstock; d. NigelSpiers / Shutterstock; e.Igor Plotnikov / Shutterstock: **657** • © From The Sustainable Development Goals Report 2020, by United Nations, • ©2020 United Nations. Reprinted with the permission of the United Nations. https://sdgs.un.org/sites/default/files/2020-07/The-Sustainable-Development-Goals-Report-2020_Page_19.png: **658 (top)** • © a. vectorplus / Shutterstock; b. Mix3r / Shutterstock: **658 (bottom)** • © Snap2Art / Alamy Stock Photo: **661 (top)** • © Volodymyr Burdiak / Shutterstock: **661 (bottom)** • © William Davies / Getty Images: **662** • © StevenK / Shutterstock: **663 (top)** • © Sipa USA / Alamy Stock Photo: **663 (bottom)** • © Amani A / Shutterstock: **664 (top)** • © CRS PHOTO / Shutterstock: **664 (bottom)** • © Raftel / Shutterstock: **665 (top)** • © Aline Tong / Shutterstock: **669** • © Map drawn by Spatial Vision. Data from • © Commonwealth of Australia, DFAT, Australian Aid Budget Summary 2017–18.: **672 (top)** • © Redrawn by Spatial Vision based on information from Mercator Institute for Chinese: **674** • © ton koene / Alamy Stock Photo: **676** • © Data source from Fairtrade International: **677** • © Karen McFarland / Shutterstock: **680** • © elenabsl / Shutterstock: **681** • © Ralph Mayhew / Shutterstock: **685** • © Marc McCormack / Newspix: **686 (top)** • © Budimir Jevtic / Shutterstock: **686 (bottom)** • © Lee Torrens / Shutterstock: **688** • © AlexRoz / Shutterstock: **689** • © Russell Shively / Shutterstock: **695** • © Laszlo66 / Shutterstock: **696** • © Zastolskiy Victor / Shutterstock: **698**

Text

The Dorothea Mackellar Estate c/- Curtis Brown Aust Pty Ltd: **72** The Australian Museum, www.dreamtime.net.au/creation: **77** The Conversation. Kimberly Wood. The 2020 Atlantic hurricane season is so intense, it just ran out of storm names. 19 September 2020. https://theconversation.com/the-2020-atlantic-hurricane-season-is-so-intense-it-just-ran-out-of-storm-names-146506: **502** Source: Australian Bureau of Statistics, Overseas Arrivals and Departures, Australia https://www.abs.gov.au/statistics/industry/tourism-and-transport/overseas-arrivals-and-departures-australia/latest-release: **536** Data source based on Australia's 12 million public transport users face a COVID19 dilemma – get back on the train or drive to work? May 28 2020. Finding No. 8421, Ryan Morgan.: **572** • © State of New South Wales Transport for NSW: **576** Department of Foreign Affairs and Trade website – www.dfat.gov.au: **619** • © 2021 World Shipping Council: **628** c University of Melbourne. William Ho. The COVID-19 shock to supply chains, retrieved from https://pursuit.unimelb.edu.au/articles/the-covid-19-shock-to-supply-chains [online resource]. Reproduced by permission under Creative Commons Attribution-No Derivatives 3.0 Australia CC BY-ND 3.0 AU https://creativecommons.org/licenses/by-nd/3.0/au/: **639**

1 The world of Geography

1.1 Overview

Numerous **videos** and **interactivities** are embedded just where you need them, at the point of learning, in your learnON title at www.jacplus.com.au. They will help you to learn the content and concepts covered in this topic.

How and why do we study Geography?

1.1.1 What is Geography?

The world around us is made up of interesting places, people, cultures and environments. Geography is the subject that you study at school to learn about different places, and how relationships between environments and people shape these places. Geographers question how environments function and why the world is the way it is. They explore geographical issues and challenges, predict their outcomes, and come up with possible solutions for the future. Geographers are active and responsible citizens, who are informed about our world and are capable of shaping the future.

FIGURE 1 Our planet is made up of a large variety of fascinating places, peoples, cultures and environments.

on Resources

eWorkbook Chapter 1 eWorkbook (ewbk-7981)

1.2 Geographical inquiry

LEARNING INTENTION

By the end of this subtopic, you will be able to describe the process of geographical inquiry and identify geographical tools.

1.2.1 The process of geographical inquiry

Have you ever visited or gone on holidays to a place other than where you live? If so, you have probably noticed that some of the features and characteristics of the people and places are similar to what you know and some are different. Studying Geography at school provides you with the skills, the knowledge and the tools to learn about and understand the relationships between the world's peoples, places and environments.

FIGURE 1 Using maps to work out locations and to plot data

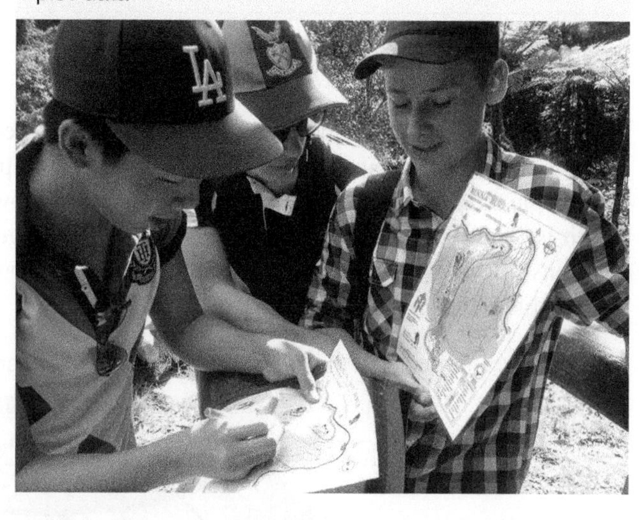

As a geographer you get to ask questions and then seek answers to them. Geographers use what is called an inquiry approach to help them learn about and understand the world around them. This could involve working individually, or as part of a group, to discover the answer to a geographical question, using a variety of geographical skills, tools and concepts.

Geographers also look at many interesting issues that face the world today; for example, different people have different viewpoints, or perspectives, about what we should do about climate change. The answer to this question might vary among individuals, within a local area or a country, or even on a world scale.

1.2.2 Inquiry skills

Have you ever noticed that young children ask many questions as they begin to learn because they are curious about the world around them? Below are some examples of questions we can call *geographical* questions:

- Why are there many different types of landscapes around the world and how are they formed?
- Where is the best place to live?
- How can we look after our water resources so we have enough for the future?
- What are the effects of tourism in different places?

Geographical inquiry skills develop your ability to collect, process and communicate information.

Acquiring geographical information

Acquiring or collecting geographical information needs to be focused and well planned. Begin a geographical inquiry by developing a problem or issue to investigate. This will be the general theme of your inquiry. Develop a few geographical questions that will help you study your issue or problem. Ensure that your questions are not so broad that they will be difficult to investigate — for example, water management in Australia — or so specific that you won't be able to find enough information to support your inquiry.

Think about how you will collect information about your inquiry. You should include both primary geographical data and information from secondary sources. Primary data is information that you have collected yourself using fieldwork. Secondary sources are data that has been collected and processed by someone else, or written by someone else. Secondary sources include websites, books and brochures. Once you have decided on the information you need, plan your investigation and carry out your fieldwork, and collate information from secondary sources.

Processing geographical information

Before you begin processing the information you have collected, you should evaluate the sources and techniques you have used to determine whether they are reliable and free from bias. Can you trust the sources of information? Did you carry out your fieldwork techniques thoroughly and with care? Present your information in a range of different forms. This might include graphs, tables, diagrams, sketch maps and annotated photographs. You might also write paragraphs explaining your results. Look at the information you have collected and reflect on your research questions. At this stage you can start to interpret the information. Did you answer your research questions? What are the answers to your research questions? Analyse the findings of your research and draw conclusions.

Communicating geographical information

You can choose to communicate your research findings in a range of ways. Consider who you will be presenting your findings to. Choose methods to communicate your information that are appropriate to your audience. Explain how you undertook your investigation and your findings. Propose actions that you think should be taken to address your problem or issue, and explain why you think this is the right course of action. If possible, take action yourself to address the geographical issue you have chosen.

1.2.3 Geographical tools

Geographers use a range of tools to help them collect information during a geographical inquiry. Geographical tools include:

- maps
- fieldwork
- graphs and statistics
- spatial technologies
- visual representations.

FIGURE 2 Collecting your own data and information

Maps

Maps are the most basic tool of the geographer as they are possibly the most effective way to locate, represent, display and record spatial information. These days, geographers are able to use and create both digital and non-digital maps.

Political maps are common; they show the boundaries of countries, states and regions, and usually show major cities and bodies of water. Topographic maps and relief maps show the shape of the land on a map. Sketch maps are hand-drawn maps that show only the most basic details. Choropleth maps or flowline maps can be used to show information about particular themes. Précis maps show a basic summary of information found on a topographic map.

It is important for geographers to develop skills in map reading to be able to use all the information that has been found. Mapping skills include being able to determine direction and use the scale of the map to determine distance between different places. Geographers use lines drawn on maps to determine and communicate the location of different places. On topographic maps, grids are used to determine the area and grid references of different places. On some maps, lines of latitude and longitude are shown to help us locate places.

FIGURE 3 Maps are a key tool for a geographer.

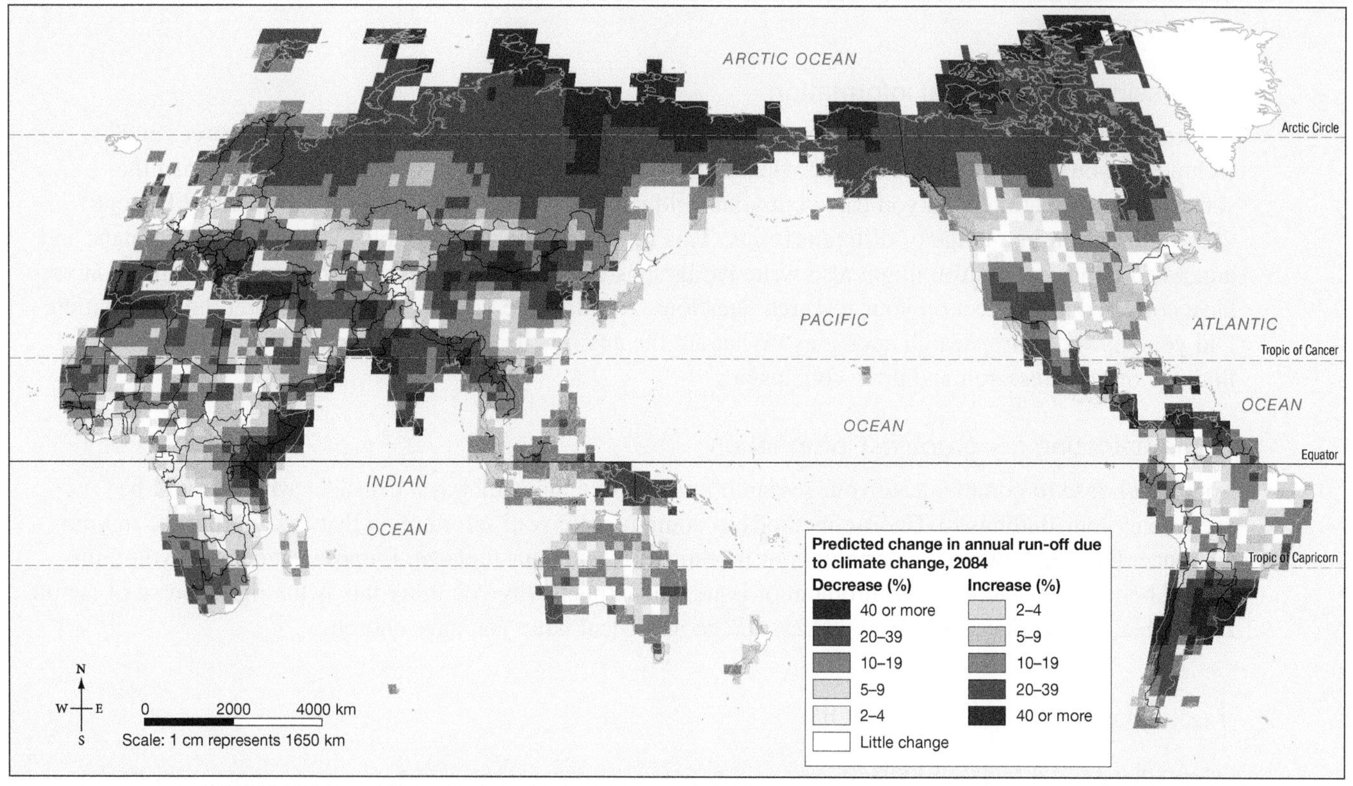

Source: Geophysical Fluid Dynamics Laboratory, National Oceanic and Atmospheric Administration, 2021

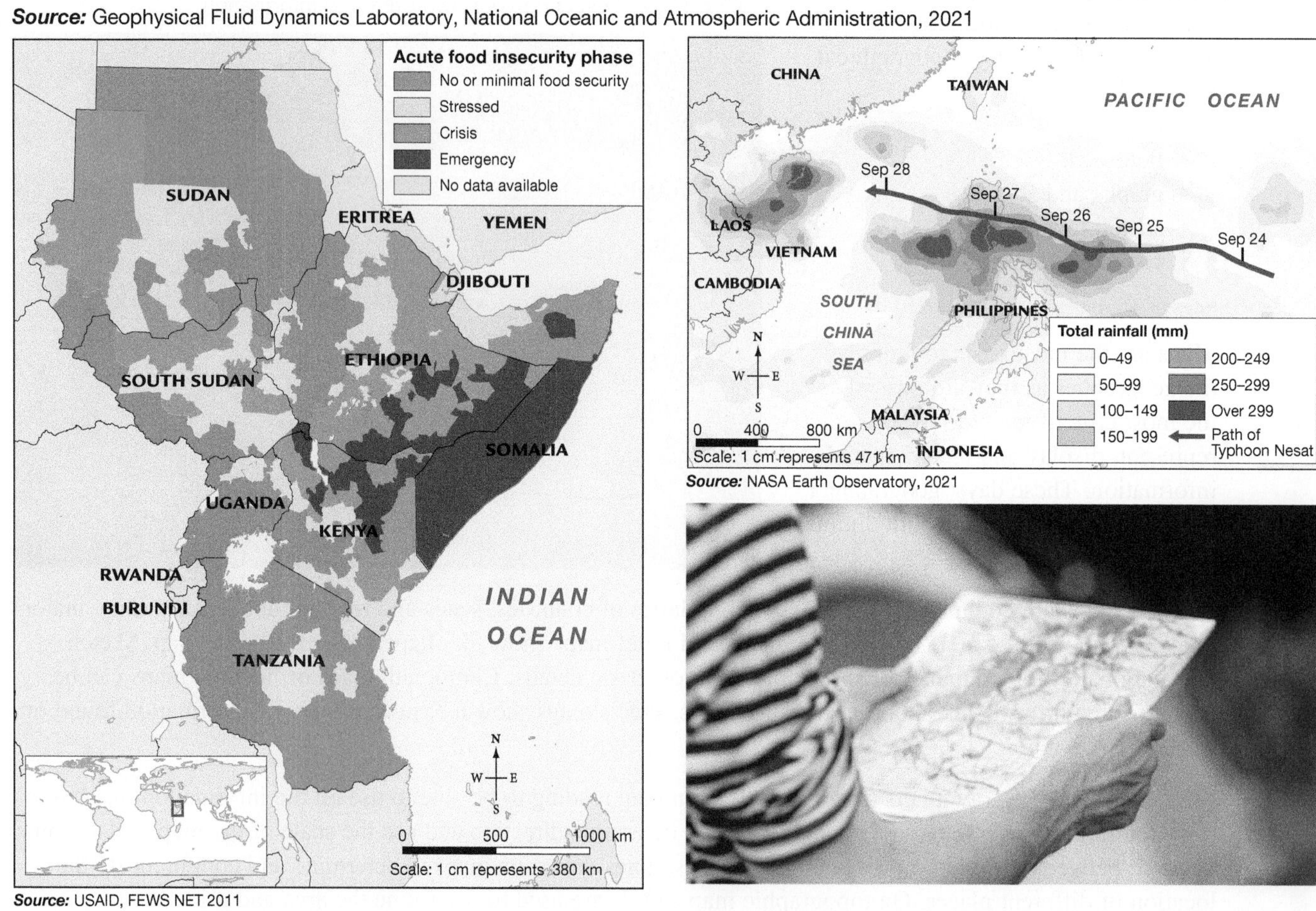

Source: USAID, FEWS NET 2011

Source: NASA Earth Observatory, 2021

FIGURE 4 A flow map shows the movement of agricultural trade around the world.

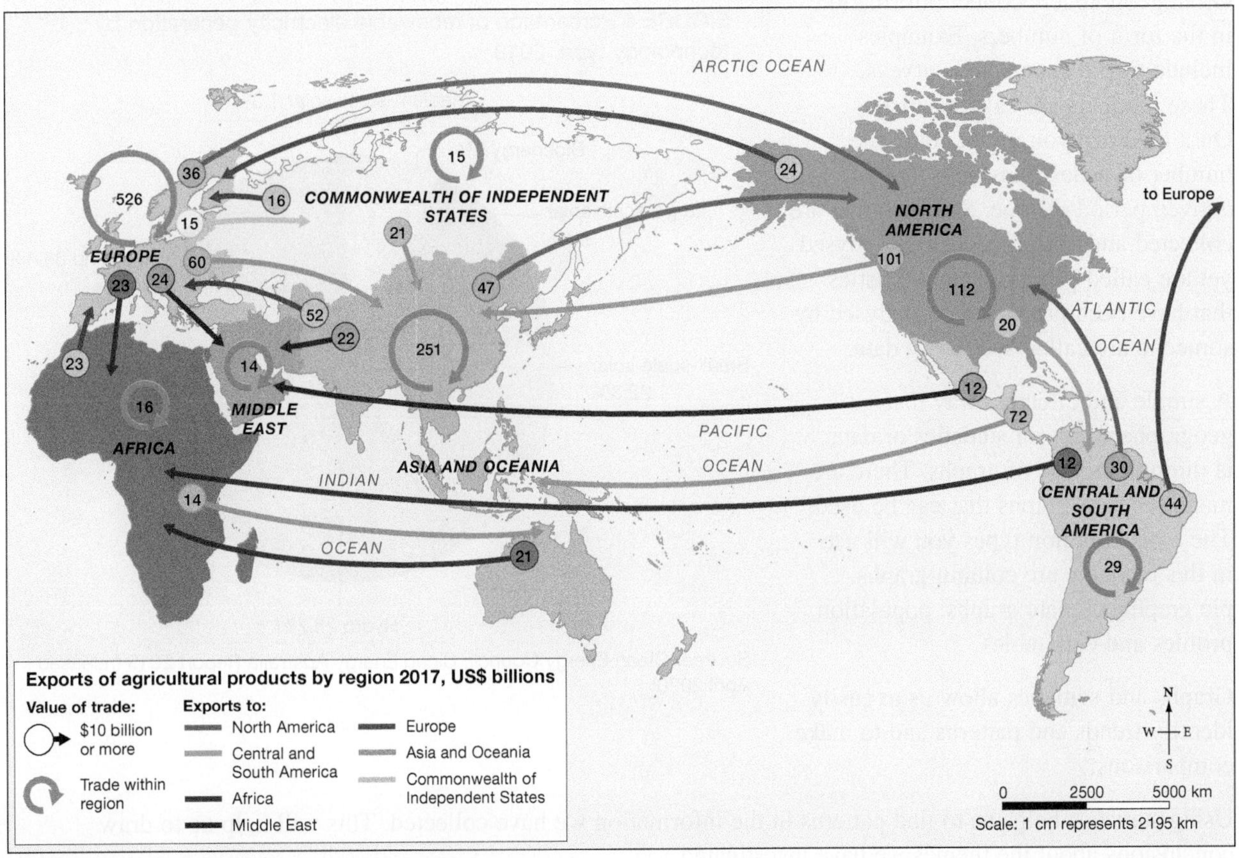

ARCTIC OCEAN
526
36
15
16
COMMONWEALTH OF INDEPENDENT STATES
15
21
24
NORTH AMERICA
to Europe
101
112
20
ATLANTIC OCEAN
EUROPE
60
23
24
52
22
47
251
14
23
16
MIDDLE EAST
AFRICA
12
72
PACIFIC OCEAN
ASIA AND OCEANIA
INDIAN OCEAN
12
30
44
CENTRAL AND SOUTH AMERICA
14
21
29
Exports of agricultural products by region 2017, US$ billions
Value of trade:
$10 billion or more
Trade within region
Exports to:
North America
Central and South America
Africa
Middle East
Europe
Asia and Oceania
Commonwealth of Independent States
N
W
E
S
0
2500
5000 km
Scale: 1 cm represents 2195 km

Source: Data from World Trade Organization, 2021

Fieldwork

There is nothing better than going into an environment, or visiting a place, that you are studying. Seeing something firsthand provides a better understanding than reading about it or looking at it in photographs. That is why fieldwork is such an important, and compulsory, part of your studies.

Fieldwork involves observing, measuring, collecting and recording information and data outside the classroom.

Fieldwork can be undertaken within the school grounds, around the local neighbouring area or at more distant locations. We can use tools such as weather instruments, identification charts, photographs and measuring devices to collect information about our environment.

Sometimes it may be necessary to use information and communication technology to undertake virtual fieldwork.

FIGURE 5 Conducting a survey in the field

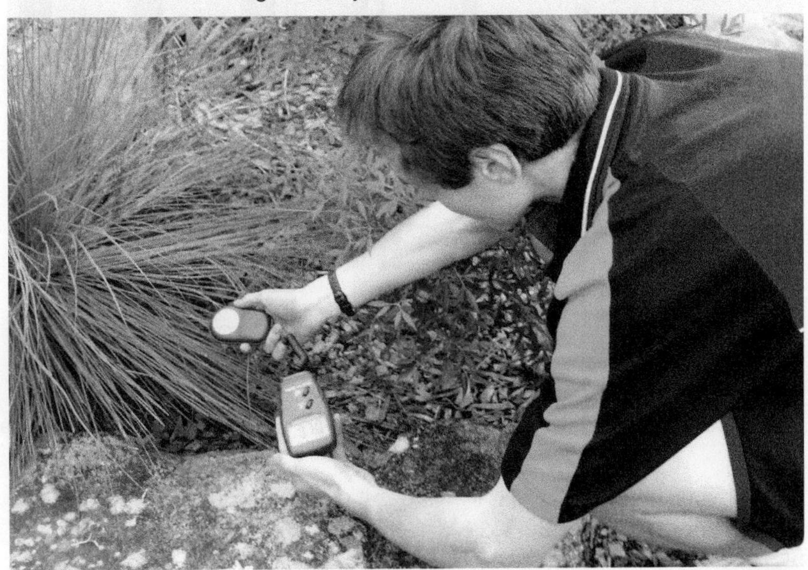

Graphs and statistics

Often geographers collect information in the form of numbers. Examples include traffic counts and surveys. These numbers are called statistics. On a field trip you might count the number of pedestrians on a footpath in a given period of time. Statistics that are collected and not processed or analysed yet are called primary data. Statistics that have been processed or analysed by someone are called secondary data.

A simple and effective way that geographers present statistics or data is through the use of graphs. There are many types of graphs that can be used. The most common types you will use in this resource are column graphs, pie graphs, climate graphs, population profiles and data tables.

Graphs and statistics allow us to easily identify trends and patterns and to make comparisons.

FIGURE 6 Percentage of renewable electricity generation by technology type, 2019

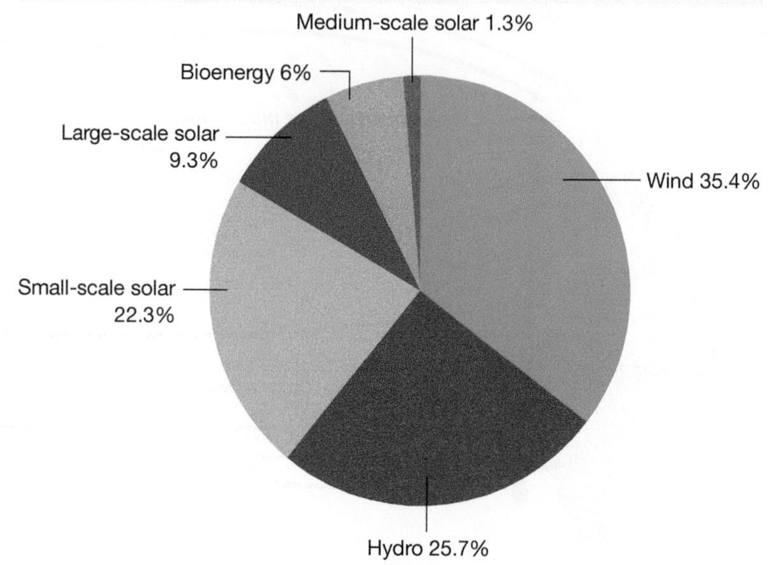

Source: Clean Energy Council, Clean Energy Australia Report 2019 (released April 2020)

Using statistics helps us to find patterns in the information we have collected. This will help us to draw conclusions about the themes we have investigated.

FIGURE 7 The growth of urban populations over time

100
90
80
70
60
% Urban
50
40
30
20
10
0

1960 1970 1980 1990 2000 2010 2020 2030* 2040*

Year

Africa Asia Europe Latin America and the Caribbean North America Oceania

(* projected)

Source: World Population Prospects: 2015 Revision, United Nations

Spatial technologies

Spatial technologies involve using satellite information and virtual maps to explore and record information. When you use Global Positioning System (GPS) or Google Earth you are using a form of spatial technology. Spatial technologies are any software or hardware that interacts with real-world locations. Geographic information systems (GIS) are another commonly used spatial technology. They help us analyse, display and record spatial data.

FIGURE 8 A false-colour satellite image of the Mt Lofty Ranges, South Australia

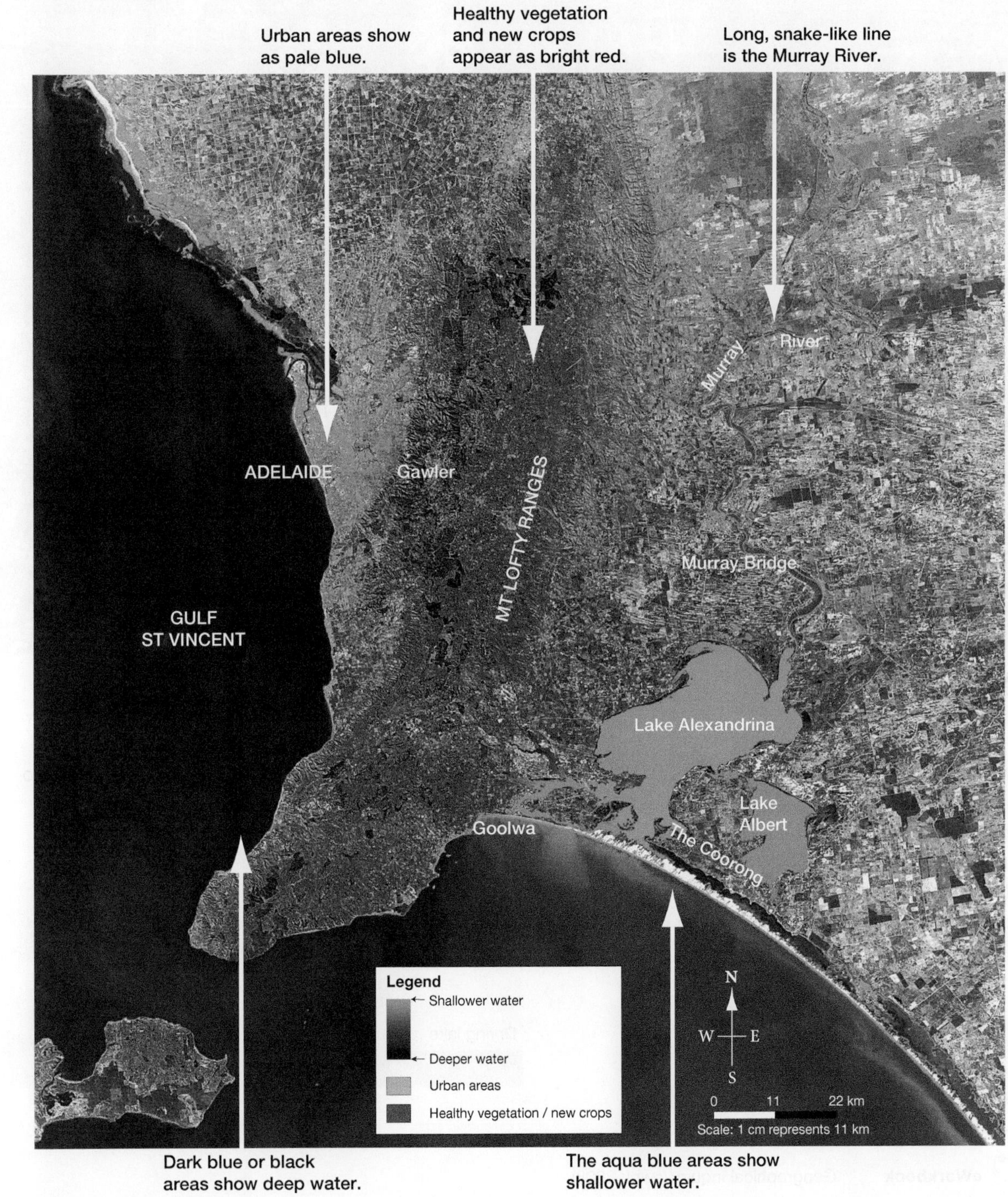

Source: © Commonwealth of Australia. Geoscience Australia 1982.

FIGURE 9 Satellite image of Canberra, by GeoEye, 26 September 2011. Satellite images show a realistic view like a photograph, providing a bird's-eye view of a place.

Source: © GeoEye

Visual representations

A visual representation is an effective way of showing complex information using pictures, symbols and diagrams. Examples of visual representations include photographs, field sketches, cartoons and infographics. They are used to display, analyse and communicate geographical data and information.

FIGURE 10 The process of erosion while Lake Mungo was drying out

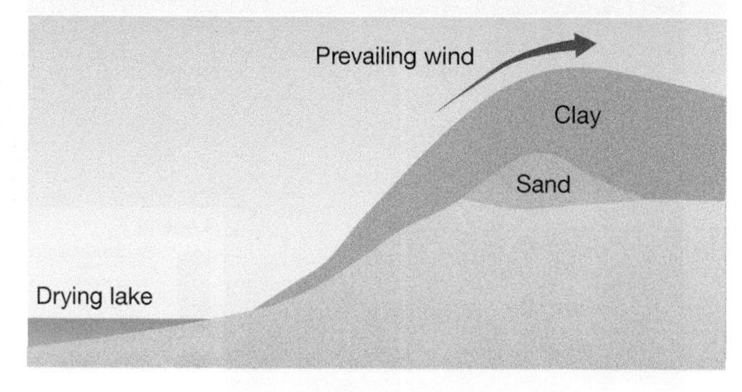

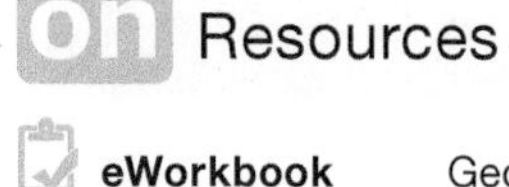

Resources

eWorkbook Geographical inquiry (ewbk-8608)

Video eLesson Geographical inquiry — Key concepts (ewbk-5080)

1.3 Geographical concepts

LEARNING INTENTION

By the end of this subtopic, you will be able to identify and summarise the seven key geographical concepts.

1.3.1 Overview

Geographical concepts help you to make sense of your world. By using these concepts you can investigate and understand the world you live in, and you can use them to imagine a different world. The concepts help you to think geographically. There are seven major concepts: *space, place, interconnection, change, environment, sustainability* and *scale*.

FIGURE 1 A way to remember these seven concepts is to think of the term SPICESS.

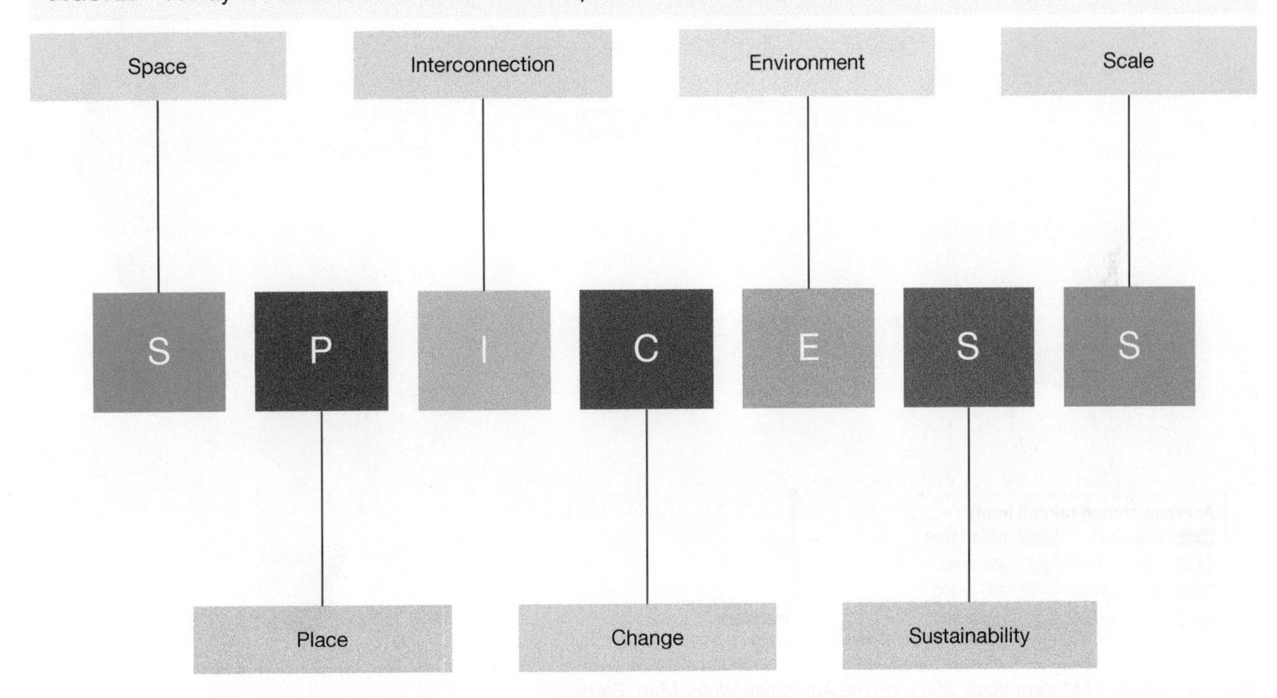

A way to remember these seven concepts is to think of the term SPICESS.

In this resource you will use and apply the seven concepts to investigate four topics: *Landscapes and landforms, Place and liveability, Water in the world* and *Interconnections*.

1.3.2 What is space?

Everything has a location on the **space** that is the surface of the Earth, and studying the effects of location, the distribution of things across this space, and how the space is organised and managed by people, helps us to understand why the world is like it is.

A place can be described by its **absolute location** (latitude and longitude) or its **relative location** (in what direction and how far it is from another place).

space geographical concept concerned with location, distribution (spread) and how we change or design places

absolute location the latitude and longitude of a place

relative location the direction and distance from one place to another

FIGURE 2 The amount of rain that falls varies across the space of Australia, as this rainfall map shows.

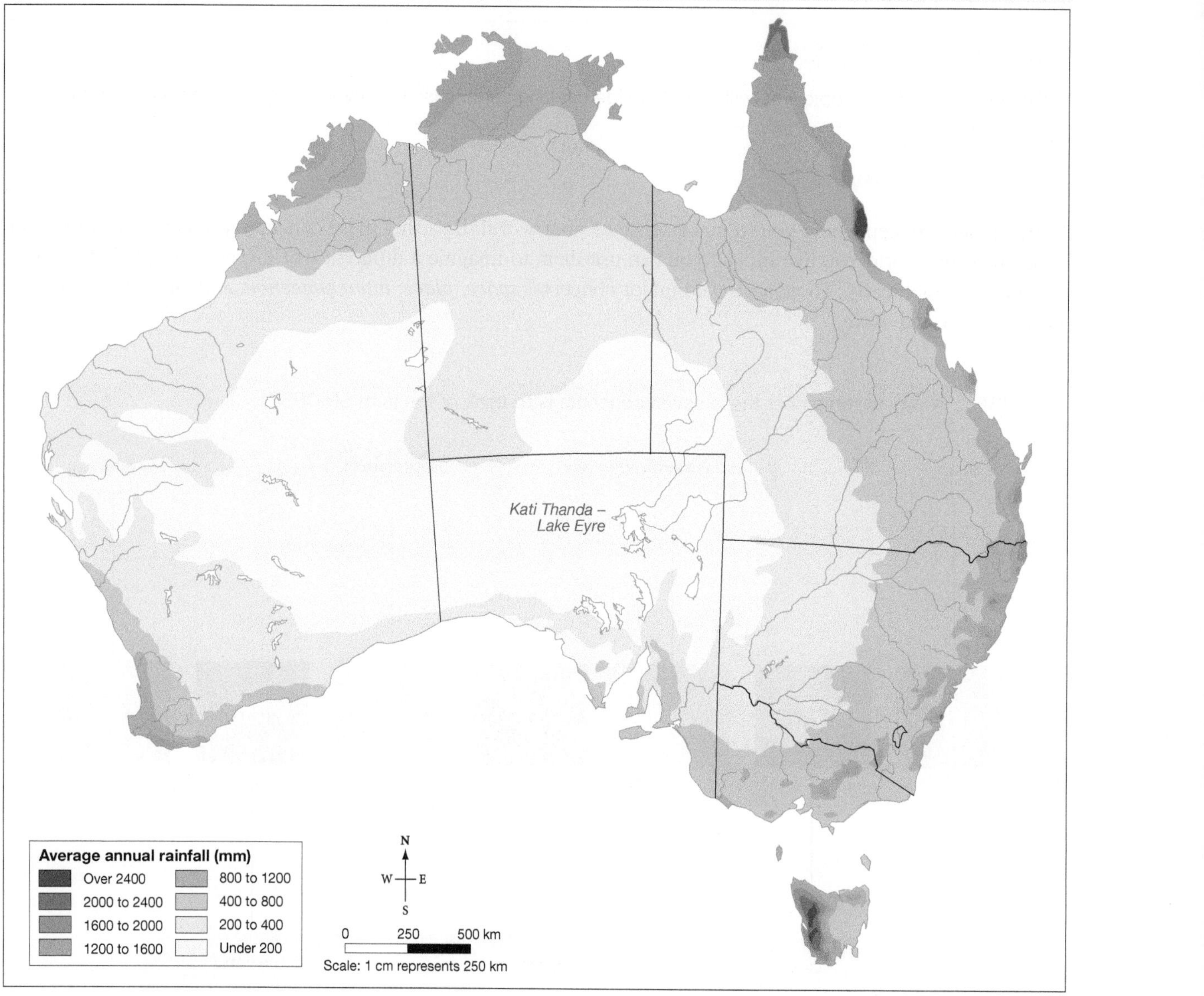

Kati Thanda – Lake Eyre

Average annual rainfall (mm)
Over 2400
2000 to 2400
1600 to 2000
1200 to 1600
800 to 1200
400 to 800
200 to 400
Under 200

N W E S

0 250 500 km
Scale: 1 cm represents 250 km

Source: Bureau of Meteorology, 2003, on the Australian Water Map, Earth

1.3.2 ACTIVITIES

Refer to **FIGURE 2** and an atlas.

1. Use an atlas to give the absolute location (latitude and longitude) of the capital city of the state/territory in which you live.
2. In which direction and how far is your capital city from Alice Springs (relative location)?
3. Describe the spatial distribution of capital cities in Australia.
4. Describe the distribution of rainfall across Australia. Why might one place have more or less rainfall than another?
5. How does rainfall (or lack of rainfall) help explain the distribution of Australia's major cities? What is the relationship between rainfall and population location?
6. Find where you live on the maps. How is the location of your place influenced by rainfall and rainfall variability?

1.3.3 What is place?

The world is made up of places, so to understand our world we need to understand its places by studying their variety, how they influence our lives and how we create and change them.

You often have mental images and perceptions of **places** — rich and poor cities, suburbs, towns or neighbourhoods — and these may be very different from someone else's perceptions of the same places.

1.3.3 ACTIVITIES

Refer to **FIGURE 3**.

1. Where in the world is this place located?
2. What effects have people had on this place?
3. List the differences you observe in the way people live on each side of this settlement.
4. How is this place similar to or different from the place where you live?
5. What decisions could be made to improve or change this place?
6. How might the environment of this place affect the people who live there?
7. How might the place where these people live affect their lives?

FIGURE 3 The Paraisópolis favela (slum), home to 60 000 people, is situated next to the gated complexes of the wealthy Morumbi district of São Paulo.

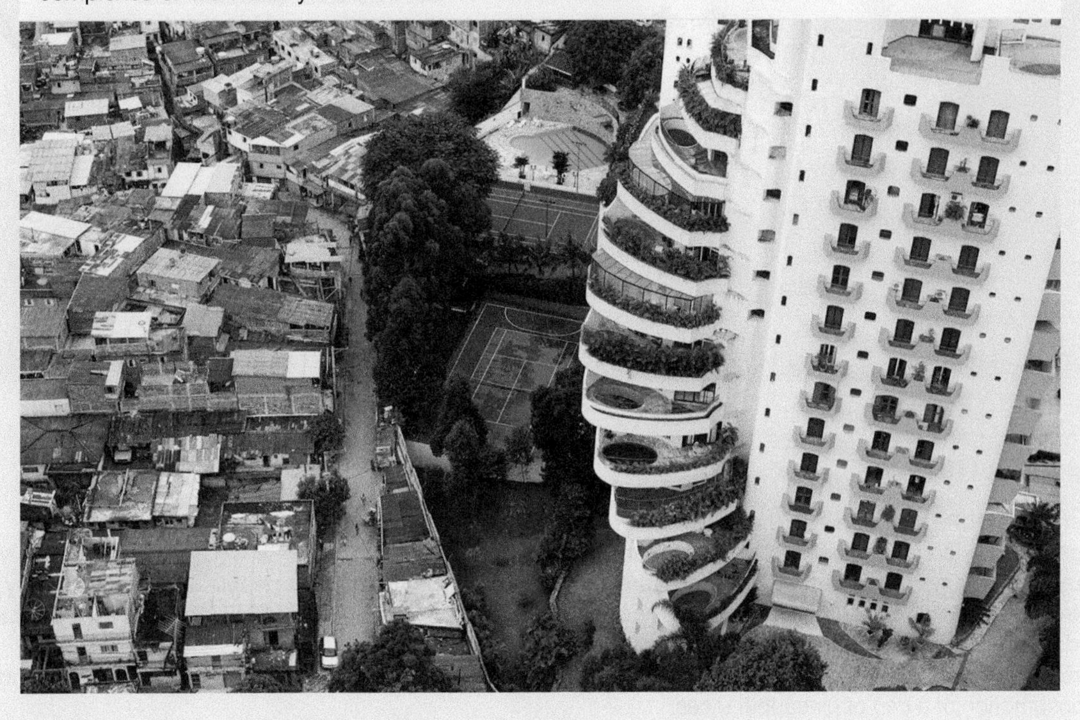

1.3.4 What is interconnection?

People and things are connected to other people and things in their own and other places, and understanding these connections helps us to understand how and why places are changing.

Individual geographical features can be **interconnected** — for example, the climate within a place or biome, such as a tropical rainforest, can influence natural vegetation, and the removal of this vegetation can affect climate. People can be interconnected to other people and other places via employment, communications, sporting events or culturally. The manufacturing of a product may create interconnections between suppliers, manufacturers, retailers and consumers.

place geographical concept concerned with why somewhere is important and the characteristics that make it unique or interesting

interconnection geographical concept concerned with how natural places, processes and features can change or affect each other and people, or be changed or affected by people

FIGURE 4 The water cycle shows many interconnections.

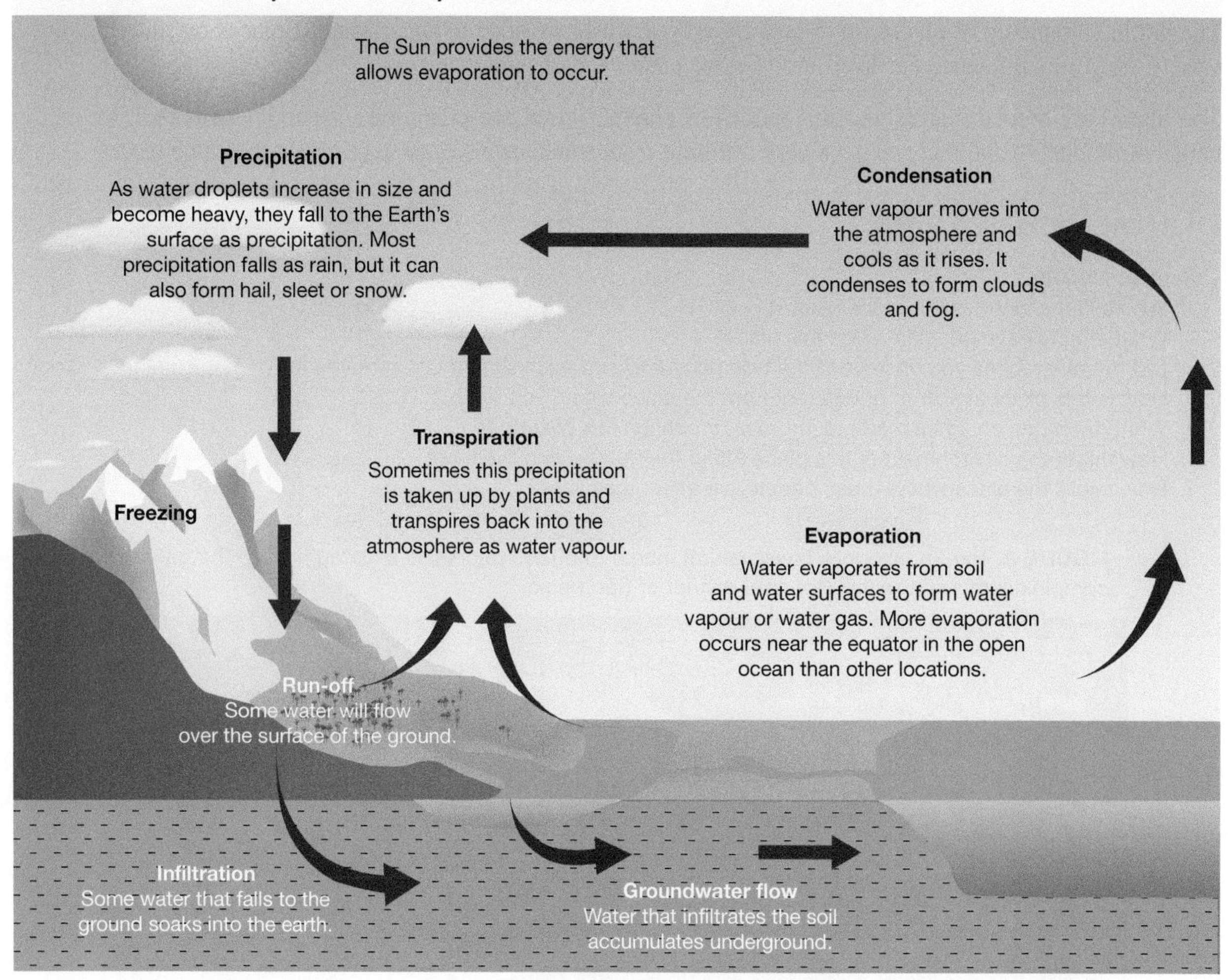

1.3.4 ACTIVITIES

1. Refer to **FIGURE 4**. Summarise the water cycle using only words.
2. Describe the interconnection between the Sun and levels of evaporation.
3. Draw a diagram to show the ways that you are interconnected with three other people, and the ways they are interconnected with each other. You might choose family, friends or people you know from school. Annotate (add notes to) your diagram, like in **FIGURE 4**, to outline each connection.
4. Create a diagram to show the interconnections that could occur for the growing, manufacturing, sales and consumption of a can of pineapple slices.

1.3.5 What is change?

The concept of change is about using time to better understand a place, an environment, a **spatial** pattern or a geographical problem.

The concept of **change** involves both time and space — change can take place over a period of time, or over an area of space. The time period for change can be very short (for example, the impact of a flash flood) or span thousands or millions of years (for example, the development of fossil fuel resources).

The use of technology can result in rapid change — think of the explosions at a mining site that reveal mineral seams.

spatial how things relate to each other in an area or location

change geographical concept concerned with examining, comparing and predicting the impact of processes over time

The degree of change occurring can be used to predict, or plan for, actual or preferred futures (the future we hope for).

FIGURE 5 Port Douglas, 60 kilometres north of Cairns, was a busy port in the 1870s, with a population of more than 10 000 people. The mining that had attracted people to this hot, wet area did not last. By the 1960s, the population was only 100. In the 1980s, road and air access to the town improved and tourist numbers to the area grew. The permanent population is now about 3500 people. During the peak holiday season this number increases by four times.

1.3.5 ACTIVITIES

Refer to **FIGURE 5**.

1. How and why has the population of Port Douglas changed over time?
2. Which economies have declined and grown in the Port Douglas area?
3. How has technology (transport links) been important in the development of this place?
4. How do you think the changes have affected the environment, businesses and economy in the area?
5. List five changes you can observe in the three photographs of Port Douglas.
6. Use evidence from the photographs to decide if the natural environment has changed faster than the human environment.
7. How would you like this place to be in 50 years? What changes need to occur for your preferred future to come about?

1.3.6 What is environment?

People live in and depend on the environment, so it has an important influence on our lives.

The **environment**, defined as the physical and biological world around us, supports and enriches human and other life by providing raw materials and food, absorbing and recycling wastes, and being a source of enjoyment and inspiration to people.

FIGURE 6 Pacific Islanders use traditional methods to fish sustainably.

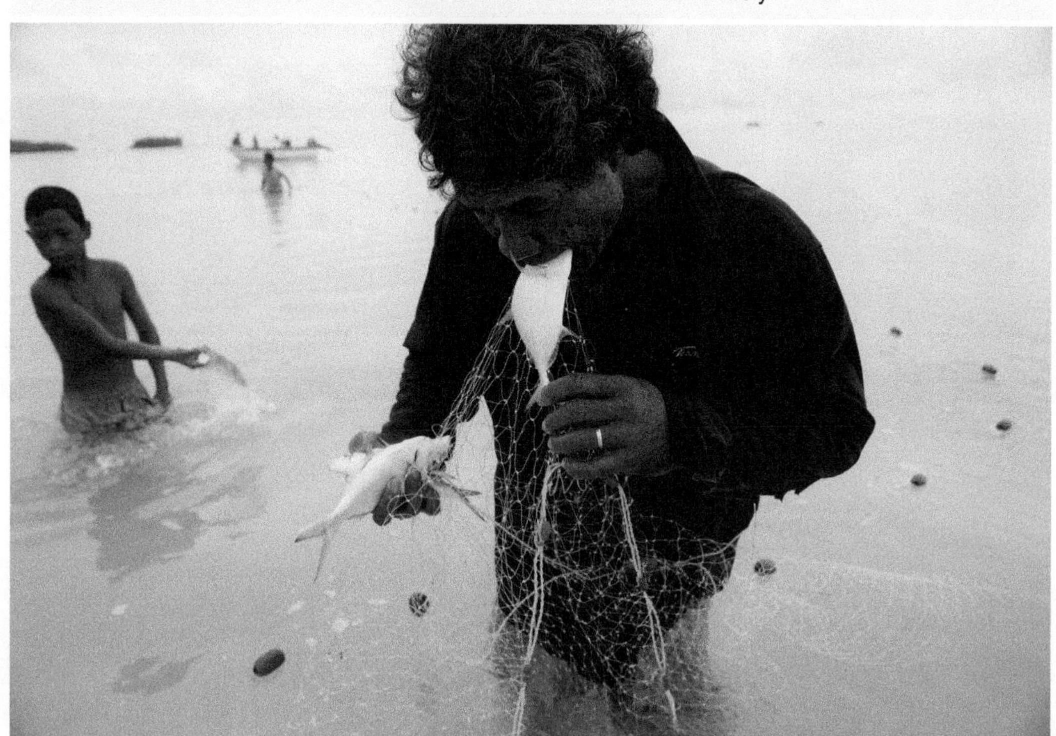

1.3.6 ACTIVITIES

Refer to **FIGURE 6**.

1. Does the photograph of Pacific Islanders fishing depict a natural environment or a human environment? Justify your response.
2. Does this environment appeal to you? Would you like to visit this place? Why? Why not?
3. Which resource/s do you think people would obtain from this environment?
4. Describe how these people are fishing. Why might this be sustainable?
5. List the impacts on this environment if a factory was built on the edge of the water.
6. How have people changed this environment (for better or worse)? What are the positive and the negative aspects of this?
7. How might technology change this environment to make it less sustainable?

1.3.7 What is sustainability?

Sustainability is about maintaining the capacity of the environment to support our lives and those of other living creatures.

Sustainability involves maintaining and managing our resources and environments for future generations. It is important to understand the causes of unsustainable situations to be able to make informed decisions about the best way to manage our natural world.

environment geographical concept concerned with why the natural world plays an important part in human life

sustainability geographical concept concerned with how to protect the environment and life on Earth

FIGURE 7 The unsustainable nature of fishing

1.3.7 ACTIVITIES

Refer to **FIGURE 7**.

1. What does this cartoon tell us about the sustainability of fishing?
2. How has modern technology helped to reduce world fish populations?
3. Suggest reasons that fishing can be considered an unsustainable practice.
4. Give reasons that some fish species are over-fished while others are not.
5. Explain why it might be more difficult to manage a resource in the ocean compared to a resource found on land.

1.3.8 What is scale?

When we examine geographical questions at different spatial levels we are using the concept of **scale** to find more complete answers. That means we are looking at issues and how they affect different ranges or sizes of place: personal, local, regional, national or global.

Local events can have global outcomes; for example, removing areas of forest at a local scale can have an impact on climate at a global scale. A policy at a national scale, such as forest protection, can have an impact at a local scale, such as the protection of an endangered species.

scale geographical concept concerned with the different levels or ranges at which issues can be examined

1.3.8 ACTIVITIES

Refer to **FIGURES 8 (a)** and **(b)**, and **FIGURE 9**.

1. Zoom in on the areas on these maps, and determine whether you would see more or less detail.
2. List the extra information and detail you see on Annette's map compared to Jayden's map. Determine which map gives more information.
3. The railway map is at a regional scale. Identify the region of Australia it is showing.
4. Use the scale to measure the longest straight stretch of railway shown on the railway map. Determine the length. Why is it significant?
5. Determine the main information each map is trying to show.
6. Refer to the railway map. Explain the relationships between the location of settlements and the location of the railway.

FIGURE 8 Mental map of Jayden's local place (a) by Jayden and (b) by Annette, Jayden's mother

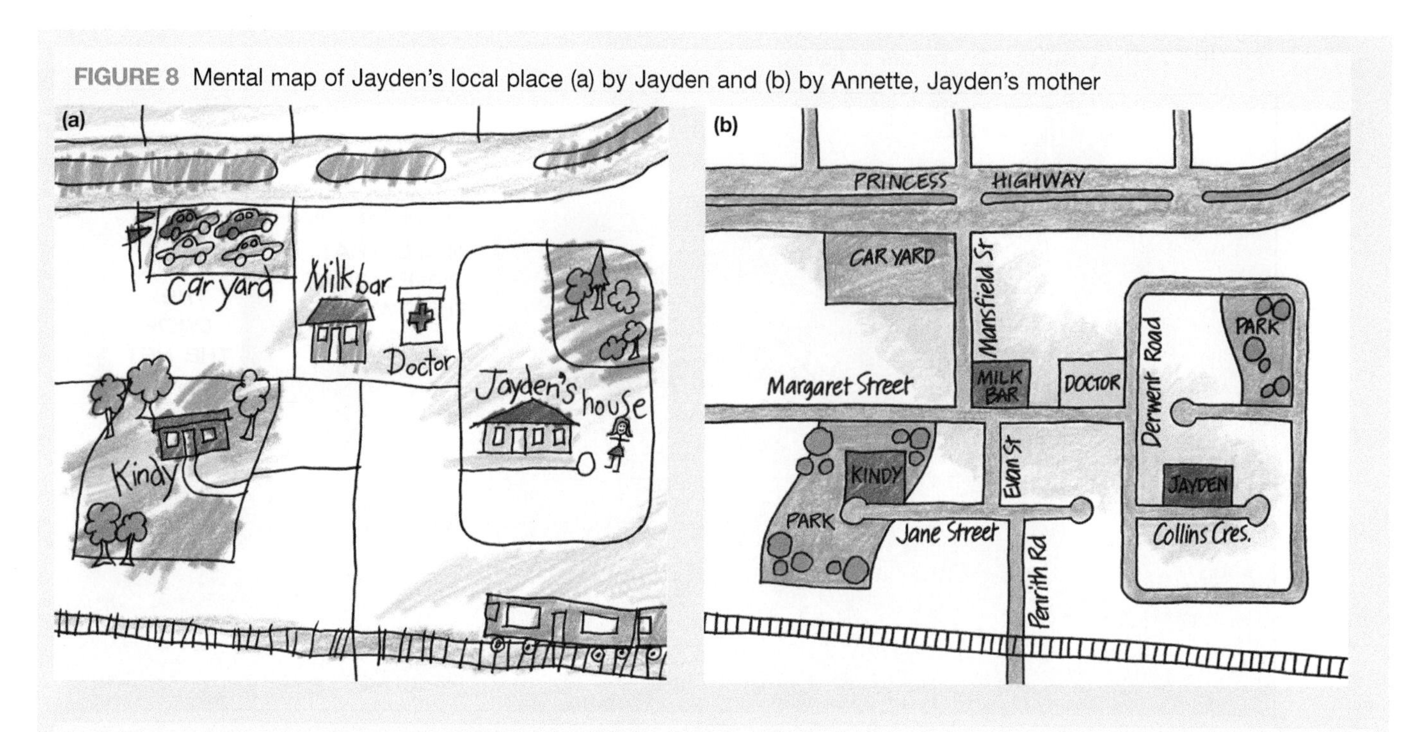

FIGURE 9 Railway route and main settlements between Sydney and Perth

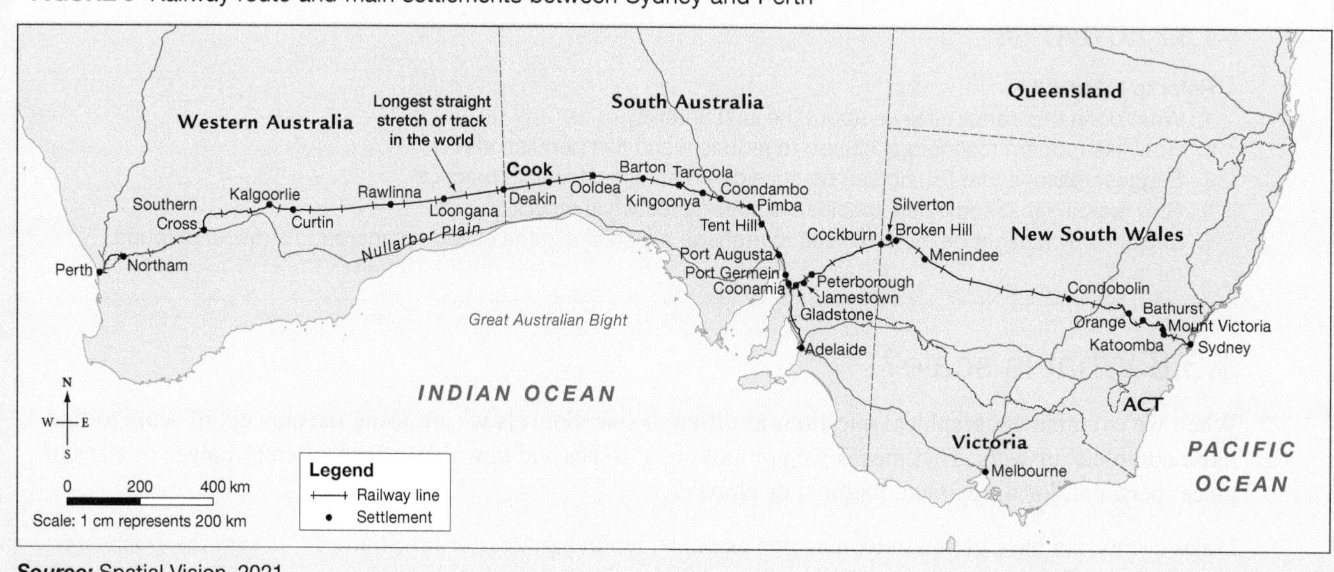

Source: Spatial Vision, 2021

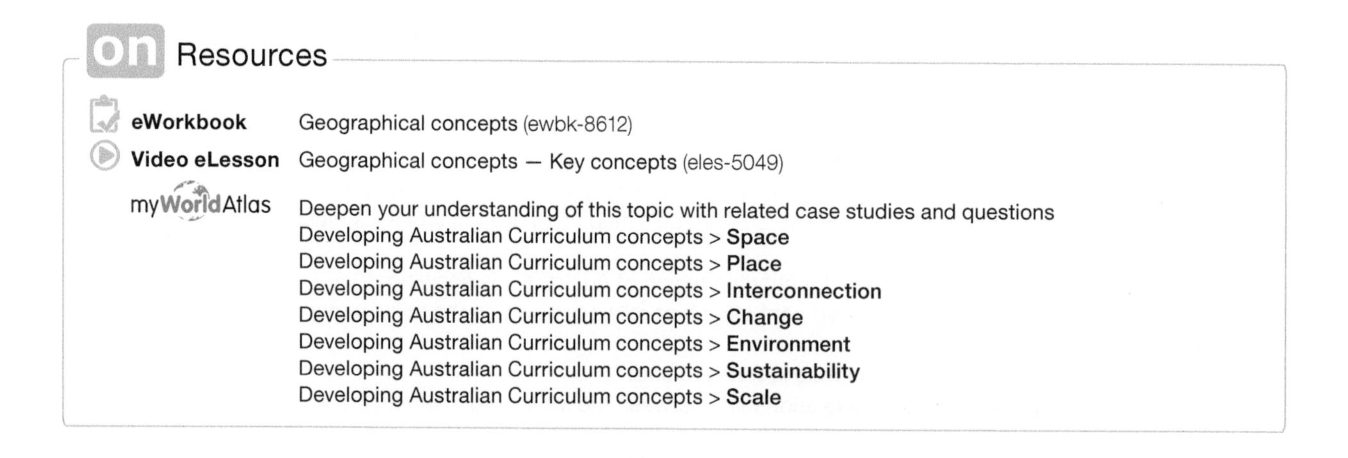

on Resources

eWorkbook Geographical concepts (ewbk-8612)

Video eLesson Geographical concepts — Key concepts (eles-5049)

myWorldAtlas Deepen your understanding of this topic with related case studies and questions
Developing Australian Curriculum concepts > **Space**
Developing Australian Curriculum concepts > **Place**
Developing Australian Curriculum concepts > **Interconnection**
Developing Australian Curriculum concepts > **Change**
Developing Australian Curriculum concepts > **Environment**
Developing Australian Curriculum concepts > **Sustainability**
Developing Australian Curriculum concepts > **Scale**

1.4 Work and careers in Geography

LEARNING INTENTION

By the end of this subtopic, you will be able to list some of the careers that Geography can lead to, and outline some of the skills these jobs require.

1.4.1 Connecting Geography to school and work

The concepts and skills that you will use will not only help you in Geography, but can also be applied to everyday situations, such as finding your way from one place to another. Studying Geography may even help you in a future career here in Australia or overseas.

Throughout the year you will be studying topics that will give you a better understanding of the world around you – both the local and global environment. You will be investigating issues that need to be addressed and options for the future.

Skills you need for a job

Many questions come up during a typical Geography class, such as the ones in **TABLE 1**. These questions need to be answered in the real world by people in a wide variety of occupations that have links with Geography.

TABLE 1 Questions to consider in Geography

What might you want to know?	Who finds the answers?
How can we protect our wildlife?	Park ranger; planner; environmental manager
Where should we build more houses?	Urban planner; demographer
How can we survive droughts and floods?	Civil engineer
Does our town really have enough water?	Coastal engineer; hydrologist; cartographer
Do we have good quality drinking water?	Chemist; hydrologist
How should we deal with air pollution problems?	Environmental scientist/manager
How do we help people in other countries?	Air force/navy/army officer; Red Cross, World Vision and other aid agencies
How do we make sure our homes are sustainable?	Architect; landscape architect; civil engineer/ construction manager; town planner; real estate salesperson

1.4.2 Geography careers

A great part of studying Geography is being able to explore the many occupations and areas that it opens up. **TABLE 2** shows some occupations that you may not have thought studying Geography could lead you into.

A wide range of exciting new jobs are developing in the spatial sciences, which use geographical tools such as GPS, GIS, satellite imaging and surveying. These tools help people make important decisions about managing and planning places and resources, such as managing water in the Middle East or designing a new housing estate here in Australia. These skills and occupations will be an important part of working as a global citizen.

TABLE 2 Careers in Geography

I would enjoy ...	I could become a ...
Working outdoors	Surveyor, mining engineer, geologist, cartographer
Helping other people	Paramedic, navy officer, firefighter, tour guide
Working indoors	Land economist, landscape designer, geoscientist
Designing communities	Urban planner, architect, demographer
Working in parks and gardens	Horticulturalist, landscape architect, park ranger
Discovering new things	Geologist, meteorologist, environmental scientist

FIGURE 1 Studying Geography can open up a range of careers

1.4 ACTIVITIES

1. Consider spatial technologies and work and enterprise careers of the future. Which geographical tools do you predict will be used by:
   - a meteorologist
   - a naval officer
   - an airline pilot
   - a farmer?
2. We can develop a better understanding of work and enterprise by exploring what others have to say about their careers. Use the **Geography careers** weblink in your online Resources panel to help you locate one male and one female geographer working as local or global citizens.
   - What career path did they follow?
   - How did they include their passion for geography in their career journey?
   - What advice did they share about their career journey?

on Resources

eWorkbook Work and careers in Geography (ewbk-10245)

Video eLesson Work and careers in Geography — Key concepts (eles-5081)

1.5 SkillBuilder — How to read a map

online only

LEARNING INTENTION

By the end of this subtopic, you will be able to recognise key features of maps.

The content in this subtopic is a summary of what you will find in the online resource.

1.5.1 Tell me

What are maps?

Maps represent parts of the world as if you were looking down from above. A mapmaker (or cartographer) simplifies the view from an aerial photograph or satellite image. They use colours and symbols on the map to show how features such as roads, rivers and towns are organised in a spatial way.

1.5.2 Show me

eles-1634

Step 1

When you read a map, check each of the six BOLTSS features to get to know what information is being presented, and where the information has come from.

- **B**order — to show the boundaries of the map
- **O**rientation — to show direction on the map, usually which way is north
- **L**egend — to explain the symbols and colours used (also called the key)
- **T**itle — to describe what the map is showing
- **S**cale — to indicate distances on the map
- **S**ource — to explain the source of the information for the map

Step 2

Read the title of the map carefully to determine what it shows and identify important factors that might affect your interpretation, such as the date of the information that has been mapped. Read the source of the map to see who has provided the information. Is the source reliable? Can you trust the information they are providing? Examine the scale and any notes underneath the map to help you interpret the information.

Step 3

Examine the legend (also called a key) to understand what is shown on the map in more detail. Read each item in the legend, then find examples on the map. For example, in **FIGURE 1**, closed forest is shown as the darkest green colour in the key. This colour is found in about half the area of Tasmania, in small linear tracts along parts of the east coast of mainland Australia and in the Otway region of Victoria.

Step 4

Identify and describe the patterns shown in the whole map, in parts of the map (e.g. states, countries or regions), and any parts that do not fit the patterns (anomalies).

1.5.3 Let me do it

int-3130

Go to learnON to access the following additional resources to help you build this skill:

- a longer explanation of this skill and its application in Geography (Tell me)
- a video showing the step-by-step process of this skill (Show me)
- an activity and interactivity for you to practise this skill (Let me do it)
- self-marking questions to help you understand and use this skill.

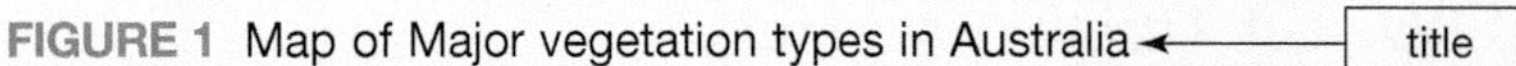

FIGURE 1 Map of Major vegetation types in Australia

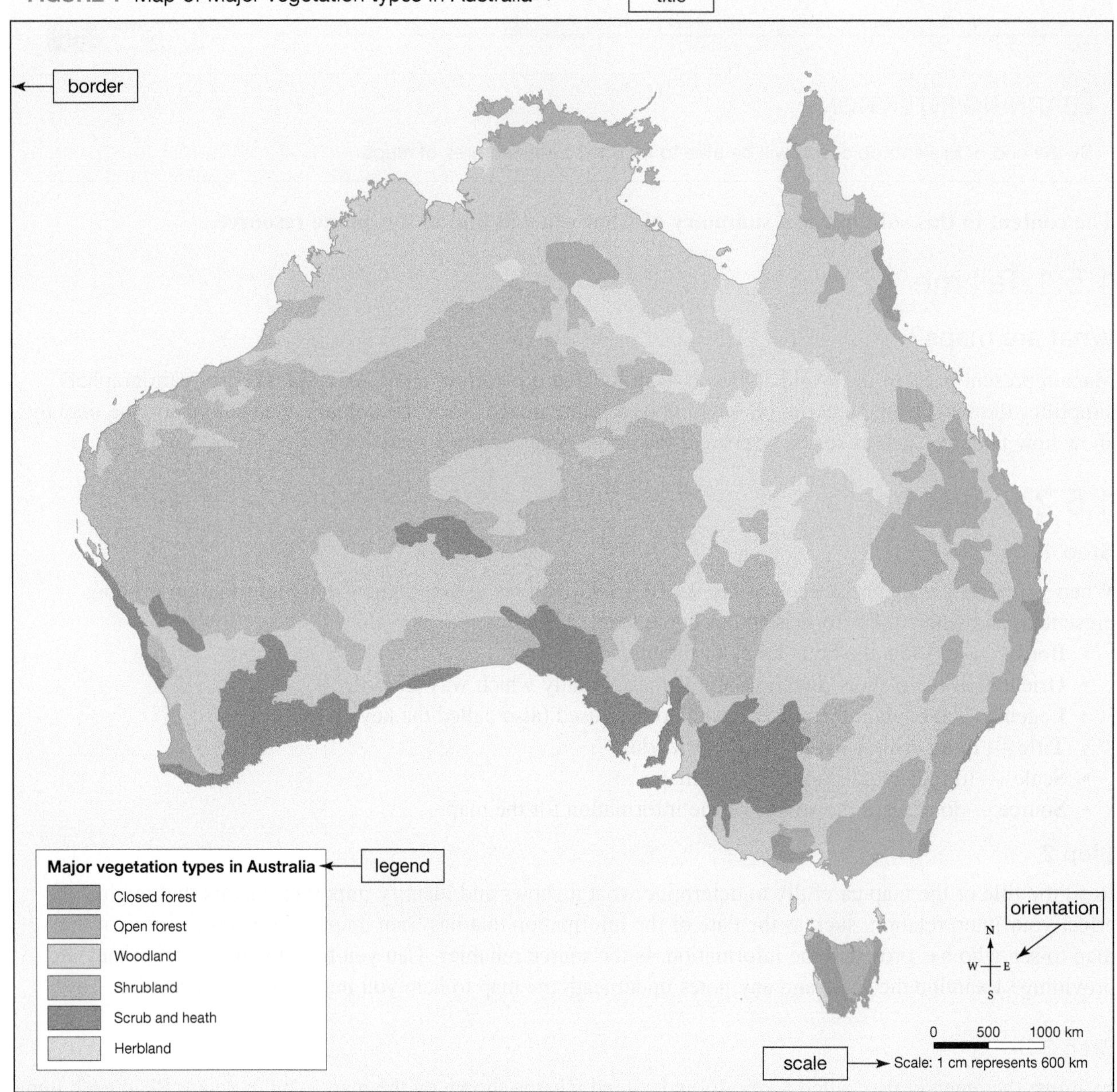

Source: Map taken from http://www.anbg.gov.au/aust-veg/veg-map.html. Reproduced with permission from Sydney University Press, 2021

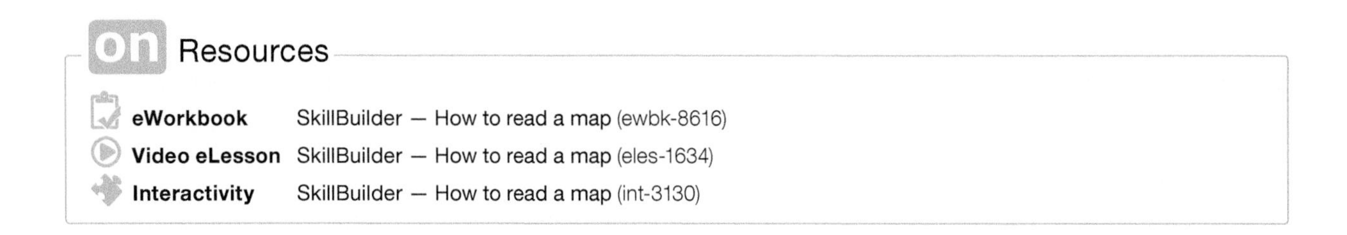

on Resources

eWorkbook SkillBuilder — How to read a map (ewbk-8616)

Video eLesson SkillBuilder — How to read a map (eles-1634)

Interactivity SkillBuilder — How to read a map (int-3130)

1.6 Writing skills in Geography

online only

LEARNING INTENTION

By the end of this subtopic, you will know where to find templates and worksheets in your online resources to help you write well in Geography.

Communicating your ideas to other people is an important part of your learning. How can your teachers assess whether you understand the ideas and skills you are learning if you can't communicate them? More importantly, there are many situations in life where you have to communicate your ideas to other people in writing. Want to apply to build an extension to your house? Need to access help from the government after a bushfire? Want to protest against a new mine in your area? Most likely, you will need to do these things in writing.

FIGURE 1 Need help with your writing in Geography?

Whether you are reading this in print or online, you have access to a range of SkillBuilder worksheets and templates online to help you improve your writing skills.

FIGURE 2 Help can be found at the click of a button — go to your online resources for writing templates and help sheets!

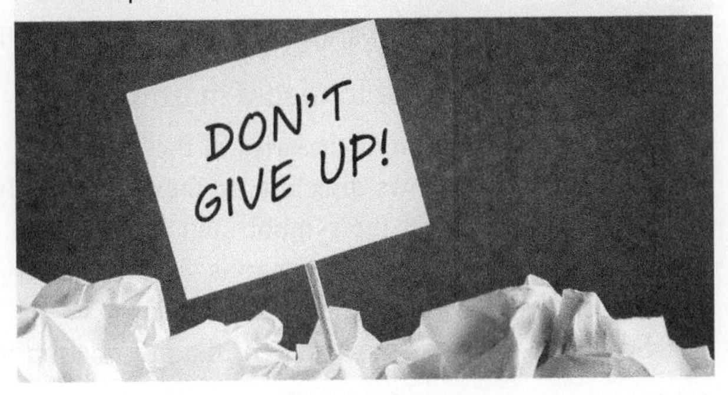

Go to your online Resources panel to download Word documents to help you:

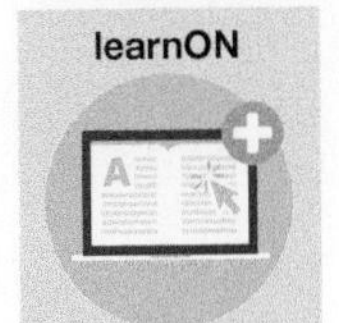

- take notes
- structure a paragraph
- interpret a short answer question
- evaluate a source
- analyse an essay question
- write an essay
- create presentation slides.

on Resources

eWorkbook SkillBuilder — Notetaking (ewbk-8622)
SkillBuilder — Structuring a paragraph (ewbk-8626)
SkillBuilder — Interpreting short answer questions (ewbk-8630)
SkillBuilder — Evaluating a source (ewbk-8634)
SkillBuilder — Analysing an essay question (ewbk-8638)
SkillBuilder — Structuring an essay (ewbk-8642)
SkillBuilder — Creating presentation slides (ewbk-8646)

Video eLessons SkillBuilder — Notetaking (eles-5042)
SkillBuilder — Structuring a paragraph (eles-5043)
SkillBuilder — Interpreting short answer questions (eles-5044)
SkillBuilder — Evaluating a point of view (eles-5045)
SkillBuilder — Analysing an essay question (eles-5046)
SkillBuilder — Structuring an essay (eles-5047)
SkillBuilder — Creating presentation slides (eles-5048)

1.7 Review

1.7.1 Key knowledge summary

1.2 Geographical inquiry

- Geography is a way of exploring, analysing and understanding this world of ours, especially its people and places.
- Geographers use what is called an 'inquiry' approach. This means that you will investigate questions by collecting, analysing and interpreting information and data in order to develop your own understanding and draw your own conclusions.
- Geography is a way of thinking and a way of looking at the world. One of the key tools geographers use is a map.
- Maps contain a lot of information about people and places. As a geographer you will produce your own maps and spatial information, by hand or digitally.

1.3 Geographical concepts

- The acronym SPICESS helps you remember the seven geographical concepts:
  - space
  - place
  - interconnection
  - change
  - environment
  - sustainability
  - scale.

1.4 Work and careers in Geography

- Many jobs in a wide variety of occupations have links with Geography.

1.5 SkillBuilder — How to read a map

- **B**order — shows the boundaries of the map
- **O**rientation — shows direction on the map, usually which way is north
- **L**egend — explains the symbols and colours used (also called the key)
- **T**itle — describes what the map is showing
- **S**cale — indicates distances on the map
- **S**ource — explains the source of the information for the map

1.6 Writing skills in Geography

- Communicating your ideas well is an important skill in Geography and in life.
- Find templates and SkillBuilders to help in your eWorkbook in the online Resources panel.

1.7.2 Key terms

absolute location the latitude and longitude of a place

change geographical concept concerned with examining, comparing and predicting the impact of processes over time

environment geographical concept concerned with why the natural world plays an important part in human life

interconnection geographical concept concerned with how natural places, processes and features can change or affect each other and people, or be changed or affected by people

place geographical concept concerned with why somewhere is important and the characteristics that make it unique or interesting

relative location the direction and distance from one place to another

scale geographical concept concerned with the different levels or ranges at which issues can be examined

space geographical concept concerned with location, distribution (spread) and how we change or design places

spatial how things relate to each other in an area or location

sustainability geographical concept concerned with how to protect the environment and life on Earth

1.7.3 Reflection

Revisit the inquiry question posed in the Overview.

How and why do we study Geography?

1. Think about everything you did this morning before you got to school. Brainstorm examples of how your morning routine might relate to the seven key concepts in Geography.
2. Describe a situation in daily life that might be made easier if you knew how to use one of the geographical tools.

Subtopic	Success criteria			
1.1	I can describe what Geographers do.			
1.2	I can explain the three main stages in the geographical inquiry process: acquiring, processing and communicating information.			
	I can name some common types of maps.			
	I can explain the importance of fieldwork to studying Geography.			
	I can name common types of graphs used in Geography.			
	I can name some types of spatial technology.			
1.3	I can list the seven geographical concepts.			
	I can explain the geographical concept of space.			
	I can explain the geographical concept of place.			
	I can explain the geographical concept of interconnection.			
	I can explain the geographical concept of change.			
	I can explain the geographical concept of environment.			
	I can explain the geographical concept of sustainability.			
	I can explain the geographical concept of scale.			
1.4	I can list some of the careers that studying Geography can lead to.			
	I can outline some of the skills that jobs in geographical fields require.			
1.5	I can recognise key features of maps.			
	I can use a map to gain meaning.			
1.6	I know where I can find templates and skills sheets to help me write well in Geography.			

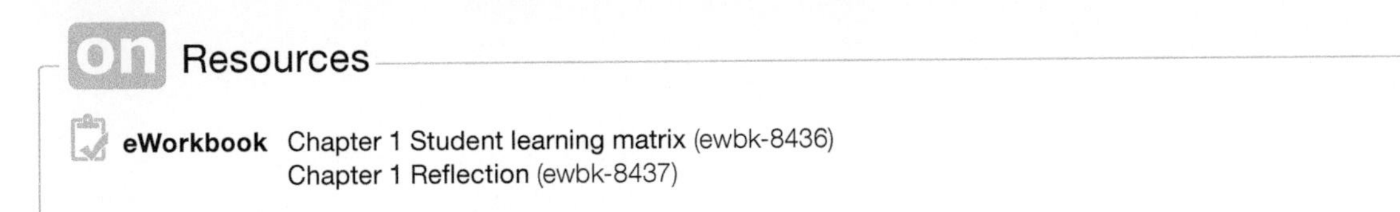

ONLINE RESOURCES

on Resources

Below is a full list of **rich resources** available online for this topic. These resources are designed to bring ideas to life, to promote deep and lasting learning and to support the different learning needs of each individual.

eWorkbook

- 1.1 Chapter 1 eWorkbook (ewbk-7981) ☐
- 1.2 Geographical inquiry (ewbk-8608) ☐
- 1.3 Geographical concepts (ewbk-8612) ☐
- 1.4 Work and careers in Geography (ewbk-10245) ☐
- 1.5 SkillBuilder — How to read a map (ewbk-8616) ☐
- 1.6 SkillBuilder — Notetaking (ewbk-8622) ☐
  - SkillBuilder — Structuring a paragraph (ewbk-8626) ☐
  - SkillBuilder — Interpreting short answer questions (ewbk-8630) ☐
  - SkillBuilder — Evaluating a source (ewbk-8634) ☐
  - SkillBuilder — Analysing an essay question (ewbk-8638) ☐
  - SkillBuilder — Structuring an essay (ewbk-8642) ☐
  - SkillBuilder — Creating presentation slides (ewbk-8646) ☐
- 1.7 Chapter 1 Student learning matrix (ewbk-8436) ☐
  - Chapter 1 Reflection (ewbk-8437) ☐

Video eLessons

- 1.2 Geographical inquiry — Key concepts (eles-5080) ☐
- 1.3 Geographical concepts — Key concepts (eles-5049) ☐
- 1.4 Work and careers in Geography — Key concepts (eles-5081) ☐
- 1.5 SkillBuilder — How to read a map (eles-1634) ☐
- 1.6 SkillBuilder — Notetaking (eles-5042) ☐
  - SkillBuilder — Structuring a paragraph (eles-5043) ☐
  - SkillBuilder — Interpreting short answer questions (eles-5044) ☐
  - SkillBuilder — Evaluating a source (eles-5045) ☐
  - SkillBuilder — Analysing an essay question (eles-5046) ☐
  - SkillBuilder — Structuring an essay (eles-5047) ☐
  - SkillBuilder — Creating presentation slides (eles-5048) ☐

Interactivity

- 1.5 SkillBuilder — How to read a map (int-3130) ☐

Weblink

- 1.2 Geography careers (web-1070) ☐

myWorld Atlas

- 1.3 Deepen your understanding of this topic with related case studies and questions ☐
  - Developing Australian Curriculum concepts > Space (mwa-1599) ☐
  - Developing Australian Curriculum concepts > Place (mwa-1601) ☐
  - Developing Australian Curriculum concepts > Interconnection (mwa-1600) ☐
  - Developing Australian Curriculum concepts > Change (mwa-1602) ☐
  - Developing Australian Curriculum concepts > Environment (mwa-1603) ☐
  - Developing Australian Curriculum concepts > Sustainability (mwa-1605) ☐
  - Developing Australian Curriculum concepts > Scale (mwa-1604) ☐

Teacher resources

There are many resources available exclusively for teachers online.

UNIT

1 Landscapes and landforms

2 The diversity and formation of landscapes and landforms

INQUIRY SEQUENCE

To access a pre-test with **immediate feedback** and **sample responses** to every question in this chapter, select your learnON format at www.jacplus.com.au.

2.1 Overview

Numerous **videos** and **interactivities** are embedded just where you need them, at the point of learning, in your learnON title at www.jacplus.com.au. They will help you to learn the content and concepts covered in this topic.

How do we identify landscapes and what landform features make them unique?

2.1.1 Introduction

Landscapes are the visible features occurring across an area of land. They are created by a combination of environmental processes and human activities. Examples include coastal, desert and mountain landscapes.

Landforms are distinctive, natural features occurring on the surface of the Earth. They are identifiable by their shape. Examples of landforms include valleys, canyons, hills and dunes. Landscapes often include a variety of landforms.

STARTER QUESTIONS

1. Identify key features of the landscape shown in **FIGURE 1**.
2. Watch the Video eLesson (see onResources box below) and make a list of the landforms shown.
3. Brainstorm different landscapes around the world. Describe your favourite place and why it appeals to you.
4. Watch the Video eLesson **Landscapes and landforms — Photo essay**. If you could travel anywhere in Australia to see a unique type of landscape or a famous landform, where would you choose to go?

eles-1623

FIGURE 1 A desert landscape — Kings Canyon in the Watarrka National Park, Northern Territory

on Resources

eWorkbook	Chapter 2 eWorkbook (ewbk-7982)
Video eLesson	World landscapes and landforms (eles-1623) Landscapes and landforms — Photo essay (eles-5306)

2.2 Different types of landscapes

LEARNING INTENTION

By the end of this subtopic, you will be able to describe a variety of landscapes, including their key features, and locate examples on a world map.

2.2.1 Global landscapes

There are many different **landscapes** across the Earth, and similarities can be observed within regions. Variations in landscapes are influenced by factors such as climate; geographical features, including mountains and rivers; latitude; the impact of humans; and where the landscapes are located.

int-3102

FIGURE 1 Selected world landscapes

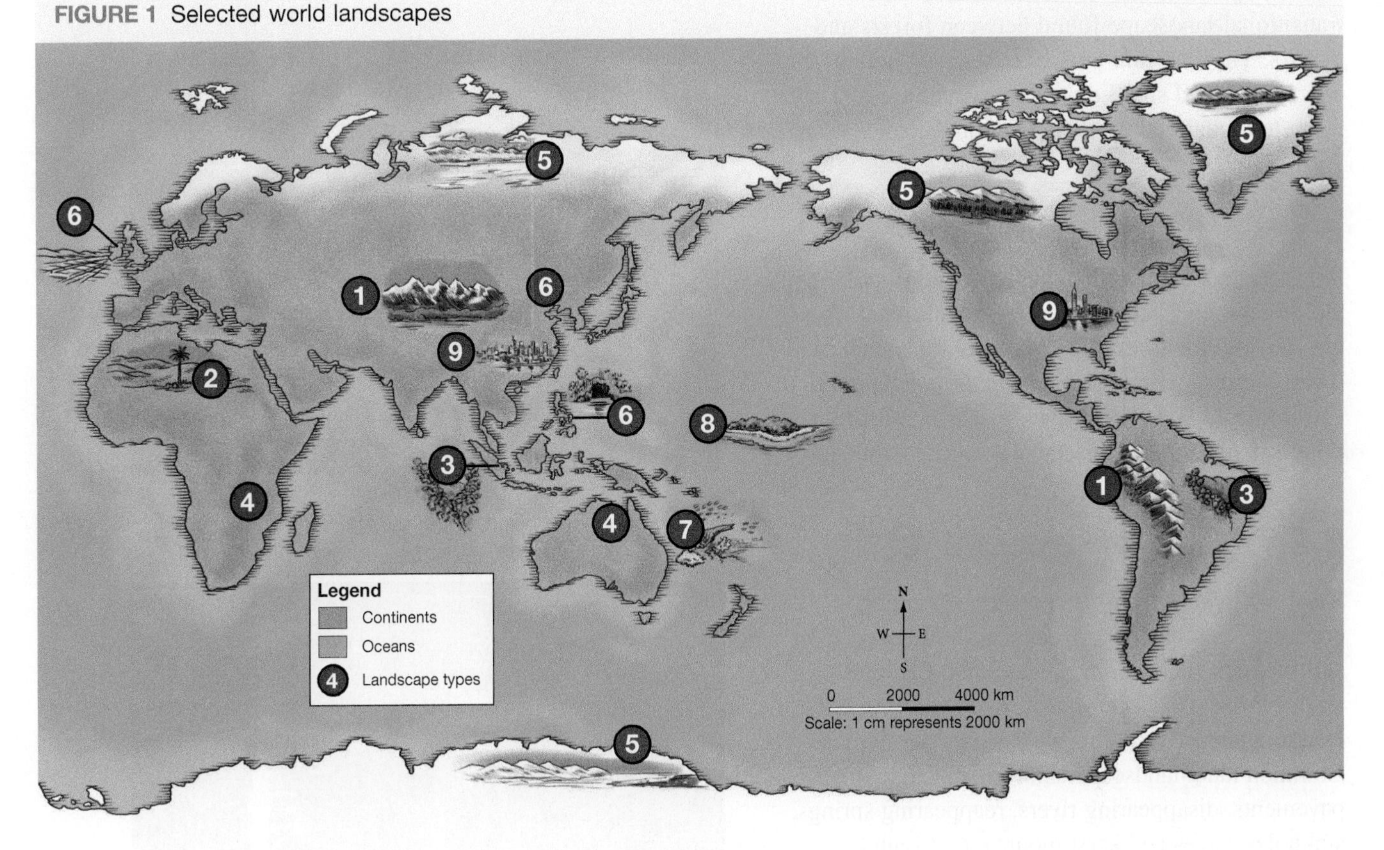

2.2.2 Mountain landscapes 1

Mountains rise above the surrounding land. They often have steep sides and high peaks and are the result of processes operating deep inside the Earth. Some reach high into the atmosphere where it is so cold that snow is found on their peaks.

2.2.3 Desert landscapes 2

Deserts are areas of low rainfall; they are an **arid** or dry environment. They can experience temperature extremes: hot by day and freezing at night. However, not all deserts are hot. Antarctica is the world's largest desert, and the Gobi Desert, located on a high **plateau** in Asia, is also a cold desert.

landscape visible features occurring across an area of land
arid somewhere dry, without much precipitation
plateau a large area of flat land that is higher than the land around it. Plateaus are sometimes referred to as tablelands.

2.2.4 Rainforest landscapes 3

Rainforests are the most diverse landscapes on Earth. They are found in a variety of climates, ranging from the hot wet tropics to the cooler temperate areas. The lush vegetation found in these regions depends on a high level of rainfall. Over 50 per cent of all known plant and animal species are found within them. In addition, many of our foods and medicines come from rainforests.

FIGURE 2 At 8848 metres, Mount Everest in the Himalayas is the highest mountain on Earth.

2.2.5 Grassland landscapes 

Grasslands, or savanna, are sometimes seen as a transitional landscape found between forests and deserts. They contain grasses of varying heights and coarseness, and small or widely spaced trees. They are often inhabited by grazing animals.

FIGURE 3 Tundra landscapes feature low vegetation and lichen, shown on the rocks in the foreground.

2.2.6 Polar landscapes 

Polar regions and **tundra** can be found in polar and alpine regions. Characterised by **permafrost**, they are too cold for trees to grow. Vegetation such as dwarf shrubs, grasses and lichens have adapted to the extreme cold and short growing season (**FIGURE 3**). **Glaciers** often carve spectacular landscape features.

2.2.7 Karst landscapes 

Karst landscapes form when water containing chemicals that make it a weak acid flows over **soluble** rock such as limestone. Small fractures or cracks form, which increase in size over time and lead to underground drainage systems developing. Common karst landscape landforms include limestone pavements, disappearing rivers, reappearing springs, sinkholes, caves and karst mountains. Around 25 per cent of the world's population obtains water from karst **aquifers**.

FIGURE 4 Rickwood Caverns, Alabama, United States, are protected karst caves.

2.2.8 Aquatic landscapes 

Aquatic landscapes cover around three-quarters of the Earth and can be classified as freshwater or marine. Marine landscapes are the saltwater regions of the world, and include oceans and coral reefs. Freshwater landscapes are found on land, and include lakes, rivers estuaries and wetlands.

tundra landscape without trees, with small shrubs and grass, found in cold or high-altitude places

permafrost a layer beneath the surface of the soil where the ground is permanently frozen

glacier a large body of ice, formed by a build-up of snow, which flows downhill under the pressure of its own weight

soluble something that can be dissolved

aquifer a body of permeable rock below the Earth's surface which contains water, known as groundwater (*permeable* means liquid or gas can pass through)

2.2.9 Island landscapes 8

Islands are areas of land that are completely surrounded by water. They can be continental or oceanic. Continental islands lie on a continental shelf — an extension of a continent that is submerged in the sea. Oceanic islands rise from the ocean floor and are generally volcanic in origin. A group or chain of islands is known as an archipelago.

2.2.10 Built landscapes 9

Human or built landscapes are those that have been created or changed by humans.

FIGURE 5 Hong Kong

on Resources

- **eWorkbook** Different types of landscapes (ewbk-8620)
- **Video eLesson** Different types of landscapes — Key concepts (eles-5023)
- **Interactivity** Landscapes galore (int-3102)
- **myWorldAtlas** Deepen your understanding of this topic with related case studies and questions > **Grasslands**

2.2 ACTIVITIES

1. The map in **FIGURE 1** shows the variety of landscapes found on the surface of the Earth. However, it does not show all locations for each landscape type. Investigate one of these landscapes and find out other places in which it is found. Show this information on a map. Annotate your map with characteristics of the landscape.
2. Copy the following table into your workbook.

Landscape characteristics	How people use it	Positive effects	Negative effects

   a. Select one of the landscape types described in this section and complete the table.
   b. Which list is larger — the positive effects or negative effects?
   c. Review the column of negative effects. Select three of these effects and suggest a way in which the environment could be used in a more sustainable way.

2.2 EXERCISE

Learning pathways

■ LEVEL 1	■ LEVEL 2	■ LEVEL 3
1, 2, 7, 8, 9, 15	3, 6, 11, 12, 16	4, 5, 10, 13, 14

Check your understanding

1. List the factors that make landscapes different.
2. Describe why people change landscapes.
3. Define the term *plateau.*
4. Examine **FIGURE 1**. Identify two types of landscapes that occur in more than one continent.
5. What is an aquifer?
6. Match the landscape type to its best description.

Landscape type	Description
Mountain	Cold place without vegetation
Desert	Place made by humans
Rainforest	Land surrounded by water
Grassland	Rock landscape formed by water
Polar	Landscape with steep, high peaks
Karst	Place with lots of thick vegetation
Aquatic	Land covered in grasses and some spread-out trees
Island	Water landscape such as the ocean, rivers and lakes
Built	Very dry place

7. What is the Earth's highest mountain?

Apply your understanding

8. Do glaciers move? Explain why/why not.
9. Explain how karst landscapes are formed.
10. Is Australia likely to have any places with permafrost? Explain why/why not.
11. In which of the nine landscape types might you need special equipment or clothing to survive? Choose one landscape and explain what you would need and why you wouldn't survive without it.
12. Explain the difference between the two types of aquatic landscapes.
13. Grasslands are sometimes called a 'transitional' landscape. Give your interpretation of what this means.

Challenge your understanding

14. Can a landscape fall into more than one of the nine categories shown in **FIGURE 1**? Give reasons to support your answer.
15. What kind of landscape would you most like to live in? Explain your answer, giving examples of what features of the landscape appeal to you and why.
16. Considering the proportion of known plant and animal species that live in rainforests, suggest what might happen if this landscape is not protected.

To answer questions online and to receive **immediate feedback** and **sample responses** for every question, go to your learnON title at www.jacplus.com.au.

2.3 Landscapes and landforms in Australia

LEARNING INTENTION

By the end of this subtopic, you will be able to explain how geomorphic processes have helped to shape Australian landscapes.

2.3.1 Processes that shaped Australia's landscapes and landforms

The tectonic forces of folding, faulting and volcanic activity have created many of Australia's major **landforms**. Other forces that work on the surface of Australia, and give our landforms their present appearance, include **weathering**, **erosion** and **deposition**. These forces are examples of geomorphic processes.

Australia is an ancient landmass. The Earth is about 4600 million years old, and parts of the Australian continent are about 4300 million years old.

Over millions of years, Australia has undergone many changes. Mountain ranges and seas have come and gone. As mountain ranges eroded, sediments kilometres thick were laid over vast areas. These formed sedimentary rocks that were then subjected to folding, faulting and uplifting. This means that the rocks that make up the Earth's crust have buckled and folded along areas of weakness, known as faults. Sometimes fractures or breaks occur, and forces deep within the Earth cause sections to be raised, or uplifted. Over time the forces of weathering and erosion have worn these raised parts down again. Erosion acts more quickly on softer rocks, forming valleys and bays. Harder rocks remain as mountains, hills and coastal headlands.

landform distinctive, natural feature occurring on the surface of the Earth

weathering the breaking down of rock through the action of wind and water and the effects of climate, mainly by water freezing and cooling as a result of temperature change

erosion the wearing away and removal of soil and rock by natural elements, such as wind, waves, rivers or ice, and by human activity

deposition the laying down of material carried by rivers, wind, ice and ocean currents or waves

FIGURE 1 Many of Queensland's mountain peaks were formed by volcanic activity around 20 million years ago. The Glasshouse Mountains, north of Brisbane, are volcanic plugs. They are made of volcanic rock that hardened in the vent of a volcano. Over millions of years, weathering and erosion have worn away the softer rock that surrounded the vent, leaving only the plugs.

Because it is located in the centre of a **tectonic plate**, rather than at the edge of one, Australia currently has no active volcanoes on its mainland, and has very little tectonic lift from below. This means its raised landforms such as mountains have been exposed to weathering forces for longer than mountains on other continents and are therefore more worn down.

tectonic plate one of the slow-moving plates that make up the Earth's crust. Volcanoes and earthquakes often occur at the edges of plates.

hotspot an area on the Earth's surface where the crust is quite thin, and volcanic activity can sometimes occur, even though it is not at a plate margin

About 33 million years ago, when Australia was drifting northwards after splitting from Antarctica, the continent passed over a large **hotspot**. Over the next 27 million years, about 30 volcanoes erupted while they were over the hotspot. The oldest eruption was 35 million years ago at Cape Hillsborough, in Queensland, and the most recent was at Macedon in Victoria around six million years ago. Over millions of years, these eruptions formed a chain of volcanoes in eastern and south-eastern Australia, which are known today as the Great Dividing Range. At present, the hotspot that caused the earlier eruptions is probably beneath Bass Strait (see **FIGURE 2**).

FIGURE 2 Relief map of the Australian continent. The Great Dividing Range stretches from north of Cairns in Queensland to Mount Dandenong, near Melbourne.

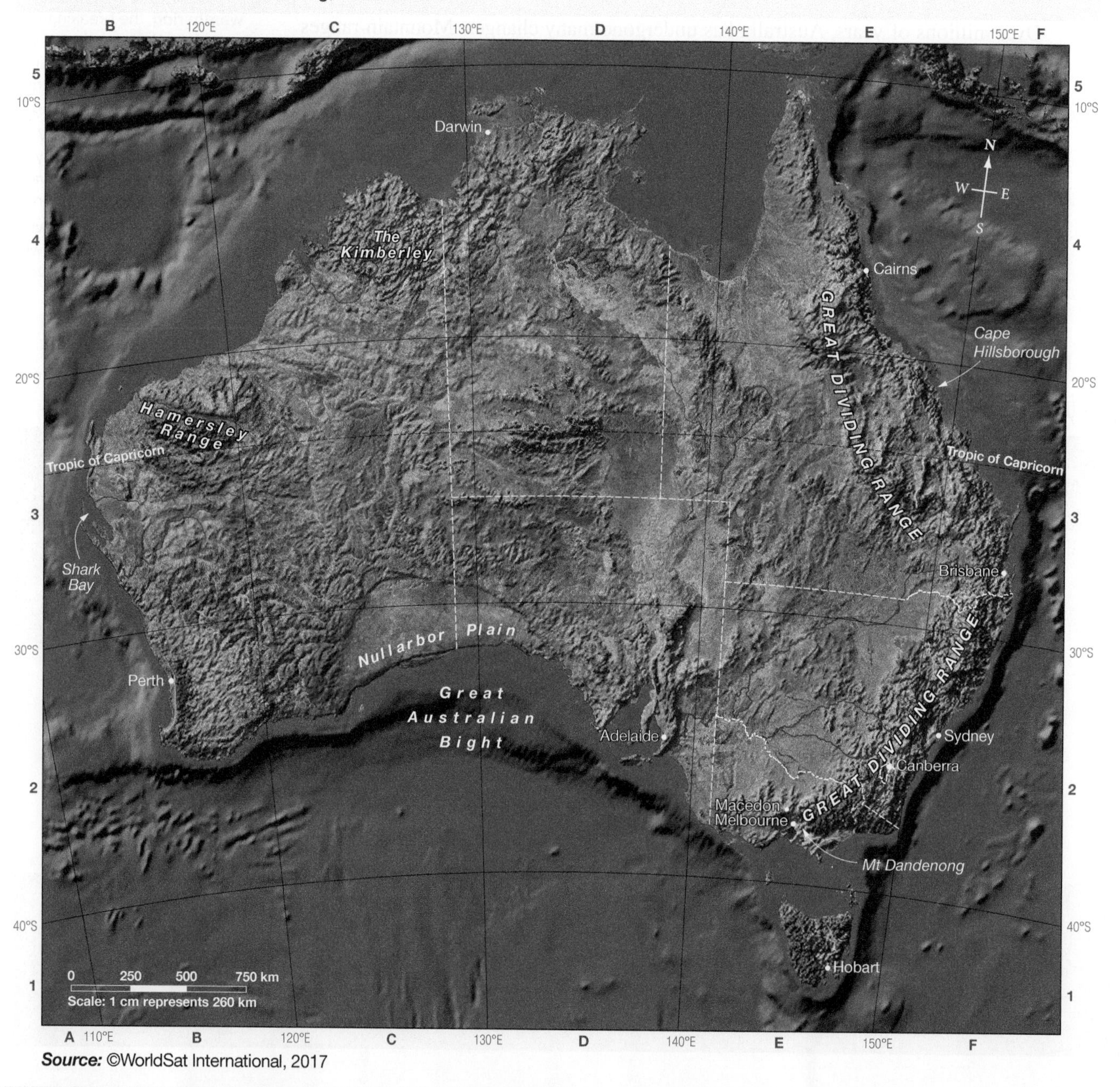

Source: ©WorldSat International, 2017

Many of the features of the Australian landscape result from erosion caused by ice. For example, about 290 million years ago a huge ice cap covered parts of Australia. After the ice melted, parts of the continent subsided (became lower) and were covered by sediment such as sand, soil and small pieces of rock. This formed sedimentary basins (a low area where sediments gather) such as the Great Artesian Basin. On a smaller scale, parts of the Australian Alps and Tasmania have also been eroded by glaciers during the last ice age.

Rivers and streams are another cause of erosion and have carved many of the valleys in Australia's higher regions.

When streams, glaciers and winds slow down, they deposit, or drop, the material they have been carrying. This is called deposition. Many broad coastal and low-lying inland valleys have been created by stream deposition. These areas are called floodplains.

int-3606

FIGURE 3 Australia's four major landform regions

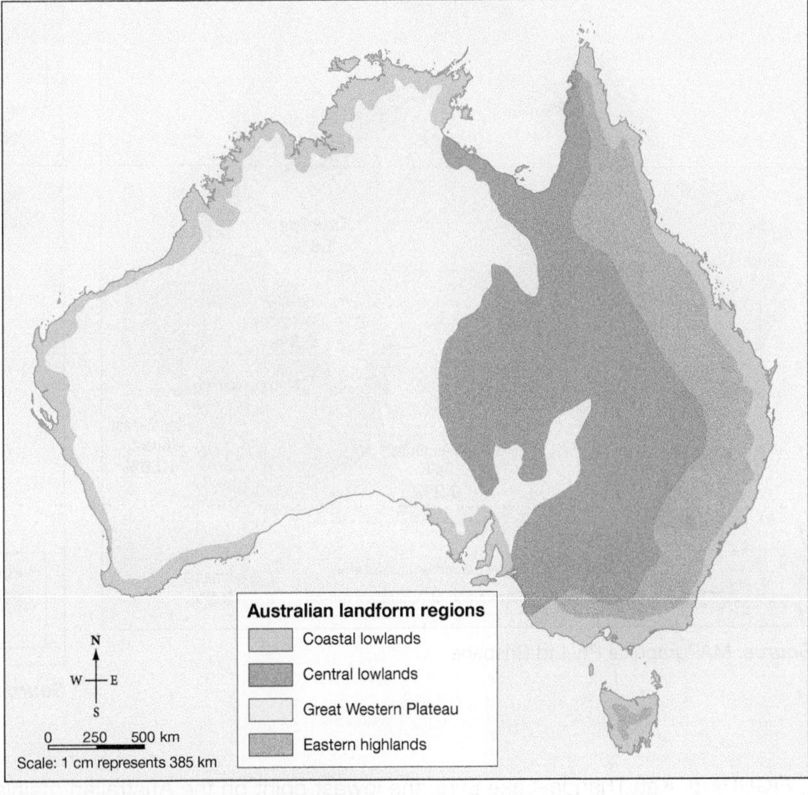

Source: MAPgraphics Pty Ltd Brisbane

2.3.2 Australia's landform regions

The topography of Australia can be divided into four major regions (see **FIGURE 3**).

- The coastal lowlands around Australia's edge are narrow and fragmented. The plains often take the form of river valleys, such as the Hunter River Valley near Newcastle.
- The eastern highlands region (which includes the Great Dividing Range) is mainly a series of tablelands and plateaus. Most of the area is very rugged because rivers have cut deep valleys. It is the source of most of Australia's largest rivers, including the Fitzroy, Darling and Murray. The highest part is in the south-east, where a small alpine area is snow-covered for more than half the year.
- The central lowlands are a vast area of very flat, low-lying land that contains three large **drainage basins**: the Carpentaria Lowlands in the north, the Kati Thanda–Lake Eyre Basin in the centre (see **FIGURE 4**) and the Murray–Darling Basin in the south.
- The Great Western Plateau is a huge area of tablelands, most of which are about 500 metres above sea level. It includes areas of gibber (or stony) desert and sandy desert. There are several rugged upland areas, including the Kimberley and the McDonnell Ranges.

2.3.3 Water flow across the land

Permanent rivers and streams flow in only a small proportion of the Australian continent. Australia is the driest of all the world's inhabited continents. It has:

- the least amount of run-off
- the lowest percentage of rainfall as run-off
- the least amount of water in rivers
- the smallest area of permanent wetlands
- the most variable rainfall and stream flow.

drainage basin the entire area of land that contributes water to a river and its tributaries

Australia has many lakes, but they hold little water compared with those found on other continents. The largest lakes are Kati Thanda–Lake Eyre (see **FIGURE 6**) and Lake Torrens in South Australia. During the dry seasons, these become beds of salt and mud. Yet an inland sea did once exist in this area. It covered approximately 100 000 square kilometres around present-day Kati Thanda–Lake Eyre and Lake Frome. South Australia is Australia's driest state and has very few permanent rivers and streams.

FIGURE 4 Australia's drainage basins

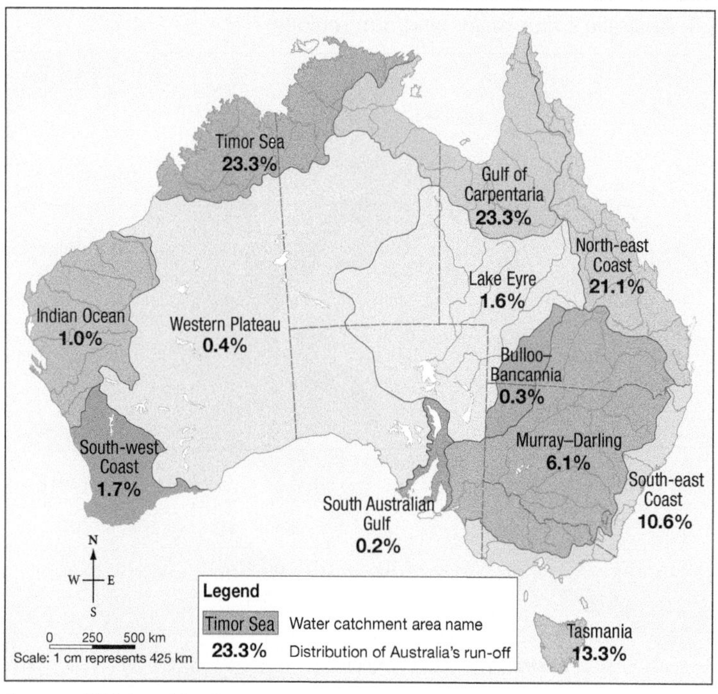

Source: MAPgraphics Pty Ltd Brisbane

FIGURE 5 Kati Thanda–Lake Eyre

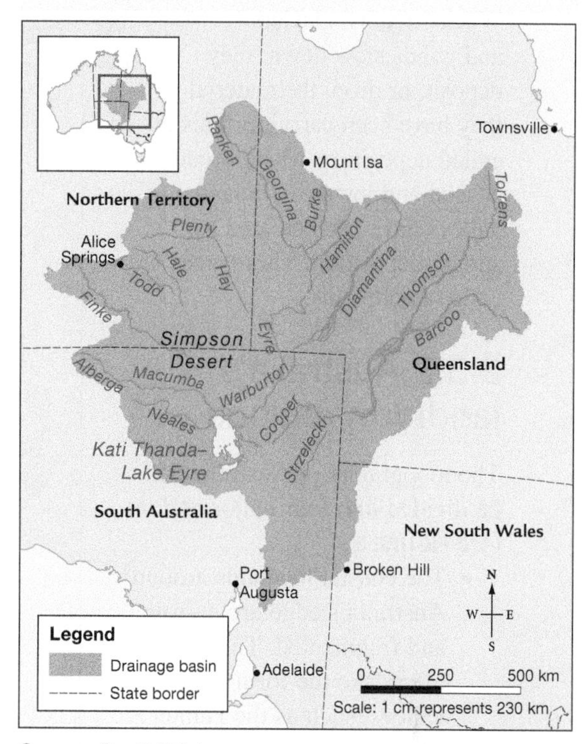

Source: Spatial Vision

FIGURE 6 Kati Thanda–Lake Eyre, the lowest point on the Australian mainland, is part of the Great Artesian Basin. It is 15 metres below sea level. Once a freshwater lake, the region is now the world's largest salt pan. The evaporated salt crust shows white in the satellite image (a) to the left. The lake fills with water only a few times each century. Deep water is shown as black in image (b) to the right.

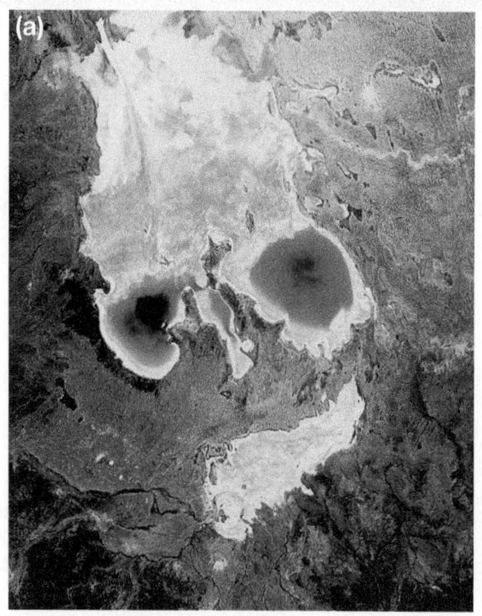

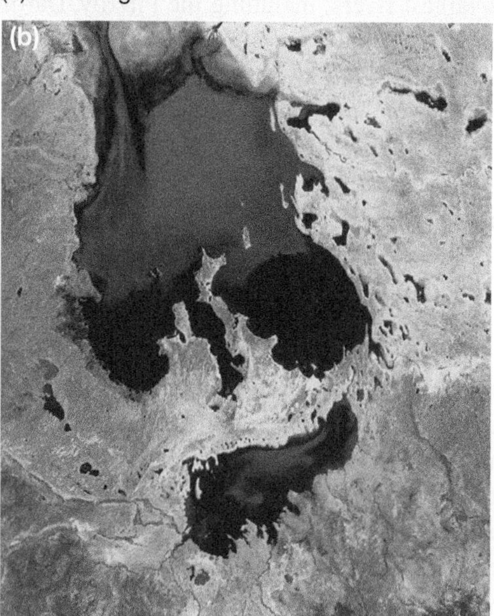

CASE STUDY

The Murray–Darling Basin

The Murray–Darling River system covers about one million square kilometres, and more than 20 major rivers flow into it. It has a wide variety of landscapes, ranging from alpine areas in the south-east to plains in the west. The basin produces 43 per cent of Australia's food and over 40 per cent of Australia's total agricultural income.

The Murray–Darling Basin is the largest and most important drainage basin in Australia, covering one-seventh of the continent. However, the amount of water flowing through it in one year is about the same as the *daily* flow of the Amazon River. The basin is facing severe problems.

- Only about 20 per cent of the water flowing through the basin ever reaches the sea. The rest is diverted for agriculture, industry and domestic use.
- The Murray supplies about 40 per cent of Adelaide's drinking water. The quality of the water continues to decline, mainly because of salinity (salt) levels.
- Approximately 50 to 80 per cent of the wetlands in the basin have been severely damaged or destroyed, and more than a third of the native fish species are threatened with extinction.
- In 2008, inflows into the river system were at their lowest levels since records began 117 years earlier.
- An estimate of weather trends shows that the flow to the Murray River mouth (where the river meets the sea) may be reduced by a further 25 per cent by 2030. However, with the added problem of climate change, it is predicted that precipitation in the Murray–Darling catchment will decrease, so that the reduction in flow to the mouth could be as high as 70 per cent.

FIGURE 7 The Murray River, where it enters the Coorong and Lake Alexandrina in South Australia

on Resources

eWorkbook Landscapes and landforms in Australia (ewbk-8624)

Video eLesson Landscapes and landforms in Australia — Key concepts (eles-5024)

Interactivities Kati Thanda–Lake Eyre (int-3605)
Australia's four major landform regions (int-3606)

myWorldAtlas Deepen your understanding of this topic with related case studies and questions > **Murray–Darling Basin**

2.3 ACTIVITIES

1. Use an atlas to complete the table below, listing the highest mountain in each Australian state and territory. Use your data to write a short description of the location of each.

State/Territory	Highest mountain	Elevation (metres)	Absolute location	Relative location
New South Wales				
Victoria				
Queensland				
Tasmania				
Western Australia				
South Australia				
Northern Territory				
Australian Capital Territory				

2. Use Google Earth to view any part of the Murray–Darling Basin. Describe the landscape.
3. Divide your class into four groups. Assign each group one of Australia's landform regions to investigate. Collectively compile a list of landforms that are found in each region. Then have each member of the group investigate a different landform and prepare a series of slides that show the following:
   a. the landform
   b. where it is located
   c. how it was formed
   d. whether it is a popular place for tourists to visit and the reasons why or why not.
4. Use your atlas to find the Cape Hillsborough and Macedon volcanoes, or refer to **FIGURE 2**.
   a. Calculate the distance between them.
   b. Use the information in this section to work out the rate at which the Australian landmass is moving.
   c. How far has Australia moved over the Bass Strait hotspot? Now calculate where under Bass Strait this hotspot might now lie.
   d. Use the information in this section to explain why this hotspot has changed its location over time.
5. It is said that the amount of water that flows down the Amazon River in a day is more than flows down the Murray in a year.
   a. What does that tell you about how dry Australia's climate is?
   b. How might this affect the way the environment around the Murray River is affected?

2.3 EXERCISE

Learning pathways

■ LEVEL 1	■ LEVEL 2	■ LEVEL 3
3, 5, 6, 11, 14	1, 2, 8, 12, 15	4, 7, 9, 10, 13

Check your understanding

1. In your own words, define the following terms:
   a. folding
   b. faulting
   c. uplift.
2. Describe three of the physical changes Australia's landmass has undergone.
3. Describe the major characteristics of Australia's four main landform regions.
4. What does the process of weathering do?
5. Identify the four major landform regions in Australia.
6. What is the lowest point on the Australian mainland?
7. Give three reasons why Australia is considered the driest inhabited continent.
8. Examine **FIGURE 4**. Which drainage basin do you live on?

Apply your understanding

9. Explain why the Murray–Darling Basin is important to Australia's agriculture industry.
10. Which of the following problems with the Murray–Darling do you think will have the worst impact on people?
    - Low water quality that makes the water unsafe for drinking
    - Destroyed wetlands
    - Low rainfall, which reduces water flow
11. Explain what a tectonic plate is and outline how they contribute to the creation of landforms.
12. What is the difference between erosion and deposition?

Challenge your understanding

13. Suggest why Australia's inland sea no longer exists.
14. Australia is an ancient landmass and has undergone many changes over millions of years. Brainstorm and compile lists under the following headings.
    - Physical changes that have taken place on the Australian landmass
    - Tectonic processes that have contributed to these changes
    - Changes caused by processes such as weathering and erosion

    Write a paragraph that explains the interconnection between these factors. (If you need help structuring your paragraphs, you will find templates in the Chapter 1 eWorkbook in your online Resources panel.)
15. Australia is so low in altitude compared with other continents. Predict whether this will change in the next 30 million years.

To answer questions online and to receive **immediate feedback** and **sample responses** for every question, go to your learnON title at www.jacplus.com.au.

2.4 SkillBuilder — Recognising land features

online only

LEARNING INTENTION

By the end of this subtopic, you will be able to identify key features and landforms on a topographic map.

The content in this subtopic is a summary of what you will find in the online resource.

2.4.1 Tell me

What are land features?

Land features are landforms with distinct shapes, such as hills, valleys and mountains. On topographic maps you recognise land features from the patterns formed by the contour lines (the word *contour* means outline or shape). By reading the contour lines, you can understand the shape of the land.

2.4.2 Show me

How to recognise a land feature

Step 1

The contour lines are the brown lines on **FIGURE 1**. You will see that sometimes they are close together and sometimes they are further apart.

int-7829

Step 2

The numbers on the contour lines show the height of that point of land above sea level. In **FIGURE 1**, each line represents a 200 metre increase.

Step 3

The patterns in the lines show how steep the land is. If the contour lines are close together then the shape of the land is steep. If the contour lines are further apart then the land is flatter.

FIGURE 1 Landforms matched to a topographic map

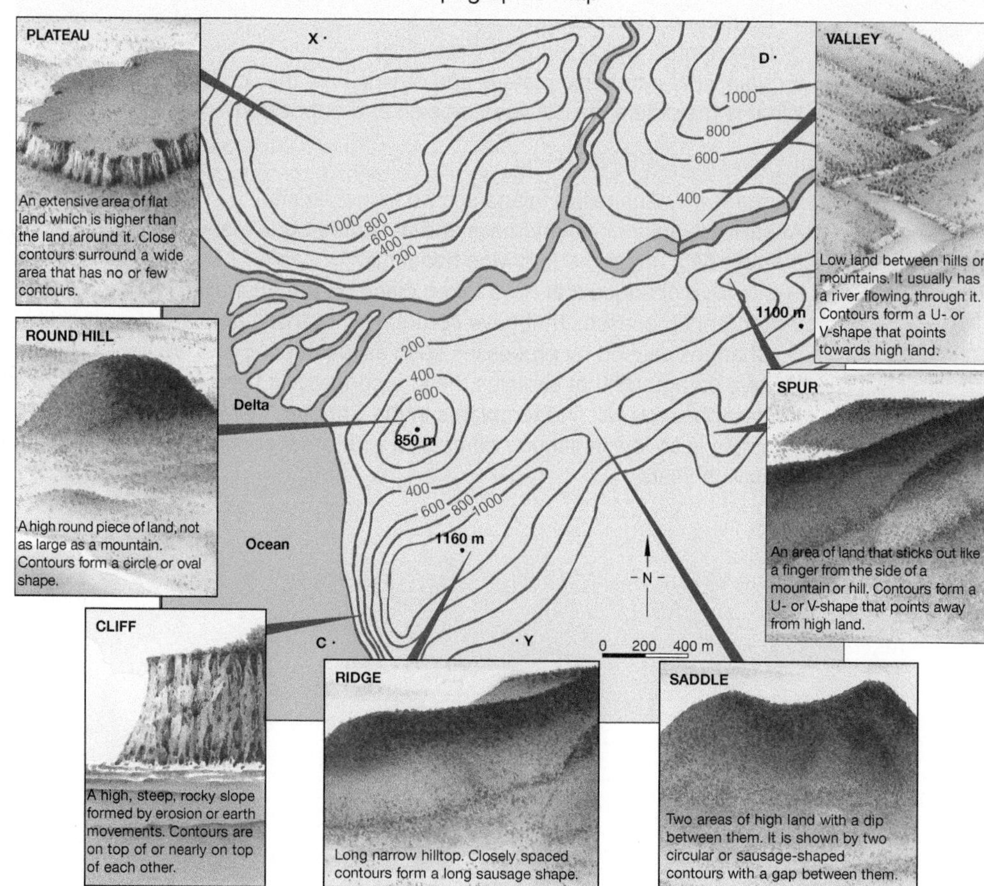

2.4.3 Let me do it

Go to learnON to access the following additional resources to help you build this skill:
- a longer explanation of this skill and its application in Geography (Tell me)
- a video showing the step-by-step process of this skill (Show me)
- an activity and interactivity for you to practise this skill (Let me do it)
- self-marking questions to help you understand and use this skill.

on Resources

eWorkbook	SkillBuilder — Recognising land features (ewbk-8628)
Digital document	Topographic map of Yarra Yarra Creek Basin (doc-36315)
Video eLesson	SkillBuilder — Recognising land features (eles-1648)
Interactivities	SkillBuilder — Recognising land features (int-3144) Landforms matched to a topographic map (int-7829)

2.5 Landscapes in the Pacific region

online only

LEARNING INTENTION

By the end of this subtopic, you will be able to identify and explain the processes that have shaped the Pacific islands.

The content in this subtopic is a summary of what you will find in the online resource.

The Pacific Ocean is the world's largest ocean, and occupies almost a third of the Earth's surface, making it larger than all the Earth's land areas combined. It stretches from the Arctic in the north to Antarctica in the south and is bordered by Australia and Asia in the west and the Americas in the east. The 25 000 Pacific islands are home to around 10 million people.

To learn more about the Pacific islands, go to your learnON resources at www.jacPLUS.com.au.

FIGURE 6 An underwater volcanic eruption in 2009 created a new island off the coast of Tonga, an island group in the South Pacific.

Source: © AFP/Getty Images

Contents

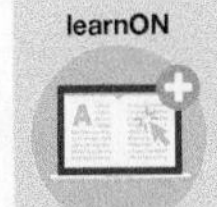

- 2.5.1 Pacific landscape
- 2.5.2 Records of the Pacific
- 2.5.3 Low islands
- 2.5.4 High islands

on Resources

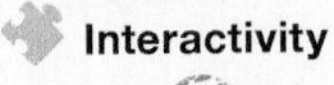

eWorkbook	Landscapes in the Pacific region (ewbk-8632)
Video eLesson	Landscapes in the Pacific region – Key concepts (eles-5025)
Interactivity	Pacific island groups (int-8358)
myWorldAtlas	Deepen your understanding of this topic with related case studies and questions > **Pacific nations**

2.6 Processes that transform landscapes

LEARNING INTENTION

By the end of this subtopic, you will be able to explain the processes that are involved in the formation of landscapes.

2.6.1 Natural and human processes

In the future, the Earth's surface will look very different from the way it looks today. There are a variety of natural processes that shape and reshape not only the surface of the Earth, but also what lies beneath it. Natural processes, also called geomorphological (or geomorphic) processes, include uplift (such as that caused by tectonic activity), erosion, deposition and weathering. People also change the landscape when they clear land for agriculture or build cities and road networks. Sometimes people alter the course of a river or trap its flow behind the walls of a dam. When people change the landscape these processes are called human processes.

2.6.2 Tectonic activity

The Earth's surface, or **crust**, is split into a number of plates, which fit together like a giant jigsaw puzzle. These plates sit on a layer of semi-molten (melted) material in the Earth's **mantle** — the layer of the Earth between the crust and the core. Heat from the Earth's core creates **convection currents** within the mantle, causing the plates to move. Most of the Earth's great mountain regions were formed as a result of this movement. The force of this movement in the Earth's crust is called **tectonic** activity or force

crust the Earth's outer layer or surface

mantle the layer of the Earth between the crust and the core

convection current a current created when a fluid is heated, making it less dense, and causing it to rise through surrounding fluid and to sink if it is cooled; a steady source of heat can start a continuous current flow

tectonic relating to force or movement in the Earth's crust

FIGURE 1 After tectonic forces cause a section of the Earth to be raised, other processes take over and re-sculpt the landscape.

1 Weathering is the breakdown of rocks due to the action of rainwater, temperature change and biological action. The material is not transported (removed).

It can be physical, chemical or biological.

2 Erosion is the process whereby soil and rocks are worn away and moved to a new location by agents such as wind, water or ice.

3 Transportation is the process that moves eroded material to a new location — examples include soil carried by the wind, and sediment or pebbles in a stream.

4 Deposition — materials moved by wind and water eventually come to a halt. Over time new landforms are built. Sand dunes and beaches are common landforms associated with deposition.

Physical — occurs where water is continually freezing and thawing. The water penetrates cracks and holes in the rocks. As water freezes it expands, making the cracks larger. Over time the rock breaks apart.

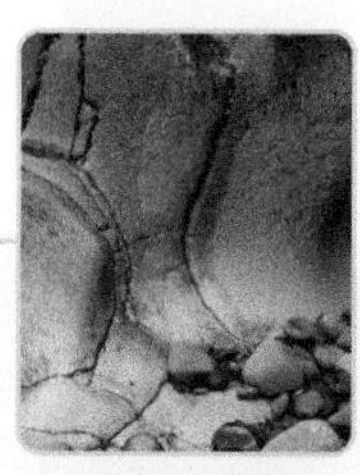

Chemical — some rocks, such as limestone, contain chemicals that react with water causing the limestone to dissolve.

Biological — living organisms such as algae produce chemicals that break down rocks. They can also be forced apart by plant

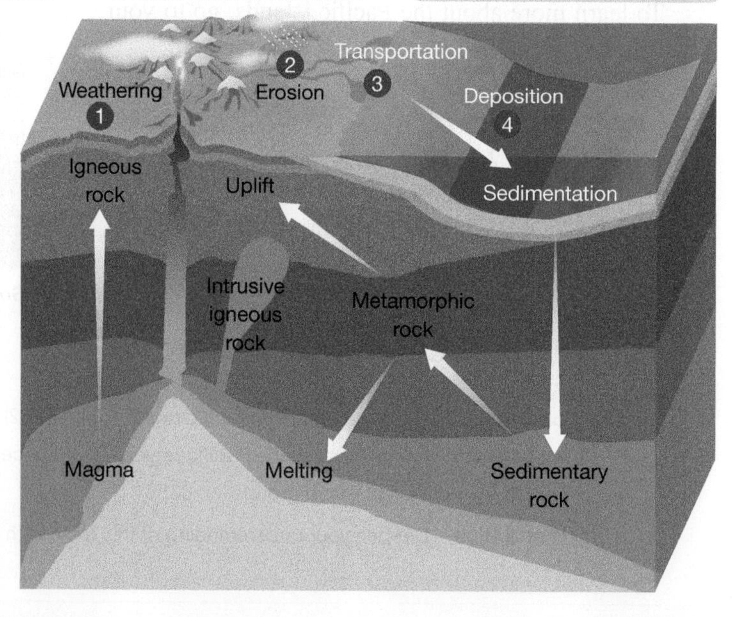

2.6.3 Weathering

Weathering is the breakdown of rock due to the impact of physical, chemical and biological actions, or the effects of climate. These might include temperature changes, rainwater, reactions with chemicals or the impact of algae or other living organisms.

2.6.4 Erosion and deposition

A torrent of gushing water can shift rocks, remove topsoil or shape river valleys. Gentle rain can change the chemical structure of any surface material, making it more likely that soil will be transported by the next heavy shower. In cold climates, frozen water in glaciers works like a slow-moving bulldozer to erode land and create unique landscape features. At ocean shorelines, the power of waves creates coastal landscape features.

FIGURE 2 Water flowing over rocks erodes the landscape over time

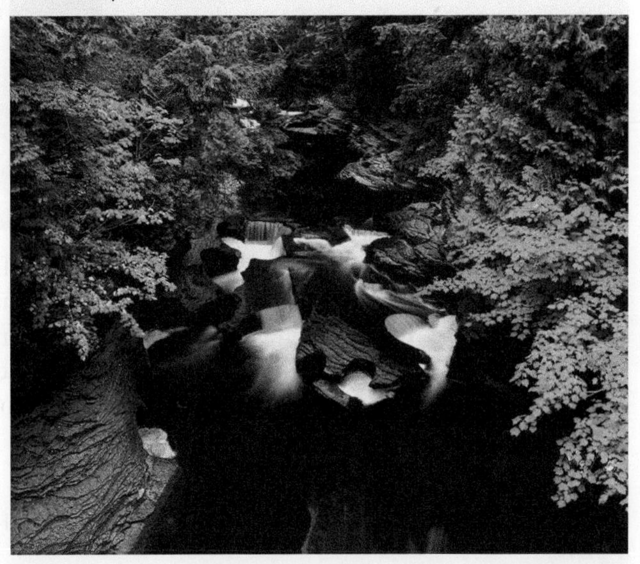

Most often, landscapes are changed or created by two processes: erosion and deposition (see section 2.6.1). Water is one of the most powerful causes of change (or agents) in the landscape, breaking up, moving and repositioning material across the Earth's surface. In **FIGURE 2** you can see the power of water as it rushes over a rockface and carves pools in its hard surface. You may have seen pools of a similar shape carved by waves in rocky coastal landforms.

Erosion

Erosion is the wearing away of the Earth's surface by natural elements such as wind, water, ice and human activity. The landscape is further eroded when agents such as wind, water and ice **transport** these materials to new locations. Eventually, transported material is deposited in a new location. Over time, this material can build up and new landforms result. The Grand Canyon in the United States (**FIGURE 3**) is an example of these elements at work. These processes work more quickly on softer rocks.

FIGURE 3 Over millions of years, the Colorado River has cut deep channels to form the Grand Canyon.

Human activity also contributes to erosion. Deforestation, agriculture, urban sprawl, logging and road construction all alter the natural balance and increase erosion by as much as 40 per cent in some areas. Vegetation not only provides valuable habitat for native animals but also is vital for binding (holding together) the soil. Once vegetation is removed, the soil is more easily broken down and carried away by wind and water. When topsoil is removed, plants are unable to obtain the nutrients they need for growth. Erosion sometimes forms wide, deep channels known as gullies (**FIGURE 4**).

transport the movement of eroded materials to a new location by agents such as wind and water

Deposition

Deposition occurs when material is carried to another place by rivers, the wind, ice or ocean currents. This process moves rock, soil, sand and other material from one place to another. This natural process can also be affected by human activity, For example, when a river is dammed, the material in the river might collect and build up behind the dam walls.

FIGURE 4 Gullies created by erosion

Change over time

As water makes contact with landscapes, it can change the shape and size of its features or landforms through combinations of natural and human processes. The coastal landscapes that you see today are not the same as they were hundreds or thousands of years ago. **FIGURE 5** is a photo of the Twelve Apostles, located on the coast of south-western Victoria. The name suggests that there may once have been 12 pillars of rock, or stacks, visible along this stretch of coastline. In the foreground you can see the remnants of two quite recently collapsed stacks. Even these stacks were once joined to the cliffs as part of the mainland. The soft rock that makes up these cliffs has been constantly altered by many years of rainfall and wave action. The resulting coastline has seen much creation and destruction of stacks over time.

FIGURE 5 The Twelve Apostles in Port Campbell National Park, Victoria.

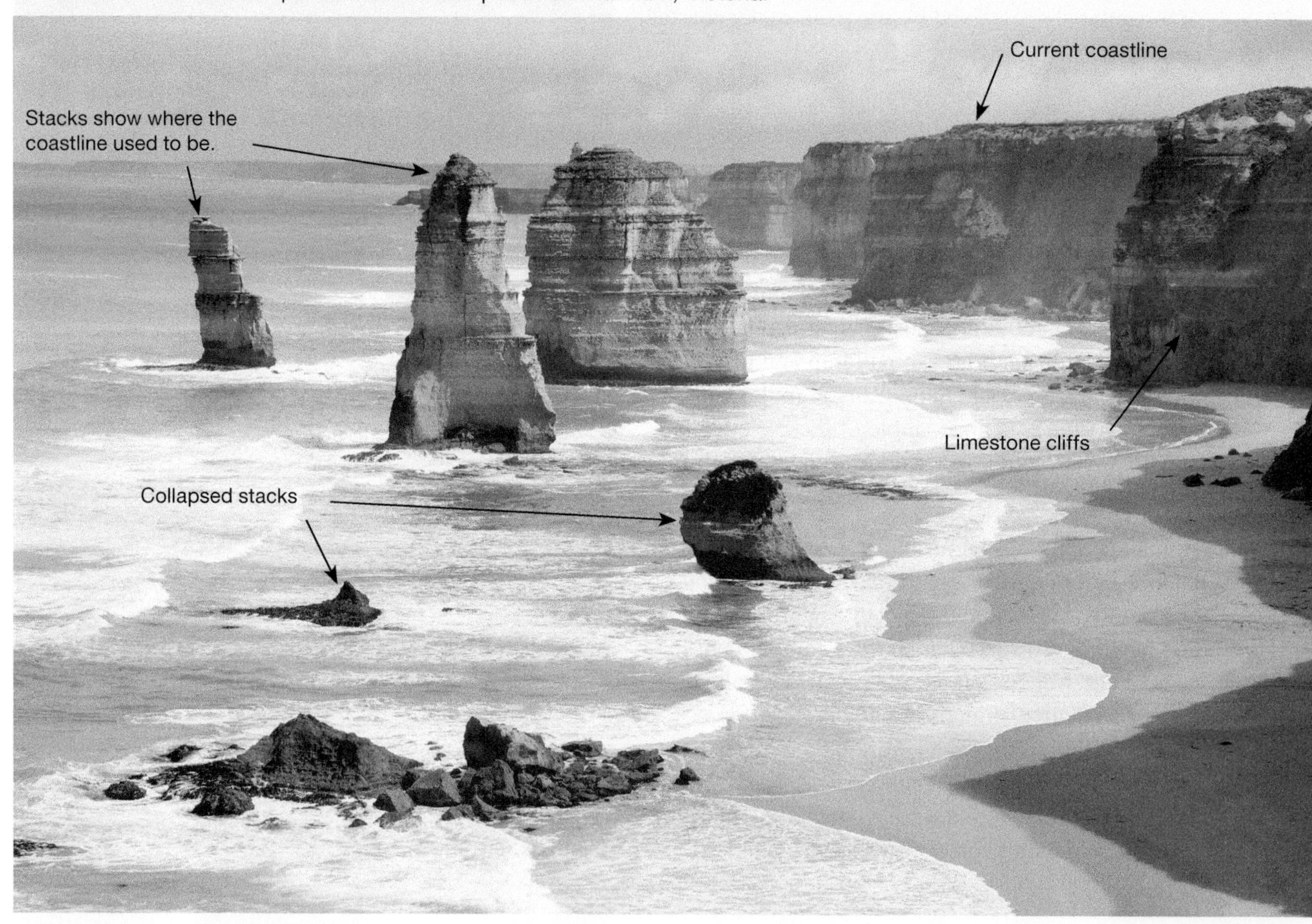

FIGURE 6 Water constantly moves over and through the Earth and through the air.

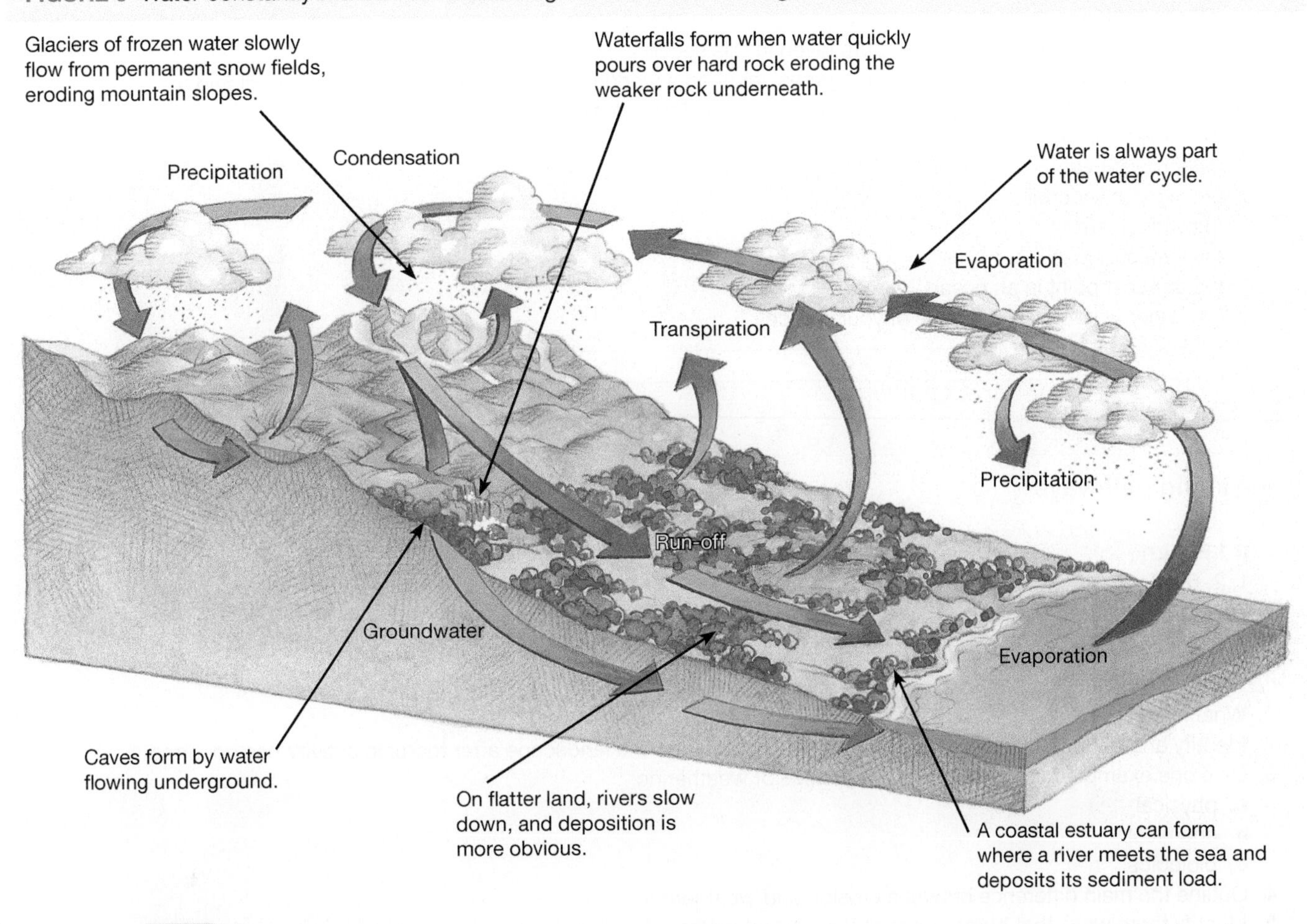

Resources

eWorkbook	Processes that transform landscapes (ewbk-8636)
Digital document	Blank world map template (doc-36260)
Video eLesson	Processes that transform landscapes — Key concepts (eles-5026)
Interactivity	Break down! (int-3101)
myWorldAtlas	Deepen your understanding of this topic with related case studies and questions > **Active Earth**

2.6 ACTIVITIES

1. Choose a landscape to investigate online.
   a. Copy and paste an image of this landform into a digital file (a document, page or image editor).
   b. Annotate the image with information about its location, formation and change over time.
   c. Create a hand-drawn or digital sketch to show how this landform might have looked in the past and how it might look in the future.
   d. Ensure that you include the source of the original image as a source line.
2. Many landscapes change rapidly; for example, the Twelve Apostles. With a partner or group, discuss another landscape that has been shaped by the power of water. Do you think the changes to the landscape have been positive or negative? To what extent should people try to stop the changes caused by water?
3. Study the environment around your home or school and find a place where there is evidence of erosion.
   a. Make a sketch and label the features of the landscape.
   b. Highlight areas where erosion is evident and add annotations to explain what you think might have caused this change and, in particular, the scale of this change.

c. Estimate the proportion of this environment that has been affected. What proportion do you think is the result of human activity?
d. Compare your estimate with the figure given in this section.
4. Using an atlas, Google Earth and other sources, label on a world map each of the following. (You will find a blank world map in the Digital documents in your online Resources.)
a. the largest glacier
b. the longest river
c. the highest waterfall
d. the widest river
e. the largest ocean
f. the deepest point in an ocean
g. one other water-related feature of your choice.

2.6 EXERCISE

Learning pathways

■ LEVEL 1	■ LEVEL 2	■ LEVEL 3
1, 2, 5, 10, 13	4, 6, 11, 12, 15	3, 7, 8, 9, 14, 16

Check your understanding

1. What is the mantle?
2. Identify and define the four types of processes that change the landscape after tectonic activity has occurred.
3. Give one example for each of the three causes of weathering:
a. physical
b. chemical
c. biological.
4. Outline the main difference between erosion and weathering.
5. Identify three ways that humans can make erosion worse.

Apply your understanding

6. Explain how and why human activity might contribute to weathering and erosion.
7. Explain how damming a river might affect deposition.
8. Using terms such as uplift, erosion, deposition, weathering and transportation, explain the interconnection between landscapes and physical processes.
9. Describe how convection currents work and explain how the process contributes to the formation of landforms.
10. Are the processes that transform the landscape always slow? Explain why/why not.
11. Describe the two natural processes powered by water, and explain how they are interconnected.
12. Explain how the water cycle and the formation of landscapes are interconnected.

Challenge your understanding

13. Is the result of erosion sometimes considered to be beautiful? Give examples to support your view.
14. The natural processes that form landscapes can remove or move topsoil, which makes it hard for plants to gain the nutrients they need. Discuss the positive and negative impacts of humans interfering in this process.
15. Water can be considered one of the most important architects of desert landscape features. After looking at the images in this section, try to explain how you think water can change the landscapes of arid or desert environments.
16. Identify three possible ways that people can change the flow of water, either across the surface of the Earth or along the coast. Predict how you believe this may alter landscape features. Examples may include the use of river water for irrigation or the construction of a marina.

To answer questions online and to receive **immediate feedback** and **sample responses** for every question, go to your learnON title at www.jacplus.com.au.

2.7 Landscapes created underground

LEARNING INTENTION

By the end of this subtopic, you will be able to explain how a karst landscape is formed.

2.7.1 Karst landscapes

Apart from rivers and streams that flow across the surface of the Earth, vast networks of rivers also exist under the ground. The result is a network of caves and channels that carve a very different landscape, known as **karst**.

Karst is formed by water dissolving bedrock (solid rock beneath soil) over hundreds of thousands of years (see **FIGURE 1**). On the surface of the Earth, sinkholes (holes in the Earth's surface), vertical shafts (tunnels), and fissures (cracks) will be evident. Rivers and streams may seem to simply disappear, but underground there are intricate drainage networks, complete with caves, rivers, **stalactites** and **stalagmites** (see **FIGURE 2**).

Karst topography makes up about 10 per cent of the Earth's surface; however, a quarter of the world's population depends on karst environments to meet its water needs.

karst underground landscapes of caves and water channels

stalactite a feature made of minerals, which forms from the ceiling of limestone caves, like an icicle.

stalagmite a feature made of minerals found on the floor of limestone caves.

FIGURE 1 Formation of a karst landscape

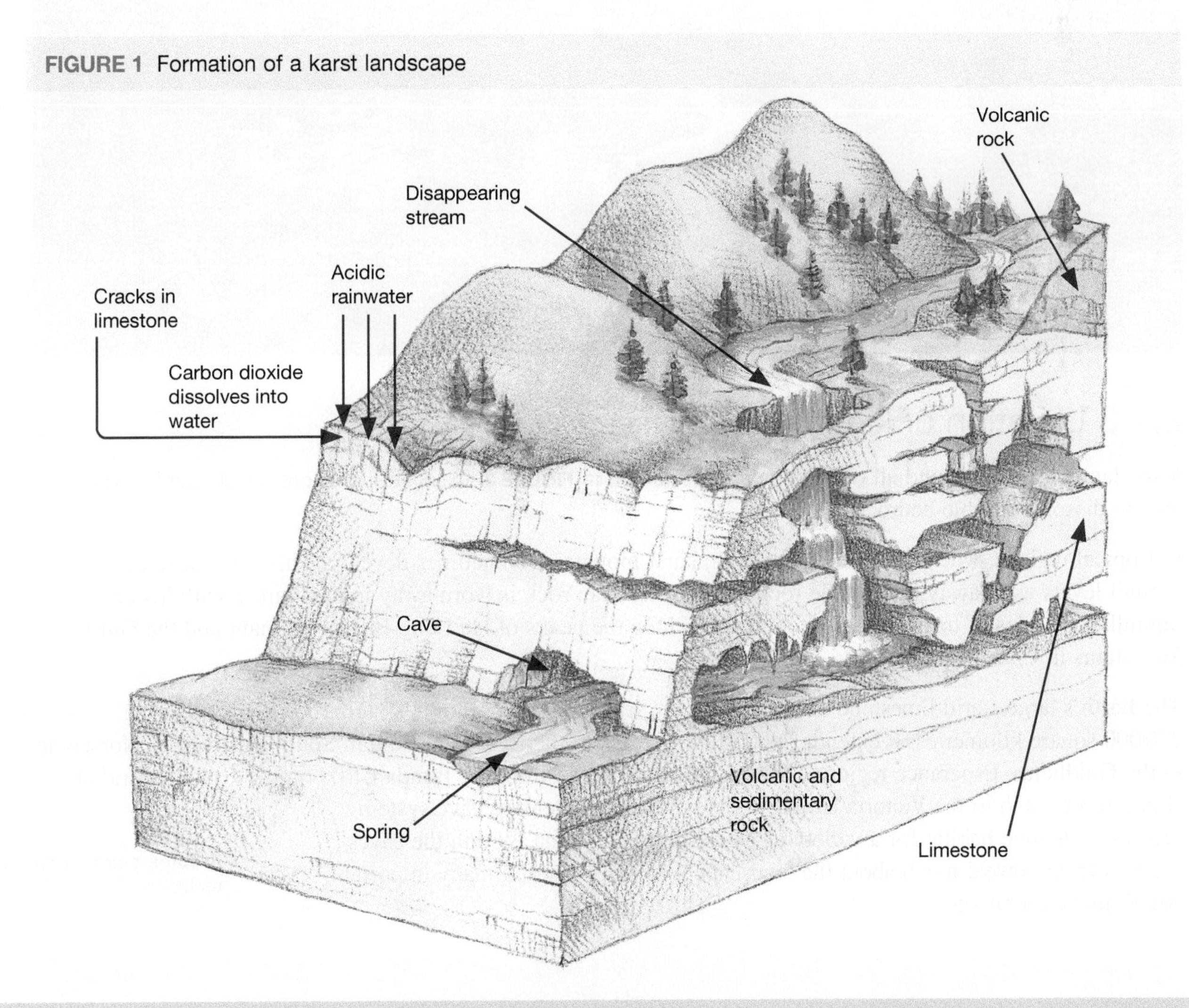

2.7.2 Karst landscape formation

Water becomes slightly acidic when it comes into contact with carbon dioxide (CO_2) in the atmosphere (as it does when raindrops form) or when it filters through organic matter in the soil and soaks into the ground. Acidic water is able to dissolve soluble bedrock, such as limestone and dolomite. This creates cracks (also called **fissures**), allowing more water to move through the rocks. When the water reaches a layer of non-dissolving rocks, it begins to erode sideways, forming an underground river or stream. As the process continues, the water creates hollows, eventually creating a cave. Some karst landscapes contain aquifers that are capable of providing large amounts of water.

FIGURE 2 Li River and karst mountains, near Yangshuo County, Guilin City, Guangxi Province, China

2.7.3 Location of karst landscapes

Karst landscapes are found all over the world, as shown in **FIGURE 3**, in locations where mildly acidic water is able to dissolve soluble bedrock.

In tropical regions, where rainfall is very high, karst mountains sometimes develop. This is because the high rainfall levels wear away the soluble rock much faster than rock is worn away in karst areas with lower rainfall. Examples of tropical karst mountains include the peaks of Ha Long Bay in Vietnam and the Guilin Mountains in China (**FIGURE 2**).

The Earth's largest arid limestone karst cave system is located on Australia's Nullarbor Plain, covering 270 000 square kilometres. It extends 2000 kilometres from the Eyre Peninsula in South Australia to Norseman in the Goldfields–Esperance region of Western Australia, and from the Bunda Cliffs on the Great Australian Bight in the south to the Victoria Desert in the north. The extensive cave system provides a unique habitat for a variety of native flora and fauna. Within the caves are fossils that reveal much about the landscape's distant past, and many important Aboriginal cultural sites.

fissures cracks, especially in rocks

FIGURE 3 Karst regions of the world

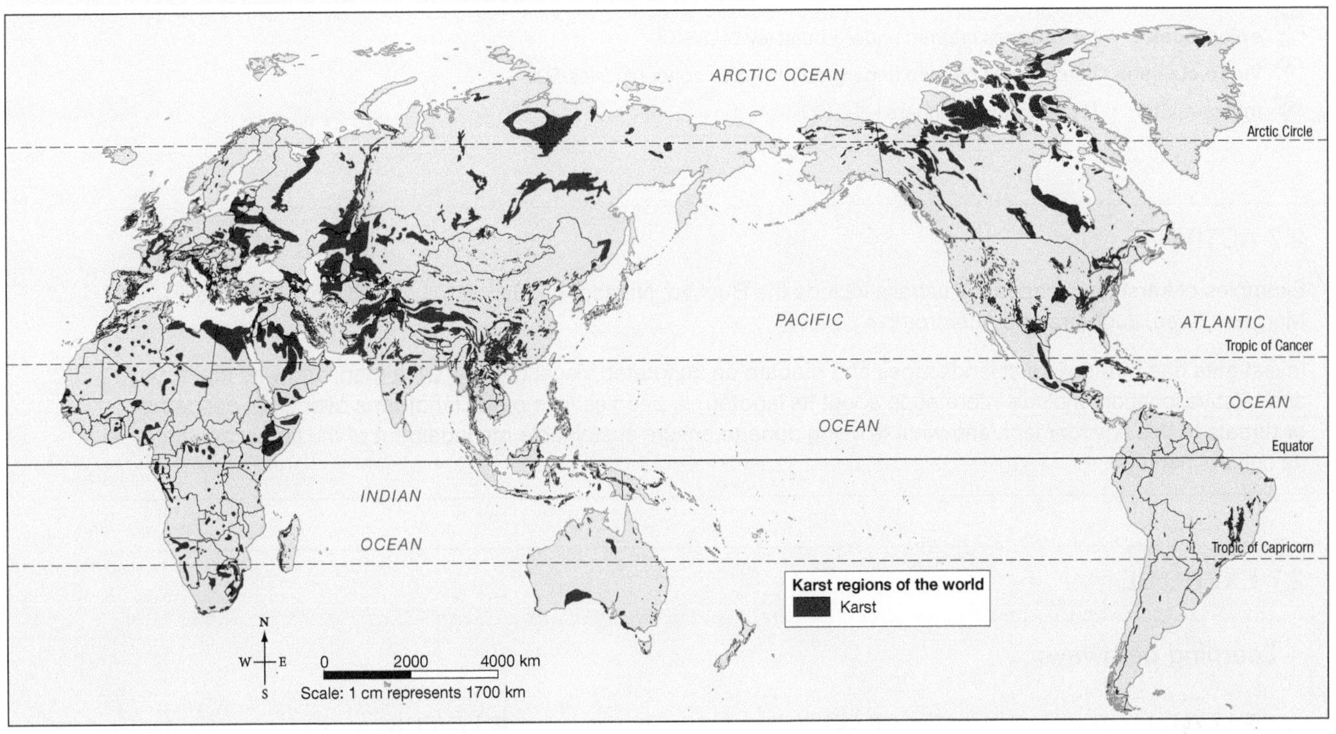

Source: Spatial Vision

FIGURE 4 Mulwaree Cave (upper levels of Wollondilly Cave), Wombeyan Caves, Wombeyan Karst Conservation Reserve, New South Wales

Resources

eWorkbook Landscapes created underground (ewbk-8640)

Video eLesson Landscapes created underground — Key concepts (eles-5027)

Interactivity Underground wonders (int-3103)

2.7 ACTIVITY

Examples of karst landscapes in Australia include the Buchan, Naracoorte, Jenolan, Labertouche, Princess Margaret Rose, Judbarra and Abercrombie caves.

Investigate one of these karst landscapes and prepare an annotated visual display. On a map, show its absolute and relative location. Include information about its landforms, peoples' use of the landforms over time, concerns or threats to this environment, and what is being done to ensure sustainable management of the landscape and its cultural heritage.

2.7 EXERCISE

Learning pathways

■ LEVEL 1	■ LEVEL 2	■ LEVEL 3
1, 3, 8, 12, 15	2, 4, 5, 7, 14	6, 9, 10, 11, 13

Check your understanding

1. In your own words, explain how a karst landscape is formed.
2. Describe the global distribution of karst landscapes.
3. How do stalactites and stalagmites differ?
4. Complete this sentence by choosing the correct options: Water becomes slightly *acidic/basic* when it comes into contact with $CO_2/H_2 0$.
5. Examine **FIGURE 1**. Describe a 'disappearing stream'.
6. Which kinds of rock are more affected by acidic rainwater?

Apply your understanding

7. Why should we preserve karst landscapes? Give at least two reasons.
8. Explain why karst landscapes might pose dangers for hikers.
9. Almost a quarter of the world's population rely on karst systems for their water supply. Would you expect many of these essential karst water supplies to be in tropical areas? Explain why or why not.
10. Refer to **FIGURE 3**. Explain how it is possible that there are karst landscapes in polar regions.
11. Based on **FIGURE 1**, explain how a disappearing stream forms.
12. Imagine you are caving, and you trip over a finger-shaped rock formation that is coming from the floor of the cave. What have you tripped over, and how did it form there?

Challenge your understanding

13. The largest limestone arid karst system is found on the Nullarbor Plain, Australia. The Nullarbor Plain is an example of a desert landscape. Suggest how an environment formed by water can occur in this location.
14. Suggest strategies for the sustainable management of karst in popular tourist areas.
15. Create a mnemonic (memory aid) to help you remember the difference between stalactites and stalagmites.

To answer questions online and to receive **immediate feedback** and **sample responses** for every question, go to your learnON title at www.jacplus.com.au.

2.8 Landscapes created by rivers

LEARNING INTENTION

By the end of this subtopic, you will be able to identify the different phases of a river and describe the landform features that form from a river's source to its mouth.

2.8.1 Moving water

Erosion transportation and deposition by running water are the main processes that create our landscapes. Some rivers, such as the Gordon River in Tasmania, are **perennial**; some, such as Coopers Creek in Queensland, are **intermittent**; others, such as the Colorado River in the United States, have eroded amazing landforms including the Grand Canyon.

Water is always on the move. It evaporates and becomes part of the water cycle; it rains and flows over the surface of the Earth and into streams that make their way to a sea, lake or ocean; and it soaks through the pores of rocks and soil into **groundwater**.

FIGURE 1 The longest river on each continent

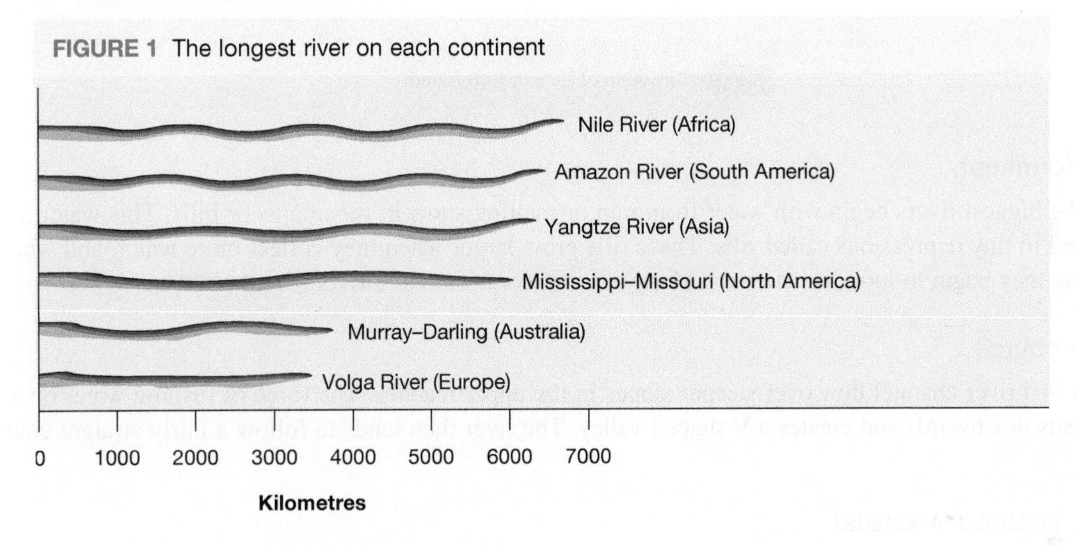

2.8.2 River systems and features

A river is a natural feature, and what we see is the result of the interaction of a range of inputs and processes. All parts of the Earth are related to the formation of river landscapes. This includes the lithosphere (rocks and soil), the hydrosphere (water), the biosphere (plants and animals) and the atmosphere (temperature and water cycle). Changes can happen both quickly and over a very long period of time. Changes at one location along a river can have an effect at other locations along the river.

Water flows downhill, and the source (the start) of a river will be at a higher altitude than its mouth (the end). As the water moves over the Earth's surface, it erodes, transports and deposits material.

The volume of water and the speed of flow will influence the amount and type of work carried out by a river. A fast-flowing flooded river will erode enormous amounts of material and transport it **downstream**. As the speed or volume of the water decreases, much of the material it carries will be deposited.

Watershed

A river gathers its water from a region known as its drainage basin, or **catchment** (**FIGURE 2**). The boundary of this region is identified by mountain tops, hilltops or any land that is slightly higher than surrounding land. This is known as the watershed, and it is the point that divides the direction of water flow.

perennial describes a stream that flows all year

intermittent describes a stream that does not always flow

groundwater water that seeps into soil and gaps in rocks

downstream nearer the mouth of a river, or going in the same direction as the current

catchment the area where the water that forms a river collects

FIGURE 2 The watershed and catchment, or drainage basin, of a river system

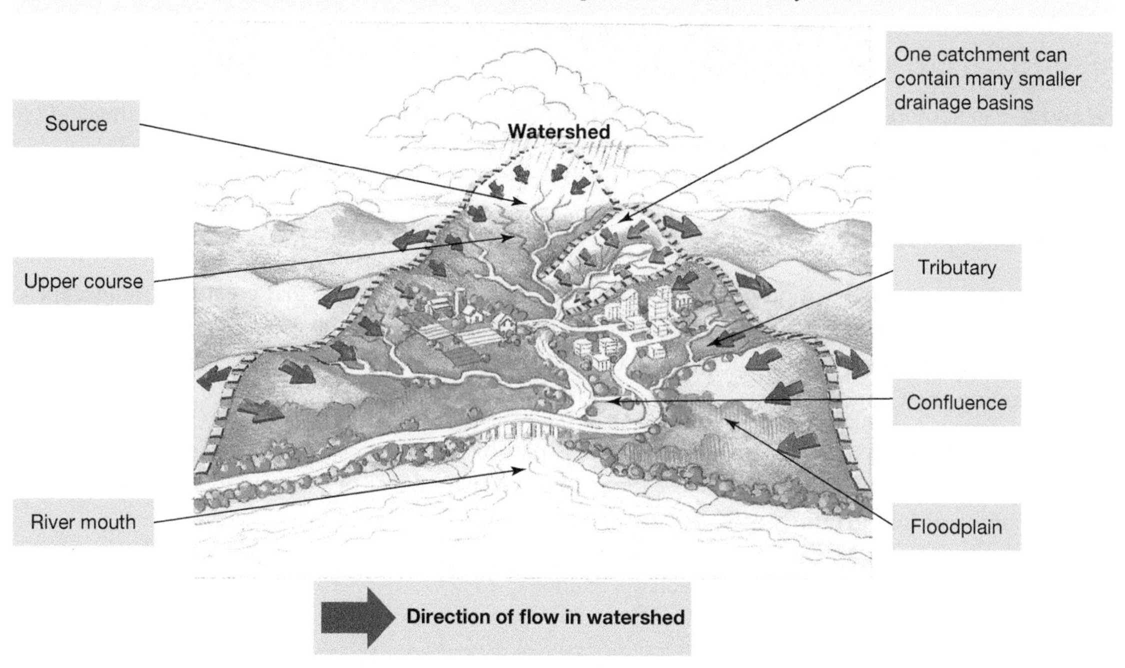

River formation

Even the biggest rivers begin with water from rain or melting snow in mountains or hills. This water has collected in tiny depressions called rills. These rills grow larger when they collect more water, and when they combine they begin to look like streams. Many streams contribute to a river (see **FIGURE 2**).

Upper course

Waters in a river channel flow over steeper slopes in the upper reaches. The force of rushing water on a steep slope cuts downwards and creates a V-shaped valley. The river then tends to follow a fairly straight course.

FIGURE 3 A waterfall

Waterfall retreats.

Hard rock

Overhang

Ridges of hard rock create an uneven slope. This creates rapids.

Steep-sided gorge develops as waterfall retreats.

Plunge pool

Soft rock

Fallen rocks

Hard rock

Waterfall

When a river meets resistant (hard) rock, a waterfall can occur (see **FIGURE 3**). If the river has to cross bands of resistant rock, rapids will form. The turbulent water flow in rapids is called white water. A plunge pool forms at the base of a waterfall when rocks and soil moved by the fast-flowing water erode the banks and base of the river.

Meanders

On flatter land, a river is wider than it is in the hills, and water added from tributaries has increased its volume (see **FIGURE 5**). Much of the erosion is in a sideways direction, and the valley of a river is much wider. Sideways erosion causes meanders (curves) along its course (see **FIGURE 4**). Over time, a meandering river will change the path it follows, as some bends become more obvious and some disappear. A meander that is cut off is called an oxbow lake.

Tributary

A river or stream that adds or contributes water to the main river is known as a **tributary** (see **FIGURE 2**). The place where two rivers join is called the **confluence**.

Floodplains

Flooding over thousands of years creates floodplains. During a flood, the water flows over the banks of the river. Once outside the river, it slows down and deposits the **alluvium** it was transporting. This alluvium is often very fertile. These regions are highly suitable for farming and settlement (see **FIGURE 5**).

River mouth

Deltas are found at the mouths of large rivers, such as the Mississippi in the USA (see **FIGURE 7**). A delta is formed when a river deposits its material faster than the sea can remove it. The material is a mix of mud, sand and clay. The river will sometimes split into smaller streams to find its way through the deposited material. These little streams are called distributaries. There are three main shapes of delta: fan shaped, arrow shaped and bird-foot shaped. The shape is influenced by tides, ocean waves and the volume of sediment and river water.

Sometimes a river will have a wide mouth, where fresh water and salt water mix. This is known as an estuary.

FIGURE 4 The formation of a meander and oxbow lake

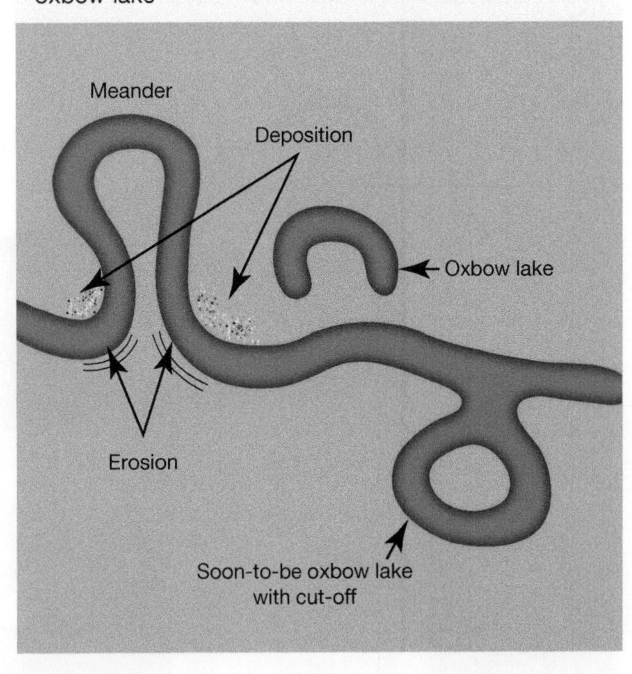

FIGURE 5 Murray River meanders

tributary river or stream that flows into a larger river or body of water

confluence where two rivers join

alluvium sand, silt, clay and gravel being transported in flowing water

FIGURE 6 (a) Long profile of a river — a view along its length. The slope of the river tends to get flatter and the riverbed smoother as it moves downstream. (b)–(d) Cross-sections showing the shape of the river channel and valley at three points along the river — arrows indicate the main direction of erosion.

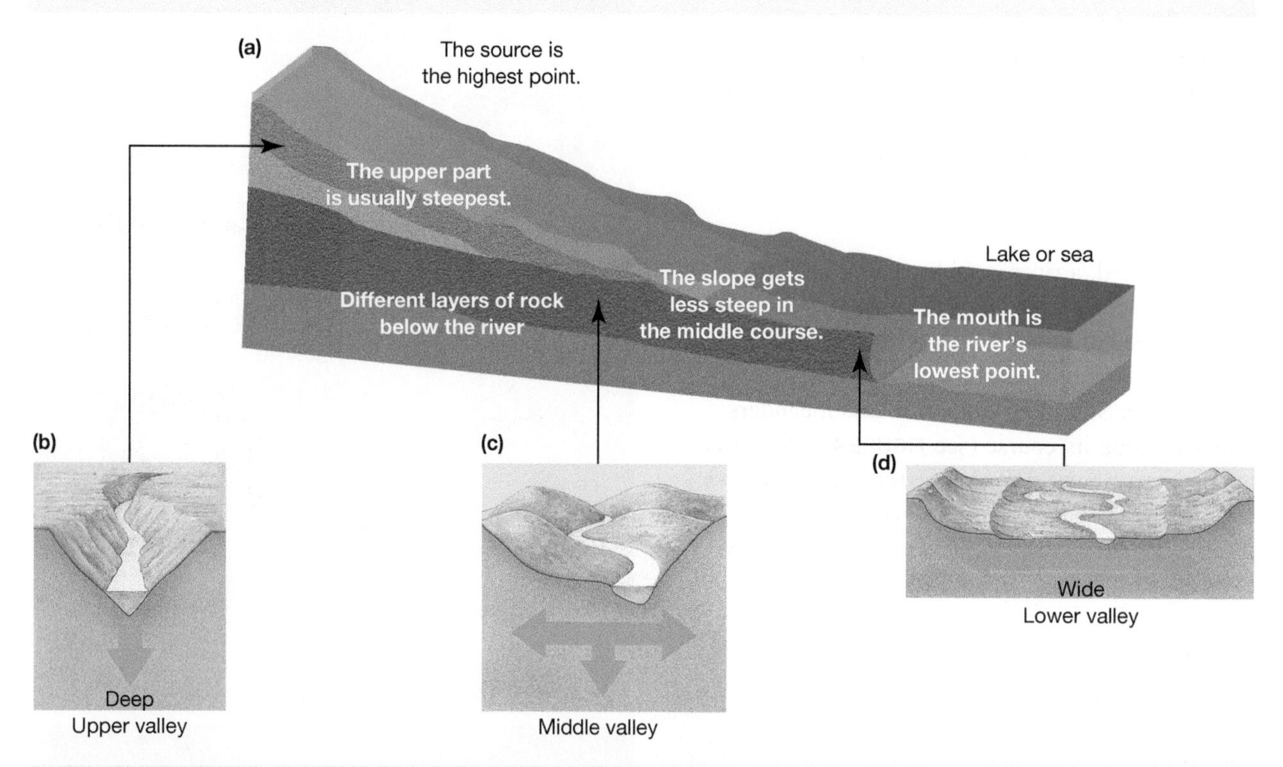

FIGURE 7 Satellite image of the Mississippi River Delta, Louisiana, USA

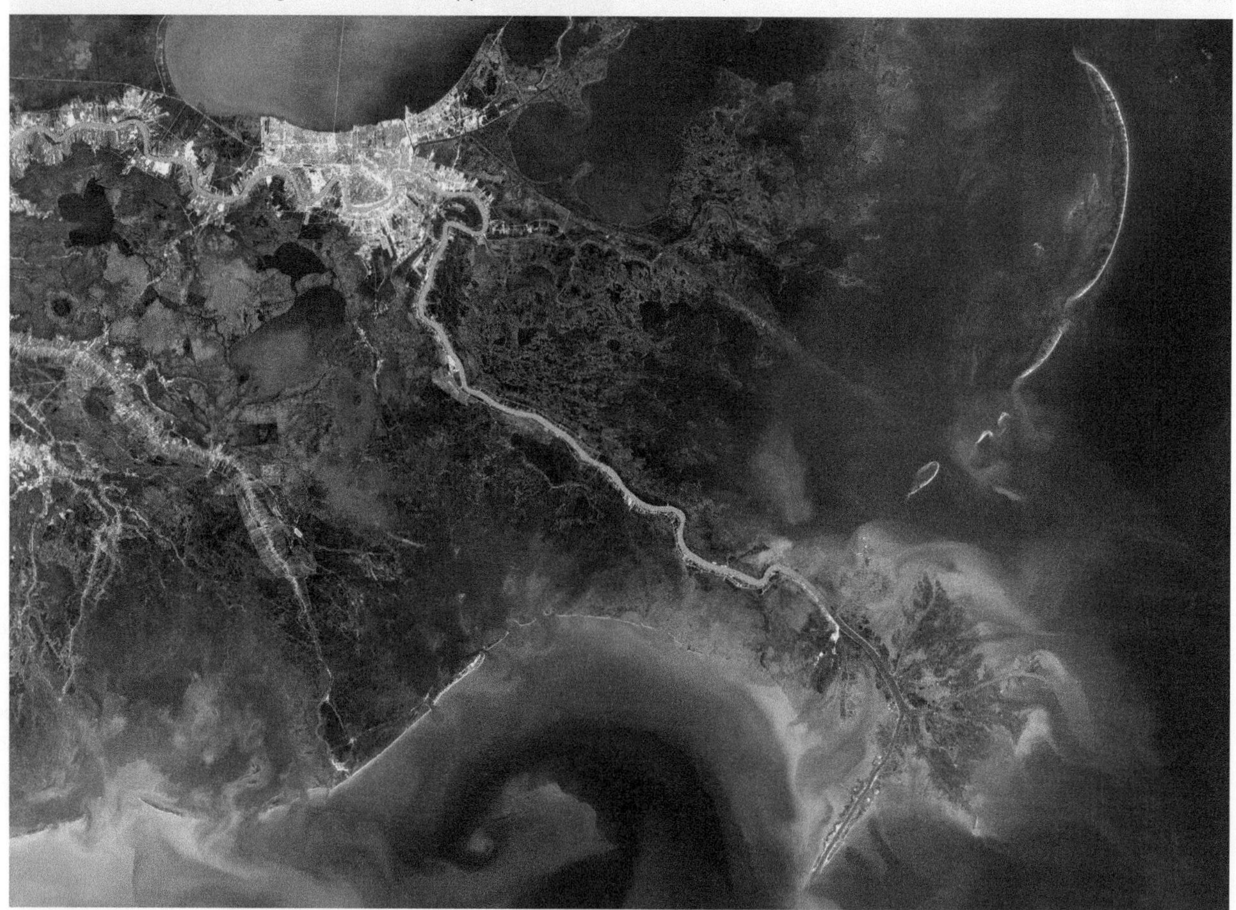

Resources

- **eWorkbook** Landscapes created by rivers (ewbk-8644)
- **Video eLesson** Landscapes created by rivers — Key concepts (eles-5028)
- **Interactivity** River carvings (int-3104)
- **myWorldAtlas** Deepen your understanding of this topic with related case studies and questions > **Fjords**

2.8 ACTIVITIES

1. Identify a river that flows through the capital city in one state or territory in Australia. Describe its source, any tributaries, and its mouth.
2. After some rain, investigate an area of bare ground on a small slope near school or home. Sketch the pattern that the rills have made. Identify the watershed and catchment for each rill.
3. Using Google Earth or an atlas, find the Nile delta, the Ebro delta and the Mississippi delta. Draw a sketch and write a short description of the shape of each delta, presenting your findings in a table.
4. Produce a flow chart or animation to explain the formation of an oxbow lake, a delta, a waterfall or rapids.
5. Refer to the paragraph about meanders and to **FIGURES 4**, **5** and **6**. Sketch a cross-section (like **FIGURE 6**) of a river at a meander. This will show the shape of the riverbank on each side of the river. What are the advantages and disadvantages of living on each side of the river?

2.8 EXERCISE

Learning pathways

■ LEVEL 1	■ LEVEL 2	■ LEVEL 3
3, 4, 7, 8, 15	1, 2, 10, 11, 14	5, 6, 9, 12, 13

Check your understanding

1. Refer to **FIGURE 1** and compare the scale of Australia's longest river with the world's longest river.
2. What feature, other than water, has to be present for waterfalls and rapids to form? Refer to **FIGURE 3**.
3. Define the following terms in your own words.
   a. Groundwater
   b. Tributary
   c. Alluvium
4. What does the term *confluence* mean?
5. In which direction does a meander erode the land?
6. Outline the process or processes that cause a river mouth to form a delta.

Apply your understanding

7. Explain how rivers are part of the water cycle.
8. Why do people settle and farm on floodplains?
9. Refer to **FIGURE 5**. Predict and label where the next oxbow lake might form.
10. What changes will occur along a river if there is unusually high rainfall along its upper course? Think in terms of erosion and deposition.
11. Sketch the long profile shown in **FIGURE 6**. Label the source, the mouth and the direction the river flows.
12. Explain why the upper course of a river creates a V-shaped valley.

Challenge your understanding

13. What do you think will happen to deltas if sea levels rise?
14. Predict the changes that will occur to the waterfall in **FIGURE 3**. Justify your answer.
15. Examine **FIGURE 3**. Predict what will eventually happen to the overhang.

To answer questions online and to receive **immediate feedback** and **sample responses** for every question, go to your learnON title at www.jacplus.com.au.

2.9 Coastal landscapes shaped by erosion

online only

LEARNING INTENTION

By the end of this subtopic, you will be able to identify coastal features created by erosion, and distinguish between the impacts of wave action and running water.

The content in this subtopic is a summary of what you will find in the online resource.

Coastal landscapes have landforms that are common to coastlines in different places around the world because they are built up or worn away in similar ways.

To learn more about erosion of coastal landscapes, go to your learnON resources at www.jacPLUS.com.au.

Contents

- 2.9.1 Erosion of coastal landscapes
- 2.9.2 Coastal landscape features created by erosion

Resources

eWorkbook	Coastal landscapes shaped by erosion (ewbk-8648)
Video eLesson	Coastal landscapes shaped by erosion — Key concepts (eles-5029)
Interactivity	Coastal sculpture (int-3124)
myWorldAtlas	Deepen your understanding of this topic with related case studies and questions > **Coastal processes**

2.10 Coastal landscapes shaped by deposition

online only

LEARNING INTENTION

By the end of this subtopic, you will be able to identify coastal features created by deposition, and distinguish between constructive and destructive waves.

The content in this subtopic is a summary of what you will find in the online resource.

Smaller, gentle waves push material such as sand and shells on to the beach, building new coastal features. This is known as deposition.

To learn more about coastal deposition, go to your learnON resources at www.jacPLUS.com.au.

Contents

- 2.10.1 Constructive waves
- 2.10.2 The role of the wind

Resources

eWorkbook	Coastal landscapes shaped by deposition (ewbk-8652)
Video eLesson	Coastal landscapes shaped by deposition — Key concepts (eles-5030)
Interactivity	Coastal landscapes shaped by deposition — Interactivity (int-7838)

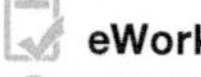

2.11 Undertaking coastal fieldwork

online only

LEARNING INTENTION

By the end of this subtopic, you will be able to undertake preparation and planning for coastal fieldwork.

The content in this subtopic is a summary of what you will find in the online resource.

2.11.1 Your fieldwork task

A fieldwork activity will allow you to put the knowledge you have gained in the classroom into practice. Any coastal landscape would be suitable to investigate. Once a fieldwork site has been identified, there is quite a lot of planning that you should do before you get there.

What is your fieldwork task?

Your task is to:

1. Identify the landforms and dynamic nature of a coastal landscape.
2. Assess the influence of people on coastal landscape.

Go to learnON to access more detailed instructions for conducting coastal fieldwork.

FIGURE 1 Lighthouse Beach, Ballina, NSW

FIGURE 2 Breakwater, East Ballina

fdw-0011

FOCUS ON FIELDWORK — Creating a labelled field sketch

Sketching the land and its features when you are doing fieldwork can help you to understand and look more closely at what you can see. To draw something, you have to focus on it carefully and pay attention to how all of the parts of the landscape work together.

Learn more about creating a field sketch in 2.12 SkillBuilder — Constructing a field sketch, then practise using the **Creating a labelled field sketch** activity in your online Resources panel.

on Resources

eWorkbook How do I undertake coastal fieldwork? (ewbk-8656)

Fieldwork Creating a labelled field sketch (fdw-0011)

2.12 SkillBuilder — Constructing a field sketch

online only

LEARNING INTENTION

By the end of this subtopic, you will be able to identify the key features of a field sketch and record geographical information in the form of a field sketch.

The content in this subtopic is a summary of what you will find in the online resource.

2.12.1 Tell me

What are field sketches?

Field sketches are drawings completed during fieldwork — geography outside the classroom. They are free-hand drawings with annotations that capture only the important or relevant features for your research. This kind of sketch helps to record and interpret environments by forcing us to look closely at the relevant information. It may be easier to take a photo, but you are then capturing the non-relevant data as well.

FIGURE 1 Field sketch of Cape Schanck

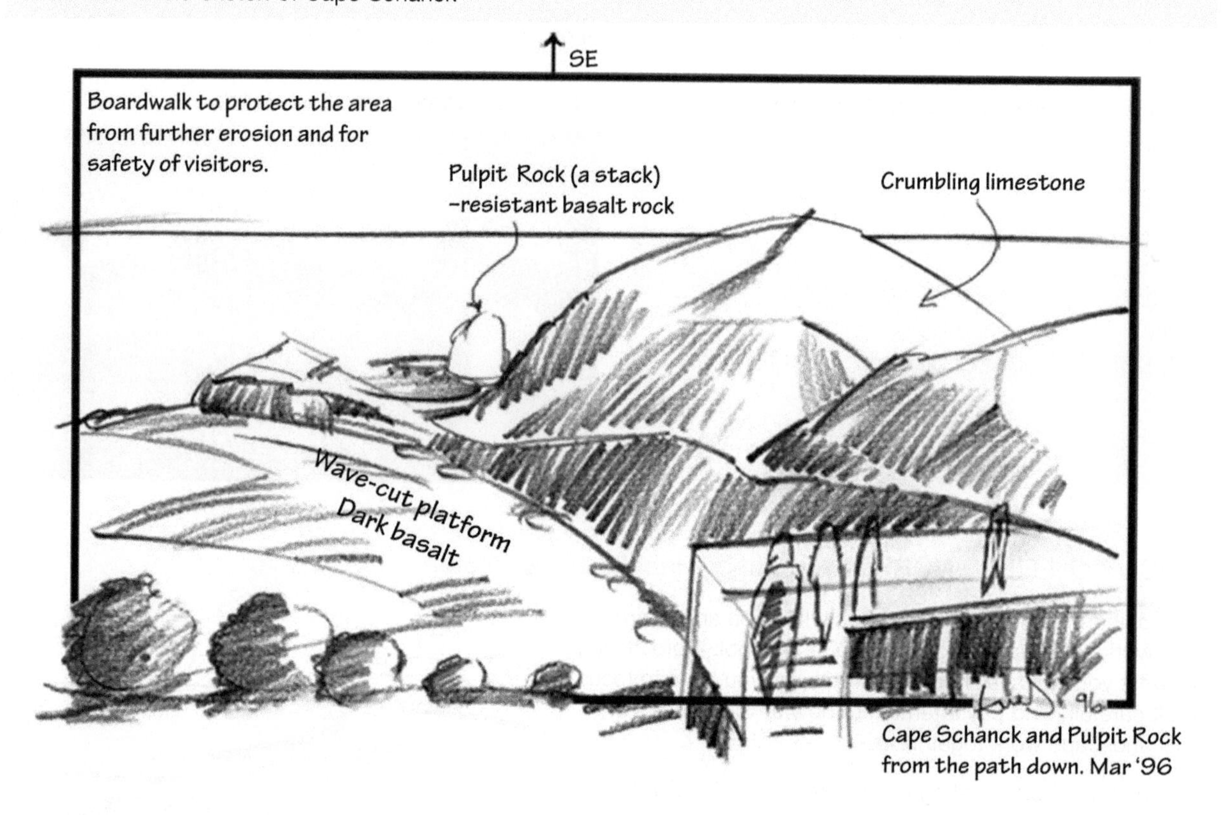

2.12.2 Show me

How to complete a field sketch

Step 1

Choose the field of view to be sketched. Make yourself comfortable so you can take your time to draw.

Step 2

Using a pencil, draw a border (frame) around your sketch area.

Step 3

Draw the horizon as a baseline; that is, where the land meets the sky.

Step 4

Divide your sketch horizontally into three portions: background, middle ground and foreground (the closest to you).

Step 5

Partly close your eyes to focus on the main outlines. Draw these shapes and outlines within your border, placing them in the correct position according to your horizon.

Step 6

Add main features in the background (most distant), then middle ground and finally foreground. There will be a few shapes on your page, but no detail.

Step 7

Identify the aspects that are the most relevant to your study; add these in with detail and shading.

Step 8

Annotate (label) your sketch to draw attention to the landscape features and how they are interconnected. (See **FIGURE 1**)

Step 9

- Add a direction indicator on the border to show which way you are looking at the landscape.
- Title your sketch (identify the place with as much detail as possible).
- Date your drawing.
- Include the direction you are facing (if known).

2.12.3 Let me do it

Go to learnON to access the following additional resources to help you build this skill:

- a longer explanation of this skill and its application in Geography (Tell me)
- a video showing the step-by-step process of this skill (Show me)
- an activity and interactivity for you to practise this skill (Let me do it)
- self-marking questions to help you understand and use this skill.

Resources

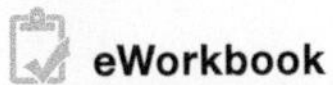
eWorkbook SkillBuilder — Constructing a field sketch (ewbk-8610)

Video eLesson SkillBuilder — Completing a field sketch (eles-1650)

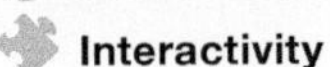
Interactivity SkillBuilder — Completing a field sketch (int-3146)

2.13 Investigating topographic maps — What is a topographic map?

LEARNING INTENTION

By the end of this subtopic, you will be able to identify each of the elements of BOLTSS on a topographic map and explain the purpose of contour lines.

2.13.1 What is a topographic map?

A topographic map is a representation of the Earth that shows the features of a landscape. The map is overlaid with a coordinate grid. You can revise the basic elements of a topographic map (contour lines) with SkillBuilder 2.4 — Recognising land features.

2.13.2 Features of a topographic map

- **Cultural features:** urban developments, buildings, borders and boundaries, and transportation infrastructure such as roads, bridges, airports and railways
- **Vegetation:** different types of vegetation such as grasslands, heathlands, peatlands, forests and woodlands
- **Water sources:** physical features related to water such as rivers, oceans, lakes and swamps
- **Relief:** the height of the land and different landforms such as mountains, slopes and depressions

2.13.3 How to interpret a topographic map

The basic features that help you read a topographic map are the border, orientation, legend, title, scale and source. This can be remembered with the acronym BOLTSS.

B Border
O Orientation
L Legend
T Title
S Scale
S Source

Border

The border defines the boundaries of the map face. Around the edges of the border you will find grid coordinates and possibly latitude and longitude references. The border outlines all the key features of the map.

Orientation

Direction is used to tell us the location of one place in relation to another. The four cardinal points are:

- north
- south
- east
- west.

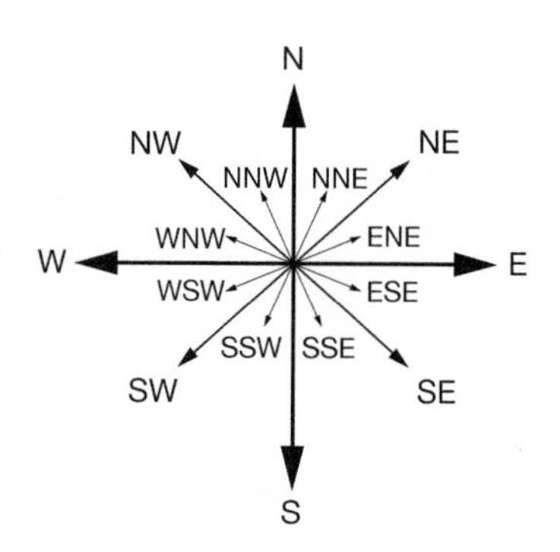

Legend

The features shown on a topographic map are explained in a key or legend (see below). The key provides symbols with explanations.

Legend

121 . Spot height
Building
Index contour
Contour (interval 25 m)
route no. A7 Highway
Major road
unsealed Minor road
Track
unsealed Private road
River
Stream
Water body
Intertidal
Vegetation
Plantation
Mine
Urban area

Title

The title tells you the location and purpose of the map.

Scale

Scale describes the relationship between the size of an object in real life and the size of an object on a map. Scale helps us to determine the distance of one location from another.

Scale can be shown in three ways.

Linear scale

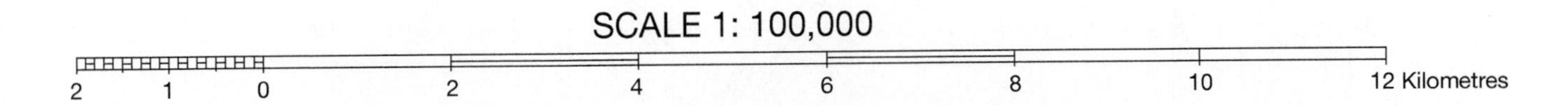

Scale statement

A scale statement expresses the map scale in words. For example:

- One centimetre represents half a kilometre.

Scale as a ratio

A map scale may be expressed as the ratio of the distance on the map to the corresponding distance on the ground. For example:

- 1 : 50 000

The most common scale is 1 : 100 000. This means that 1 centimetre represents 100 000 centimetres, or 1000 metres or 1 kilometre; so, if you measure 4 centimetres on the map it represents 4 kilometres on the ground.

Source

For Australian locations, many topographic maps are created by the state and federal governments of Australia. Others can be made by private companies. It is always important to examine the source of a map to ensure it is from a trusted source. For example:

- Map supplied by MAPLAND, Environmental and Geographic Information, Department for Environment and Heritage, South Australia.

2.13.4 Contour lines

Topographic maps enable us to understand the shape of the land even though they are shown in two dimensions (on a flat piece of paper).

Contour lines and **spot heights** help us to determine the altitude of different places on the map.

contour line a line joining places of equal height above sea level

spot height a point on a map that shows the exact height above sea level (in metres) at that place

FIGURE 1 Topographic map extract, Tamar Valley, Tasmania, 2021

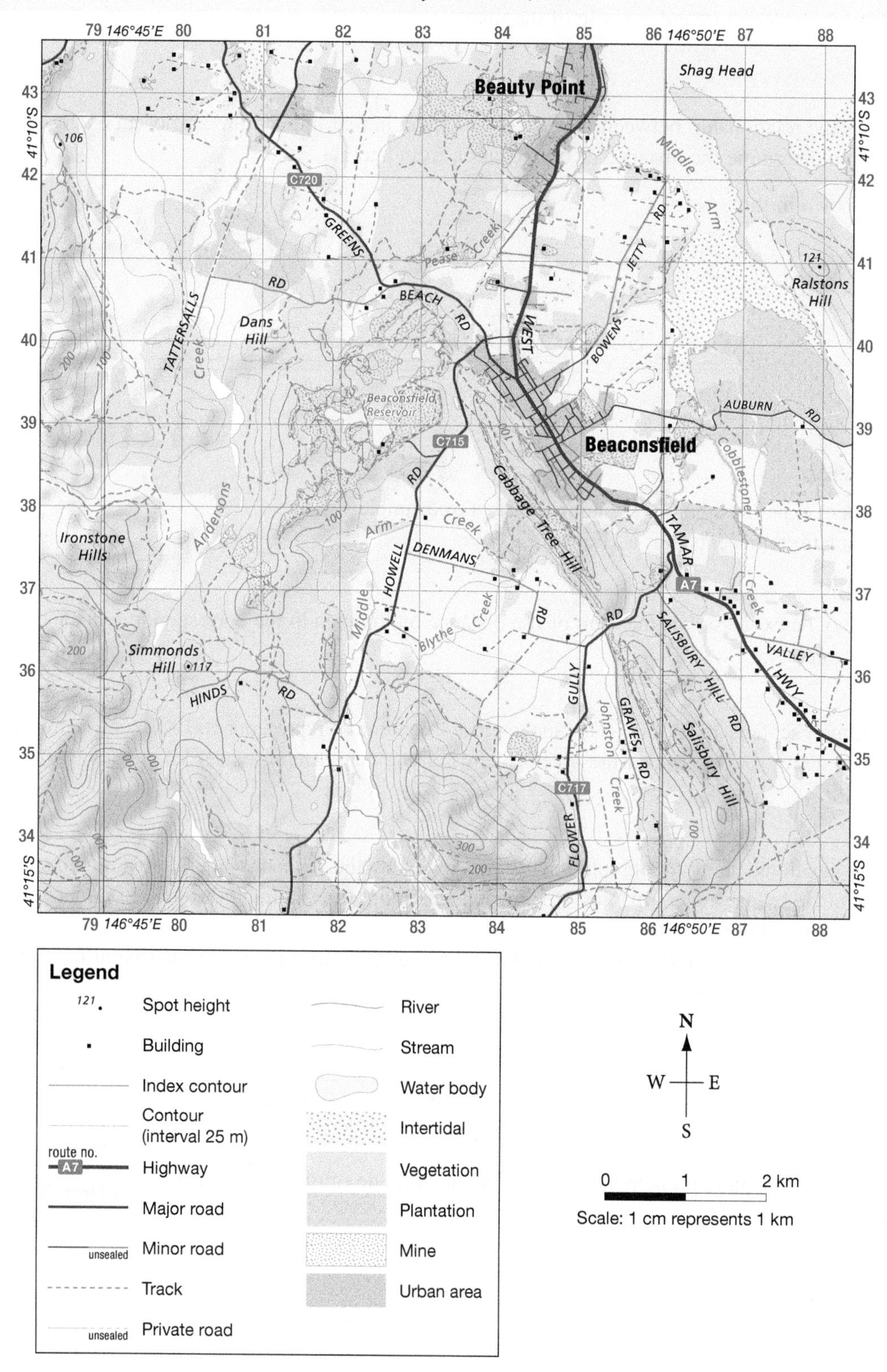

Source: Address Points, Coastline, Contours, Hydrographic Areas, Hydrographic Lines, Spot Heights, Tasmania 25m DEM, TASVEG 3.0, Transport Nodes, Transport Segments from www.theLIST.tas.gov.au © State of Tasmania

on Resources

eWorkbook	Investigating topographic maps (ewbk-8672)
Digital document	Topographic map of Tamar Valley, Tasmania (doc-36187)
Video eLesson	Investigating topographic maps — Key concepts (eles-5032)
Interactivity	Investigating topographic maps (int-8360)
Google Earth	Tamar Valley, Tasmania (gogl-0132)

2.13 EXERCISE

Learning pathways

■ LEVEL 1	■ LEVEL 2	■ LEVEL 3
3, 5, 6, 7, 8, 12	1, 2, 4, 9, 13	10, 11, 14, 15

Check your understanding

1. Find two examples of each of the following in **FIGURE 1**:
   a. roads that run through areas of vegetation
   b. fresh water sources
   c. hills or mountains.
2. What important information sits around the border of a topographic map?
3. List the four cardinal points.
4. Use the orientation arrow of the map to answer the following.
   a. If you travelled from Beaconsfield to Beauty Point on the West Tamar Highway, what direction would you be travelling?
   b. Which hill is further west: Simmonds Hill, Ralstons Hill, Salisbury Hill or Dans Hill?
5. How many water sources are included in the legend?
6. Describe how plantations are shown differently to other types of vegetation.
7. Describe how unsealed roads are shown differently to sealed roads.
8. What information does the title of **FIGURE 1** tell us about this map?
9. Identify the scale statement in **FIGURE 1**.
10. Based on the source line of **FIGURE 1**, has any of the information used in the map come from a government source?
11. Locate Ralstons Hill. What is the height of this landform?

Apply your understanding

12. What is the purpose of the scale on a map? Explain your answer.
13. Why would it be difficult to interpret a map that did not have a key or legend?
14. **FIGURE 1** has a scale statement. How would you express this as a ratio?
15. If you were going to build a new housing development at Beauty Point, would you plan to build it to the north, south, east or west of town? Justify your answer by giving examples of why the three directions you did not choose would not be good options.

To answer questions online and to receive **immediate feedback** and **sample responses** for every question, go to your learnON title at www.jacplus.com.au.

2.14 Thinking Big research project — Coastal erosion animation

online only

The content in this subtopic is a summary of what you will find in the online resource.

Scenario

The erosive power of water is one of the strongest forces of nature. The continued pounding of waves upon rocks, cliffs and beaches has sculpted our coastal landscapes into the diverse environments that we see today. From towering cliffs to lonely sea stacks and deep, dark caves, all across the world, there are thousands of examples of the effect of erosion on coastal landscapes.

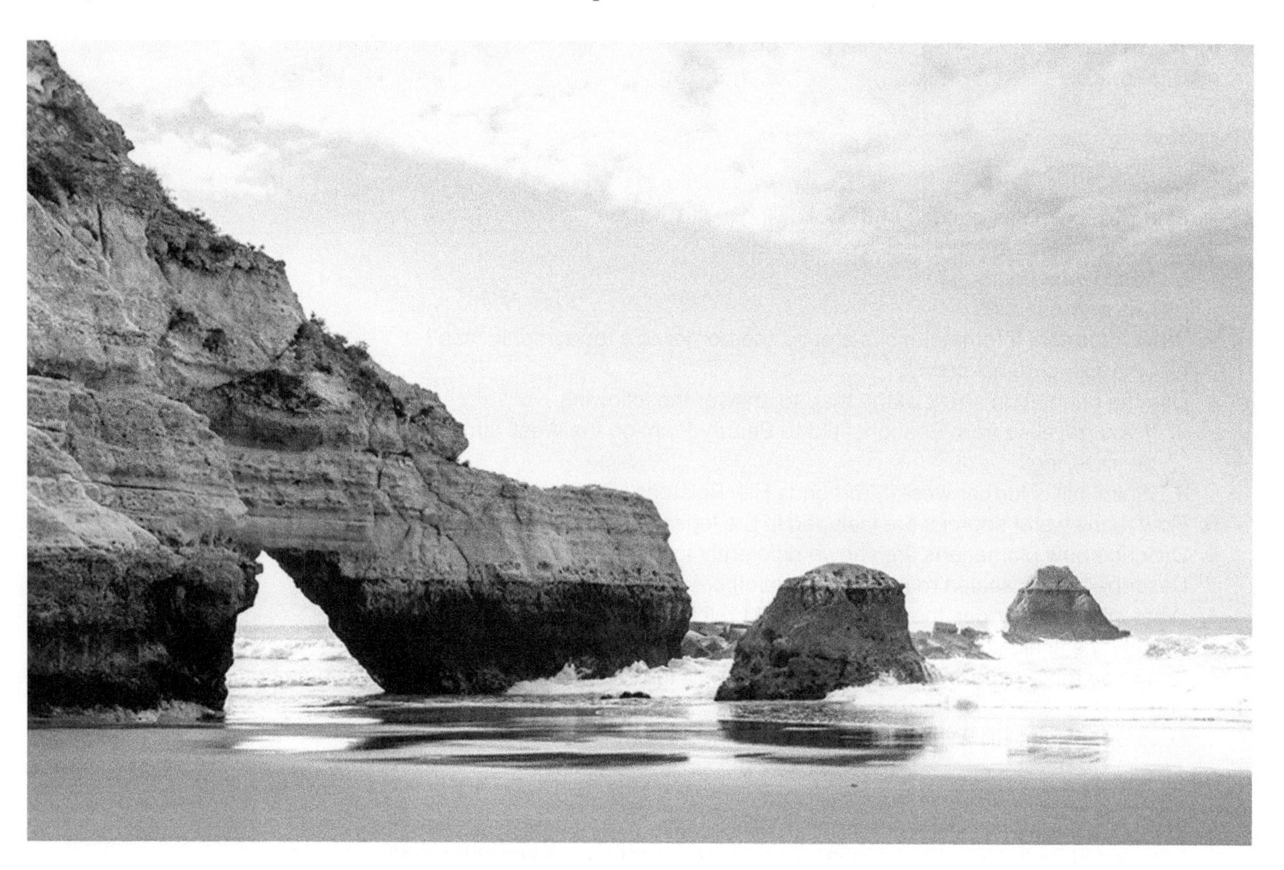

Task

Unless you are lucky enough to be watching at the exact moment that a sea stack tumbles into the ocean, it can be difficult to catch erosion in action. Although wave action is continuous, it can often take an extremely long time for us to see the effects of erosion. In this task, you will do what few people before you have achieved — you will capture the effect of erosion on film! You will complete this task by creating an animation (either hand-drawn or using a computer) that shows how a coastal landform is created. Your animation should be accompanied by written annotations that describe each step of the process. Go to your Resources tab to access all of the resources you need to complete this research project.

 Resources

 **ProjectsPLUS** Thinking Big research project — Coastal erosion animation (pro-0169)

2.15 Review

2.15.1 Key knowledge summary

2.2 Different types of landscapes

- There are a variety of landscapes such as mountain, desert, rainforest, grassland, polar, karst, aquatic, island and built.

2.3 Landscapes and landforms in Australia

- Folding, faulting and volcanic activity are tectonic forces that have shaped Australia's major landforms.
- Australia is an ancient landmass and parts of the continent are about 4300 million years old.
- Australia's landform regions are the coastal lowlands, the eastern highlands, the central lowlands and the Great Western Plateau.

2.5 Landscapes in the Pacific region

- The Pacific islands are divided into three main groups: Melanesia, Micronesia and Polynesia.
- Low islands are the remains of volcanoes that have eroded over time, or are coral reefs or atolls built on top of reefs, and are common in the Pacific.
- Volcanoes have created a range of larger, higher islands in the Pacific, including Tahiti, Fiji, Vanuatu and Rarotonga.

2.6 Processes that transform landscapes

- Geomorphic processes include uplift, erosion, deposition and weathering.
- The Earth's surface is split into a number of plates that sit on a layer of semi-molten material. Convection currents move the plates around. Plate movement can create mountains.
- Erosion is the wearing away of the Earth's surface by natural elements such as wind, water and ice. Deposition is the laying down of material carried by rivers, wind, ice and ocean currents or waves.
- Rushing water can shift rocks, remove topsoil and shape river valleys.
- Gentle rain can change the chemical structure of rocks so that they break up and dissolve.
- Frozen water can erode and scrape away rocks and soil.

2.7 Landscapes created underground

- Water can dissolve rock underground, creating networks of caves, channels, cracks (fissures), sinkholes and vertical shafts.
- Water becomes acidic when exposed to carbon dioxide in the atmosphere and when it filters through organic material in soils. The acidic water is then able to dissolve soluble bedrock.

2.8 Landscapes created by rivers

- Water flows downhill, and the source (the start) of a river will be at a higher altitude than its mouth (the end). As the water moves over the Earth's surface, it erodes, transports and deposits material.
- A fast-flowing flooded river will erode enormous amounts of material and transport it downstream.
- As water slows, it deposits much of the material it carries along the river.

2.9 Coastal landscapes shaped by erosion

- Powerful ocean waves (destructive waves) crash onto rocky coastlines, wearing away the cliff base between the high and low tide marks.
- Coastal erosion is mostly caused by waves moving sand and other material and energy to and from the beach.
- Features such as cliffs, headlands, bays, caves and stacks are all landforms found along an eroding coastline.

2.10 Coastal landscapes shaped by deposition

- Smaller, gentler waves that carry less energy than destructive waves are known as constructive waves. They push material such as sand and shells and deposit them on the beach, building new coastal features.
- The coastal features created by deposition are created by materials such as sand, shells, coral and pebbles.
- The construction material may come from eroding cliffs, from an offshore source, or from rivers which, when they enter the sea, dump any material they were transporting.

2.15.2 Key terms

alluvium sand, silt, clay and gravel being transported in flowing water

aquifer a body of permeable rock below the Earth's surface which contains water, known as groundwater (*permeable* means liquid or gas can pass through)

archipelago a chain or line of islands

arid somewhere dry, without much precipitation

backwash the movement of water from a broken wave as it runs down a beach returning to the ocean

catchment the area where the water that forms a river collects

confluence where two rivers join

constructive wave a gentle backwash that leads to material being deposited on land

contour line a line joining places of equal height above sea level

convection current a current created when a fluid is heated, making it less dense, and causing it to rise through surrounding fluid and to sink if it is cooled; a steady source of heat can start a continuous current flow

coral reef (or atoll) reef that partially or completely encircles a lagoon

crust the Earth's outer layer or surface

deposition the laying down of material carried by rivers, wind, ice and ocean currents or waves

destructive wave a large powerful storm wave that has a strong backwash

downstream nearer the mouth of a river, or going in the same direction as the current

drainage basin the entire area of land that contributes water to a river and its tributaries

erosion the wearing away and removal of soil and rock by natural elements, such as wind, waves, rivers or ice, and by human activity

field sketch a diagram with geographical features labelled or annotated

fissures cracks, especially in rocks

glacier a large body of ice, formed by a build-up of snow, which flows downhill under the pressure of its own weight

groundwater water that seeps into soil and gaps in rocks

hotspot an area on the Earth's surface where the crust is quite thin, and volcanic activity can sometimes occur, even though it is not at a plate margin

human features structures built by people

intermittent describes a stream that does not always flow

islet a very small island

karst underground landscapes of caves and water channels

lagoon a shallow body of water separated by islands or reefs from a larger body of water, such as a sea

landform distinctive, natural feature occurring on the surface of the Earth

landscape visible features occurring across an area of land

longshore drift backwash returning to the sea by the shortest possible route

mantle the layer of the Earth between the crust and the core

perennial describes a stream that flows all year

permafrost a layer beneath the surface of the soil where the ground is permanently frozen

plateau a large area of flat land that is higher than the land around it. Plateaus are sometimes referred to as tablelands

prevailing wind the main direction from which the wind blows

Ring of Fire the area around the Pacific Ocean where most of the world's earthquakes occur and volcanoes are found

soluble something that can be dissolved

spot height a point on a map that shows the exact height above sea level (in metres) at that place

stalactite a feature made of minerals, which forms from the ceiling of limestone caves, like an icicle

stalagmite a feature made of minerals found on the floor of limestone caves

swash the movement of water in a wave as it breaks onto a beach

tectonic relating to force or movement in the Earth's crust

tectonic plate one of the slow-moving plates that make up the Earth's crust. Volcanoes and earthquakes often occur at the edges of plates.

transport the movement of eroded materials to a new location by agents such as wind and water

tributary river or stream that flows into a larger river or body of water

tundra landscape without trees, with small shrubs and grass, found in cold or high-altitude places

weathering the breaking down of rock through the action of wind and water and the effects of climate, mainly by water freezing and cooling as a result of temperature change

2.15.3 Reflection

Complete the following to reflect on your learning.

Revisit the inquiry question posed in the Overview:

How do we identify landscapes and what landform features make them unique?

1. Now that you have completed this topic, describe the variety of landscapes and landforms on Earth. Discuss with a partner. Has your learning in this topic changed your understanding of landscapes and landforms? If so, how?
2. Write a paragraph in response to the inquiry question outlining your views.

Subtopic	Success criteria			
2.2	I can identify different landscapes and their key features.			
	I can locate different landscapes on a world map.			
2.3	I can describe the processes that have shaped the Australian landscape and explain how these processes work.			
	I can identify the key features of the Australian landscape.			
2.4	I can recognise key features of topographic maps.			
	I can identify landforms on a topographic map.			
2.5	I can identify landscapes and landforms in the Pacific region and explain the processes that shaped them.			
2.6	I can explain the processes that form landscapes.			
	I can explain how geomorphic processes change landscapes, including weathering, erosion and deposition.			
2.7	I can describe the features of karst landscapes and explain how they are formed.			
2.8	I can identify the different phases of a river.			
	I can describe the landform features of a river.			
2.9	I can identify coastal features created by erosion.			
	I can explain the impact of wave action and running water.			
2.10	I can identify coastal features created by deposition.			
	I can distinguish between constructive and destructive waves.			
2.11	I can undertake preparation and planning for coastal fieldwork.			
2.12	I can identify the key features of a field sketch.			
	I can create a field sketch to record geographical information.			
2.14	I can recognise and interpret orientation, legends, scale and contour lines on a topographic map.			

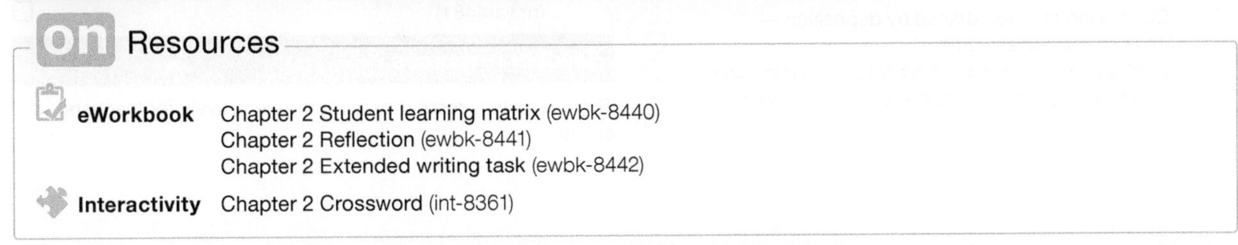
on Resources

eWorkbook Chapter 2 Student learning matrix (ewbk-8440)
Chapter 2 Reflection (ewbk-8441)
Chapter 2 Extended writing task (ewbk-8442)

Interactivity Chapter 2 Crossword (int-8361)

ONLINE RESOURCES

Below is a full list of **rich resources** available online for this topic. These resources are designed to bring ideas to life, to promote deep and lasting learning and to support the different learning needs of each individual.

eWorkbook

- 2.1 Chapter 2 eWorkbook (ewbk-7982) ☐
- 2.2 Different types of landscapes (ewbk-8620) ☐
- 2.3 Landscapes and landforms in Australia (ewbk-8624) ☐
- 2.4 SkillBuilder — Recognising land features (ewbk-8628) ☐
- 2.5 Landscapes in the Pacific region (ewbk-8632) ☐
- 2.6 Processes that transform landscapes (ewbk-8636) ☐
- 2.7 Landscapes created underground (ewbk-8640) ☐
- 2.8 Landscapes created by rivers (ewbk-8644) ☐
- 2.9 Coastal landscapes shaped by erosion (ewbk-8648) ☐
- 2.10 Coastal landscapes shaped by deposition (ewbk-8652) ☐
- 2.11 How do I undertake coastal fieldwork? (ewbk-8656) ☐
- 2.12 SkillBuilder — Constructing a field sketch (ewbk-8610) ☐
- 2.14 Investigating topographic maps (ewbk-8672) ☐
- 2.15 Chapter 2 Student learning matrix (ewbk-8440) ☐
- Chapter 2 Reflection (ewbk-8441) ☐
- Chapter 2 Extended writing task (ewbk-8442) ☐

Sample responses

- 2.1 Chapter 2 Sample responses (sar-0137) ☐

Digital documents

- 2.4 Topographic map of Yarra Yarra Creek Basin (doc-36315) ☐
- 2.6 Blank world map template (doc-36260) ☐
- 2.13 Topographic map of Tamar Valley, Tasmania (doc-36187) ☐

Video eLessons

- 2.1 World landscapes and landforms (eles-1623) ☐
- Landscapes and landforms — Photo essay (eles-5306) ☐
- 2.2 Different types of landscapes — Key concepts (eles-5023) ☐
- 2.3 Landscapes and landforms in Australia — Key concepts (eles-5024) ☐
- 2.4 SkillBuilder — Recognising land features (eles-1648) ☐
- 2.5 Landscapes in the Pacific region – Key concepts (eles-5025) ☐
- 2.6 Processes that transform landscapes — Key concepts (eles-5026) ☐
- 2.7 Landscapes created underground — Key concepts (eles-5027) ☐
- 2.8 Landscapes created by rivers — Key concepts (eles-5028) ☐
- 2.9 Coastal landscapes shaped by erosion — Key concepts (eles-5029) ☐
- 2.10 Coastal landscapes shaped by deposition — Key concepts (eles-5030) ☐
- 2.12 SkillBuilder — Completing a field sketch (eles-1650) ☐
- 2.13 Investigating topographic maps — Key concepts (eles-5032) ☐

Interactivities

- 2.2 Landscapes galore (int-3102) ☐
- 2.3 Kati Thanda–Lake Eyre (int-3605) ☐
- Australia's four major landform regions (int-3606) ☐
- 2.4 SkillBuilder — Recognising land features (int-3144) ☐
- Landforms matched to a topographic map (int-7829) ☐
- 2.5 Pacific island groups (int-8358) ☐
- 2.6 Break down! (int-3101) ☐
- 2.7 Underground wonders (int-3103) ☐
- 2.8 River carvings (int-3104) ☐
- 2.9 Coastal sculpture (int-3124) ☐
- 2.10 Coastal landscapes shaped by deposition (int-7838) ☐
- 2.12 SkillBuilder — Completing a field sketch (int-3146) ☐
- 2.13 Investigating topographic maps (int-8360) ☐
- 2.15 Chapter 2 Crossword (int-8361) ☐

ProjectsPLUS

- 2.14 Thinking Big research project — Coastal erosion animation (pro-0169) ☐

Weblinks

- 2.9 Cliffed coast (web-0069) ☐
- Stack formation (web-1036) ☐

Google Earth

- 2.14 Tamar Valley, Tasmania (gogl-0132) ☐

Fieldwork

- 2.11 Creating a labelled field sketch (fdw-0011) ☐

myWorld Atlas

- 2.2 Deepen your understanding of this topic with related case studies and questions > Grasslands (mwa-7335) ☐
- 2.3 Deepen your understanding of this topic with related case studies and questions > Murray–Darling Basin (mwa-4538) ☐
- 2.5 Deepen your understanding of this topic with related case studies and questions > Pacific nations (mwa-4438) ☐
- 2.6 Deepen your understanding of this topic with related case studies and questions > Active Earth (mwa-4475) ☐
- 2.8 Deepen your understanding of this topic with related case studies and questions > Fjords (mwa-7333) ☐
- 2.9 Deepen your understanding of this topic with related case studies and questions > Coastal processes (mwa-4481) ☐

Teacher resources

There are many resources available exclusively for teachers online.

3 Valuing, managing and protecting landscapes and landforms

INQUIRY SEQUENCE

To access a pre-test with **immediate feedback** and **sample responses** to every question in this chapter, select your learnON format at www.jacplus.com.au.

3.1 Overview

Numerous **videos** and **interactivities** are embedded just where you need them, at the point of learning, in your learnON title at www.jacplus.com.au. They will help you to learn the content and concepts covered in this topic.

From oceans to deserts to cities, what exactly are landscapes and how is each one unique?

3.1.1 Introduction

There are a variety of reasons why people value landscapes and landforms. They are central to many cultures' identities, they inspire because of their natural beauty and they play a role in the economic activities of humans. Whatever the reason, landscapes and landforms are important to us and we should manage and protect them for current and future generations to use and appreciate.

STARTER QUESTIONS

1. Use the **Iguazu Falls** weblink in your Resources tab to watch a video of the Iguazu Falls.
   a. List as many adjectives as you can to describe this landscape.
   b. As a class, use an online word cloud generator to create a word cloud of your class's adjectives. Ask each person in turn to call out their adjectives to be entered into the word cloud generator. Do include repeated words, so that they are bigger in the word cloud.
2. Identify and describe two ways in which Iguazu Falls has the potential to make money for its local community.
3. Humans need water to survive. Most Australians live near the coast, and our inland places also rely on access to water. Watch the **Landscapes sculpted by water** Video eLesson. How does water affect the value people place on where you live? Does it make it beautiful? Attract tourists? Create jobs? Represent an important spiritual or cultural idea? Create fun activities? Or something else?
4. Watch the **Valuing, managing and protecting landscapes and landforms — Photo essay**. What places have a special importance to you and your family, culture or town? Why are they considered to be special?

FIGURE 1 Iguazu Falls sits on the border of Argentina and Brazil, in South America (the falls are viewed here from Argentina).

on Resources

eWorkbook	Chapter 3 eWorkbook (ewbk-7983)
Video eLessons	Landscapes sculpted by water (eles-1624) Valuing, managing and protecting landscapes and landforms — Photo essay (eles-5307)

3.2 Valuing landforms and landscapes

LEARNING INTENTION

By the end of this subtopic, you will be able to explain the aesthetic value of landscapes and landforms.

3.2.1 Aesthetic value of landscapes and landforms

Landscapes and landforms are valuable because they play an important part in the way the physical environment functions. However, humans have always been interconnected with landscapes and believe they are important for their aesthetic, economic, cultural and spiritual values.

Many landscapes and landforms are valuable to humans because of their natural beauty. This is also known as **aesthetic** value. All sorts of landscapes can inspire people, from vibrant and colourful coral reef aquatic ecosystems to beautiful mountain landscapes.

Many recreational activities are taken up because of the aesthetic value of landscapes and landforms (**FIGURE 1**) and the enjoyment that being in a beautiful place brings. People participate in outdoor recreations such as surfing, scuba diving, bushwalking, rock sports and photography.

The aesthetic value of the beach plays an important part in Australia's national identity. Globally, Australia is well known for its beautiful beaches and its leisure lifestyle based around the coast (**FIGURE 2**). Spending time at the beach and surf lifesavers are a part of that culture. Many international tourists will visit one or more of Australia's famous beaches, such as Bondi Beach in Sydney, Broadbeach on the Gold Coast or Bells Beach in Victoria. Australia is a large island continent with 10 685 beaches, and about 85 per cent of the population lives within 50 kilometres of the coastline, making access to the beach relatively easy. Important coastal places, such as the Great Barrier Reef in Queensland, Kangaroo Island in South Australia and Ningaloo Reef in Western Australia are also valued around the world for their natural beauty.

FIGURE 1 Outdoor recreational activities take advantage of the natural beauty of landscapes and landforms.

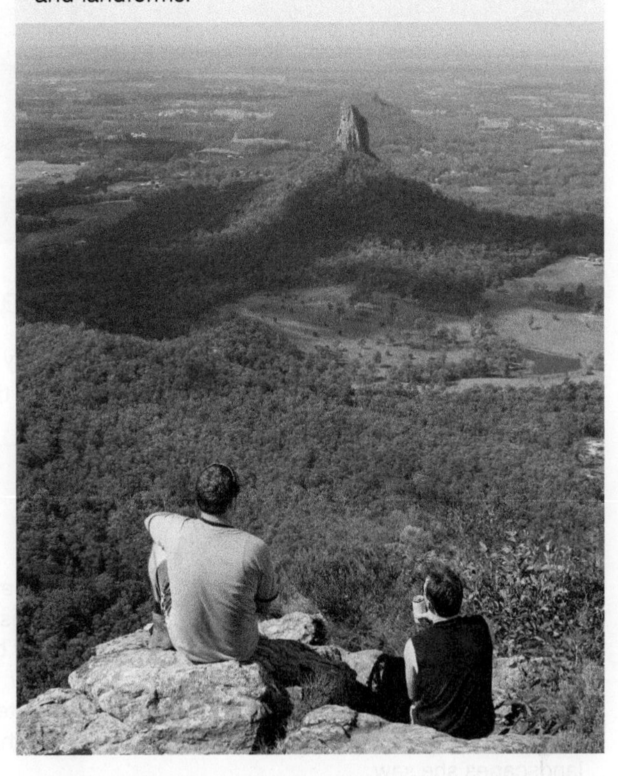

FIGURE 2 The beach and lifesavers are recognised as part of Australia's cultural identity.

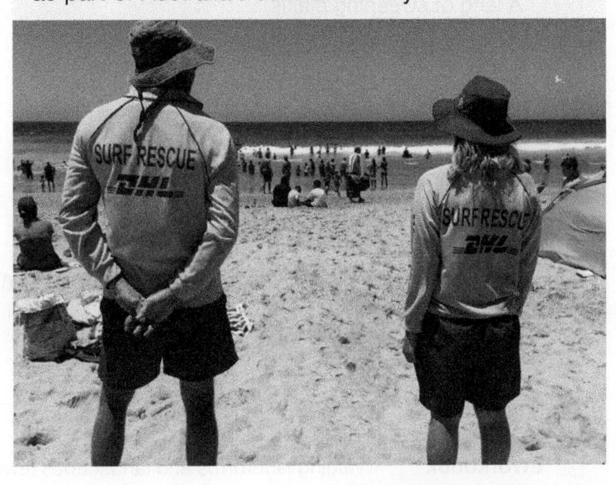

aesthetic relating to something beautiful or pleasing to look at

FIGURE 3 The Blue Mountains in NSW are known for their aesthetic value.

3.2.2 Landscapes and landforms as creative inspiration

Artists, poets and musicians have also been inspired by the aesthetics of landscapes, and have created some of the most famous pieces of poetry, art and songs about them.

CASE STUDY

'My Country'

Dorothea Mackellar (shown, right) was an Australian poet who wrote 'My Country', one of the most famous poems about the Australian landscape. She grew up in the city but her family owned properties in the Hunter Valley of New South Wales, and near Gunnedah to the north. She visited these places frequently and was heavily influenced by the landscapes she saw.

> I love a sunburnt country,
> A land of sweeping plains,
> Of ragged mountain ranges,
> Of droughts and flooding rains.
> I love her far horizons,
> I love her jewel-sea,
> Her beauty and her terror –
> The wide brown land for me!

Source: The Dorothea Mackellar Estate c/- Curtis Brown (Aust) Pty Ltd

on Resources

eWorkbook Valuing landforms and landscapes (ewbk-8676)

Video eLesson Valuing landforms and landscapes — Key concepts (eles-5033)

Interactivity Valuing landforms and landscapes (int-8363)

3.2 ACTIVITIES

1. Numerous songs and artists have been inspired by landscapes. Research either a song or an artwork that is well known. Outline how the artist or songwriter described or portrayed the beauty of the landscape.
2. Think of a landscape or landform you have visited. Write a poem or a song using as many adjectives as you can to describe what the landscape/landform looks like and how it made you feel.
3. Conduct a survey in your class to count how many people visit the beach and how often (frequency).
   a. Create a table that will tally your results like this:

TABLE 1 Visits to the beach — frequency

Daily	Weekly	Monthly	Yearly	Never

   b. Based on your class results, do you think the beach is an important part of the culture where you live? Explain possible factors that might have influenced the results.

3.2 EXERCISE

Learning pathways

■ LEVEL 1	■ LEVEL 2	■ LEVEL 3
1, 2, 7, 12, 14	3, 5, 9, 10, 15	4, 6, 8, 11, 13

Check your understanding

1. Define the word *aesthetic*.
2. Identify and describe two recreational activities you can do in landscapes.
3. Approximately how far from a beach does the majority of Australia's population live?
4. How would people's recreational activities change if they lived inland?
5. Name three famous Australian beaches.
6. Recall the proportion of the Australian population that lives within 50 kilometres of the coast. If Australia's population is about 26 million, about how many people live further than 50 kilometres from the ocean?

Apply your understanding

7. Explain why the beach is considered part of Australia's cultural identity.
8. Evaluate the importance of beach culture for those people who do not live on the coast.
9. In what ways does the coast influence lifestyles beyond the beach?
10. Do you think that the beach is an important part of what makes Australia a beautiful place to live? Discuss the arguments in favour of and against its importance.
11. Present a case to argue that the beach is not the most beautiful landscape in Australia.

Challenge your understanding

12. Discuss the changes that may arise from more people living in coastal areas.
13. Provide two recommendations for how coastal communities can preserve the aesthetic value of their beaches.
14. Predict how someone who has never seen the ocean would react to seeing it for the first time.
15. Explore, in a well-structured paragraph, why people are often moved to write or create artwork about landscapes. (For help structuring your paragraphs, use the SkillBuilder template in your Chapter 1 eWorkbook.)

To answer questions online and to receive **immediate feedback** and **sample responses** for every question, go to your learnON title at www.jacplus.com.au.

3.3 The cultural and spiritual value of landscapes and landforms

LEARNING INTENTION

By the end of this subtopic, you will be able to explain the cultural and spiritual value of landscapes and landforms, and appreciate the aesthetic value of landscapes to Aboriginal and Torres Strait Islander Cultures.

3.3.1 The Australian experience

Humans have always been interconnected with the physical environment, and landscapes and landforms have helped shape our beliefs and way of life. Many of the world's indigenous peoples continue to have a special relationship with their physical surroundings, and these landscapes are an important part of their cultures and spirituality (religious beliefs).

Aboriginal Peoples and Torres Strait Islander Peoples inhabited Australia for more than sixty thousand years before European colonisation. Their cultures and spirituality are linked to the landscapes that they live in. Both Aboriginal Peoples and Torres Strait Islander Peoples recount how their ancestors created the landforms as they travelled across the landscape during a time that is commonly referred to by the English term the **Dreaming**. The Dreaming relates to past, present and future. It is a never-ending cycle that is embedded in **Country**.

Dreaming in Aboriginal culture, the time when the Earth took on its present form, and cycles of life and nature began; also known as the Dreamtime. Dreaming Stories pass on important knowledge, laws and beliefs.

Country the land and its features, which is bound to the concept of belonging to a place that is fundamental to Aboriginal Peoples' sense of identity and Culture

FIGURE 1 Wandjina Spirits are important spiritual symbols for people of the Mowanjum language groups — Worrorra, Ngarinyin and Wunumbal — of the Kimberley in Western Australia

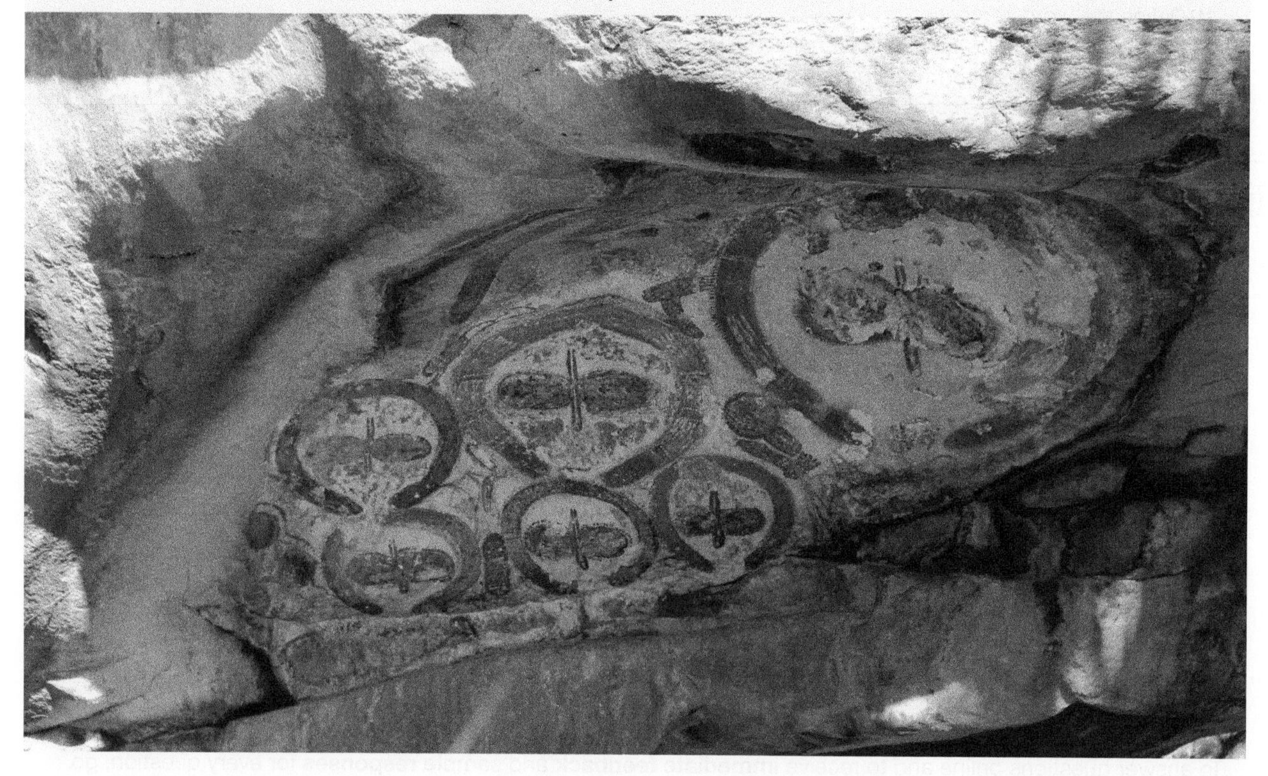

FIGURE 2 Exploring Indigenous relationships to the land

Indigenous people have a strong attachment to their land. They had — and retain — a strong spiritual relationship with it, as well as relying on it for their food, shelter and livelihood. Indigenous societies didn't entertain concepts of ownership commonplace among European cultures. Their relationship to the land was more one of custodianship than of ownership. In other words, the relationship of Indigenous people to their traditional land was defined by their responsibilities to that land. Indigenous people believed they had responsibility to look after their country, to conduct particular ceremonies and to ensure that the ecosystems were kept in balance. And these responsibilities still exist today.

For Indigenous cultures across Australia, land means more than just an essential for physical survival, with spiritual life strongly linked to it too. In this way, ancestral land had personal significance, with each person obliged to look after it . . . and conflicts over the boundaries of where one nation's land ended and another's began were rare. Some land was shared between different groups because of its spiritual or ecological importance but, generally, a nation or clan had sole responsibility for each area of country.

The relationship where Indigenous people believe they are guardians of their country is very different from European concepts of ownership of land, where it can be bought and sold, transferring it from one owner to the next.

Source: Larissa Behrendt, *Indigenous Australia for Dummies* (Second Edition), Wiley Publishing Australia, Milton, 2021 p. 38

Evidence of human presence in Australia is found across the continent in Aboriginal Peoples' rock art (for example, as shown in **FIGURE 1**), in **archaeological** records, and through the **cultural heritage** that is passed down through generations, where Country is a core aspect.

The perspective of Aboriginal Peoples and Torres Strait Islander Peoples is one of belonging to the landscape. This is very different to the European perspective, which is based on the idea of owning land. Europeans arrived in 1788 and occupied areas of Australia. They had a very different view of the landscape, based on ideas they brought with them from Britain. They sought to change the landscape and adapt it to meet their needs. They established wide-spreading settlements and depended on farming introduced species.

3.3.2 Connection to Country

The Dreaming has different meanings to the various Aboriginal Nations across Australia. Generally, Dreaming Stories tell of a time when the Aboriginal Ancestors travelled across the land and waters, and created the variety of landscapes, landforms, plants and animals by either carving them out or becoming the landforms themselves (see the case study in this subtopic).

This makes all elements of the physical environment vital to Aboriginal Peoples and their cultures and heritage. All aspects of Aboriginal culture are centred around the Dreaming and are expressed through art, dance, song, ceremony and stories/lore. These landscapes and landforms are their sacred spaces and spiritual homes. Additionally, many of these sacred and/or significant sites are linked through **songlines**, which identify an interconnectedness between places and peoples.

The spiritual connection Aboriginal Peoples have with Country is highly important as it contributes to an individual's sense of belonging and is a key factor in their identity. Being at one with their traditional homelands also strengthens connections and cultural ties with the land, people and place. The term Country is much more than just the physical land we live, walk and play on. It also includes:

- rivers, waterways, oceans, lakes
- animals and plants
- weather patterns
- the Sun, the Moon and the stars
- the sky
- mountains, hills, forests.

archaeological concerning the study of past civilisations and cultures by examining the evidence left behind, such as graves, tools, weapons, buildings and pottery

cultural heritage beliefs and ways of living passed down through generations

songlines paths across the land that follow the Ancestor's creation journeys that are used to teach about Culture and Country; also known as Dreaming Tracks

Therefore, connection with Country is much more than being attached to physical land; it is a sense of being a part of and in tune with land and the elements of the environment that are listed above.

Torres Strait Islander People are the Indigenous people living on the tip of Cape York and the 17 islands in the Torres Strait between Australia and Papua New Guinea (**FIGURE 3**). They are culturally different to mainland Aboriginal Peoples and their stories are set around the Tagai, a warrior who is visible in the stars. These stories help them navigate the islands and explain their close connection with the landscapes and oceans around them.

For Aboriginal Peoples and Torres Strait Islander Peoples the landscapes and landforms are crucial to their culture and spirituality. They are not simply reminders of what to do and how to act, but are a vital part of a person's understanding of themselves, their Culture, language, past, present, future and identity.

FIGURE 3 Location map of Torres Strait Islands

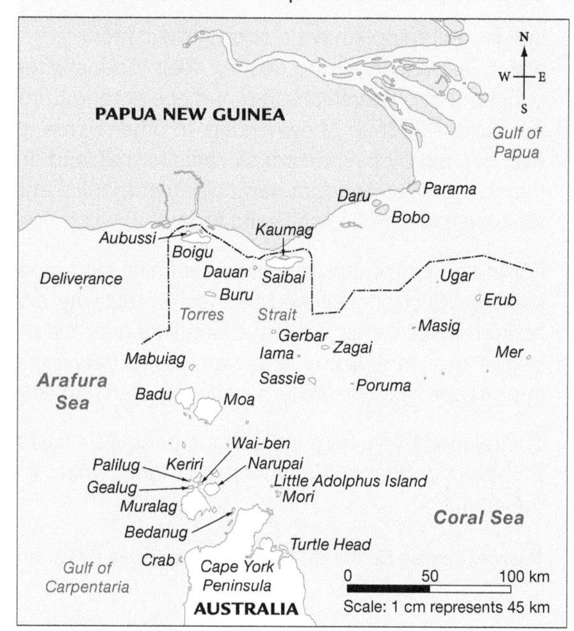

Source: Geoscience Australia

FIGURE 4 The view from Wai-ben (also sometimes called Thursday Island) in the Torres Strait — Muralag peoples are the traditional owners of the land and seas surrounding the island.

CASE STUDY

Ngiyaampaa Dreaming

The following Dreaming Story is told by Aunty Beryl Carmichael, from the Ngiyaampaa people of western New South Wales. It tells the story of Ngiyaampaa Country and how the Darling River was created (**FIGURE 5**).

FIGURE 5 The meandering landscape of the Darling River in Menindee, NSW, where Ngiyaampaa people tell of Weowie the water serpent who travelled and made the water holes and depressions.

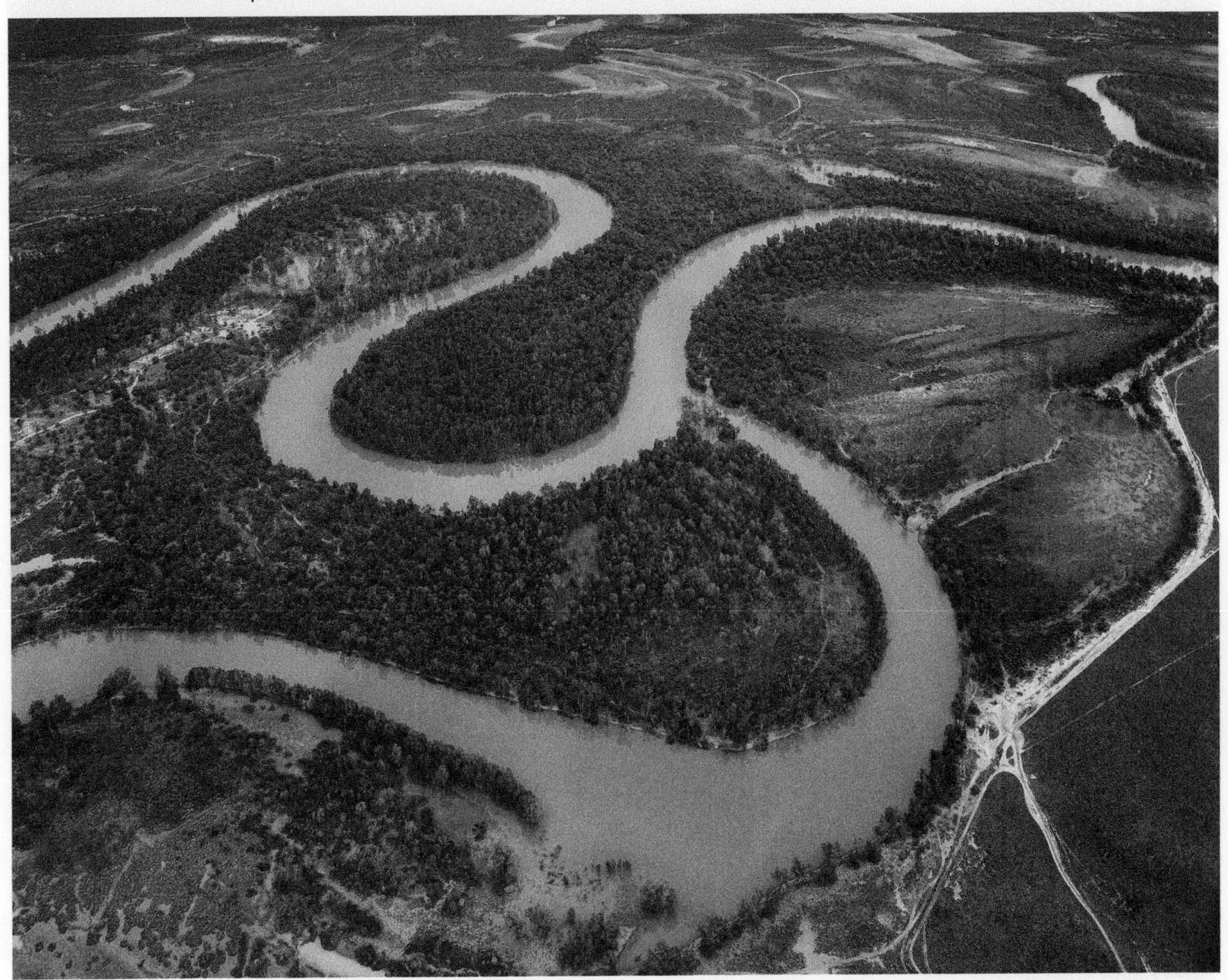

Now long, long time ago of course, in the beginning, when there was no people, no trees, no plants whatever on this land, 'Guthi-guthi', the spirit of our ancestral being, he lived up in the sky. So he came down and he wanted to create the special land for people and animals and birds to live in. So Guthi-guthi came down and he went on creating the land for the people — after he'd set the borders in place and the sacred sites, the birthing places of all the Dreamings, where all our Dreamings were to come out of.

Guthi-guthi put one foot on Gunderbooka Mountain and another one at Mt Grenfell. And he looked out over the land and he could see that the land was bare. There was no water in sight, there was nothing growing. So Guthi-guthi knew that the water serpent, Weowie, was trapped in a mountain — Mt Minara. So Guthi-guthi called out to him, 'Weowie, Weowie', but because Weowie was trapped right in the middle of the mountain, he couldn't hear him. Guthi-guthi went back up into the sky and he called out once more, 'Weowie', but once again Weowie didn't respond. So Guthi-guthi came down with a roar like thunder and banged on the mountain and the mountain split open. Weowie the water serpent came out. And where the water serpent travelled he made waterholes and streams and depressions in the land. So once all that was finished, of course, Weowie went back into the mountain to live and that's where Weowie lives now, in Mount Minara. Old Pundu, the cod, it was his duty to drag and create the river known as the Darling River today. So what I'm telling you — the stories that were handed down to me — all come from within this country.

Resources

eWorkbook — The cultural and spiritual value of landscapes and landforms (ewbk-8680)

Video eLesson — The cultural and spiritual value of landscapes and landforms — Key concepts (eles-5034)

Interactivity — The cultural and spiritual value of landscapes and landforms (int-8364)

3.3 ACTIVITIES

1. Research and identify the name/s of the Aboriginal nation/s where you live and go to school.
2. Research an Aboriginal sacred site in your area or region. Describe its landscape or landforms. Locate the site using an online mapping engine such as Google Maps, Bing Maps or OpenStreetMap.
3. Organise for an Aboriginal Elder from your local area to visit your class or to share a Dreaming Story (that you are allowed to hear) that describes the creation of a local landscape or landform.
4. Not all stories belonging to Aboriginal Peoples can be shared with people from other cultures, or even with everyone who is part of that specific group. Discuss as a class what other cultures have stories, activities or rules related to the land that can't be shared with other people from different cultures or communities.

3.3 EXERCISE

Learning pathways

■ LEVEL 1	■ LEVEL 2	■ LEVEL 3
1, 7, 8, 9, 14	2, 3, 6, 10, 11, 15	4, 5, 12, 13, 16

Check your understanding

1. What is the Dreaming? Define the term in your own words.
2. Describe where the Torres Strait Islands are located in relation to the Australian mainland.
3. Which aspects of Aboriginal cultures are centred around the Dreaming?
4. In what ways is Culture expressed by Aboriginal Peoples? Describe two.
5. How is the European perspective of land different to Aboriginal Peoples' view?
6. Outline why the physical environment is an important part of Aboriginal Peoples' identities.
7. According to the Ngiyaampaa Dreaming, summarise how the Darling River was created.

Apply your understanding

8. Read the Dreaming Story as told by Aunty Beryl Carmichael. What landscapes and landforms are being created?
9. Why do Aboriginal Peoples and Torres Strait Islander Peoples believe land is important?
10. What is a sacred site?
11. Explain the importance of Country to Aboriginal Peoples' sense of belonging. Contrast this with a European understanding of the land.
12. Account for why a specific landform might be important to Aboriginal Peoples or Torres Strait Islander Peoples.
13. What is the relationship between the land and Aboriginal Peoples' identity?

Challenge your understanding

14. Predict what problems may arise for Aboriginal Peoples' cultures if Aboriginal Dreaming sites are destroyed.
15. How can non-Aboriginal people protect and respect sites that are sacred for Aboriginal Peoples? Suggest two actions that individuals could take. (An action might also include something that people should not do.)
16. Propose a strategy or plan that will help to maintain all aspects of Country while also allowing access for all Australians.

To answer questions online and to receive **immediate feedback** and **sample responses** for every question, go to your learnON title at www.jacplus.com.au.

3.4 The economic value of landscapes and landforms

LEARNING INTENTION

By the end of this subtopic, you will be able to identify some of the ways that landscapes have economic value.

3.4.1 Tourism

'Economic' is a word we use to describe how we, as humans, make money and spend money. Humans have used landscapes and landforms by digging out their natural resources, but they can also be a valuable source of money in their natural state. As outlined in 3.2, landscapes and landforms hold aesthetic value and people are willing to spend money to enjoy them.

Tourism is when people travel to other places from where they live for recreation, pleasure, relaxation or education. Many landscapes and landforms are excellent tourist destinations and are important for local communities because of the money and jobs created for the local economy.

3.4.2 Adventure tourism

A type of tourism that attracts people to landscapes and landforms is adventure tourism. An adventure tourism trip is one that has at least two of the following features:

- includes a physical activity
- occurs in the natural environment
- involves cultural immersion.

Many landscapes and landforms are a great place for people to enjoy adventure tourism by participating in activities such as rock climbing, mountain biking and hiking (**FIGURE 1**). Adventure tourism activities are classified as 'soft' when most people can participate and 'hard' when significant skills and equipment are needed. Examples of soft activities include backpacking, bird watching, camping, canoeing, hiking, horse riding and scuba diving. Examples of hard activities include caving, climbing, kayaking, whitewater rafting and mountain biking. Economically, adventure tourism makes money at a local scale by directly benefiting local communities through the money spent on accommodation, eating out or guides. Adventure tourism often requires specialised equipment, such as skis, bikes or ropes so, at a larger scale, it contributes to the national and global economy. For example, the hard adventure activity of climbing the landform of Mount Everest, the world's highest mountain, costs on average US$48 000 per person. The permit to climb is US$11 000 and there are the additional costs of buying training, specialised equipment, airfare and tour guides.

FIGURE 1 Mountain biking is a form of adventure tourism that allows people to enjoy landscapes and landforms and helps local economies.

CASE STUDY

Grand Canyon National Park, United States

The Grand Canyon is a famous, steep-sided canyon in the United States that was created by the processes of weathering and erosion caused by the Colorado River. It is a distinctive landform that is 446 kilometres long and in some parts 18 kilometres wide. This landform has a large economic impact on the local communities. In 2019, the Grand Canyon National Park attracted 6 million visitors. They spent more than $890 million and supported 11 806 jobs. This generated $1.1 billion total economic output.

FIGURE 2 **Pictorial map** showing tourist activities in the Grand Canyon

Most visitor spending within the park is for accommodation, followed by food and drinks, fuel, admissions and fees, and souvenirs. Local communities up to 100 kilometres away also benefit from tourist spending as tourists travel to the Grand Canyon National Park and purchase food and other supplies on the way. Apart from guided tours, visitors participate in a variety of activities, including hiking and walking tours, river rafting, helicopter tours and tandem skydiving, which provide employment opportunities for local people.

pictorial map a map using illustrations to represent information

on Resources

eWorkbook	The economic value of landscapes and landforms (ewbk-8684)
Video eLesson	The economic value of landscapes and landforms — Key concepts (eles-5035)
Interactivity	The economic value of landscapes and landforms (int-8365)
myWorldAtlas	Deepen your understanding of this topic with related case studies and questions > **Grand Canyon** Deepen your understanding of this topic with related case studies and questions > **Himalayas video**

3.4 ACTIVITIES

1. Consider the economic value of Watarrka National Park.
   a. Locate Watarrka National Park (Kings Canyon) in the Northern Territory on a map.
   b. Research the tourist activities available in or near to the park. Choose the three that you would most like to do if you visited the area.
   c. Write a list of what you would need to buy if you wanted to do these activities safely, and research how much each of these would cost.
   d. Write a list of the services you would need to pay for while you were in the area for three days, and research how much each of these would cost.
   e. Estimate how much (dollar value) your visit would contribute to the economy of the area.
   f. Consider the economic impact of travel bans on tourist destinations like Watarrka. Might they have suffered more from the COVID-19 shutdowns and travel restrictions than places in easily accessible areas?
   g. Imagine you are a business owner. Identify other activities that you could promote in this area.

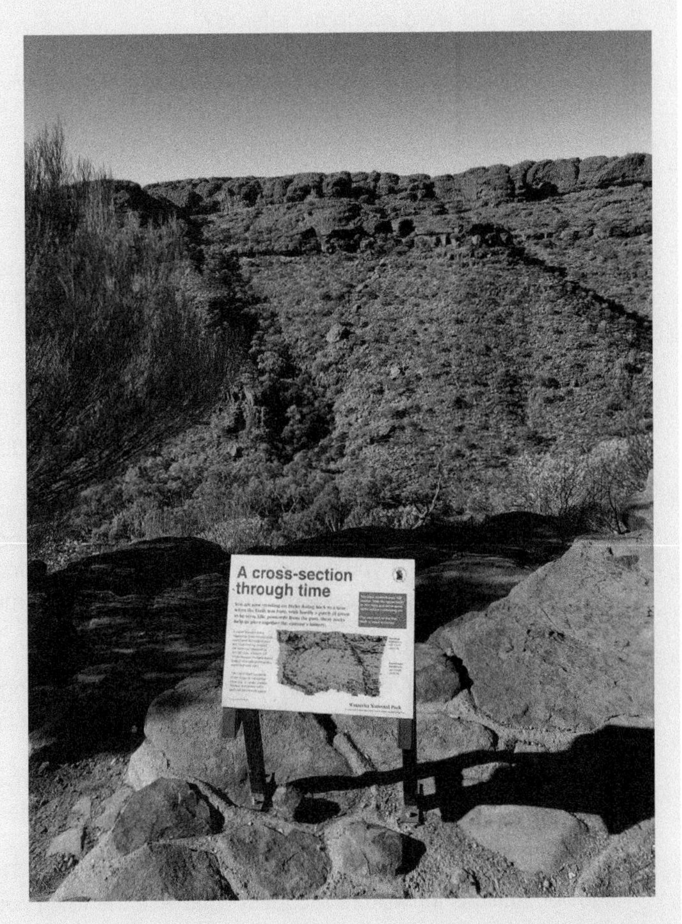

2. Conduct research online to complete the following table to compare the economic benefits of tourism in the Great Barrier Reef and the Grand Canyon.

	Great Barrier Reef	Grand Canyon
Continent and country		
Visitors		
Money spent by visitors		
Jobs created		
Activities		

3.4 EXERCISE

Learning pathways

■ LEVEL 1	■ LEVEL 2	■ LEVEL 3
1, 3, 7, 11, 13	2, 4, 6, 8, 12, 14	5, 9, 10, 15

Check your understanding

1. Define the word *tourism*.
2. What does the word *economic* refer to?
3. Describe two features of adventure tourism.
4. Categorise the following activities as either soft or hard adventure tourism.
   a. hiking
   b. sea kayaking
   c. mountain-bike riding
   d. cross-country skiing
   e. surfing.
5. Give two examples of how adventure tourism helps to boost the economy of a place.
6. Describe the location and dimensions of the Grand Canyon.

Apply your understanding

7. Explain the difference between hard and soft adventure tourism.
8. Explain why adventure tourism relies on landscapes and landforms.
9. How does adventure tourism make money directly and indirectly for the economy? Give examples to support your explanation.
10. It costs around A$65 000 to climb Mount Everest. Outline some of the people who would benefit economically from this, and explain how they might feel about increasing tourist numbers.
11. Explain two ways that you would like to contribute to the local economy if you visited the Grand Canyon. Give specific examples, and explain who in the local community would make money from your activities.

Challenge your understanding

12. Examine **FIGURE 2**. Are there any businesses that you recognise as big companies that operate around the world? What are the advantages and disadvantages of setting up a hotel or restaurant at a place like the Grand Canyon?
13. Suggest why so many tourists want to visit the Grand Canyon every year.
14. Predict some of the economic impacts of COVID-19 pandemic travel bans on communities that usually rely on tourists for their income.
15. Derby in Tasmania is known for its mountain-biking trails. The town offers free camping sites and cheap shower facilities for visitors. Suggest what else the town could provide to encourage more mountain-bike riders to visit. Consider both free and paid facilities or services.

To answer questions online and to receive **immediate feedback** and **sample responses** for every question, go to your learnON title at www.jacplus.com.au.

3.5 SkillBuilder – Using positional language

online only

LEARNING INTENTION

By the end of this subtopic, you will be able to use geographical terminology to identify direction and position.

The content in this subtopic is a summary of what you will find in the online resource.

3.5.1 Tell me

What is positional language?

Positional language is the use of compass points to find and give directions between places. It is based on the directions on magnetic compasses, which always point to north. An eight-point compass is standard in most geography books and atlases — it shows north, north-east, east, south-east, south, south-west, west and north-west.

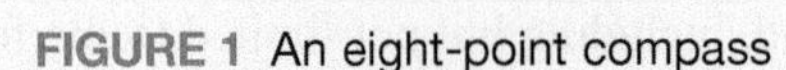

FIGURE 1 An eight-point compass

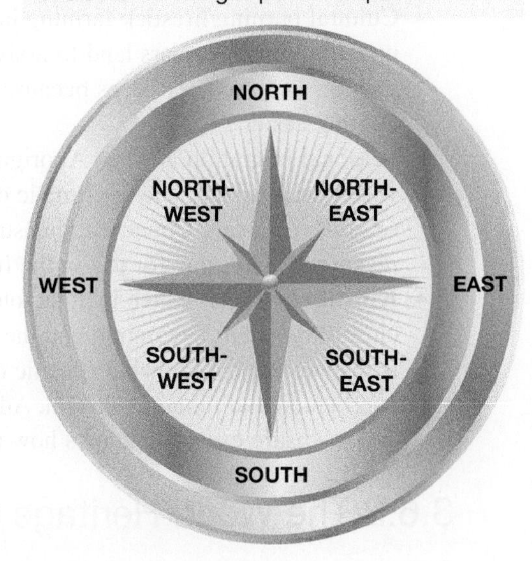

3.5.2 Show me

How to use positional language

Step 1

On the piece of tracing paper draw a simple 16-point compass based on that shown in FIGURE 1. Mark the centre of the compass with a dot.

Step 2

Place the centre of your 16-point compass (the dot) on the point of origin from which a direction is being given. (North should be pointing to the top of the map.)

Step 3

Read the compass direction from the centre dot to the place identified and write down that direction.

Step 4

The placement of the centre of the compass must be moved for each individual direction required.

3.5.3 Let me do it

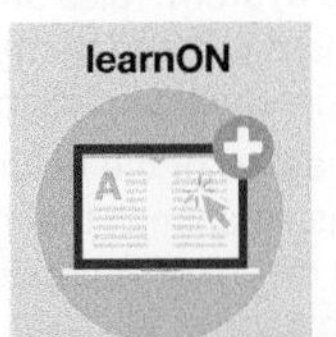

Go to learnON to access the following additional resources to help you build this skill:

- a longer explanation of this skill and its application in Geography (Tell me)
- a video showing the step-by-step process of this skill (Show me)
- an interactivity for you to practise this skill (Let me do it)
- self-marking questions to help you understand and use this skill.

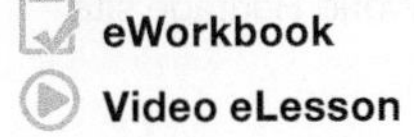

Resources

eWorkbook	SkillBuilder — Using positional language (ewbk-8688)
Video eLesson	SkillBuilder — Using positional language (eles-1649)
Interactivity	SkillBuilder — Using positional language (int-3145)

3.6 Protecting and managing landscapes

LEARNING INTENTION

By the end of this subtopic, you will be able to describe ways that landscapes can be protected at a range of scales.

3.6.1 Aboriginal land management practices

Aboriginal Peoples have been caring for Country for tens of thousands of years. They have and continue to use sophisticated and dynamic ways to manage the land that have sustained land and water resources. An important part of Aboriginal cultures is only taking what is necessary for survival and therefore maintaining a highly productive and respectful relationship with Country. There are many ways that Aboriginal Peoples managed the land, including but not limited to the following examples.

- Cultural burning/firestick farming involves the careful use of hot and cold fires, which manages the undergrowth and clears land to assist in minimising uncontrollable fires. This is also a tool for hunting animals and food resources because the smoke moves the animals to clearer areas, which makes hunting easier.
- Seasonal migration is when Aboriginal groups stay in particular areas when resources such as food and water are abundant due to climatic conditions and patterns. Seasonal migration occurs with the changing seasons to sustain Country and ensures that resources are not overused. It also ensures that resources can regenerate and grow for use in the future.
- **Kinship systems**, which include totemic systems, are a highly sophisticated land management practice closely linked to Dreaming, and the roles and responsibilities derived from (given by) **totems**. A person is responsible for knowing specific details about their totems and for ensuring that their totems are never overhunted and remain plentiful. Additionally, it is their role to teach others about their specific totems and to ensure everyone knows how to protect and use them sustainability.

3.6.2 The World Heritage Convention

Over time, people have realised the value of landscapes and landforms, and the need to protect and manage them more effectively. This is so that they can ensure the landscapes and landforms continue to function in their natural state, as they are habitats for a variety of plants and animals, and so that future generations are able to enjoy them too. There are international agreements in place at the global scale to help protect these places, and at a local scale governments have created areas that can be managed and protected by special laws that ensure the sustainable use of landscapes and landforms.

The World Heritage Convention is an international agreement that has been signed by 190 countries since 1972. This gives recognition at a global scale that there are important parts of the physical and human environments that needed to be protected. The convention is overseen by UNESCO (United Nations Educational, Scientific and Cultural Organization). The process begins by a country nominating a place inside its territories as a World Heritage site, then the World Heritage Committee assesses if it matches at least one of the ten set criteria. If it adequately addresses the criteria, it is inscribed on to the **World Heritage List** and it will become a World Heritage site (**FIGURE 1**). It is expected that the country the site is located in will take all precaution to protect and manage it. In some countries, once a place is put on the World Heritage List, it is protected by that country's national and local laws.

kinship system rules and relationships that determine how people interact with each other

totem a part of the natural world (such as an animal) that has a special meaning or importance for specific people

World Heritage List a list of human and natural sites that have been approved by UNESCO as meeting the criteria of being a World Heritage site

int-8727

FIGURE 1 The World Heritage List includes 1031 sites of significance.

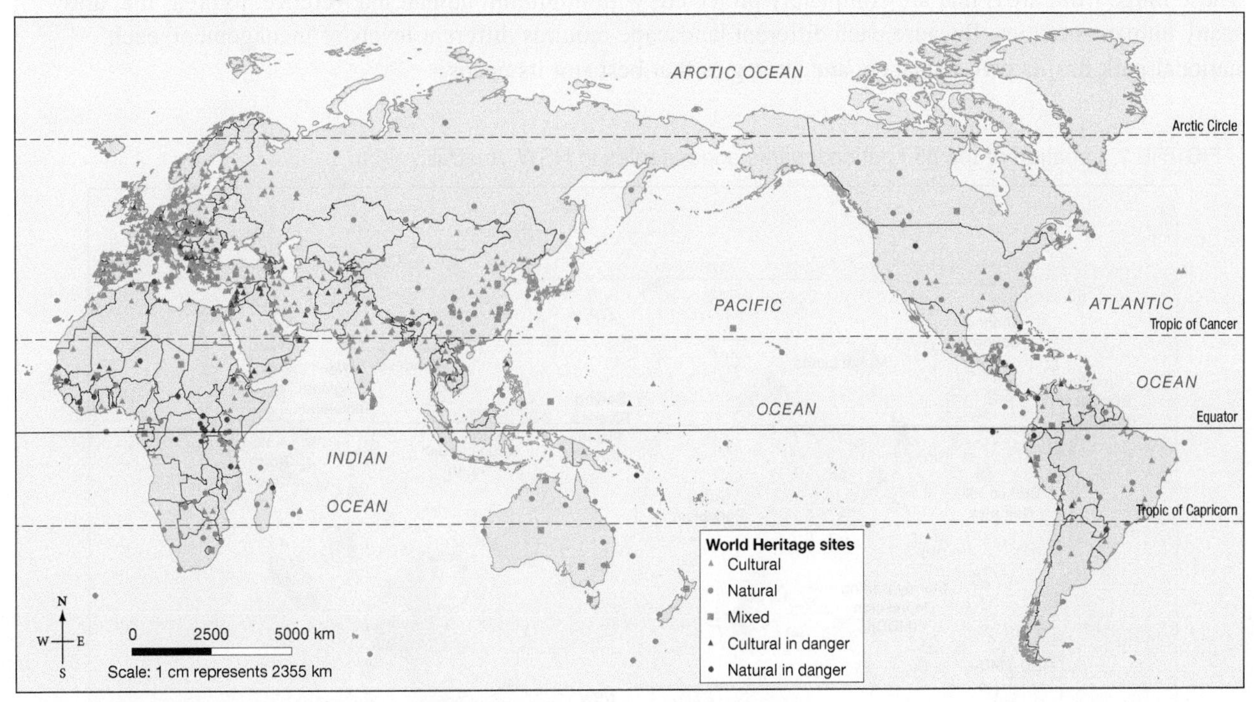

The UNESCO criteria for World Heritage sites

The ten criteria for the selection of outstanding and/or unique World Heritage sites are summarised below.

Cultural sites

1. Human creative genius
2. Exchange of human values, over time or within a culture
3. Culture or civilisation (current or from the past)
4. Building or landscape that illustrates significant stage(s) in human history
5. Human settlement representing human interaction with the environment, especially if vulnerable to extinction
6. Events, traditions, ideas or beliefs

Natural sites

7. Exceptional natural beauty and importance
8. Major stages of the Earth's natural processes such as landforms
9. Ongoing processes in the evolution of living things and developing ecosystems
10. Important natural habitats for conservation of biodiversity

Source: UNESCO

3.6.3 National parks and reserves

At a national and state level, governments have created special places called national parks and reserves. These are places that are protected by law and have special rules for how they can be accessed and used. In NSW, the National Parks and Wildlife Service (NPWS) manages 850 national parks and reserves that cover over 7 million hectares of land (**FIGURE 2**). This includes four World Heritage sites, a number of Australian National Heritage sites and 17 Ramsar wetlands (these are wetlands that have been officially recognised as being important by the International Convention on Wetlands).

There are different types of parks with different purposes that are managed by National Parks and Wildlife. These range from areas that are completely protected, with minimum human interference, to areas that allow many human activities. Because each different landscape requires different levels of management, each national park has its own managers and strategies that best suit its needs.

int-8735

FIGURE 2 Location of the 850 national parks and reserves in NSW, Australia

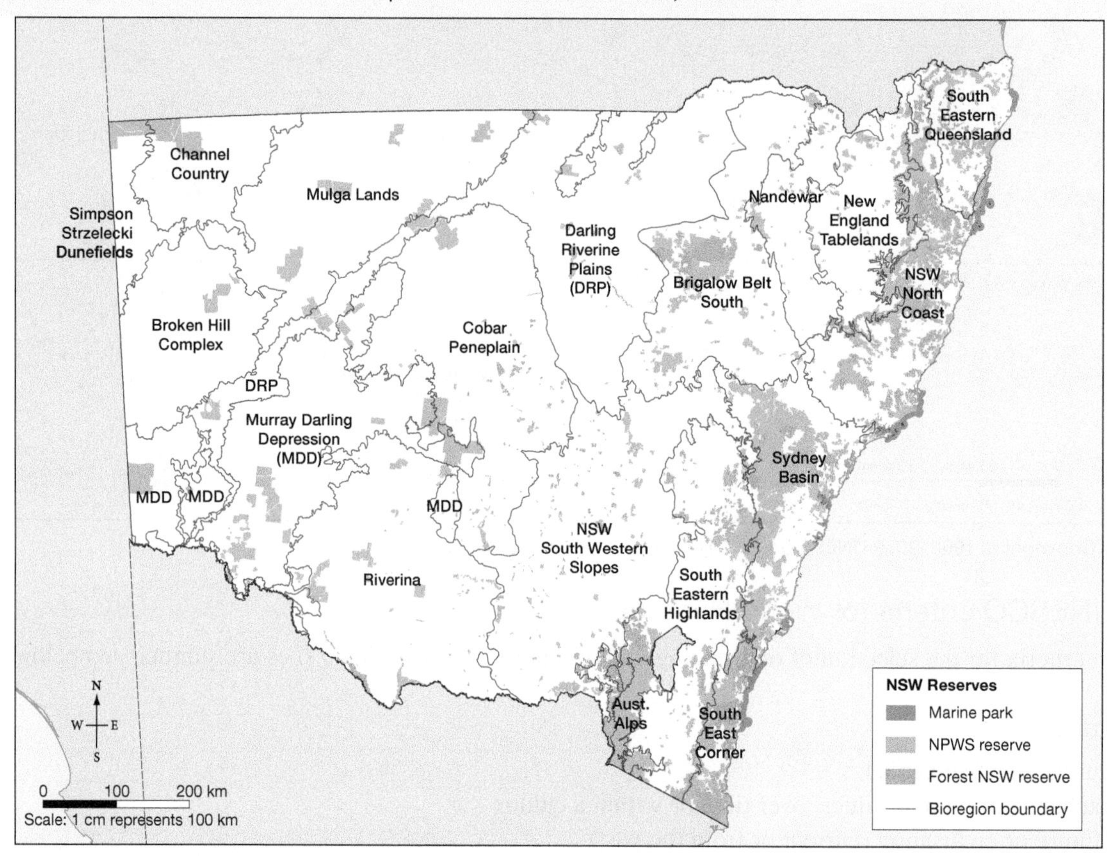

Source: Commonwealth Dept. of Environment, CAPAD 2018 (terrestrial and marine), IBRA7, Forestry Corporation NSW

fdw-0001

fdw-0006

FOCUS ON FIELDWORK

Using a GPS

GPS stands for Global Positioning System. You may have used or heard of GPS being used to locate people, find where images were taken or to navigate unfamiliar places in a car. But did you know that a GPS is a valuable tool used in fieldwork by geographers?

GPS devices use geostationary satellites. These are satellites that orbit with the Earth's rotation and appear to be stationary in the sky. This makes finding the exact location of something on the surface of the Earth using longitude and latitude more accurate. Lines of longitude go around the Earth from north to south (they go 'longways'). Lines of latitude go around the Earth (they go 'laterally', or sideways). GPS coordinates give the longitude and latitude of an exact location. In Geography, these are known as waypoints.

Practise how to use GPS and plot points of interest on a map using the **GPS Home** or **GPS Sydney** activity in the Fieldwork section of your online Resources panel.

on Resources

eWorkbook	Protecting and managing landscapes (ewbk-8692)
Video eLesson	Protecting and managing landscapes — Key concepts (eles-5036)
Interactivities	The World Heritage List sites of significance (int-8727) Location of the 850 national parks and reserves in NSW, Australia (int-8735)
Fieldwork	GPS Home (fdw-0001) GPS Sydney (fdw-0006)
myWorldAtlas	Deepen your understanding of this topic with related case studies and questions > **World Heritage sites**

3.6 ACTIVITIES

1. Research NSW national parks. Create a table with the column headings shown below and research the different types of parks within the National Parks and Wildlife Service of NSW.

Type of park	Description	Examples	Example nearest to my home

2. Research a landscape or landform that is on the World Heritage List and write an explanation of the criteria that were used to put it on the list. Begin your research with the **World Heritage List** weblink in your online Resources panel.
3. Choose a landscape or landform that is not on the World Heritage List that you think should be included. Prepare a proposal to the World Heritage Committee, either through a report, slide presentation or multimedia presentation, trying to persuade them, by using the appropriate criteria, to place your chosen site on the list.

3.6 EXERCISE

Learning pathways

■ LEVEL 1	■ LEVEL 2	■ LEVEL 3
1, 5, 9, 10, 13	2, 3, 8, 11, 15	4, 6, 7, 12, 14

Check your understanding

1. Define the word *sustainable*.
2. Identify three methods used by Aboriginal Peoples to sustainably manage Country.
3. Identify two reasons why it is important to protect sites with natural or cultural significance.
4. Use **FIGURE 2** to name three areas in NSW that includes examples of all three kinds of reserves.
5. List the criteria a site needs to meet to be considered a UNESCO World Heritage site of natural significance.
6. Give an example of a global-scale strategy and an example of a local-scale strategy to protect and manage landscapes and landforms.
7. Describe how Aboriginal land management practices work together to maintain Country. Give specific examples.

Apply your understanding

8. Explain how Aboriginal land management balances human needs with the needs of the environment.
9. Create a flow diagram that shows how a site becomes a World Heritage area.
10. Why do different national parks and reserves have their own managers and management strategies?
11. Discuss whether it might be difficult for Aboriginal Peoples living in urban or built-up areas in Australia to protect their Country.
12. Explain the interconnection that Aboriginal and Torres Strait Islander Peoples have with the landscape.

Challenge your understanding

13. Suggest two ways that a community can show that it values the Aboriginal heritage of the area.
14. Recommend one place in your local community that might meet at least three of the criteria for becoming a UNESCO World Heritage site of cultural importance. List the criteria and explain how your chosen place fits the description.
15. Recommend two ways that people could be encouraged to more actively protect important Australian landscapes.

To answer questions online and to receive **immediate feedback** and **sample responses** for every question, go to your learnON title at www.jacplus.com.au.

3.7 Protecting and managing Kakadu

online only

LEARNING INTENTION

By the end of this subtopic, you will be able to assess the sustainability of the use and management of an important Australian landscape.

The content in this subtopic is a summary of what you will find in the online resource.

Kakadu is an important part of Australia's human and natural heritage. It is rich with the historical records and Culture of its Aboriginal Peoples. In addition, it supports a treasure trove of native plant and animal species and provides a temporary home to a large number of migratory birds. More than 170000 tourists visit Kakadu every year. In 1981 a large part of Kakadu became Australia's first World Heritage site.

However, Kakadu also has large deposits of uranium ore, which some regard as a valuable export for Australia. How do we protect and manage the resources in such important places?

To learn more about protecting and managing Kakadu, go to your learnON resources at www.jacPLUS.com.au.

FIGURE 6 Kakadu National Park is an important cultural, historical, environmental and economic place.

Contents

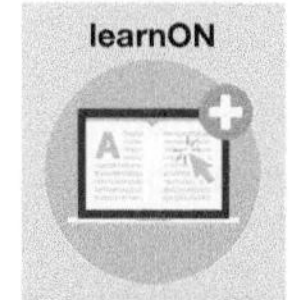

- 3.7.1 Kakadu's landscape
- 3.7.2 World Heritage listing
- 3.7.3 Kakadu and its resources

on Resources

eWorkbook	Protecting and managing Kakadu (ewbk-8696)
Video eLesson	Protecting and managing Kakadu — Key concepts (eles-5037)
Interactivity	Protecting and managing Kakadu (int-3609)

3.8 Protecting and managing Uluṟu

online only

LEARNING INTENTION

By the end of this subtopic, you will be able to provide examples to demonstrate the contribution of Aboriginal Peoples' knowledge to the use and management of an Australian landform.

The content in this subtopic is a summary of what you will find in the online resource.

Uluṟu is one of Australia's most famous and recognisable landforms. It is located very close to the centre of Australia in the south-west corner of the Northern Territory.

Uluṟu is a sacred place to the Aṉangu People, who have lived in the area for hundreds of generations. Uluṟu's uniqueness and beauty have also made it a popular tourist destination and an important source of money for the small community that lives in the area. Today, Aṉangu People and National Parks Australia Rangers work together to help protect the landscape and unique landforms by using both scientific and Aboriginal knowledge.

To learn more about protecting and managing Uluṟu, go to your learnON resources at www.jacPLUS.com.au.

Contents

FIGURE 1 The beauty of Uluṟu's red rock face at sunrise

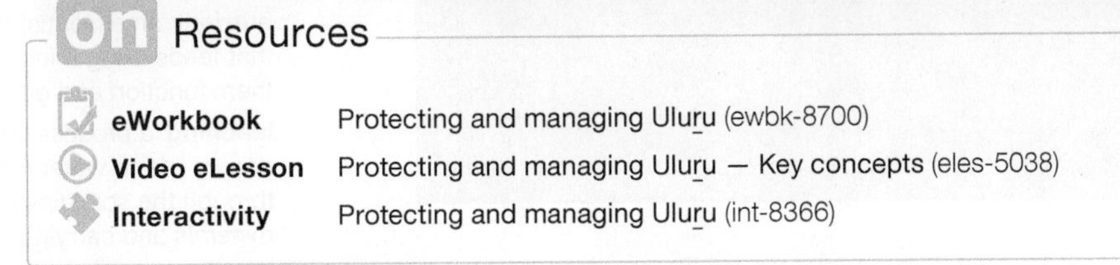

on Resources

eWorkbook Protecting and managing Uluṟu (ewbk-8700)

Video eLesson Protecting and managing Uluṟu — Key concepts (eles-5038)

Interactivity Protecting and managing Uluṟu (int-8366)

3.9 Changed and degraded landscapes

LEARNING INTENTION

By the end of this subtopic, you will be able to describe human causes of land degradation and its effects.

3.9.1 The impact of human activities

Humans depended on landscapes for their survival for thousands of years and lived harmoniously within them. At the end of the last **ice age**, humans began to settle in one place and made changes to the landscape. Some scientists have named the period of time since then the **Anthropocene**. People built farms and settlements, such as towns and cities, and they used spaces for other industrial purposes such as mines. As a consequence, the built environments humans created have often resulted in the **degradation** of landscapes.

Farming

Human communities, generally, moved from being **nomadic** to settling in one place. This happened because people discovered that they could grow crops and **domesticate** animals such as cattle, sheep, goats and pigs. Since humans no longer needed to migrate during different seasons to find food, they could change the environment around them to farm more effectively. Farms need areas of clear flat land with good soil and access to water. Clear land does not often occur naturally, so humans removed many trees and other vegetation. Landscapes that often look natural now were altered by humans thousands of years ago (**FIGURE 1**). The roots of trees help hold soil together and falling organic material from plants and animals replaces the **nutrients** in the soil. But removing the vegetation increases erosion and reduces the nutrients in soil. Ploughing and irrigation also speed up the process of **leaching** nutrients out of the soil.

FIGURE 1 Once covered in vegetation, this Irish landscape in County Kerry was changed by the removal of trees for farming thousands of years ago.

ice age a geological period during which the Earth is colder, glaciers and ice sheets expand and sea levels fall. The last ice age was approximately 13 000 years ago.

Anthropocene a period of time when humans made significant changes to the Earth's environments

degradation a decline in quality caused by time or improper use

nomadic a lifestyle in which people do not live in one fixed spot but move according to the seasons from place to place in search of food, water and grazing land

domesticate to tame a wild animal or plant so it can live with or be looked after by people

nutrient an essential element that feeds living things to help them function and grow

leaching a process that occurs where water runs through the soil, dissolving minerals and carrying them into the subsoil

3.9.2 Settlements

Since farming allowed humans to stay in one place they started to build large **settlements** to create communities with other people. Settlements began as small villages and some eventually turned into large towns or cities. To make way for housing, streets and other utilities, landscapes needed to be cleared of vegetation and the ground excavated (**FIGURE 2**). Cities and towns were very compact for many centuries, as most people had to walk around them. Since the invention of transport technologies, such as trains and motor vehicles, cities have expanded outwards to create suburbs. Where once landscapes were dominated by natural elements, settlements have created large areas of built features that changed the landscape and reduced the size of habitats (natural homes and environments) of many plant and animal species.

FIGURE 2 Impacts of spreading settlements

1. A new treeline is created after trees have been removed to clear land.
2. New surfaces change the pattern of water run-off and absorption.
3. Earth is excavated and the landscape reshaped.
4. New suburban houses are built.

settlement place where humans live together, e.g. houses, villages, towns and cities

CASE STUDY

Open-cut mining

Mining for minerals, such as iron ore and coal, has been an activity that humans have been involved in for thousands of years, and that has provided many economic benefits and resources to communities. Open-cut mining is an activity in which humans dig large holes in the earth to access the minerals below the surface (**FIGURE 3**). To access the minerals, miners remove the top layers of soil and dirt (overburden) and pile them up (**FIGURE 4**). The mineral being mined is removed, but it is often mixed with other substances that are not valuable, so the waste (tailings) is separated using water and chemicals, and stored in large ponds that are engineered to keep the contaminated water away from the local environment. In the past, miners pumped tailings straight into local waterways, causing significant contamination and environmental degradation. Once all the valuable minerals have been removed from the mine a large pit is left called a final void. Unfortunately, in many parts of the world final voids are not backfilled (filled in) because it is very expensive to do. Final voids that are not backfilled experience high rates of erosion, and may sometimes damage water tables. Also, many believe that modern, large-scale open-cut mining ruins the beauty (aesthetics) of the landscapes because it significantly changes the topography of the area.

FIGURE 3 Mt Tom Price Mine is a huge iron ore mine in the Pilbara region in Western Australia.

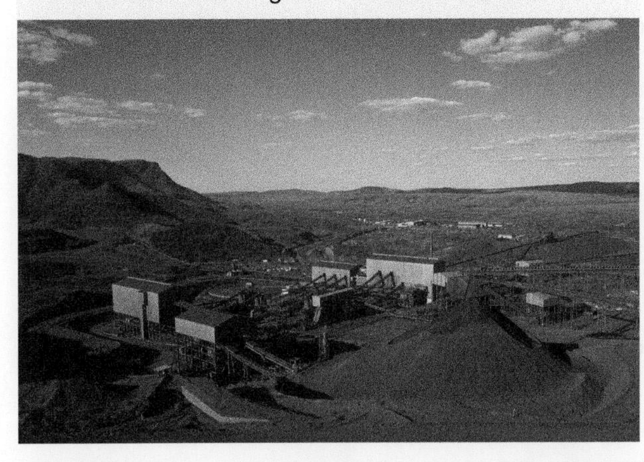

FIGURE 4 Overburden piled up from an open-cut mine

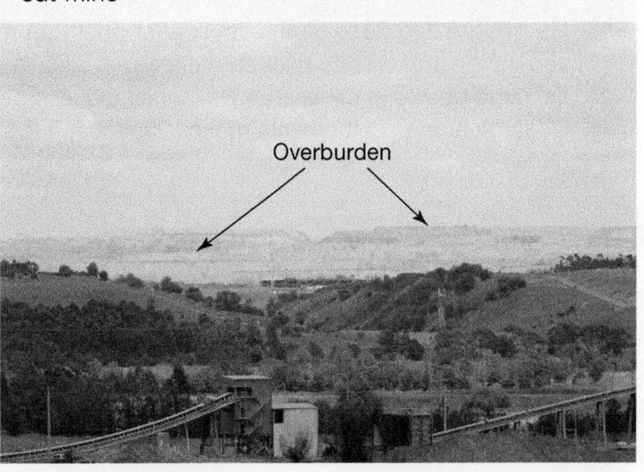

Resources

eWorkbook Changed and degraded landscapes (ewbk-8704)

Video eLesson Changed and degraded landscapes — Key concepts (eles-5039)

Interactivity Changed and degraded landscapes (int-8367)

3.9 ACTIVITIES

1. Research the term 'Anthropocene' and write a paragraph to explain what this period of time is known for.
2. Research open cut mining and draw an annotated diagram that identifies and describes the basic features of an open cut mine.
3. Research why people move from a city to a suburb. What are the positives and negatives of creating new suburbs?

3.9 EXERCISE

Learning pathways

■ LEVEL 1	■ LEVEL 2	■ LEVEL 3
1, 6, 7, 9, 14	2, 4, 5, 10, 15	3, 8, 11, 12, 13

Check your understanding

1. Outline, in your own words, what the term *degradation* means.
2. Define the term *Anthropocene*.
3. What was the name of the large climate event that occurred before humans started to have large impacts on landscapes?
4. Give two examples of ways that farming can degrade soil.
5. Identify two kinds of animals that humans have domesticated.
6. List three types of settlements.
7. Describe some of the 'built features' near your school.
8. What are the tailings from a mine?

Apply your understanding

9. Explain the reasons why many human groups stopped being nomadic.
10. Discuss how settlements change landscapes.
11. Explain how the roots of trees help to stop land degradation.
12. How might tailings from a mine be a threat to the environment near a mine site?

Challenge your understanding

13. How might degraded land recover after open-cut mining? Suggest what people could do to encourage vegetation and wildlife back to the area.
14. Do large open-cut mines make a landscape less beautiful? Give reasons why or why not.
15. Do large human settlements make a landscape less beautiful? Give reasons why or why not.

To answer questions online and to receive **immediate feedback** and **sample responses** for every question, go to your learnON title at www.jacplus.com.au.

3.10 Geomorphic hazards — Landslides

LEARNING INTENTION

By the end of this subtopic, you will be able to outline the natural and human causes of landslides and describe their potential impacts.

3.10.1 Landslides

FIGURE 1 Travellers look at a landslide-affected road in Manali, India. Landslides regularly occur in this region.

Geographers who study geomorphology examine the processes that shape and create the natural features of the Earth's crust. The layer of the Earth's crust they are interested in is called the **lithosphere**. **Natural hazards** are often events that happen in the natural environment that pose a threat to humans. A **geomorphic** hazard is any natural hazard that occurs in the lithosphere such as **mass movements**, hazards caused by glaciers moving and soil erosion.

Landslides are a geomorphic hazard in which loose rock, debris or earth moves down a steep slope. Landslides happen when the materials that make up the slope give way and move because of gravity (**FIGURE 1**). Landslides are also known as landslips, slumps or slope failure.

lithosphere the Earth's crust, including landforms, rocks and soil

natural hazard an extreme event that is the result of natural processes and has the potential to cause serious material damage and loss of life

geomorphic describes a process that occurs in the lithosphere

mass movement large areas of earth that move

3.10.2 Types of Landslides

FIGURE 2 The different types of landslides

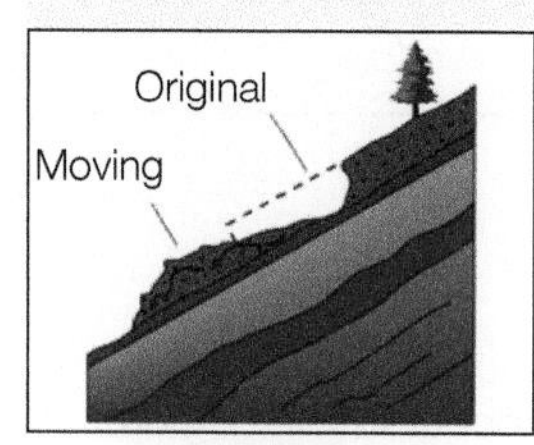

FLOW

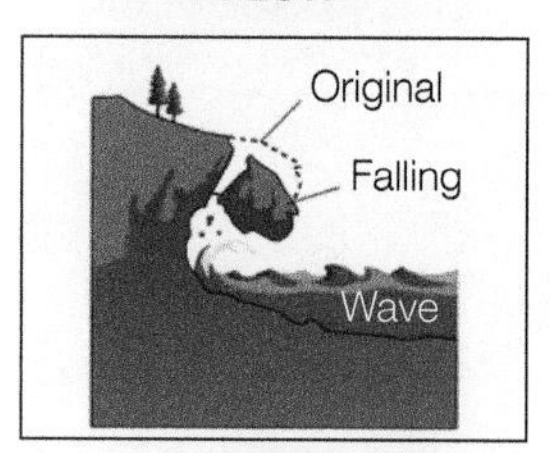

FALL

TOPPLE

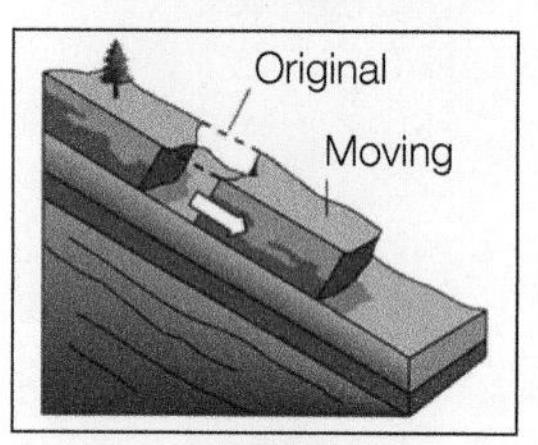

SLIDE

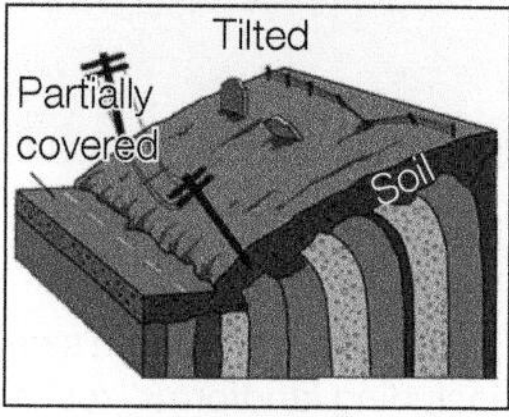

SPREAD

There are five categories of landslides (**FIGURE 2**). The most dangerous type is a *flow*. They are called a flow because the rock and debris move like water down the slope very quickly. A *fall* is where rock or debris free-falls before it rolls or slides down a slope. A *topple* is when rock tilts and does not collapse; often caused by the fissure pattern in a rock, similar to dominos toppling. A *slide* is when large intact pieces of earth slip down another layer of a slope and often move slowly. *Spread* is when large areas of land creep slowly sideways over very gentle or flat terrain.

3.10.3 Causes of landslides

Landslides can either occur naturally or be caused by humans, or be caused and made worse by a combination of the two factors.

Natural causes of landslides

Landslides that are caused naturally have three main triggers: water, earthquakes and volcanic activity. Each of these triggers can occur independently or can work in combination with each other to cause a landslide.

Water

Slopes becoming saturated with water are one of the main causes of landslides. This can happen after lots of rain, as snow melts, or as a result of changes to water levels. Flooding and landslides are often related: as floodwater saturates the ground, loosening the earth, it creates the trigger for a landslide to happen.

Earthquakes

Areas that experience earthquakes often have landslide events. When the ground starts shaking it often loosens the material on steep slopes, causing a landslide or making the soil less compact, meaning water can saturate the ground quickly. Also, some materials when they shake can turn almost liquid, a process called **liquefaction**, and then they flow down a slope rapidly.

Volcanoes

When volcanoes erupt, the lava can melt snow that then causes debris such as ash, rock, soil and water to flow down the slope. Volcanic debris is called *lahar*, an Indonesian word, and it flows very rapidly, often leaving a trail of destruction. Also, the large volcanic explosion can cause earthquakes that then loosen the rock and create landslides.

Human causes

When humans move into an area they modify the landscape significantly. The roots of grass, plants and trees on slopes hold the soil together, but when vegetation is removed the earth is loosened, creating conditions for a landslide. Digging out the bottom of a slope (undercutting) can change how steep a slope is and also make it top heavy. Humans' use of water also contributes to landslides as the ground can be saturated by watering from farming (irrigation), from drain pipes that leak, and from changing the way water flows when hard surfaces such as roads are created. All these factors, if not properly addressed by good engineering, can cause landslides.

liquefaction when solids are shaken heavily and act like liquids

3.10.4 Impacts of landslides

Landslides can happen anywhere in the world, and most countries experience them. Many landslides occur with minimal damage and no loss of life, but some have had disastrous impacts. In Australia, there have been 100 disastrous landslides recorded since 1842. These have killed 105 people and injured 129, and were mainly caused by human action (**FIGURE 3**). One such disaster was the Thredbo landslide on 30 July 1997. It killed 18 people, injured one person, and cost the government $40 million in compensation. This disastrous landslide occurred on a steep slope in the Thredbo village in the Kosciuszko National Park in NSW. Poor drainage was a contributing factor when the road was created above the village (**FIGURE 4**).

Overseas, many poor communities are built on the sides of steep hills due to the land being of poor quality and, therefore, the only place they can afford to live. Those who have built housing on steep slopes are at risk of being in a landslide, especially when heavy rains occur (see the case study in this subtopic).

FIGURE 3 Locations of major landslides in Australia

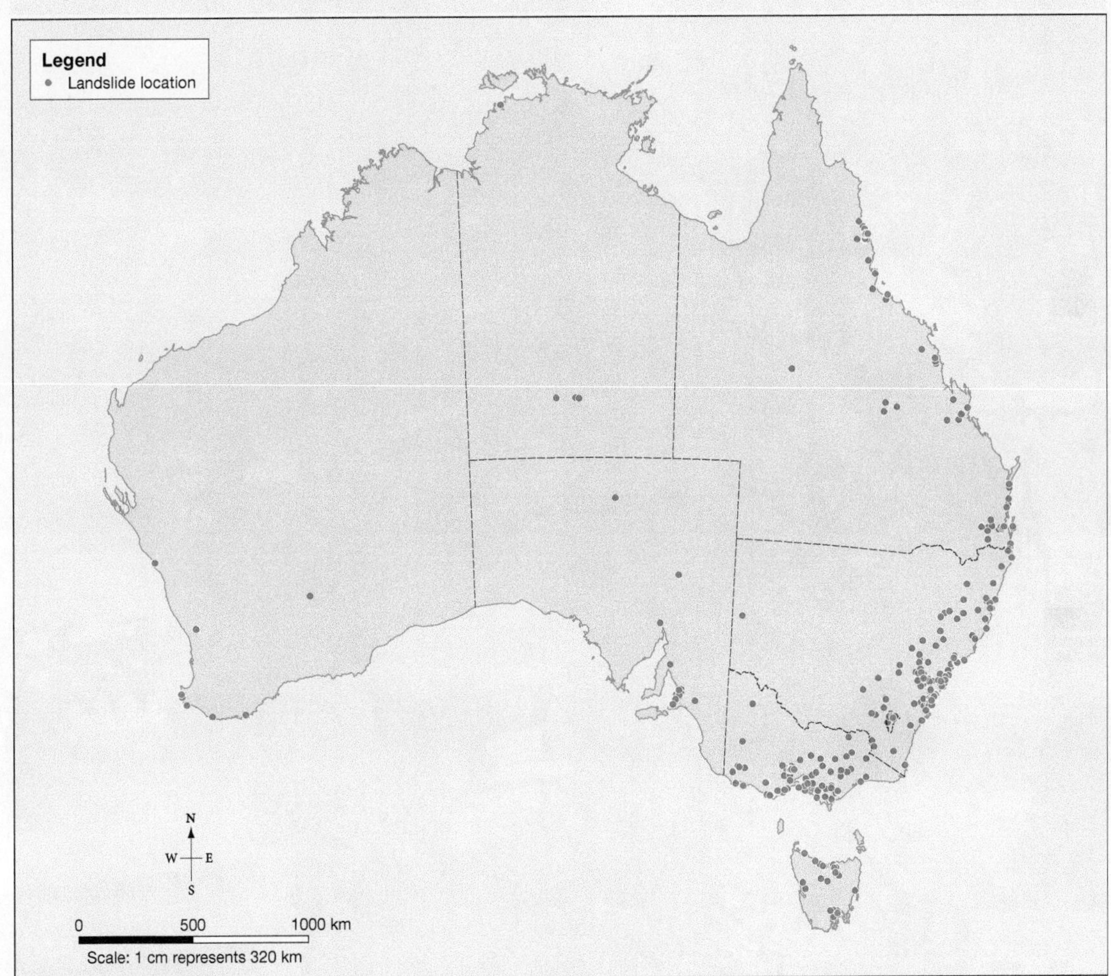

Source: Geoscience Australia

FIGURE 4 The destructive path of the Thredbo landslide 1997

3.10.5 Responding to landslides

After landslides occur, emergency services help with rescue and recovery. However, it is a better option to try to minimise the impacts of landslides or to reduce the likelihood that they will occur at all.

Ideally, houses should not be built on land that is prone to landslides. If an area is prone to landslides, meshing can be used on the hillside to catch falling rocks, retaining walls can be built where a hill has been undercut, and signage can be used to warn people of the dangers of landslides (**FIGURE 5**).

FIGURE 5 Signage warning of possible landslides

CASE STUDY

Brazil landslides, March 2020

In 2020, heavy rains, landslides and flooding impacted three Brazilian states. Extreme rainfall events are becoming more common, and result in the ground becoming saturated and waterlogged. *Favela* is the name given to the slum-style housing that poorer people live in in Brazil, and many favelas were built on steep slopes on the outskirts of cities. The saturated ground, the stresses put on the slope by removing the vegetation, and undercutting when building the houses caused massive flow landslides in heavily populated areas (**FIGURE 6**). These landslides killed more than 150 people, injured many more and left countless already poor people homeless.

FIGURE 6 Damage to homes from a mudslide caused by heavy rains in Guarujá, south-eastern Brazil, in March 2020

on Resources

eWorkbook	Geomorphic hazards — Landslides (ewbk-8708)
Video eLesson	Geomorphic hazards — Landslides — Key concepts (eles-5040)
Interactivity	Geomorphic hazards — Landslides (int-8368)
myWorldAtlas	Deepen your understanding of this topic with related case studies and questions > **Thredbo landslide**

3.10 ACTIVITIES

1. Using an online mapping tool, such as Google Maps, locate a site where landslides could occur in your area. Explain why this site might be at risk.
2. Refer to **FIGURE 3** and an atlas showing the physical characteristics of Australia. Using positional language, describe where most of Australia's landslides occur. What reasons can you give as to why landslides are occurring in these places?

3.10 EXERCISE

Learning pathways

■ LEVEL 1	■ LEVEL 2	■ LEVEL 3
1, 4, 10, 12, 15	2, 5, 7, 8, 13	3, 6, 9, 11, 14

Check your understanding

1. Define the term *geomorphology*.
2. Describe what a natural hazard is.
3. Describe what a geomorphic hazard is.
4. List and describe the different types of landslides.
5. Refer to **FIGURE 3**. Describe the distribution of landslides that have occurred in Australia.
6. Outline the key factors that contributed to the Thredbo landslide.

Apply your understanding

7. Explain why a landslide is considered a geomorphic hazard.
8. Account for why changes to landscapes can cause landslides.
9. Explain why a flow landslide is considered the most dangerous.
10. Explain what caused the landslide in Brazil.
11. Identify the different types of landslides and their causes. Then sketch a diagram showing how they occur.
12. Explain one way that a community can protect itself from landslides.

Challenge your understanding

13. Suggest one way that a government could help to protect people who live in areas that are at risk of landslides.
14. Refer to **FIGURE 1**. Using evidence from the photo, explain why this part of India has many landslides. List possible strategies to ensure landslides are minimised in the area.
15. Predict how a landslide in your local area might impact the community.

To answer questions online and to receive **immediate feedback** and **sample responses** for every question, go to your learnON title at www.jacplus.com.au.

3.11 SkillBuilder — Using topographic maps

online only

LEARNING INTENTION

By the end of this subtopic, you will be able to use a topographic map to determine the location of a place using area and grid references.

The content in this subtopic is a summary of what you will find in the online resource.

3.11.1 Tell me

What are topographic maps?

Topographic maps show the height and shape of the land, features of the natural environment (such as forests and lakes) and features of human environments (such as roads and settlements).

A 1-kilometre-square grid is overprinted on a topographic map. These grid lines are numbered with two-digit numbers in the map's margins. Lines that run up and down the map (north–south) are called **eastings**. Lines that run horizontally across the map (east–west) are called **northings** The example in **FIGURE 3** also has an additional grid laid over the top showing how to work out six-figure grid references (see step 3).

FIGURE 3 This map extract shows a single square kilometre of a map (AR8513).

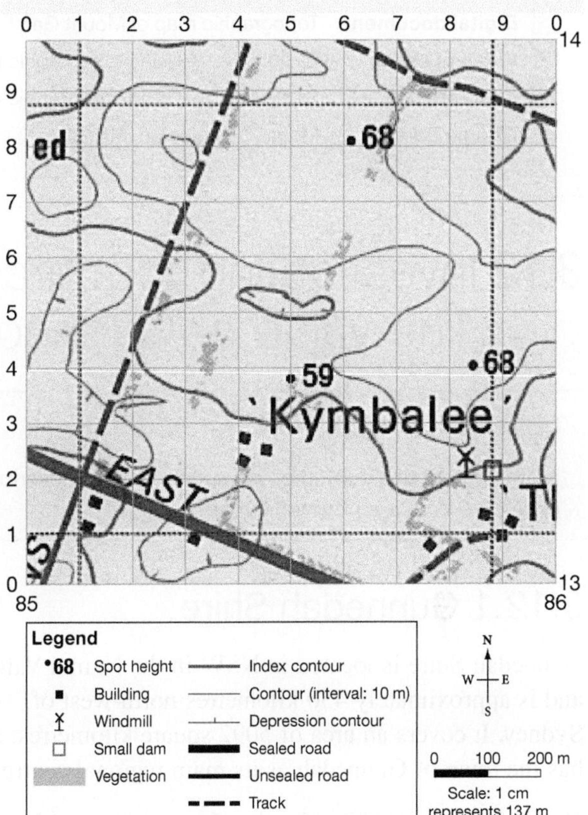

3.11.2 Show me

How to read grid references

Step 1

Find the place that you want to provide a location for.

Step 2

Four-figure grid reference: Go to the bottom left corner of the square, where the grid lines join. Write down the number labelling the horizonal line first, and then write the number labelling the vertical line. This is the area reference. Place the letters AR in front of the number to show it is an area reference (e.g. AR8513 is Kymbalee in **FIGURE 3**).

Step 3

Six-figure grid reference: Find an exact point in the grid square you referenced in step 2. The six-figure grid reference adds two more numbers to the AR.

Imagine a 10 x 10 grid over your area reference square (as shown in **FIGURE 3**). For your particular place, estimate or measure the tenths across the grid square from the side and top.

eastings lines that run up and down a map (north–south)

northings lines that run horizontally across a map (east–west)

Write the number of tenths from the side after the first two numbers in your area reference. Write the number of tenths from the top after the last two numbers in your area reference. Write the letters GR in front of the six-figure grid reference (e.g. GR854132 is Kymbalee in **FIGURE 3**).

3.11.3 Let me do it

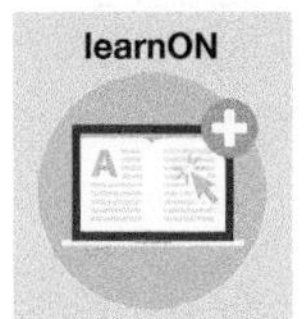

Go to learnON to access the following additional resources to help you build this skill:

- a longer explanation of this skill and its application in Geography (Tell me)
- a video showing the step-by-step process of this skill (Show me)
- an activity and interactivity for you to practise this skill (Let me do it)
- self-marking questions to help you understand and use this skill.

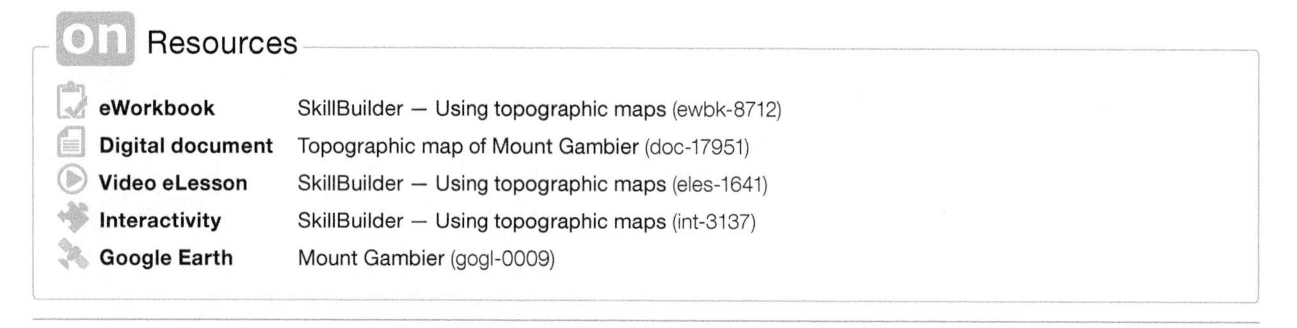

on Resources

eWorkbook	SkillBuilder — Using topographic maps (ewbk-8712)
Digital document	Topographic map of Mount Gambier (doc-17951)
Video eLesson	SkillBuilder — Using topographic maps (eles-1641)
Interactivity	SkillBuilder — Using topographic maps (int-3137)
Google Earth	Mount Gambier (gogl-0009)

3.12 Investigating topographic maps — The value of Gunnedah Shire

LEARNING INTENTION

By the end of this subtopic, you will be able to describe how a place can be valued in different ways, giving examples from a topographic map.

3.12.1 Gunnedah Shire

Gunnedah Shire is located in NSW in the Namoi Valley, and is approximately 450 kilometres north-west of Sydney. It covers an area of 5092 square kilometres, and has the town of Gunnedah as its main regional centre.

Gunnedah was inhabited by the *Gunn-e-dar* people of the Kamilaroi nation and its name means 'place of white stones'. This area holds many special places for the various Aboriginal clans within this area, from the Mooki River, which is still home today to a small Aboriginal community called Walhallow, to the Doona State Forest, the site of a large section of grinding stones and albino kangaroos. Bindea, also known as Porcupine Lookout, Bee Burra and Boonalla are also significant places that the Kamilaroi have maintained connection with following colonisation.

FIGURE 1 Location of Gunnedah

Source: Spatial Vision

Across the whole Gunnedah Shire and within the geographical area of the Liverpool Plains there is an abundance of flora and fauna which the Kamilaroi Peoples relied on for medicine, food, shelter and cultural practices. This area is one of the most highly populated area of koalas in Australia and both Aboriginal and non-Aboriginal locals are focused on ensuring this important native species is well looked after.

The area is situated on the Liverpool Plains so it is very flat. The plains are 264 metres above sea level and the hills are only about 200 metres above the flat plains. As the area is flat, Gunnedah's main industry is agriculture, with approximately 403 000 hectares of land used to grow crops such as wheat, sorghum, barley, cotton, sunflowers, soya beans and vegetables. Gunnedah also has a strong coal mining industry with 10 million tonnes of coal exported every year.

FIGURE 2 The Namoi Valley in the flat Liverpool Plains is good agricultural land.

FIGURE 3 Aerial view of the Namoi Valley

FIGURE 4 Topographic map extract of Gunnedah Shire

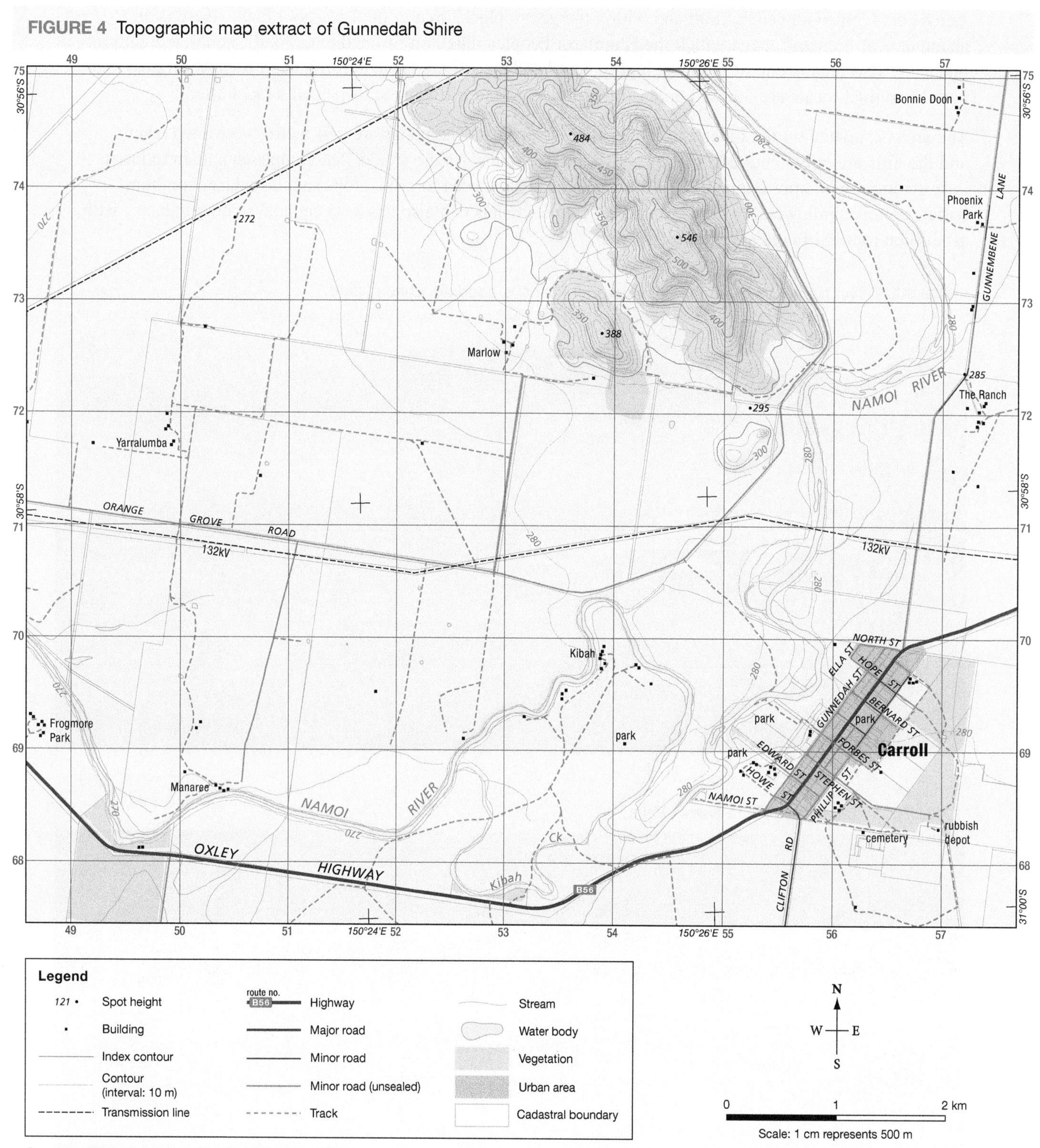

Source: Data based on Spatial Services 2019

Resources

eWorkbook The value of Gunnedah Shire (ewbk-8720)
Digital document Topographic map of Gunnedah Shire (doc-36188)
Video eLesson The value of Gunnedah Shire — Key concepts (eles-5041)
Interactivity The value of Gunnedah Shire (int-8369)
Google Earth Gunnedah Shire (gogl-0080)

3.12 EXERCISE

Learning pathways

■ LEVEL 1	■ LEVEL 2	■ LEVEL 3
1, 7, 8, 12, 13	2, 3, 4, 5, 10, 14, 17	6, 9, 11, 15, 16

Check your understanding

Refer to **FIGURE 4** and use your topographic map reading skills for the following exercises.

1. State the contour interval of this map.
2. State the ratio scale of the map.
3. Name the road in AR5671 and identify the type of road it is in the key.
4. How many buildings are located in AR5069?
5. What is the grid reference of the cemetery?
6. What is the height of the highest point on this map?
7. Name the main river evident on this map.
8. Name the major road evident on this map.
9. Find the name of the town located in the south-east quadrant of this topographic map.

Apply your understanding

10. Refer to the text and **FIGURES 2**, **3** and **4**, and explain why farming is the main activity that occurs in this landscape.
11. Why do you think the town in the south-east quadrant was built in this space?
12. What natural hazard might the town be under threat from?

Challenge your understanding

13. Choose an area on this map where you would establish a farm. Give the area reference and outline three reasons why you would locate a farm there.
14. Propose a solution for the protection of significant Aboriginal sites that may be located on farming land.
15. Construct a plan that includes the maintenance of Aboriginal sites and ensures sustainable farming.
16. List the advantages and disadvantages of farming in Gunnedah Shire. Based on your list, which groups of people might believe that the advantages outweigh the disadvantages? Who might believe the disadvantages are more significant?
17. How might the negative effects of agriculture be managed in Gunnedah Shire? Suggest one strategy that could be put in place.

To answer questions online and to receive **immediate feedback** and **sample responses** for every question, go to your learnON title at www.jacplus.com.au.

3.13 Thinking Big research project — Karst landscape virtual tour

online only

The content in this subtopic is a summary of what you will find in the online resource.

Scenario

Karst environments feature in some Aboriginal Peoples' creation stories. They feature in ceremonies and are thought to have curative powers. Part of their mystique is that they predominantly exist underground and are difficult to access. The Australian government is creating a display at the National Museum and wants to include a diorama that showcases this environment and its connection to creation stories. Your job is to create it!

Task

Create a virtual tour of a karst landscape that:

- showcases how the landscape was created from both the geographical perspective and based on Aboriginal Dreaming stories
- includes a guided tour of a unique karst environment within Australia.

Go to your Resources tab to access all of the resources you need to complete this project.

ProjectsPLUS Thinking Big research project — Karst landscape virtual tour (pro-0168)

3.14 Review

3.14.1 Key knowledge summary

3.2 Valuing landforms and landscapes

- Landscapes and landforms have aesthetic, cultural, spiritual and economic value.
- Aesthetic value refers to the importance of a landscape or landform being pleasing to look at.
- In Australia, the beach is an example of a landscape that has aesthetic value as well as cultural value.

3.3 The cultural and spiritual value of landscapes and landforms

- Landscapes and landforms can shape our beliefs and way of life.
- Aboriginal Peoples and Torres Strait Islander Peoples have strong connections with Country, which plays a central role in Culture, their sense of belonging and identity.
- Dreaming Stories and Tagai provide examples of a cultural and spiritual value of landscapes.

3.4 The economic value of landscapes and landforms

- Landscapes and landforms can have economic value through mining and tourism.
- The Grand Canyon is an example of a landscape that has economic value because it attracts many tourists.
- Adventure tourism is a type of tourism that involves people engaging in activities such as mountain biking, climbing and hiking.

3.6 Protecting and managing landscapes

- Aboriginal Peoples managed and cared for Country with methods such as cultural burning, seasonal migration and kinship systems.
- The World Heritage Convention is an international agreement that provides protection for significant landscapes and landforms.
- National parks and reserves allow for protection of landscapes and landforms at a national and state level.

3.7 Protecting and managing Kakadu

- Kakadu National Park covers 20 000 square kilometres in the Northern Territory.
- A large part of Kakadu National Park was listed as a World Heritage site in 1981.
- Kakadu has many significant cultural and heritage sites, and large deposits of uranium.

3.8 Protecting and managing Uluṟu

- Uluṟu's colour and dramatic shapes mean that Uluṟu has aesthetic value.
- Uluṟu has deep spiritual (sacred) value for the Aṉangu People.
- Uluṟu has a large number of visitors each year who generate spending on tour guides, airfares, accommodation and restaurants. It has cultural value to many Australians.
- Uluṟu has been managed over thousands of generations by the Aṉangu People. It has been a national park since 1950, which has been jointly managed with the Aṉangu People since 1985.
- Uluṟu is recognised as a World Heritage site.

3.9 Changed and degraded landscapes

- Humans have used and modified landscapes and landforms for thousands of years.
- Farming, including domesticating animals and growing crops, has altered environments.
- Large settlements and the development of cities and towns have enabled humans to dominate landscapes, and resulted in landscapes and landforms becoming degraded.

3.10 Geomorphic hazards—Landslides

- Geomorphic hazards are extreme events that occur in the lithosphere.
- Landslides are an example of geomorphic hazards, and involve loose rock, debris or earth moving down a steep slope.
- Landslides can be caused by natural and human causes (or a combination of both natural and human causes) such as water, earthquakes, volcanoes or by land clearing.

3.14.2 Key terms

aesthetic relating to something beautiful or pleasing to look at
Anthropocene a period of time when humans made significant changes to the Earth's environments
archaeological concerning the study of past civilisations and cultures by examining the evidence left behind, such as graves, tools, weapons, buildings and pottery
Country the land and its features, which is bound to the concept of belonging to a place that is fundamental to Aboriginal Peoples' sense of identity and Culture
cultural heritage beliefs and ways of living passed down through generations
degradation a decline in quality caused by time or improper use
domesticate to tame a wild animal or plant so it can live with or be looked after by people
Dreaming in Aboriginal culture, the time when the Earth took on its present form, and cycles of life and nature began; also known as the Dreamtime. Dreaming Stories pass on important knowledge, laws and beliefs.
eastings lines that run up and down a map (north–south)
geomorphic describes a process that occurs in the lithosphere
ice age a geological period during which the Earth is colder, glaciers and ice sheets expand and sea levels fall. The last ice age was approximately 13 000 years ago.
inselberg an isolated hill, knob, ridge, outcrop or small mountain that rises abruptly from the surrounding landscape
kinship system rules and relationships that determine how people interact with each other
leaching a process that occurs where water runs through the soil, dissolving minerals and carrying them into the subsoil
lithosphere the Earth's crust, including landforms, rocks and soil
liquefaction when solids are shaken heavily and act like liquids
mass movement large areas of earth that move
natural hazard an extreme event that is the result of natural processes and has the potential to cause serious material damage and loss of life
nomadic a lifestyle in which people do not live in one fixed spot but move according to the seasons from place to place in search of food, water and grazing land
northings lines that run horizontally across a map (east–west)
nutrient an essential element that feeds living things to help them function and grow
pictorial map a map using illustrations to represent information
plateau an extensive area of flat land that is higher than the land around it. Plateaus are sometimes referred to as tablelands.
settlement place where humans live together, e.g. houses, villages, towns and cities
songlines paths across the land that follow the creation journeys of Aboriginal Peoples' ancestors, which are used to teach about Culture and Country; also known as Dreaming Tracks
totem a part of the natural world (such as an animal) that has a special meaning or importance for specific people
World Heritage List a list of human and natural sites that have been approved by UNESCO as meeting the criteria of being a World Heritage site

3.14.3 Reflection

Complete the following to reflect on your learning.

Revisit the inquiry question posed in the Overview:

From oceans to deserts to cities, what exactly are landscapes and how is each one unique?

1. Now that you have completed this topic, what is your view on the question? Discuss with a partner. Has your learning in this topic changed your view? If so, how?
2. Write a paragraph in response to the inquiry question outlining your views.

Subtopic	Success criteria	(light)	(medium)	(dark)
3.2	I can explain the aesthetic value of landscapes and landforms.			
3.3	I can explain the cultural and spiritual value of landscapes and landforms.			
	I can describe the cultural value of landscapes to Aboriginal Peoples and Torres Strait Islander Peoples.			
3.4	I can identify and give examples of the economic value of landscapes or landforms.			
3.5	I can use geographical terminology to identify direction and position.			
3.6	I can describe the ways that landscapes can be protected at a range of scales.			
3.7	I can assess whether the use and management of an important Australian landscape is sustainable.			
3.8	I can appreciate the contribution of Aboriginal Peoples' knowledge to the use and management of an Australian landform.			
3.9	I can describe human causes of land degradation.			
	I can give examples of the effects of land degradation.			
3.10	I can outline the natural and human causes of geomorphic hazards.			
	I can describe how geomorphic hazards can affect people and the landscape.			
3.11	I can use a topographic map to determine the location of a place using area and grid references.			
3.12	I can describe how a place can be valued in different ways, giving examples from a topographic map.			

on Resources

eWorkbook Chapter 3 Student learning matrix (ewbk-8444)
Chapter 3 Reflection (ewbk-8445)
Chapter 3 Extended writing task (ewbk-8446)

Interactivity Chapter 3 Crossword (int-8370)

ONLINE RESOURCES

on Resources

Below is a full list of **rich resources** available online for this topic. These resources are designed to bring ideas to life, to promote deep and lasting learning and to support the different learning needs of each individual.

eWorkbook

- 3.1 Chapter 3 eWorkbook (ewbk-7983) ☐
- 3.2 Valuing landforms and landscapes (ewbk-8676) ☐
- 3.3 The cultural and spiritual value of landscapes and landforms (ewbk-8680) ☐
- 3.4 The economic value of landscapes and landforms (ewbk-8684) ☐
- 3.5 SkillBuilder — Using positional language (ewbk-8688) ☐
- 3.6 Protecting and managing landscapes (ewbk-8692) ☐
- 3.7 Protecting and managing Kakadu (ewbk-8696) ☐
- 3.8 Protecting and managing Uluṟu (ewbk-8700) ☐
- 3.9 Changed and degraded landscapes (ewbk-8704) ☐
- 3.10 Geomorphic hazards — Landslides (ewbk-8708) ☐
- 3.11 SkillBuilder — Using topographic maps (ewbk-8712) ☐
- 3.12 Investigating topographic maps — The value of Gunnedah Shire (ewbk-8720) ☐
- 3.14 Chapter 3 Student learning matrix (ewbk-8444) ☐
- Chapter 3 Reflection (ewbk-8445) ☐
- Chapter 3 Extended writing task (ewbk-8446) ☐

Sample responses

- 3.1 Chapter 3 Sample responses (sar-0138) ☐

Digital documents

- 3.11 Topographic map of Mount Gambier (doc-17951) ☐
- 3.12 Topographic map of Gunnedah Shire (doc-36188) ☐

Video eLessons

- 3.1 Landscapes sculpted by water (eles-1624) ☐
- Valuing, managing and protecting landscapes and landforms — Photo essay (eles-5307) ☐
- 3.2 Valuing landforms and landscapes — Key concepts (eles-5033) ☐
- 3.3 The cultural and spiritual value of landscapes and landforms — Key concepts (eles-5034) ☐
- 3.4 The economic value of landscapes and landforms — Key concepts (eles-5035) ☐
- 3.5 SkillBuilder — Using positional language (eles-1649) ☐
- 3.6 Protecting and managing landscapes — Key concepts (eles-5036) ☐
- 3.7 Protecting and managing Kakadu — Key concepts (eles-5037) ☐
- 3.8 Protecting and managing Uluṟu — Key concepts (eles-5038) ☐
- 3.9 Changed and degraded landscapes — Key concepts (eles-5039) ☐
- 3.10 Geomorphic hazards — Landslides — Key concepts (eles-5040) ☐
- 3.11 SkillBuilder — Using topographic maps (eles-1641) ☐
- 3.13 Investigating topographic maps — The value of Gunnedah Shire — Key concepts (eles-5041) ☐

Interactivities

- 3.2 Valuing landforms and landscapes (int-8363) ☐
- 3.3 The cultural and spiritual value of landscapes and landforms (int-8364) ☐
- 3.4 The economic value of landscapes and landforms (int-8365) ☐
- 3.5 SkillBuilder — Using positional language (int-3145) ☐
- 3.6 World Heritage List sites of significance (int-8727) ☐
- Location of the 850 national parks and reserves in NSW, Australia (int-8735) ☐
- 3.7 Protecting and managing Kakadu (int-3609) ☐
- 3.8 Protecting and managing Uluṟu (int-8366) ☐
- 3.9 Changed and degraded landscapes (int-8367) ☐
- 3.10 Geomorphic hazards Landslides (int-8368) ☐
- 3.11 SkillBuilder — Using topographic maps (int-3137) ☐
- 3.12 Investigating topographic maps — The value of Gunnedah Shire (int-8369) ☐
- 3.14 Chapter 3 Crossword (int-8370) ☐

Fieldwork

- 3.6 GPS Home (fdw-0001) ☐
- GPS Sydney (fdw-0006) ☐

ProjectsPLUS

- 3.13 Thinking Big research project — Karst landscape virtual tour (pro-0168) ☐

Weblinks

- 3.1 Iguazu Falls (web-0042) ☐
- 3.6 World Heritage list (web-0065) ☐

Google Earth

- 3.11 Mount Gambier (gogl-0009) ☐
- 3.12 Gunnedah Shire (gogl-0080) ☐

myWorld Atlas

- 3.4 Deepen your understanding of this topic with related case studies and questions > Grand Canyon (mwa-4532) ☐
- Deepen your understanding of this topic with related case studies and questions > Himalayas video (mwa-4493) ☐
- 3.6 Deepen your understanding of this topic with related case studies and questions > World Heritage sites (mwa-4434) ☐
- 3.10 Deepen your understanding of this topic with related case studies and questions > Thredbo landslide (mwa-4472) ☐

Teacher resources

There are many resources available exclusively for teachers online.

4 Desert landscapes

INQUIRY SEQUENCE

To access a pre-test with **immediate feedback** and **sample responses** to every question in this chapter, select your learnON format at www.jacplus.com.au.

4.1 Overview

Numerous **videos** and **interactivities** are embedded just where you need them, at the point of learning, in your learnON title at www.jacplus.com.au. They will help you to learn the content and concepts covered in this topic.

Hot and sandy? Cold and windy? What are the features of a landscape that make it a desert?

4.1.1 Introduction

Have you ever stood on a hill, looking at the landscape and variety of landforms such as hills, valleys, volcanoes and plains, and wondered how they are created? Despite people valuing our planet's landscapes, some landscapes and landforms have been degraded so we need to manage and protect them for the future.

Approximately one-third of the Earth's land surface is desert — arid land with little rainfall. These arid regions may be hot or cold. The actions of wind and, sometimes, water shape the rich variety of landscapes found there.

STARTER QUESTIONS

1. Have you ever visited a desert? Where was it? Why did you go?
2. Make a list of the geographical characteristics of desert areas. What is the most common feature of a desert?
3. Find a description of a desert online. Copy the text and use it to create a wordcloud, using an online wordcloud creator. What is the most common word in your cloud. Why do you think this is the case?
4. Use the **Sahara** weblink in your Resources tab to watch a video of desert sandstorms.
   a. Describe the sandstorm you see. Would it be difficult for people to cope with such an event?
   b. Describe how some animals have adapted to cope with desert sandstorms.
   c. How do the wind and sand shape the desert?
   d. How do communities survive in the desert?
5. Watch the **Deserts — Photo essay** Video eLesson. Is this somewhere you would like to live? What would you enjoy? What would you find challenging about living in that kind of environment?

FIGURE 1 (a) and (b) Arches National Park, Utah, USA

(a)

(b)

Resources

eWorkbook Chapter 4 eWorkbook (ewbk-7984)

Video eLesson Desert landscapes — Desertscapes (eles-1625)
Deserts — Photo essay (eles-5334)

4.2 Deserts around the world

LEARNING INTENTION

By the end of this subtopic, you will be able to explain some characteristics of deserts and describe where they are located in the world.

4.2.1 Defining a desert

A desert is a hot or cold region with little or no rainfall. Around one-third of the Earth's surface is desert and these areas are home to about 300 million people. Although they receive little rainfall, most deserts receive some form of precipitation. When it does rain, it is usually during a few heavy storms that last a short time.

TABLE 1 Types of deserts

Rainfall (mm/year)	Type of desert	Examples
<25	Hyper-arid	Namib, Arabian
25–200	Arid	Mojave
200–500	Semi-arid	Parts of Sonoran Desert

Hot deserts

Most of the world's hot deserts are located between the Tropic of Cancer and the Tropic of Capricorn (see **FIGURE 3**). They have very hot summers and warm winters. Temperature extremes are common because cloud cover is rare and humidity is very low; this means there is nothing to block the heat of the Sun during the day or prevent its loss at night. Temperatures can range between around 45 °C and –15 °C in a 24-hour period.

Examples of hot deserts include the Sahara Desert in northern Africa, the Kalahari Desert in southern Africa, and the many deserts found in Australia including the Great Victoria Desert, the Great Sandy Desert, the Tanami Desert and the Simpson Desert.

FIGURE 1 The Sahara is an example of a hot desert.

Cold deserts

Cold deserts lie on high ground generally north of the Tropic of Cancer and south of the Tropic of Capricorn (see **FIGURE 3**). They include the polar deserts. Any precipitation falls as snow. Winters are very cold and often windy; summers are dry and cool to mild.

Examples of cold deserts include Antarctica; the Gobi Desert, which sits on the border of Mongolia and China; and the Great Basin Desert in North America.

FIGURE 2 The Gobi is an example of a cold desert.

4.2.2 Deserts of the world

FIGURE 3 The distribution of the world's deserts

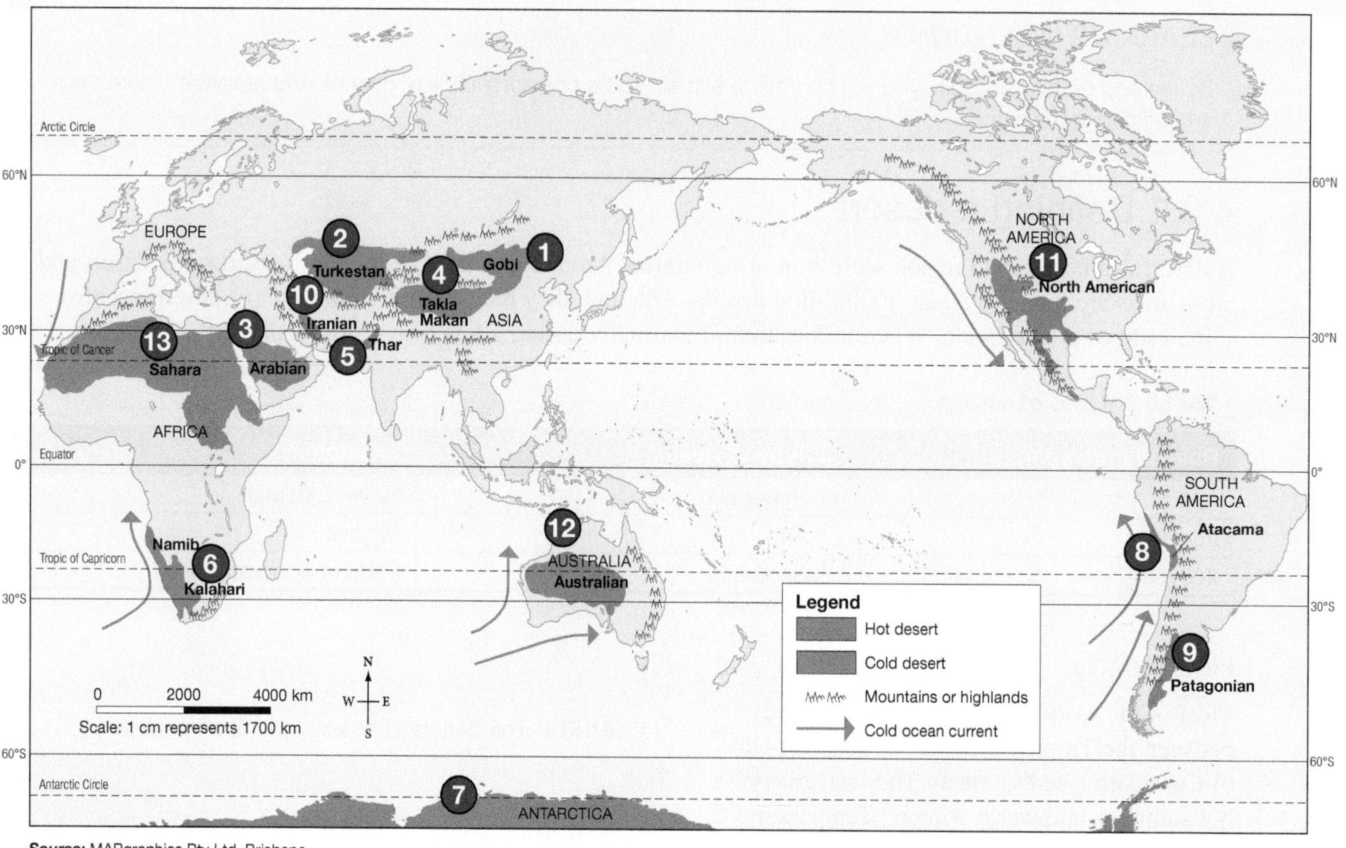

Source: MAPgraphics Pty Ltd, Brisbane

1 Gobi Desert

Asia's biggest desert, the Gobi, is a cold desert. It sits about 900 metres above sea level and covers an area of approximately 1.2 million square kilometres. Its winters can be freezing.

2 Turkestan Desert

The cold Turkestan Desert covers parts of south-western Russia and the Middle East.

3 Arabian Desert

This hot desert is as big as the deserts of Australia. Towards its south is a place called Rub' al-Khali (meaning 'empty quarter'), which has the largest area of unbroken sand dunes, or erg, in the world.

4 Takla Makan Desert

The Takla Makan Desert is a cold desert in western China. Its name means 'place of no return'. The explorer Marco Polo crossed it around 800 years ago.

5 Thar Desert

The Thar Desert is a hot desert covering north-western parts of India and Pakistan. Small villages of around 20 houses dot the landscape.

6 Namib and Kalahari deserts

The Namib Desert extends for 1200 kilometres down the coast of Angola, Namibia and South Africa. It seldom rains there, but an early-morning fog often streams across the desert from the ocean. The dew it leaves behind provides moisture for plants and animals. It joins the Kalahari Desert, which is about 1200 metres above sea level.

7 Antarctic Desert

The world's biggest and driest desert, the continent of Antarctica is another cold desert. Only snow falls there, equal to about 50 millimetres of rain per year.

8 Atacama Desert

The Atacama Desert is the driest hot desert in the world. Its annual average rainfall is only 0.1 millimetre.

9 Patagonian Desert

The summer temperature of this cold desert rarely rises above 12 °C. In winter, it is likely to be well below zero, with freezing winds and snowfalls.

10 Iranian Desert

Two large deserts extend over much of central Iran. The Dasht-e-Lut is covered with sand and rock, and the Dasht-e-Kavir is mainly covered in salt. Both have virtually no human populations.

11 North American deserts

The desert region in North America is made up of the Mojave, Sonoran and Chihuahuan deserts (all hot deserts) and the Great Basin (a cold desert). The Great Basin's deepest depression, Death Valley, is the lowest point in North America.

12 Australian deserts

After Antarctica, Australia is the driest continent in the world. Its deserts are hot, are generally flat lands and are often vibrant in colour.

13 Sahara Desert

The largest hot desert in the world, the Sahara stretches some 9 million square kilometres across northern Africa over 12 countries. Only a small part is sandy. It is the sunniest place in the world (see **FIGURE 1**).

on Resources

eWorkbook	Deserts around the world (ewbk-8724)
Video eLesson	Deserts around the world — Key concepts (eles-5082)
Interactivities	Great deserts of the world (int-3106) The distribution of the world's deserts (int-3611)

4.2 ACTIVITIES

1. Use the information in this subtopic to design a quiz of 10 questions entitled 'Deserts of the world'. Test your friends and family.
2. Draw up and complete a table like the one below to show your understanding of the locations and features of desert environments. Look for photos online. (Many people who have visited deserts have uploaded images to Google Maps, and for some locations you might even be able to access Street View images.)

FIGURE 4 Paracas National Reserve meets the Pacific Ocean, Paracas Peninsula, Peru

Name of desert	Type of desert	Continent	Features	Photos

3. Research how the following deserts are formed:
   - subtropical deserts
   - coastal deserts.

4.2 EXERCISE

Learning pathways

■ LEVEL 1	■ LEVEL 2	■ LEVEL 3
1, 3, 4, 10, 14	5, 6, 11, 12, 13	2, 7, 8, 9, 15

Check your understanding

1. Define the term *desert*.
2. What climate conditions are needed for:
   a. hot deserts to form?
   b. cold deserts to form?
3. Where is the sunniest place in the world?
4. Name three deserts in the Asia–Pacific region.
5. On what major line of latitude are Australian deserts located?
6. Identify two deserts in each of the following continents, and indicate if each example is a hot or a cold desert:
   a. Africa
   b. Asia
7. Describe key differences between arid, semi-arid and hyper-arid deserts.
8. Describe the spatial distribution of the world's deserts and provide reasons for their location.

Apply your understanding

9. Look carefully at the map in **FIGURE 3**.
   a. Which continent has the largest area of hot desert?
   b. Which continent has the largest area of cold desert?
10. Look carefully at the map in **FIGURE 3**.
    a. What is the largest hot desert in the world?
    b. What is the largest hot desert in the Asia–Pacific region?
11. Look carefully at the map in **FIGURE 3**.
    a. Which is the driest continent in the world?
    b. Which continent contains the driest hot desert?
12. Look carefully at the map in **FIGURE 3**. Which North American desert contains the lowest land on the continent?

Challenge your understanding

13. Suggest why the only precipitation that falls in a polar desert is snow.
14. If you had to live in one of the desert regions labelled in **FIGURE 3**, which would you choose and why?
15. Create a list of equipment and supplies that you would need to take to survive for 24 hours in a hot desert. Suggest how many of these things would also be useful for a similar expedition to a cold desert.

To answer questions online and to receive **immediate feedback** and **sample responses** for every question, go to your learnON title at www.jacplus.com.au.

4.3 SkillBuilder — Using latitude and longitude

online only

LEARNING INTENTION

By the end of this subtopic, you will be able to interpret and calculate latitude and longitude readings on a map.

The content in this subtopic is a summary of what you will find in the online resource.

4.3.1 Tell me

What are latitude and longitude?

Latitude and longitude are imaginary grid lines encircling the Earth. They can be drawn over a map to help us locate a place. The lines that run parallel to the equator are called parallels of latitude. Lines of longitude run north to south from the North Pole to the South Pole.

FIGURE 1 The parallels of latitude

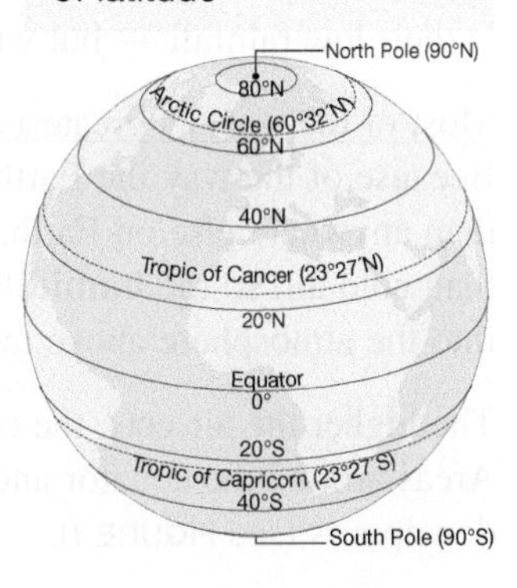

4.3.2 Show me

How to use latitude and longitude

Step 1

Determine the place for which you want to give a latitude and longitude reading.

Step 2

Begin with the parallels of latitude. Determine the degrees on the line closest to the location. Each degree on the grid is made up of 60 minutes. It is likely that a place is not situated exactly on the degree line, so you will need to determine a minute reading as well. Calculate the minutes for the place you are identifying. Combine the readings to obtain a precise latitude for place.

Step 3

Longitude is determined in a similar manner. Find the north–south line (meridian) closest to the place. When combining the grid readings, latitude always comes first.

FIGURE 2 The meridians of longitude

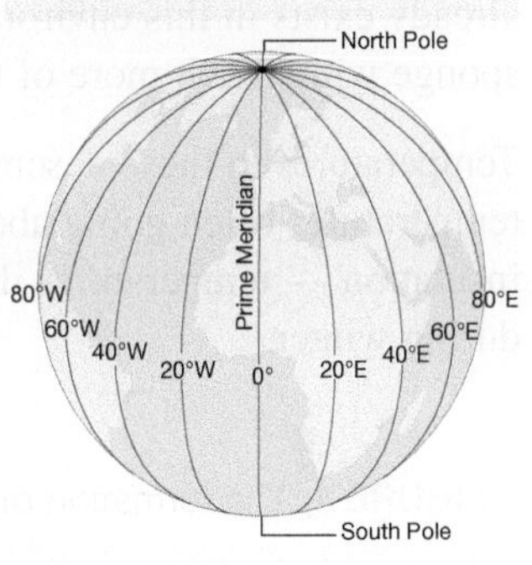

4.3.3 Let me do it

learnON

Go to learnON to access the following additional resources to help you build this skill:

- a longer explanation of this skill and its application in Geography (Tell me)
- a video showing the step-by-step process of this skill (Show me)
- an activity and interactivity for you to practise this skill (Let me do it)
- self-marking questions to help you understand and use this skill.

on Resources

eWorkbook SkillBuilder — Using latitude and longitude (ewbk-8728)

Video eLesson SkillBuilder — Using latitude and longitude (eles-1652)

Interactivity SkillBuilder — Using latitude and longitude (int-3148)

4.4 How the climate forms deserts

LEARNING INTENTION

By the end of this subtopic, you will be able to explain the climatic processes that form and transform deserts.

4.4.1 The subtropics

Deserts form in many different parts of the globe: the **subtropics**; continental interior areas at middle latitudes; on the **leeward** side of mountain ranges; along coastal areas; and in the polar regions. The only common factor is their low rainfall — but why do these areas experience low rainfall?

Most of the world's greatest deserts are found in the subtropics near the Tropics of Cancer and Capricorn. Because of the way the Earth rotates around the Sun, areas around the equator receive more direct sunlight than anywhere else on Earth. This means the air there is always very hot. Hot air can hold much more moisture than cold air, so the **humidity** in these areas is always very high. Hot air also rises. As the air heads upwards into the atmosphere above the equator, it drifts away, heading north and south.

The higher the air gets, the cooler it becomes. Cool air can't hold as much moisture, so it releases it as rain. Areas around the equator and to the immediate north and south of it (the tropics) receive frequent heavy downpours (see **FIGURE 1**).

With its moisture gone, the cool, dry air continues moving north and south away from the equator until it meets zones of high air pressure around the tropics. Here, it is forced downwards. The further the dry air descends, the warmer it gets. This means it can hold more moisture and is likely to absorb any moisture that already exists in this environment. It is like using a sponge to wipe up water on the kitchen bench; a dry sponge will absorb more of the spill than a wet sponge. This is how the subtropical deserts form.

Temperatures in these deserts are usually high all year round. In summer the heat is extreme, with daytime temperatures often going above 38 °C and sometimes as high as 49 °C. At night — with no clouds to provide insulation — temperatures drop quickly to an average of 21 °C in summer and sometimes below freezing during winter.

FIGURE 1 The formation of subtropical deserts in Africa

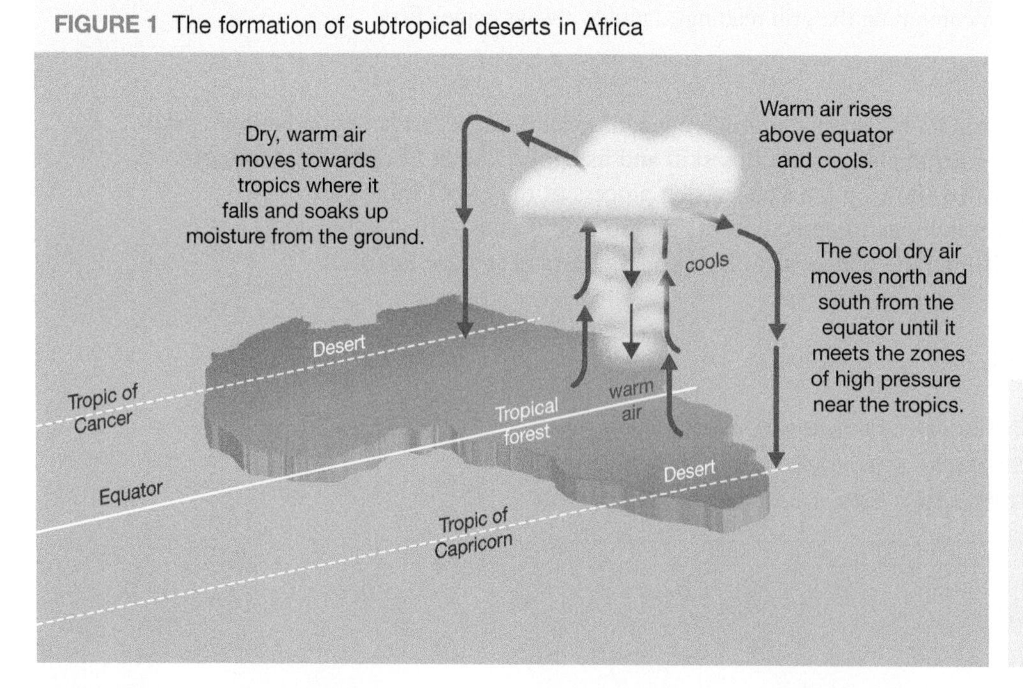

subtropics the areas of Earth just north of the Tropic of Cancer and just south of the Tropic of Capricorn

leeward the side of mountains that faces away from rainbearing winds

humidity the amount of water vapour in the atmosphere

4.4.2 Rain shadow deserts

Rain shadows form on the leeward side of a mountain range (opposite the windward side that faces rainbearing winds). Deserts commonly form in rain shadows in the following conditions.

- Moist air blowing in from the ocean is forced to rise when it hits a range of mountains. This cools the air. As cool air cannot hold as much moisture, it releases it as precipitation, which falls on the **windward** side of the mountain (see FIGURE 3).
- By the time the air moves over the top of the range and down the other side, it is likely to have lost most, if not all, of its moisture. It will therefore be fairly dry.
- The more the air descends on the other side of the range, the more it warms up. Hence, it can hold more moisture. So, as well as not bringing any rain to the land, the air absorbs what little moisture the land contains.
- In time, as this pattern continues, the country in the rain shadow of the mountain range is likely to become arid (see FIGURE 3).

FIGURE 2 Satellite image of the coast of California, USA

An example of this effect is the Great Dividing Range in Australia: cool moist air produces rainfall on the eastern side of these mountains and desert to the west.

Another example is the Mojave Desert in the south-western United States. It is located on the leeward side of the Sierra Nevada mountain range. In FIGURE 2, the dry landscape of the Mojave Desert (top right) can be seen to the north (here, the leeward side) of the mountains created by the San Andreas Fault (centre), which separate the desert from the Pacific Ocean.

rain shadow the drier side of a mountain range, cut off from rain-bearing winds

windward the side of mountains that face rain-bearing winds

FIGURE 3 The formation of rain shadow deserts

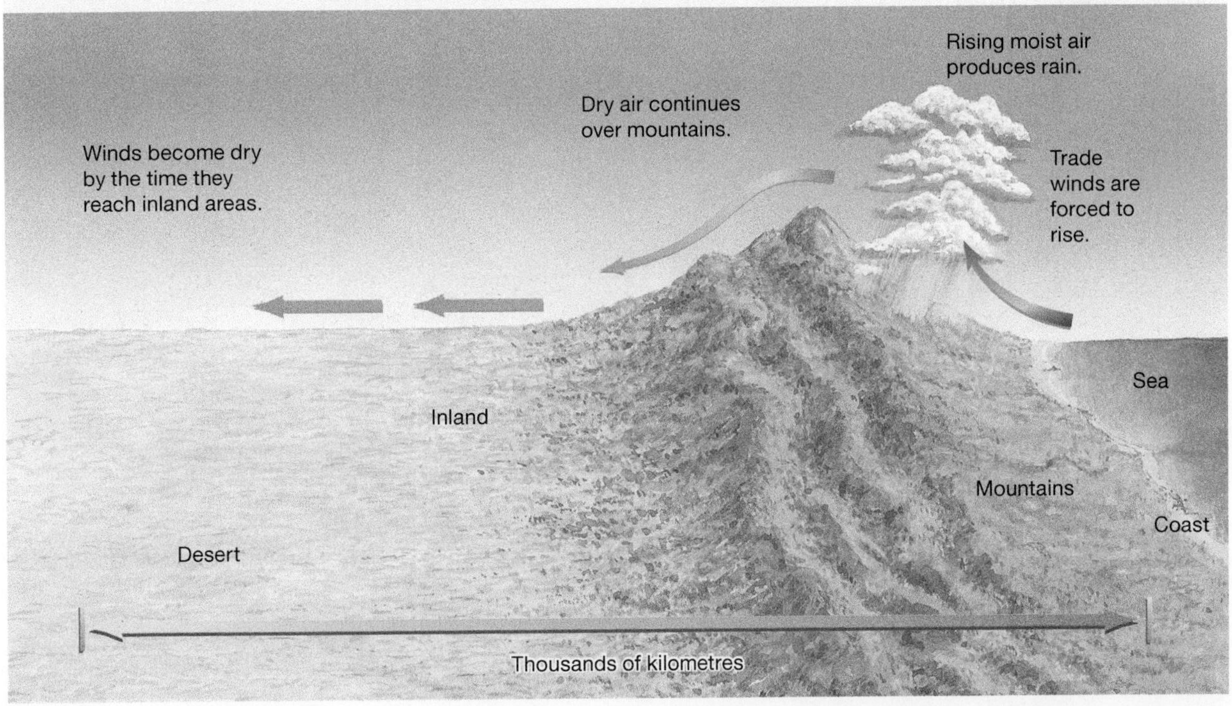

Rising moist air produces rain.
Dry air continues over mountains.
Winds become dry by the time they reach inland areas.
Trade winds are forced to rise.
Sea
Inland
Mountains
Coast
Desert
Thousands of kilometres

4.4.3 Coastal deserts

Currents in the oceans are both warm and cold, and are always moving. Cold currents begin in polar and temperate waters (with moderate temperatures), and drift towards the equator. They flow in a clockwise pattern in the northern hemisphere, and in an anticlockwise pattern in the southern hemisphere. As they move, they cool the air above them (see **FIGURE 4**).

If cold currents flow close to a coast, they can contribute to the creation of a desert. This occurs because cold ocean currents cause the air over the coast to become stable, which stops cloud formation. If the cool air the currents create blows in over warm land, the air warms up; it can then hold more moisture (see **FIGURE 4**). It is therefore not likely to release any moisture it contains unless it is forced up by a mountain range. Large coastal deserts, including the Atacama Desert in Chile and the Namib Desert in Namibia, are formed in this way (see **FIGURE 5**).

FIGURE 4 The formation of coastal deserts

FIGURE 5 The coastal Namib Desert

4.4.4 Inland deserts

Some deserts form because they are so far inland that they are beyond the range of any rainfall. By the time winds reach these dry centres, they have dumped any rain they were carrying or have become so warm they cannot release any moisture they still hold. The air that enters such areas is usually extremely dry and the skies are cloudless for most of the year. Summer daytime temperatures can rise as high as those of subtropical deserts. In winter, however, temperatures are much lower. Average daily temperatures below freezing are common during winter.

FIGURE 6 The Simpson Desert in central Australia

Examples of inland deserts are the central deserts of Australia (see **FIGURE 6**), the Thar Desert in north-west India and the vast Gobi and Takla Makan deserts of Central Asia.

4.4.5 Polar deserts

Polar deserts are areas with a precipitation rate of less than 250 millimetres per year and an average temperature lower than 10 °C during the warmest month of the year. Polar deserts cover almost 5 million square kilometres of our planet and consist mostly of rock or gravel plains. Snow dunes may be present in areas where precipitation occurs. Temperatures in polar deserts often alternate between freezing and thawing, a process that can create patterned textures on the ground as much as 5 metres across.

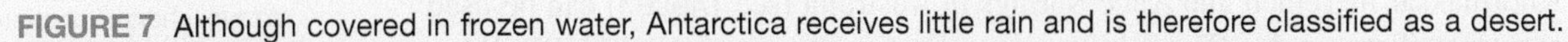

FIGURE 7 Although covered in frozen water, Antarctica receives little rain and is therefore classified as a desert.

4.4.6 Desert climate

Temperature

One geographical characteristic of many deserts is the high temperature, which quickly evaporates any water that might be around. The Earth's highest recorded temperature was long considered to be 58 °C as measured at El Azizia in the Libyan Desert of northern Africa on 13 September 1922. However, recently meteorologists have questioned this reading suggesting it was an error, and suggest the highest verifiable temperature is actually 56.7 °C, recorded in Death Valley in the US in 1913.

During the 1920s, the semi-arid town of Marble Bar in Western Australia (average rainfall 361 mm per year) experienced temperatures of more than 37.8 °C for 160 days in a row, from 31 October 1923 to 7 April 1924. This set an Australian temperature record, and possibly a world record. During that summer the temperature at Marble Bar peaked at 47.5 °C on 18 January 1924.

FIGURE 8 Temperature and rainfall data can be displayed clearly in a climograph, such as this one for Yuma, Arizona.

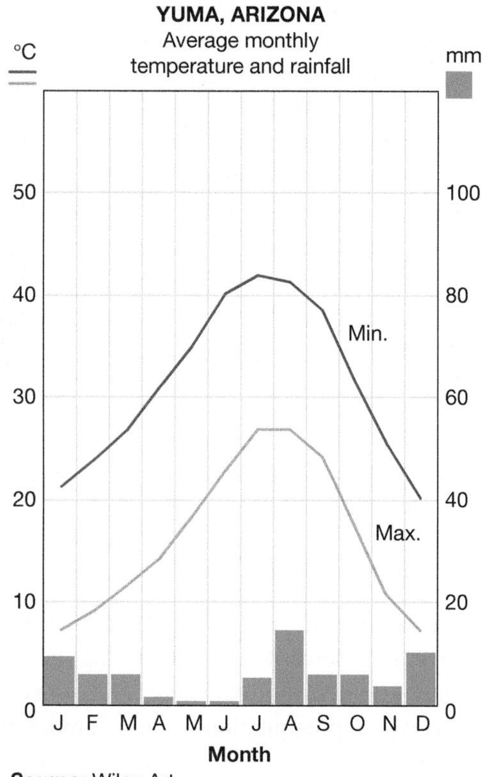

Source: Wiley Art

Rainfall

Although low rainfall is a characteristic of deserts, rain does fall and violent storms can sometimes occur. A record 44 millimetres of rain once fell within three hours in the Sahara. Large Saharan storms may deliver up to 1 millimetre of rain per minute. Normally dry stream channels, called **arroyos** or wadis, can quickly fill after heavy rains, and flash floods make these channels dangerous.

Monthly data for rainfall and temperature can be used to create climographs (see **FIGURE 8**). Data for other desert locations such as Khormaksar in Yemen and Alice Springs in Australia (see **TABLE 1**) can be used to construct climographs. (For help creating climographs, go to 10.7 SkillBuilder—Drawing a climate graph.)

arroyo channel or stream that is normally dry but fills quickly with heavy rain (also called a wadi)

TABLE 1 Climate data for (a) Khormaksar, Yemen, and (b) Alice Springs, Australia

(a) Khormaksar, Yemen

	Jan	Feb	Mar	Apr	May	Jun	Jul	Aug	Sep	Oct	Nov	Dec	Total
Average temperature (°C)	25.0	25.5	27.0	28.5	30.5	33.0	32.0	32.0	32.0	28.5	26.5	25.5	
Average rainfall (mm)	5.0	0.0	5.0	0.0	0.0	0.0	5.0	3.0	0.0	0.0	0.0	5.0	23.0

(b) Alice Springs, Australia

	Jan	Feb	Mar	Apr	May	Jun	Jul	Aug	Sep	Oct	Nov	Dec	Total
Average temperature (°C)	28.5	27.7	24.8	20.0	15.4	12.4	11.5	14.3	18.3	22.8	25.8	27.7	
Average rainfall (mm)	40.5	41.5	34.7	16.6	17.0	16.7	12.1	10.0	9.0	20.0	25.3	37.2	281.2

Source: Wiley Art

fdw-0002

FOCUS ON FIELDWORK

Investigating your local climate — Collecting and using rainfall data

Setting up a weather station at your school is a great way to record one aspect of your local climate. A weather station typically includes a rain gauge and a thermometer in an unobstructed area. It's easy to create a rain gauge to collect your own data, but you can also access rainfall data through the Bureau of Meteorology.

Learn how to investigate and graph one aspect of your locate climate using the **Recording rainfall** fieldwork activity in your Resources tab.

on Resources

eWorkbook	How the climate forms deserts (ewbk-8732)
Video eLesson	How the climate forms deserts — Key concepts (eles-5051)
Interactivities	How to make a desert (int-3107) Formation of coastal deserts (int-3629) Formation of rain shadow deserts (int-3628) Formation of subtropical deserts in Africa (int-3627) Yuma, Arizona climograph (int-3612) Climate graphs (int-0780)
Fieldwork	Recording rainfall (fdw-0002)
Google Earth	Alice Springs (gogl-0047) Khormaksar, Yemen (gogl-0048)

4.4 ACTIVITIES

1. Use the **Desert rain** weblink in the Resources tab, and then answer the following questions.
   a. What is a flash flood?
   b. What happens to water as it flows over sand? Think of what happens to water at the beach.
   c. How do animals and plants respond to these rare water events?
   d. Describe how the landscape quickly changes once there is water in the desert.
2. As a class discuss the following: Climate change is already leading to increasing areas of desertification. How important is it for Australians to consider the impact of their high carbon-producing lifestyle on the impact of such landscapes?
3. There is some debate about what the highest temperatures and hottest places in the world have been. How would you judge the title of 'hottest and driest place on Earth'? Would you award the title to the place with the least precipitation or access to water? The hottest single temperature? The longest run of hot days? Create a set of rules to determine which place on Earth would win this title, and investigate which place would win.

FIGURE 9 Arches in a desert landscape, Utah, United States

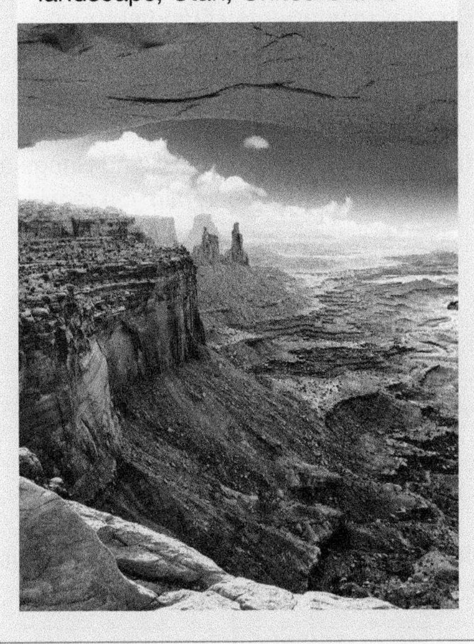

4.4 EXERCISE

Learning pathways

■ LEVEL 1	■ LEVEL 2	■ LEVEL 3
1, 2, 3, 9, 13	5, 6, 7, 11, 14	4, 8, 10, 12, 15

Check your understanding

1. Decide whether the following statements are true or false. Rewrite the false statements to make them true.
   a. The cooler the air, the more moisture it can hold.
   b. Rain shadows often contain dry areas of land.
   c. Cold ocean currents cool the air above them.
   d. Deserts do not form along coastlines.
2. Identify the highest and lowest temperatures that have been recorded in hot deserts.
3. Define the term *humidity*.
4. Complete the following table.

	Polar desert	Inland desert
Reasons for formation of desert		
Example		
Average temperature		
Average precipitation		

5. What does a climograph show?
6. Describe the characteristics of polar deserts.
7. On which side of a mountain range does a rain shadow form?
8. Use **FIGURE 8** to complete these questions.
   a. Which month received the most and least rainfall?
   b. Which season received the most and least rainfall?

Apply your understanding

9. Explain why temperatures in deserts drop so much at night after being so high during the day.
10. Use **FIGURE 1** to explain why deserts form around areas near the tropics but not at the equator.
11. Explain why deserts tend to form in rain shadows.
12. Explain how cold ocean currents influence the formation of a desert environment.

Challenge your understanding

13. Use **TABLES 1a** and **1b** to draw climate graphs for Khormaksar, Yemen, and Alice Springs, Australia.
14. Propose a system by which people living in or near desert areas that experience storms with heavy rain could collect the rainfall, even if they don't know the storm is coming.
15. How does the Great Dividing Range affect the rainfall where you live? Suggest whether and how it might influence how much rain you receive and when it falls during the seasons.

To answer questions online and to receive **immediate feedback** and **sample responses** for every question, go to your learnON title at www.jacplus.com.au.

4.5 The processes that shape desert landforms

LEARNING INTENTION

By the end of this subtopic, you will be able to explain how deserts are shaped by wind and water.

4.5.1 Shaping the desert

Although most people imagine a sea of sand when they think of deserts, sand covers only about 20 per cent of the world's deserts. The landforms and patterns of a desert are created by a number of natural processes. The unprotected land surfaces are prone to erosion. After heavy rain, often a long distance from the desert flood plains, erosion of ancient river channels can be major. Extreme temperatures, along with strong winds and the rushing water that can follow a desert rainstorm, cause rocks to crack and break down into smaller fragments. This process is called weathering.

FIGURE 1 Desert landforms

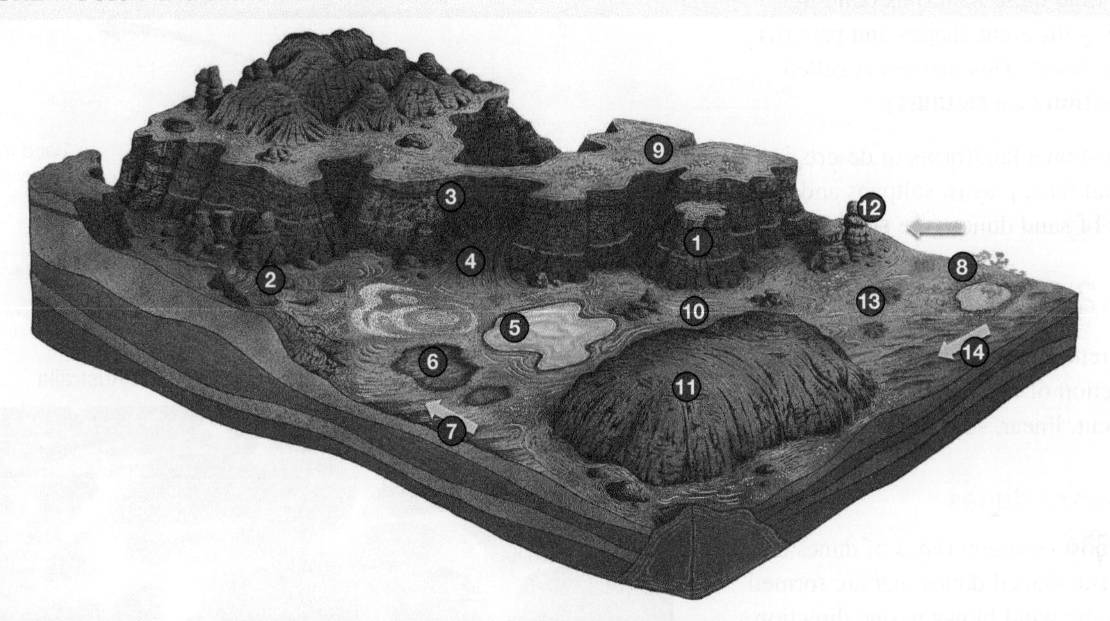

1. A **butte** is the remaining solid core of what was once a mesa. It often is shaped like a castle or a tower.
2. Crescent-shaped **barchan dunes** are produced when sand cover is fairly light.
3. An **arch,** or window, is an opening in a rocky wall that has been carved out over millions of years by erosion.
4. An **alluvial fan** is the semicircular build-up of material that collects at the base of slopes and at the end of wadis after being deposited there by water and wind.
5. A **playa** lake may cover a wide area, but it is never deep. Most water in it evaporates, leaving a layer of salt on the surface. These salt-covered stretches are called saltpans.
6. **Clay pans** are low-lying sections of ground that may remain wet and muddy for some time.
7. The rippled surface on **transverse dunes** is the result of a gentle breeze blowing in the one direction.
8. An **oasis** is a fertile spot in a desert. It receives water from underground supplies.
9. A **mesa** is a plateau-like section of higher land with a flat top and steep sides. The flat surface was once the ground level, before weathering and erosion took their toll.
10. **Sand dunes** often start as small mounds of sand that collect around an object such as a rock. As they grow larger, they are moved and shaped by wind.
11. An **inselberg** is a solid rock formation that was once below ground level. As the softer land around it erodes, it becomes more and more prominent. Uluru is an inselberg.
12. A **chimney rock** is the pillar-like remains of a butte.
13. **Star dunes** are produced by wind gusts that swirl in from all directions.
14. Strong winds blowing in one direction form **longitudinal dunes**.

Erosional landforms in the desert

The process of **erosion** removes material such as weathered rock. Most erosion in deserts is caused by wind and, at times, running water. During heavy rainfall, water carves channels in the ground. Fast-flowing water can carry rocks and sand, which help to scour the sides of the channel. As vegetation is usually sparse or non-existent, there are few roots to hold the soil together. Eventually, deep gullies called wadis can form.

Erosion can also result from the action of wind and from chemical reactions. Some rock types, such as limestone, contain compounds that react with rainwater and then dissolve in it. Wind is a very important agent of transport and deposition, and can change the shape of land by abrasion — the wearing down of surfaces by the grinding and sandblasting action of windborne particles. Erosional landforms in deserts include buttes, mesas and inselbergs (see **FIGURE 1**).

erosion the wearing away and removal of soil and rock by natural elements, such as wind, waves, rivers or ice, and by human activity

deposition the laying down of material carried by rivers, wind, ice and ocean currents or waves

Depositional landforms in the desert

Materials carried along by rushing water and wind must eventually be put down. Over time these materials build up, forming different shapes and patterns in the desert. This process is called **deposition** (see **FIGURE 2**).

Depositional landforms in deserts include alluvial fans, playas, saltpans and various types of sand dunes (see **FIGURE 1**).

FIGURE 2 Transport and deposition of sand creates and moves dunes.

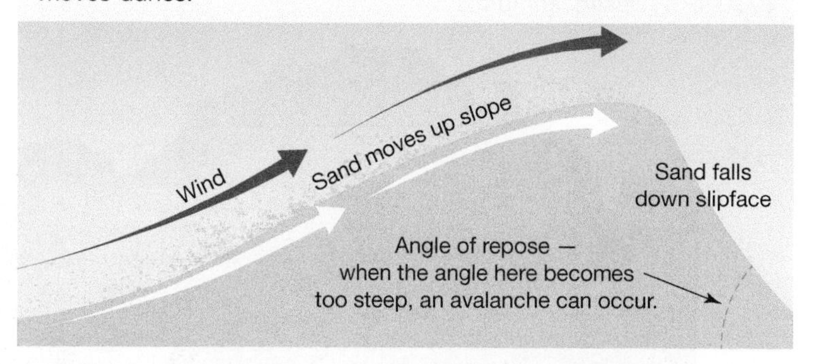

4.5.2 Sand dunes

Different dune shapes are created by the action of the wind. These include crescent, linear, star, dome and parabolic.

Crescent dunes

The most common types of dunes are the crescent-shaped dunes that are formed when the wind blows in one direction. The dunes face the direction of the prevailing winds. These are known as crescent dunes or barchan (or barkhan) dunes. They are usually wider than they are long and can move very quickly across desert landscapes (see **FIGURE 3**).

FIGURE 3 A series of crescent dunes in Western Australia

Linear dunes

Linear dunes are a series of dunes running parallel to each other (running side-by-side as shown in **FIGURE 4**).

They can vary in length from a few metres to over 100 kilometres. It appears that winds blowing in opposite directions help create these dunes. The Namib Desert in Namibia has linear dunes (see **FIGURE 4**).

FIGURE 4 Linear dunes, Namib Desert, Namibia

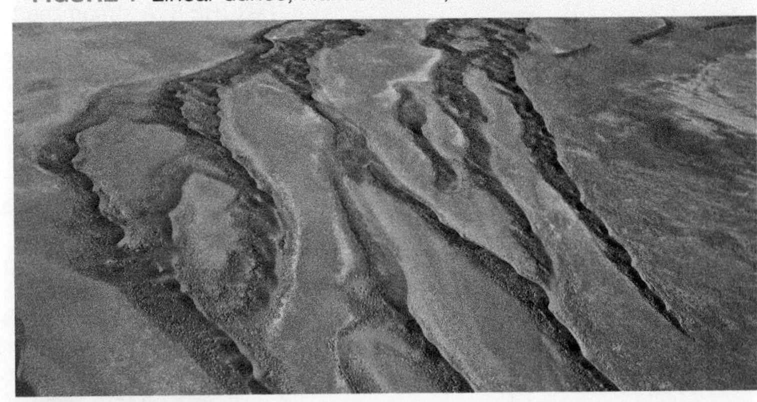

Star dunes

Star dunes have 'arms' that radiate from a high central pyramid-shaped mound. They form in regions that have winds blowing in many different directions and can become very tall rather than wide — some are up to 500 metres high (see **FIGURE 5**). Star dunes are found in many deserts including the Namib, the Grand Erg Oriental of the Sahara, and the southeast Badain Jaran Desert of China.

FIGURE 5 Star dunes

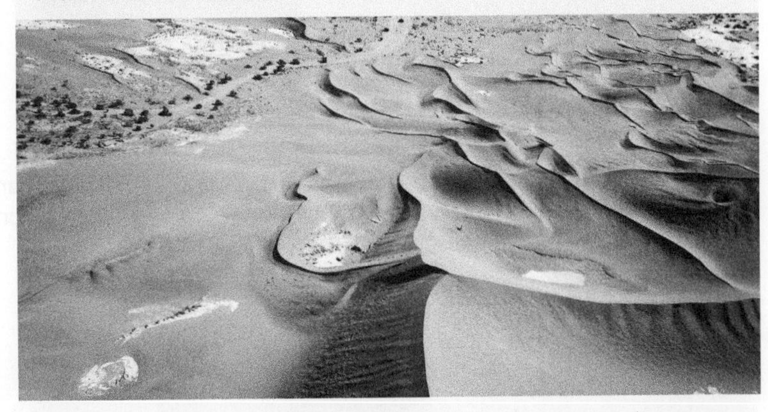

Dome dunes

Dome dunes are made up of fine sand. They have gentle sloping sides, rather than steep sides. These rounded structures tend to be only one or two metres high and are very rare.

In **FIGURE 6**, you can compare the dome dune with the larger dunes in the background that have steeper sides that have slipped away.

FIGURE 6 A dome dune in the Chihuahuan Desert, New Mexico, USA

Parabolic dunes

Parabolic dunes have a U shape and do not grow very high (see **FIGURE 7**). They often occur in coastal deserts. The longer section follows the 'head' of the dune (the opposite process to the formation of crescent dunes) because vegetation has anchored them in place. The arms can be long — in one case, a dune arm was measured at 12 kilometres.

FIGURE 7 Formation of a parabolic dune

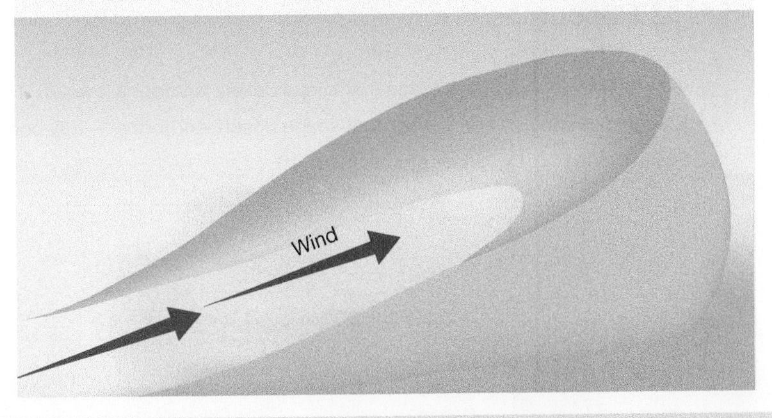

4.5.3 Playas and pans

A desert basin may fill with water after heavy rains to form a shallow lake, but for the majority of the time the often salt-encrusted surface is hard and dry. Such expanses of land are known as playas, saltpans or hardpans. The flat terrains of pans and playas make them excellent race tracks and natural runways for aeroplanes and spacecraft. Ground-vehicle speed records are commonly established on Bonneville Speedway, a race track on the Great Salt Lake hardpan. Space shuttles used to land on Rogers Lake Playa at Edwards Air Force Base, California.

FIGURE 8 A streamlined racing car, Bonneville Salt Flats, Utah, USA

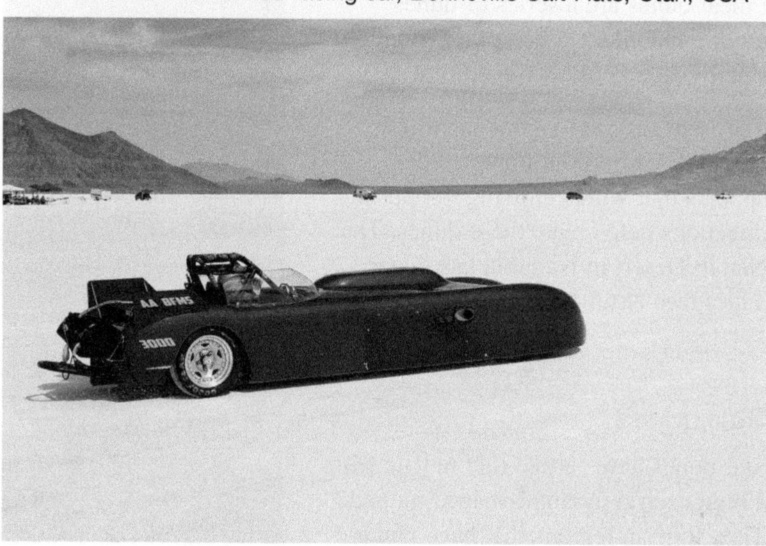

FIGURE 9 (a) and (b) The salt-encrusted surface of Lake Ballard, near Menzies in Western Australia, is home to 'Inside Australia' by British artist Antony Gormley. The art installation consists of 51 sculptures set over 10 square kilometres of the salt pan.

on Resources

eWorkbook	The processes that shape desert landforms (ewbk-0717)
Video eLesson	The processes that shape desert landforms — Key concepts (eles-5052)
Interactivity	Desert landforms (int-7841)

4.5 ACTIVITIES

1. Locate all the desert places named in this subtopic. Use Google Maps to create your own map of these locations, and add some interesting facts and images of each location. Save and send or upload a link to your completed map to your teacher.
2. Draw up a table like the one below.

Name of landform	Picture of landform	Location	Type of erosion (wind or water)	Type of deposition (wind or water)
Butte				
Mesa				
Inselberg				

Add the rest of the landforms shown in **FIGURE 1** to your table. Add examples of other desert landforms that you have found when researching this topic.
3. Work in small groups to create a model of a desert (using plasticine or playdough, for example) that contains a number of desert forms and patterns. Use **FIGURE 1** as a guide. Show your completed model to the other groups, explaining the reasons why the landscapes you have sculpted are created, then provide and respond to constructive feedback.

4.5 EXERCISE

Learning pathways

■ LEVEL 1	■ LEVEL 2	■ LEVEL 3
1, 2, 8, 9, 13	3, 4, 7, 10, 14	5, 6, 11, 12, 15

Check your understanding

1. Name two erosional and two depositional landforms in a desert.
2. Name the most common dune shapes that are formed in deserts.
3. Explain the difference between a mesa and a butte.
4. What wind conditions are needed to create a:
   a. star dune
   b. longitudinal dune
   c. parallel dune?
5. Which desert in Australia contains many linear dunes? Describe what conditions create these dunes.
6. On which landforms are land-speed records often held and why?

Apply your understanding

7. Explain how erosion and weathering processes cause change in a desert.
8. What do chimney rocks and arches have in common?
9. Explain what playa lakes and saltpans have in common.
10. What causes a sand dune avalanche?
11. Explain why some sand dunes are slow moving.
12. Why do you think dome dunes are very rare?

Challenge your understanding

13. Study the landforms labelled 1, 3 and 9 in **FIGURE 1**. Sketch what each of these may look like in the future as erosion and weathering continue to occur.
14. How does vegetation help to prevent erosion in a desert?
15. Suggest why oases are such fertile places.

To answer questions online and to receive **immediate feedback** and **sample responses** for every question, go to your learnON title at www.jacplus.com.au.

4.6 Australia's desert areas

LEARNING INTENTION

By the end of this subtopic, you will be able to describe where Australia's deserts are located and their characteristics, and to explain why Lake Mungo is a significant place.

4.6.1 The location of Australia's deserts

Australia is the world's driest inhabited continent. Over 70 per cent of the country receives between 100 and 350 millimetres of rainfall annually, which makes most of Australia arid or semi-arid.

Australia's deserts are subtropical and are located mainly in central and western Australia, making up about 18 per cent of the country (see **FIGURE 1**). They are hot deserts, which mean that they are areas of little rainfall and extreme temperatures — rainfall can be less than 250 millimetres per year, and temperatures can rise to over 50 °C. The average humidity (moisture in the air) is between 10 and 20 per cent. The desert terrain is very diverse and can range from red sand dunes to the polished stones of the gibber plains — the term *gibber* comes from the Dharug (Sydney region) word for stone (*gibba*).

FIGURE 1 The location and distribution of Australia's deserts

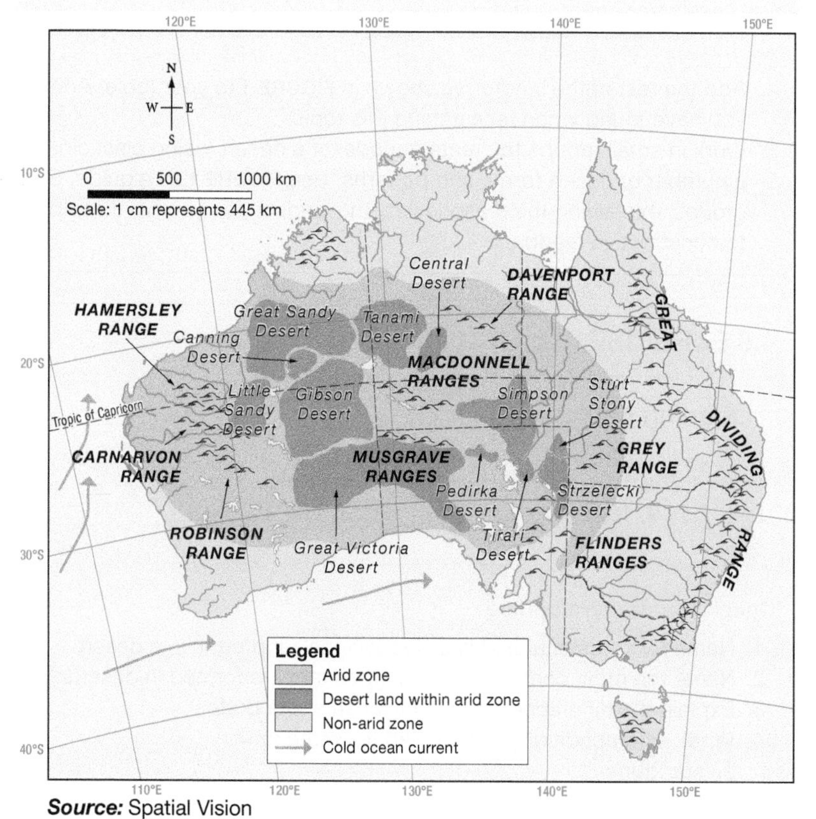

Source: Spatial Vision

The Australian deserts listed below are located on traditional lands of various Aboriginal nations. Within these nations there are multiple language groups and Aboriginal communities. There have been many native title claims concerning these special places and many have been identified as Indigenous Protected Areas.

Great Victoria Desert

The Great Victoria Desert is located on the lands of a number of Aboriginal groups including the Pila Nguru Peoples of Western Australia and the A̲nangu Pitjantjatjara Yankunytjatjara Peoples of South Australia.

This is Australia's largest desert, covering 424 400 square kilometres. It is not a desert of dunes, but has some desert-adapted plants including marble gums, mulga and spinifex grass. Part of this desert has been named a Biosphere Reserve by UNESCO and is one of the largest arid-zone biospheres in the world.

Great Sandy Desert

The Great Sandy Desert is located on the lands of a number of Aboriginal groups including the Martu Peoples and the Pintupi Peoples.

It makes up 3.5 per cent of Australia, covering just under 285 000 square kilometres. The red sands of this desert reach almost to the Western Australia coast, where they join the white sand of Eighty Mile Beach south of Broome. The limited vegetation is mostly spinifex. The Great Sandy Desert averages more rainfall than many arid areas because parts are close to the tropics that receive heavy wet-season rains, but the very high average temperatures mean that water quickly evaporates.

Simpson Desert

The Simpson Desert is located on the lands of a number of Aboriginal groups including the Arrernte Peoples and Wangkangurru Yarluyandi Peoples.

It is in one of the driest areas of Australia, with rainfall of less than 125 millimetres per year. It is located near the geographical centre of Australia. Dunes make up nearly three-quarters of the desert. Long parallel dunes (see **FIGURE 2**, and **FIGURE 4** in subtopic 4.5) form in a north–north-west/south–south-east direction; some can be straight and unbroken for up to 300 kilometres and can be 40 metres high. The space between the dunes can vary from 100 metres to 1000 metres.

FIGURE 2 Sand dunes and vegetation in the Simpson Desert

Strzelecki Desert

The Strzelecki Desert is located on the lands of the Yandruwandha Peoples, and within three Australian states — far northern South Australia, south-west Queensland and western New South Wales. The dunes support vegetation such as sandhill wattle, needlebush and hard spinifex.

Tanami Desert

The Tanami Desert is located on the lands of a number of Aboriginal groups including the Warlpiri Peoples and the Kukatja Peoples. It lies to the east of the Great Sandy Desert. This desert is mostly characterised by red sand plains with hills and ranges.

Little Sandy Desert

The Little Sandy Desert is located on the lands of a number of Aboriginal groups including the Martu Peoples. It is in Western Australia and borders three other deserts. Its landforms are similar to those in the Great Sandy Desert. It includes a vast salt lake called Lake Disappointment.

Sturt Stony Desert

The Sturt Stony Desert is located on the lands of a number of Aboriginal groups including the Wangkumara Peoples and the Yandruwandha Peoples.

This harsh gibber desert in north-eastern South Australia is covered in closely spaced glazed stones (**FIGURE 3**). These are left behind when the wind blows away the loose sand between the dense covering of pebbles. The desert also contains some dunes and hills that are resistant to weathering.

FIGURE 3 Gibber landscape in the Sturt Stony Desert in South Australia

Pedirka Desert

The Pedirka Desert (**FIGURE 4**) is on the lands of the Arabana Peoples, in South Australia. It is Australia's smallest desert, located north-east of Oodnadatta (see **FIGURE 1**). The lines of parallel red dunes run north-east to south-west, and the space between the dunes can be up to 1 kilometre. Hamilton Creek is located in this desert and its banks are home to river red gums, coolabah, mulga, prickly wattle and dead finish.

FIGURE 4 Pedirka Desert between Oodnadatta and William Creek, South Australia

Tirari Desert

The Tirari Desert is located on a number of the lands of a number of Aboriginal groups including the Dieri Peoples. This small desert covers almost 1600 square kilometres of far northern South Australia, east of Kati Thanda–Lake Eyre. It contains many linear (parallel) dunes and salt lakes. Cooper Creek runs through the centre of the desert, as do many other **intermittent creeks**. Where there is enough water — usually in waterholes — river red gums and coolabah gums will grow. Tall, open shrubland also occurs in some areas.

intermittent creek a creek that flows for only part of the year following rainfall

Gibson Desert

The Gibson Desert is located on the traditional homelands of the Ngaanyatjarra Peoples. The fifth largest desert in Australia, the Gibson is in Western Australia and borders three other deserts. It consists of sand plains and dunes plus some low, rocky ridges. Some small salt-water lakes are also present in the south-western part of the desert.

CASE STUDY

Lake Mungo and the Willandra Lakes

Lake Mungo, in Mungo National Park, is just one of 13 ancient dry lake beds in a section of the Willandra Lakes Region World Heritage area in semi-arid New South Wales. There is no water there now, yet the lakes were once full of water and teeming with life, supporting the Paakantji, Ngiampaa, and Mutthi Mutthi Peoples for more than 47 000 years — archaeological records show this continuous human presence. What happened to change this environment into the semi-arid landscape it is today?

FIGURE 5 The 'Walls of China' at Lake Mungo. The dry lake bed is covered by low bushes and grasses.

The Willandra Lakes are located in far south-western New South Wales and the region is part of the Murray–Darling River Basin. Lake Mungo is 110 kilometres north-east of Mildura, Victoria. The lakes were originally fed by water from Willandra Creek (see **FIGURE 6**), which was a branch of the Lachlan River. The average rainfall in this area is 325 millimetres per year, making it a semi-arid desert region.

How Mungo has changed over time

40 000 years ago

During the last ice age, huge amounts of water filled the shallow lake. At its fullest, Lake Mungo was 6–8 metres deep and covered 130 square kilometres (more than twice the area of Sydney Harbour). The lakes were rich with life, including water birds, freshwater mussels, yabbies, and fish such as golden perch and Murray cod. Giant kangaroos, giant wombats, large emus and the buffalo-sized *Zygomaturus* — all now extinct — grazed around the water's edge. Remains of more than 55 species have been found in the area and identified — 40 of these are no longer found in the region, and 11 are extinct.

Aboriginal Peoples lived here in large numbers — evidence for this has been found in more than 150 human fossils, including 'Mungo Lady' discovered in 1968 and 'Mungo Man' in 1974, both believed to be over 40 000 years old.

30 000–19 000 years ago

A west wind blows across this landscape. During low-water years, red dust and clay was blown across the plains to the eastern side of the lake and mixed with sand dunes on the edge of the lake (formed when the lake was full; see **FIGURE 7**). This began the formation of lunettes (crescent-shaped dunes) on the east side — called 'the Walls of China' in Lake Mungo. Vegetation covered the dunes, protecting them.

FIGURE 6 Location of the Willandra Lakes World Heritage Area, including Lake Mungo

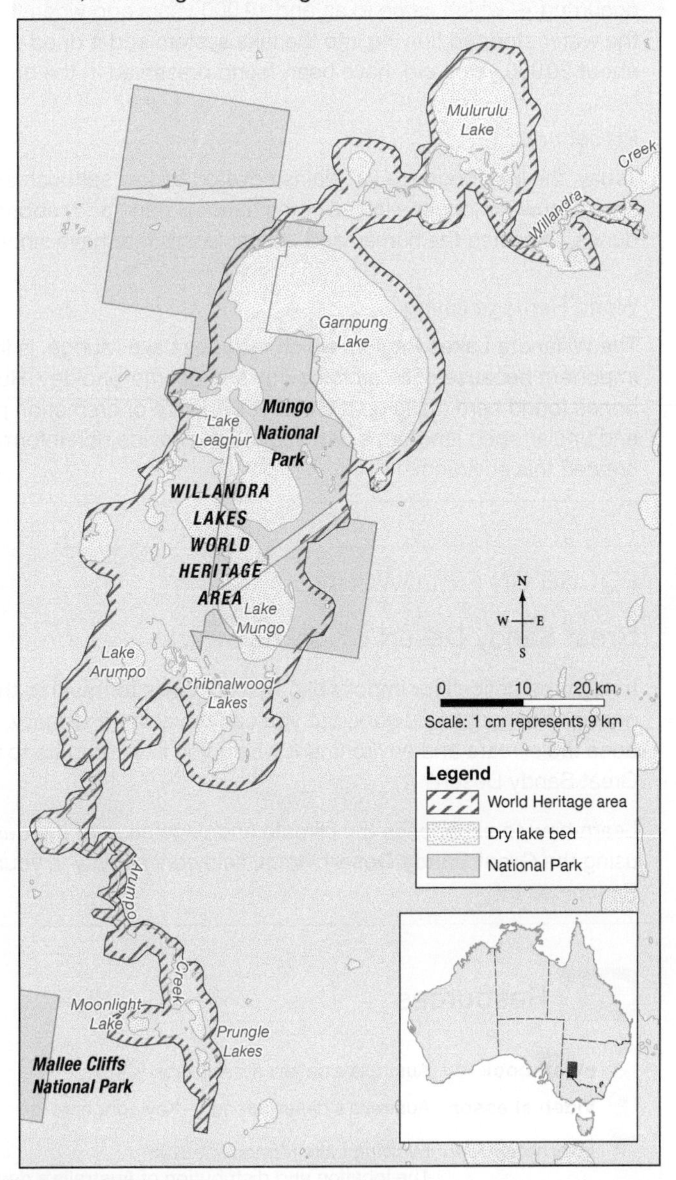

Source: Spatial Vision

FIGURE 7 The process of erosion while Lake Mungo was full

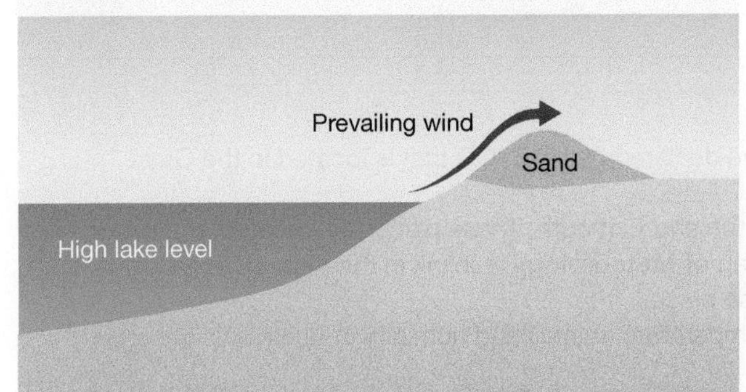

FIGURE 8 The process of erosion while Lake Mungo was drying out

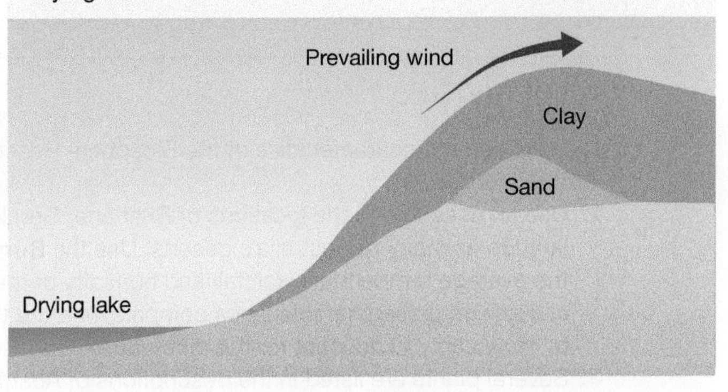

19 000 years ago

The lakes were full of deep, relatively fresh water for a period of 30 000 years — with cycles of wet and dry occurring — which came to an end 19 000 years ago when the climate became drier and warmer. Eventually, the water stopped flowing into the lake system and it dried out (see **FIGURE 8**). Ancient footprints, thought to be about 20 000 years old, have been found preserved in the dried mud of the former lake.

Present day

Today, the lake beds are flat plains covered by low saltbush and bluebush as well as grasses. Grazing cattle and sheep (now no longer allowed in the national park) and rabbits have caused erosion of the lunettes and sand dunes, exposing the human and animal fossils that have since been discovered.

World Heritage listing

The Willandra Lakes Region, which includes Lake Mungo, is listed as a World Heritage area. This region is important because of its archaeology and geomorphology. Human skeletons, tools, shell middens and animal bones found here make up the oldest evidence of cremation practices and burial places in the world. The ancient and undisturbed landforms and sediments provide rich information about the geomorphological processes that created this environment.

fdw-0003

FOCUS ON FIELDWORK

Great Sandy Desert virtual fieldwork

It might be difficult (or impossible) for your class to travel to the desert to conduct fieldwork, but you can travel there virtually. How does the climate and environment where you live compare to the Great Sandy Desert?

Learn how to experience the climate and vegetation of the desert using the **Great Sandy Desert** virtual fieldwork activity in your Resources tab.

on Resources

eWorkbook	Australia's desert areas (ewbk-8740)
Video eLesson	Australia's desert areas — Key concepts (eles-5053)
Interactivities	Evolving Lake Mungo (int-3108) The location and distribution of Australia's deserts (int-3614) Location of Willandra Lakes, including Lake Mungo (int-3615)
Google Earth	Mungo National Park (gogl-0084) The Great Sandy Desert (gogl-0121)
Fieldwork	Great Sandy Desert virtual fieldwork (fdw-0003)

4.6 ACTIVITIES

1. Research the characteristics of the Biosphere Reserve declared by UNESCO that is located in the Great Victorian Desert.
2. Use an atlas to find the locations of Brisbane, Geraldton and Exmouth. These places are located at the same latitude as many of Australia's deserts. Use the **Bureau of Meteorology** weblink in the Resources tab to find the average temperature, rainfall and humidity of these places.
   a. How do these characteristics compare with the temperature, rainfall and humidity in Australia's deserts?
   b. How can you account for the differences?
3. Several plants are listed in the descriptions of Australia's deserts in this subtopic. Choose two plant types (for example, a grass and a tree) and research how they are adapted to desert conditions.

4. As a class, discuss the following claim: 'It is right for Lake Mungo to be protected under a World Heritage listing because of its significant cultural characteristics.' How does this type of protection reflect the cultural values of a society?
5. Work in small groups to create an identification brochure with pictures and facts about these three extinct animals that once lived at Lake Mungo.
   a. *Genyornis newtoni* (giant emu)
   b. *Protemnodon goliah* (giant short-faced kangaroo)
   c. *Zygomaturus trilobus* (Zygomaturus)
6. Use the **Timeline maker** weblink in the Resources tab and images you find through online research to create your own colourful electronic timeline of the changes that occurred at Lake Mungo.
7. Research what a World Heritage listing means in terms of protecting Lake Mungo. Why is this place culturally important to Aboriginal Peoples? To which groups is it important? Research the history and consequences of native title. What kinds of protections does native title provide a place? How do these protections compare with World Heritage protection?

4.6 EXERCISE

Learning pathways

■ LEVEL 1	■ LEVEL 2	■ LEVEL 3
1, 2, 10, 12, 14, 18	3, 4, 5, 6, 11, 13, 15, 20	7, 8, 9, 16, 17, 19

Check your understanding

1. Name the deserts bordered by:
   a. the Gibson Desert
   b. the Little Sandy Desert.
2. What is a gibber desert?
3. What percentage of Australia is arid or semi-arid?
4. List the types of vegetation that can be found in the Strzelecki Desert.
5. Where is the Great Victoria Desert located?
6. Name Australia's smallest desert. Where is it located?
7. Define the term *intermittent creek*.
8. Which desert contains a UNESCO Biosphere Reserve?
9. Use **FIGURE 6** to describe where Lake Mungo is located.
10. When did Lake Mungo dry up?
11. What is a lunette?

Apply your understanding

12. Explain why Lake Mungo and the Willandra Lakes region are considered a semi-desert.
13. Look at **FIGURE 1** showing the distribution of Australia's deserts. Where are they located in terms of the tropics?
14. Which desert is Australia's driest and what are its characteristics?
15. Explain why Lake Mungo received World Heritage listing.
16. Use the scale in **FIGURE 6** to measure the north–south extent of Mungo National Park.
17. Use the Lake Mungo case study and explain how Lake Mungo changed over time.

Challenge your understanding

18. Outline the evidence that shows that many Aboriginal Peoples lived in this area.
19. Suggest what human activity caused the lunettes to erode and explain what the erosion unearthed.
20. Suggest why Lake Mungo dried out and whether it might fill with water in the future.

To answer questions online and to receive **immediate feedback** and **sample responses** for every question, go to your learnON title at www.jacplus.com.au.

4.7 SkillBuilder — Calculating distance using scale

online only

LEARNING INTENTION

By the end of this subtopic, you will be able to use the scale of a map to calculate distance.

The content in this subtopic is a summary of what you will find in the online resource.

4.7.1 Tell me

What does it mean to calculate distance using scale?

Calculating distance using scale involves working out the actual distance from one place to another using a map. The scale on a map allows you to convert distance on a map or photograph to distance in the real world — what it represents on Earth's surface.

FIGURE 4 Measuring curved distances with a scale using a paper edge

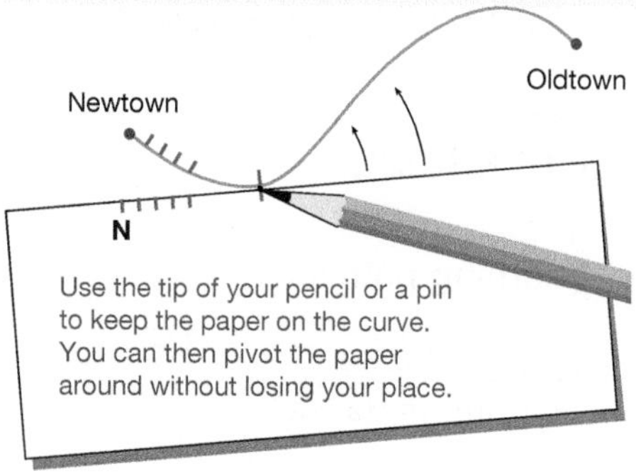

Newtown
Oldtown
N
Use the tip of your pencil or a pin to keep the paper on the curve. You can then pivot the paper around without losing your place.

4.7.2 Show me

How to calculate distance using scale

Step 1

Find the two places you want to know the distance between. If you need to know the distance in a straight line, use a ruler. If the distance is winding, use paper or use a piece of string.

Step 2

Method 1: Using a ruler Place your ruler between two places to find the distance in centimetres. Find out what 1 cm equals, either on the linear scale or in the scale statement. Multiple the distance you measured by the distance 1 cm represents on the map.

Method 2: Using paper Place the edge of the paper next to the starting point on your map. Mark the edge with a small line and the place name. Move the paper carefully so the edge follows the curve on the map. When you move the paper's edge around the curve, use your pencil to hold the paper on your last marking place. Mark and label the end point on your paper. Place the paper along the linear scale and read the distance between your two places.

Method 3: Using string Place one end of the string at your starting point, and carefully bend the string around the winding distance. Mark the total distance. Place the string against the linear scale to calculate the total distance between the two places.

4.7.3 Let me do it

learnON

Go to learnON to access the following additional resources to help you build this skill:

- a longer explanation of this skill and its application in Geography (Tell me)
- a video showing the step-by-step process of this skill (Show me)
- an activity and interactivity for you to practise this skill (Let me do it)
- self-marking questions to help you understand and use this skill.

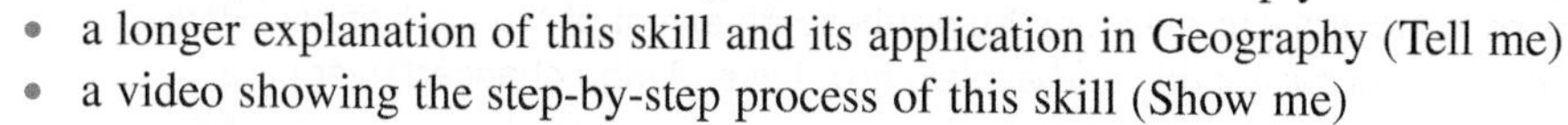

on Resources

- **eWorkbook** SkillBuilder — Calculating distance using scale (ewbk-8744)
- **Video eLesson** SkillBuilder — Calculating distance using scale (eles-1653)
- **Interactivity** SkillBuilder — Calculating distance using scale (int-3149)

4.8 How people use deserts

LEARNING INTENTION

By the end of this subtopic, you will be able to explain how deserts are used and valued, and identify human causes of land degradation in desert landscapes.

4.8.1 Traditional livelihoods

Although not many people live in deserts, these environments have been important to traditional communities for many years. People either adapt to living in deserts or transform deserts to suit their needs. People are also attracted to desert regions to mine resources.

FIGURE 1 (a) and (b) The San people now live in permanent settlements, rather than living a seasonally nomadic lifestyle; however, some do earn a living sharing their culture and customs with tourists.

There are communities who live in deserts, many using a mixture of traditional and modern ways of living. Examples include the Bedouin of the Middle East and Sahara, the Tuareg of the Sahara in North Africa, the Topnaar in the Namib Desert and the San peoples of the Kalahari in southern Africa (see **FIGURE 1**), the Timbisha Shoshone of the Mohave Desert in the USA, and the communities from the Atacama Desert in South America.

Some of these communities are **nomadic** and live traditional lives, moving with the seasons and obtaining all their needs from the land or herding animals and trading with people in settlements. It is important to understand that not all people from traditional desert cultures are still desert-dwellers. In some deserts the traditional ways have adapted and evolved as greater contact has been made with other cultures. Many previously nomadic groups now choose or have been forced to live in permanent settlements or towns.

The San people

The San people, also commonly known as the 'Bushmen of the Kalahari', are the First Nations people of Southern Africa. San were traditionally semi-nomadic, moving seasonally within certain defined areas based on the availability of resources. The San adapted to living in desert conditions through their excellent survival and hunting tactics, involving poison arrows and sucking water from wet soil through hollow reeds. The last of the Kalahari San were forcibly removed and resettled into camps by the Botswana government in 2002 to make way for diamond mining. The San people that remain have adopted many strategies for political, economic and social survival but still pass along many of their traditional cultural practices.

nomadic a group that moves from place to place depending on the food supply, or on pastures for animals

4.8.2 Desert resources

Many of the changes in deserts have been brought about by developments in technology. These changes have resulted in water being extracted and used to grow crops, and minerals being mined and used in many ways.

Water in the desert

Drilling equipment and pumps have allowed bores to tap into groundwater in aquifers deep below the desert surface. This has transformed some deserts in northern Africa and the Middle East into a series of circular irrigation fields — some of these can be up to 3 kilometres in diameter. In Australia, groundwater from the Great Artesian Basin (a huge underground water source) has enabled desert communities to exist and grazing to take place. Unfortunately, the groundwater in many areas is being pumped out far more quickly than it is being replaced and may be in danger of running out.

Desalination plants have also provided water to desert communities in many areas, especially the Middle East, including large cities such as Dubai.

desalination the process of removing salt from sea water

FIGURE 2 (a) Aerial photograph of a circular irrigation field in the USA; (b) False-colour Landsat satellite image of circular irrigation fields near Riyadh, Saudi Arabia

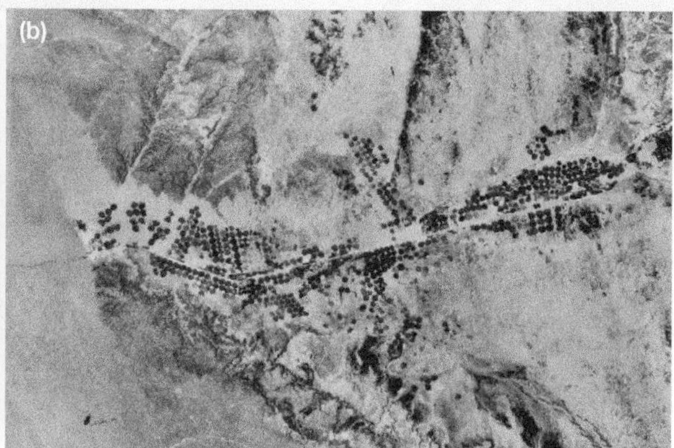

Mining in deserts

Many deserts contain valuable mineral deposits that were formed in the arid environment or have been exposed by erosion.

Examples of mining resources include:
- iron and lead-zinc ore — mined in Australian deserts
- phosphorus (used to make fertilisers) — mined in the Sahara region
- borates (used to make glass, ceramics, enamels and agricultural chemicals) mined in the deserts of the United States
- copper, iron ore and nitrates (used in fertiliser) — mined in the Atacama Desert
- precious metals such as gold, silver and platinum — mined in the deserts of Australia, America and central Asia
- uranium (used in nuclear power generation) — mined in Australia and the United States
- diamonds — mined in the Kalahari and Namib deserts of Africa
- oil — found in the desert regions of the Middle East, mainly in Kuwait, Iraq, Iran and Saudi Arabia.

FIGURE 3 A uranium mine next to the Colorado River in the United States

Resources

eWorkbook How people use deserts (ewbk-0723)

Video eLesson How people use deserts — Key concepts (eles-5054)

Interactivity How people use deserts (int-8403)

4.8 ACTIVITIES

1. Investigate one of the desert communities mentioned in this subtopic. Conduct research to identify where these communities generally live, and find examples of their traditional ways of life, including living conditions and shelter. Present your information in an interesting way, such as a Prezi, Keynote or PowerPoint presentation, and use images and maps where possible.
2. Make a sketch of **FIGURE 3**. Label your sketch to include the river, the mine site and buildings/roads.

4.8 EXERCISE

Learning pathways

■ LEVEL 1	■ LEVEL 2	■ LEVEL 3
3, 4, 10, 13, 16	1, 6, 7, 9, 14	2, 5, 8, 11, 12, 15

Check your understanding

1. List the sources of water in a desert that can be used to grow crops and provide water for people.
2. Outline two ways that technology has enabled water to be used in deserts.
3. Where do the following peoples traditionally live: Bedouin, Tuareg, Topnaar, San and Timbisha?
4. Define the term nomadic.
5. Outline why the lives of peoples who traditionally lived in the desert have changed over time.
6. Identify three types of mining resources found in Australian deserts, and outline what they are used for.

Apply your understanding

7. Why is it important to use groundwater sustainably?
8. Do you think most desert people adapt to live in the desert environment, or adapt the environment to live in the desert? Give two examples to support your reasoning.
9. What do you think are the similarities and differences in the lifestyles of the Bedouin, Tuareg, Topnaar, San and Timbisha peoples? Explain two similarities and two differences.
10. Study **FIGURE 2 (b)**. Explain what the small red circles in the image show.
11. Study **FIGURE 3**.
    a. How has mining changed this environment?
    b. What issues could arise due to the location of this mine?
    c. What might happen to this area when mining stops?
12. Why might a desert environment suit the needs of nomadic peoples?
13. Where is the Great Artesian Basin located and why is it important?

Challenge your understanding

14. Suggest specific ways that desert mining might affect desert-dwelling people.
15. Predict whether the numbers of people living in deserts will increase or decrease in the future. Give reasons for your view.
16. If you were required to move to a desert location in central Australia, what do you think would be the greatest difficulty you would face? Give a detailed explanation of how you made your decision.

To answer questions online and to receive **immediate feedback** and **sample responses** for every question, go to your learnON title at www.jacplus.com.au.

4.9 Antarctica — A cold desert

online only

LEARNING INTENTION

By the end of this subtopic, you will be able to explain why Antarctica is classified as a desert, describe how Antarctica is used by people and outline how it has changed as a result.

The content in this subtopic is a summary of what you will find in the online resource.

Antarctica and the seas that surround it contain valuable natural resources. The landscape and unique wildlife also attract growing numbers of tourists who have the potential to cause significant damage. How do we protect and manage such an important place?

FIGURE 1 Emperor penguins on Coulman Island

To learn more about protecting and managing Antarctica, go to your learnON resources at www.jacPLUS.com.au.

Contents

on Resources

eWorkbook	Antarctica — A cold desert (ewbk-8752)
Video eLesson	Antarctica — A cold desert — Key concepts (eles-5055)
Interactivity	Antarctica — A cold desert (int-7842)
myWorldAtlas	Deepen your understanding of this topic with related case studies and questions > Exploring places > Antarctica > **Antarctica: human features**

4.10 Investigating topographic maps — The Sahara Desert

LEARNING INTENTION

By the end of this subtopic, you will be able to locate and describe key features on a topographic map of a desert region, and use your observations as evidence of how deserts affect settlements.

4.10.1 The expanding Sahara

Nouakchott, the capital city of Mauritania in West Africa, is slowly being engulfed by the Sahara Desert. Years of drought, deforestation and poor farming methods have all contributed to the desertification (the process of an area becoming a desert). The Mauritanian government has planted corridors of vegetation (greenbelts) to act as a barrier; however, not all of these have worked. (View **FIGURE 1** full-size in this topic online.)

FIGURE 1 Colour satellite image of Nouakchott, Mauritania, 2017

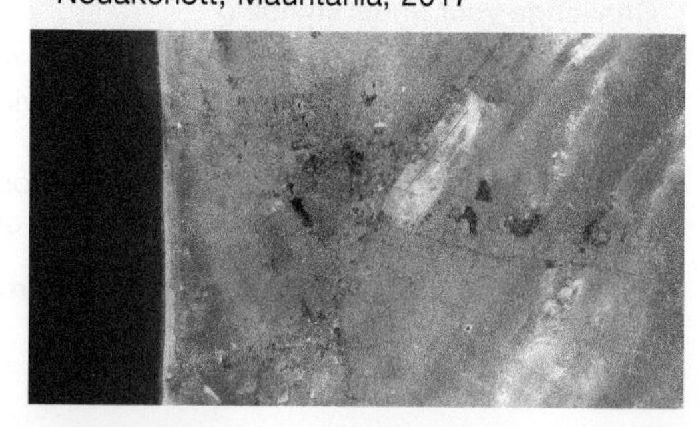

FIGURE 2 Topographic map of Nouakchott, Mauritania, 2021

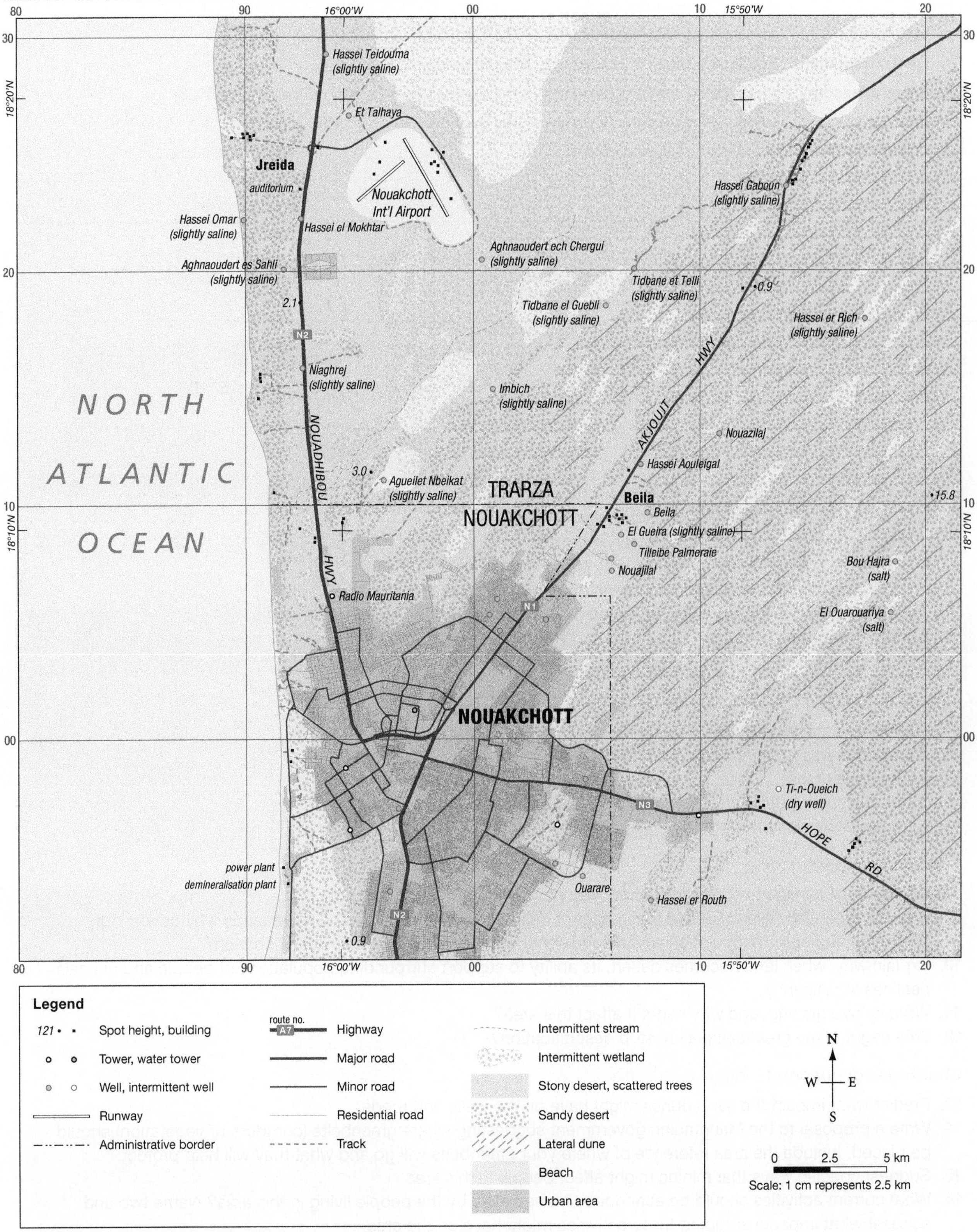

Legend

Symbol	Meaning	Symbol	Meaning	Symbol	Meaning
121 • ▪	Spot height, building	route no. A7	Highway		Intermittent stream
o o	Tower, water tower		Major road		Intermittent wetland
● o	Well, intermittent well		Minor road		Stony desert, scattered trees
	Runway		Residential road		Sandy desert
	Administrative border		Track		Lateral dune
					Beach
					Urban area

N, W, E, S

0 2.5 5 km

Scale: 1 cm represents 2.5 km

Source: Map data © OpenStreetMap contributors, https://openstreetmap.org. Data is available under the Open Database Licence, https://opendatacommons.org/licenses/odbl/

Resources

eWorkbook	Investigating topographic maps — The Sahara Desert (ewbk-8760)
Digital document	Topographic map of Nouakchott, Mauritania (doc-20643)
Video eLesson	Investigating topographic maps — The Sahara Desert — Key concepts (eles-5056)
Interactivity	Investigating topographic maps — The Sahara Desert (int-8404)
Google Earth	Nouakchott, Mauritania (gogl-0131)

4.10 EXERCISE

Learning pathways

■ LEVEL 1	■ LEVEL 2	■ LEVEL 3
2, 5, 7, 10, 13	1, 3, 8, 9, 14	4, 6, 11, 12, 15, 16

Check your understanding

1. Refer to the topographic map of Nouakchott to answer the following.
   a. List four landform features that are shown on the map.
   b. Identify the area reference of the city of Nouakchott.
   c. Using the scale, how far is Nouakchott from the coastline?
   d. How close are the sand and the dunes to the built-up areas?
2. Decide whether the following wells provide a good drinking water source for humans.
   a. Hassei Omar
   b. Ti-n-Oueich
   c. El Ouarouariya
3. Describe how the map confirms that desertification is a problem in Nouakchott.
4. What is an 'intermittent' well?
5. Identify the area reference of:
   a. the power plant
   b. the demineralisation plant.
6. Approximately what proportion of this map area is sand?

Apply your understanding

7. Explain what is meant by 'desertification' and how it is affecting Nouakchott.
8. Would a greenbelt help to reduce the impact of desertification in this area? Give reasons why or why not.
9. Why might increasing populations in arid and semi-arid areas contribute to desertification?
10. Explain why, when land becomes desert, its ability to support surrounding populations of people and animals declines significantly.
11. What is 'overgrazing' and why might it affect this area?
12. Why might some greenbelts fail to stop desertification?

Challenge your understanding

13. Predict what impact the sand dunes might have on the wells and roads.
14. Write a proposal to the Mauritanian government suggesting where greenbelts (corridors of vegetation) should be placed. Include the area reference of where your greenbelts will go and what they will help protect.
15. Suggest specific ways that mining might affect people in this area.
16. What current activities should be abandoned immediately by the people living in this area? Name two and suggest what impacts stopping these activities might have on the area.

To answer questions online and to receive **immediate feedback** and **sample responses** for every question, go to your learnON title at www.jacplus.com.au.

4.11 Thinking Big research project — Desert travel brochure

online only

The content in this subtopic is a summary of what you will find in the online resource.

Scenario

Your graphic design business is applying for a contract to design travel brochures to amazing locations. The tourism company offering the contract would like you to create a sample brochure to showcase your ideas for the project.

Task

Your team must create a trifold tourist brochure featuring one desert.

Go to your Resources tab to access all of the resources you need to complete this project.

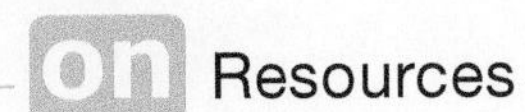

ProjectsPLUS Thinking Big research project — Desert travel brochure (pro-0170)

4.12 Review

4.12.1 Key knowledge summary

4.2 Deserts around the world

- Deserts can be hot or cold and are defined by the amount of rainfall they receive.
- Deserts are located on every continent except Europe.

4.4 How the climate forms deserts

- Different climate types are responsible for the formation of deserts in a variety of places in the world.
- Latitude and longitude, mountain ranges, ocean currents, hot interiors and polar locations can all contribute to desert formation.

4.5 The processes that shape desert landforms

- There are many different landscapes in deserts — sand dunes, claypans, alluvial fans and mesas are examples.
- Desert landscapes are formed by a combination of erosion and deposition.

4.6 Australia's desert areas

- Australia is the world's driest inhabited continent and over 70 per cent is arid or semi-arid.
- The deserts in Australia are hot deserts with low rainfall and high temperatures. Most are located in central and western Australia; some are sandy, others are stony and many have shrubs, trees and intermittent streams.
- Lake Mungo in New South Wales is in a semi-arid environment. Over 40 000 years ago Lake Mungo was a shallow lake teeming with fish and birds that supported a large human population. As a result of a drying climate over thousands of years, Lake Mungo became dry and is now protected for its cultural and natural importance.

4.8 How people use deserts

- There are many communities around the world who live in deserts.
- Many important minerals are found in deserts, creating significant industries in some countries.
- Human activities are increasing rates of desertification.

4.9 Antarctica – A cold desert

- Antarctica is a polar desert where the coldest temperature on Earth was recorded in 2018. It receives so little precipitation that it is even drier than the Sahara Desert.
- The Antarctic Treaty was formulated by many countries to protect Antarctica.
- Tourism and scientific research are the main activities in Antarctica.

4.12.2 Key terms

arroyo channel or stream that is normally dry but fills quickly with heavy rain (also called a wadi)

blizzard a strong and very cold wind containing particles of ice and snow that have been whipped up from the ground

deposition the laying down of material carried by rivers, wind, ice and ocean currents or waves

desalination the process of removing salt from sea water

erosion the wearing away and removal of soil and rock by natural elements, such as wind, waves, rivers or ice, and by human activity

humidity the amount of water vapour in the atmosphere

intermittent creek a creek that flows for only part of the year following rainfall

katabatic wind very strong winds that blow downhill

leeward the side of mountains that faces away from rainbearing winds

nomadic a group that moves from place to place depending on the food supply, or on pastures for animals

rain shadow the drier side of a mountain range, cut off from rain-bearing winds

sastrugi parallel wave-like ridges caused by winds on the surface of hard snow, especially in polar regions

subtropics the areas of Earth just north of the Tropic of Cancer and just south of the Tropic of Capricorn
treaty a formal agreement between two or more countries
windward the side of mountains that face rain-bearing winds

4.12.3 Reflection

Complete the following to reflect on your learning.

Revisit the inquiry question posed in the Overview:

Hot and sandy? Cold and windy? What are the features of a landscape that make it a desert?

1. Now that you have completed this topic, what is your view on the question? Discuss with a partner. Has your learning in this topic changed your view? If so, how?
2. Write a paragraph in response to the inquiry question, outlining your views.

Subtopic	Success criteria			
4.2	I can explain the characteristics of deserts.			
	I can describe where the world's deserts are located.			
4.3	I can interpret latitude and longitude readings on a map.			
4.4	I can explain the climatic processes that form and change deserts.			
4.5	I can explain how deserts are shaped by wind and water.			
4.6	I can describe where Australia's deserts are located and outline their characteristics.			
	I can explain the cultural and natural significance of Lake Mungo.			
4.7	I can use the scale of a map to calculate distance.			
4.8	I can explain how deserts are used and valued.			
	I can identify human causes of land degradation.			
4.9	I can explain why Antarctica is classified as a desert.			
	I can explain how Antarctica is used and changed by people.			
4.10	I can describe features on a topographic map of a desert region.			
	I can use my observations as evidence of how deserts affect settlements, and how human activities affect deserts.			

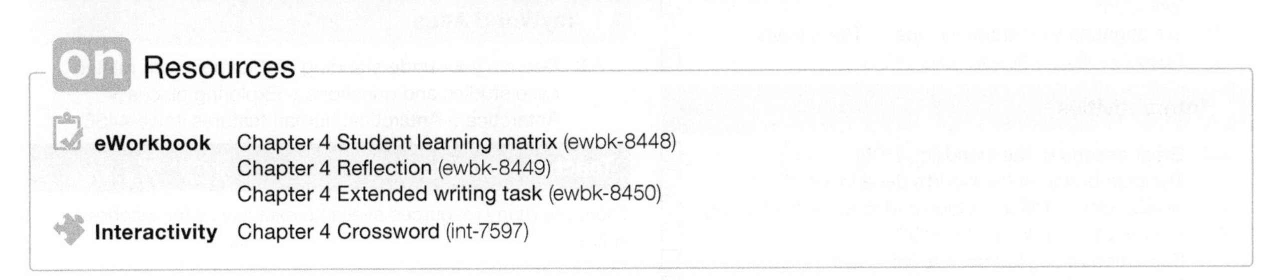

on Resources

eWorkbook Chapter 4 Student learning matrix (ewbk-8448)
Chapter 4 Reflection (ewbk-8449)
Chapter 4 Extended writing task (ewbk-8450)

Interactivity Chapter 4 Crossword (int-7597)

ONLINE RESOURCES

on Resources

Below is a full list of **rich resources** available online for this topic. These resources are designed to bring ideas to life, to promote deep and lasting learning and to support the different learning needs of each individual.

eWorkbook

- 4.1 Chapter 4 eWorkbook (ewbk-7984) ☐
- 4.2 Deserts around the world (ewbk-8724) ☐
- 4.3 SkillBuilder — Using latitude and longitude (ewbk-8728) ☐
- 4.4 How the climate forms deserts (ewbk-8732) ☐
- 4.5 The processes that shape desert landforms (ewbk-0717) ☐
- 4.6 Australia's desert areas (ewbk-8740) ☐
- 4.7 SkillBuilder — Calculating distance using scale (ewbk-8744) ☐
- 4.8 How people use deserts (ewbk-0723) ☐
- 4.9 Antarctica — A cold desert (ewbk-8752) ☐
- 4.10 Investigating topographic maps — The Sahara Desert (ewbk-8760) ☐
- 4.12 Chapter 4 Student learning matrix (ewbk-8448) ☐
  - Chapter 4 Reflection (ewbk-8449) ☐
  - Chapter 4 Extended writing task (ewbk-8450) ☐

Sample responses

- 4.1 Chapter 4 Sample responses (sar-0139) ☐

Digital document

- 4.10 Topographic map of Nouakchott, Mauritania (doc-20643) ☐

Video eLessons

- 4.1 Desert landscapes — Desertscapes (eles-1625) ☐
  - Deserts — Photo essay (eles-5334) ☐
- 4.2 Deserts around the world — Key concepts (eles-5082) ☐
- 4.3 SkillBuilder — Using latitude and longitude (eles-1652) ☐
- 4.4 How the climate forms deserts — Key concepts (eles-5051) ☐
- 4.5 The processes that shape desert landforms — Key concepts (eles-5052) ☐
- 4.6 Australia's desert areas — Key concepts (eles-5053) ☐
- 4.7 SkillBuilder — Calculating distance using scale (eles-1653) ☐
- 4.8 How people use deserts — Key concepts (eles-5054) ☐
- 4.9 Antarctica — A cold desert — Key concepts (eles-5055) ☐
- 4.10 Investigating topographic maps — The Sahara Desert — Key concepts (eles-5056) ☐

Interactivities

- 4.2 Great deserts of the world (int-3106) ☐
  - The distribution of the world's deserts (int-3611) ☐
- 4.3 SkillBuilder — Using latitude and longitude (int-3148) ☐
- 4.4 How to make a desert (int-3107) ☐
  - Formation of coastal deserts (int-3629) ☐
  - Formation of rain-shadow deserts (int-3628) ☐
  - Formation of subtropical deserts in Africa (int-3627) ☐
  - Yuma, Arizona climograph (int-3612) ☐
  - Climate graphs (int-0780) ☐
- 4.5 Desert landforms (int-7841) ☐
- 4.6 Evolving Lake Mungo (int-3108) ☐
  - The location and distribution of Australia's deserts (int-3614) ☐
  - Location of Willandra Lakes, including Lake Mungo (int-3615) ☐
- 4.7 SkillBuilder — Calculating distance using scale (int-3149) ☐
- 4.8 How people use deserts (int-8403) ☐
- 4.9 Antarctica — A cold desert (int-7842) ☐
- 4.10 Investigating topographic maps — The Sahara Desert (int-8404) ☐
- 4.12 Chapter 4 Crossword (int-7597) ☐

ProjectsPLUS

- 4.11 Thinking Big research project — Desert travel brochure (pro-0170) ☐

Fieldwork

- 4.4 Recording rainfall (fdw-0002) ☐
- 4.6 Great Sandy Desert virtual fieldwork (fdw-0003) ☐

Google Earth

- 4.4 Alice Springs (gogl-0047) ☐
  - Khormaksar, Yemen (gogl-0048) ☐
  - Mungo National Park (gogl-0084) ☐
- 4.6 The Great Sandy Desert (gogl-0121) ☐
- 4.10 Nouakchott, Mauritania (gogl-0131) ☐

Weblinks

- 4.1 Sahara (web-1039) ☐
- 4.4 Desert rain (web-4070) ☐
- 4.6 Bureau of Meteorology (web-4071) ☐
  - Timeline maker (web-4072) ☐
- 4.9 Life in Antarctica (web-4066) ☐
  - Biosecurity fears (web-4068) ☐
  - Antarctic weather (web-4069) ☐
  - Antarctic tourism dangers (web-6034) ☐

myWorld Atlas

- 4.9 Deepen your understanding of this topic with related case studies and questions > Exploring places > Antarctica > Antarctica: human features (mwa-4436) ☐

Teacher resources

There are many resources available exclusively for teachers online.

5 Mountain landscapes

INQUIRY SEQUENCE

To access a pre-test with **immediate feedback** and **sample responses** to every question in this chapter, select your learnON format at www.jacplus.com.au.

5.1 Overview

Numerous **videos** and **interactivities** are embedded just where you need them, at the point of learning, in your learnON title at www.jacplus.com.au. They will help you to learn the content and concepts covered in this topic.

Magma, water and tectonic plates — can they really move mountains?

5.1.1 Introduction

Mountains occupy 24 per cent of the Earth's landscape, and are characterised by many different landforms. The forces that form and shape mountains come from deep within the Earth, and have been shaping landscapes for millions of years. The Earth is a very active planet — every day, many volcanoes are erupting somewhere on the planet, and even more tremors are occurring.

STARTER QUESTIONS

1. What is the highest mountain you have visited? What did you do there?
2. Have you ever seen a volcano? Where was it? If you haven't, what do you imagine a volcano to look like?
3. Is there a mountain or mountain range you would like to visit? Why?
4. Have you ever felt an earthquake or earth tremor? Where were you when it happened? How did you react?
5. Watch the video **Majestic mountains** to describe how movements in the Earth can transform and destroy the landscape.
6. Watch the **Mountain landscapes — Photo essay**. How do mountains change where we live? How might your life change if you lived in the mountains?

FIGURE 1 The eruption of Kilauea volcano, Hawaii, 2018

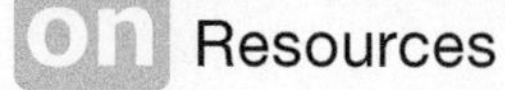

Resources

eWorkbook	Chapter 5 eWorkbook (ewbk-7985)
Video eLessons	Majestic mountains (eles-1626) Mountain landscapes — Photo essay (eles-5335)

5.2 Mountain ranges of the world

LEARNING INTENTION

By the end of this subtopic, you will be able to describe the location of the world's main mountain ranges and explain the effects of altitude on mountain environments.

5.2.1 The world's mountains and ranges

A mountain is a landform that rises high above the surrounding land. Most mountains have certain characteristics in common, although not all mountains have all these features. Many have steep sides and form a peak at the top, called a **summit**. Some mountains located close together have steep valleys between them known as **gorges**.

Mountains make up a quarter of the world's landscape. They are found on every continent and in three-quarters of all the world's countries. Only 46 countries have no mountains or high plateaus, and most of these are small island nations.

Some of the highest mountains are found beneath the sea. Some islands are actually mountain peaks emerging from the water. Even though the world's highest peak (from sea level) is Mount Everest in the Himalayas (8850 metres high), Mauna Loa in Hawaii is actually higher when measured from its base on the ocean floor (9170 metres high). Long chains or groups of mountains located close together are called a mountain range.

summit the peak at the top of a mountain

gorge a deep valley between mountains

FIGURE 1 The world's main mountains and mountain ranges

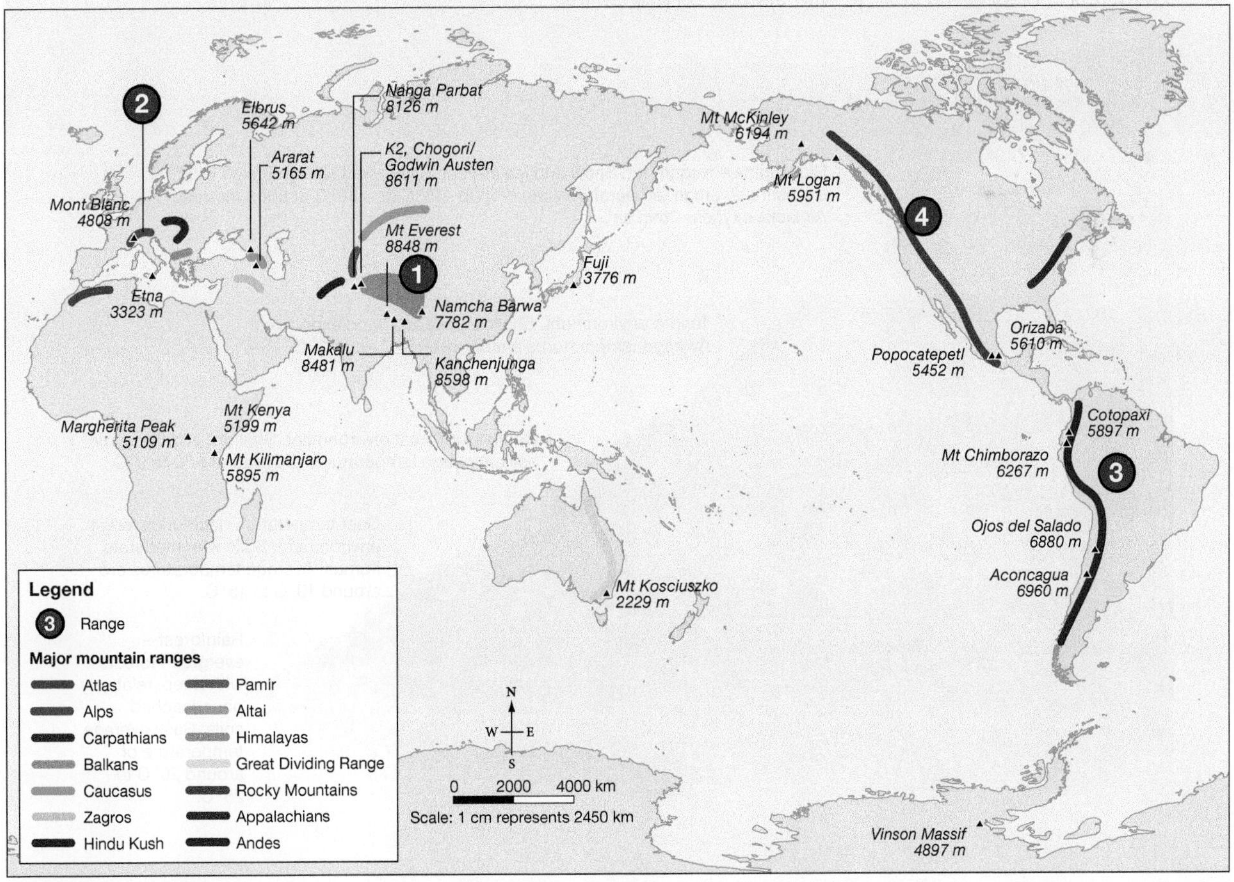

Source: Spatial Vision

1 The Himalayas

Located in Asia, the Himalayas are the highest mountain range in the world. They extend from Bhutan and southern China in the east, through northern India, Nepal and Pakistan, and to Afghanistan in the west.

2 The Alps

The Alps, located in south central Europe, extend 1200 kilometres from Austria and Slovenia in the east, through Italy, Switzerland, Liechtenstein and Germany, to France in the west. The highest mountain in the Alps is Mont Blanc in France, which is 4808 metres high.

3 The Andes

The Andes Mountains are located in South America, extending north to south along the western coast of the continent. The Andes has many mountains over 6000 metres and, at 7200 kilometres long, it is the longest mountain range in the world.

4 The Rocky Mountains

The Rocky Mountains in western North America extend north–south from Canada to New Mexico, a distance of around 4800 kilometres. The other large mountain range in North America is the Appalachian Mountains, which extend 2400 kilometres from Alabama in the south to Canada in the north.

5.2.2 Mountain climate and weather

It is usually colder at the top of a mountain than at the bottom, because air gets colder with an increase in **altitude**. Air becomes thinner and is less able to hold heat. For every 1000 metres you climb, the temperature drops by 6 °C. The effect of this on vegetation is shown in **FIGURE 2**.

altitude height above sea level

FIGURE 2 Ecosystems change with altitude on mountains.

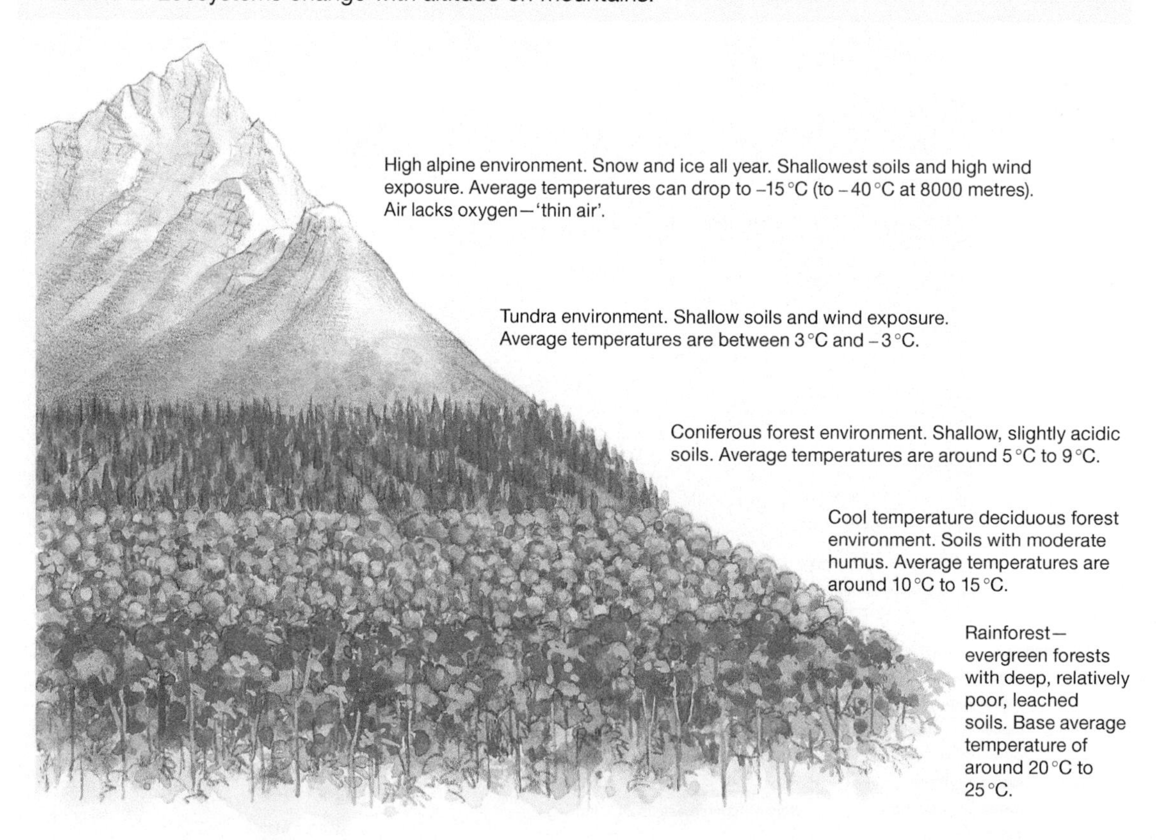

fdw-0004

FOCUS ON FIELDWORK

Recording temperature data

One of the key ways to map how our climate is changing is to examine how average temperatures are changing over long periods of time (decades and centuries). Recording temperature data yourself, and knowing how to display it, can help you to better interpret data that has been recorded over longer periods of time. It's also an important part of describing a specific environment. Learn how to create identification charts using the **Recording temperature data** fieldwork activity in your Resources tab.

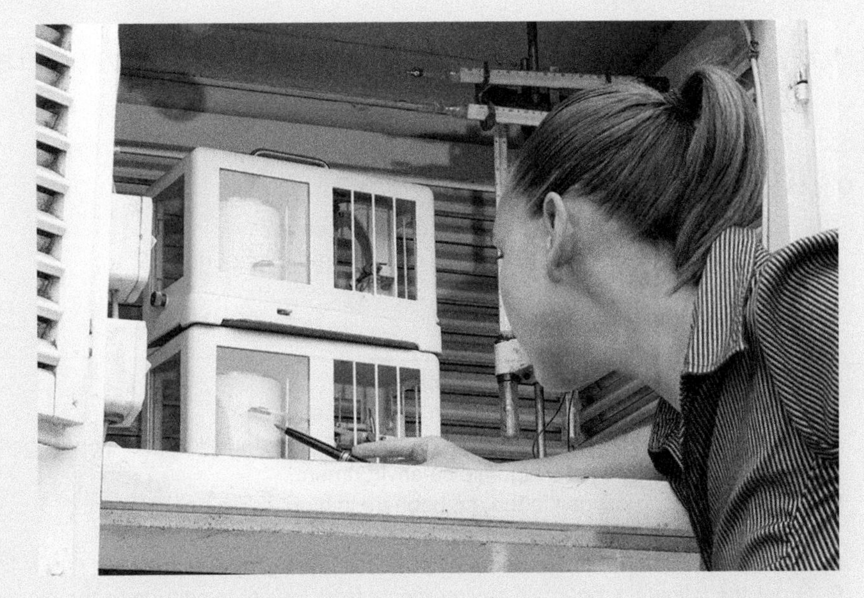

on Resources

eWorkbook	Mountain ranges of the world (ewbk-8764)
Video eLesson	Mountain ranges of the world — Key concepts (eles-5057)
Interactivity	Mountain ranges of the world (int-7844)
 **Fieldwork**	Recording temperature data (fdw-0004)
myWorldAtlas	Deepen your understanding of this topic with related case studies and questions > **Himalayas** Deepen your understanding of this topic with related case studies and questions > **Mountain environments**

5.2 ACTIVITY

Work in groups of four to six to investigate some of the following mountain ranges. Each student should choose a different range.

- Antarctica — Antarctic Peninsula, Transantarctic Mountains
- Africa — Atlas Mountains, Eastern African Highlands, Ethiopian Highlands
- Asia — Hindu Kush, Himalayas, Taurus, Elburz, Japanese Mountains
- Australia — MacDonnell Ranges, Great Dividing Range
- Europe — Pyrenees, Alps, Carpathians, Apennines, Urals, Balkan Mountains
- North America — Appalachians, Sierra Nevada, Rocky Mountains, Laurentians
- South America — Andes, Brazilian Highlands

Complete the following tasks. Present your information in Google Maps.

a. Map the location of the range in its region.
b. Describe the climate experienced throughout the range.
c. Name and provide images of a selection of plants and animals found in the range.
d. Suggest how some of the plants and animals might have adapted over time to survive in this region.

5.2 EXERCISE

Learning pathways

■ LEVEL 1	■ LEVEL 2	■ LEVEL 3
1, 2, 3, 12, 13, 16	4, 5, 7, 10, 11, 17	6, 8, 9, 14, 15

Check your understanding

1. Is a mountain considered a landform or a landscape?
2. What percentage of the Earth's surface is covered by mountains?
3. Name the:
   a. highest mountain range in the world
   b. longest mountain range in the world
   c. highest mountain in Western Europe
   d. second-highest mountain range in North America.
4. What name is given to long chains or groups of mountains located close together?
5. Describe the features of a high alpine environment.
6. What happens to oxygen in the atmosphere in high alpine environments?
7. What is the 'summit' of a mountain?
8. What is the name for the deep valleys that run between mountains?
9. Order the following mountain environments from lowest in altitude to the highest: coniferous forest, deciduous forest, rainforest, alpine, tundra.

Apply your understanding

10. Refer to **FIGURE 2**. How does vegetation change on a mountain?
11. Refer to **FIGURE 1**. How does the scale of the world's mountains vary across the continents?
12. List the countries across which the European Alps extend.
13. Where are the Appalachian Mountains located?
14. Explain why it is colder at the top of a mountain.

Challenge your understanding

15. If global warming leads to a rise in average global temperatures, suggest how this might affect mountain environments.
16. Based on the location of mountain ranges and the heights of mountains shown in **FIGURE 1**, do you think Australia could be considered a mountainous country? Provide evidence for your view.
17. Imagine you are a mountaineer, climbing to the top of Mont Blanc.
    a. Suggest the type of clothing you need to wear for such a climb.
    b. When you begin your climb at 1500 metres, the weather is perfect; it is sunny and clear and the temperature is 8 °C. You climb 2200 metres before you set up camp. What is the elevation? What is the temperature at this elevation?
    c. The next day the weather holds, and you climb to the summit. How far did you climb to reach the top of the mountain?
    d. What is the temperature?

FIGURE 3 Aiguille du Midi, Mont Blanc, France

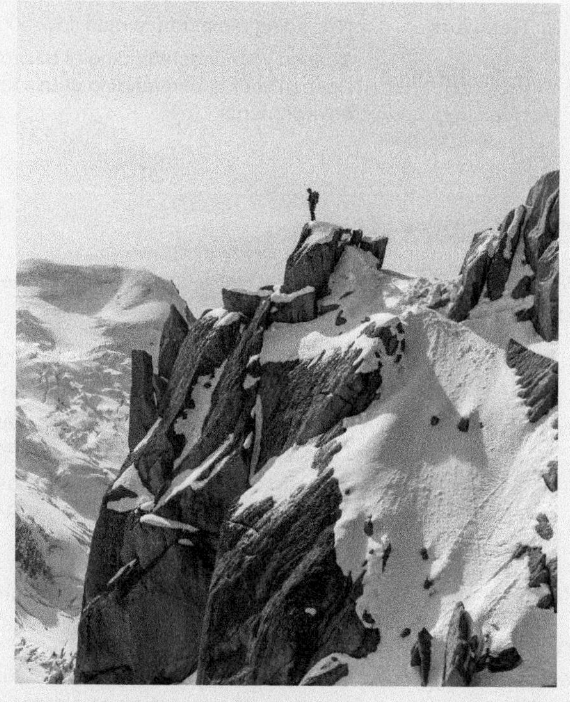

To answer questions online and to receive **immediate feedback** and **sample responses** for every question, go to your learnON title at www.jacplus.com.au.

5.3 The value of mountains

LEARNING INTENTION

By the end of this subtopic, you will be able to explain the different ways that mountains are valued.

5.3.1 Mountain people and cultures

Around 12 per cent of the world's people live in mountain regions. About half of those live in the Andes, the Himalayas, and various African mountain regions.

Usually, **population density** is very low in mountain areas. One reason for this is that mountains are very difficult to cross, as they are often rugged, steep and covered with forests and wild animals. They can also be hard to climb and may have ice, snow or glaciers that make travel dangerous.

Sacred and special places

Mountain landscapes often have special meaning to specific groups of people. This might be because the location includes sacred sites or religious symbols; it might also be because people want to be close to nature or to feel spiritually inspired or renewed.

Mountaineers who take great risks, climbing alone or in small groups, often find a special meaning in mountain environments. These areas may hold deep spiritual, **cultural** and aesthetic (relating to beauty) values and ideas, and these will often inspire people to care for and protect mountain environments.

The following list gives examples of beliefs that are connected to mountains.

- Mount Kailash in the Himalayas is sacred for Hindus, Buddhists and Jains.
- Hindus in Bali, Indonesia, have a special connection with Gunung Agung.
- Tibetan Buddhists revere Chomolungma (Mount Everest).
- Nanda Devi in the Himalayas is a sacred site for both Sikh and Hindu communities, and is a UNESCO World Heritage site.
- Mt Fuji is a place of spiritual and cultural importance for Japanese people.
- St Catherine Protectorate in Egypt, is holy to Jews, Christians and Muslims, and includes Mount Catherine and Mount Moses (Mount Horeb).
- Jabal La'lam is a mountain that is sacred to the people of northern Morocco.

population density the number of people within a given area, usually per square kilometre

cultural relating to the ideas, customs and social behaviour of a society

FIGURE 1 Mount Everest

FIGURE 2 Mount Fuji

Mountain landscapes in the Dreaming

There are many examples of Aboriginal Peoples' and Torres Strait Islander Peoples' stories that are linked to mountain landscapes. These creation stories help explain the formation and importance of each landscape and landform.

Aboriginal Peoples are guided by elders who know the local Dreaming stories and customs and cultural practices that are linked to them. Dreaming stories are passed on through lore and ceremonial practices across generations. They explain the origin of the world around them and all that it encompasses.

FIGURE 3 A storm over Mount Yengo, NSW

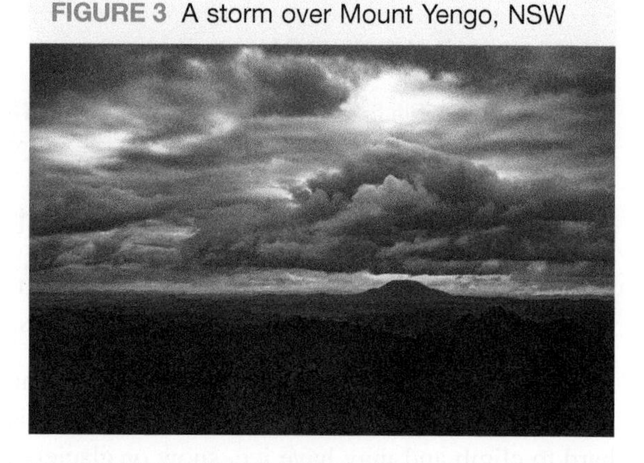

Mount Yengo

Mount Yengo is a significant landform for the Wonnarua, Awabakal, Worimi and Darkinjung Aboriginal Peoples. It holds significant spiritual and cultural places that have direct links to Dreaming stories of their creator spirit Baiame. It is also a central space that links neighbouring nations and clans to each other as well as being located on an Aboriginal songline that runs through other states and territories. It is listed as World Heritage site due to its cultural heritage for Aboriginal Peoples.

FIGURE 4 Mount Kaputar National Park, NSW

Mount Kaputar

Mount Kaputar is a significant landform on the lands of the Gamilaroi Peoples of north-western New South Wales. It holds many sites of cultural and spiritual importance where ancient stories of the Gamilaroi Peoples are embedded in the landscape in the forms of rock engravings and rock paintings, scarred trees and engravings in the landscape.

fdw-0005

FOCUS ON FIELDWORK

Using clinometers to measure slopes

Clinometers measure the angle of a slope, indicating how steep the incline or decline is. Measuring slopes is a useful fieldwork tool to help work out the gradient of the various slopes around us. Clinometers can come in many forms and are easily made or used through an app on a phone or other device. Additionally, clinometers can give us the height of tall objects by using simple trigonometry calculations.

Measure and record the slopes around your school area or the peaks in your local landscape. Refer to the **Using clinometers to measure slopes** fieldwork activity in your Resources tab.

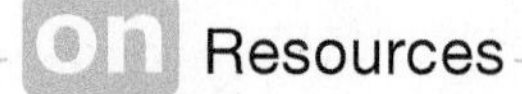

Resources

eWorkbook	The value of mountains (ewbk-8768)
Video eLesson	The value of mountains — Key concepts (eles-5058)
Interactivity	The value of mountains — Interactivity (int-8405)
Fieldwork	Using clinometers to measure slopes (fdw-0005)

5.3 ACTIVITIES

1. From this section, choose one of the Hindu or Buddhist beliefs linked to mountains. Research this connection. Present your information as a print or electronic brochure.
2. Research Mount Kaputar, Mount Yengo or a different mountain with cultural significance to local Aboriginal Peoples in your area, and complete the following:
   a. Find a map of the mountain and describe its location. View the place in Google Earth or Maps. Describe the shape of the landform.
   b. List plants and animals that are found in the landscape of the mountain.
   c. Describe how animals and plants have adapted for their environment.
   d. List the Aboriginal Peoples for whom this is a special place, and explain why it is important to them.
3. Find photographs online of the Himalayas in Nepal. Use these photographs to examine the ways that the Nepalese people use the mountain environment. Complete further research on the traditional Sherpa lifestyle and describe the connection that Sherpas have with the land.
4. Research where your water supply comes from. How do mountains determine your water availability?

5.3 EXERCISE

Learning pathways

■ LEVEL 1	■ LEVEL 2	■ LEVEL 3
1, 2, 4, 8, 13	3, 5, 7, 9, 14	6, 10, 11, 12, 15

Check your understanding

1. List the geographical characteristics of mountains that limit the number of people who live there.
2. What type of work and recreation can people undertake in mountain regions?
3. Outline the role that Aboriginal Elders play in passing on Dreaming stories about mountains.
4. List five mountains and identify the people for whom they are special.
5. What does the term *cultural value* mean?
6. Why is Mount Yengo listed as a World Heritage site?

Apply your understanding

7. How can spiritual or religious beliefs linked to mountain landscapes help in protecting them?
8. Describe how two different groups of people value mountainous places.
9. Why is population density generally low in mountain areas?
10. What physical evidence exists that shows the importance of Mount Kaputar to the Gamilaroi Peoples?
11. Explain why a mountaineer might feel they have a strong connection with a mountain they have climbed.
12. How might the cultural importance of a place also generate economic benefits for a community?

Challenge your understanding

13. Imagine you work as a park ranger in the Yengo National Park. Suggest one way that visiting tourists could be informed about the Aboriginal heritage of the area.
14. Suggest why so many cultures and religious traditions consider mountains to be important spiritual places.
15. Many mountains in the world are spiritually important to more than one religion. Suggest why you think this might be the case.

FIGURE 5 Camping at Yengo National Park

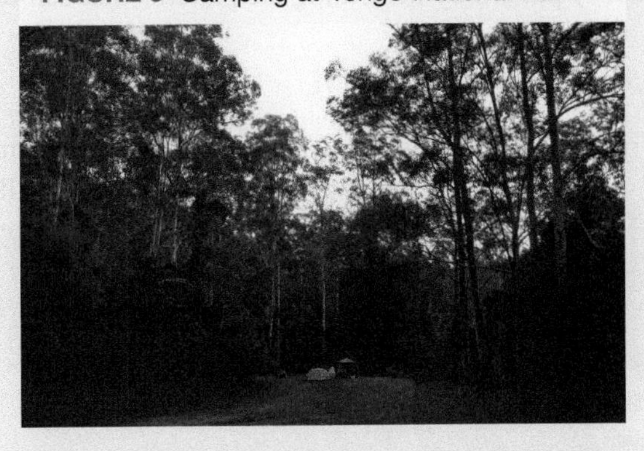

To answer questions online and to receive **immediate feedback** and **sample responses** for every question, go to your learnON title at www.jacplus.com.au.

5.4 SkillBuilder — Interpreting an aerial photograph

online only

LEARNING INTENTION

By the end of this subtopic, you will be able to identify and describe features shown in an aerial photo.

The content in this subtopic is a summary of what you will find in the online resource.

5.4.1 Tell me

What are aerial photos?

Aerial photographs are those that are taken from above the Earth from an aircraft. They record how a place looks at a particular moment in time. Vertical aerial photos are taken from directly above; that is, looking straight down on objects. Oblique aerial photos are taken from an angle.

FIGURE 1 Cartographers use different types of photographs.

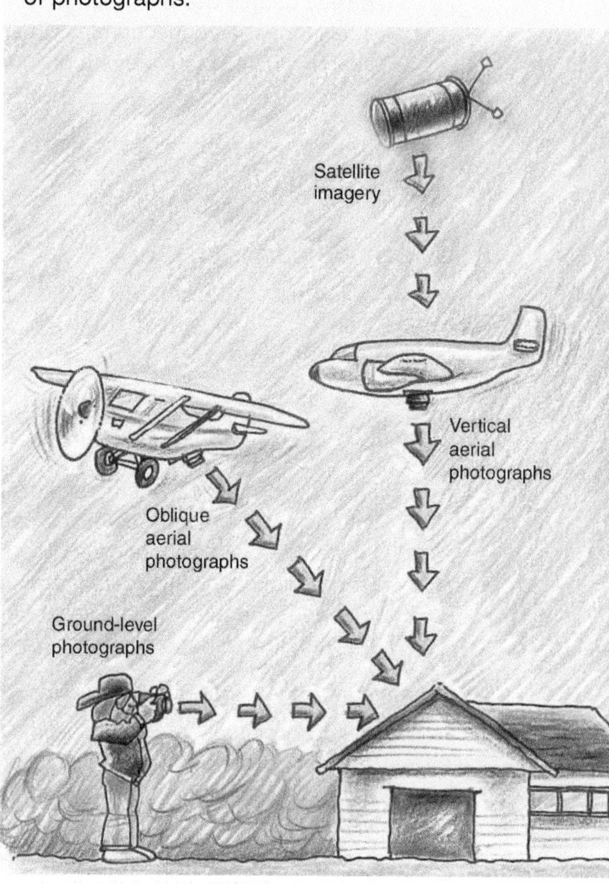

5.4.2 Show me

How to interpret a vertical aerial photo

eles-1654

Interpreting an aerial photograph involves identifying key features by recognising elements (shapes, colours, patterns and textures), and describing the main aspects in detail.

Step 1

Identify shapes and their size. Consider what this means for how the area is used.

Step 2

Identify texture and tone. Texture indicates whether the object has a degree of smoothness or whether it is rough. Tone is the reflection of light from objects that is picked up by the camera; for example, water glistens when clear, but appears brown when in flood.

Step 3

Identify pattern by describing the patterns that can be observed in the photo.

5.4.3 Let me do it

int-3150

learnON

Go to learnON to access the following additional resources to help you build this skill:

- a longer explanation of this skill and its application in Geography (Tell me)
- a video showing the step-by-step process of this skill (Show me)
- an activity and interactivity for you to practise this skill (Let me do it)
- self-marking questions to help you understand and use this skill.

on Resources

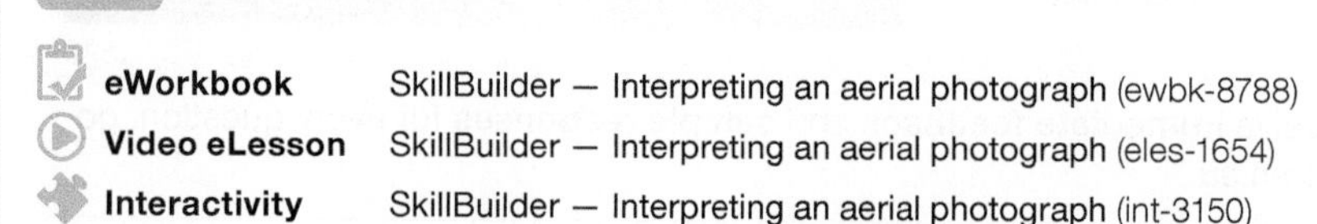

eWorkbook	SkillBuilder — Interpreting an aerial photograph (ewbk-8788)
Video eLesson	SkillBuilder — Interpreting an aerial photograph (eles-1654)
Interactivity	SkillBuilder — Interpreting an aerial photograph (int-3150)

5.5 Altering mountain landscapes

LEARNING INTENTION

By the end of this subtopic, you will be able to identify the impact of human activities on mountain landscapes.

5.5.1 Modifying mountains

It can be hard to make a living in mountain regions. People living in small, isolated mountain communities have learned to use the land and resources sustainably. Many practise **shifting cultivation**, migrate with grazing herds, and have terraced fields.

shifting cultivation process of growing crops in different places (or rotating where crops are grown)

Terracing mountain slopes

Some of the world's oldest rice terraces (see **FIGURE 1**) are over 2000 years old. Rice and vegetables could be grown quite densely on the terraces. This enabled people to survive in a region with very steep slopes and high altitude. Terracing can reduce soil erosion by slowing down water run-off as well as allowing for greater water-holding capacity. This meant that farmers would have consistent and productive crops year-round. Around the world, terraces are being abandoned due to a move away from agriculture. Italy's Cinque Terre region (see **FIGURE 2**) was once a landscape of terraces filled with viticulture (wine growing) and vegetables. As generations have moved away and no longer maintain the agricultural lifestyle, many terraces were abandoned and deteriorated. During a significant rainfall event in October 2011, the deteriorated terraces in Cinque Terre significantly contributed to large-scale landslides in the area.

FIGURE 1 The Longsheng rice terraces in China show how a mountainside can be changed to grow food.

FIGURE 2 Terraces above the village of Corniglia in Cinque Terre, Italy

Grazing mountain slopes

Not every human modification to mountain slopes has been sustainable. Areas such as the Australian Alps were once used for sheep and cattle grazing, which resulted in erosion, soil structure damage and damage to alpine vegetation such as mosses. The impacts on the alpine environments have been significant. Hard-hooved animals, such as sheep, cattle and horses, compact soils and increase soil erosion because new plants cannot establish root systems. Additionally, wetlands with naturally peaty soils (with a lot of decaying organic matter) that remain wet all year round are easily trampled and have become significantly degraded. This, in turn, impacts water supply and quality in the catchment.

Mountains supply 60 to 80 per cent of the world's fresh water. This is due to orographic rainfall (caused by warm, moist air rising and cooling when passing over high ground, such as a mountain; as the air cools, the water vapour condenses and falls as rain). Where precipitation falls as snow, water is stored in snowfields and glaciers. When these melt, they provide water to people when they need it most.

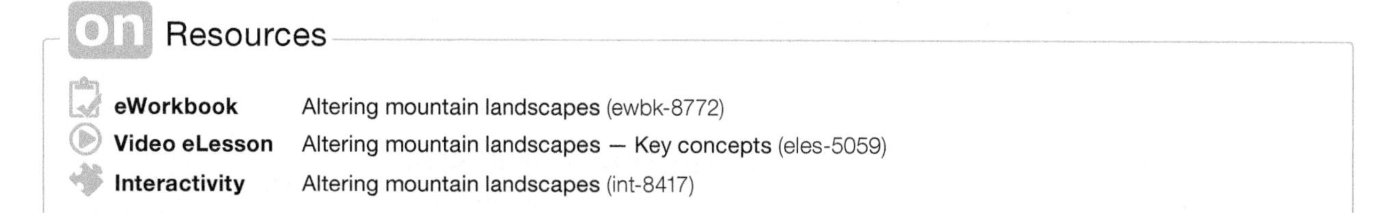

5.5 ACTIVITIES

1. Research where your water supply comes from. Which mountains, if any, are located near your water supply?
2. Draw a consequence chart to show how and why mountains are important for water supply. Now add information to your chart about what might happen if this was reduced for some reason; for example, through climate change. Use the **Climate change and water storage** weblink in the Resources tab to help you with this task.

5.5 EXERCISE

Learning pathways

LEVEL 1	LEVEL 2	LEVEL 3
1, 2, 7, 10, 15	3, 5, 9, 12, 13	4, 6, 8, 11, 14

Check your understanding

1. Identify one reason it is hard for people to live in mountainous regions.
2. What is terracing?
3. How recent is terracing as a farming technique?
4. What can happen when terraces become abandoned?
5. How does grazing cause soil erosion on mountain slopes?
6. How much of the world's fresh water lands in mountains as precipitation?

Apply your understanding

7. Explain how the natural environment in **FIGURE 1** has been changed by people.
8. Explain why mountains are important for global water supply.
9. Draw and annotate a diagram to demonstrate your understanding of orographic rainfall.
10. Why is terracing used on mountain slopes?
11. Discuss the advantages and disadvantages of terrace farming.
12. Explain one reason why a farming family might move away from their land.

Challenge your understanding

13. Predict what will happen if mountain environments are not managed properly.
14. Do you think terracing is a sustainable practice? Give reasons for your answer.
15. Suggest one strategy to rehabilitate the old terraces of Italy's Cinque Terre.

To answer questions online and to receive **immediate feedback** and **sample responses** for every question, go to your learnON title at www.jacplus.com.au.

5.6 How mountains are formed

> **LEARNING INTENTION**
>
> By the end of this subtopic, you will be able to explain how continental plates and continental drift have formed mountains and landscapes over billions of years. You should also be able to describe different mountain types and identify where some of these mountains are located.

5.6.1 The forces that form mountains

Mountains and mountain ranges have formed over billions of years from tectonic activity; that is, movement in the Earth's crust. The Earth's surface is always changing — sometimes very slowly and sometimes dramatically.

Continental plates

The Earth's crust is cracked, and is made up of many individual moving pieces called continental plates, which fit together like a jigsaw puzzle (see **FIGURE 1**). These plates float on the semi-molten rocks, or magma, of the Earth's mantle. Enormous heat from the Earth's core, combined with the cooler surface temperature, creates convection currents in the magma. Convection currents are created when a fluid is heated, making it less dense and causing it to rise through surrounding fluid and to sink if it is cooled. A steady source of heat can start a continuous current flow. These currents can move the plates by up to 15 centimetres per year. Plates beneath the oceans move more quickly than plates beneath the continents.

Continental drift

Scientific evidence shows that, about 225 million years ago, all the continents were joined.

FIGURE 1 World map of plates, volcanoes and hotspots

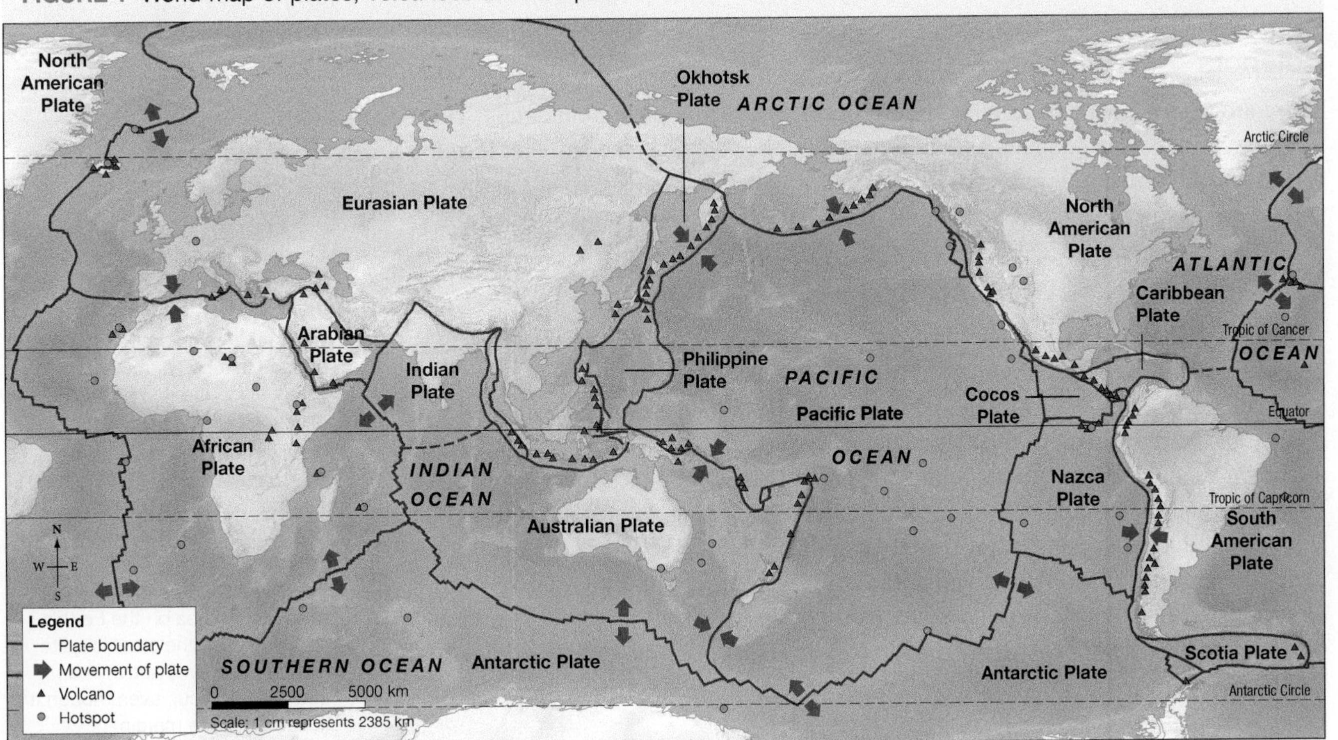

Source: Spatial Vision

Convergent plates

When two continental plates of similar density collide, the pressure of the **converging plates** can push up land to form mountains. The Himalayas were formed by the collision of the Indian subcontinent and Asia. The European Alps were formed by the collision of Africa and Europe.

Oceanic and continental plates are different densities, and when they collide the thinner oceanic plate is subducted, meaning it is forced down into the mantle. Heat melts the plate and pressure forces the molten material back to the surface. This can produce volcanoes and mountain ranges. The Andes in South America were formed this way.

Subduction can also occur when two oceanic plates collide. This forms a line of volcanic islands in the ocean about 70–100 kilometres past the subduction line. The islands of Japan have been formed in this way. Deep oceanic trenches are also formed when this occurs. The Mariana Trench in the Pacific Ocean is 2519 kilometres long and 71 kilometres wide, and is the deepest point on Earth — 10.911 kilometres deep.

FIGURE 2 The Earth's core is very hot, while its surface is quite cool. This causes hot material within the Earth to rise until it reaches the surface, where it moves sideways, cools, and then sinks.

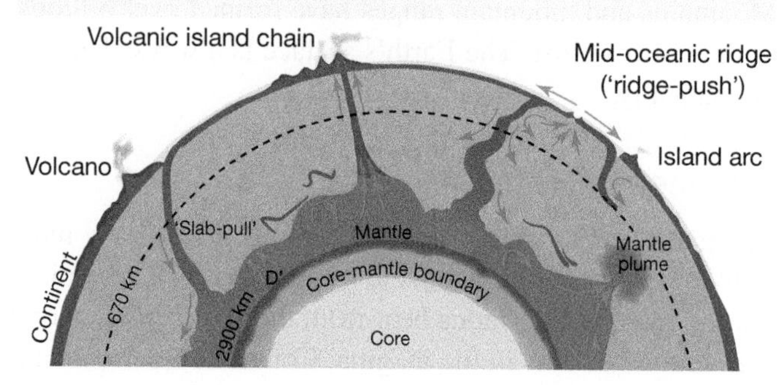

Lateral plate slippage

Convection currents can sometimes cause plates to slide, or slip, past one another, forming **fault** lines. The San Andreas Fault, in California in the western United States, is an example of this.

Divergent plates

In some areas, plates are moving apart, or diverging, from each other (for example, the Pacific Plate and Nazca Plate). As the **divergent plates** separate, magma can rise into the opening, forming new land. Underwater volcanoes and islands are formed in this way.

Hotspots

There are places where volcanic eruptions occur away from plate boundaries. This occurs when there is a weakness in the oceanic plate, allowing magma to be forced to the surface, forming a volcano. As the plate drifts over the **hotspot**, a line of volcanoes is formed. Hotspots are found in the ocean and on continents. Examples include the Hawaiian Islands and many of Australia's extinct volcanoes.

The Pacific Ring of Fire

The most active volcanic region in the world is the Pacific Ring of Fire. It is located on the edges of the Pacific Ocean and is shaped like a horseshoe. The Ring of Fire is a result of the movement of tectonic plates. For example, the Nazca and Cocos plates are being subducted beneath the South American Plate, while the Pacific and Juan de Fuca plates are being subducted beneath the North American Plate. The Pacific Plate is being subducted under the North American Plate on its east and north sides, and under the Philippine and Australian plates on its west side. The Ring of Fire is an almost continuous line of volcanoes and earthquakes. Most of the world's earthquakes occur here, and 75 per cent of the world's volcanoes are located along the edge of the Pacific Plate.

converging plates a tectonic boundary where two plates are moving towards each other

fault an area on the Earth's surface that has a fracture, along which the rocks have been displaced

divergent plates a tectonic boundary where two plates are moving away from each other and new continental crust is forming from magma that rises to the Earth's surface between the two

hotspot an area on the Earth's surface where the crust is quite thin, and volcanic activity can sometimes occur, even though it is not at a plate margin

5.6.2 How different types of mountains form

The different movements and interactions of the **lithosphere** plates result in many different mountain landforms. Mountains can be classified into five types, based on what they look like and how they were formed. The five types are fold, fault-block, dome, plateau and volcanic mountains. (Volcanic mountains are discussed in subtopic 5.10.)

lithosphere the crust and upper mantle of the Earth

FIGURE 3 Selected world mountains and range types

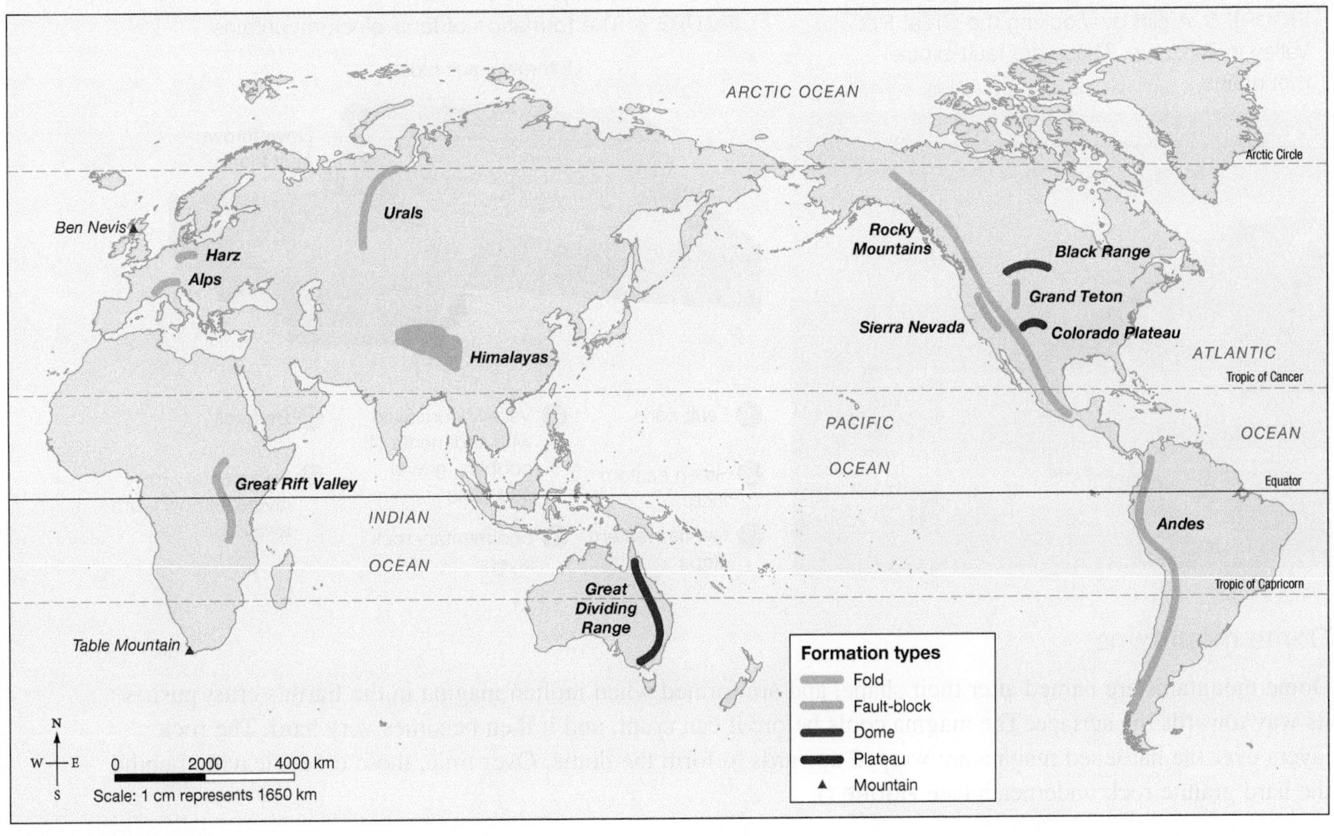

Source: Spatial Vision

Fold mountains

The most common type of mountain, and the world's largest mountain ranges, are fold mountains. The process of folding occurs when two continental plates collide, and rocks in the Earth's crust buckle, fold and lift. The upturned folds are called anticlines, and the downturned folds are synclines (see **FIGURE 4**). These mountains usually have pointed peaks.

Examples of fold mountains include:

- the Himalayas in Asia
- the Alps in Europe
- the Andes in South America
- the Rocky Mountains in North America
- the Urals in Russia.

FIGURE 4 The formation of fold mountains

Thrust fault
Anticline valley
Anticline valley
Syncline ridge
Syncline ridge
Anticline Syncline Anticline Syncline Anticline

Fault-block mountains

Fault-block mountains form when pressure within the Earth forces some parts of rock up and other parts to collapse, along a fault line (or crack) in the Earth's crust. Instead of folding, the crust fractures (pulls apart) and breaks into blocks. The exposed parts then begin to erode and shape mountains and valleys (see **FIGURE 6**).

Fault-block mountains usually have a steep front side and then a sloping back. The Sierra Nevada and Grand Tetons in North America, the Great Rift Valley in Africa and the Harz Mountains in Germany are examples of fault-block mountains. Another name for the uplifted blocks is *horst*, and the collapsed blocks are *graben*.

FIGURE 5 A cliff overlooking the Great Rift Valley in Tanzania. These are fault-block mountains.

FIGURE 6 The formation of fault-block mountains

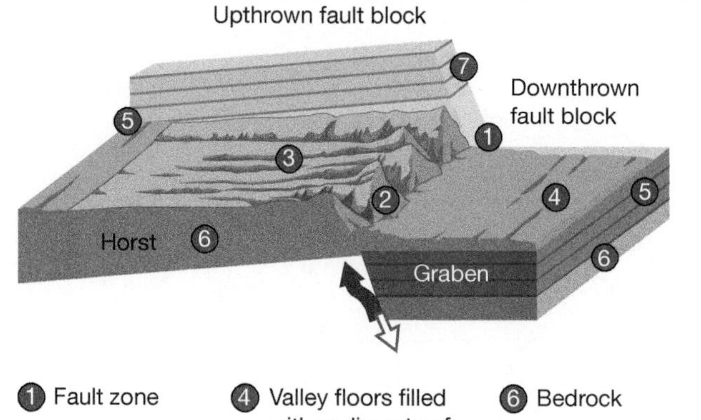

Dome mountains

Dome mountains are named after their shape, and are formed when molten magma in the Earth's crust pushes its way towards the surface. The magma cools before it can erupt, and it then becomes very hard. The rock layers over the hardened magma are warped upwards to form the dome. Over time, these erode, leaving behind the hard granite rock underneath (see **FIGURE 7**).

FIGURE 7 Very hot magma pushes towards the surface to form dome mountains.

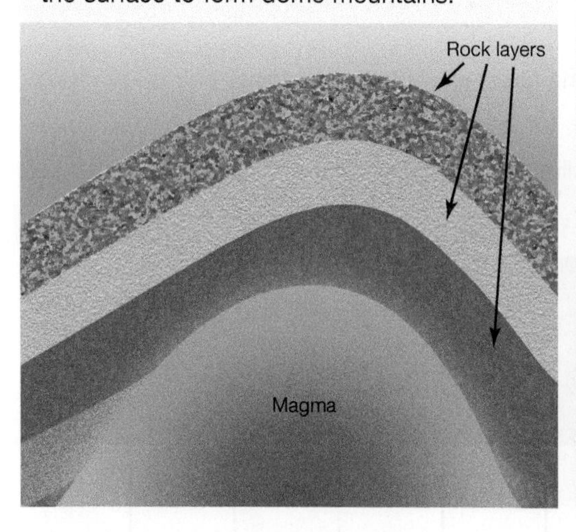

FIGURE 8 Ben Nevis, the highest peak in Scotland, is an example of a dome mountain.

Plateau mountains

Plateaus are high areas of land that are large and flat. They have been pushed above sea level by tectonic forces or have been formed by layers of lava. Over billions of years, streams and rivers cause erosion, leaving mountains standing between valleys. Plateau mountains are sometimes known as erosion mountains.

Examples of plateau mountains include Table Mountain in South Africa (see **FIGURE 9**), the Colorado Plateau (see **FIGURE 10**) in the United States, and parts of the Great Dividing Range in Australia.

FIGURE 9 The plateau of Table Mountain towers over the city of Cape Town in South Africa.

FIGURE 10 The Colorado Plateau in the United States was raised as a single block by tectonic forces. As it was uplifted, streams and rivers cut deep channels into the rock, forming the features of the Grand Canyon.

FIGURE 11 The movement of the Indian landmass

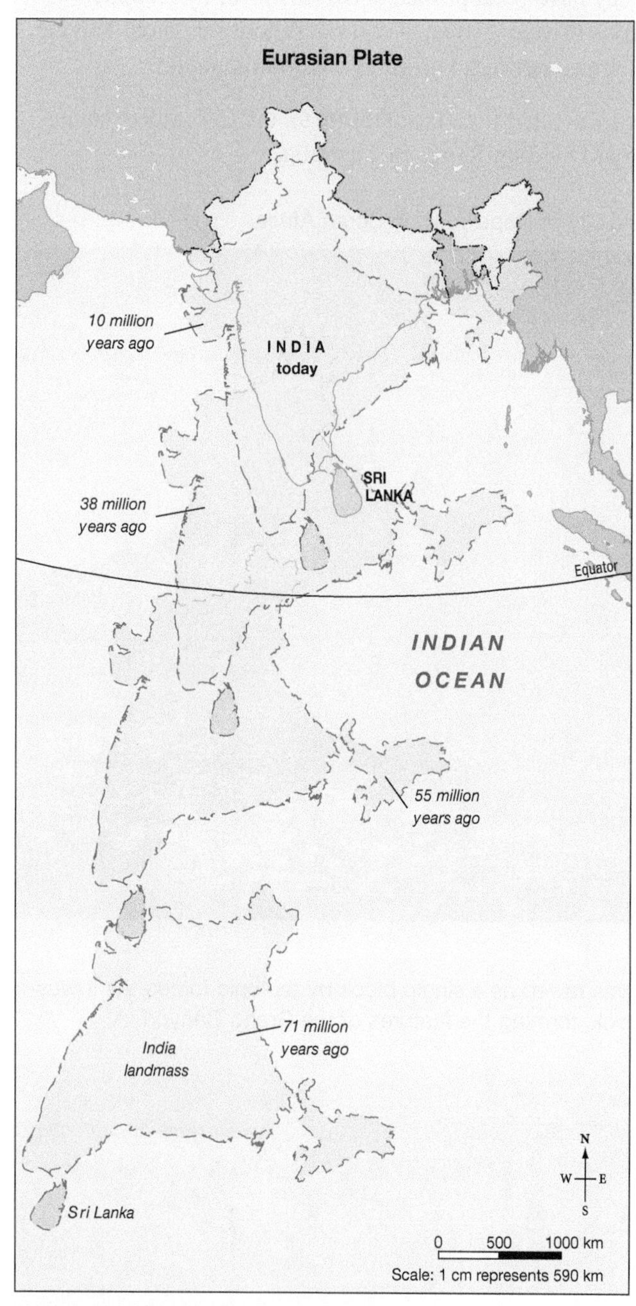

Source: MAPgraphics Pty Ltd Brisbane

CASE STUDY 1

Formation of the Himalayas

Before the theory of tectonic plate movement, scientists were puzzled by findings of fossilised remains of ancient sea creatures near the Himalayan peaks. Surely these huge mountains could not once have been under water?

Since understanding plate movements, the mystery has been solved. About 220 million years ago, India was part of the ancient supercontinent we call **Pangaea**. When Pangaea broke apart, India began to move northwards at a rate of about 15 centimetres per year. About 200 million years ago, India was an island separated from the Asian continent by a huge ocean.

When the plate carrying India collided with Asia 40 to 50 million years ago, the oceanic crust (carrying fossilised sea creatures) slowly crumpled and was uplifted, forming the high mountains we know today. It also caused the uplift of the Tibetan Plateau to its current position. The Bay of Bengal was also formed at this time.

The Himalayas were thus formed when India crashed into Asia and pushed up the tallest mountain range on the continents.

The Himalayas are known as young mountains, because they are still forming. The Indo-Australian plate is still moving northwards at about 2 centimetres each year, making this boundary very active. It is predicted that over the next 10 million years it will travel more than 180 kilometres into Tibet and that the Himalayan mountains will increase in height by about 5 millimetres each year. Old mountains are those that have stopped growing and are being worn down by the process of erosion.

Pangaea the name given to all the landmass of the Earth before it split into Laurasia and Gondwana

FIGURE 12 Panoramic view of Kathmandu city and the Himalayas from Patan, Nepal

CASE STUDY 2

Formation of the Sierra Nevada range, United States

The Sierra Nevada range began to rise about 5 million years ago. As the western part of the block tilted up, the eastern part dropped. As a result there is a long, gentle slope towards the west and a steep slope to the east.

FIGURE 13 Yosemite Valley in the Sierra Nevada mountains

FIGURE 14 The Sierra Nevada range was formed by fault-block tilting.

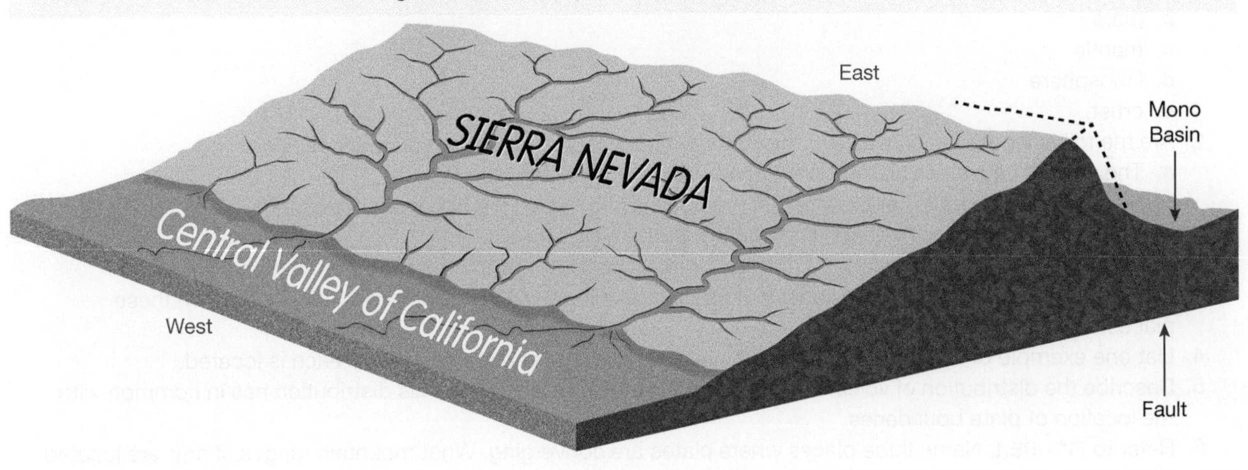

Resources

eWorkbook	How mountains are formed (ewbk-8776)
Video eLesson	Drifting continents (eles-0129)
Interactivities	Grand peaks (int-3110) Mountain builders (int-3109)
Google Earth	Sierra Nevada mountains (gogl-0058) Table Mountain (gogl-0059) Grand Canyon (gogl-0002) Ben Nevis (gogl-0049)
myWorldAtlas	Deepen your understanding of this topic with related case studies and questions > **Investigate additional topics > Earthquakes and volcanoes > Active Earth**

5.6 ACTIVITIES

1. Use different coloured strips of plasticine to make models showing how a collision of continental and oceanic plates differs from a collision of two continental plates. Use the **Fold mountains** weblink in the online Resources to explain the formation of fold mountains and fault-block mountains.
2. Sketch **FIGURE 5** and annotate where erosion has taken place. Label areas of hard and weak rocks.

3. Draw a sketch of **FIGURE 13**, noting the plateau and areas of erosion and weathering.
4. Use an atlas to locate the Sierra Nevada range. Describe where it is. Identify two national parks in this mountain range. Choose one and investigate its geographical characteristics. Present your findings to the class.
5. Using the **Anticline and syncline** weblink in your online Resources as inspiration, create a video or animation to demonstrate the ways that different mountains form.

5.6 EXERCISE

Learning pathways

■ LEVEL 1	■ LEVEL 2	■ LEVEL 3
1, 4, 8, 9, 13, 15	2, 3, 5, 11, 14	6, 7, 10, 12, 16

Check your understanding

1. Define the following terms related to the ways mountains form:
   a. hotspot
   b. plate
   c. mantle
   d. lithosphere
   e. crust.
2. Are the following statements true or false? Rewrite any statements that you mark as false.
   a. The world's volcanoes are randomly scattered over the Earth's surface.
   b. Most of the world's volcanoes are concentrated along the edges of certain continents.
   c. Island chains are closely linked with the location of volcanoes.
   d. There is a weak link between the distribution of volcanoes and the location of continental plates.
3. Name two locations where plates are moving apart. Outline what is happening to the sea floor in these places.
4. List one example of fold, fault, dome and plateau mountains, and identify where each is located.
5. Describe the distribution of volcanoes shown in **FIGURE 1**. Outline what this distribution has in common with the location of plate boundaries.
6. Refer to **FIGURE 1**. Name three places where plates are converging. What mountain ranges, if any, are located in these places?
7. What is the difference between converging and diverging?
8. In which type of mountain range can a 'horst' and a 'graben' be found?

Apply your understanding

9. Explain, in your own words, the meaning of subduction when referring to plate movements.
10. Look at **FIGURE 2**. How do convection currents help explain plate tectonics?
11. Use **FIGURES 7** and **8** to explain the formation of dome mountains.
12. How does the shape of each of the mountains shown in this subtopic provide clues as to how they were formed? How have the effects of erosion changed these mountains?
13. Mountains form in many ways. Explain which is the quickest.

Challenge your understanding

14. What you think the world's continents will look like millions of years into the future, based on the way continents move and change? Justify your decisions.
15. Apart from changing its location on the Earth's surface, suggest one other way that the movement of a tectonic plate might change the environment on a particular landmass over millions of years.
16. Suggest why the movement of continental plates is referred to as the *theory* of tectonic plate movement.

To answer questions online and to receive **immediate feedback** and **sample responses** for every question, go to your learnON title at www.jacplus.com.au.

5.7 Geomorphic processes that cause earthquakes and tsunamis

LEARNING INTENTION

By the end of this subtopic, you will be able to explain the geomorphic processes that cause earthquakes and tsunamis.

5.7.1 Earthquakes

Earthquakes and **tsunamis** are frightening events and often strike with little or no warning. An earthquake can shake the ground so violently that buildings and other structures collapse, causing damage to infrastructure and harming people. If an earthquake occurs at sea, it may cause a tsunami, which produces waves of water that move to the coast and further inland, sometimes with devastating effects.

FIGURE 1 What happens in an earthquake?

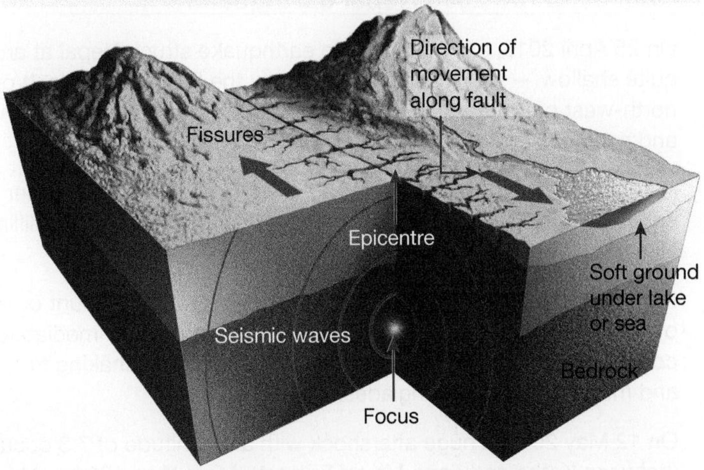

Earthquakes occur every day somewhere on the planet, usually on or near the boundaries of tectonic plates. The map at **FIGURE 1** in subtopic 5.5 shows a strong relationship between the location of plate boundaries and the occurrence of earthquakes. Earthquakes occur when there is sudden movement at a weak point or crack in the Earth's crust near these plate boundaries, or faults.

The point where this earthquake movement begins is called the **focus** (see **FIGURE 1**). Earthquakes can occur near the surface or up to 700 kilometres below. The shallower the focus, the more powerful the earthquake will be. Energy travels quickly from the focus point in powerful **seismic waves**, radiating like ripples in a pond. The seismic waves decrease in strength as they travel away from the **epicentre**.

The energy released at the focus can be immense, and it travels in seismic waves through the mantle and crust of the Earth. Primary waves, or P-waves, are the first waves to arrive, and are felt as a sudden jolt. Depending on the type of rock or water in which they are moving, these waves travel at speeds of up to 30 000 kilometres an hour.

Secondary waves, or S-waves, arrive a few seconds after P-waves and travel at about half the speed of P-waves. These waves cause more sustained up-and-down movement.

Surface waves radiate from the epicentre and arrive after the main P-waves and S-waves. These move the ground either from side-to-side, like a snake moving, or in a circular movement.

Even very strong buildings can collapse with these stresses. The energy that travels in waves across the Earth's surface can destroy buildings many kilometres from the epicentre.

tsunami large wave that can be triggered by an earthquake under the ocean

focus the point where the sudden movement of an earthquake begins

seismic waves waves of energy that travel through the Earth as a result of an earthquake, explosion or volcanic eruption

epicentre the point on the Earth's surface directly above the focus of an earthquake

Measuring earthquakes

Earthquakes are measured according to their magnitude (size) and intensity. Magnitude is measured on the Richter scale, which shows the amount of energy released by an earthquake. The scale is open-ended as there is no upper limit to the amount of energy an earthquake might release. An increase of one in the scale is 10 times greater than the previous level. For example, energy released at the magnitude of 7.0 is 10 times greater than the energy released at 6.0.

Earthquake intensity is measured on the Modified Mercalli scale. This indicates the amount of damage caused. Intensity depends on the nature of buildings, time of day and other factors.

CASE STUDY 1

Causes of the 2015 Nepal earthquake

On 25 April 2015, a 7.8-magnitude earthquake struck Nepal at around midday. The focus of this earthquake was quite shallow — only 15 kilometres below the Earth's surface. It occurred approximately 80 kilometres to the north-west of Kathmandu, Nepal's capital. During the Nepal earthquake event, nearly 9000 people were killed and nearly 18 000 were injured.

At this location, the Indian Plate to the south is subducting under the Eurasian Plate to the north (see **FIGURE 1** in subtopic 5.5). This is occurring at a rate of approximately 45 millimetres per year and is causing the uplift of the Himalayas (see Case study 1 in subtopic 5.5).

FIGURE 3 shows that the earthquake released a large amount of energy and caused large slips of up to 4 metres of the Earth's surface. There were severe aftershocks immediately after the main earthquake and the aftershocks continued for many weeks — up to 100 in total. The shaking from this earthquake was felt in China, India, Bhutan and much of western Bangladesh.

On 12 May 2015, a huge aftershock with a magnitude of 7.3 occurred near the Chinese border with Nepal (between Kathmandu and Mount Everest). More than 160 people died and more than 2500 were injured as a result of this aftershock.

FIGURE 2 The shake intensity and the tectonic plate boundary involved in the Nepal earthquake

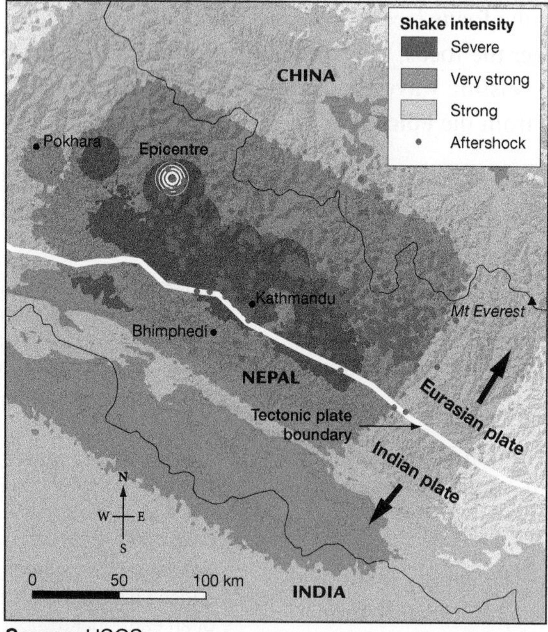

Source: USGS

FIGURE 3 Magnitudes of earthquake and aftershocks in Nepal, 2015

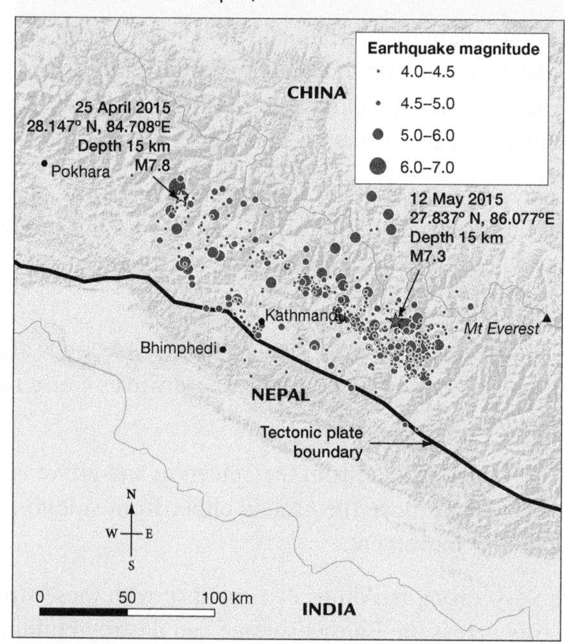

Source: USGS

5.7.2 Tsunamis

A tsunami is a large ocean wave that is caused by sudden motion on the ocean floor. The sudden motion could be caused by an earthquake, a volcanic eruption or an underwater landslide. About 90 per cent of tsunamis occur in the Pacific Ocean, and most are caused by earthquakes that are over 6.0 on the Richter scale.

A tsunami at sea will be almost undetectable to ships or boats. The reasons for this are that the waves travel extremely fast in the deep ocean (about 970 kilometres per hour — as fast as a large jet) and the wavelength is about 30 kilometres, yet the wave height is only 1 metre.

When tsunamis reach the continental slope, several things happen. First, the wave slows down and, as it does, the wave height increases and the wavelength decreases; in other words, the waves get higher and closer together. Sometimes, the sea may recede quickly, very far from shore, as though the tide has suddenly gone out. If this happens, the best course of action is to head to higher ground as quickly as possible.

A tsunami is not a single wave. There may be between five and twenty waves altogether. Sometimes the first waves are small and then become larger; at other times there is no apparent pattern. Tsunami waves will arrive at fixed periods between ten minutes and two hours.

FIGURE 4 The earthquake and subsequent tsunami in the Indian Ocean in 2004 occurred along the boundary between tectonic plates.

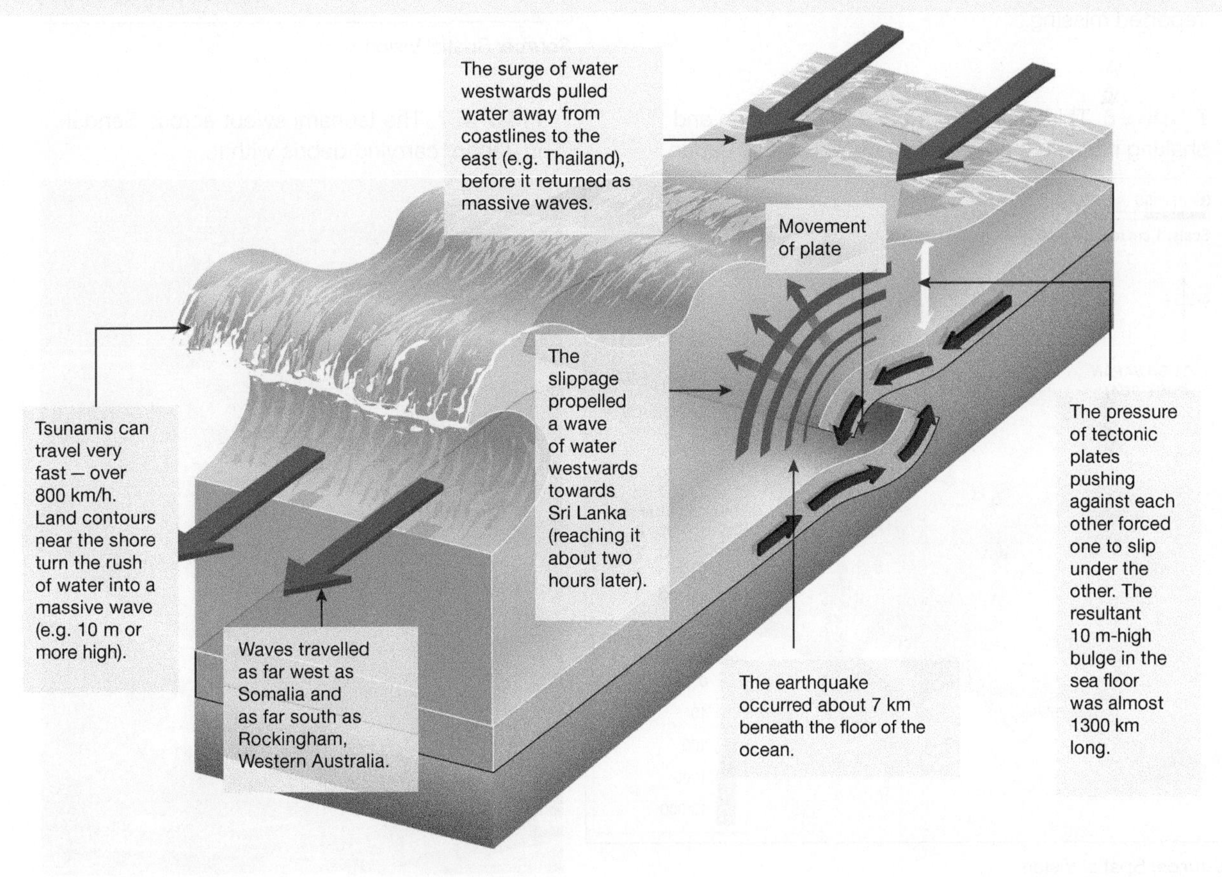

CASE STUDY 2

The 2011 Japanese tsunami

The region of Japan is seismically active because four plates meet there: the Eurasian, Philippine, Pacific and Okhotsk (see **FIGURE 5**). Many landforms in this region are influenced by the collision of oceanic plates. Chains of volcanic islands called island arcs are formed, and an ocean trench is located parallel to the island arc.

On 11 March 2011, an 8.9-magnitude earthquake struck near the coast of Japan. The earthquake was caused by movement between the Pacific Plate and the North American Plate (see **FIGURE 5**). It occurred about 27 kilometres below the Earth's surface along the Japan Trench, where the Pacific Plate moves westwards at about 8 centimetres each year. The sudden upward movement released an enormous amount of energy and caused huge displacement of the sea water, causing the tsunami. When the tsunami reached the Japanese coast, waves more than 6 metres high moved huge amounts of water inland. Strong aftershocks were felt for a number of days. Nearly 16 500 people were killed and 4800 were reported missing.

FIGURE 5 The location and magnitude of the earthquake that caused the Japanese tsunami

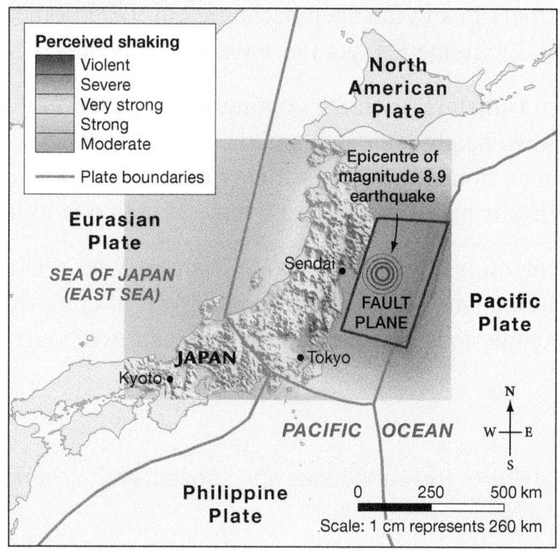

Source: Spatial Vision

FIGURE 6 This map shows the ground motion and shaking intensity from the earthquake across Japan.

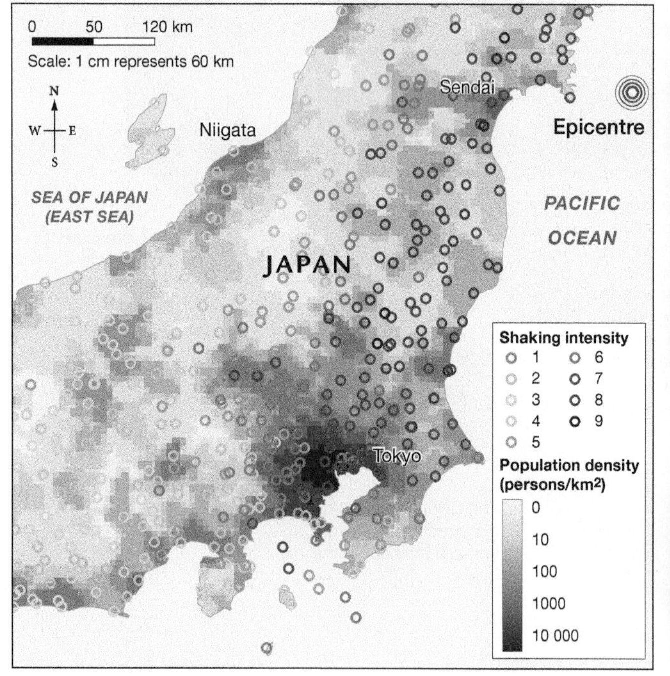

Source: Spatial Vision

FIGURE 7 The tsunami swept across Sendai in Japan, carrying debris with it.

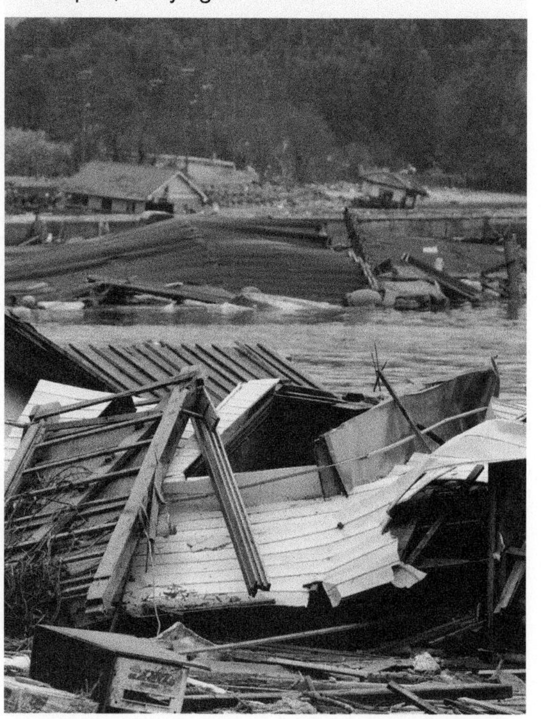

FIGURE 8 An area of Sendai, Japan, on 14 April 2011 (top) and 3 March 2014 (bottom) — the extent of damage and time taken to recover after such an event is significant.

on Resources

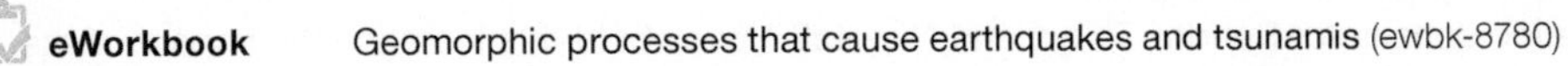

eWorkbook	Geomorphic processes that cause earthquakes and tsunamis (ewbk-8780)
Video eLesson	Geomorphic processes that cause earthquakes and tsunamis — Key concepts (eles-5060)
Interactivities	Anatomy of a tsunami (int-3111) Magnitudes of earthquake and aftershocks in Nepal, 2015 (int-7846) The location and magnitude of the earthquake that caused the Japanese tsunami (int-7847) The shake intensity and the tectonic plate boundary involved in the Nepal earthquake (int-7845) Map of ground motion and shaking intensity from the earthquake across Japan (int-7848)
myWorldAtlas	Deepen your understanding of this topic with related case studies and questions > **Haiti earthquake** Deepen your understanding of this topic with related case studies and questions > **Banda Aceh tsunami**

5.7 ACTIVITIES

1. 'The strongest earthquakes result in the worst disasters.' Work in pairs or groups of three to decide: do you agree, partially agree, or disagree with this statement. Use the data in this subtopic and particular examples in your response.
2. Use the **P- and S-waves** weblink in the online Resources. What is the difference between the waves? How fast do they travel? How is damage caused by the waves?
3. Use an atlas or Google Earth to find the location of Lituya Bay. Draw a map to show the location. Use the **World's biggest tsunami** weblink in the online Resources to listen to eyewitness accounts of the event. How does this help give you a sense of the scale of this event?
4. Use the **Sendai tsunami** weblink in the online Resources to look at satellite images showing areas before and after the 2011 Japanese tsunami. Choose two locations to draw sketches of before and after, and annotate your sketches to record the changes that have taken place.
5. Use the **Nepal earthquake — before and after photos** weblink in the online Resources to look at more before and after images. Choose one of the before/after images and sketch the after image, providing annotations that show the impact on people and/or the environment.
6. Research the Richter scale and the Modified Mercalli scale. Create an infographic to demonstrate how each scale works.

5.7 EXERCISE

Learning pathways

■ LEVEL 1	■ LEVEL 2	■ LEVEL 3
3, 5, 7, 11, 14	2, 4, 8, 9, 15	1, 6, 10, 12, 13

Check your understanding

1. Describe the difference between the focus and epicentre of an earthquake.
2. Outline the steps that lead to an earthquake occuring.
3. What does the Richter scale measure?
4. How much more powerful is the magnitude of an earthquake at 7.0 than at 5.0?
5. What is a seismic wave?
6. Describe the difference between a P-wave and an S-wave.

Apply your understanding

7. Explain why an earthquake is likely to create more damage to buildings when its focus is closer to the surface of the Earth than an earthquake of the same intensity that occurs deeper in the Earth.
8. Study **FIGURE 5**. Describe where the most violent shaking occurred as a result of the earthquake. How many plates meet in this region? What impact does this have?
9. Study **FIGURE 6**.
   a. Where in Japan was the greatest intensity felt?
   b. What is the population density for Sendai, Tokyo and Niigata? How would this increase the impact of the earthquake?
10. Study **FIGURE 2**.
    a. In which direction is the Indian Plate moving? Is it moving under or over the Eurasian Plate?
    b. Describe the location of the highest intensity shaking. How close was it to the epicentre? To the tectonic plate boundary?
    c. What type of mountain ranges would you expect to see in this region?
11. Study **FIGURE 3**. Are the following statements true or false? If they are false, rewrite them to make them true.
    a. The earthquake and aftershocks were between 4.0 and 6.0 in magnitude.
    b. The furthest earthquake and aftershocks were 100 kilometres apart.
    c. The earthquake on 12 May was the same intensity as the earthquake on 25 April.
    d. Most of the aftershocks were felt to the east of the main earthquake on 25 April.
12. Which waves do you think will cause more destruction: P-waves or S-waves? Explain your answer.

FIGURE 9 Tsunami damage, Sendai, Japan, 2011

Challenge your understanding

13. How does the earthquake event in Nepal support the idea that the Himalayas are a young mountain range that is still forming?
14. Study the photo of the Japanese tsunami in **FIGURES 7, 8,** and **9.**
    a. Imagine you are a radio news reporter. Describe what you see and what might be happening to people in the area.
    b. Imagine you were a Sendai resident. Describe what you would have done to take care of yourself.
15. Why would most Australians not know what to do if an earthquake occurred?

To answer questions online and to receive **immediate feedback** and **sample responses** for every question, go to your learnON title at www.jacplus.com.au.

5.8 Impacts of earthquakes and tsunamis

LEARNING INTENTION

By the end of this subtopic, you will be able to identify impacts of earthquakes and tsunamis and explain how people respond to these hazards.

5.8.1 Measuring the impact

Earthquakes and tsunamis can have an enormous impact. The degree of impact can be affected by several factors: the size of the quake; its location; the density of the population near the epicentre; and whether there are any densely populated areas nearby. Poverty also plays a role, because it can increase the vulnerability of a country or region to such disasters. Measuring the event by the impact can be difficult. Should it be measured by the number of people killed and made homeless (social impact); the cost of recovery (economic impact); or the effect on the surroundings (environmental impact)?

5.8.2 Impact on people

The data in the map in **FIGURE 1**, and in **TABLES 1** and **2**, show some of the worst earthquake and tsunami disasters that have occurred. The amount of damage and death they cause does not always relate to the magnitude of the earthquake. Some smaller magnitude earthquakes can have a devastating impact. Likewise, to measure the impact of a tsunami, we have to look at its effect on people, not at the magnitude of the earthquake (or volcano) that caused it, and not at the size of the waves, which are difficult to measure.

FIGURE 1 Map of the ten biggest earthquakes and tsunamis ever recorded

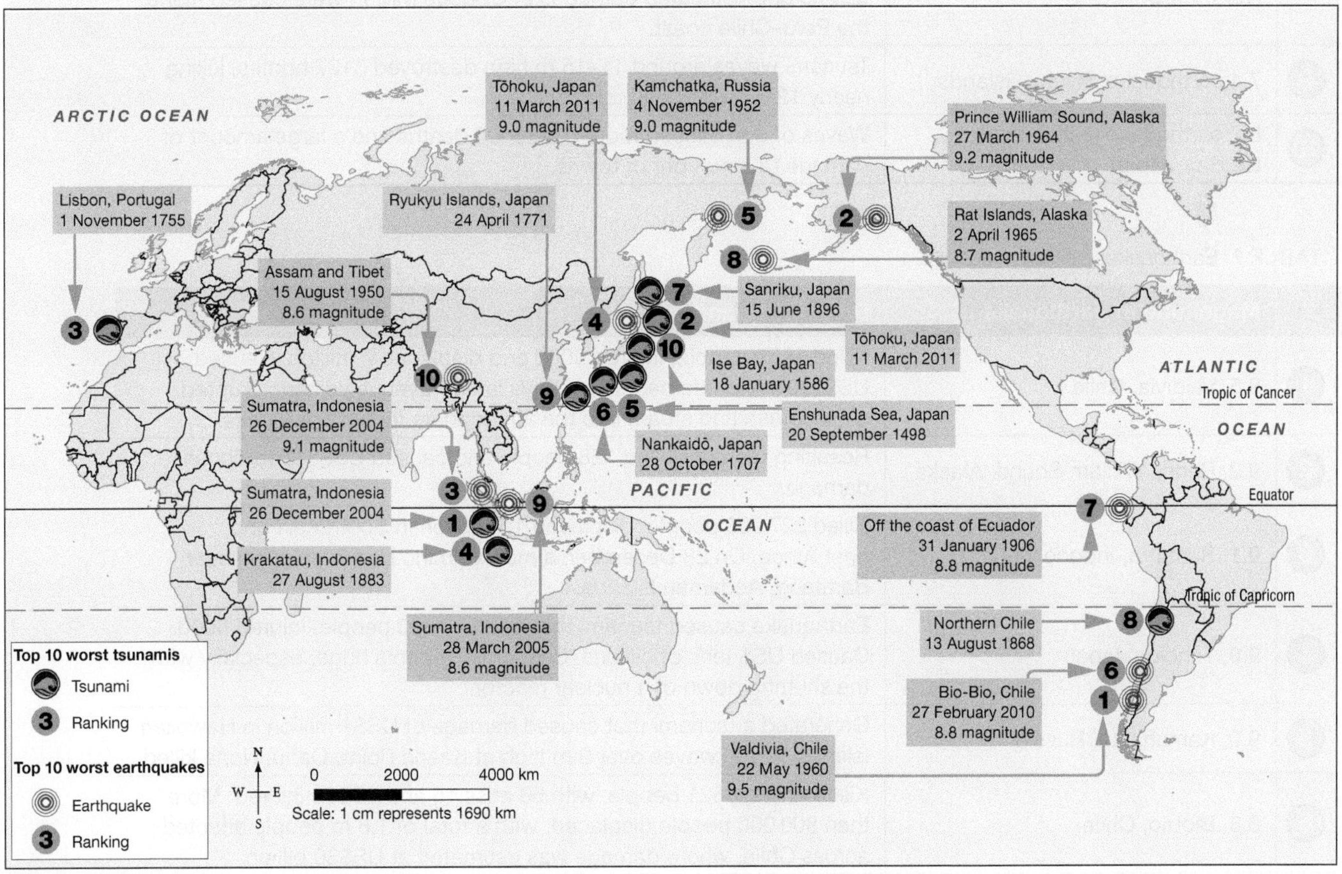

Source: Spatial Vision

TABLE 1 Tsunamis shown in **FIGURE 1**

	Cause of tsunami	Description and impact
1	9.1 earthquake, Sumatra, Indonesia	Tsunami 50 m high, reaching 5 km inland near Meubolah. 230 000 people died. Estimated damages of US$10 billion.
2	9.0 earthquake, Tōhoku, Japan	Waves of 10 m swept over the east coast of Japan. 19 000 people died. Caused nuclear emergency at Fukushima Daiichi nuclear power plant. $235 billion damages.
3	8.5 earthquake, Lisbon, Portugal	Waves up to 30 m high struck towns along western Portugal and southern Spain. Earthquake and tsunami killed 60 000 in Portugal, Morocco and Spain.
4	Volcano, Krakatau, Indonesia	Tsunami linked to the explosion of the Krakatau volcano. Waves as high as 37 m demolished the towns of Anjer and Merak. Killed 40 000 people, with 2000 deaths caused by the volcanic eruptions rather than the tsunami.
5	8.3 earthquake, Enshu-nada Sea, Japan	Homes were flooded and swept away; 31 000 people killed.
6	8.4 earthquake, Nankaidō, Japan	Waves up to 25 m high struck the Pacific coasts of Kyushyu, Shikoku and Honshin. Nearly 30 000 buildings were damaged in the affected regions and about 30 000 people were killed.
7	7.6 earthquake (estimated), Sanriku, Japan	Tsunami was reported to have reached a height of 38.2 m, causing damage to more than 11 000 homes and killing around 22 000 people. Reports were also found of a corresponding tsunami hitting the east coast of China, killing around 4000 people and doing extensive damage to local crops.
8	Two 8.5 earthquakes, Northern Chile	Waves up to 21 m high affected the entire Pacific Rim for two or three days. Tsunami registered as far away as Sydney, Australia. 25 000 deaths and estimated damages of US$300 million were caused along the Peru–Chile coast.
9	7.4 earthquake, Ryukyu Islands	Tsunami waves around 11–15 m high destroyed 3137 homes, killing nearly 12 000 people in total.
10	8.2 earthquake (estimated), Ise Bay, Japan	Waves of 6 m caused more than 8000 deaths and a large amount of damage to a number of towns.

TABLE 2 Earthquakes shown in **FIGURE 1**

	Magnitude of earthquake	Description and impact
1	9.5, Valdivia, Chile	Killed 1655 people, injured 3000 and displaced 2 million. Caused US$550 million damage. Two days later, Puyehue volcano erupted, sending ash and steam into the atmosphere for several weeks.
2	9.2, Prince William Sound, Alaska	Resulting tsunami killed 128 people and caused US$311 million in damage.
3	9.1, Sumatra, Indonesia	Killed 227 900 people, displaced 1.7 million in South Asia and East Africa. On 28 December, a mud volcano began erupting near Baratang, Andaman Islands.
4	9.0, Tōhoku, Japan	Earthquake caused tsunami that killed 19 000 people, injured 6000. Caused US$ tens of billions. Economic impacts huge, especially with the shutting down of a nuclear reactor.
5	9.0, Kamchatka, Russia	Generated a tsunami that caused damage of US$1 million in Hawaiian Islands. Some waves over 9 m high at Kaena Point, Oahu. None killed.
6	8.8, Biobío, Chile	Killed at least 521 people, with 56 missing and 12 000 injured. More than 800 000 people displaced, with a total of 1.8 m people affected across Chile, where damage was estimated at US$30 billion.

(continued)

	Magnitude of earthquake	Description and impact
7	8.8, off the coast of Ecuador	Earthquake caused tsunami reported to have killed between 500 and 1500 people in Ecuador and Colombia.
8	8.7, Rat Islands, Alaska	Generated a tsunami about 10 m high that caused damage on Shemya Island, plus US$10 000 in property damage from flooding on Amchitka Island. No deaths or injuries reported.
9	8.6, Sumatra, Indonesia	Killed 1313, with more than 400 people injured by the tsunami as far away as Sri Lanka.
10	8.6, Assam and Tibet	This inland earthquake caused widespread damages to buildings as well as large landslides. 780 people were killed in eastern Tibet.

Less developed countries often do not have the resources to prepare adequately for an earthquake. Often, many people are housed in badly constructed buildings in densely populated areas on poor land which does not withstand an earthquake or tsunami event. When a disaster strikes, poorer countries often do not have the resources to act quickly and get help for relief efforts.

Developed countries have strict building codes and better infrastructure to withstand disasters. They have warning systems and better communication. Usually, help is quick to arrive, with army and police personnel sent in to help with rescue efforts.

Studies have shown that a higher proportion of people will die when a disaster strikes a low-income country or region, in comparison with the number who will die from a similar size and type of disaster in a similarly populated high- or middle-income region. This is partly because low-income countries are not as well prepared or as well equipped to manage after the disaster has occurred.

FIGURE 2 Landslide caused by an earthquake, June 2008, Honshu, Japan.

5.8.3 Impact on the environment

The impact of an earthquake or tsunami on a human environment can be catastrophic. It can damage and destroy entire settlements. **Landslides** can be triggered by earthquakes, permanently changing the landscape.

Liquefaction

Liquefaction occurs when soil suddenly loses strength and, mixed with groundwater, behaves like a liquid. This usually occurs as a result of ground shaking during a large earthquake. The types of soils that can liquefy include loose sands and silts that are below the water table, so all the space between the grains is filled with water. Dry soils above the water table will not liquefy.

Once a soil liquefies, it cannot support the weight of dry soil, roads, concrete floors and buildings above it. The liquefied soil comes to the surface through cracks, and widens them.

landslide a rapid movement of rocks, soil and vegetation down a slope, sometimes caused by an earthquake or by excessive rain

liquefaction transformation of soil into a fluid, which occurs when vibrations created by an earthquake, or water pressure in a soil mass, cause the soil particles to lose contact with one another and become unstable; for this to happen, the spaces between soil particles must be saturated or near saturated

FIGURE 3 Cars swallowed by liquefied soil on a road in Christchurch, New Zealand, 2011

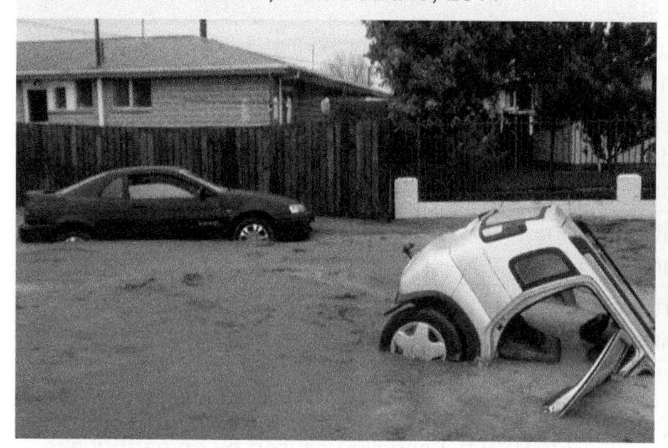

FIGURE 4 Liquefaction brings sand, silt and water to the surface, Kaiapoi, New Zealand, 2010

CASE STUDY

Impact of the 2015 Nepal earthquake

Nearly 9000 people were killed during the 25 April and 12 May earthquakes in Nepal, with more than 17 800 injured. Nearly 400 people were never found. In addition, more than 500 000 houses were destroyed and nearly 270 000 were damaged. Nepal's historic Dharahara Tower collapsed, killing 180 people, and an avalanche at the Mount Everest Base Camp killed 21 people and injured 120. A huge avalanche also occurred in the Langtang Valley, where all the homes were destroyed and 250 people were reported missing. Hundreds of thousands of people were made homeless after buildings were destroyed or seriously damaged.

A further 78 people were killed and more than 500 injured in India. In China 25 people were killed and more than 380 people injured, and 2500 homes were destroyed and 24 700 damaged. The earthquake occurred during working hours so many people were outdoors. Had it occurred at night, with more people at home, the number of dead and injured would have been higher.

FIGURE 5 (a) and (b) Before and after images of Dunbar Square, in Nepal

(a)

(b)

The economic costs of such an event are also huge — damage costs are estimated to be between US$4 and $5 billion as a result of the earthquake and aftershocks. This is disastrous for a very poor country like Nepal.

The region around Kathmandu is known as one of the most dangerous places in the world in terms of earthquake risk. Apart from earthquakes, other geophysical hazards that occur in Nepal include landslides, avalanches and flash flooding.

In addition to its location, Nepal is extremely vulnerable because of its poverty. This means Nepal has poor building standards (many of the buildings were quickly reduced to rubble) and inadequate public health emergency response and infrastructure systems resources to support its people in times of crisis, when compared with wealthier nations. Without this support, clean water, safe food and effective disposal of sewerage cannot be guaranteed.

FIGURE 6 Compared to buildings in some wealthier countries, such as Japan and the United States, very few buildings in Nepal are earthquake-proof.

 Resources

eWorkbook	Impacts of earthquakes and tsunamis (ewbk-8784)
Video eLesson	Impacts of earthquakes and tsunamis — Key concepts (eles-5061)
Interactivity	Map of the 10 biggest earthquakes and tsunamis in recorded history (int-8728)

5.8 ACTIVITIES

1. Research how Japan has recovered from the 2011 tsunami and how Nepal has recovered from the 2015 earthquake. How can you account for any differences in the recovery process?
2. Use the **Liquefaction** weblink in the online Resources to view a video of liquefaction occurring. Write a paragraph describing what liquefaction is and why it occurs.
3. In groups, research the liquefaction that occurred in the Sulawesi earthquake in 2018. Discuss whether the earthquake, tsunami or liquefaction was the most dangerous part of this event.
4. Use the **Earthquakes — vulnerable cities** weblink in the online Resources to read more about cities that are most at risk from earthquakes. Use an atlas or online map to locate these cities. Where are they located in relation to plate boundaries and, in particular, to the Pacific Ring of Fire?
5. Earthquake engineers often say earthquakes don't kill people, collapsing buildings do. Discuss this statement in relation to poor and rich countries. What role should people in rich countries play in helping those in poor countries at risk of these events?
6. Research the location of Kamchatka, Russia. Why do you think a major earthquake in this place did not cause any deaths?

5.8 EXERCISE

Learning pathways

■ LEVEL 1	■ LEVEL 2	■ LEVEL 3
2, 5, 6, 7, 15	1, 4, 8, 10, 13	3, 9, 11, 12, 14

Check your understanding

1. Rewrite the following statement in your own words: 'When one compares the number of people killed per disaster against the income of the country, one finds a steep rise in mortality with decreasing income'.
2. List two impacts that earthquakes and tsunamis can have within the categories of social, environmental and economic impacts.
3. List the factors that combine to cause an earthquake or tsunami to turn into a disaster.
4. Describe what happens when *liquefaction* occurs.
5. Using **FIGURE 1** and **TABLES 1** and **2**, identify the place and time of the following:
   a. The highest tsunami waves shown.
   b. The highest magnitude earthquake shown.
   c. The tsunami shown that occurred closest to Australia.
   d. The event shown that caused the greatest number of deaths.

Apply your understanding

6. What is a landslide?
7. Geophysicists and other experts warned for decades that Nepal was vulnerable to a deadly earthquake. Explain why Nepal was not better prepared for this event.
8. How does poverty in Nepal increase vulnerability to disasters?
9. What is the relationship between income and disaster risk? Why is the risk of earthquakes and tsunamis higher in poor countries?
10. Study **FIGURE 1** and consider what you have learnt about tectonic plates. What is the interconnection between the distribution of earthquakes and tsunamis and the distribution of tectonic plates?
11. Explain one way in which technology can help to reduce the numbers of people killed in a disaster.
12. Explain how a small magnitude earthquake can have a devastating impact.

Challenge your understanding

13. Why might Japan experience so many destructive earthquakes and tsunamis?
14. Suggest why some of the earthquakes and tsunamis listed in **TABLES 1** and **2** might have been larger, but caused fewer deaths.
15. Consider **FIGURES 3** and **7**. How might the authorities in New Zealand have removed the cars from the liquefied soil?

FIGURE 7 A classic car caught by liquefaction, Christchurch, New Zealand, 2011

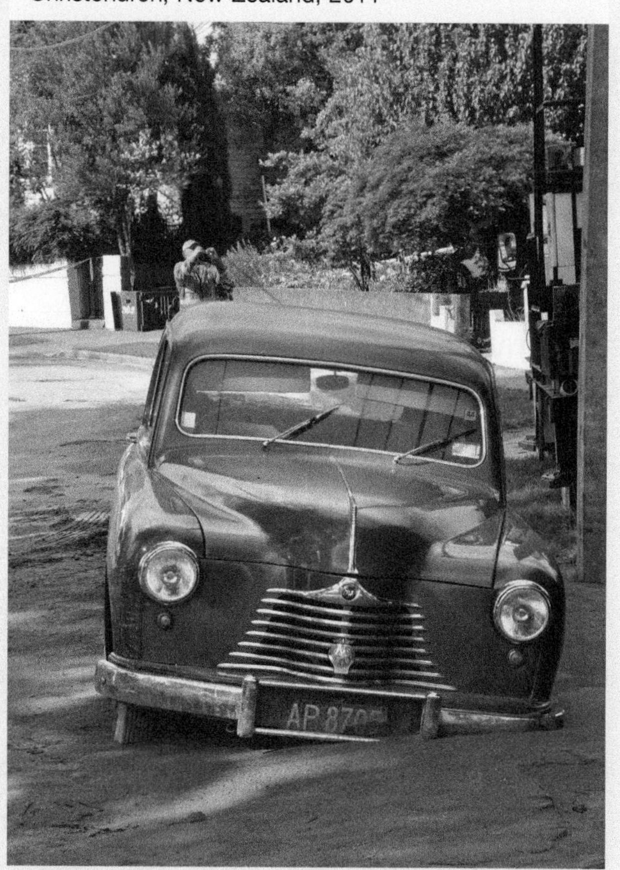

To answer questions online and to receive **immediate feedback** and **sample responses** for every question, go to your learnON title at www.jacplus.com.au.

5.9 SkillBuilder — Drawing simple cross-sections

online only

LEARNING INTENTION

By the end of this subtopic, you will be able to draw a cross-section of a landscape from a map.

The content in this subtopic is a summary of what you will find in the online resource.

5.9.1 Tell me

What are cross-sections?

A cross-section is a side-on, or cut-away, view of the land as if it had been sliced through by a knife. It is like taking a vertical slice of the landscape and looking at it side-on.

FIGURE 3 Marking up the paper edge where each contour touches the page

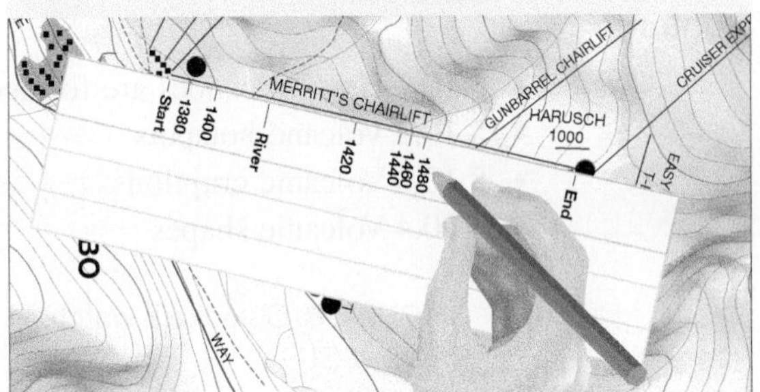

5.9.2 Show me

eles-1655

Step 1

Determine the two points between which you want to create a cross-section. Place the straight edge of a piece of paper between the two points. Mark the two points on the paper. Mark where each contour line touches the edge and write beside the mark the height of the contour line.

Step 2

On another sheet of paper, use your ruler to draw an axis onto which to transfer your markings. The horizontal (base) line should be as long as your cross-section from 'start' to 'end'. The vertical scale needs to give a realistic impression of the slopes and landforms.

Step 3

Place the marked edge of the paper along the base axis. At each contour marking, find the appropriate height according to the vertical scale and put a small dot directly above the contour marked on the edge of the paper. Join the dots with a smooth line to show the slope of the land.

Step 4

Complete the cross-section with the geographic conventions of a title and labelling of the axis. Shade the area below the line of your cross-section.

5.9.3 Let me do it

int-3151

Go to learnON to access the following additional resources to help you build this skill:

- a longer explanation of this skill and its application in Geography (Tell me)
- a video showing the step-by-step process of this skill (Show me)
- an activity and interactivity for you to practise this skill (Let me do it)
- self-marking questions to help you understand and use this skill.

on Resources

eWorkbook SkillBuilder — Drawing simple cross-sections (ewbk-8792)

Video eLesson SkillBuilder — Drawing simple cross-sections (eles-1655)

Interactivity SkillBuilder — Drawing simple cross-sections (int-3151)

5.10 Volcanic mountains

LEARNING INTENTION

By the end of this subtopic, you will be able to explain the processes that create volcanic mountains and describe the features of a volcano.

The content in this subtopic is a summary of what you will find in the online resource.

Volcanic mountains are formed when magma pushes its way to the Earth's surface and then erupts as lava, ash, rocks and volcanic gases. These erupting materials build up around the vent through which they erupted. A volcanic eruption can be slow or spectacularly fast, and can result in a number of different displays.

To learn more about volcanoes, go to your learnON resources at www.jacPLUS.com.au.

Contents

- 5.10.1 How volcanoes are formed
- 5.10.2 Volcano hotspots
- 5.10.3 Volcanic eruptions
- 5.10.4 Volcanic shapes

FIGURE 8 Onlookers watch an eruption of Mt Fagradalsfjall, Iceland, in April 2021

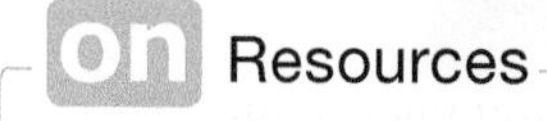

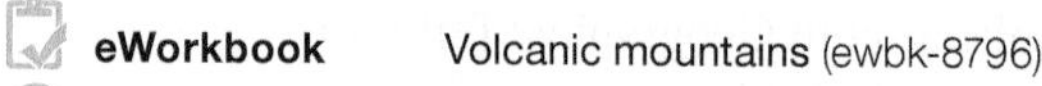

eWorkbook	Volcanic mountains (ewbk-8796)
Video eLesson	Volcanic mountains — Key concepts (eles-5062)
Interactivities	The Great Rift Valley, Africa (int-7850) Four volcanic landforms (int-7851) The anatomy of a volcano (int-3620)
Google Earth	Iceland (gogl-0057) Great Rift Valley, Africa (gogl-0055)
myWorldAtlas	Deepen your understanding of this topic with related case studies and questions > **Hawaii's hotspot** Deepen your understanding of this topic with related case studies and questions > **Lahars**

5.11 Impacts of volcanoes

online only

LEARNING INTENTION

By the end of this subtopic, you will be able to identify the impacts of volcanoes and explain how people respond to volcanic eruptions.

The content in this subtopic is a summary of what you will find in the online resource.

Volcanic eruptions both create and destroy landscapes. How can the worst volcanoes be measured? Should it be based on the number of people killed or the cost of the damage and destruction? Or should it be based on the size of the explosion?

To learn more about the impacts of volcanoes, go to your learnON resources at www.jacPLUS.com.au.

Contents

- 5.11.1 The worst volcanic eruptions
- 5.11.2 Living near volcanoes
- 5.11.3 How to prepare for volcanic eruption

FIGURE 1 Aerial photograph of the Sidoarjo 'mud volcano' in Indonesia, which has been erupting since May 2006

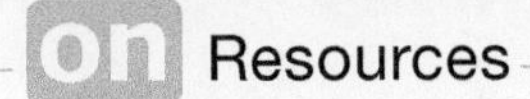

on Resources

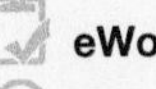

eWorkbook	The impact of volcanic eruptions (ewbk-0745)
Video eLesson	Impacts of volcanoes — Key concepts (eles-5063)
Interactivity	Impacts of volcanoes — Interactivity (int-8407)
myWorldAtlas	Deepen your understanding of this topic with related case studies and questions > **Mount Vesuvius**

5.12 Investigating topographic maps — Mount Taranaki

LEARNING INTENTION

By the end of this subtopic, you will be able to describe the landscape around a volcano using data from a topographic map.

New Zealand's Mount Taranaki is named after the Maori terms *tara* meaning 'mountain peak' and *ngaki* meaning 'shining' (because the mountain is covered with snow in winter).

Mount Taranaki is 2518 metres high and is the largest volcano on New Zealand's mainland. It is located in the south-west of the North Island (see **FIGURE 1**).

Mount Taranaki was formed 135 000 years ago by subduction of the Pacific Plate below the Australian Plate. It is a stratovolcano — a conical volcano consisting of layers of pumice, lava, ash and tephra. Mount Taranaki is symmetrical, looking the same on both sides of a central point. It is the only active volcano in a chain in this region. The other volcanoes were once very large but have been eroded over time.

The summit of Mount Taranaki is a lava dome in the middle of a crater that is filled with ice and snow. The mountain is considered likely to erupt again. There are significant potential hazards from lahars, avalanches and floods. A circular plain of volcanic material surrounding the mountain was formed from lahars (see **FIGURE 3**) and landslides. Some of these flows reached the coast in the past. The volcano's lower flanks are covered in forest, and are part of a national park. There is a clear line between the park boundary and surrounding farmland.

FIGURE 1 Location of Mount Taranaki on the North Island of New Zealand

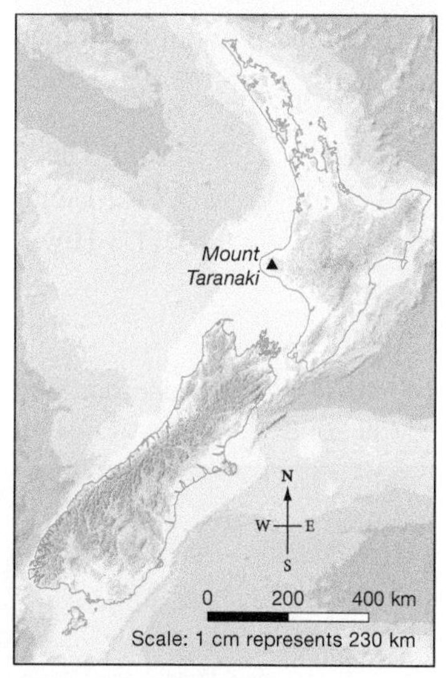

Source: Spatial Vision

Use the **Mt Taranaki live** weblink in the Resources tab to view a live webcam of Mount Taranaki.

FIGURE 2 Mount Taranaki — a near-perfect cone

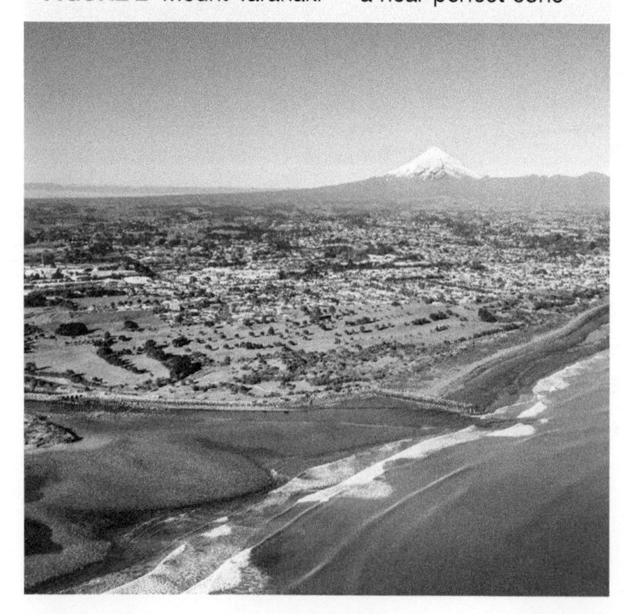

FIGURE 3 Satellite photo of Mount Taranaki

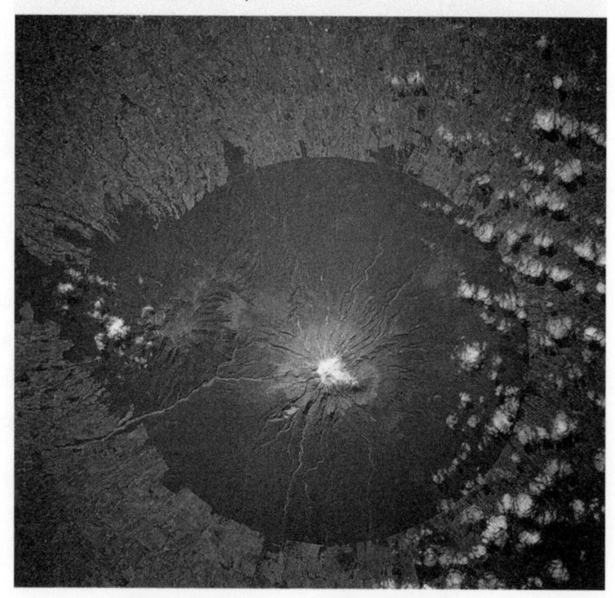

FIGURE 4 Topographic map of Mount Taranaki

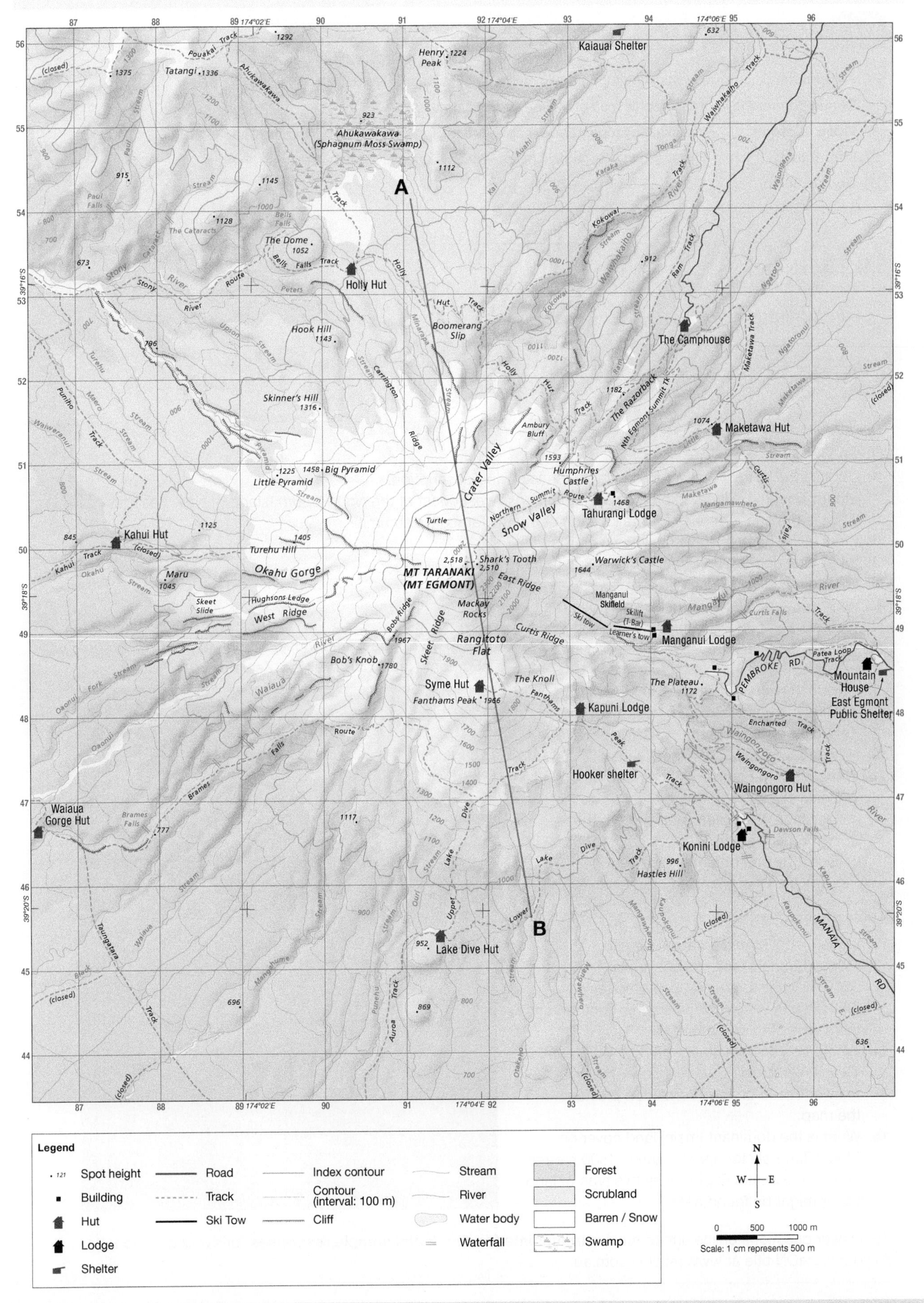

on Resources

eWorkbook	Investigating topographic maps—Mount Taranaki (ewbk-10403)
Digital document	Topographic map of Mount Taranaki (doc-36253)
Video eLesson	Investigating topographic maps—Mount Taranaki — Key concepts (eles-5064)
Interactivity	Investigating topographic maps—Mount Taranaki (int-8408)
Google Earth	Mount Taranaki (gogl-0134)

5.12 EXERCISE

Learning pathways

■ LEVEL 1	■ LEVEL 2	■ LEVEL 3
1, 2, 8, 10, 12	3, 4, 5, 7, 15	6, 9, 11, 13, 14

Check your understanding

1. Choose the correct answer to complete the following description.
   Mount Taranaki is located on the *north/south/east/west* coast of the *North/South* Island of New Zealand.
2. What is the grid reference for the location of Mount Taranaki's summit?
3. What is the highest elevation shown on the map?
4. What direction is Ouri Stream flowing?
5. What direction is Waiaua flowing?
6. What is the difference in height between Big Pyramid and Little Pyramid?
7. What is the distance between Big and Little Pyramid?
7. How many huts can you see on the map?
9. What is the distance of the track between Kapuni Lodge and Syme Hut?
10. How did Mount Taranaki get its name?
11. What might the consequences be if Mount Taranaki erupted?
    a. Describe one potential change to the human environment.
    b. Describe one potential change to the natural environment.

FIGURE 5 The Camphouse, Mount Taranaki, Egmont National Park, New Zealand

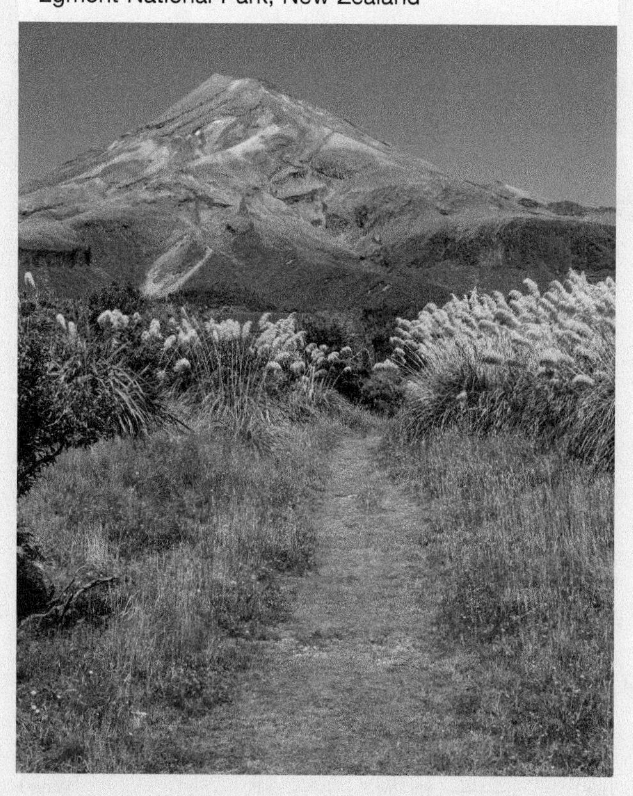

Apply your understanding

12. Explain how Mount Taranaki formed.
13. Mount Taranaki receives between 3200 millimetres and 6400 millimetres of rainfall each year. How would this contribute to the shape of this landform?
14. Where do you think lava would flow if Mount Taranaki erupted? Give reasons for your answer based on the information provided in the map.
15. What is the dominant (main) land cover on Mount Taranaki for approximately 1500 metres around the peak? Explain why this type of land cover might be found here.

To answer questions online and to receive **immediate feedback** and **sample responses** for every question, go to your learnON title at www.jacplus.com.au.

5.13 Thinking Big research project — Earthquakes feature article

online only

The content in this subtopic is a summary of what you will find in the online resource.

Scenario

Congratulations! You have been promoted to feature writer of the *Weekly Rattle*, a leading geographical magazine. Your readers are very keen geographers and earth scientists who like to have detailed and up-to-date information about earthquakes.

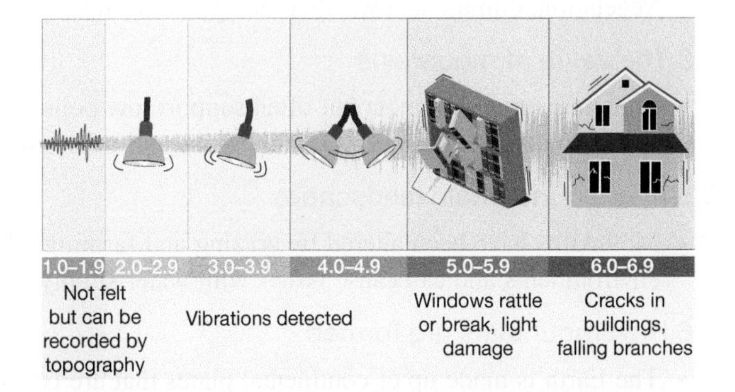

Task

Your first assignment is to write a feature article about the strongest earthquakes that occur in the world over a one-week period.

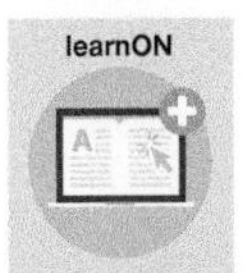

Go to your Resources tab to access all of the resources you need to complete this project.

on Resources

ProjectsPLUS Thinking Big research project — Earthquakes feature article (pro-0171)

5.14 Review

5.14.1 Key knowledge summary

5.2 Mountain ranges of the world

- Mountains are found on every continent on Earth. There are major chains of mountains — mountain ranges — on all continents.
- Vegetation, climate and weather change as altitude in mountains increases.

5.3 The value of mountains

- Mountains can be remote but often support low population densities. Specific mountain landforms are special places to indigenous communities and other groups of people.

5.5 Altering mountain landscapes

- Mountains have been altered by grazing and farming. These alterations have impacted mountain environments and can cause issues with water supply.

5.6 How mountains are formed

- The Earth is made up of continental plates that are constantly moving slowly. Some plates converge, others diverge and others slide past one another. This tectonic activity (moving plates) is a process for forming mountains.
- Mountains are classified by what they look like and how they were formed. The most common formations are fold mountains. Other mountain formations include fault-block, dome and plateau.

5.7 Geomorphic processes that cause earthquakes and tsunamis

- Earthquakes are a common occurrence each day across the Earth.
- There is a strong relationship between the location of plate boundaries (weaknesses in the Earth's crust) and the location of earthquakes.
- A tsunami can result if a large earthquake occurs on the ocean floor.

5.8 Impacts of earthquakes and tsunamis

- Earthquakes and tsunamis can affect people and result in deaths, injuries and damage to homes and infrastructure. The impact of the same magnitude earthquake can vary depending on a country's level of income.
- The environment can be affected through landslides, erosion and liquefaction.

5.10 Volcanic mountains

- Volcanoes are formed when molten magma in the Earth's mantle is forced through an opening in the Earth's surface.
- Volcanoes can be formed in rift valleys and over hotspots. The shapes and sizes of volcanic landscapes depend on the type of lava, the amount of ash and the speed of the eruption.

5.11 Impacts of volcanoes

- Volcanic eruptions can destroy landscapes and buildings and kill people.
- Large numbers of people across the world live near volcanoes because of the location of fertile soils, ore deposits and geothermal energy.

5.14.2 Key terms

altitude height above sea level
converging plates a tectonic boundary where two plates are moving towards each other
cultural relating to the ideas, customs and social behaviour of a society
divergent plates a tectonic boundary where two plates are moving away from each other and new continental crust is forming from magma that rises to the Earth's surface between the two
epicentre the point on the Earth's surface directly above the focus of an earthquake
fault an area on the Earth's surface that has a fracture, along which the rocks have been displaced
focus the point where the sudden movement of an earthquake begins
geyser a hot spring that erupts, sending hot water and steam into the air
gorge a deep valley between mountains
geothermal energy energy derived from the heat in the Earth's interior
hotspot an area on the Earth's surface where the crust is quite thin, and volcanic activity can sometimes occur, even though it is not at a plate margin
landslide a rapid movement of rocks, soil and vegetation down a slope, sometimes caused by an earthquake or by excessive rain
liquefaction transformation of soil into a fluid, which occurs when vibrations created by an earthquake, or water pressure in a soil mass, cause the soil particles to lose contact with one another and become unstable; for this to happen, the spaces between soil particles must be saturated or near saturated
lithosphere the crust and upper mantle of the Earth
Pangaea the name given to all the landmass of the Earth before it split into Laurasia and Gondwana
population density the number of people within a given area, usually per square kilometre
rift zone a large area of the Earth in which plates of the Earth's crust are moving away from each other, forming an extensive system of fractures and faults
seismic waves waves of energy that travel through the Earth as a result of an earthquake, explosion or volcanic eruption
shifting cultivation process of growing crops in different places (or rotating where crops are grown)
summit the peak at the top of a mountain
tsunami large wave that can be triggered by an earthquake under the ocean
volcanic loam a volcanic soil composed mostly of basalt, which has developed into a crumbly mixture

5.14.3 Reflection

Complete the following to reflect on your learning.

Revisit the inquiry question posed in the Overview:

Magma, water and tectonic plates — can they really move mountains?

1. Now that you have completed this topic, what is your view on the question? Discuss with a partner. Has your learning in this topic changed your view? If so, how?
2. Write a paragraph in response to the inquiry question, outlining your views.

on Resources

eWorkbook Chapter 5 Student learning matrix (ewbk-8452)
Chapter 5 Reflection (ewbk-8453)
Chapter 5 Extended writing task (ewbk-8454)

Interactivity Chapter 5 Crossword (int-7598)

Subtopic	Success criteria	●	●	●
5.2	I can describe where the world's main mountain ranges are.			
	I can explain the impact that altitude has on mountain environments.			
5.3	I can explain the different ways that mountains are valued.			
5.4	I can interpret an aerial photo.			
5.5	I can identify the impact of human activities on mountain environments.			
5.6	I can explain how mountains form.			
	I can describe the different mountain types and identify where some of these mountains are located.			
5.7	I can explain how earthquakes and tsunamis occur.			
5.8	I can describe the impacts of earthquakes and tsunamis.			
	I can explain how people respond to earthquakes and tsunamis.			
5.9	I can draw a cross-section of a landscape.			
5.10	I can explain how volcanic mountains are formed.			
	I can describe the features of a volcano.			
5.11	I can describe the impacts of volcanoes.			
	I can explain how people respond to volcanoes.			
5.12	I can describe the landscape around a volcano using data from a topographic map.			

ONLINE RESOURCES

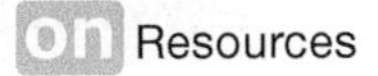

Below is a full list of **rich resources** available online for this topic. These resources are designed to bring ideas to life, to promote deep and lasting learning and to support the different learning needs of each individual.

eWorkbook

- 5.1 Chapter 5 eWorkbook (ewbk-7985) ☐
- 5.2 Mountain ranges of the world (ewbk-8764) ☐
- 5.3 The value of mountains (ewbk-8768) ☐
- 5.4 SkillBuilder — Interpreting an aerial photograph (ewbk-8788) ☐
- 5.5 Altering mountain landscapes (ewbk-8772) ☐
- 5.6 How mountains are formed (ewbk-8776) ☐
- 5.7 Geomorphic processes that cause earthquakes and tsunamis (ewbk-8780) ☐
- 5.8 Impacts of earthquakes and tsunamis (ewbk-8784) ☐
- 5.9 SkillBuilder — Drawing simple cross-sections (ewbk-8792) ☐
- 5.10 Volcanic mountains (ewbk-8796) ☐
- 5.11 The impact of volcanic eruptions (ewbk-0745) ☐
- 5.12 Investigating topographic maps — Mount Taranaki (ewbk-10403) ☐
- 5.14 Chapter 5 Student learning matrix (ewbk-8452) ☐
  - Chapter 5 Reflection (ewbk-8453) ☐
  - Chapter 5 Extended writing task (ewbk-8454) ☐

Sample responses

- 5.1 Chapter 5 Sample responses (sar-0140) ☐

Digital document

- 5.12 Topographic map of Mount Taranaki (doc-36253) ☐

Video eLessons

- 5.1 Majestic mountains (eles-1626) ☐
  - Mountain landscapes — Photo essay (5335) ☐
- 5.2 Mountain ranges of the world — Key concepts (eles-5057) ☐
- 5.3 The value of mountains — Key concepts (eles-5058) ☐
- 5.4 SkillBuilder — Interpreting an aerial photograph (eles-1654) ☐
- 5.5 Altering mountain landscapes — Key concepts (eles-5059) ☐
- 5.6 Drifting continents (eles-0129) ☐
- 5.7 Geomorphic processes that cause earthquakes and tsunamis — Key concepts (eles-5060) ☐
- 5.8 Impacts of earthquakes and tsunamis — Key concepts (eles-5061) ☐
- 5.9 SkillBuilder — Drawing simple cross-sections (eles-1655) ☐
- 5.10 Volcanic mountains — Key concepts (eles-5062) ☐
- 5.11 Impacts of volcanoes — Key concepts (eles-5063) ☐
- 5.12 Investigating topographic maps — Mount Taranaki — Key concepts (eles-5064) ☐

Interactivities

- 5.2 Mountain ranges of the world (int-7844) ☐
- 5.3 The value of mountains (int-8405) ☐
- 5.4 SkillBuilder — Interpreting an aerial photograph (int-3150) ☐
- 5.5 Altering mountain landscapes (int-8417) ☐
- 5.6 Mountain builders (int-3109) ☐
  - Grand peaks (int-3110) ☐
- 5.7 Anatomy of a tsunami (int-3111) ☐
  - Magnitudes of earthquake and aftershocks in Nepal, 2015 (int-7846) ☐
  - The location and magnitude of the earthquake that caused the Japanese tsunami (int-7847) ☐
  - The shake intensity and the tectonic plate boundary involved in the Nepal earthquake (int-7845) ☐
  - Map of ground motion and shaking intensity from the earthquake across Japan (int-7848) ☐
- 5.8 Map of the 10 biggest earthquakes and tsunamis in recorded history (int-8728) ☐
- 5.9 SkillBuilder — Drawing simple cross-sections (int-3151) ☐
- 5.10 The Great Rift Valley, Africa (int-7850) ☐
  - Four volcanic landforms (int-7851) ☐
  - The anatomy of a volcano (int-3620) ☐
- 5.11 Impacts of volcanoes (int-8407) ☐
- 5.12 Investigating topographic maps — Mount Taranaki (int-8408) ☐
- 5.14 Chapter 5 Crossword (int-7598) ☐

Fieldwork

- 5.2 Recording temperature data (fdw-0004) ☐
- 5.3 Using clinometers to measure slopes (fdw-0005) ☐

ProjectsPLUS

- 5.13 Thinking Big research project — Earthquakes feature article (pro-0171)

Weblinks

- 5.4 Climate change and water storage (web-4810) ☐
- 5.5 Anticline and syncline (web-0092) ☐
  - Fold mountains (web-1085) ☐
- 5.6 World's biggest tsunami (web-0083) ☐
  - P- and S-waves (web-0087) ☐
  - Sendai tsunami (web-1087) ☐
  - Nepal earthquake — before and after photos (web-4078) ☐
- 5.7 Earthquakes — vulnerable cities (web-4079) ☐
  - Liquefaction (web-4080) ☐
- 5.10 Hawaii's hotspot (web-4081) ☐
  - Iceland webcams (web-6035) ☐
- 5.11 Timeline (web-4072) ☐
- 5.12 Mt Taranaki live (web-4082) ☐

Google Earth

- 5.6 Sierra Nevada mountains (gogl-0058) ☐
  - Table Mountain (gogl-0059) ☐
  - Grand Canyon (gogl-0002) ☐
  - Ben Nevis (gogl-0049) ☐
- 5.10 Iceland (gogl-0057) ☐
  - Great Rift Valley, Africa (gogl-0055) ☐
- 5.12 Mount Taranaki (gogl-0134) ☐

myWorld Atlas

- 5.2 Himalayas (mwa-4493) ☐
- Mountain environments (mwa-4492) ☐
- 5.6 Active Earth (mwa-4475) ☐
- 5.7 Haiti earthquake (mwa-4478) ☐
- Banda Aceh tsunami (mwa-4479) ☐
- 5.10 Hawaii' s hotspot (mwa-4477) ☐
- Lahars (mwa-7334) ☐
- 5.11 Mount Vesuvius (mwa-4476) ☐

Teacher resources

There are many resources available exclusively for teachers online.

6 Rainforest landscapes

INQUIRY SEQUENCE

To access a pre-test with **immediate feedback** and **sample responses** to every question in this chapter, select your learnON format at www.jacplus.com.au.

6.1 Overview

Numerous **videos** and **interactivities** are embedded just where you need them, at the point of learning, in your learnON title at www.jacplus.com.au. They will help you to learn the content and concepts covered in this topic.

We can plant new trees any time and anywhere. What makes the world's rainforests so special?

6.1.1 Introduction

What do you know about rainforests? Did you know they have the greatest biodiversity of any forest environment? They contain complex layers that support thousands of species of plants and animals. The rainforest has supplied resources to all people, including indigenous communities. People are concerned that clearing large areas of rainforest is creating negative impacts that are unsustainable.

STARTER QUESTIONS

1. Create a mind map of all the words that come to your mind when you hear the word *rainforest*. Use categories such as *animals, plants, colours* or *structure.*
2. Watch the **Rainforest — Photo essay**. Using magazines, newspapers, the internet and other resources, create a collage that represents a rainforest to you. What emotions does this image give you?
3. Refer to **FIGURE 1** and the chapter opener image. Why do we call rainforests a 'green landscape'?
4. Try to list things in your home that might originally have come from a rainforest. Have they changed from their original form? How do you think they got here?
5. Think of your favourite place. How would you feel if that place was changed or destroyed and you could not visit or use it in the ways you used to? What could you do about this?

FIGURE 1 Tortuguero National Park, in Costa Rica, Central America

on Resources

eWorkbook Chapter 6 eWorkbook (ewbk-7986)

Video eLesson Protecting our landscapes — Rainforest (eles-1627)
Rainforest – Photo essay (eles-5336)

6.2 Rainforest characteristics

LEARNING INTENTION

By the end of this subtopic, you will be able to describe the different types of rainforest and where they are located, and explain the characteristics of rainforests.

6.2.1 Introducing rainforests

Forests that grow in constantly wet conditions are called rainforests. A rainforest is an example of a biome (a community of plants and animals spread over a large natural area). Rainforests are located wherever the annual rainfall is more than 1300 millimetres and is evenly spread throughout the year. While tropical rainforests are the best known, there are also other types of rainforest.

Rainfall

Rainforests thrive in the hot and wet conditions experienced in the **equatorial** regions of the world. Most of the time the rainfall in these regions is extremely heavy, caused by thunderstorms. These heavy downpours, called **convectional rainfall**, result in very hot and humid conditions.

equatorial near the equator, the line of latitude around the Earth that creates the boundary between the northern and southern hemispheres

convectional rainfall heavy rainfall as a result of thunderstorms, caused by rapid evaporation of water by the Sun's rays

int-8736

FIGURE 1 World rainforest types

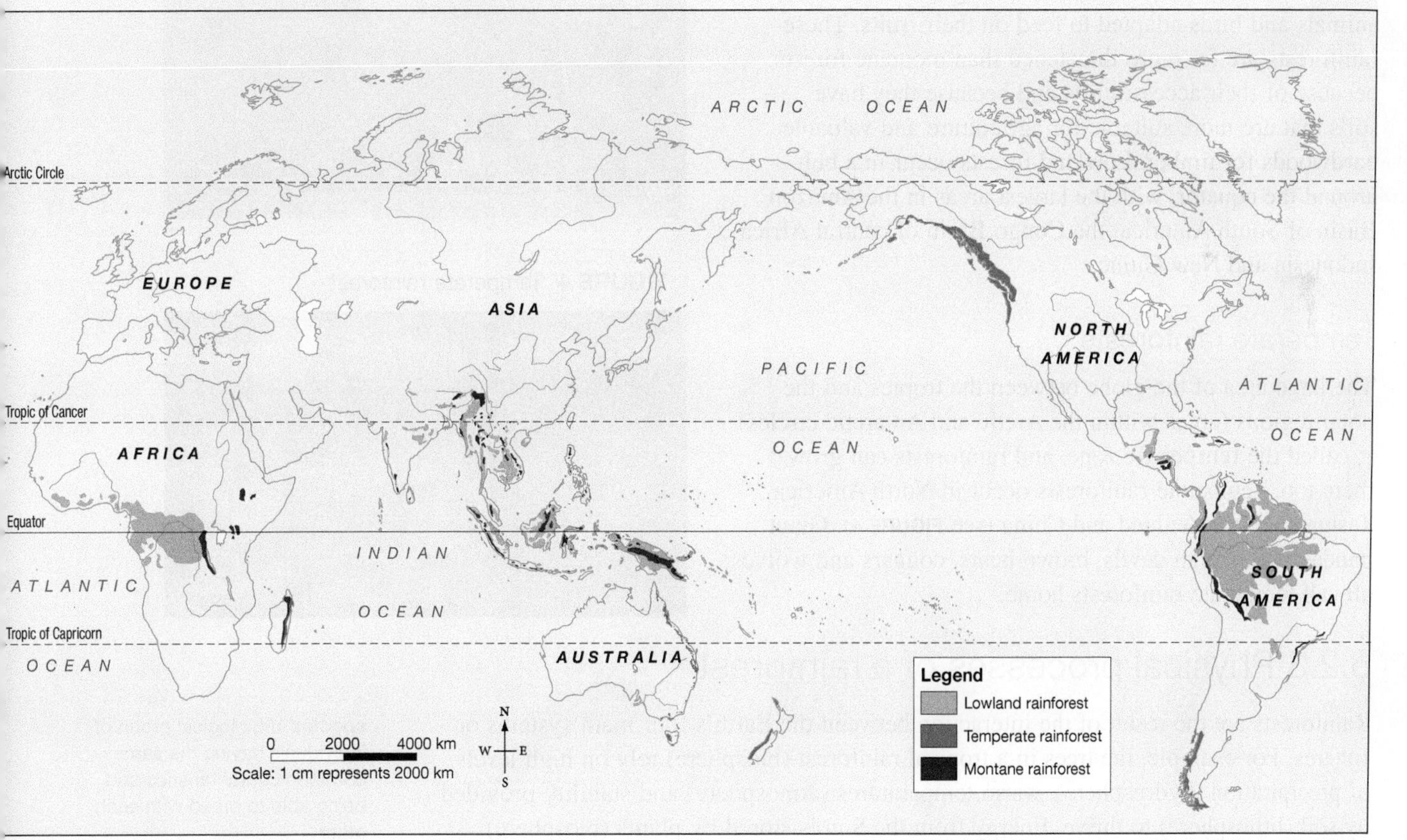

Source: MAPgraphics Pty Ltd Brisbane

6.2.2 Types of rainforests

Tropical rainforests

Tropical rainforests are located around the equator (see **FIGURE 1**) where the temperatures are warm all year round and there is a lot of rainfall. Plants flourish in humid, moist tropical rainforests, which support a huge number of plants and animals — perhaps as many as 90 per cent of all known **species**. Poison-dart frogs, birds of paradise, piranha, tarantulas, anacondas, Komodo dragons and vampire bats are all found in tropical rainforests. Tropical rainforests can occur at different altitudes; they are called montane rainforests and lowland tropical forests.

FIGURE 2 Montane rainforest

Montane rainforests

Tropical rainforests that occur in the mountains, 1000 metres or more above sea level, are called montane rainforests. Montane rainforests have shorter trees than lowland tropical forests and are almost always covered by clouds, making the leaves drip with moisture (see **FIGURE 2**).

FIGURE 3 Lowland rainforest

Lowland tropical forest

Lowland tropical rainforests form the majority of the world's tropical rainforests (see **FIGURE 3**). They grow at elevations generally below 1000 metres. Trees in lowland forests have a great diversity of fruiting trees, attracting animals and birds adapted to feed on their fruits. These rainforests are far more threatened than montane forests because of their accessibility, and because they have soils that are more suitable for agriculture and valuable hardwoods for timber. Lowland forests occur in a belt around the equator, with the largest areas in the Amazon Basin of South America, the Congo Basin of central Africa, Indonesia and New Guinea.

FIGURE 4 Temperate rainforest

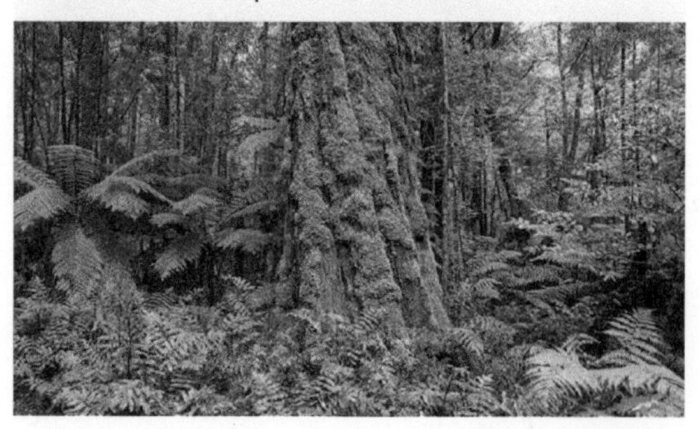

Temperate rainforests

The large area of the globe between the tropics and the polar regions (areas within the Arctic and Antarctic circles) is called the **temperate** zone, and rainforests can grow there too. Temperate rainforests occur in North America, Tasmania, New Zealand and China (see **FIGURE 4**). Giant pandas, Tasmanian devils, brown bears, cougars and wolves all call temperate rainforests home.

6.2.3 Physical processes of a rainforest

Rainforests are the result of the interaction between the Earth's four main systems or spheres. For example, the trees in a tropical rainforest (biosphere) rely on high levels of precipitation (hydrosphere), warm temperatures (atmosphere) and stability provided by soil (lithosphere) to thrive. Energy from the Sun is stored by plants (biosphere). When humans or animals (biosphere) eat the plants, they acquire the energy originally captured by the plants.

species a biological group of individuals having the same common characteristics and being able to breed with each other

temperate describes the relatively mild climate experienced in the zones between the tropics and the polar circles

6.2.4 Rainforest characteristics

Rainforests are unique **ecosystems** consisting of four layers — the emergent, canopy and understorey layers and the forest floor. Each layer can be identified by its distinct characteristics. Rainforests are actually a community of plants and animals working together to survive, linked in a food web.

FIGURE 5 Layers in a tropical rainforest

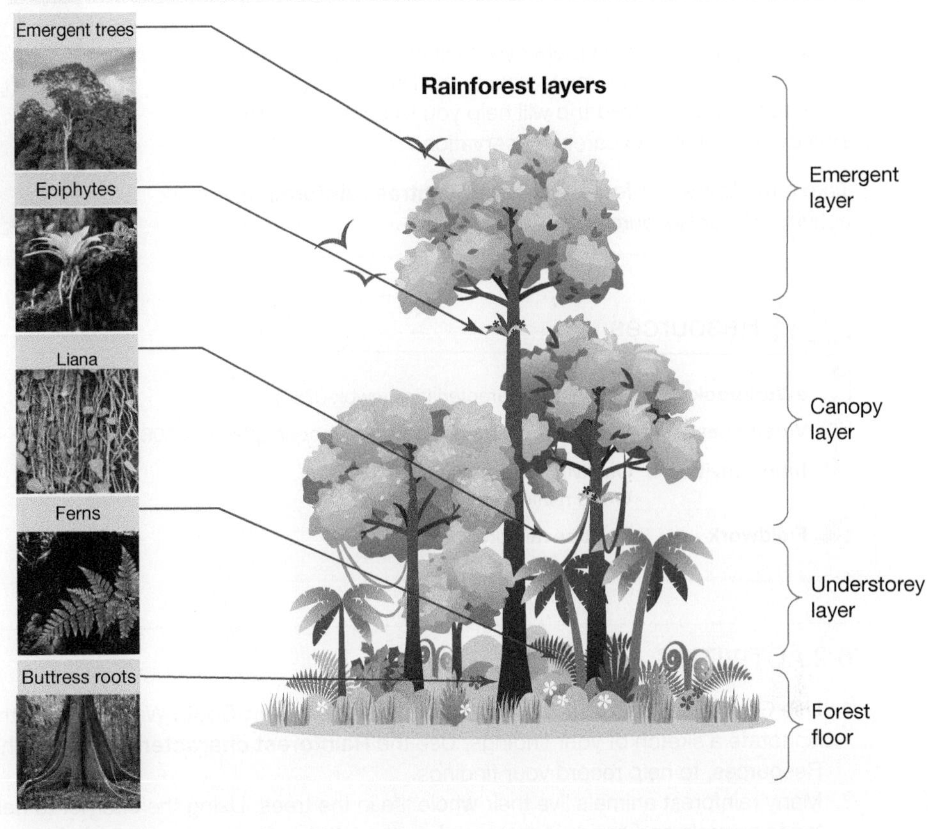

Emergents

These are the tallest trees, ranging in height from 30 to 50 metres. They are so named because they rise or emerge out of the forest canopy. Huge crowns of leaves and abundant animal life thrive on plenty of available sunlight.

Canopy

This describes the array of treetops that form a barrier between the sunlight and the underlying layers. Their height can vary from 20 to 45 metres. This layer contains a distinct **microclimate** and supports a variety of plants and animals. The taller trees host special vines called lianas that intertwine the branches. Other plants called **epiphytes** use the tree trunks and branches as anchors in order to capture water and sunlight.

Understorey

This layer contains a mixture of smaller trees and ferns that receive only about 5 per cent of the Sun's energy. Many animals move around in the darkness and humidity, using the vines as highways.

Forest floor

This bottom layer is dominated by a thick carpet of leaves, fallen trees and huge buttress roots that support the giant trees above. Rainforest soils give the impression of being fertile because they support an enormous number of trees and plants. However, this impression is wrong, as the soil in rainforests is generally poor. Leaves and other matter are recycled by the many organisms to create a living **compost**. The roots of trees must 'snatch' these nutrients from the soil before heavy rains wash them away and they are lost through a process called leaching. Larger animals also roam through this layer in search of food.

ecosystem an interconnected community of plants, animals and other organisms that depend on each other and on the non-living things in their environment

microclimate specific atmospheric conditions within a small area

epiphyte a plant that grows on another plant but does not use it for nutrients

compost a mixture of various types of decaying organic matter such as dung and dead leaves

fdw-0007

FOCUS ON FIELDWORK

Virtual visit to the Daintree

Not everyone lives close to a rainforest, but that doesn't mean you can't visit one virtually. Immerse yourself in the sights of one of Australia's most famous rainforests in the Daintree National Park in Queensland. This field trip will help you to understand rainforest environments through careful observation.

Take a trip to the rainforest using the **Daintree rainforest** fieldwork activity in your Resources tab.

on Resources

eWorkbook	Rainforest characteristics (ewbk-0750)
Video eLesson	Rainforest characteristics — Key concepts (eles-5065)
Interactivities	Our living green dinosaurs (int-3112) World rainforest types (int-8736)
Fieldwork	Daintree rainforest (fdw-0007)

6.2 ACTIVITIES

1. Use Google Earth to visit **Dorrigo National Park** in New South Wales to view the layers of the rainforest. Annotate a sketch of your findings. Use the **Rainforest characteristics** worksheet, found in the online Resources, to help record your findings.
2. Many rainforest animals live their whole life in the trees. Using the internet to help you, give some examples of these animals and conduct research into the habits of one animal.
3. Use the **Rainforest layers** weblink in the online Resources to explore the layers of the rainforest and the plants and animals that inhabit them.

6.2 EXERCISE

Learning pathways

■ LEVEL 1	■ LEVEL 2	■ LEVEL 3
2, 5, 6, 11, 15	1, 3, 4, 12, 14	7, 8, 9, 10, 13

Check your understanding

1. What conditions do rainforest environments thrive in?
2. Create and complete a table with the headings shown below that summarises the features of a rainforest environment.

Layer	Height	Amount of light	Features

3. Identify key characteristics of a tropical rainforest.
4. Why are montane forests often called 'cloud forests'?
5. How many layers are there in a rainforest environment?
6. What are the tallest trees in the rainforest called?
7. Describe how conditions in the canopy layer differ from those on the forest floor.

8. Study **FIGURE 1**. Describe the distribution of rainforests around the world. (Think about in which continents and between which latitudes they are found, the size and scale of them, and whether they are continuous or scattered.)

Apply your understanding

9. What are the differences between montane and lowland rainforest environments? What causes these changes in rainforest type?
10. Why are lowland rainforest environments more threatened by human activity than montane rainforests?
11. Refer to **FIGURE 5**. List the layers of a rainforest. Give several examples of features in each layer in a rainforest and explain one way in which they interact.
12. Why are rainforest environments able to support a large range of animals and plants?

Challenge your understanding

13. What change might you expect in the success of plant growth if rainforest trees are removed and crops are planted instead? Why?
14. Imagine you are a raindrop. Recreate your journey through a rainforest, passing through each of the forest layers.
15. How do you interact with some of the Earth's spheres in your daily activities? Refer to at least two spheres in your answer.

To answer questions online and to receive **immediate feedback** and **sample responses** for every question, go to your learnON title at www.jacplus.com.au.

6.3 SkillBuilder — Creating and describing complex overlay maps

online only

LEARNING INTENTION

By the end of this subtopic, you will be able to create an overlay map to show relationships between factors in a landscape.

The content in this subtopic is a summary of what you will find in the online resource.

6.3.1 Tell me

What is a complex overlay map?

A complex overlay map is created when one or more maps of the same area are laid over one another to show similarities and differences between the mapped information. Complex overlay maps are analysed to show relationships between factors — the similarities and the differences in a pattern.

6.3.2 Show me

eles-1656

How to create and describe a complex overlay map

Step 1

Select the base map — this will show information that is unlikely to vary.

Step 2

Trace each of the other maps onto separate sheets of tracing paper. Don't forget to include BOLTSS on each map.

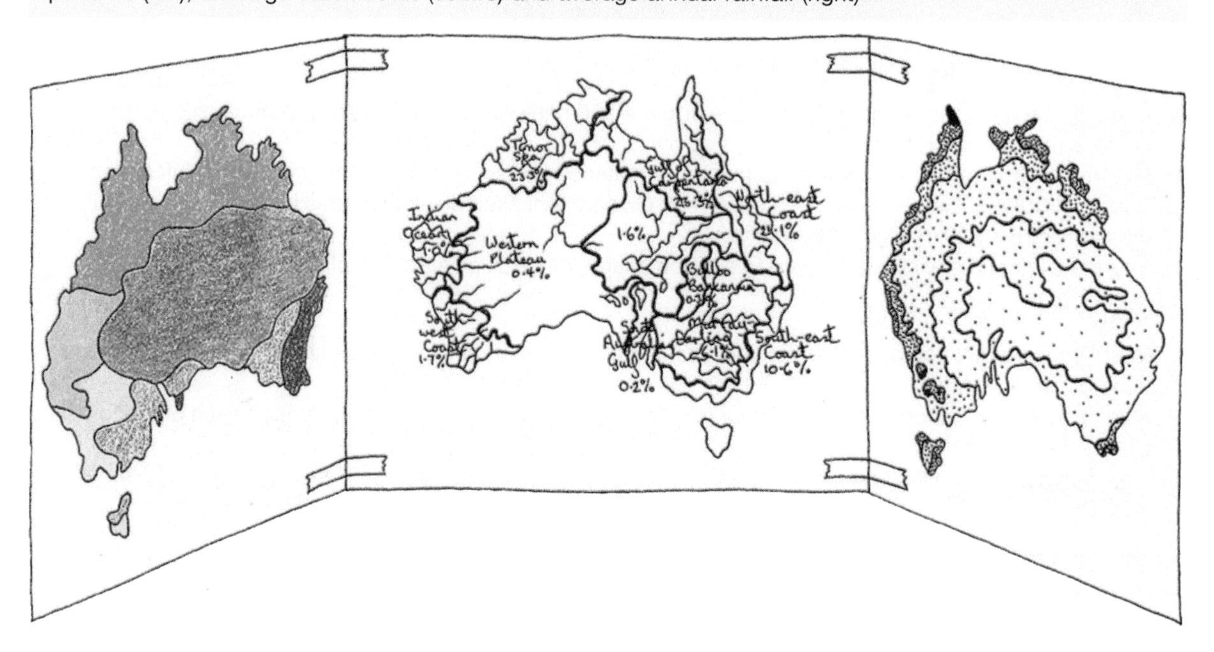

FIGURE 1 An illustration of a completed complex overlay map showing Australia's seasonal rainfall patterns (left), drainage catchments (centre) and average annual rainfall (right)

Step 3

Using adhesive tape, hinge the maps to fold on top of each other so that the map outlines (e.g. coastlines) match up.

Step 4

You are now able to lift each map separately from the others to see the information individually, or view two or more maps combined.

Step 5

To analyse the information that the overlay maps show, write a sentence about each of the following:

- Where is there a relationship or interconnection of features?
- What are the significant differences across two or more maps?
- What unusual occurrences can you see? (For example, where things appear to be random and show no interconnection.)

6.3.3 Let me do it

int-3152

Go to learnON to access the following additional resources to help you build this skill:

- a longer explanation of this skill and its application in Geography (Tell me)
- a video showing the step-by-step process of this skill (Show me)
- an activity and interactivity for you to practise this skill (Let me do it)
- self-marking questions to help you understand and use this skill.

Resources

eWorkbook SkillBuilder — Creating and describing complex overlay maps (ewbk-9042)

Video eLesson SkillBuilder — Creating and describing complex overlay maps (eles-1656)

Interactivity SkillBuilder — Creating and describing complex overlay maps (int-3152)

6.4 The value of rainforest environments

LEARNING INTENTION

By the end of this subtopic, you will be able to outline the various ways that rainforests are valued.

6.4.1 Amazing rainforests

More than 7000 modern medicines are made from rainforest plants. They can be used to treat problems from headaches to killer diseases such as malaria. They are used by people who suffer from multiple sclerosis, Parkinson's disease, leukaemia, asthma, acne, arthritis, diabetes, dysentery and heart disease, among many other illnesses.

Even animals can be used to cure human diseases. Tree frogs from Australia give off a chemical that can heal sores, and a similar chemical from a South American frog is used as a powerful painkiller. The poisonous venom from an Amazonian snake is used to treat high blood pressure.

Only 1 per cent of the known plants and animals of the rainforest have been properly analysed for their medicinal potential. Perhaps the greatest benefits to medicine and our own health are yet to come.

FIGURE 1 Rainforests play a vital role in controlling the world's climate and oxygen supply. Scientists believe that half of all the world's oxygen is produced by the Amazon rainforest alone.

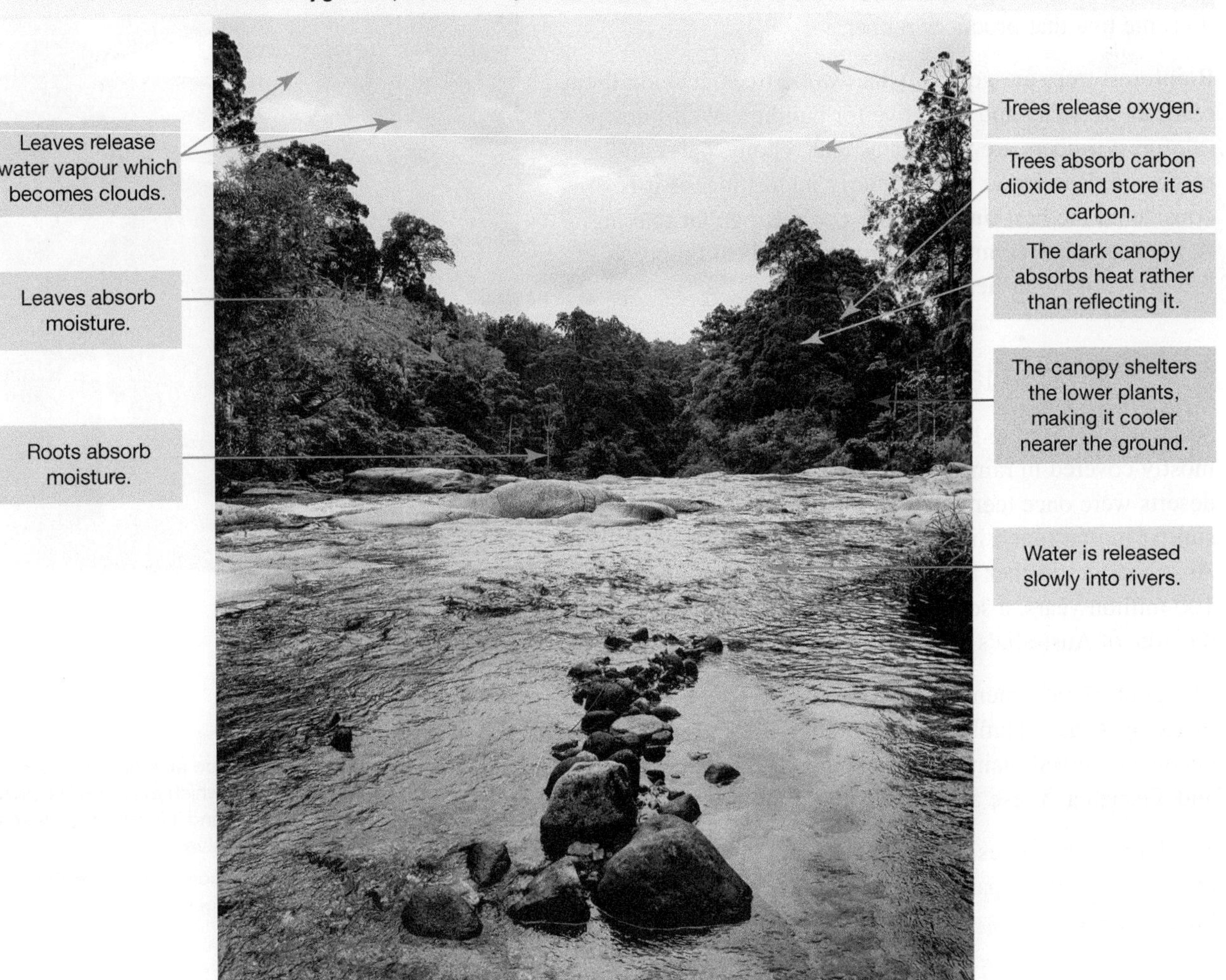

Rainforests are home to the greatest profusion of life on the planet: at least half of all known plants and animals live in rainforests.

At least 50 million indigenous peoples live in rainforests worldwide. From the Kuna people of Panama and the Yanomami of Brazil to the Baka people of Cameroon and the Penan of Borneo (Indonesia), these people traditionally lived in harmony with their forest home.

The people who live in or near the rainforests gain much of their food from the forest. But rainforests also supply the supermarkets of the world with food. Most of these fruits and nuts are now grown by farmers rather than harvested directly from the forest, but it was in the rainforests that they originated.

Chocolate first came from cacao trees native to the Amazon rainforest. Today the cocoa in chocolate is most likely to have come from huge cacao plantations in West Africa. Similarly, brazil and cashew nuts, cinnamon, ginger, pepper, vanilla, bananas, pineapples, coconuts, paw-paws, mangoes and avocados were all originally rainforest plants. Even the gum used in chewing gum comes from a rainforest plant, as does the tree that produces rubber.

Rainforest trees are generally hardwood trees, making them resistant to decay and attractive for building. Well-known rainforest timbers are mahogany, teak, ebony, balsa and rosewood. Rosewood is particularly interesting, as it is considered the best timber in the world for guitar making. In many tropical countries, people also collect timber as fuel for cooking or heating.

FIGURE 2 Skin secretions from frogs such as the Waxy Monkey Treefrog (*Phyllomedusa bicolor*) contain powerful painkillers.

FIGURE 3 Food products such as chocolate and chewing gum are made from ingredients that originally came from the rainforest.

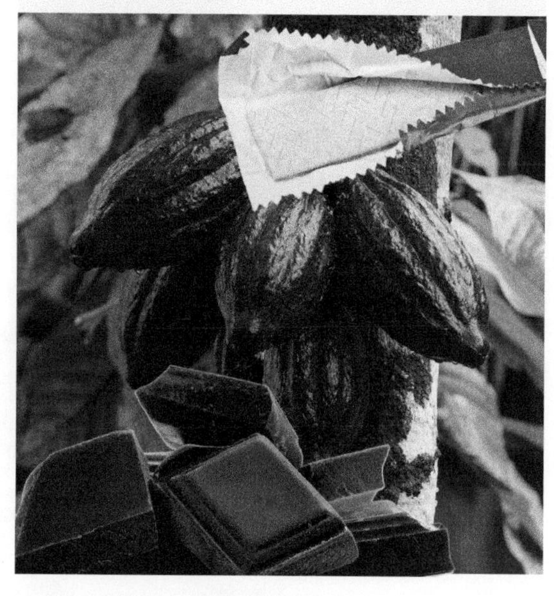

6.4.2 Australian rainforests

Although it's hard to believe now, Australia was once mostly covered in rainforest. Even areas that today are deserts were once teeming with plant and animal life similar to that found in the Amazon. This is because Australia was further north than it is today. Over the past 100 million years, a series of events has gradually reduced the area of Australia's rainforests (see **FIGURE 4**).

The gradual movement of Australia as it separated from Gondwanaland, and a series of **ice ages**, have combined to make it a drier place (see **FIGURE 5**). Rainforests have become confined mainly to the mountains and **gorges** of the Great Dividing Range and Tasmania. These areas have higher rainfall and fewer fires.

The farming practices of European immigrants have reduced much of the remaining rainforest — in the past 200 years, more than 70 per cent of these forests have been cleared.

ice age historical period during which the Earth is colder, glaciers and ice sheets expand and sea levels fall

gorge narrow valley with steep rocky walls

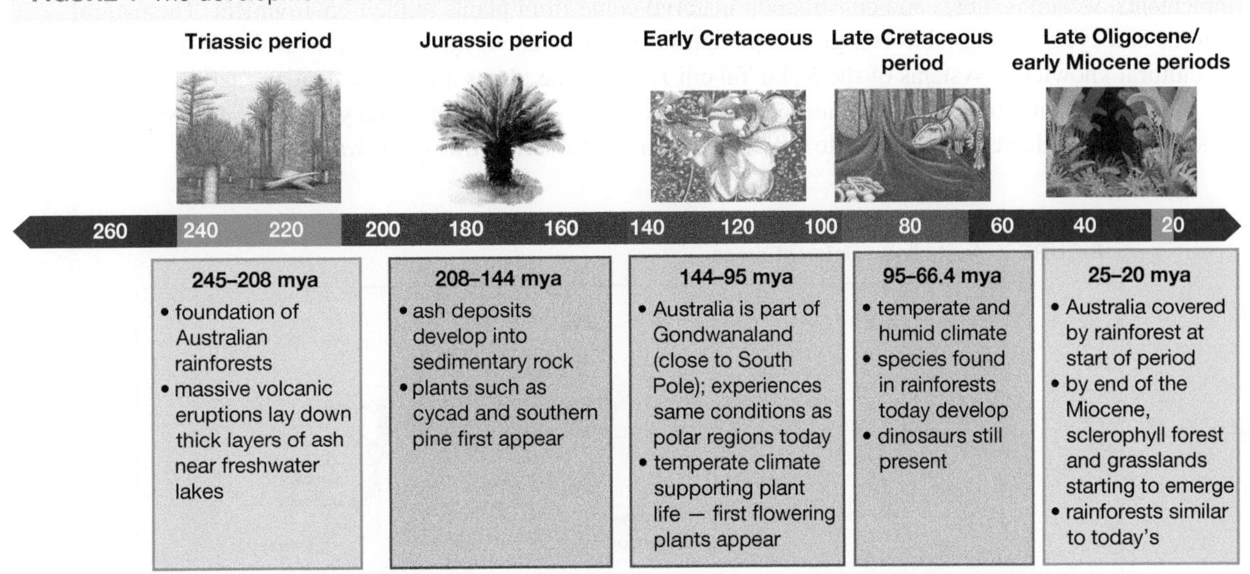

FIGURE 4 The development of Australian rainforests over time

*Dates are expressed in *mya* — million years ago.

Scientists have identified three major types of rainforest in Australia (see **FIGURE 7**), all three of which exist in Queensland.

Much of Australia's tropical rainforests are now World Heritage areas. This means they have been listed by UNESCO as being of global importance. The Wet Tropics of Queensland are a World Heritage area containing some of the oldest rainforests in the world. They have the world's highest concentration of flowering plants, and have records that show Aboriginal Peoples of the area are the world's oldest rainforest cultures.

The Aboriginal People of the Daintree rainforest in North Queensland are the Kuku Yalanji, who are believed to have lived in this area for more than 50 000 years. Their culture is uniquely adapted to the rainforest environment.

For the Kuku Yalanji, the natural world is often thought of in human terms and is closely linked to the people. Any changes to the environment are seen as changes to themselves. Due to powerful properties being attributed to most story places (sites with links to Dreamings and Creation stories) of the Daintree, the Kuku Yalanji regard damage and destruction to the environment as disrespectful and unacceptable.

FIGURE 5 Difference in the location of Australia when it formed part of Gondwanaland compared with its location today

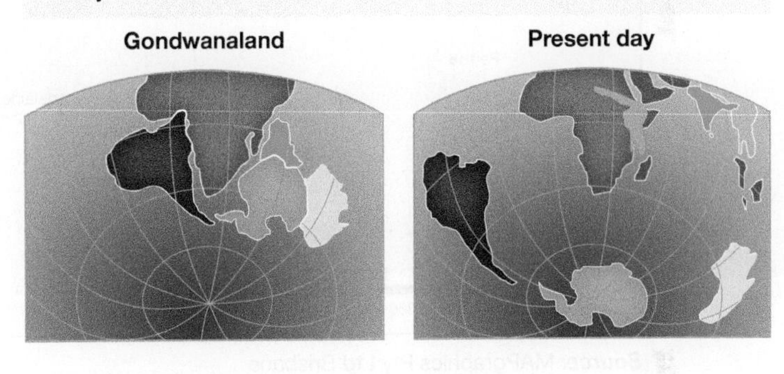

FIGURE 6 Rainforest plants are used to make goods such as these baskets — used for storage, food collection, carrying personal possessions, and leaching poisons from seeds in fresh running water.

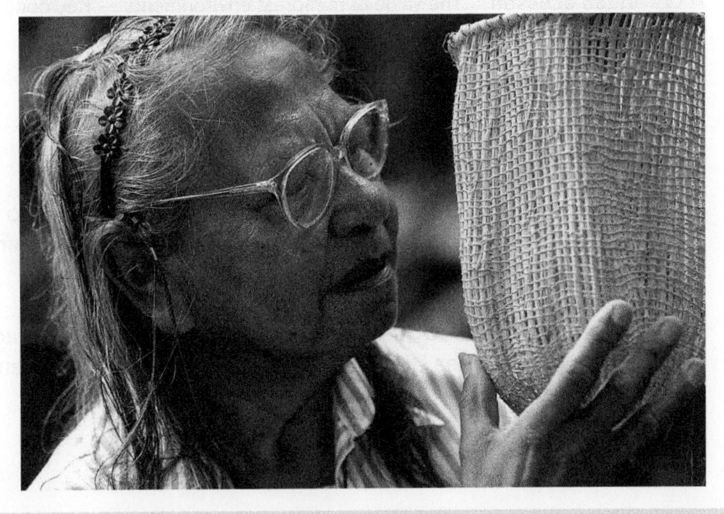

The Kuku Yalanji People use resources from the environment for their food and medicine, and many of their implements, weapons, fibres and construction material come from plants in their environment. The natural patterns and cycles of the rainforest give important information about the food that is available and is reliant on the cultural knowledge systems of the Kuku Yalanji People. The plants are their calendar, marking the seasons. For example, when blue ginger (*jun jun*) is fruiting it is time to catch scrub turkey (*diwan*), and when mat grass (*jilngan*) is flowering it is time to collect the eggs of the scrub fowl (*jarruka*).

FIGURE 7 Australia's rainforest areas

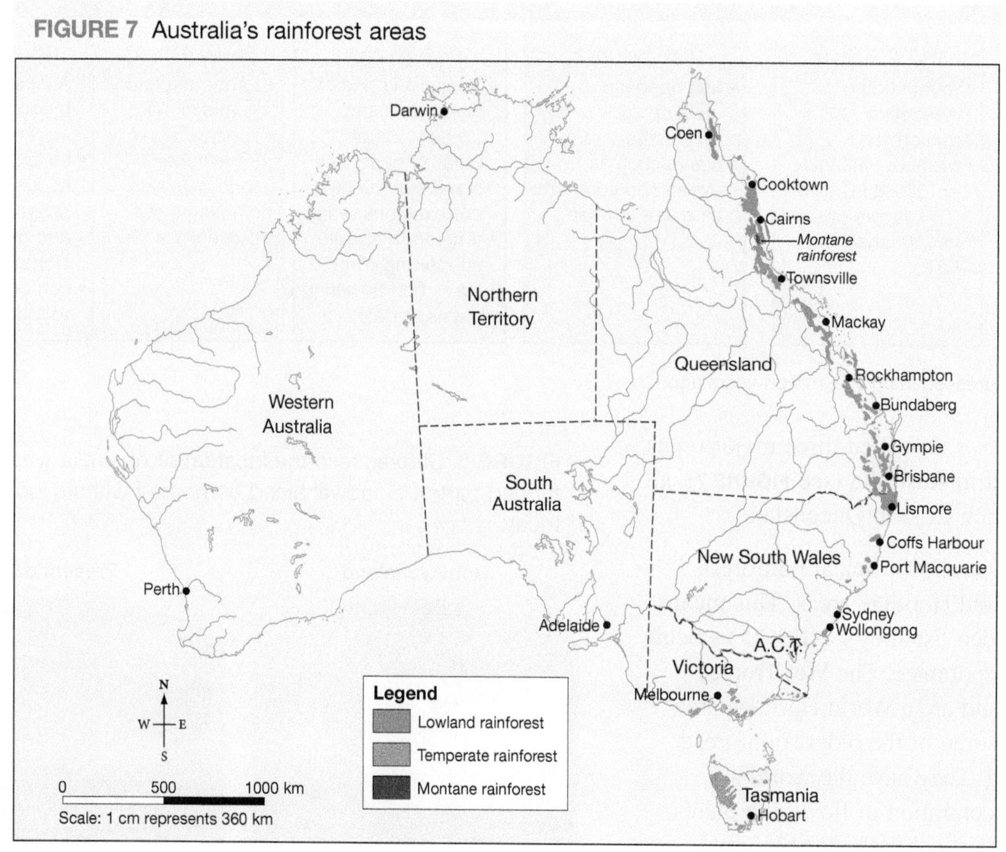

Source: MAPgraphics Pty Ltd Brisbane

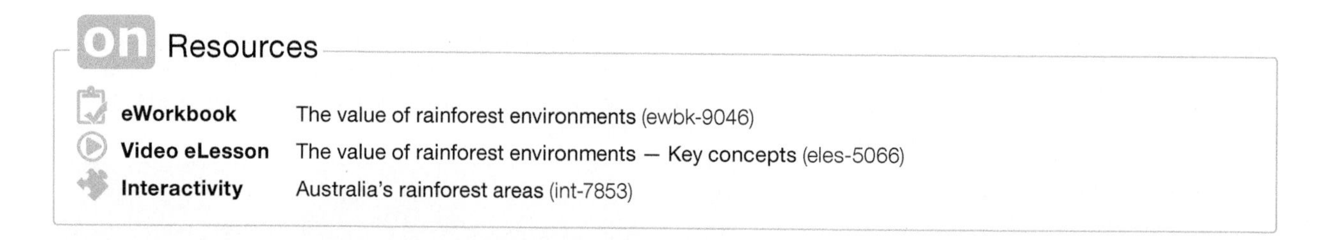

Resources

eWorkbook The value of rainforest environments (ewbk-9046)

Video eLesson The value of rainforest environments — Key concepts (eles-5066)

Interactivity Australia's rainforest areas (int-7853)

6.4 ACTIVITIES

1. Use the **UNESCO Heritage** weblink in the online Resources to complete the following.
   a. On a map of Australia, locate and label Australia's World Heritage sites.
   b. Which three sites have been added most recently?
   c. Which two sites protect Australian rainforests?
   d. The Wet Tropics of Queensland border another World Heritage site. What is this other site?
   e. What criteria does UNESCO use to determine whether a natural region should be placed on its list?

2. A hotel chain has applied to the Queensland government for permission to build a resort in the Daintree. Assess this proposal from the perspectives of the developers, government, local residents, environmentalists and Kuku Yalanji People. Try to make a decision as to whether this project should be approved.
3. Use the **Rainforest foods** weblink in the Resources tab to learn which foods come from the rainforest.
4. Make a list of things in your home that may come from the rainforest environment. Remember to look in the medicine cupboard and the pantry as well as at the furniture. Perhaps you could bring some examples to school and your class could set up a display.
5. This subtopic lists only a few of the products we use from rainforests. List the value of these and other rainforest products according to whether they are valued for their economic, aesthetic or cultural characteristics.
6. Research and share Dreamtime stories about Australia's rainforests from different Aboriginal Peoples and Torres Strait Islander Peoples. Do other cultures from rainforests around the world share similar culturally important stories about their rainforests?

FIGURE 8 Dorrigo National Park

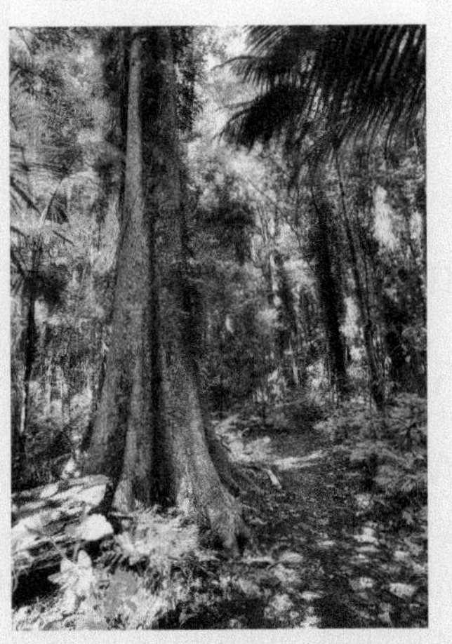

6.4 EXERCISE

Learning pathways

■ LEVEL 1	■ LEVEL 2	■ LEVEL 3
1, 2, 3, 10, 15	4, 6, 7, 8, 12	5, 9, 11, 13, 14

Check your understanding

1. Give three examples of diseases that are treated by medicines made from rainforest plants.
2. List five foods that were originally from rainforest plants.
3. List the reasons for the gradual disappearance of Australia's rainforest environments.
4. What resources does the rainforest provide for the Kuku Yalanji People?
5. How has the scale of Australian rainforest environments changed over time?
6. Why do the Kuku Yalanji People regard damage to the Daintree rainforest as unacceptable?

Apply your understanding

7. Describe the location and distribution of Australia's remaining rainforest environments. What factors have contributed to their survival in these places?
8. Suggest the impacts that the built environment could have on the Daintree.
9. Refer to **FIGURE 1**.
   a. Explain the role of the rainforest environment in relation to the climate.
   b. Why are rainforests sometimes called 'the lungs of the Earth'?
10. Why are rainforest trees favoured for building?
11. Detail three ways that rainforest trees help to regulate the world's climate and oxygen levels.

Challenge your understanding

12. There are three major types of rainforest environments found in Australia. What makes Queensland's rainforests unique? Why is this possible?
13. Refer to **FIGURE 7**. Why are there no rainforest environments on the western side of Australia?
14. Based on the history of Australia's rainforests and the protection now in place for the remaining forests, what do you think the future holds for this important resource?
15. Which of the present uses of the rainforest do you think is the most sustainable for the forest's future? Explain your answer.

To answer questions online and to receive **immediate feedback** and **sample responses** for every question, go to your learnON title at www.jacplus.com.au.

6.5 SkillBuilder — Drawing a précis map

online only

LEARNING INTENTION

By the end of this subtopic, you will be able to create a précis map to highlight a feature of a landscape.

The content in this subtopic is a summary of what you will find in the online resource.

6.5.1 Tell me

What is a précis map?

A précis map is a summary of an area. There may be just one feature shown, such as rainforest. Sometimes more features are shown, such as vegetation, urban areas and roads.

6.5.2 Show me

eles-1657

How to draw a précis map

The précis map in **FIGURE 1** shows four aspects — the height of the land, the major towns, the rivers and the areas where the Penan people live.

Step 1

Determine the area that you want to use to create a précis map. In **FIGURE 1** this has been done by removing details for surrounding countries, so that only Sarawak is detailed.

FIGURE 1 A précis map showing where the Penan people of Borneo live

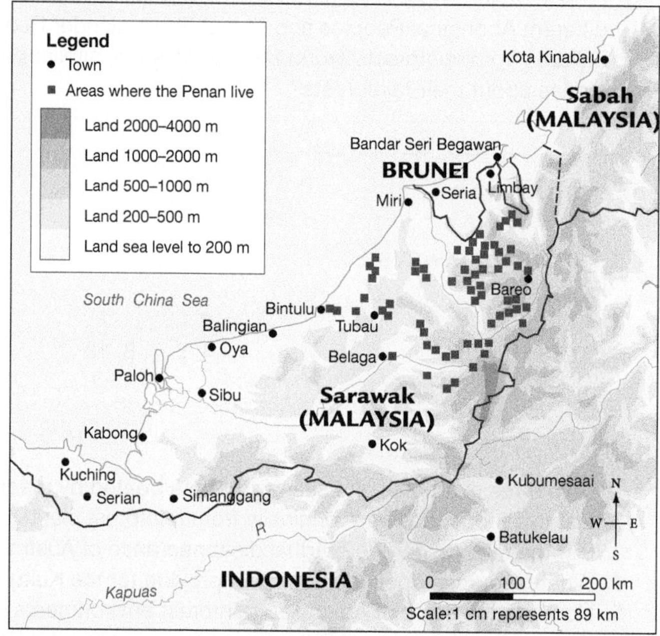

Source: MAPgraphics Pty Ltd, Brisbane

Step 2

Rule a border on your page. Make this the same size as the original map to avoid having to scale your drawing.

Step 3

Identify the features that you are going to include on your précis map. In **FIGURE 1**, the cartographer has chosen to leave in land heights, rivers and towns, and has chosen to leave out roads and vegetation.

Step 4

Create a colour-coded key/legend for each feature and place it next to or below the map.

Step 5

Within the border that you created in Step 2, draw an outline of the area that is to be mapped; keep the scale of the original map that you are using.

Step 6

Individually, take each of the features that you identified in Step 3 and mark onto your map, in a generalised way, the area that it covers. When you have completed one feature, colour it before moving to the next feature and mark your key/legend appropriately.

Step 7

Complete the précis map with BOLTSS.

6.5.3 Let me do it

int-3153

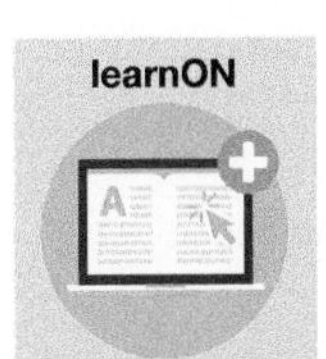

Go to learnON to access the following additional resources to help you build this skill:
- a longer explanation of this skill and its application in Geography (Tell me)
- a video showing the step-by-step process of this skill (Show me)
- an activity and interactivity for you to practise this skill (Let me do it)
- self-marking questions to help you understand and use this skill.

on Resources

eWorkbook	SkillBuilder — Drawing a précis map (ewbk-9050)
Video eLesson	SkillBuilder — Drawing a précis map (eles-1657)
Interactivity	SkillBuilder — Drawing a précis map (int-3153)

6.6 Living in the rainforest

online only

LEARNING INTENTION

By the end of this subtopic, you will be able to explain and provide examples of how some groups of Indigenous peoples value and use rainforests around the world.

The content in this subtopic is a summary of what you will find in the online resource.

It is difficult to accurately count all the people around the world who live in rainforests, but some estimates put the number as high as 150 million. While these people are usually described as living a traditional **subsistence** way of life, this is usually combined with selling and buying items such as their labour, their land and assorted forest products.

FIGURE 2 The Huli people of Papua New Guinea

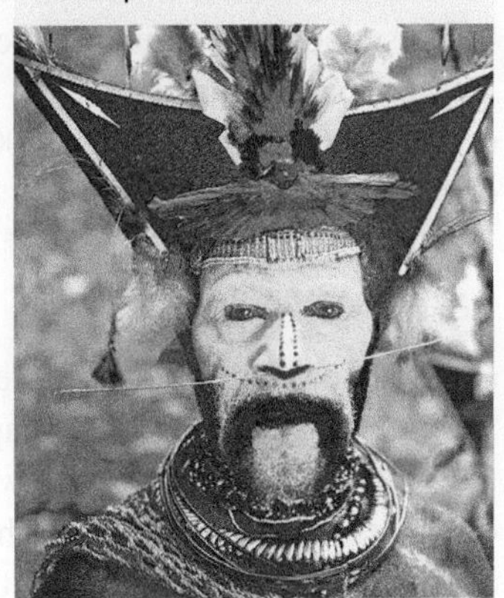

To learn more about the peoples who live in the world's rainforests and how their lives are changing, go to your learnON resources at www.jacPLUS.com.au.

subsistence producing only enough crops and raising only enough animals to feed yourself and your family or community

Contents

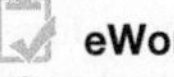

on Resources

eWorkbook	Living in the rainforest (ewbk-9054)
Video eLesson	Living in the rainforest — Key concepts (eles-5067)
Interactivity	Living in the rainforest — Interactivity (int-8409)

6.7 Disappearing rainforests

LEARNING INTENTION

By the end of this subtopic, you will be able to explain why rainforests are being destroyed, and outline the consequences of this.

6.7.1 Reasons rainforests are being cleared

Rainforests can provide a wide variety of useful resources, many of which are economically valuable. Mining and farming these areas, or selling other rainforest resources, can earn a lot of money and can lift the standard of living in a community if it remains in control of the profits. If large companies or governments mine or clear land for farming, this is not always the case. As a result, all around the world, rainforests are being destroyed for economic gain. The main reasons for rainforests being cleared are shown in **FIGURE 1** and described below.

FIGURE 1 Causes of deforestation in the Amazon, 2000–2015

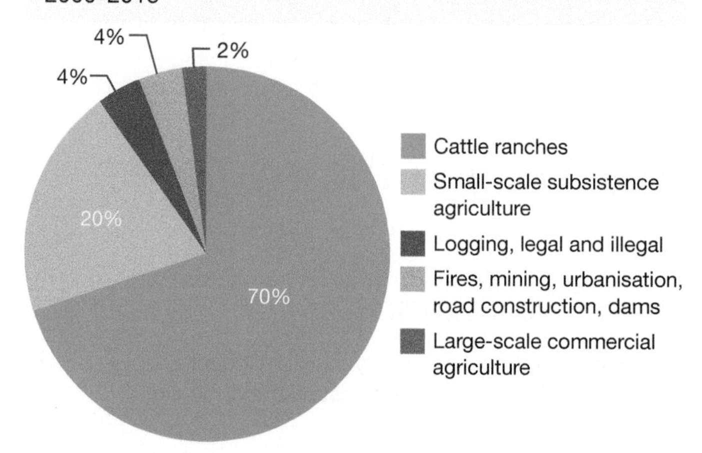

Commercial logging

There are two main types of logging: **clearfelling** and **selective logging**. When a forest is clearfelled, all trees are removed either by chainsaw or with heavy machinery such as bulldozers. In selective logging, only the best and most valuable trees are cut down. But in clearing forest to reach those trees, it is estimated that a hectare (10 000 square metres) of forest is destroyed for each log removed.

FIGURE 2 It is thought that up to 70 per cent of logging in Brazil and Indonesia could be illegal.

Farming

Rainforests grow in many developing countries where populations are often rapidly increasing and standards of living are low. In these countries, the land on which the forest grows is seen as more valuable than the forest itself because it provides a way for people to earn a living to support their families.

New farms are often started where new highways open up parts of the rainforest that were once almost impossible to reach. Soon after the roads are built, settlers (called homesteaders) arrive, claiming a piece of the forest that borders the road. Homesteaders often chop down a few trees as timber for fencing or a house, and then set fire to the rest so that the land can easily be cleared for farming (see **FIGURE 3**).

Once all the land beside the roads has been claimed, tracks and roads leading from the highways are built to provide deeper access into the forest. Soon these access roads also provide an easy way for more homesteaders to claim land in the forest. These might be small farms or large-scale commercial farms that raise beef or crops for export to more developed countries.

clearfelling a forestry practice in which most or all trees and forested areas are cut down

selective logging a forestry practice in which only selected trees are cut down

FIGURE 3 Blocks of rainforest in Peru burned to clear the area for agricultural use

FIGURE 4 Deforestation and pollution from gold mining, Matto Grosso Peixoto de Azevedo River, Brazil

FIGURE 5 An aerial view of a gold mining area in the Amazon forest region, Para state, Brazil

Mining

Many rainforests are growing on land that also contains large energy and mineral deposits such as oil, gold, silver, bauxite, iron ore, copper and zinc (see **FIGURE 4**). Mining companies build roads to the deposits and set up large-scale mining and processing plants. These plants require large amounts of electricity, and this is often supplied by burning trees to create charcoal or by constructing vast **hydroelectric dams**.

Deep in the Brazilian rainforest, hydroelectric dams are being constructed to provide electricity for mining and related operations. The dams flood the lands of local peoples, and are so large that they can alter the climate in the area, making it drier.

Another problem created by mining is the pollution of nearby rivers and streams from chemicals used in the processing plants. Rivers downstream from a vast goldmine in Papua New Guinea have been found to contain four times the safe limit of cyanide. Cyanide is used to extract gold from rock. Dam tailings (left-over material from the mining process) also cause significant damage if they burst or leak.

6.7.2 The Amazon rainforest

The world's largest remaining rainforest is in the Amazon Basin in South America. This truly remarkable forest is under increasing threat from forestry, mining and farming. The loss may cause severe problems worldwide. Most of us use rainforest products every day. More importantly, however, rainforests help control the world's climate and our oxygen supply. So the next time you eat chocolate, treat your asthma, play a guitar or even take a deep breath, you should thank the Amazon rainforest.

hydroelectric dam a dam that harnesses the energy of falling or flowing water to generate electricity

FIGURE 6 The Amazon Basin

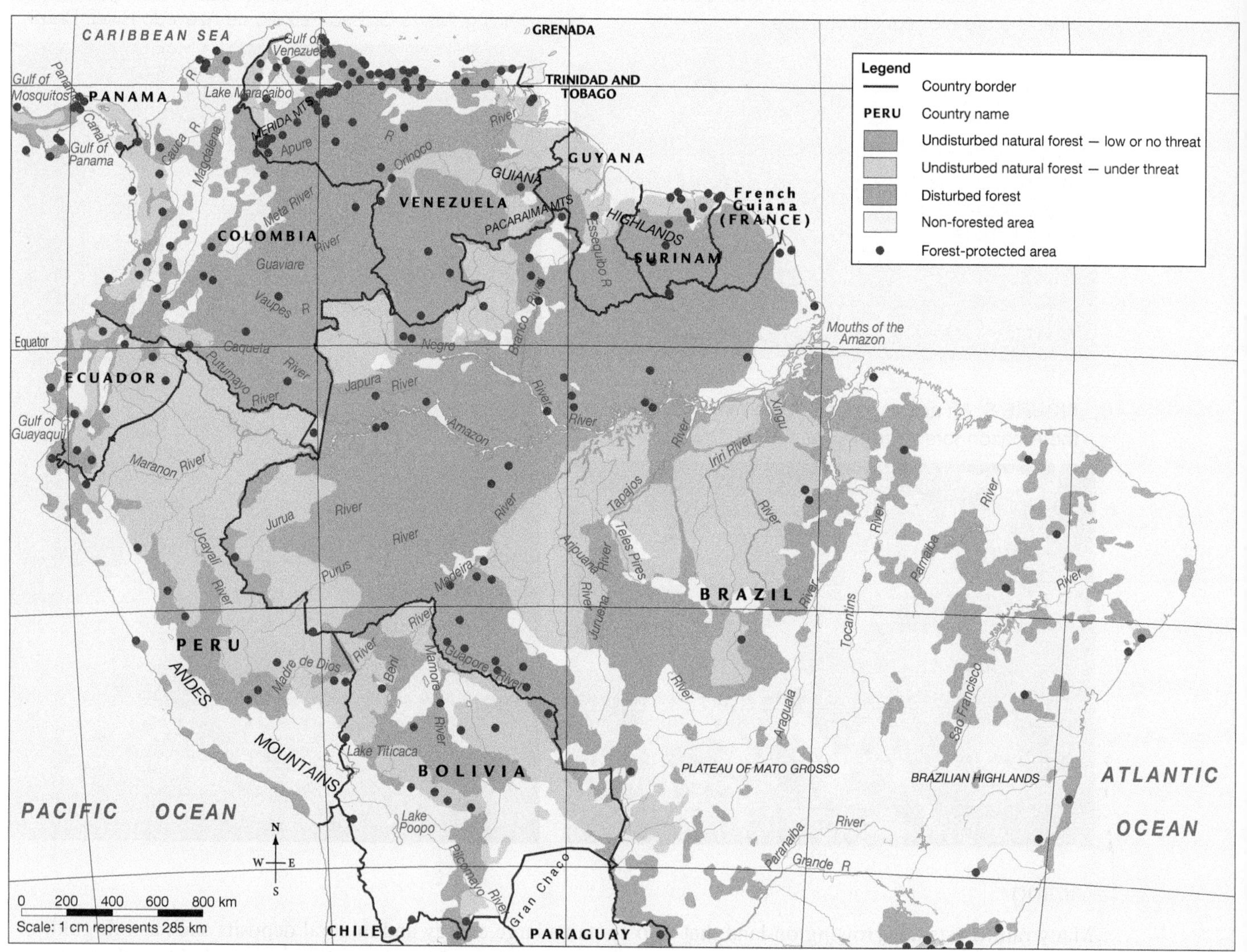

Source: MAPgraphics Pty Ltd, Brisbane

The ecosystem services that the Amazon provides

The Amazon River and the rivers that feed into it (tributaries) contain one-fifth of the world's fresh water, and more than 2000 species of fish — more than in the Atlantic and Pacific oceans combined. There are many other significant facts about the Amazon:

- The mouth of the Amazon River is approximately 325 kilometres wide and contains an island the size of Switzerland!
- The Amazon forest is home to more than 40 000 species of plants, 1300 bird species, 430 different mammals and 2.5 million different insects.
- Approximately 1.3 million tons of sediment is transported by the Amazon River to the sea daily.
- No bridges cross the main trunk of the Amazon River, which locals call the Ocean River.
- Since 2000, the Amazon rainforest has been facing deforestation at an average rate of 50 football fields per minute. This rate reportedly rose to a 12-year high in 2020, with an estimated 11 088 square kilometres of rainforest destroyed between August 2019 and August 2020.
- The Amazon is the second longest river in the world, but it carries more water than the next six largest rivers combined.
- The Amazon River drains nearly 40 per cent of South America.
- There are official plans for 412 dams to be in operation in the Amazon River and its headwaters. Since 1900, more than 90 indigenous groups have disappeared in Brazil alone.

FIGURE 7 The brown waters of the Amazon show that it is carrying a lot of sediment.

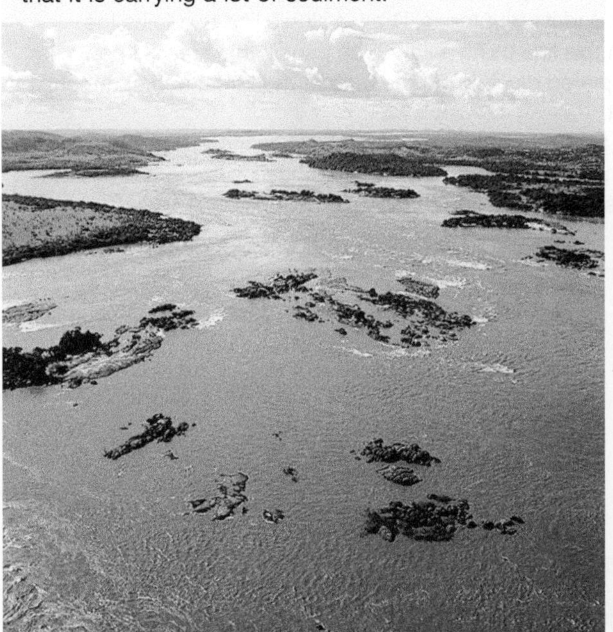

FIGURE 8 Area cleared for ranching in the Amazon rainforest

FIGURE 9 Deforestation patterns in the rainforest in 2001 and 2019 in Rondônia in Brazil

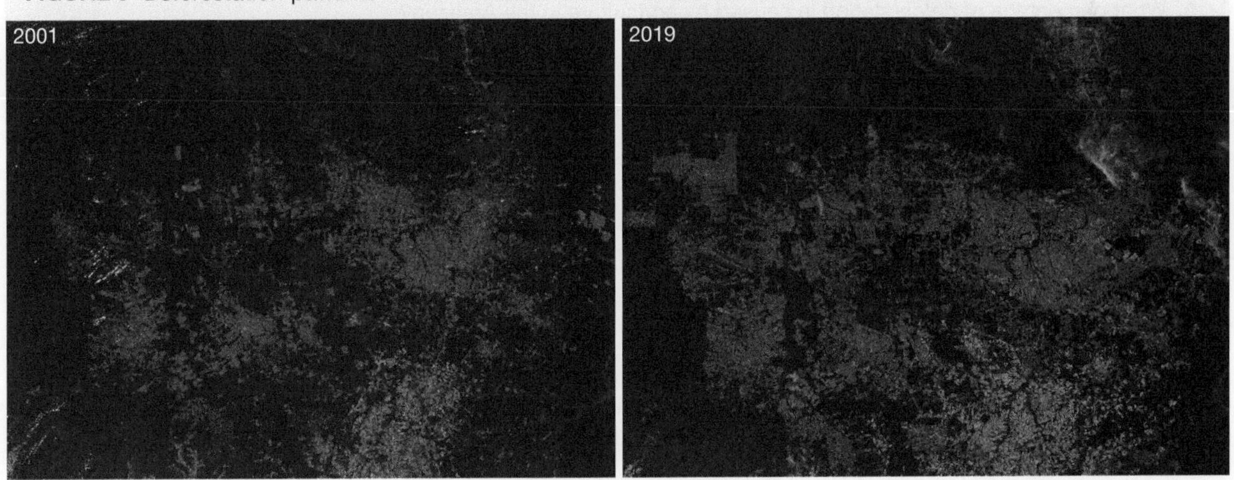

Amazon fires

Each year during the dry season, slash-and-burn methods of clearing rainforest cause many fires to be lit in the Amazon. During the 2019 dry season, there was a 77 per cent increase in fires from the previous year for the same period. In July 2020, the amount of fires increased by a further 28 per cent. The loss of 906 thousand hectares of rainforest in 2019 not only significantly impacted biodiversity, but also the indigenous peoples that live within the forest and the global climate. As agricultural demands increase, the Amazon is put under more and more stress each year.

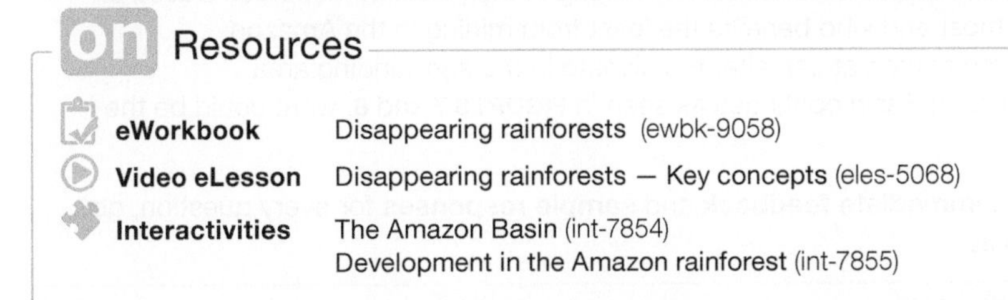

on Resources

eWorkbook Disappearing rainforests (ewbk-9058)

Video eLesson Disappearing rainforests — Key concepts (eles-5068)

Interactivities The Amazon Basin (int-7854)
Development in the Amazon rainforest (int-7855)

6.7 ACTIVITIES

1. Using a piece of tracing paper, trace the Amazon River and its tributaries. Draw a single line that joins the source of each of the tributaries. Shade the area within this line using a light blue pencil: this area is known as the catchment, or basin, of the river. Overlay your completed diagram on the map of the forest and comment on the interconnection between the river and the forest.
2. Use the **Amazon tour** weblink in the Resources tab to take a tour through an Amazon rainforest slideshow.
3. As a class, discuss the potential long-term problems that could result from the continued commercial use of rainforest environments around the world. Develop a list of the top five potential problems.

6.7 EXERCISE

Learning pathways

■ LEVEL 1	■ LEVEL 2	■ LEVEL 3
1, 3, 9, 11, 13	2, 5, 7, 10, 15	4, 6, 8, 12, 14

Check your understanding

1. What is the difference between clearfelling and selective logging?
2. List four changes/problems caused by mining operations in rainforests.
3. What is the major cause of rainforest destruction?
4. What is hydroelectricity?
5. What percentage of deforestation in caused by agriculture in the Amazon?

Apply your understanding

6. 'Many homesteaders are unable to make a good living from the poor tropical soils.' Explain the reasoning behind this statement.
7. Explain how producing hydroelectricity can lead to destruction of the rainforest.
8. Why might the clearing and change in the Amazon appear to occur in straight lines when viewed from an aerial photograph?
9. Looking at **FIGURES 7, 8** and **9** for interconnections, what do you think could be contributing to the high levels of sediment in the Amazon River? Why?
10. Look carefully at **FIGURE 6**.
    a. List the countries of South America into which the Amazon rainforest extends.
    b. Which country contains most of the Amazon rainforest?
    c. Why do you think there are so few large cities in the rainforest?
    d. Estimate the percentage of the rainforest that can be considered:
        i. under low or no threat
        ii. under threat
        iii. disturbed.
    e. Describe in your own words what each of these terms means.
11. Many rainforest environments are located in less economically developed countries. Why does this make the problem of rainforest destruction harder to solve?
12. 'With the population of the world increasing, we have no choice — cutting down the rainforest is in the best interests of both people and the environment.' Write a paragraph for or against this statement.

Challenge your understanding

13. Create a table to weigh up the benefits and problems caused by mining in the Amazon rainforest. Based on your table, suggest who benefits the most and who benefits the least from mining in the Amazon.
14. Suggest how a dam built in the Brazilian rainforest can alter the climate in the surrounding area.
15. If development and clearing in the Amazon Basin continues as seen in **FIGURES 7** and **8**, what could be the consequences locally and globally?

To answer questions online and to receive **immediate feedback** and **sample responses** for every question, go to your learnON title at www.jacplus.com.au.

6.8 Social and environmental impacts of deforestation

LEARNING INTENTION

By the end of this subtopic, you will be able to explain the social and environmental impacts of deforestation.

6.8.1 Impacts of deforestation

Deforestation is the major cause of problems in the rainforest ecosystem. The loss of unique **habitats** is the primary reason species are becoming endangered. Clearing creates smaller islands of vegetation, making it more difficult for animals to communicate and breed. People are also affected by the removal of the rainforest. While indigenous people may feel the effects first, others also experience negative consequences.

Speed and impacts of deforestation:

- About 1 hectare of rainforest is destroyed every second across the world: this is about twice the size of a soccer pitch.
- Scientists estimate that 137 plants and animals are made extinct daily: that's 50 000 each year (including some that haven't been discovered yet).
- It is believed that in the year 1500 up to 9 million people lived in the Amazon rainforest. The number of indigenous people living in the Amazon is now lower than 200 000.
- The world loses about 2 per cent of its rainforest each year. Rates differ between countries.

6.8.2 Environmental impacts

Many forests are cleared using fire. These fires release millions of tonnes of carbon dioxide into the air, contributing to global warming. At the same time, destroying the trees robs the planet of the natural system that helps regulate the amount of carbon dioxide in the air.

In many areas where forests are cleared, it has become a practice to leave behind 'islands' of rainforest. This is meant to assist in the natural regeneration of the forest and also to leave areas of the natural habitats of plants and animals that live in the rainforest. But is this working?

The islands that are left are often not big enough to ensure the survival of the large numbers of species that live there. For example, the endangered Queen Alexandra's Birdwing (the world's largest butterfly) is facing extinction as its distribution is being condensed into seven isolated blocks of rainforest measuring approximately 1–2 square kilometres in northern Papua New Guinea (see **FIGURE 1**). These remaining refuges are threatened by surrounding palm oil plantations.

FIGURE 1 The wingspan of the Queen Alexandra's Birdwing can reach 30 centimetres.

habitat the total environment where a particular plant or animal lives, including shelter, access to food and water, and all of the right conditions for breeding

FIGURE 2 Leftover pockets of rainforest are at risk from reduced rainfall and cannot survive drought conditions.

(a) Rainforest trees are cleared, with 'islands' left for regeneration.

(b) There is less evapotranspiration and less rain on forest 'islands'.

There are other problems. When the forest is cleared, the exposed earth can quickly erode as the tree roots no longer hold the soil together, making the regrowth of vegetation slow. On steep slopes this can increase the risk of landslides, and sediments can flow into rivers.

During drought, the bare ground can become hot and barren. With the removal of the forest cover there is little moisture stored in the ground and a much lower rate of **evapotranspiration**. This in turn affects the water cycle, reducing the amount of rain that falls on the remaining islands of rainforest, which quickly dry out (see **FIGURE 2**). With the removal of the trees, no nutrients are returned to the soil through decaying leaf litter, which alters the nutrient cycle.

evapotranspiration the process by which water is transferred to the atmosphere from surfaces such as the soil and plants

CASE STUDY

Deforestation in Indonesia and the orangutan

Nearly 10 per cent of the world's rainforests and 40 per cent of all Asian rainforests are found in Indonesia. Less than half of Indonesia's original rainforest area remains. Much of this is in Kalimantan, on the island of Borneo. Forests have been cleared for timber, for plantation crops such as palm-oil trees, and to make way for Indonesia's growing population, which is now more than 200 million. Fires lit to clear land in 1982 and 1997 have severely damaged large areas of rainforest in Kalimantan. Orangutans, Sumatran tigers and Javan hawk-eagles may disappear from Indonesia as their natural habitats disappear.

Orangutans are the largest tree-living mammals and the only great ape that lives in Asia. They survive only on the islands of Borneo and Sumatra. Current estimates are that orangutans have lost 80 per cent of their habitat in the last 20 years. In 1997–98, wildfires burned through nearly 2 million hectares of land in Indonesia, killing up to 8000 orangutans.

It is estimated that orangutan numbers have declined by more than 50 per cent in the last 60 years. The current orangutan population is believed to range between 45000 and 69000.

FIGURE 3 Mother and baby orangutan

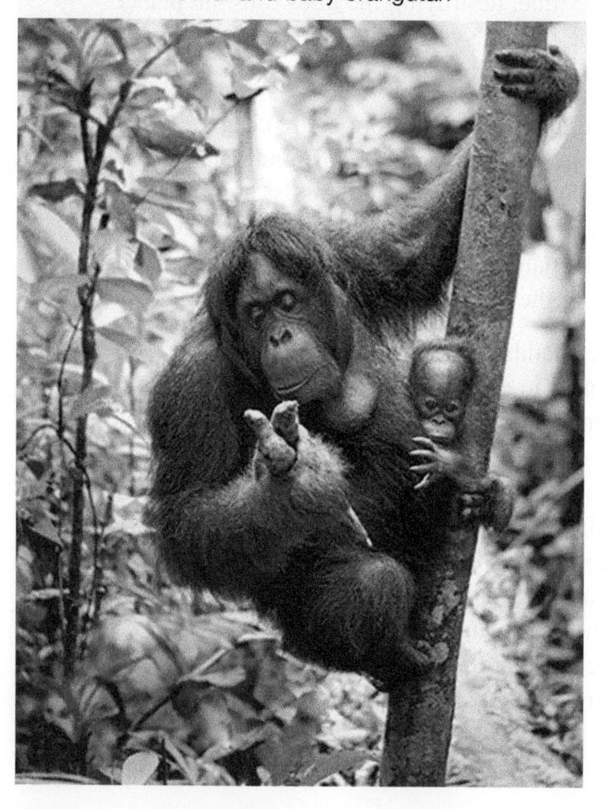

FIGURE 4 Orangutan distribution in Borneo, 1930–2015

Source: IUCN Red List

FIGURE 5 Rainforest distribution in Borneo, 1950–2020

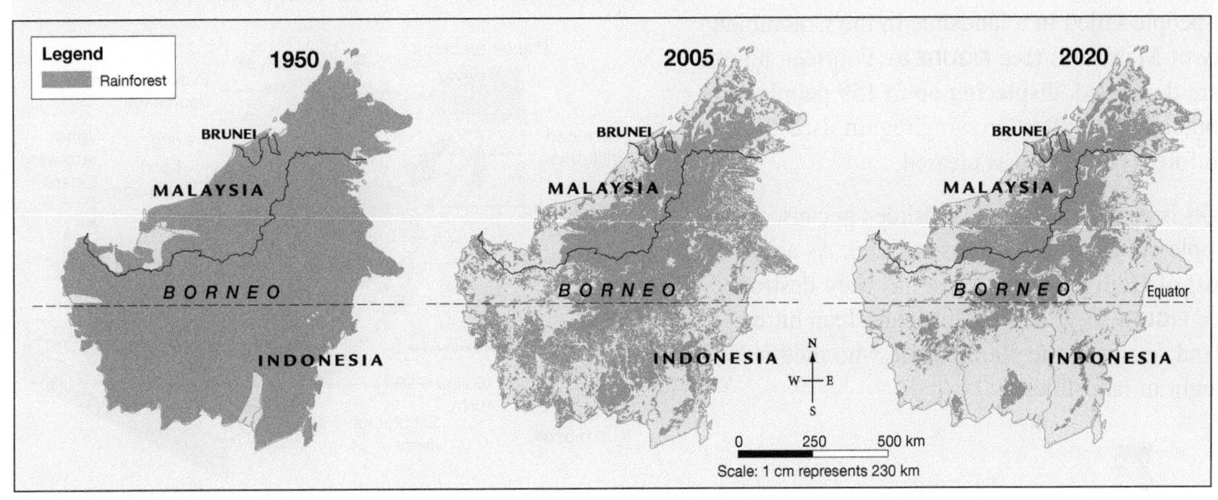

Source: Spatial Vision

6.8.3 Social impacts

Indigenous Peoples

As forests are cleared and new settlers move into a region, the Indigenous Peoples of the area are often displaced and their culture may disappear. The homesteaders bring new diseases to which Indigenous people have no natural immunity. One tribe, the Nambiquara of Brazil, lost half its population to illness when a road was placed through its land.

Indigenous Peoples aren't often given a choice about 'progress' coming to their section of the rainforest. As a result, tension can be created between these tribes and the government. In 1999 the Bakun Dam Project began in Malaysia, resulting in the eviction of approximately 10 000 people from their homeland. As compensation they were resettled, but the land provided was too small to support their traditional methods of hunting and agriculture and many failed to adapt to their new lifestyles.

Landslides

A landslide, the downward movement of earth and rocks on a slope, occurs in the lithosphere (see subtopic 6.2). It can be caused by natural physical processes such as rainfall and earthquakes, or by human activities such as deforestation and road building. Usually, the roots of rainforest plants keep the soil together and add stability to mountainous areas. This is especially important during times of heavier rainfall. However, sometimes the ground becomes so waterlogged that the roots can't keep the soil in place and it slips downhill, creating a landslide. The risk of this increases if deforestation has taken place on the hillside, as there are no tree roots to provide added stability (see **FIGURE 6**).

Therefore, when these hills are cleared and settled by communities, the danger of property damage, and even death, increases. November 2011 saw 35 people killed in a landslide in the Colombian city of Manizales (see **FIGURE 8**). Fourteen houses were destroyed, displacing up to 159 people. This mountainous, coffee-growing region used to be rainforest before it was cleared.

Haiti is at a high risk of landslides because its people cut down trees to use as fuel. As a result, most of Haiti's natural forest has been destroyed (see **FIGURE 9**). In 2004, Hurricane Jean hit the island; many of the 3000 people who died were caught in landslides.

FIGURE 6 A forested hillside (a) before and (b) after deforestation

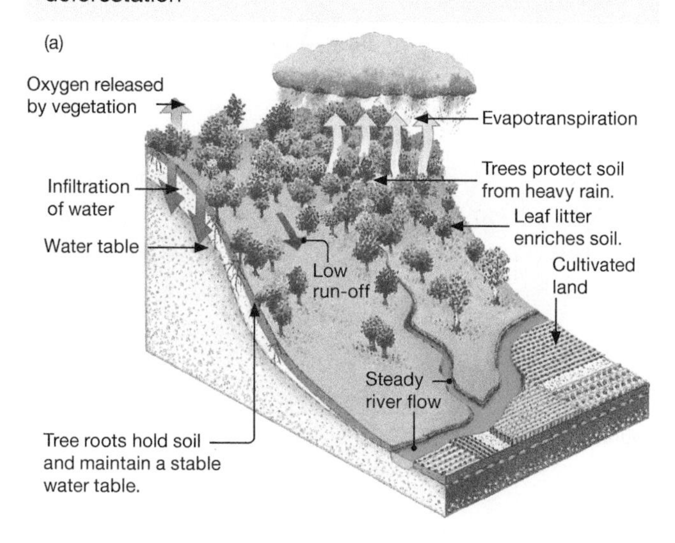

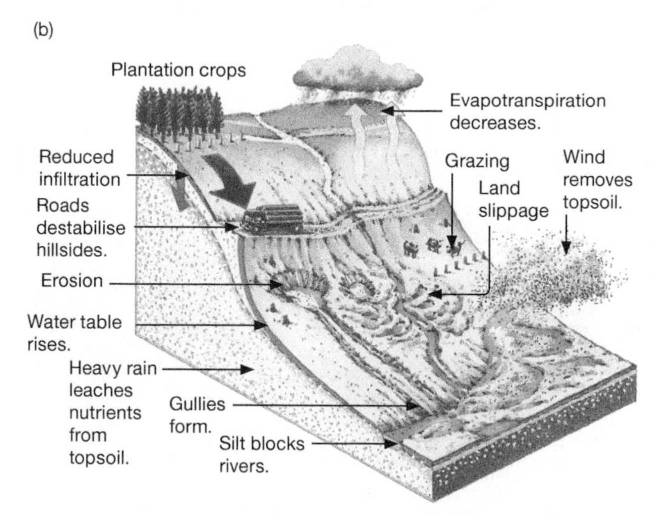

FIGURE 7 Landslides in intact forest in the Mata Atlantica rainforest, Brazil

FIGURE 8 Landslide in Manizales, Colombia, in November 2011

Disease

The arrival of new tropical diseases is a less obvious result of deforestation. As animal **hosts** disappear and new human settlers move into previously inaccessible areas, 'new' disease-causing microorganisms are transferred to the human population. The frequency of mosquito-borne diseases such as malaria has increased due to the creation of more water puddles — for example, in ditches and tyre treads — which are an excellent breeding ground for the mosquito. It is estimated that malaria is responsible for the deaths of 20 per cent of the Yanomami people in Brazil and Venezuela. Today, more than 99 per cent of malaria cases in Brazil occur in the Amazon Basin region, even though the mosquitoes that carry the disease are found across 80 per cent of the country.

The outbreak of such diseases doesn't affect only the local area. The impact can also spread into other countries via people who visit these areas, unknowingly contract an illness and then travel home, spreading the disease along the way.

The spread of COVID-19 in Brazil also prompted communities to protest the government's response to the pandemic — especially the lack of help provided to remote rainforest communities. In Novo Progresso, Brazil, members of the Kayapó people blocked a major highway in protest.

host an organism that supports another organism

FIGURE 9 Deforestation visible at the Haiti–Dominican Republic border: Haiti is on the left and the Dominican Republic is on the right

FIGURE 10 Members of the Kayapó people block the BR-163 motorway in Novo Progresso, Brazil, August 2020.

fdw-0008

FOCUS ON FIELDWORK

Interview techniques — What should I ask?

One activity that can help you better understand a place when you are conducting fieldwork is interviewing local people or representatives of companies or organisations who operate in the area. But it's not always easy to know what to ask, or how to ask a question to get the specific information you need.

Build your ability to ask great interview questions with the **Designing interview questions** fieldwork activity in your Resources tab.

Resources

eWorkbook	Social and environmental impacts of deforestation (ewbk-9062)
Video eLesson	Social and environmental impacts of deforestation — Key concepts (eles-5069)
Interactivity	Deforestation dilemma (int-3113)
Fieldwork	Designing interview questions (fdw-0008)
myWorldAtlas	Deepen your understanding of this topic with related case studies and questions. **Orangutans**

6.8 ACTIVITIES

1. Research and create a list of 10 other animal species threatened by deforestation around the world. Choose one of these animals and report back to the class on its current location, the remaining population level and the main causes of deforestation. Present your report as a poster, PowerPoint presentation, movie (documentary), poem, song or drama performance.
2. Using the internet, investigate two management strategies, policies or laws that have been implemented around the world to try to conserve the rainforest environment. Note the positive and negative aspects of these strategies. Comment on their ability to support the sustainable use of rainforests. Discuss your results as a class. Create a summary on the board to evaluate all the options that are shared.
3. Produce an A4-sized poster designed to publicise the rate and consequences of rainforest destruction. Your poster must include a colourful diagram and a short slogan based on the facts and figures presented in this subtopic.
4. Research the ways that indigenous peoples from throughout the Amazon have protested and resisted deforestation, mining, new settlements and their effects. As part of your research, investigate whether there have been different strategies used in different countries, and what each country is doing to protect the rainforest and its peoples. (If you need help getting started, investigate responses to the Belo Monte dam in Brazil.)

6.8 EXERCISE

Learning pathways

■ LEVEL 1	■ LEVEL 2	■ LEVEL 3
1, 2, 3, 9, 13	4, 6, 8, 11, 14	5, 7, 10, 12, 15

Check your understanding

1. Name three species threatened by deforestation in Indonesia.
2. Define the term *habitat*.
3. List the main threats to orangutans.
4. What is the interconnection between deforestation and the impact of disease on indigenous peoples?
5. Why were people in remote rainforest areas of Brazil protesting their government's actions in response to the COVID-19 pandemic?
6. Outline the negative impacts of clearing a rainforest area with fire.
7. What is transferred between the atmosphere and plants during evapotranspiration?

Apply your understanding

8. Explain how deforestation can affect the:
   a. lithosphere
   b. atmosphere
   c. biosphere.
9. Why does having separate small islands of vegetation make it difficult for animals to communicate and breed?

10. Refer to **FIGURES** 4 and 5. Describe the interconnection between the two sets of data.
11. Refer to **FIGURE** 6. Write a paragraph that explains how deforestation results in the consequences and changes illustrated in the diagram.
12. Explain why it is important to save species from extinction.

Challenge your understanding

13. Indonesia recently granted a licence to a pulp paper producer to clear 50 000 hectares of forest near an orangutan sanctuary in Sumatra. What impact might this have on the orangutan population?
14. What could the consequences be if the rainforest continues to be cleared at its current rate?
15. Do you think there are species of plants and animals in Australian rainforests that have not been 'discovered' (named and catalogued by scientists)? Explain why this is/is not likely.

To answer questions online and to receive **immediate feedback** and **sample responses** for every question, go to your learnON title at www.jacplus.com.au.

6.9 Saving and conserving rainforests

online only

LEARNING INTENTION

By the end of this subtopic, you will be able to explain reasons why rainforests should be protected and identify ways that this can be done.

The content in this subtopic is a summary of what you will find in the online resource.

As people begin to realise the importance of rainforests, many have started to work towards preserving these valuable 'green dinosaurs'. Some methods of conservation are relevant only to governments and large companies, but some are relevant to you and the choices you make.

While most of us do not have rainforests growing in our backyards, the choices we make each day can and do make a difference to the way resources are used around the world. There are many organisations that aim to conserve the world's remaining rainforests.

To learn more about saving and conserving rainforests, go to your learnON resources at www.jacPLUS.com.au.

FIGURE 3 The keel-billed toucan inhabits the rainforests of Costa Rica.

Contents

on Resources

- **eWorkbook** Saving and conserving rainforests (ewbk-9066)
- **Video eLesson** Saving and conserving rainforests — Key concepts (eles-5070)
- **Interactivity** Protecting or plundering rainforests (int-3114)

6.10 Investigating topographic maps — Features of the Daintree rainforest

LEARNING INTENTION

By the end of this subtopic, you will be able to locate and describe the features of the Daintree rainforest on a topographic map.

6.10.1 Daintree rainforest

The Daintree National Park is home to one of the oldest surviving rainforests on Earth. The Daintree rainforest, located in far north Queensland, is part of the Wet Tropics of Queensland, which stretches along the north-east coast of Australia for 450 kilometres. This area was recognised as a World Heritage area in 1988 for its significance to Australian geological history and biological diversity.

FIGURE 1 The Daintree area of northern Queensland: (a) A mangrove on the mudflats of Cooya Beach; (b) Newell beach; (c) Mossman Gorge

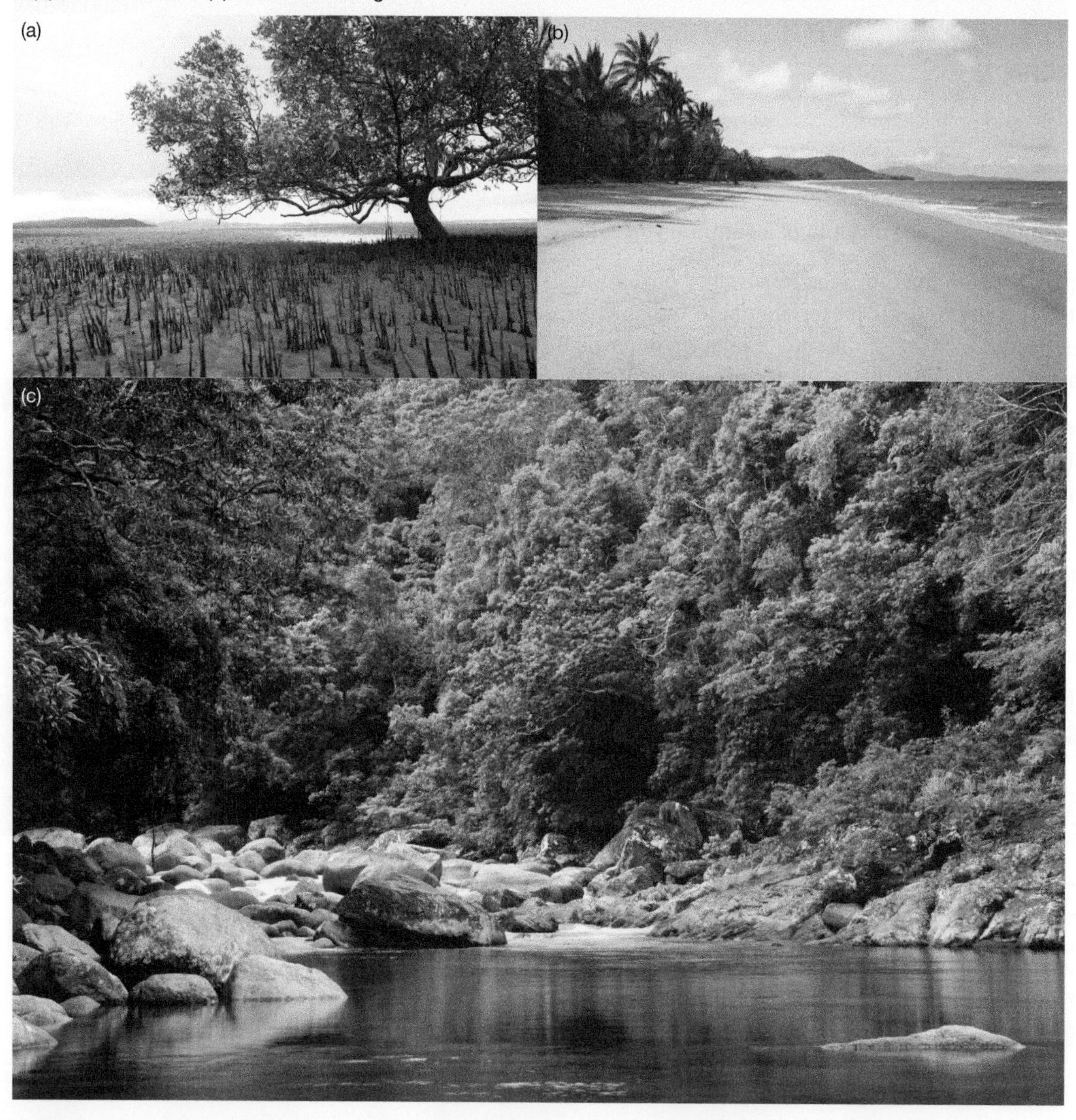

FIGURE 2 Topographic map extract of the south-eastern section of the Daintree River National Park

Legend

- 121 • Spot height
- ▪ Building
- Index contour
- Contour (interval: 25 m)
- route no. B56 Highway
- Major road
- Minor road
- Track
- Rail
- Stream
- River
- Water body
- Intertidal zone
- Vegetation
- Mangrove
- Urban area
- National park

N, W, E, S

0 2 4 km

Scale: 1 cm represents 1 km

Source: Data based on QSpatial, State of Queensland (Department of Natural Resources, Mines and Energy, Department of Environment and Science), http://qldspatial.information.qld.gov.au/catalogue/

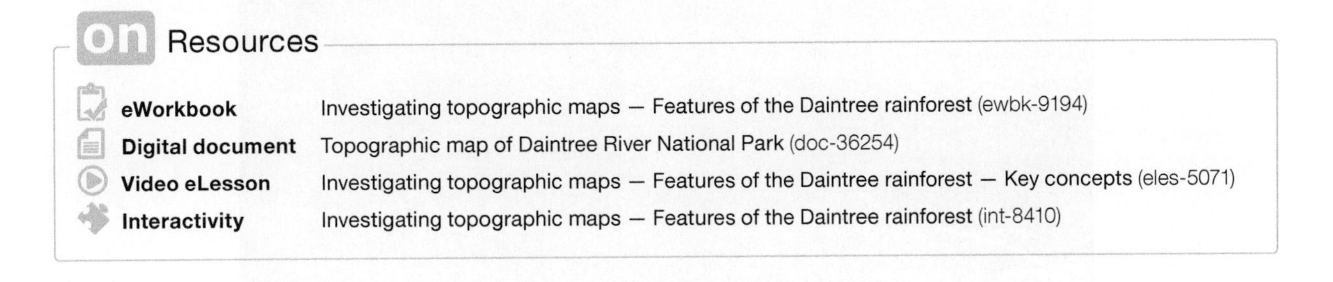

6.10 EXERCISE

Learning pathways

■ LEVEL 1	■ LEVEL 2	■ LEVEL 3
1, 2, 8, 11	3, 6, 7, 10	4, 5, 9, 12

Check your understanding

1. List the types of vegetation that can be seen on the map.
2. What types of farming occur in this area?
3. List the features found at the following locations.
   a. GR298835
   b. GR298807
   c. GR268831
   d. GR307760
4. Describe what the landscape would look like if you were at the following locations.
   a. GR195768
   b. GR335765
   c. GR272867
5. Which of these locations is the highest: GR252763 or GR301759?
6. Give the six-figure grid references for:
   a. Dayman Point
   b. Port of Mossman
   c. Miallo.

Apply your understanding

7. What features shown on the map provide evidence that this is a rainforest landscape?
8. What impact might extending the settlement of Mossman to the west have on the rainforest?
9. Explain three ways that humans have had an impact on this rainforest environment.

Challenge your understanding

10. Predict what would happen to the agricultural land in this area if it was left without any human intervention for ten years.
11. Suggest what negative impacts agriculture might be having on the rainforest in this area.
12. Propose one way that the economy of this area could be boosted without clearing additional forest for agriculture or settlements.

FIGURE 3 Daintree National Park, Cape Tribulation, Queensland

To answer questions online and to receive **immediate feedback** and **sample responses** for every question, go to your learnON title at www.jacplus.com.au.

6.11 Thinking Big research project — Rainforest display

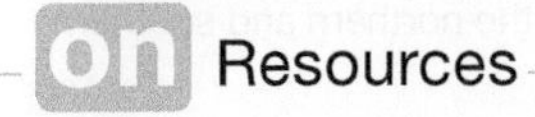

The content in this subtopic is a summary of what you will find in the online resource.

Scenario

Over the past 10 000 years, human activities have reduced the Earth's forest by about one-third. Trees have been removed to make way for urban development and agriculture, and to obtain fuel and building materials. Today, approximately 34 per cent of the world's land area is covered by forest.

Rainforests are one of the most complicated environments on Earth. Mostly found in warm, moist areas near the equator, rainforests contain nearly three-quarters of all varieties of life on Earth and perform several important functions on our planet. Australia's rainforests are very important, as they contain half of our plants and one-third of our native animals in a very small area (20 000 square kilometres).

Task

You have been commissioned by the Department of Natural Resources and Environment to complete an in-depth study of rainforests and present your information on their website, as part of an ongoing educational program.

Go to your Resources tab to access all of the resources you need to complete this project.

on Resources

ProjectsPLUS Thinking Big research project — Rainforest display (pro-0172)

6.12 Review

6.12.1 Key knowledge summary

6.2 Rainforest characteristics

- There are different types of rainforest: montane, temperate and lowland.
- Rainforests have similar characteristics and a similar structure with distinct layers.

6.4 The value of rainforest environments

- Australia was once covered by rainforests. Over time the gradual movement of the continent southwards has resulted in a dramatic decrease in the amount of rainforest found in Australia.
- Australia's tropical rainforests have World Heritage status. Aboriginal Peoples and Torres Strait Islander Peoples have lived in Australia's rainforests for thousands of years.
- Many products found in our pantries and medicine cabinets have their origins in the rainforest.

6.6 Living in the rainforest

- Indigenous Peoples rely on the rainforest to supply all their needs, but many now use a combination of traditional and modern ways of living.

6.7 Disappearing rainforests

- The major issue facing rainforests is deforestation, mainly due to commercial logging, farming and mining activities. Rainforests in developing nations are most at risk. Here the population is expanding rapidly, and the people are poor. Using rainforests is viewed as more valueable than preserving them.
- The Amazon Basin is the world's largest remaining rainforest and plays an important role in controlling the world's climate and oxygen supply.

6.8 Social and environmental impacts of deforestation

- Deforestation has a dramatic impact on the environment. The regulating effect on the planet of forests is lost and carbon dioxide is released into the atmosphere, accelerating global warming. The land becomes more prone to landslides once vegetation is removed, posing a threat to the inhabitants of the region.
- As the rainforest becomes fragmented, animal species such as the orangutan lose their habitat and become isolated. Entire species are threatened with extinction.
- Indigenous Peoples can no longer live in traditional ways and many lose their lands. Outsiders bring diseases putting entire populations at risk.

6.9 Saving and conserving rainforests

- Only a small proportion of rainforests is protected, the smallest proportion in developing countries.
- Sustainable development, finding alternatives to timber products and educating the public are some of the measures being used to manage and preserve rainforests.

6.12.2 Key terms

clearfelling a forestry practice in which most or all trees and forested areas are cut down

compost a mixture of various types of decaying organic matter such as dung and dead leaves

convectional rainfall heavy rainfall as a result of thunderstorms, caused by rapid evaporation of water by the Sun's rays

ecosystem an interconnected community of plants, animals and other organisms that depend on each other and on the non-living things in their environment

ecotourist a tourist who travels to threatened ecosystems in order to help preserve them

epiphyte a plant that grows on another plant but does not use it for nutrients

equatorial near the equator, the line of latitude around the Earth that creates the boundary between the northern and southern hemispheres

evapotranspiration the process by which water is transferred to the atmosphere from surfaces such as the soil and plants

gorge narrow valley with steep rocky walls

habitat the total environment where a particular plant or animal lives, including shelter, access to food and water, and all of the right conditions for breeding
host an organism that supports another organism
hydroelectric dam a dam that harnesses the energy of falling or flowing water to generate electricity
ice age historical period during which the Earth is colder, glaciers and ice sheets expand and sea levels fall
microclimate specific atmospheric conditions within a small area
nomadic a group that moves from place to place depending on the food supply, or pastures for animals
selective logging a forestry practice in which only selected trees are cut down
shifting agriculture process of moving gardens or crops every couple of years because the soils are too poor to support repeated sowing
species a biological group of individuals having the same common characteristics and being able to breed with each other
subsistence producing only enough crops and raising only enough animals to feed yourself and your family or community
sustainable development economic development that causes a minimum of environmental damage, thereby protecting the interest of future generations
temperate describes the relatively mild climate experienced in the zones between the tropics and the polar circles

6.12.3 Reflection

Complete the following to reflect on your learning.

Revisit the inquiry question posed in the Overview:

We can plant new trees any time and anywhere. What makes the world's rainforests so special?

1. Now that you have completed this topic, what is your view on the question? Discuss with a partner. Has your learning in this topic changed your view? If so, how?
2. Write a paragraph in response to the inquiry question, outlining your views.

Subtopic	Success criteria			
6.2	I can describe types of rainforest and identify where they are located.			
	I can describe the characteristics of rainforests.			
6.3	I can create an overlay map to show relationships between factors in a landscape.			
6.4	I can outline the ways that rainforests are valued and provide examples.			
6.5	I can create a précis map to highlight a feature of a landscape.			
6.6	I can explain how Indigenous Peoples value and use rainforests.			
6.7	I can explain and give examples of reasons why rainforests are disappearing.			
	I can explain some of the consequences of the loss of rainforests.			
6.8	I can identify and explain the social impacts of deforestation.			
	I can identify and explain the environmental impacts of deforestation.			
6.9	I can explain the reasons why rainforests should be conserved.			
	I can identify ways to conserve rainforests.			
6.10	I can locate and describe the features of the Daintree rainforest on a topographic map.			

on Resources

eWorkbook Chapter 6 Student learning matrix (ewbk-8456)
Chapter 6 Reflection (ewbk-8457)
Chapter 6 Extended writing task (ewbk-8458)

Interactivity Chapter 6 Crossword (int-7599)

ONLINE RESOURCES

Below is a full list of **rich resources** available online for this topic. These resources are designed to bring ideas to life, to promote deep and lasting learning and to support the different learning needs of each individual.

eWorkbook

- 6.1 Chapter 6 eWorkbook (ewbk-7986) ☐
- 6.2 Rainforest characteristics (ewbk-0750) ☐
- 6.3 SkillBuilder — Creating and describing complex overlay maps (ewbk-9042) ☐
- 6.4 The value of rainforest environments (ewbk-9046) ☐
- 6.5 SkillBuilder — Drawing a précis map (ewbk-9050) ☐
- 6.6 Living in the rainforest (ewbk-9054) ☐
- 6.7 Disappearing rainforests (ewbk-9058) ☐
- 6.8 Social and environmental impacts of deforestation (ewbk-9062) ☐
- 6.9 Saving and conserving rainforests (ewbk-9066) ☐
- 6.10 Investigating topographic maps — Features of the Daintree rainforest (ewbk-9194) ☐
- 6.12 Chapter 6 Student learning matrix (ewbk-8456) ☐
  Chapter 6 Reflection (ewbk-8457) ☐
  Chapter 6 Extended writing task (ewbk-8458) ☐

Sample responses

- 6.1 Chapter 6 Sample responses (sar-0141) ☐

Digital document

- 6.10 Topographic map of Daintree River National Park (doc-36254) ☐

Video eLessons

- 6.1 Protecting our landscapes — Rainforest (eles-1627) ☐
  Rainforest — Photo essay (eles-5336) ☐
- 6.2 Rainforest characteristics — Key concepts (eles-5065) ☐
- 6.3 SkillBuilder — Creating and describing complex overlay maps (eles-1656) ☐
- 6.4 The value of rainforest environments — Key concepts (eles-5066) ☐
- 6.5 SkillBuilder — Drawing a précis map (eles-1657) ☐
- 6.6 Living in the rainforest — Key concepts (eles-5067) ☐
- 6.7 Disappearing rainforests — Key concepts (eles-5068) ☐
- 6.8 Social and environmental impacts of deforestation — Key concepts (eles-5069) ☐
- 6.9 Saving and conserving rainforests — Key concepts (eles-5070) ☐
- 6.10 Investigating topographic maps — Features of the Daintree rainforest — Key concepts (eles-5071) ☐

Interactivities

- 6.2 Our living green dinosaurs (int-3112) ☐
  World rainforest types (int-8736) ☐
- 6.3 SkillBuilder — Creating and describing complex overlay maps (int-3152) ☐
- 6.4 Australia's rainforest areas (int-7853) ☐
- 6.5 SkillBuilder — Drawing a précis map (int-3153) ☐
- 6.6 Living in the rainforest (int-8409) ☐
- 6.7 The Amazon Basin (int-7854) ☐
  Development in the Amazon rainforest (int-7855) ☐
- 6.8 Deforestation dilemma (int-3113) ☐
- 6.9 Protecting or plundering rainforests (int-3114) ☐
- 6.10 Investigating topographic maps — Features of the Daintree rainforest (int-8410) ☐
- 6.12 Chapter 6 Crossword (int-7599) ☐

Fieldwork

- 6.2 Daintree rainforest (fdw-0007) ☐
- 6.8 Designing interview questions (fdw-0008) ☐

ProjectsPLUS

- 6.11 Thinking Big research project — Rainforest display (pro-0172) ☐

Weblinks

- 6.2 Rainforest layers (web-4812) ☐
- 6.4 UNESCO Heritage (web-0095) ☐
  Rainforest foods (web-4811) ☐
- 6.6 Forest guardians (web-4084) ☐
- 6.7 Amazon tour (web-4083) ☐

myWorld Atlas

- 6.8 Orangutans (mwa-4488) ☐

Teacher resources

There are many resources available exclusively for teachers online.

7 Fieldwork inquiry — How waterways change from source to sea

7.1 Overview

Numerous **videos** and **interactivities** are embedded just where you need them, at the point of learning, in your learnON title at www.jacplus.com.au. They will help you to learn the content and concepts covered in this topic.

LEARNING INTENTION

By the end of this subtopic, you will be able to plan fieldwork related to waterways, identify relevant fieldwork techniques to investigate an inquiry question and analyse your data, and communicate your findings in an appropriate way.

7.1.1 Scenario and task

Task: Produce a report and presentation about the way in which your local catchment changes from the upper to the lower reaches.

Everybody lives in a catchment and its health is influenced by the activities in all areas within it. Your local water authority has received contradictory reports about the current state and health of your local catchment. As the reports are contradictory and the local water authority is not sure which is valid and which is not, they need to undertake a detailed study of the natural and built environments in the local catchment area. This will put them in an expert position to question and quash statements made by non-experts.

Go to learnON to access the following additional resources to complete this fieldwork task:

- fieldwork data recording templates
- a report template
- a presentation template
- a selection of images and audio and video files to add richness to your presentation
- weblinks to sites to assist in your catchment research and to provide sample presentations
- an assessment rubric.

FIGURE 1 In this inquiry, you will evaluate the health of a water catchment.

Your task

Your team has been commissioned by the local water authority to compile and present a report evaluating the current state of your local catchment. Your team must gather data to investigate how the catchment changes from the upper to the lower reaches. Your investigations will cover river characteristics such as depth, width and other channel characteristics, the fauna and flora in the area, and the land use in the catchment. In order to ensure that your report is accurate, your team can gather data about a local waterway and its immediate catchment by observing, collecting, interpreting and presenting your findings.

7.2 Inquiry process

7.2.1 Process

Open the ProjectsPLUS application for this chapter located in your Resources tab. Watch the introductory video lesson and then click the 'Start Project' button and set up your project group. You can complete this project individually or invite members of your class to form a group. Save your settings and the project will be launched.

Planning: Navigate to your Research Forum. You will need to research the characteristics of your local catchment area. In order to complete sufficient research, you will need to visit a number of sites within the catchment, comparing different locations upstream and downstream of one creek or river. Research topics have been loaded in the system to provide a framework for your research:

- **What** sort of data and information will you need to collect at your fieldwork sites?
- **How** will you collect and record this information?
- **Where** would be the best locations to obtain data? You can determine this once you know which waterway(s) you will be visiting.
- **How** will you record the information you are collecting? Consider using GPS, video recorders, cameras and mobile devices (laptop computer, tablet, mobile phone).

Before going out into the field, examine topographic maps and aerial photos or satellite images of the relevant area to identify key landmarks (such as the location of your school and the location of the waterway relative to the school). Locate the catchment boundary, the path of the waterway and the watercourse it contributes to. Construct a sketch map of the waterway — this map should show the catchment boundary/watershed, the river channel and the direction in which the water is flowing. Clearly note compass directions on the map. Gather spatial (mapped data) information about the region (using, for example, street directories, topographic maps, aerial photos and satellite images from sources such as Google Earth) and information about planning, population, land use, and flora and fauna. Refer to your Media Centre for potential sources of information.

Discuss with your group what you might already know about your catchment and then divide the research tasks between you. Discuss the information you will be looking for and where you might find it. Choose land use categories that you will be able to recognise and a mapping symbol to be used for each. The weblinks in your Media Centre will help you get started. You can view and comment on other group members' articles and rate the information they have entered.

7.2.2 Collecting and recording data

Depending on the catchment you visit, you could investigate some or all of the following:

- channel depth at various points across the stream
- channel width or cross-section
- stream flow velocity (how fast the water is flowing)
- flora transects
- fauna surveys
- land-use surveys.

Other relevant observations may include:

- condition of the waterway banks
- native and exotic vegetation
- cleared land
- evidence of erosion
- land-use zones
- potential pollution sources (including stormwater drains entering the waterway and sewage overflow points)
- building sites, industrial and residential areas
- pollution control devices
- erosion control.

Ensure that you take relevant measuring equipment into the field, and that several measurements are taken at each site. It is useful to divide tasks among groups and then share data when you are back at school. Use a copy of your map to record the information at each site.

7.2.3 Processing and analysing your information and data

Once you have collected, collated and shared your data, you will need to decide what information to include in your report and the most appropriate way to show your findings. If using spreadsheet data, make total and percentage calculations. Some measurements are best presented in a table; others in graphs or on maps. If you have used a spreadsheet, you may like to produce your graphs electronically. Use photographs as map annotations (either scanned and attached to your electronic map or attached to your hand-drawn map) to show features recorded at each site. You may also like to annotate each photograph to show the geographical features you observed. Describing and interpreting your data is important.

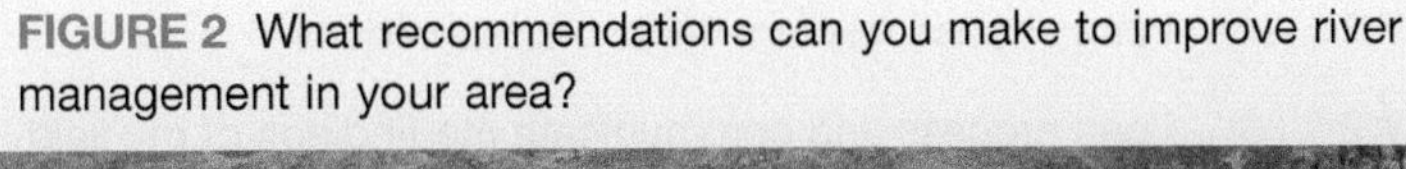

FIGURE 2 What recommendations can you make to improve river management in your area?

Visit your Media Centre and download the report template and the presentation planning template to help you complete this project. Your Media Centre also includes images, videos and audio files to help bring your presentation to life. Use the report template to create your report. Use the presentation template to create an engaging presentation that showcases all of your important findings.

7.2.4 Communicating your findings

You will now produce a fieldwork report and presentation of your findings. Your report should include all of the research that you completed and all evidence to support your findings. Ensure that your report includes a title, an aim, a hypothesis (what you think you will find, which is written before you go into the field), your findings and a conclusion. You will also need to recommend some type of action that needs to be taken to improve river management at the creek or river you visited.

7.3 Review

7.3.1 Reflecting on your work

Think back over how well you worked with your group on the various tasks for this inquiry. Determine strengths and weaknesses and recommend changes you would make if you were to repeat the exercise. Identify one area where you were pleased with your performance, and an area where you would like to improve. Write two sentences outlining how you might be able to do this.

Print your Research Report from ProjectsPLUS and hand it in with your fieldwork report and presentation, and reflection notes.

7.3.2 Reflection

Subtopic	Success criteria			
7.2	I can plan fieldwork related to waterways.			
7.2	I can choose relevant fieldwork techniques.			
7.2	I can process and communicate the findings of my fieldwork.			

on Resources

ProjectsPLUS Fieldwork inquiry — How waterways change from source to sea (pro-0145)

FIGURE 3 Allyn River near Gresford, NSW

UNIT

2 Place and liveability

8 How liveability differs around the world

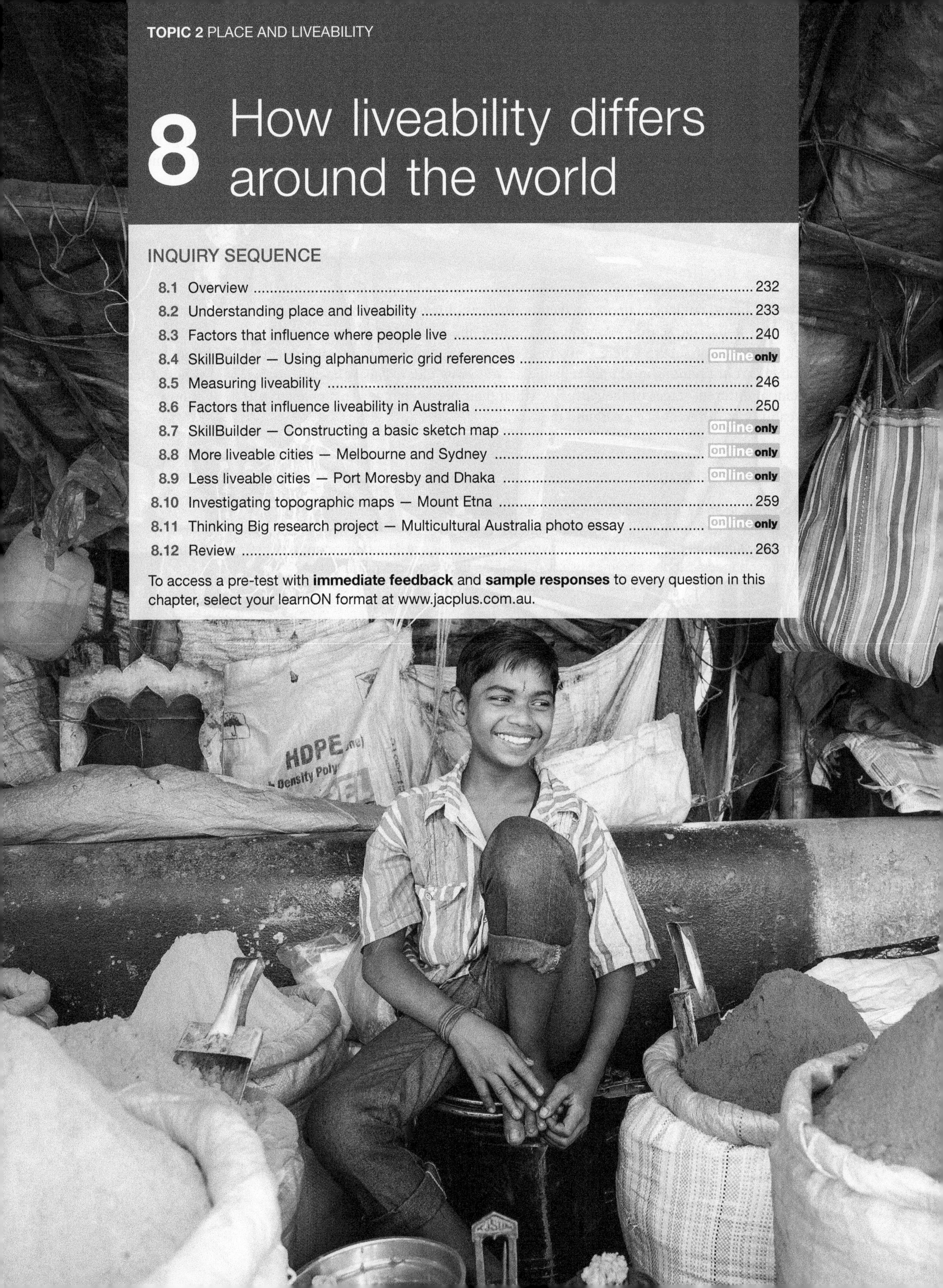

INQUIRY SEQUENCE

To access a pre-test with **immediate feedback** and **sample responses** to every question in this chapter, select your learnON format at www.jacplus.com.au.

8.1 Overview

Numerous **videos** and **interactivities** are embedded just where you need them, at the point of learning, in your learnON title at www.jacplus.com.au. They will help you to learn the content and concepts covered in this topic.

What does a place need to have to make it liveable? Is everyone's idea of liveability the same?

8.1.1 Introduction

Your quality of life is influenced by many factors, such as climate, landscape, community facilities, the location of your home, the sense of community identity and links to other settlements. You might have an idea of a street, town, city or suburb where you would like to live, and your opinions may be quite different from those of others. Your opinions might also change as you grow older. This is because people see different factors as important at different stages of their lives or when their needs change. This chapter will look at how people define liveability in different places around the world.

STARTER QUESTIONS

1. Why do people enjoy living in different kinds of places?
2. How can the natural environment, environmental quality and access to services affect people's wellbeing?
3. Watch **How liveability differs around the world — Photo essay**. How important are feeling like you belong and being connected to the community when it comes to choosing a place to live?
4. How can people improve the liveability of places?

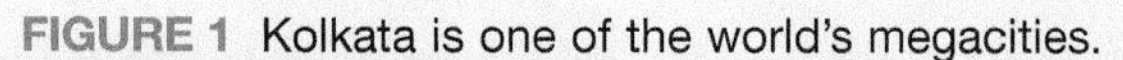

FIGURE 1 Kolkata is one of the world's megacities.

on Resources

eWorkbook Chapter 8 eWorkbook (ewbk-7988)

Video eLessons Choosing a place to live (eles-1619)
How liveability differs around the world — Photo essay (eles-5337)

8.2 Understanding place and liveability

LEARNING INTENTION

By the end of this subtopic, you will be able to define the term *liveability* and suggest reasons why people's perceptions of place and liveability differ.

8.2.1 Creating a sense of place

Places are central to the study of Geography. This is because geographers are interested in where things are found on Earth and why they are there. But what exactly is a place?

To understand what a place is, think about **location** and **region**. Each place has a unique identity that makes it different from other places. A combination of characteristics is specific to that place, making it individual.

The characteristics of a place can come from:

1. natural features
2. human features (features built by people)
3. a combination of natural and human features.

Eventually, one or more of these features becomes a symbol of that place in people's minds. Consider the natural and human features in **FIGURES 1–10**. How do the characteristics of each place affect the way you think about that place? What feature might be a symbol of that place?

place specific area of the Earth's surface that has been given meaning by people

location a point on the surface of the Earth where something is to be found

region any area of varying size that has one or more characteristics in common

FIGURE 1 (a) Hawkesbury River near Sydney (b) New York City (c) Mount Buller in Victoria

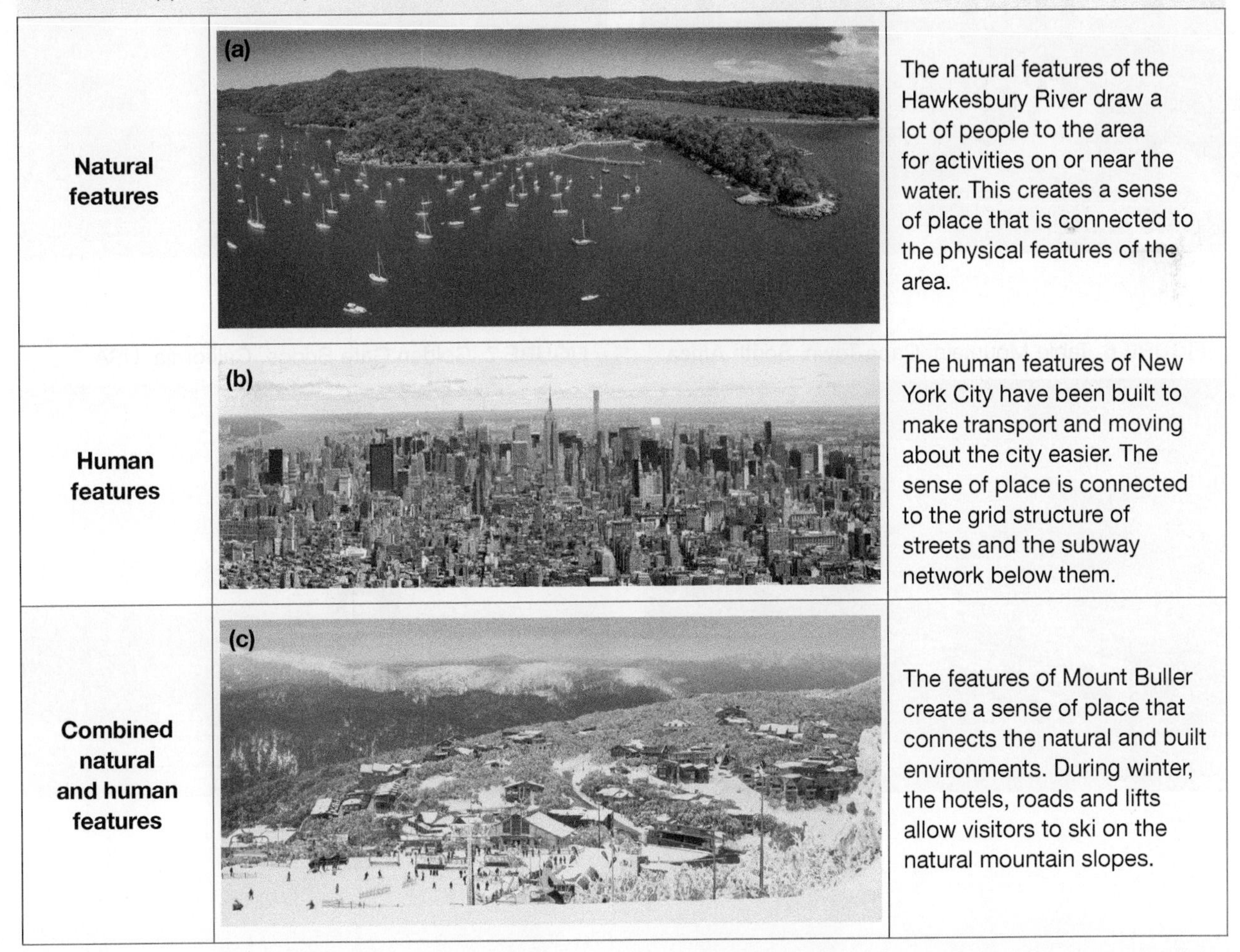

Natural features	(a)	The natural features of the Hawkesbury River draw a lot of people to the area for activities on or near the water. This creates a sense of place that is connected to the physical features of the area.
Human features	(b)	The human features of New York City have been built to make transport and moving about the city easier. The sense of place is connected to the grid structure of streets and the subway network below them.
Combined natural and human features	(c)	The features of Mount Buller create a sense of place that connects the natural and built environments. During winter, the hotels, roads and lifts allow visitors to ski on the natural mountain slopes.

FIGURE 2 Rio de Janeiro, Brazil

FIGURE 3 Disney World, Orlando, Florida, USA

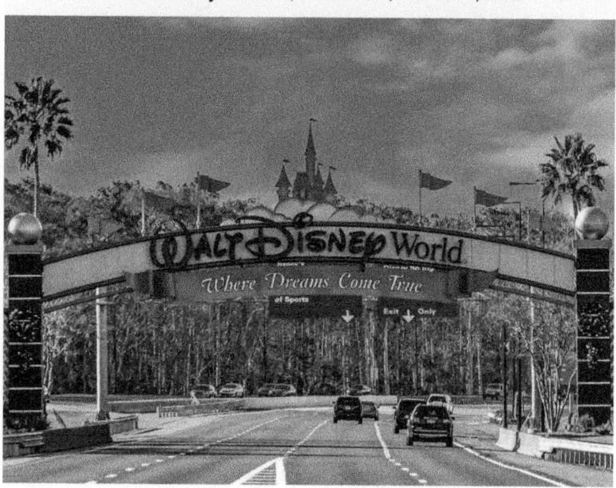

FIGURE 4 Taj Mahal, Agra, India

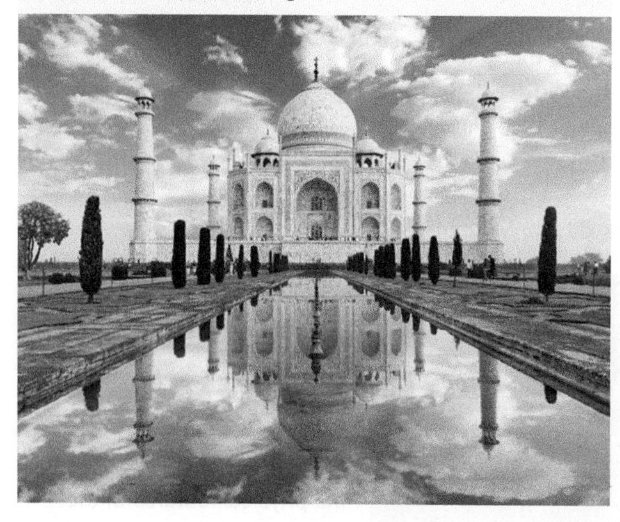

FIGURE 5 Grand Canyon, Arizona, USA

FIGURE 6 Table Mountain, Cape Town, South Africa

FIGURE 7 Golden Gate Bridge, California, USA

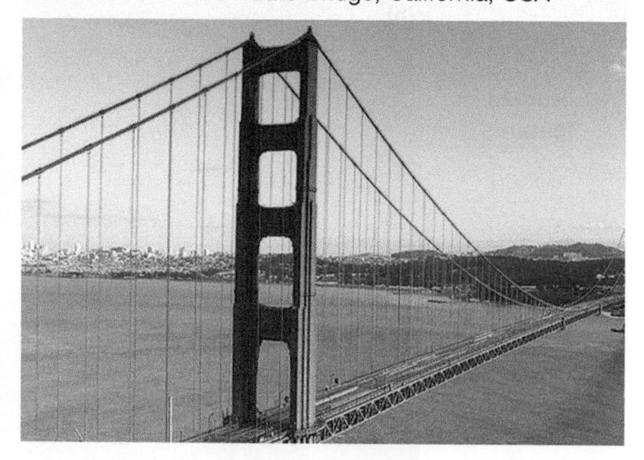

FIGURE 8 Great Wall of China

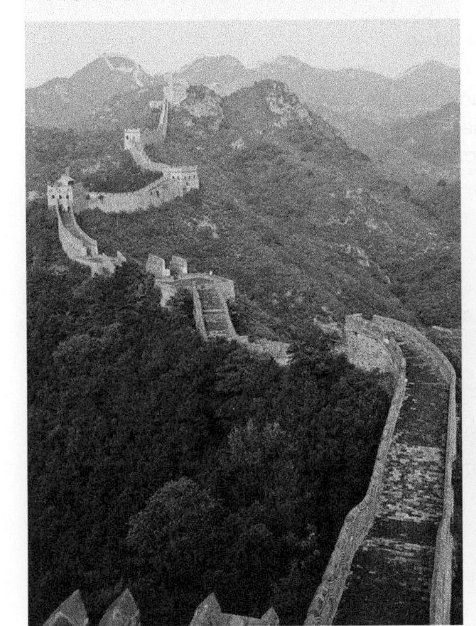

FIGURE 9 Purnululu, Western Australia

FIGURE 10 Machu Picchu, Peru

8.2.2 What people think about liveability

If you were told that Vancouver or Melbourne was the world's best place to live in, or the world's most liveable city, what would you think this means? Do city councils just brag about how good their city is or can **liveability** be measured? Is liveability the quality of life experienced by a city's residents? How do you measure quality? Is it based on wealth, happiness, opportunities or something else?

Liveability, or how liveable a place is, measures the quality of life people in that place experience. There are many factors that might be considered when assessing liveability, such as the cleanliness of the environment, how safe a place is, whether there are good jobs available, how expensive it is to live there, whether there are interesting things to do, or how easily you can access important services such as healthcare or education. Different people often come to different conclusions about how liveable a place is because they value different things.

FIGURE 11 shows examples of how different people live around the world, with about the same monthly income. Each family picture shows happy people, but what's inside each home varies. Consider how the term *liveability* might mean different things for each member of these families, or whether there might be similarities in what they value as being important to their quality of life.

liveability an analysis of what it is like to live somewhere, based on a set of criteria

FIGURE 11 How might liveability differ around the world?

The Rayez family

Country: France; **Income:** US$875 (month)
Electricity: In-house, reliable; **Water:** In-house, safe to drink
Food: Electricity and biogas cooking; all food bought
Nathalie (54) works in sales. Her daughter, Cliea (18), and her son, Edeban (14), are both students. The family lives in a three-bedroom house that they own and have been living in for 14 years. The next big thing they plan to buy is a new car.

Kitchen	Living room

The Hilgenstieler family

Country: Brazil; **Income:** US$956 (month)
Electricity: In-house, reliable; **Water:** In-house, safe to drink
Food: LPG cooking; all food bought
Schelina works (37) in sales. Her husband, Vitor (28), is a mechanic. They have a daughter, Helena (4), who is a student. The family owns a three-bedroom house, which they have been living in for one year. The family is hardly saving any money; however, their next big plan is to start their own car workshop.

Kitchen	Living room

The Prayapati family

Country: Nepal; **Income:** US$1019 (month)
Electricity: In-house, unstable reliable; **Water:** In-house, safe for drinking but filtered
Food: LPG cooking; all food bought
Nava Raj (46) is a shop owner. His wife, Bijaya Laxmi (40), is a housewife. They have a son, Pujam (16), who is a student. The family lives in a two-bedroom house, which they own. They moved there after losing their previous house in the 2015 Nepal earthquake. They also own some agricultural land. They are planning to buy a TV and a fridge.

Kitchen	Living room

The Yanvar family

Country: Indonesia; **Income:** US$1017 (month)
Electricity: In-house, unstable; **Water:** In-house, safe for drink
Food: Propane cooking; 90% of food is bought (the rest is given or grown)
Yuli (33) is a teacher. Her husband, Angga (30), is a social worker. They live with their daughter, Luce (6), and their son, Rado (5), in a rented two-room home, which they have lived in for three years. The family is saving money to buy a house.

Kitchen	Living room

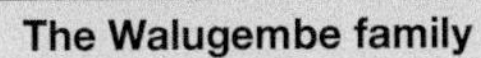

The Walugembe family

Country: Tanzania; **Income:** US$1051 (month)
Electricity: In-house, unstable; **Water:** In-house but also collected for drinking
Food: Biogas and charcoal cooking; all food bought
Francis (35) is a university teacher and environmental consultant. His wife, Penina (32), is a university lecturer. They live with their two daughters, Prathana (5) and Paloma (3), and their maid, Loise (17). They have lived in their two-bedroom house for two years; it is provided free of rent by their employer. They own another house. Their next big purchase will be an iPad.

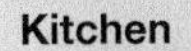

Kitchen	Living room

on Resources

eWorkbook	Understanding place and liveability (ewbk-9198)
Video eLesson	Understanding place and liveability — Key concepts (eles-5072)
Interactivity	Understanding place and liveability (int-8411)
Google Earth	Disney World (gogl-0001) Grand Canyon (gogl-0002) Table Mountain (gogl-0004) Taj Mahal (gogl-0005) Purnululu (gogl-0113)
myWorldAtlas	Deepen your understanding of this topic with related case studies and questions > Investigating Australian Curriculum topics > Year 7: Place and liveability > **New York City**

8.2 ACTIVITIES

1. a. Ask a much older person to describe the living conditions in the community they lived in as a teenager. Record or write down their memories.
   b. Ask this older person how they would have measured liveability when they were young.
   c. Do you think the current liveability of your community is better than that described by the older person? Provide examples to support your view.
2. Conduct a survey of your class to find out each person's top five places they would like to visit.
   a. Collate the results in a table similar to the following.

Place	Student A	Student B	Student C
New York City	✓		✓
Uluru		✓	✓
My cousin's house near Narrabri, NSW		✓	

   b. Conduct the survey again, but for the top five places to live rather than visit.
   c. Graph the results to show the ranking of the places by class percentage; for example, perhaps 45 per cent of the class named Uluru in their top five places to visit in Australia, but no one wanted to live there.
   d. As a class, discuss the patterns shown by the survey data. Suggest reasons why people like or dislike certain places, and why great places to visit might not always be great places to live.
3. Use the **Gapminder, Dollar Street** weblink in the online Resources to compare more places around the world. Is liveability always linked to income?

8.2 EXERCISE

Learning pathways

■ LEVEL 1	■ LEVEL 2	■ LEVEL 3
2, 3, 9, 10, 11, 14	1, 5, 7, 8, 13	4, 6, 12, 15, 16

Check your understanding

1. Define the following terms in your own words.
   a. Place
   b. Location
   c. Region
2. Study **FIGURES 2-10**. Describe five characteristics in each place that creates a sense of place.
3. Study **FIGURES 2-10**. For each figure, identify whether the characteristics are natural features, human features or both.
4. Study **FIGURES 2-10**. Suggest reasons these places have become famous around the world.
5. In your own words, define the term *community*.
6. Describe what liveability means for you.

Apply your understanding

7. Do you think that people's favourite places would vary with the age of the individual? Explain your answer.
8. No matter where we live, we all live in the one place: Earth. From what you have learned so far, define what a place is in your own words. What do you think would be the characteristics of a place that would appeal to anyone, wherever they come from? (*Hint:* What feelings do you have when you are in a place that you like?)
9. Identify a place you like because of its natural characteristics. Explain how these characteristics affect how you feel about this place.
10. Identify a place you like because of its human characteristics. Explain how these characteristics affect how you feel about this place.

11. Identify a place you like that is made up of both natural and human characteristics. Explain how these characteristics affect how you feel about this place.
12. Would the liveability of your community be different if you were vision or hearing impaired, elderly or unable to speak English? Write a community liveability statement for two such residents of your community.

Challenge your understanding

13. Think about your community 50 years from now. How will the characteristics of this place be different? For example, think about the type of houses, the spread of houses, the amount and type of traffic, the age of the population, the community facilities and other characteristics you think will be significant.
14. What type of inventions might improve liveability? Propose one device or machine that doesn't exist (as far as you know) and explain how it would make life more liveable.
15. Sometimes living conditions can change quite quickly. Suggest how natural events, political events and economic events can influence living conditions, providing examples of each to support your opinion.
16. What do you believe is the most important characteristic that makes a place liveable? Explain your decision.

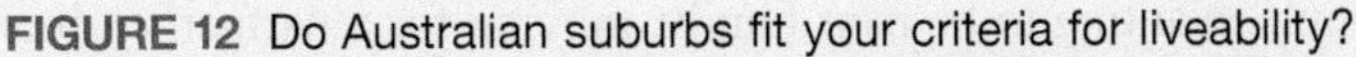

FIGURE 12 Do Australian suburbs fit your criteria for liveability?

To answer questions online and to receive **immediate feedback** and **sample responses** for every question, go to your learnON title at www.jacplus.com.au.

8.3 Factors that influence where people live

LEARNING INTENTION

By the end of this subtopic, you will be able to explain the reasons that people live in certain places and describe how places change over time.

8.3.1 Push and pull

People choose to live in specific places for a wide range of reasons. Many factors influence the liveability of places or why people decide to live there: some are human factors and some are natural. They include:

- available financial resources (money)
- employment opportunities
- relationships with other people (for example, wanting to be near family, moving for a partner's job or moving nearer a specific community of people)
- lifestyle (for example, wanting to live in a specific environment, such as by the beach, or close to services, such as a good school).

Many of these factors change throughout a person's life. For example, where a 20-year-old single person wants to live is often quite different from where someone in their forties or someone with a partner and two teenage children may want to live. Where you enjoy living now might be very different to where you would enjoy living in 20 years. For example, now you might want to live near the cinema and your friends but in 20 years you might want to live near a good childcare centre for your children.

Within these factors are **pull factors** and **push factors**. Pull factors are those that attract or 'pull' a person to a place, whereas push factors are those that discourage or 'push' people away from a place. Common pull factors are jobs, family, recreational facilities, healthcare and services. Common push factors are crime, lack of jobs, cost of living, lack of services and poor healthcare. The combination of reasons varies from person to person and what one person may see as a positive, another person may see as a negative.

pull factor positive aspect of a place; reason that attracts people to come and live in a place

push factor reason that encourages people to leave a place and live somewhere else

infrastructure the basic physical and organisational structures and facilities that help a community run, including roads, schools, sewage and phone lines

8.3.2 Towns that have been influenced by push and pull factors

Kanowna, Western Australia

In some situations, the reason for living in a place may disappear over time. This is the case for a number of towns in Western Australia. The township of Kanowna is 22 kilometres east of Kalgoorlie, and was established after gold was discovered in 1893. The chance of finding gold was a significant pull factor as it could make someone very rich instantly. Within a short period of time, its population had grown to more than 12 000 and **infrastructure** such as a hospital, railway line, school, post office and at least ten hotels had been

FIGURE 1 Kanowna in 1900

constructed. As the gold discoveries eventually dried up, and the pull factor no longer existed, the population drifted away and the township was abandoned in the mid-1950s. Interest in the geology of the area returned in the 1970s, and after new gold deposits were uncovered in the late 1980s, the Kanowna Belle Gold Mine began operation in 1993. With the mine located only 19 kilometres from Kalgoorlie, the majority of workers choose to live there, as the convenience of living in town was a strong pull factor. Some, however, work as **fly in, fly out (FIFO)** employees as the pull factors were not strong enough to convince them to move. As of the 2016 census, only ten people still lived in the original gold mining township of Kanowna.

Cossack, Western Australia

Cossack has a similar story to Kanowna. Established in the late 1860s, around 200 kilometres west-south-west of Port Hedland, it was renamed after the HMS *Cossack* in 1871 and became a base for the pearling industry in Western Australia. The development of the pastoral industry in the Pilbara region and the discovery of gold were significant pull factors that also attracted people to the port town.

In 1898, a cyclone destroyed a significant portion of the town, and after the size of pearling and transport ships began to increase in the early 1900s, the harbour was no longer suitable, as it could not accommodate them. In light of these challenges, the pearling industry was relocated to Broome, a new port was opened at Port Samson, and the town of Cossack was eventually abandoned in the 1950s. Most of the historic buildings in the town were built in the 1880s and have been redeveloped as tourist attractions; tourists can either walk or drive the 5-kilometre Cossack Heritage Trail.

fly in, fly out (FIFO) workers who fly to work in remote places, work 4-, 8- or 12-day shifts and then fly home

FIGURE 2 The restored courthouse is one of the attractions on the Cossack Heritage Trail.

FIGURE 3 The Customs House and Store is another attraction on the Cossack Heritage Trail.

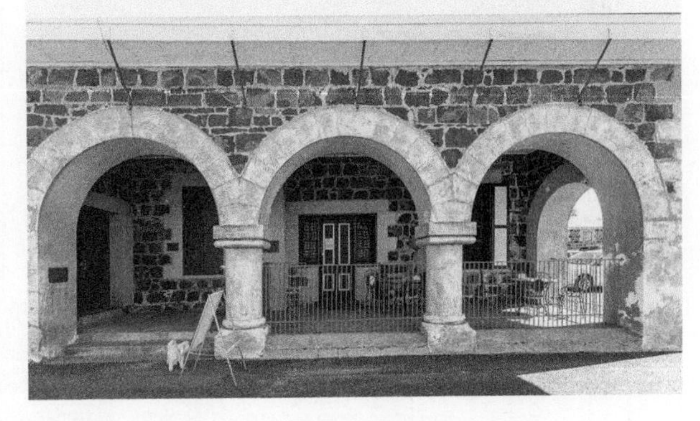

Baltimore, United States of America

Many of the towns in the north-eastern United States were established as manufacturing towns. At first they were located near major ports or iron ore and coal deposits, and some closed down when these resources ran out. In more recent times, factories such as the one shown in **FIGURE 4**, which is near Baltimore, have closed down because the owners could no longer compete with the goods produced at a lower cost in China and other South-East Asian countries. Lack of jobs was a strong push factor that drove many people from the area and it fell into a state of urban decay. Baltimore is now thriving again, especially with pull factors such as new technology jobs attracting people to the area.

FIGURE 4 A disused factory near Baltimore

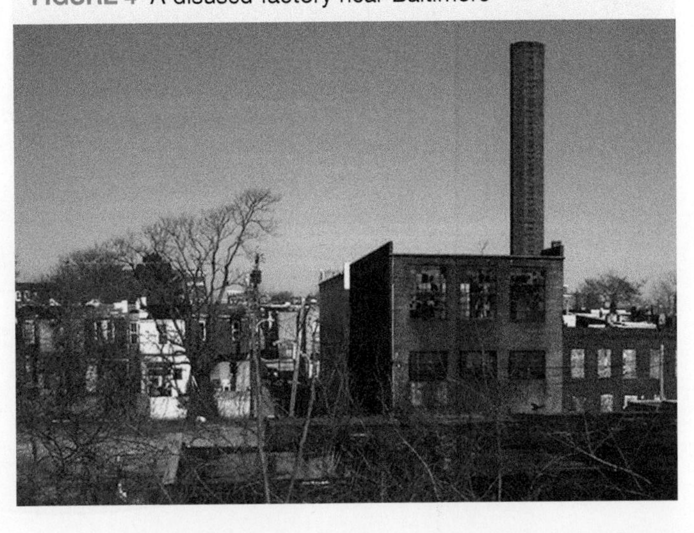

8.3.3 Services and facilities

Beyond the basic criteria of access to food, water, housing and employment, people might also choose to live in a specific place or change their view about its liveability based on whether the **services** and **facilities** they want are available. Services and facilities might be provided by the government or by community groups, religious groups or charities, such as the Red Cross. For example, a woman who is raising her first child as a single parent might consider an area close to health care facilities with a community group for young single parents and a free toy library to be more liveable than a community with older families where she has to travel long distances for medical advice and support. Every person will have their own list of services and facilities that make a place more liveable.

service program or organisation that helps people

facility place or thing that helps people

The diagram below shows some of the services and facilities that different types of people might consider as improving the liveability of a place.

FIGURE 5 Selected services and facilities that can improve liveability: examples of government services are shown in the green ring; examples of non-government services are shown in the purple ring

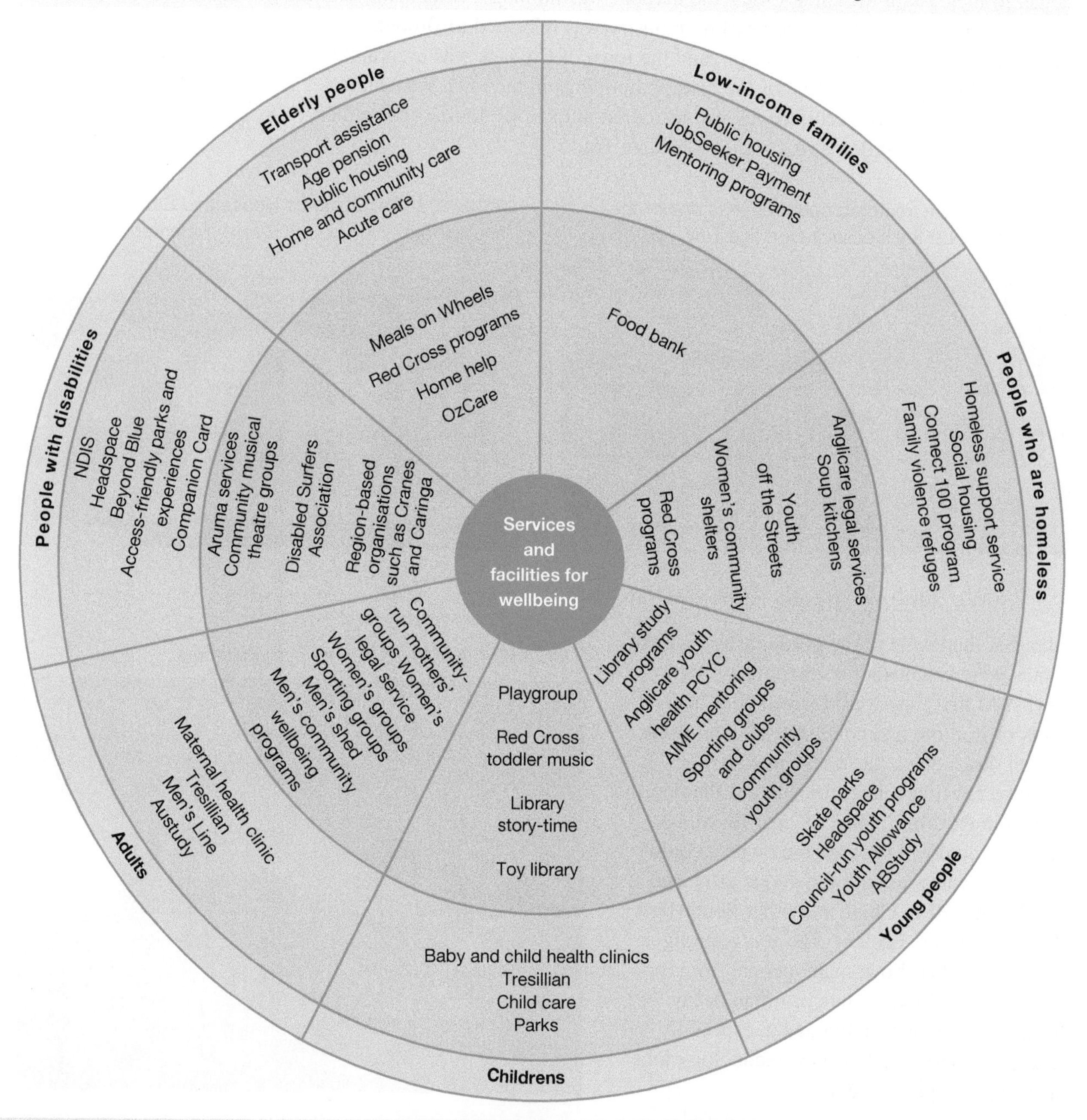

fdw-0009

FOCUS ON FIELDWORK

Suburban survey

There are many characteristics that make a place liveable, but sometimes they can be hard to define and compare, especially when you know a place well. Collecting data about a suburb and building a picture of what it is like to live there based on others' points of view can help you make a more informed assessment.

Learn some strategies to help you draw conclusions about the liveability of places using the **Suburban survey** fieldwork activity in your Resources tab.

on Resources

eWorkbook	Factors that influence where people live (ewbk-9202)
Video eLesson	Factors that influence where people live — Key concepts (eles-5073)
Interactivity	Push/pull factors (int-3089)
Fieldwork	Suburban survey (fdw-0009)
Google Earth	Baltimore, USA (gogl-0006) Kanowna (gogl-0114) Cossack (gogl-0115)
myWorldAtlas	Deepen your understanding of this topic with related case studies and questions > Investigate additional topics > Population > **Population of Australia** Investigate additional topics > Energy > **Energy in Australia**

8.3 ACTIVITIES

1. Use Google Earth to find your town or suburb.
   a. Use the historical imagery icon to display the range of images available for your location.
   b. Move the slider along the timeline to show aerial views of changes to a place.
   c. Select two times along the timeline to screen clip an image.
   d. Annotate the changes that have occurred to the environment.
2. Use the **Investigating why a place is special** link in the Resources tab to create a survey to investigate why people choose where they live.
3. Research the government and community services and facilities that are available in your community or area.
   a. Create a diagram like **FIGURE 5** to show which parts of the community received assistance from each.
   b. Share your results as a class or in small groups. Discuss your findings for each part of the community, including the balance of whether there are more government or non-government services.
   c. Based on your research, which parts of your community might find where you live the most liveable?

FIGURE 6 What services help members of your community?

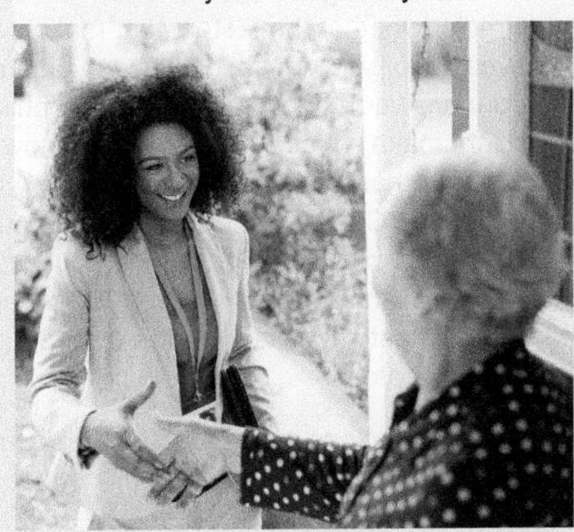

8.3 EXERCISE

Learning pathways

■ LEVEL 1	■ LEVEL 2	■ LEVEL 3
1, 2, 8, 11, 15	3, 4, 6, 7, 14	5, 9, 10, 12, 13

Check your understanding

1. Explain the difference between push and pull factors.
2. What is a FIFO worker?
3. What are four factors that influence the liveability of places and why people decide to live where they do?
4. Identify three push and three pull factors for people thinking about moving to Baltimore today.
5. Define the term *urban decay* in your own words based on how it is used in this example: 'Lack of jobs was a strong push factor that drove many people from the area and it fell into a state of urban decay'.
6. Study **FIGURE 4**. Identify some of the specific signs that indicate an area is in urban decay.

Apply your understanding

7. Why might some people stay living in decaying urban environments, and why might others choose to move?
8. Study **FIGURES 2** and **3** and think about the changes over time that have occurred in Cossack. Was the decline of the township due to push or pull factors? How did these influence people's choice of where they would live?
9. How have the types of employment opportunities changed for people in Baltimore? What might be some of the flow-on effects of this change?
10. Consider the information on Kanowna and imagine that you have been employed at the Kanowna Belle Gold Mine. You have been given the option of housing within Kalgoorlie or within the original Kanowna township. Consider the potential advantages and disadvantages of both options and explain what your final decision would be.
11. Explain one reason why someone might choose to be a FIFO worker, rather than move nearer to their job. Identify whether this reason is a push or pull factor.
12. Are natural or human features of a place more likely to be pull factors? Give reasons for your answer.

FIGURE 7 Inner harbour, Baltimore

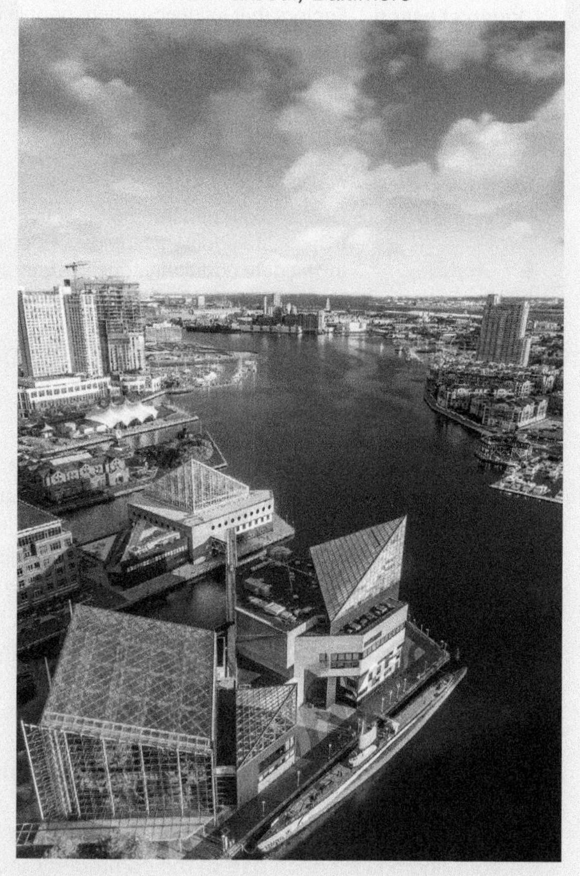

Challenge your understanding

13. Suggest what might be done with empty buildings in towns where people have moved away. Consider possible uses for heritage buildings and for former factory buildings.
14. Predict whether the factors that make a place liveable for you will change when you are older. In your answer, explain what push and pull factors might make a place liveable for you now, and whether these might change and why/why not.
15. Kanowna, Cossack and Baltimore all found ways to attract people: a new gold mine, tourism and tech developments. Predict which of these factors would be the most successful in drawing new people to live in your area.

To answer questions online and to receive **immediate feedback** and **sample responses** for every question, go to your learnON title at www.jacplus.com.au.

8.4 SkillBuilder — Using alphanumeric grid references

online only

LEARNING INTENTION

By the end of this subtopic, you will be able to locate a place on a map using alphanumeric grid references.

The content in this subtopic is a summary of what you will find in the online resource.

8.4.1 Tell me

What are alphanumeric grid references?

Alphanumeric grid references are a combination of letters and numbers that help us locate specific positions on a map. They are linked to the lines that form a grid over certain kinds of maps. The letters and numbers are placed alongside the gridlines, just outside the map. The grid, letters and numbers allow you to pinpoint a place or feature by stating its alphanumeric grid reference.

FIGURE 2 Aerial photo of elephants in Botswana

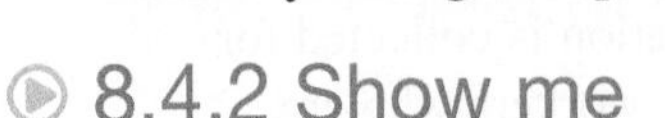

8.4.2 Show me

eles-1642

How to use alphanumeric grid references

Step 1

We will demonstrate this skill using the photo in FIGURE 1. In this **aerial photograph** of elephants in Botswana, we want to locate the veterinarian, who is at point X.

aerial photograph a photograph taken of the ground from an aeroplane or satellite

Step 2

Place your finger in the grid square that contains the X. By running your finger up the column of squares, you can see that the vet is in column D. Write this in your notebook.

Step 3

Now run your finger across the row of squares and you can see that the vet is in row 4. Write this to the right of the letter D in your notebook. This gives us the grid reference for the vet's location: D4. Now write the name of the feature (in this case, Veterinarian location) beside the grid reference.

8.4.3 Let me do it

int-3138

learnON

Go to learnON to access the following additional resources to help you build this skill:

- a longer explanation of this skill and its application in Geography (Tell me)
- a video showing the step-by-step process of this skill (Show me)
- an interactivity for you to practise this skill (Let me do it)
- self-marking questions to help you understand and use this skill.

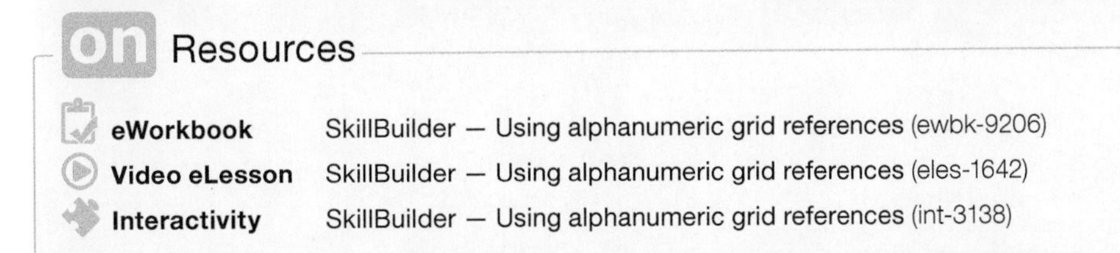

on Resources

eWorkbook SkillBuilder — Using alphanumeric grid references (ewbk-9206)

Video eLesson SkillBuilder — Using alphanumeric grid references (eles-1642)

Interactivity SkillBuilder — Using alphanumeric grid references (int-3138)

8.5 Measuring liveability

LEARNING INTENTION

By the end of this subtopic, you will be able to identify factors that are considered when determining a city's liveability ranking, and be able to describe some of the characteristics of the most liveable and least liveable cities.

8.5.1 Understanding liveability

Everyone likes to be able to tell you they are the best, or in the top ten of some category. The people who manage and live in cities are no different. If you look at the official websites for many cities around the world, they will tell you that they are the safest, wealthiest or fastest-growing or have the best events calendar. Being able to boast that your city is the world's most liveable is great publicity.

Liveability can be defined as 'the features that create a place that people want to live in and are happy to live in'. It is usually measured by factors such as safety, health, comfort, community facilities and freedom.

8.5.2 Measuring liveability

Several international organisations have created lists of the world's most liveable cities. These organisations each compare data and produce a table that ranks the liveability of cities. This information is collected for workers considering overseas transfers or for companies that may need to compensate workers who are transferred to a low-ranked city. The figures can also be used to attract migrants or investment. The various rankings compare a large number of cities; however, not all cities in the world are included in each survey.

The criteria used to produce the rankings include:

- stability or personal safety (crime, terror threats and civil unrest)
- healthcare
- culture and environment (religious tolerance, corruption, climate and potential natural disasters)
- education
- infrastructure (transport, housing, energy, water and communication)
- economic stability
- recreational and sporting facilities
- availability of consumer goods (food, cars and household items).

FIGURE 1 Vienna, Austria, regularly ranks well in liveability ratings

int-3095

FIGURE 2 City liveability rankings

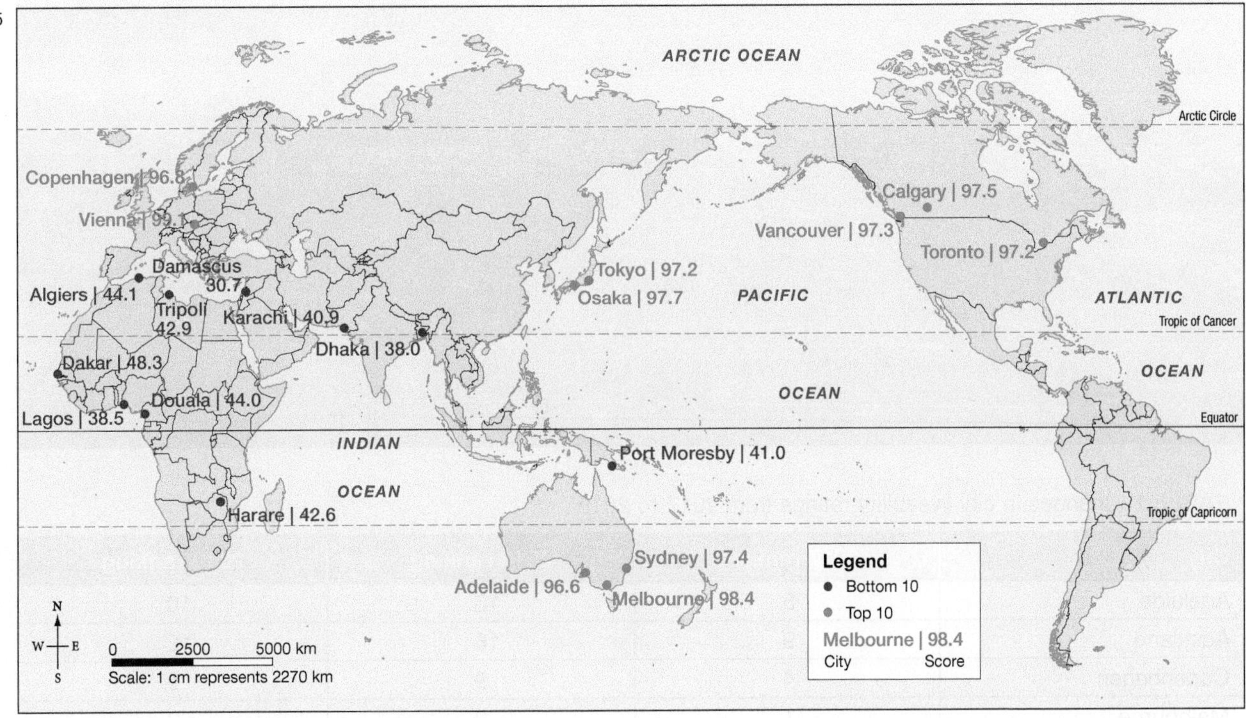

Source: Economist Intelligence Unit (EIU) 2019

FIGURE 2 shows the top ten and bottom ten in the global cities liveability rankings, as released by the Economist Intelligence Unit (EIU) in 2019. These rankings are released each year, so it is possible for you to log on (use the **Economist Intelligence Unit** weblink in the Resources tab) to get the most recent update. This survey ranks 140 cities based on 30 criteria, and a score of 100 is given to the best or ideal city to live in. In previous years Vienna, Melbourne and Vancouver have shared the top ranking as the world's most liveable cities. In 2018 Vienna took out the number one ranking, with Melbourne pushed to second. Vancouver's ranking has fallen to sixth. In 2019, Vienna took the top place once again, followed by Melbourne in second and Sydney in third.

The map shows that many of the world's top cities have scores that are very similar. The difference in score between the top ten cities is only 2.5 points.

Between 2008 and 2018, the average global liveability score has increased by almost one percentage point. Of the 140 cities included in the liveability survey, half have improved their overall status. Competition at the top of the list is fierce. Four cities were forced out of the top ten between 2017 and 2018.

Things the top ten liveable cities have in common

Looking at the locations of the most liveable cities, you can see that most are found in Australia and Canada, which have three each; followed by Europe and Japan, with two each. They are all mid-size cities, have quite low **population density**, low crime rates and infrastructure that copes quite well with the needs of the local community. They are found in places where there is a **temperate climate**, perhaps with the exception of Toronto and Calgary, which do have very cold winters.

The top cities also tend to be modern, not much more than 300 years old. They have been planned so that people can travel around them by both public and private transport. They are also found in some of the world's wealthiest or most developed nations.

Australian and Canadian cities perform better than cities in the United States due to US cities' higher crime and congestion rates. The highest ranked United States city is Honolulu, at 23.

population density the number of people living in a square kilometre

temperate climate climate with generally warm summers and cool winters, without extremes

FIGURE 3 Auckland city centre and Waitematā Harbour, New Zealand

TABLE 1 Changes in city liveability ratings from 2017 to 2019

City	2017 ranking	2018 ranking	2019 ranking
Adelaide	5	10	10
Auckland	9	16	12
Copenhagen	4	9	9
Melbourne	1	2	2
Sydney	11	5	3
Vancouver	3	6	6

on Resources

eWorkbook Measuring liveability (ewbk-9210)

Video eLesson Measuring liveability — Key concepts (eles-5074)

Interactivity My most liveable country (int-3095)

myWorldAtlas Deepen your understanding of this topic with related case studies and questions > Investigating Australian Curriculum topics > Year 7: Place and liveability > **Polluted cities**

8.5 ACTIVITIES

1. Work with a partner or in a group to find the most recent population figures for each of the cities shown on the map in **FIGURE 1**.
   a. List your findings.
   b. Write one sentence to describe the population of the most liveable cities.
   c. Write one sentence to describe the population of the least liveable cities.
2. Draw up a table or use a spreadsheet to collect at least five sets of information to compare the top ten and bottom ten in the liveable cities ranking. Use the population data you collected for the previous question as your first set of information. Other possible data sets are number of universities, number of hospitals, population density, any recent violence, traffic issues, the availability of public transport, housing types, presence of slums and water supply and sanitation. Comment on the differences between the most liveable and least liveable cities. Write at least three sentences.
3. Research which cities ranked well in the most recent liveability rankings published by the EIU in their Global Liveability Index.
   a. How might the COVID-19 pandemic have affected rankings in 2021? Discuss which criteria would have been the most affected by the pandemic.
   b. Consider the Australian cities in the top ten of the latest rankings. Create a graph to show how each of these cities has changed in the rankings over the last 10 years.

8.5 EXERCISE

Learning pathways

■ LEVEL 1	■ LEVEL 2	■ LEVEL 3
1, 3, 4, 8, 14	2, 6, 7, 9, 13	5, 10, 11, 12, 15

Check your understanding

1. How many cities are ranked in the EIU liveability ranking?
2. What does the term 'population density mean'?
3. What were the differences in ranking between the scores of the following cities in the 2018 and 2019 liveability rankings?
   a. Adelaide
   b. Auckland
   c. Vancouver
   d. Copenhagen
4. Name the three lowest-ranked cities in the 2019 liveability ranking shown in **FIGURE 1**.
5. In which type of climatic region are most of the liveable cities?
6. Analyse the information in **FIGURE 1**.
   a. How many of the top ten most liveable cities are found on each continent?
   b. How many of the most liveable cities are found in the northern hemisphere?
   c. Describe the distribution of the least liveable cities in the world.
   d. How many of the least liveable cities are found on each continent?
   e. How many of the least liveable cities are found in each hemisphere?

Apply your understanding

7. London and New York have a similar low ranking for liveability. Why do you think these well-known cities are ranked so low?
8. Why might a city suddenly fall down the liveability rankings? Explain two reasons that might account for a slip in the rankings of ten or more places.
9. What do you think could be done to improve a city's liveability ranking? Explain two changes that might improve the rankings of a city generally.
10. Suggest some reasons why Vienna might have taken the number one ranking in 2019 after sharing it with both Melbourne and Vancouver for several years.
11. The top cities in the liveability ranking tend to be modern cities, but Vienna has been an important and influential city for many centuries. How might you explain its liveability considering it is far older than cities ranking lower?
12. Explain why congestion might have a significant negative impact on a city's liveability.

Challenge your understanding

13. Suggest two reasons why cities in Australia and Canada outperform cities in the United States in the liveability rankings.
14. Make a list of what you think are the ten most important criteria to judge a liveable city.
15. How do you think the trends in the spread of COVID-19 during the pandemic affected 2020 liveability rankings? Suggest two reasons why the data from 2020 might not match previous years.

To answer questions online and to receive **immediate feedback** and **sample responses** for every question, go to your learnON title at www.jacplus.com.au.

8.6 Factors that influence liveability in Australia

LEARNING INTENTION

By the end of this subtopic, you will be able to describe environmental and human factors that contribute to liveability in Australia and apply factors of liveability to your neighbourhood.

8.6.1 The environmental factors that make Australia so liveable

Where is your favourite place in Australia? Have you been to a holiday paradise, one that you think would be the perfect place to live? Is the climate perfect, the scenery spectacular? Is it safe, fun and the place for adventure? Is this place in a city, in the **wilderness** or in the next street? Is it paradise because your friends or family live there or because of the natural or **built environment**? Many environmental factors can influence liveability, including the climate, natural surroundings, landscape and natural resources. Australia has many popular destinations where liveability is high due to their environmental factors.

wilderness a natural place that has been almost untouched or unchanged by the actions of people
built environment a place that has been constructed or created by people
remote a place that is distant from major population centres

Among the most popular and beautiful tourist destinations in Australia are the Great Barrier Reef, Uluru, Melbourne, Sydney, the Gold Coast, the Great Ocean Road, Monkey Mia, Kakadu, the Tasmanian wilderness, the Blue Mountains, Port Arthur, Byron Bay, Kangaroo Island and Ningaloo Reef. Many of these places have unique landscapes, located within naturally stunning environments. Four of these are predominantly built environments: Sydney, Melbourne, the Gold Coast and Port Arthur. The remaining ten places are best known for their natural, often **remote**, and almost wilderness environments.

These places are often perfect for a holiday but they may also be a good place to live. Is it mostly the excitement of a big city, the natural beauty, or some other factor that makes you decide which place is the most liveable? The following provides some further information on the five locations from **FIGURE 1**.

FIGURE 1 Five popular tourist destinations in Australia

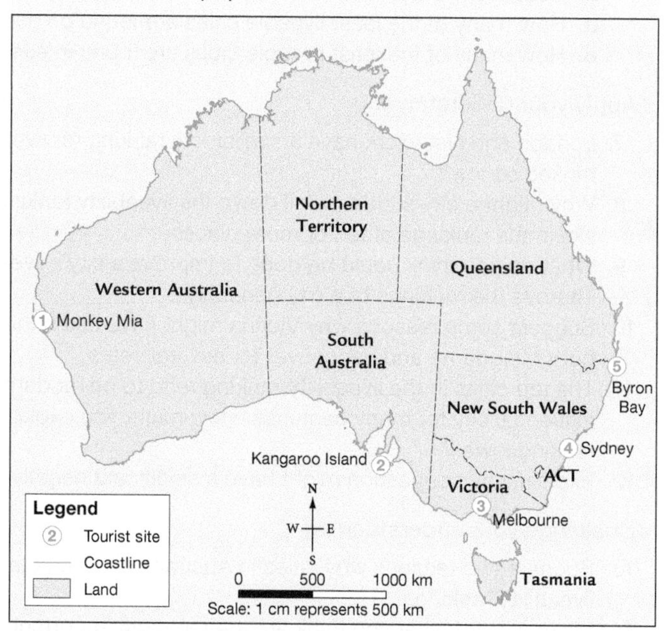

Source: Spatial Vision

FIGURE 2 Wild dolphins attract tourists to Monkey Mia.

① Monkey Mia is an environment where you can experience natural wildlife by interacting with dolphins. It is located in Shark Bay on the coast of Western Australia, 850 kilometres north of Perth. For over 40 years, a small pod of dolphins has come ashore to connect with beachgoers. The Department of Environment and Conservation provides staff who supervise the feeding of fish to these dolphins each day. It is an unusual opportunity for people to see wild dolphins up close, quite near to the shore. Monkey Mia is a place of great natural beauty. It is an important stop on the around-Australia tourist trail. At the 2016 census, fewer than 800 residents lived near the Monkey Mia Resort, but about 100 000 tourists visit each year.

(2) Kangaroo Island is a place of natural beauty. It is Australia's third largest island and is found about 160 kilometres south of Adelaide. It is a wildlife lover's paradise, home to many native Australian animals in their natural habitats, including koalas, kangaroos, seals and penguins. It has remote, unspoiled beaches and interesting rocky outcrops. Although it was first settled by Europeans in the late 1830s, its present population of over 4200 is the highest it has ever been. It was used by European settlers as a fishing and farming community but today is better known as a tourist destination, with about 170 000 people visiting for an overnight stay each year, and about 55 000 people every year travelling to the island for the day. Tourism to Kangaroo Island plummeted in 2020. Bushfires devastated the island in January and the COVID-19 pandemic prevented all but South Australians from visiting for significant parts of the year. In most years, 42 per cent of the island's visitors are from overseas or interstate.

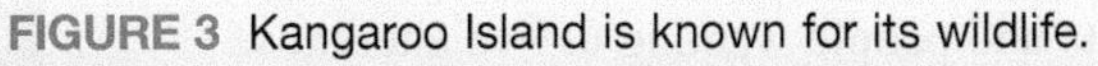
FIGURE 3 Kangaroo Island is known for its wildlife.

(3) Melbourne is the second-largest city in Australia, and was ranked the most liveable city in the world (2011–2017; second to Vienna in 2018 and 2019) according to the Economist Intelligence Group. It is the capital of Victoria and was home to more than 5.2 million residents in 2021. Melbourne is expected to overtake Sydney as Australia's largest city by 2026. It is an attractive destination for tourists, who enjoy visiting its major sporting and cultural events, shops, restaurants and theatres. About 40 million tourists visited Melbourne in 2019. Melbourne is located beside Port Phillip Bay and on the Yarra River. It is not a city known for its beautiful natural environment, but it has become known for its distinctive built environment including its laneways, bars and café culture.

FIGURE 4 Albert Park Lake and the Melbourne CBD

(4) Sydney is a built environment in a beautiful natural setting and is Australia's largest and oldest city. It is often called the 'Harbour City'. Sydney is popular with both domestic and international tourists, with more than 40 million visiting in 2019; it was home to nearly 5.4 million residents in 2021. It has many attractions, including restaurants, beautiful beaches, theatres, galleries and iconic landmarks. It has a beautiful natural environment with varied experiences provided by the built environment. This makes it an extremely popular destination.

FIGURE 5 Sydney Harbour

⑤ Byron Bay is a beachside town in northern New South Wales, located 160 kilometres south of Brisbane. It is known as a very relaxed place with a local community that includes many artists and people wanting to live alternative lifestyles. It is an important surfing place, with easy access to offshore reefs and stunning beaches. It has become a popular place for 'tree changers' from Sydney, for 'schoolies' end-of-year celebrations and for backpackers. About 2.2 million people visit Byron Bay every year. In 2021, Byron Bay Shire had a population of nearly 36 000 people (9500 in the township of Byron Bay), who rely heavily on tourism and agriculture for their income. In recent years, many residents of Byron Bay have become concerned about the negative impacts of tourism on their lives because more backpackers and budget travellers in campervans have been illegally camping in the area. The number of wealthy 'tree changers' moving into the area has also made houses more expensive, so it is harder for other local people to buy or rent homes.

FIGURE 6 Byron Bay's natural beauty and culture attract residents and tourists.

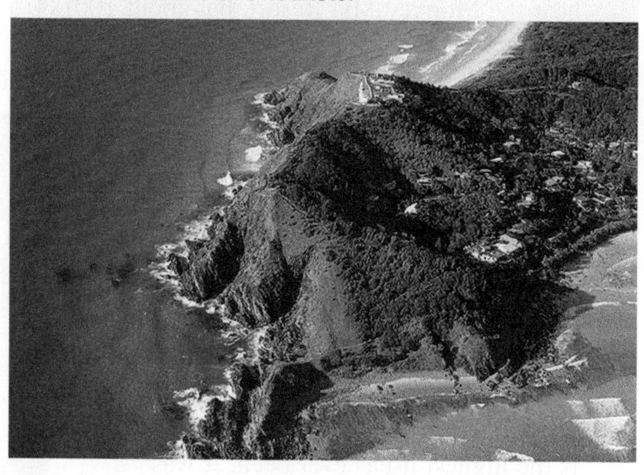

8.6.2 The human factors that make modern Australia liveable

What is your **neighbourhood** or local place like? All of us live in a **community**, and these are often centred on the place where we live, or go to school or work. Human factors, such as culture, income, employment and community, all influence levels of liveability in a place.

Teenagers have different types of local places that have special meaning for them, each one at a different scale: their bedroom, home and neighbourhood. With nearly 90 per cent of Australians living in towns and cities, most people likely live in a street that is part of a suburb, town or city, which itself is part of a state or territory. On the other hand, many Australians do not live in urban areas, but still live in their own communities that are just as distinctive as neighbourhoods in towns and cities. There are many ways we describe where our local place is and what it is like.

When you live in a neighbourhood, you become familiar with all the things that help to create the character of the place. Sometimes a neighbourhood is made up of people who have similar interests and beliefs, whether these are cultural, sporting, environmental or job-related. Other neighbourhoods have a mixture of people from different backgrounds, creating a vibrant, multicultural community identity. The fact that Australian neighbourhoods can be so different is what makes Australia such an interesting place to live in.

8.6.3 Neighbourhoods in Australia before 1788

Neighbourhoods have always existed in Australia. The Aboriginal Peoples belonging to a specific **Country** often share a common language or similar languages. There are about 29 Aboriginal groups whose Country lies within the environment and space that the Sydney metropolitan area occupies today. Each group belonged to a region that had specific boundaries (**FIGURE 7**). Within each language group, there were different dialects that overlapped. These dialects were spoken by different groups of related families. Thus, these nations see their neighbourhood as the region in which people spoke the same language and had the same customs, such as marriage rituals. People were, and are, socially connected.

Before European colonisation, Aboriginal Peoples lived in specific nations (groups) to which they belonged, with their lives intrinsically linked to the resources, customs

neighbourhood a region in which people live together in a community

community a group of people who live and work together, and generally share similar values; a group of people living in a particular region

Country the place where an Aboriginal and Torres Strait Islander Australian comes from and where their ancestors lived; it includes the living environment and the landscape and cultures embedded in Country. Regardless of the landscape and what it provided, these specific areas were the responsibility of the Aboriginal Peoples who were born to that land. These geographical areas were further divided into two **moieties** and then into further clan and family groups. As part of this, highly sophisticated social structures commonly known as kinship systems determined who was responsible for what. This contributed to a collective 'neighbourhood' or 'community' that was reliant on everyone performing certain roles and understanding expectations of behaviour to that the environment — and communities as part of that environment — were looked after and worked well together.

int-8468

FIGURE 7 Aboriginal language groups in NSW

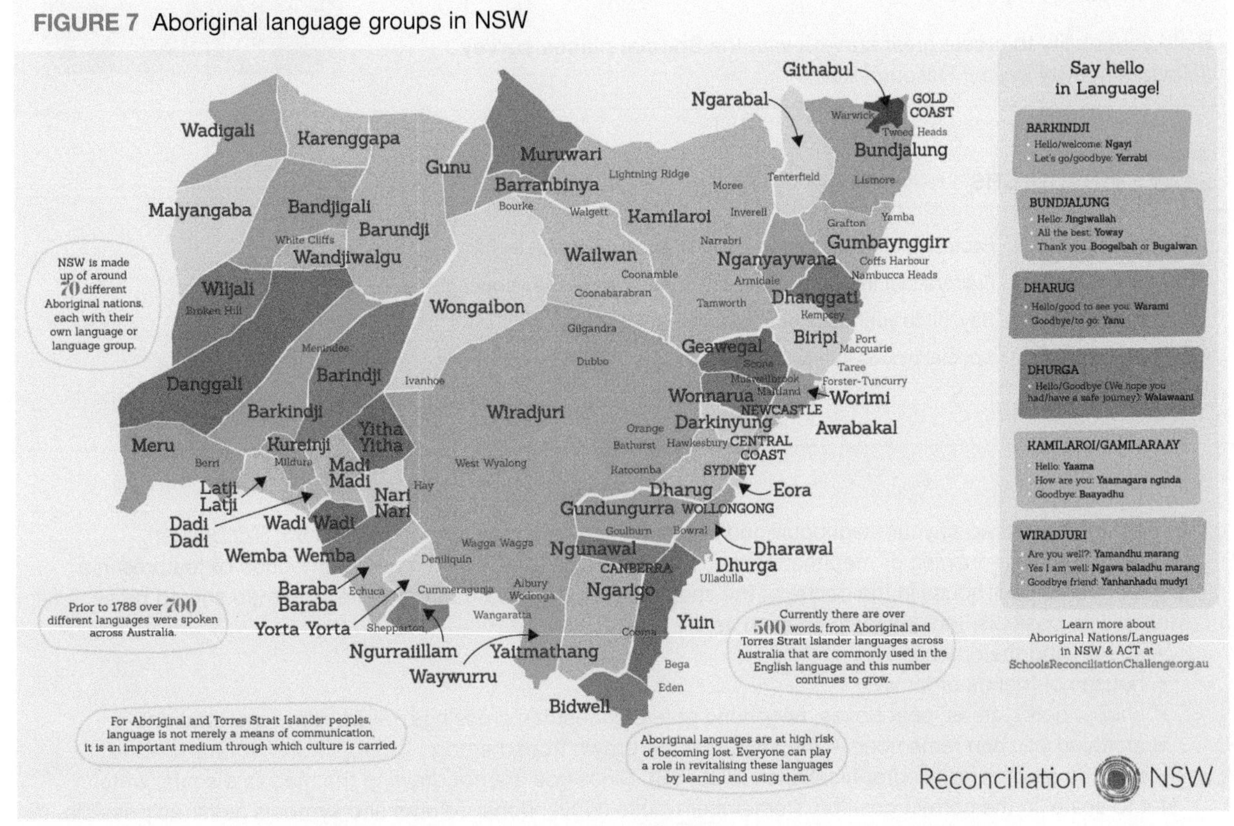

Source: Aboriginal Languages/Nations in NSW & ACT, © Reconciliation NSW, www.reconciliationnsw.org.au
Note: To look at this map in more detail, click on the int-8468 icon in learnON.

Additionally, language was a key factor in these particular areas and underpinned where a person belonged and their part in these dynamic societies. To ensure that trade could occur with neighbouring nations, specific individuals and/or groups would be required to speak the languages of the Aboriginal Peoples who shared a common boundary. This demonstrates the diversity and collective understanding that occurs when living in close proximity with others.

Aboriginal neighbourhoods in Australia after 1788

After 1788, Aboriginal Peoples still lived in the specific geographical areas outlined above. But as more colonial settlements were established, they were dislocated or removed from their traditional homelands, which had devastating impacts on their ability to maintain social structures on their Country. With the introduction of government policies in the late 1800s and early 1900s, Aboriginal Peoples were restricted from having physical connection to Country. Restrictions were placed on where people could live, what language they could speak, and cultural practices and custom were also forbidden. This had devastating impacts on Aboriginal Peoples and their cultures and societies, which further reduced the opportunity for collective communities to continue and thrive. However, despite being removed from many of the places and people that created a sense of community, Aboriginal Peoples still have a strong sense of connection to Country and Culture. Even in modern cities where the land and people living on it are dramatically changed, the spiritual connection to place is still deeply felt.

moieties the two halves that a nation were divided into

fdw-0010

FOCUS ON FIELDWORK

Why are some places considered special?

What is the difference between an interview and a survey? An interview is a geographical tool that allows the interviewer to ask detailed questions about a topic. On the other hand, a survey tends to be quicker to complete, and the questions are standardised (all the same for everyone) so they can be asked of many people.

Build your ability to create great surveys with the **Special places survey** fieldwork activity in your Resources tab.

on Resources

eWorkbook	Factors that influence liveability in Australia (ewbk-9214)
Video eLesson	Factors that influence liveability in Australia — Key concepts (eles-5075)
Interactivity	Say 'hi' to your neighbourhood (int-8468)
Fieldwork	Special places survey (fdw-0010)

8.6 ACTIVITIES

1. Create a sketch map of your neighbourhood or local place (you may wish to refer to 8.7 SkillBuilder). A sketch map is a drawing or map that contains your memory of the layout and distribution of features in a place. Locate your house in the centre of the sheet and work outwards from there. The map should be as detailed as possible. Include features such as:
   - streets and their names
   - houses of friends or family
   - shops, parks, trees, post boxes, telephone poles, pedestrian crossings, railway lines and stations
   - anything you can remember — the map must be drawn from memory.

   Present the map using geographical rules (BOLTSS). Since you are not drawing the map to a scale, write 'Not to scale' in the correct position. Remember to use conventional colours and symbols as far as possible. Compare your mental map to an actual map of your neighbourhood.
   a. In what ways was your map accurate?
   b. Which features did you not mark on your map?
   c. Which parts of your neighbourhood did you know well and which did you not know well?
   d. Think of reasons to explain your answers to (c).
2. Design a map of your most liveable place. Consider the natural and built environments; distance to a city, services, job and recreational opportunities; climate; and lifestyle. Annotate your map to explain why this is where you would like to live. Use the **Nothing like Australia** weblink in the Resources tab to help find your ideal location.

8.6 EXERCISE

Learning pathways

■ LEVEL 1	■ LEVEL 2	■ LEVEL 3
4, 5, 6, 10, 12, 14	1, 2, 3, 7, 11, 15	8, 9, 13, 16, 17

Check your understanding

1. What is a neighbourhood?
2. What percentage of Australians live in cities and towns?

3. In your own words, define the term *liveable city*.
4. Classify each of the following places in Australia as a natural environment, built environment or both.
   a. Monkey Mia
   b. Kakadu
   c. Sydney
   d. Blue Mountains
   e. Melbourne
   f. Great Barrier Reef
   g. Gold Coast
   h. Port Arthur
5. About how many different Peoples lived in the area that is now metropolitan Sydney before 1788?
6. About how many languages were spoken across Australia before 1788?
7. Describe one way in which Aboriginal Cultures establish a sense of neighbourhood.
8. Describe one way in which the British reduced Aboriginal Peoples' ability to maintain their social structures.
9. Based on the data and information provided about Monkey Mia, Kangaroo Island, Melbourne, Sydney and Byron Bay, create an infographic that summarises the number of visitors, the resident population and the key natural and built attractions of the area for tourists and residents.
10. Describe your most liveable place.

Apply your understanding

11. Explain how language was used as a way for people from different Aboriginal communities to build understanding between groups before 1788.
12. Is your most liveable place in a natural or a built environment or a mixture of the two? Explain why.
13. Look at **FIGURE 7** closely. Explain why the map has been made with the lines between each group.

Challenge your understanding

14. Which of these places is most similar to your most liveable place: Byron Bay, Sydney, Melbourne, Kangaroo Island or Monkey Mia? Suggest whether it is a place you would rather visit or live. Explain your answer.
15. If you wished to work as a national park ranger, which of the places in **FIGURE 1** would be best and why?
16. If you were planning a career in the theatre, which of the places in **FIGURE 1** would be best and why?
17. If you wished to live in a relaxed coastal environment close to a capital city, which of the places in **FIGURE 1** would be best and why?

To answer questions online and to receive **immediate feedback** and **sample responses** for every question, go to your learnON title at www.jacplus.com.au.

8.7 SkillBuilder — Constructing a basic sketch map

online only

LEARNING INTENTION

By the end of this subtopic, you will be able to construct a basic sketch map from an aerial photograph.

The content in this subtopic is a summary of what you will find in the online resource.

8.7.1 Tell me

eles-1661

What is a basic sketch map?

A basic sketch map is a map drawn from an aerial photograph or developed during fieldwork that identifies the main features of an area. It is different from a précis map, in which the cartographer opts to include or leave out certain features.

8.7.2 Show me

How to construct a basic sketch map

Step 1

Determine the relevant area of the aerial photograph that you want to use to make a basic sketch map.

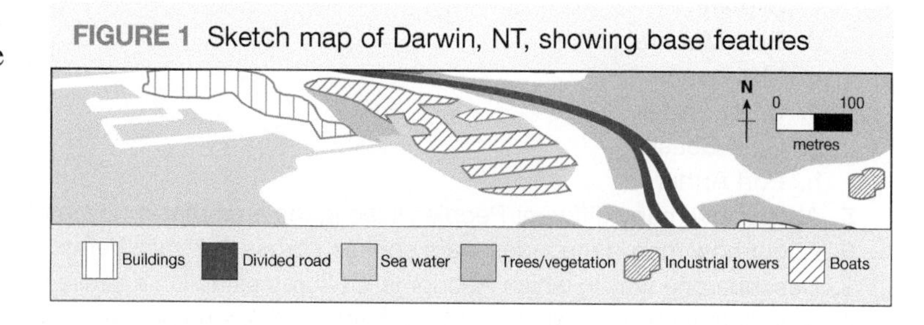

FIGURE 1 Sketch map of Darwin, NT, showing base features

Step 2

Rule a border on your page within which to create your map. Keep the border the same size as the area of the photograph you are planning to draw, to avoid scale issues.

Step 3

Identify the feature(s) that you will transfer onto your basic sketch map. Look for natural and human features. Create a colour coded key/legend for each feature and decide on any symbols too.

Step 4

Inside the border, draw an outline of the base features of the area, such as rivers, coastlines and major roads.

Step 5

Mark the locations of individual features onto your base map. Label any significant features of the sketch map. Do this neatly and horizontally.

Step 6

Complete the simplified sketch map with BOLTSS.

8.7.3 Let me do it

int-3157

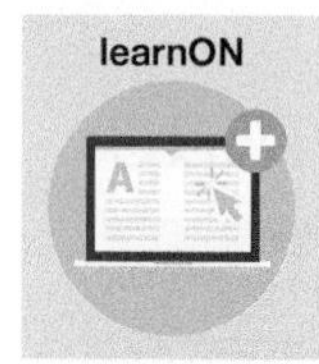

Go to learnON to access the following additional resources to help you build this skill:

- a longer explanation of this skill and its application in Geography (Tell me)
- a video showing the step-by-step process of this skill (Show me)
- an activity and interactivity for you to practise this skill (Let me do it)
- self-marking questions to help you understand and use this skill.

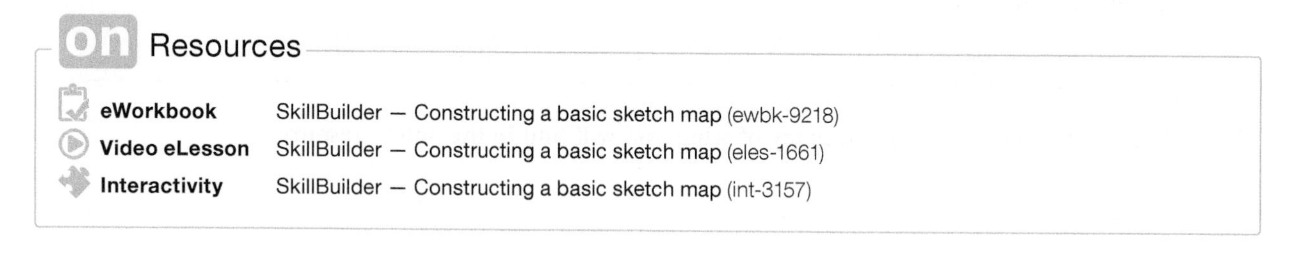
on Resources

eWorkbook SkillBuilder — Constructing a basic sketch map (ewbk-9218)

Video eLesson SkillBuilder — Constructing a basic sketch map (eles-1661)

Interactivity SkillBuilder — Constructing a basic sketch map (int-3157)

8.8 More liveable cities — Melbourne and Sydney

online only

LEARNING INTENTION

By the end of this subtopic, you will be able to identify the factors that make Melbourne and Sydney liveable cities.

The content in this subtopic is a summary of what you will find in the online resource.

What makes one Australian city more or less liveable than another Australian city? Factors such as climate, infrastructure, health care options, public safety, access to education and work opportunities all play a role. Many people also have their own personal reasons for liking one city more than another.

In this subtopic, you will be able to compare these factors for Australia's two largest cities: Sydney and Melbourne.

To learn more about liveability in Australia, go to your learnON resources at www.jacPLUS.com.au.

FIGURE 3 Central Melbourne

FIGURE 7 Aerial view of Royal Botanic Gardens

Contents

learnON

- 8.8.1 Melbourne
- 8.8.2 Sydney

on Resources

eWorkbook More liveable cities — Melbourne and Sydney (ewbk-9222)

Video eLesson More liveable cities — Melbourne and Sydney — Key concepts (eles-5076))

Interactivity More liveable cities — Melbourne and Sydney (int-8412)

8.9 Less liveable cities — Port Moresby and Dhaka

online only

LEARNING INTENTION

By the end of this subtopic, you will be able to identify factors that lead to a city being described as less liveable, and describe the impact of these factors on people and places.

The content in this subtopic is a summary of what you will find in the online resource.

What is it like living in a city that is often judged to have very low levels of liveability? Factors such as climate, infrastructure, health care options, public safety, access to education and work opportunities all play a role in how liveable a city is in a 'liveability index' but that certainly doesn't mean that everyone living in these cities hates where they live!

In this subtopic, you will be able to compare liveability in Port Moresby in Papua New Guinea with Dhaka in Bangladesh. You will be provided with data and information related to factors that are included in liveability rankings and a range of images to help you consider what life might be like in these places.

To learn more about liveability in Port Moresby and Dhaka, go to your learnON resources at www.jacPLUS.com.au.

FIGURE 5 Vegetable stall Boroko market, Port Moresby, Papua New Guinea

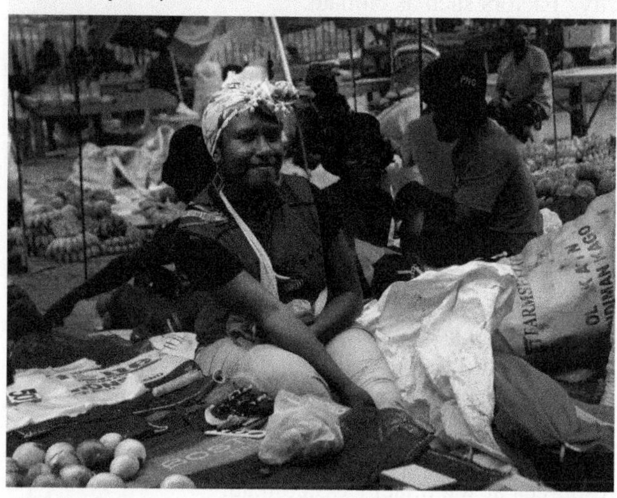

FIGURE 11 Viqarunnisa Noon School and College students celebrating their exam results, Dhaka

Contents

on Resources

eWorkbook	Less liveable cities — Port Moresby and Dhaka (ewbk-9226)
Video eLesson	Less liveable cities — Port Moresby and Dhaka — Key concepts (eles-5077)
Interactivities	Environmental quality (int-3096) Safe as houses (int-3097)
Google Earth	Dhaka (gogl-0032) Port Moresby (gogl-0034)

8.10 Investigating topographic maps — Mount Etna

LEARNING INTENTION

By the end of this subtopic, you will be able to outline some of the risks of living near a natural hazard.

8.10.1 Natural hazards

Natural hazards threaten the environmental quality and, in turn, the liveability of a place. The risks associated with natural hazards are potentially very dangerous to people and can include destruction of housing, infrastructure, environments and loss of life.

A natural hazard refers to the risk of a natural event that has the potential to cause catastrophic damage to both physical and human environments. Drought, floods, tropical storms, cyclones, volcanoes, fires, tsunamis, earthquakes and landslides are all examples of natural hazards. The actual destruction of the landscape due to a natural hazard is called a natural disaster. The risk of natural hazards and previous natural disasters reduce the environmental quality of a place.

8.10.2 Mount Etna, Italy

Mount Etna is Europe's most active volcano. Between 16 February and 2 April 2021, Etna had 17 separate small lava fountain eruptions. It is approximately 3323 metres high and situated on the island of Sicily, Italy (37°N 15°E). The slopes of Mount Etna are home to 5 million people (25 per cent of the island's population). The city of Catania is located only 35 kilometres from Mount Etna.

Why Mount Etna is dangerous

Mount Etna poses a direct threat to the populations living on its slopes — it is a very active volcano. To improve liveability and to decrease the threats to the population, Italian authorities have tried several methods, including dams and channels, to redirect lava flows away from the communities.

FIGURE 1 Mount Etna

int-8413

FIGURE 2 Topographic map extract of Mount Etna

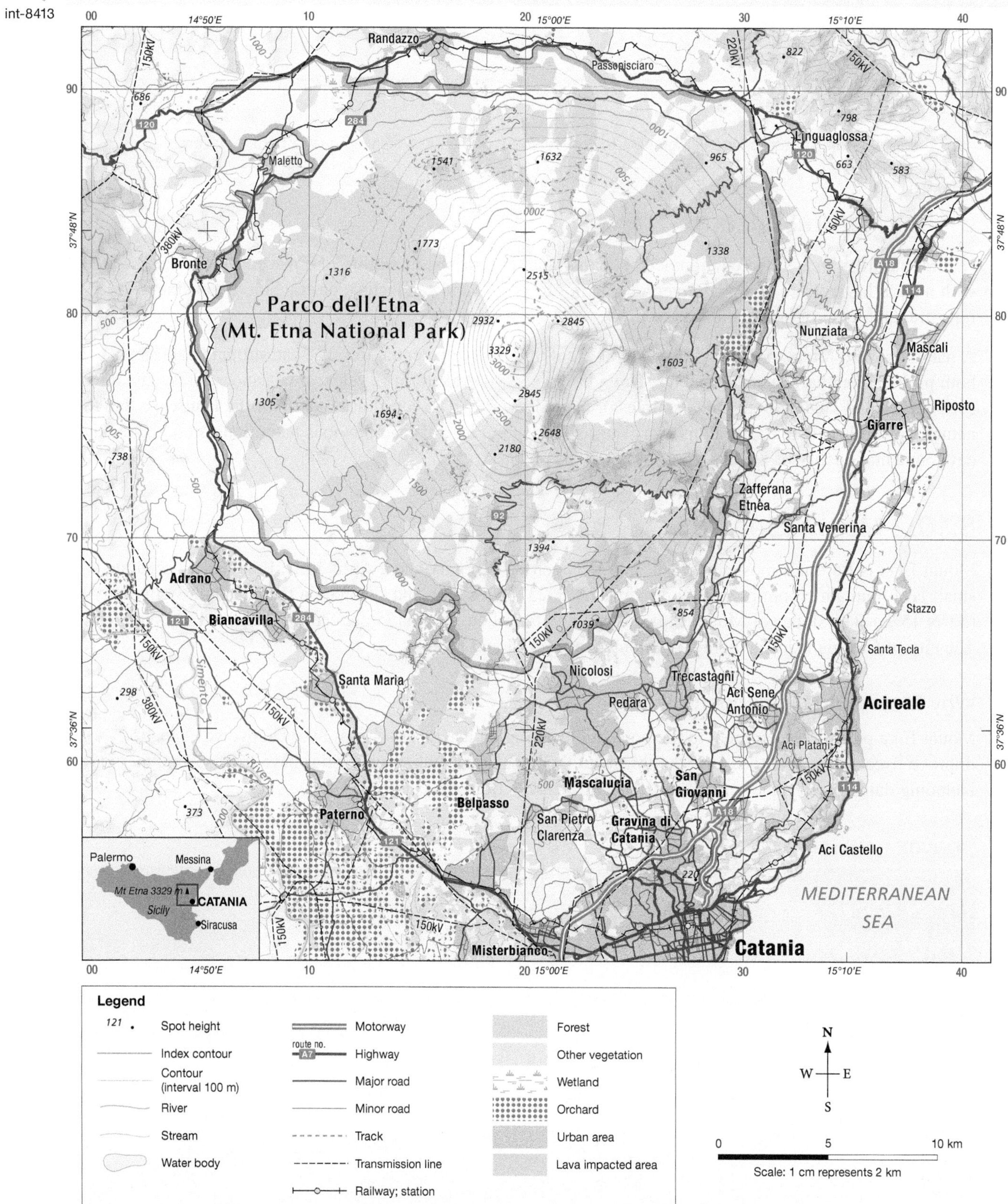

Source: Map data © Copernicus Land Monitoring Service, European Environment Agency (EEA), European Union (2018); NASA JPL (2013); © OpenStreetMap contributors, https://openstreetmap.org. Data is available under the Open Database Licence, https://opendatacommons.org/licenses/odbl/.

on Resources

eWorkbook	Investigating topographic maps — Mount Etna (ewbk-9230)
Digital document	Topographic map of Mount Etna (doc-36255)
Video eLesson	Investigating topographic maps — Mount Etna — Key concepts (eles-5078)
Interactivity	Investigating topographic maps — Mount Etna (int-8413)
Google Earth	Mount Etna (gogl-0135)

8.10 EXERCISE

Learning pathways

■ LEVEL 1	■ LEVEL 2	■ LEVEL 3
1, 2, 3, 12, 13	4, 7, 8, 10, 14	5, 6, 9, 11, 15

Check your understanding

For all of these exercises, refer to **FIGURE 2**, a topographic map extract of Mount Etna.

1. State the altitude of Mount Etna.
2. Write the scale of the map as a ratio.
3. State the contour interval of the map.
4. What are the two types of vegetation that surround Mount Etna?
5. Describe the shape of Mount Etna.
6. Describe the location and direction of the highway.
7. Give the area reference for the following locations:
   a. Misterbianco
   b. Acireale
   c. Bronte.
8. Give the grid reference for the following locations:
   a. Mount (Monte) Piniteddu
   b. spot height 1316 to the east of Bronte
   c. where the highway crosses the railway in the south-east quadrant.

Apply your understanding

9. Which direction was the photographer facing when they took the photo in **FIGURE 1**?
10. If Mount Etna erupted, which places would be affected by the lava flow? How could people plan to protect their homes from lava flow?
11. Evaluate one the methods Italian authorities have tried to protect the people living near Mount Etna from lava flows. What was it intended to do, and what do you think the likely outcome would be in the event of an eruption?
12. Explain one of the likely effects if Mount Etna erupted and the lava reached Catania.

Challenge your understanding

13. To what extent would living so close to a volcano reduce the liveability of that place for you? Explain your answer by giving examples of which aspects of liveability would be reduced, and which would be unaffected.
14. Do you think some people might find living near or visiting Europe's most active volcano a positive thing? Give reasons for your answer.
15. Apart from the immediate danger of lava flows, suggest what other dangers Mount Etna poses to the natural and built environments.

To answer questions online and to receive **immediate feedback** and **sample responses** for every question, go to your learnON title at www.jacplus.com.au.

8.11 Thinking Big research project — Multicultural Australia photo essay

online only

The content in this subtopic is a summary of what you will find in the online resource.

Scenario

Australia is celebrating a new national holiday — Multicultural Australia Day — to acknowledge the fact that Australia is made up of people from many backgrounds and origins. You are entering the inaugural photo essay competition, which aims to show aspects of Australia's rich multicultural heritage.

Task

Create a photo essay (a story told through a series of photographs with some accompanying text). The purpose of this photo essay is to inform people of the rich and diverse cultures that make up Australian society.

Go to your ProjectsPLUS Resources tab to access all of the resources you need to complete this research project.

Resources

ProjectsPLUS Thinking Big research project — Multicultural Australia photo essay (pro-0173)

8.12 Review

8.12.1 Key knowledge summary

8.2 Understanding place and liveability

- A sense of place is personal and will vary from person to person.
- Natural and human features influence a sense of place.
- Different people have different perceptions of liveability.
- Perceptions of liveability are coloured by a person's background, stage in life and expectations.

8.3 Factors that influence where people live

- Push and pull factors influence where people choose to live.
- These factors include environment, economic (work, housing, universities), social (family) and lifestyle.

8.5 Measuring liveability

- Cities around the world are ranked against a set of criteria to create the liveable city index.
- Melbourne, Vienna and Vancouver have for several years shared the number one ranking; however, in 2018 Melbourne dropped to second place and Vancouver to sixth, and Vienna has taken the top spot.
- Changes to the top ten rankings are largely the result of improvements made in cities that are pushing to move up the rankings. There is more than one ranking index, but one that is widely used and relied on for this subtopic is produced by the Economist Intelligence Unit.

8.6 Factors that influence liveability in Australia

- Local neighbourhoods can have special meaning to some people.
- Aboriginal and Torres Strait Islander Peoples often identify with their local Country or Place.
- Holiday/tourism locations often promote a positive sense of place.
- Identifying the most liveable places differs from person to person.

8.8 More liveable cities — Melbourne and Sydney

- Melbourne has featured in the top ten most liveable cities for many years.
- Some factors, such as weather, work both for and against Melbourne as a liveable city.
- Melbourne is a city of great contrasts.

8.9 Less liveable cities — Port Moresby and Dhaka

- Less liveable cities tend to be in developing countries.
- Extreme poverty is an issue in these places, with many living below the poverty line.
- Literacy levels are low, though are showing some improvement.

FIGURE 12 Kawran Bazar, business district in Dhaka

8.12.2 Key terms

aerial photograph a photograph taken of the ground from an aeroplane or satellite
built environment a place that has been constructed or created by people
community a group of people who live and work together, and generally share similar values; a group of people living in a particular region
Country the place where an Aboriginal and Torres Strait Islander Australian comes from and where their ancestors lived; it includes the living environment and the landscape
facility place or thing that helps people
fly in, fly out (FIFO) workers who fly to work in remote places, work 4-, 8- or 12-day shifts and then fly home
formal describes an event or venue that is organised or structured
informal sector jobs that are not recognised by the government as official occupations and that are not counted in government statistics
infrastructure the basic physical and organisational structures and facilities that help a community run, including roads, schools, sewage and phone lines
literacy rate the proportion of the population aged over 15 who can read and write
liveability an analysis of what it is like to live somewhere, based on a set of criteria
location a point on the surface of the Earth where something is to be found
moieties the two halves that a nation were divided into
neighbourhood a region in which people live together in a community
place specific area of the Earth's surface that has been given meaning by people
population density the number of people living in a square kilometre
pull factor positive aspect of a place; reason that attracts people to come and live in a place
push factor reason that encourages people to leave a place and live somewhere else
region any area of varying size that has one or more characteristics in common
remote a place that is distant from major population centres
service program or organisation that helps people
temperate climate climate with generally warm summers and cool winters, without extremes
wilderness a natural place that has been almost untouched or unchanged by the actions of people

8.12.3 Reflection

Complete the following to reflect on your learning.

Revisit the inquiry question posed in the Overview:

What does a place need to have to make it liveable? Is everyone's idea of liveability the same?

1. Now that you have completed this topic, what is your view on the question? Discuss with a partner. Has your learning in this topic changed your view? If so, how?
2. Write a paragraph in response to the inquiry question, outlining your views.

FIGURE 1 Port Moresby is a mixture of high-rise urbanised landscapes and village landscapes.

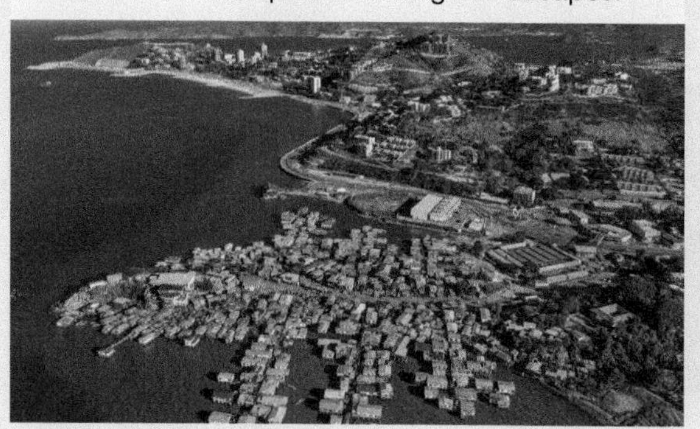

FIGURE 6 The skyline of Dhaka, Bangladesh

Subtopic	Success criteria			
8.2	I can define liveability.			
	I can outline why people's perceptions of place and liveability differ.			
8.3	I can explain the reasons that people live in certain places.			
	I can describe how places change over time.			
8.4	I can locate a place on a map using alphanumeric grid references.			
8.5	I can identify factors that determine a city's liveability.			
	I can describe the location of the most liveable and least liveable cities.			
8.6	I can describe aspects of liveability in Australia.			
	I can apply aspects of liveability to my neighbourhood.			
8.7	I can construct a basic sketch map from an aerial photograph.			
8.8	I can identify factors that make Melbourne a liveable city.			
	I can identify factors that make Sydney a liveable city.			
8.9	I can identify the factors that might make a city less liveable.			
	I can describe the impact of these factors on people and places.			
8.10	I can outline the risks of living near a natural hazard.			

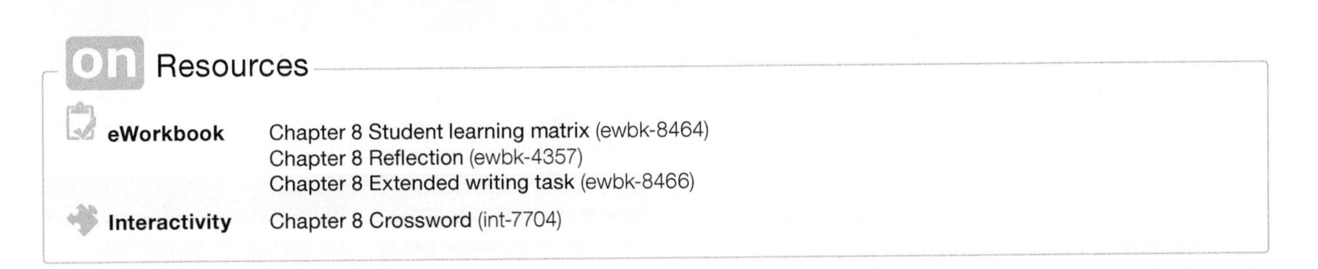
on Resources

eWorkbook Chapter 8 Student learning matrix (ewbk-8464)
Chapter 8 Reflection (ewbk-4357)
Chapter 8 Extended writing task (ewbk-8466)

Interactivity Chapter 8 Crossword (int-7704)

ONLINE RESOURCES

on Resources

Below is a full list of **rich resources** available online for this topic. These resources are designed to bring ideas to life, to promote deep and lasting learning and to support the different learning needs of each individual.

eWorkbook

- 8.1 Chapter 8 eWorkbook (ewbk-7988) ☐
- 8.2 Understanding place and liveability (ewbk-9198) ☐
- 8.3 Factors that influence where people live (ewbk-9202) ☐
- 8.4 SkillBuilder — Using alphanumeric grid references (ewbk-9206) ☐
- 8.5 Measuring liveability (ewbk-9210) ☐
- 8.6 Factors that influence liveability in Australia (ewbk-9214) ☐
- 8.7 SkillBuilder — Constructing a basic sketch map (ewbk-9218) ☐
- 8.8 More liveable cities — Melbourne and Sydney (ewbk-9222) ☐
- 8.9 Less liveable cities — Port Moresby and Dhaka (ewbk-9226) ☐
- 8.10 Investigating topographic maps — Mount Etna (ewbk-9230) ☐
- 8.12 Chapter 8 Student learning matrix (ewbk-8464) ☐
- Chapter 8 Reflection (ewbk-4357) ☐
- Chapter 8 Extended writing task (ewbk-8466) ☐

Sample responses

- 8.1 Chapter 8 Sample responses (sar-0142) ☐

Digital document

- 8.10 Topographic map of Mount Etna (doc-36255) ☐

Video eLessons

- 8.1 Choosing a place to live (eles-1619) ☐
- How liveability differs around the world — Photo essay (eles-5337) ☐
- 8.2 Understanding place and liveability — Key concepts (eles-5072) ☐
- 8.3 Factors that influence where people live — Key concepts (eles-5073) ☐
- 8.4 SkillBuilder — Using alphanumeric grid references (eles-1642) ☐
- 8.5 Measuring liveability — Key concepts (eles-5074) ☐
- 8.6 Factors that influence liveability in Australia — Key concepts (eles-5075) ☐
- 8.7 SkillBuilder — Constructing a basic sketch map (eles-1661) ☐
- 8.8 More liveable cities — Melbourne and Sydney — Key concepts (eles-5076) ☐
- 8.9 Less liveable cities — Port Moresby and Dhaka — Key concepts (eles-5077) ☐
- 8.10 Investigating topographic maps — Mount Etna — Key concepts (eles-5078) ☐

Interactivities

- 8.2 Understanding place and liveability (int-8411) ☐
- 8.3 Push/pull factors (int-3089) ☐
- 8.4 SkillBuilder — Using alphanumeric grid references (int-3138) ☐
- 8.5 My most liveable country (int-3095) ☐
- 8.6 Say 'hi' to your neighbourhood (int-8468) ☐
- 8.7 SkillBuilder — Constructing a basic sketch map (int-3157) ☐
- 8.8 More liveable cities — Melbourne and Sydney (int-8412) ☐
- 8.9 Environmental quality (int-3096) ☐
- Safe as houses (int-3097) ☐
- 8.10 Investigating topographic maps — Mount Etna (int-8413) ☐
- 8.12 Chapter 8 Crossword (int-7704) ☐

Fieldwork

- 8.3 Suburban survey (fdw-0009) ☐
- 8.6 Special places survey (fdw-0010) ☐

ProjectsPLUS

- 8.11 Thinking Big research project — Multicultural Australia photo essay (pro-0173) ☐

Weblinks

- 8.2 Gapminder, Dollar Street (web-6036) ☐
- 8.5 Economist Intelligence Unit (web-6037) ☐
- 8.6 Nothing like Australia (web-0002) ☐
- 8.9 Slum life (web-4055) ☐

Google Earth

- 8.2 Disney World (gogl-0001) ☐
- Grand Canyon (gogl-0002) ☐
- Table Mountain (gogl-0004) ☐
- Taj Mahal (gogl-0005) ☐
- Purnululu (gogl-0113) ☐
- 8.3 Baltimore, USA (gogl-0006) ☐
- Kanowna (gogl-0114) ☐
- Cossack (gogl-0115) ☐
- 8.9 Dhaka (gogl-0032) ☐
- Port Moresby (gogl-0034) ☐
- 8.10 Mount Etna (gogl-0135) ☐

myWorld Atlas

- 8.2 New York City (mwa-7329) ☐
- 8.3 Population of Australia (mwa-4505) ☐
- Energy in Australia (mwa-4525) ☐
- 8.5 Polluted cities (mwa-7331) ☐

Teacher resources

There are many resources available exclusively for teachers online.

9 Choosing a place to live

INQUIRY SEQUENCE

To access a pre-test with **immediate feedback** and **sample responses** to every question in this chapter, select your learnON format at www.jacplus.com.au.

9.1 Overview

Numerous **videos** and **interactivities** are embedded just where you need them, at the point of learning, in your learnON title at www.jacplus.com.au. They will help you to learn the content and concepts covered in this topic.

Rural, urban, remote or central — the world is full of places. Are any two places the same?

9.1.1 Introduction

We all live in different places. Places are important to people, whether those places are rural or urban, remote or central, permanent or temporary. But no two places are alike; they differ in aspects such as their appearance, size and features. In your mind's eye, try to picture the similarities and the differences between places such as a popular tourist destination, a remote village overseas, a scientific base in Antarctica, and a mining town. You may think of others to add to this list.

STARTER QUESTIONS

1. Brainstorm, as a class or individually, a list of other types of places in which people live around the world.
2. Choose three examples of places on your list that have been strongly influenced by the quality of the environment. Explain how the place is influenced by the environment.
3. Other than environmental qualities, what influences the characteristics of a place?
4. How true is it that 'no two places are the same'? Use images on this page and the previous page or places you know, to support your opinion.
5. Watch the **Choosing a place to live — Photo essay.** If you could live anywhere in Australia, where would you choose?

FIGURE 1 Homes and neighbourhoods differ around the world.

on Resources

eWorkbook Chapter 9 eWorkbook (ewbk-7989)

Video eLessons Why we live where we do (eles-1620)
Choosing a place to live — Photo essay (eles-5344)

9.2 Australians living in remote and rural places

LEARNING INTENTION

By the end of this subtopic, you will be able to classify places as either rural or remote, identify why people live in rural or in remote places, and discuss problems faced by people living in remote places including how these problems might be solved sustainably.

9.2.1 Settling inland Australia

For over 100 years, a small percentage of Australians have been moving away from large cities and coastal regions to live in more remote locations. They are often searching for new farmland or the mineral resources of the inland areas. Why do some people choose to live in places where their nearest neighbour is 50 kilometres away and it takes six hours to get to the closest supermarket? Why do they find remote places more liveable?

FIGURE 1 Stages in European land occupation in Australia

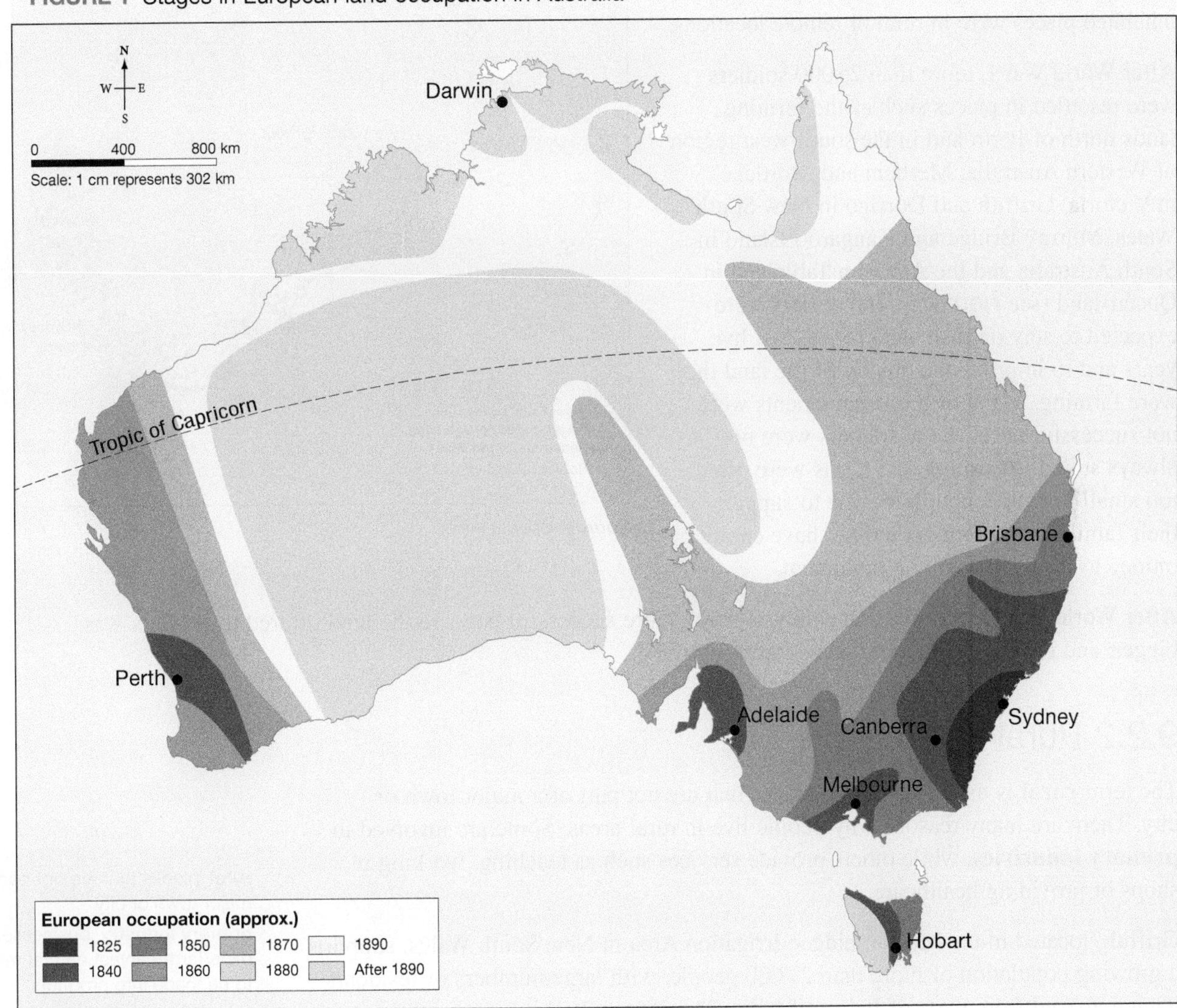

Source: Spatial Vision

The means to relocate people inland have never been faster or easier. The interconnection provided by modern transport and the high-speed communication provided by phone and internet should mean that technology has reduced remoteness. However, the general shift of Australia's population for the last 100 years has been towards the major cities and away from the country. In 2016, the average age of farmers in Australia was 56 years and getting older (it was 39 in 2010). Most children of farmers leave the country for education and work opportunities in large cities.

Over the past 100 years, governments and private industry have tried to encourage more people to move to the more remote places of Australia. Soldier settlement programs and mining developments are two examples of such schemes.

Soldier settlement schemes

After both World War I and World War II, the state and federal governments of Australia began a program of providing land to returned soldiers. This was to give these soldiers work, but it was also seen as a way of attracting people to places with few people living there. These sparsely inhabited places were in rural or remote locations.

After World War I, more than 25 000 soldiers were resettled in places such as the farming lands north of Perth and in the south-west region of Western Australia, Merbein and Mortlake in Victoria, Griffith and Dorrigo in New South Wales, Murray Bridge and Kangaroo Island in South Australia and the Atherton Tableland in Queensland (see **FIGURE 2**). The settlers were expected to stay on their land for at least five years and to improve the quality of the land they were farming. Many of these settlements were not successful because the soldiers were not always suited to farming, the farms were often too small to make enough income to support their families, and farmers did not have enough money to invest in stock or equipment.

FIGURE 2 Location of soldier settlement areas, 1917

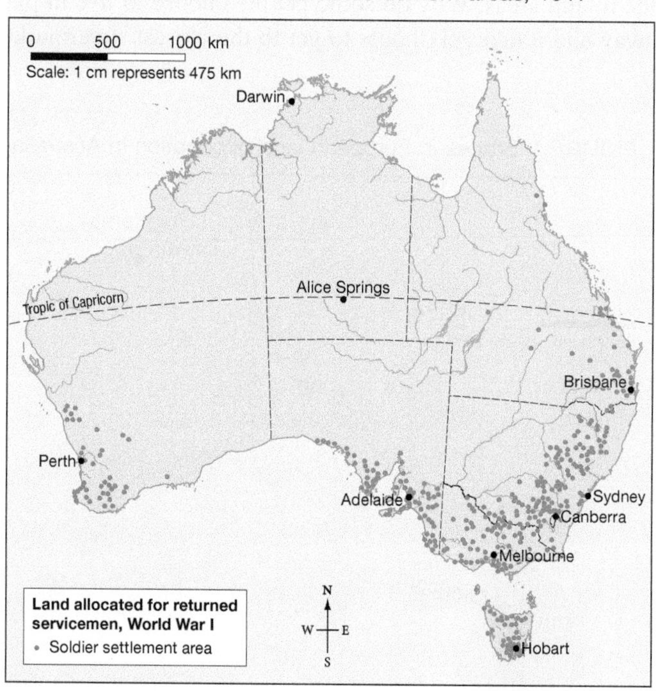

Source: Spatial Vision

After World War II, a similar scheme was much more successful because the land offered to soldiers was bigger, and roads, housing and fences were supplied.

9.2.2 Rural settlements

The term **rural** is used to describe places that are not part of a major town or city. There are many reasons why people live in rural areas. Some are involved in **primary industries**, while others provide services such as teaching, working in shops or providing healthcare.

Griffith, located in the Murrumbidgee Irrigation Area in New South Wales, supports a growing population of more than 27 000 people, with large numbers of residents having arrived from parts of Italy and India. The climate in this area is semi-arid (warm, with unreliable rainfall). The land became productive farmland after **irrigation** was provided in 1912. Reliable water and available farmland attracted many people to this area.

rural places that are not part of a major town or city

primary industry businesses that mine/farm/harvest natural resources to be made into products

irrigation water provided to crops and orchards by hoses, channels, sprays or drip systems in order to supplement rainfall

The area supports a variety of agricultural industries, and the picking, processing and distribution of locally grown products provides jobs and training opportunities for **seasonal workers** and the local population during harvesting periods.

FIGURE 3 Farms in Griffith at the edge of town

Two main types of farm are in this area.

- Type A farms are usually about 220 hectares (a hectare is 10 000 square metres). Each year they grow a combination of rice, corn, wheat, vegetables and pasture, and graze beef cattle. Irrigation water is usually used.
- Type B farms are **horticulture** farms, and are usually about 20 hectares. They grow a combination of permanent crops that may include grapes, peaches, plums, and citrus fruit such as oranges. Many of these plants last for many years, and irrigation is always needed.

Remote communities

The word **remote** is used to describe a place, town or settlement that is a long way from major population centres (towns or cities). Karratha and Tom Price in Western Australia, and Roxby Downs in South Australia are examples of current mining towns in remote areas. In some mining communities, the entire town was built and is managed by the mining company.

Today it takes only one and a half hours to fly from Perth to Tom Price in Western Australia, yet it can be difficult to attract workers to mines in this region because it seems so remote. Wages are high: workers in the mining and construction industries in these locations can earn between $90 000 and $120 000 per year. Fewer jobs are now available because the mining boom has passed, but skilled workers are still attracted to these remote places. Some workers fly in and fly out (FIFO) for their shifts. In 2017, it was estimated that between 75 000 and 90 000 Australians fly in for a shift that may last several weeks, before flying home for their days off. Some remote mining communities, such as Moomba in South Australia, do not have a permanent population. Everyone who lives and works there does so as a FIFO worker.

seasonal worker person who moves to a place for a short time to do a specific job (e.g. fruit picking)
horticulture the growing of garden crops such as fruit, vegetables, herbs and nuts
remote a place that is a long way from major towns and cities

FIGURE 4 Mount Tom Price iron ore mine

FIGURE 5 Remote mining regions

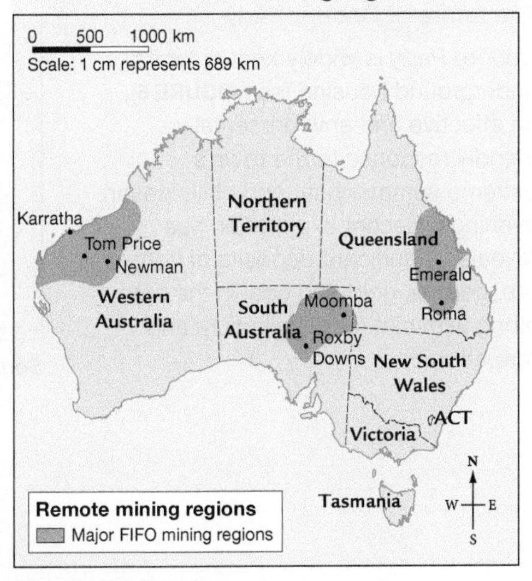

Source: Spatial Vision

9.2.3 Challenges facing rural communities

Rural and remote communities are an important part of Australia's social identity, but many are facing significant change and challenges in maintaining their population. Many are experiencing a decline because young people are leaving in search of education and employment.

CASE STUDY

Coober Pedy, South Australia

Coober Pedy is a vibrant multicultural town in the far north of South Australia, 850 kilometres north of Adelaide and 700 kilometres south of Alice Springs. The town is located in one of the most **arid** environments of Australia.

The traditional custodians of the land are the Antakirinja Matu and Yankunytjatjara Peoples. They have lived in the area around present-day Coober Pedy for thousands of years.

Opal was discovered in the area in February 1915 and, after several cycles of boom and bust, the town expanded rapidly during the 1960s. Opal developed into a multi-million dollar industry, and the town is sometimes called the opal capital of the world.

Opal continues to be important to Coober Pedy's identity and economy, but the town now draws its income from mining services, tourism and public services. The town's population had declined to an estimated 1762 people in the 2016 census.

FIGURE 6 Coober Pedy is known for its opal mining and 'dugout' houses that are built into the ground.

FIGURE 7 Coober Pedy location map

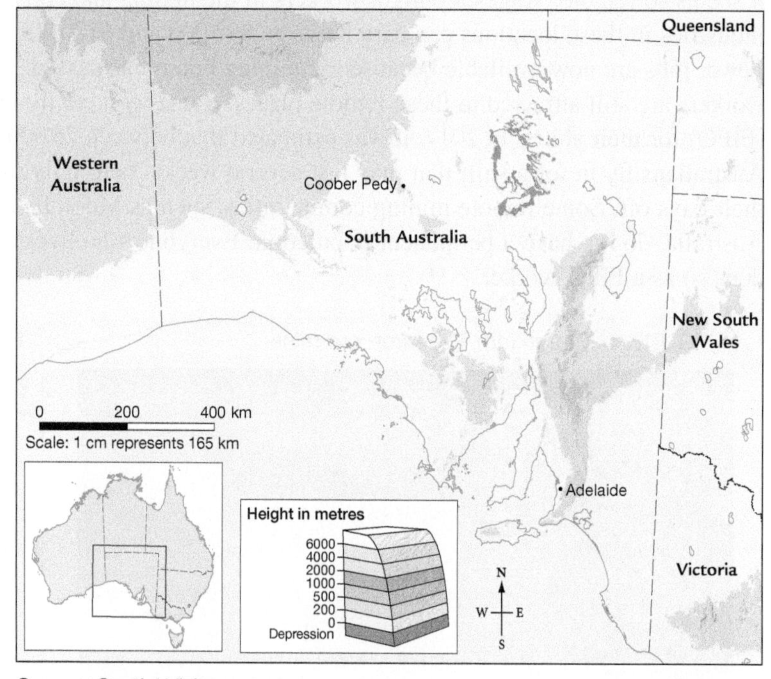

Source: Spatial Vision

The future of Coober Pedy

Coober Pedy is widely known for its underground housing (see **FIGURE 6**), an effective and environmentally friendly response to the town's extreme summer heat and chilly desert evenings. Recent exploration has revealed significant deposits of iron ore, copper, gold and coal in the area, along with platinum, palladium and rare earths.

arid lacking moisture; especially having insufficient rainfall to support trees or plants

Yet in 2014, the Cairn Hill iron ore/copper/gold mine was closed due to low iron ore prices.

The location of the town makes it an ideal centre for mining services, and a base for the delivery of state and federal government services in the region. This presents an opportunity for the town to reverse its steady population decline and again see growth in its economy and population.

Coober Pedy has hospital and medical services, primary and secondary schooling, a TAFE campus, childcare services and a police force. However, these services are under some pressure, and there is a continuing problem with the recruitment and retention of medical professionals. This place is extremely remote, so many of the pastoral properties in the region have only been linked to telecommunication services since 1987. The Stuart Highway provides the main transport and service route for the town.

FIGURE 8 Climate graph for Coober Pedy, South Australia

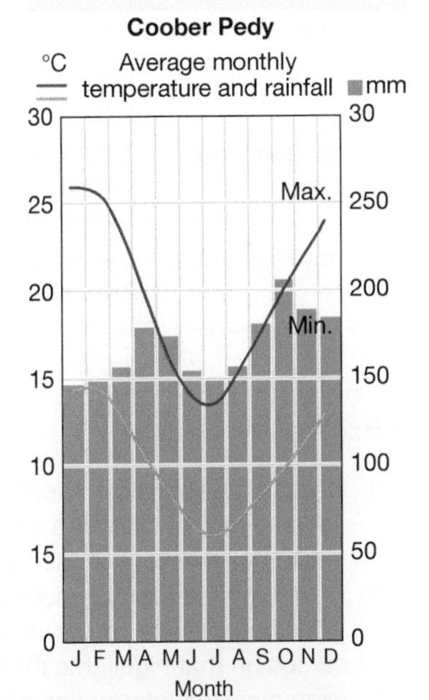

Reproduced by permission of Bureau of Meteorology, © 2021 Commonwealth of Australia.

FIGURE 9 Living underground provides cooler and more stable temperatures.

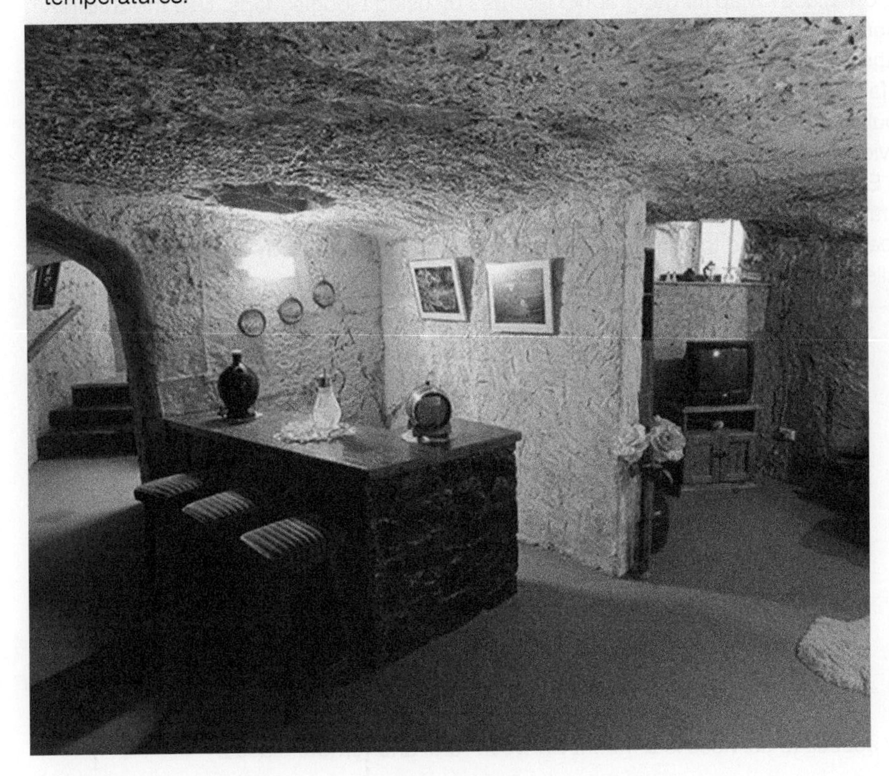

FIGURE 10 Mullock heaps create Coober Pedy's distinctive landscape.

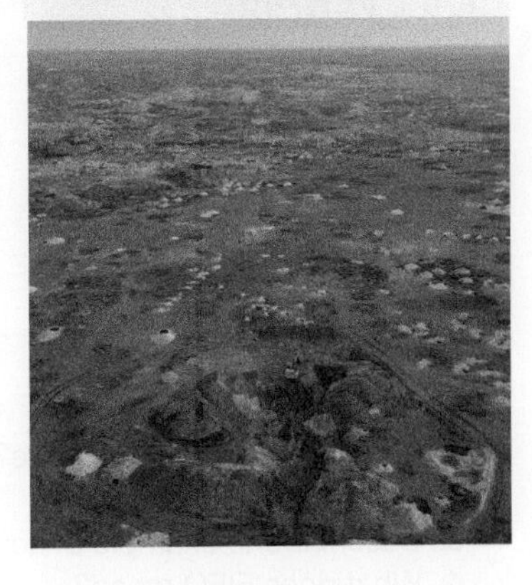

9.2.4 A question of survival

Many rural communities are facing global pressures, such as more overseas competition and fluctuations in the Australian dollar, which can affect the prices of commodities (such as minerals, wool and beef). Climate change and resultant droughts and floods also affect these rural communities. The rural communities that are not experiencing the trend of people moving to urban areas all have one thing in common: they have discovered another source of income. They may have shifted their focus to growing olives or grapes, or perhaps made use of a natural environment resource such as a nearby national park or provide adventure tourism activities like mountain biking or four-wheel driving.

In some cases, a rural community is unable to reinvent itself or tackle the problem effectively. The loss of an industry such as mining may have terrible effects on employment, leaving the resident population with lower incomes and few job prospects.

on Resources

eWorkbook	Australians living in remote and rural places (ewbk-9238)
Video eLesson	Australians living in remote and rural places — Key concepts (eles-5083)
Interactivities	Remote living (int-3090) Climate graph for Coober Pedy, South Australia (int-5292)
Google Earth	Tom Price township (gogl-0008) Griffith (gogl-0010) Coober Pedy (gogl-0122)
myWorldAtlas	Deepen your understanding of this topic with related case studies and questions > Investigate additional topics > Population > **Population of Australia**

9.2 ACTIVITIES

1. Research a local soldier settlement scheme. When was it established? How successful was it? How did this scheme help to populate a remote place? Map its geographic features by using Google Maps. Use the **Soldier settlement** weblink in the Resources tab to help with your research.
2. How might people be encouraged to move from the coastal fringe to the more remote places of Australia? What could make you or your family move or relocate? Produce a short film, a slide show or an advertising campaign that highlights the pull factors that might make people change the place where they live. Use the examples and information provided in the **Roxby Downs** weblink in your online Resources as a starting point.
3. Use the **Coober Pedy** Google Earth link in your Resources tab to explore Coober Pedy. What clues can you see that tell you about the climate of Coober Pedy? Can you see the mullock heaps, as shown in **FIGURE 10**?
4. Discuss strategies that could be implemented to entice more people to live in Coober Pedy and reverse its population decline. Develop a list of possible strategies that could be implemented.

9.2 EXERCISE

Learning pathways

■ LEVEL 1	■ LEVEL 2	■ LEVEL 3
1, 2, 3, 11, 12, 13	4, 7, 8, 10, 15	5, 6, 9, 14

Check your understanding

1. What makes a place remote?
2. What makes a place rural?
3. Describe one example of how access to services is different between rural and remote places.
4. What does FIFO mean?
5. Describe the location of Coober Pedy. Include information about direction and distance from Adelaide in your answer.
6. Outline two reasons why rural communities are under threat.

Apply your understanding

7. How does FIFO reduce the impacts of remoteness?
8. Explain two reasons why rural communities are under threat.
9. Describe the speed of settlement of inland Australia that is illustrated by **FIGURE 1**.
10. The Griffith region has many farms, which means there are many people in the area to support shops, businesses, schools and cultural activities. However, in some parts of Australia, farms are very big and it is a long way to the nearest neighbours. Anna Creek, a beef cattle property in northern South Australia, for example, is 24 000 square kilometres (2 400 000 hectares). This property is in a semi-arid region, where

vegetation is sparse and the nearest town for supplies is 170 kilometres away. Explain which location, Griffith or Anna Creek, would satisfy the following wishes:
   a. to play in a sports team every week
   b. to regularly buy clothes
   c. to collect data about arid-area lizards
   d. to grow a lush lawn
   e. to safely learn how to drive
   f. to have a private airstrip.
11. How might people who live on large farms in remote places have different perceptions of liveability than people who live in towns or city areas?
12. Explain two reasons why someone might choose to work as a FIFO worker.

Challenge your understanding

13. Predict how living in a remote community might affect liveability for elderly people.
14. Justify the link between access to services and facilities, and levels of liveability. Make reference to specific groups of people in your response.
15. Suggest ways that rural and remote communities can ensure sustainable futures.

To answer questions online and to receive **immediate feedback** and **sample responses** for every question, go to your learnON title at www.jacplus.com.au.

9.3 SkillBuilder — Interpreting population profiles

online only

LEARNING INTENTION

By the end of this subtopic, you will be able to identify and describe the age and gender spread shown in a population profile.

The content in this subtopic is a summary of what you will find in the online resource.

9.3.1 Tell me

What is a population profile?

A population profile, sometimes called a population pyramid, is a bar graph that provides information about the age and gender of a population. A population profile is used to show us the structure of a population.

9.3.2 Show me

How to compare population profiles

eles-1704

Step 1

To complete a comparison of population profiles, you must have two or more population profiles for the same place at different times, or for different places at the same time.

Step 2

Populations can be broadly grouped into three categories according to the level of dependence of the age groups:

- children (0–14 years) — dependent population
- adults (15–64 years) — economically productive and independent
- aged (65 years and over) — economically inactive and dependent.

A population is considered to be old when less than 30 per cent of the population is younger than 15 years and more than 6 per cent is aged 65 years and over. A population is considered to be young when more than 30 per cent of the population is younger than 15 years and less than 6 per cent is aged 65 years and over.

For each population profile, calculate the percentage of males and females in each of the three categories described above. What does this tell you about the population in each of the population profiles?

Step 3

Look for patterns revealed by each population profile. Write a statement about the balance of the population profiles, providing reasons for any imbalances (war, migration, government policies etc).

Step 4

Consider any unusual aspects. Are there any indents (places where the graph narrows unexpectedly) or extended age groupings? Can you suggest why these may occur? You will need to research the background of a country to gain information that will allow you to make an accurate interpretation of its population figures.

FIGURE 1 Population profiles of (a) Indonesia and (b) Vanuatu

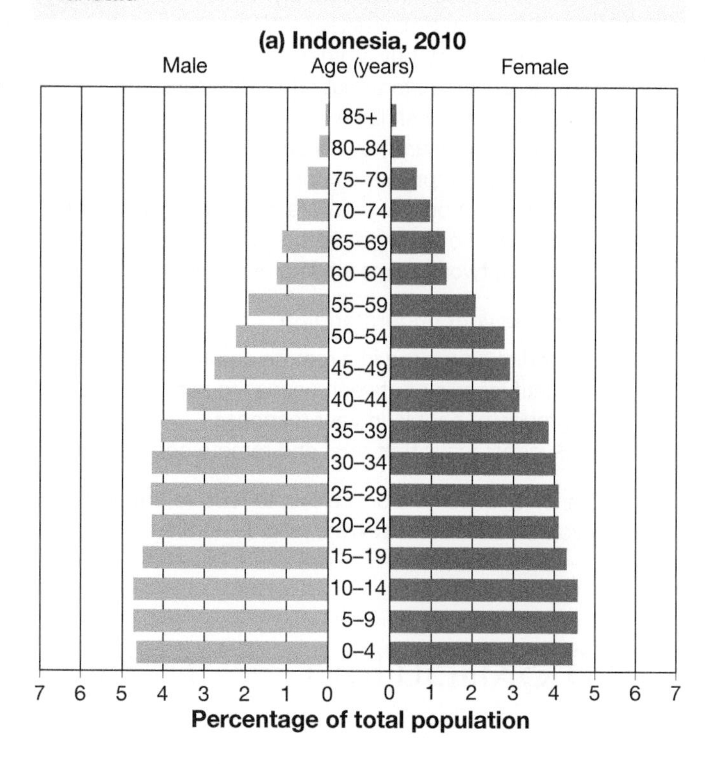

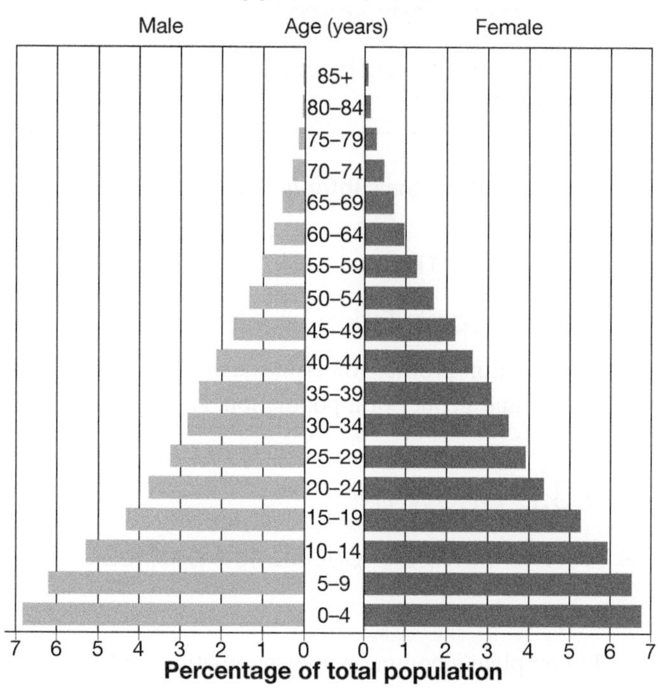

9.3.3 Let me do it

int-3284

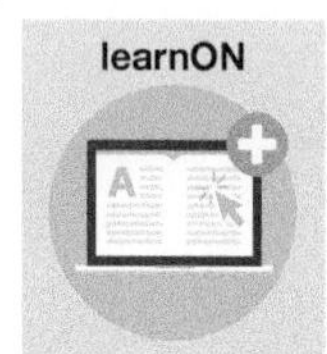

Go to learnON to access the following additional resources to help you build this skill:

- a longer explanation of this skill and its application in Geography (Tell me)
- a video showing the step-by-step process of this skill (Show me)
- an activity and interactivity for you to practise this skill (Let me do it)
- self-marking questions to help you understand and use this skill.

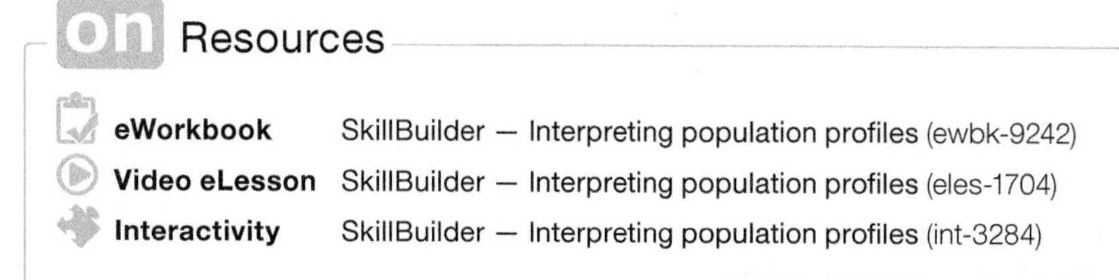

on Resources

eWorkbook SkillBuilder — Interpreting population profiles (ewbk-9242)

Video eLesson SkillBuilder — Interpreting population profiles (eles-1704)

Interactivity SkillBuilder — Interpreting population profiles (int-3284)

9.4 Life in a country town

LEARNING INTENTION

By the end of this subtopic, you will be able to identify the characteristics of country towns that make them popular places, and explain why country towns are becoming more popular places to live.

9.4.1 The attraction of the country

Country towns come in all shapes and sizes. They can be small centres with a post office and general store or they can be substantial towns. Because most of Australia's population and businesses are concentrated in the capital cities, even people who live in quite large towns outside the capital cities can see themselves as living in the country.

Even though most Australians live in large urban centres, the rural or country regions are very important because this is where food is grown, **natural resources** are extracted and **ecosystems** can flourish. Many Australians travel to country places for holidays and many dream of moving to the country. The attractions of country places include cheaper housing, less traffic, a greater sense of safety, and the allure of living within and around natural environments.

9.4.2 Demography

The **demographic** characteristics of country places are influenced by location and activities in the surrounding area.

For instance, Forbes is located approximately 380 kilometres from Sydney. It receives regular rainfall and has fertile soil. It is part of the NSW wheat belt, an area that produces large quantities of wheat and cereal crops. This is the largest employment sector in the town.

Another town, Bega, is located approximately 421 kilometres south of Sydney. It has moderate temperatures and high annual rainfall. The dairy industry is the dominant source of employment. Bega is famous for its cheese factory.

natural resources resources (such as landforms, minerals and vegetation) that are provided by nature rather than people

ecosystem an interconnected community of plants, animals and other organisms that depend on each other and on the non-living things in their environment

demographic describes statistical characteristics of a population

FIGURE 1 (a) Bega and surrounding farmland, NSW (b) Lachlan Street, Forbes, NSW

FIGURE 2 Population pyramids for (a) Bega and (b) Forbes, 2014 and 2019

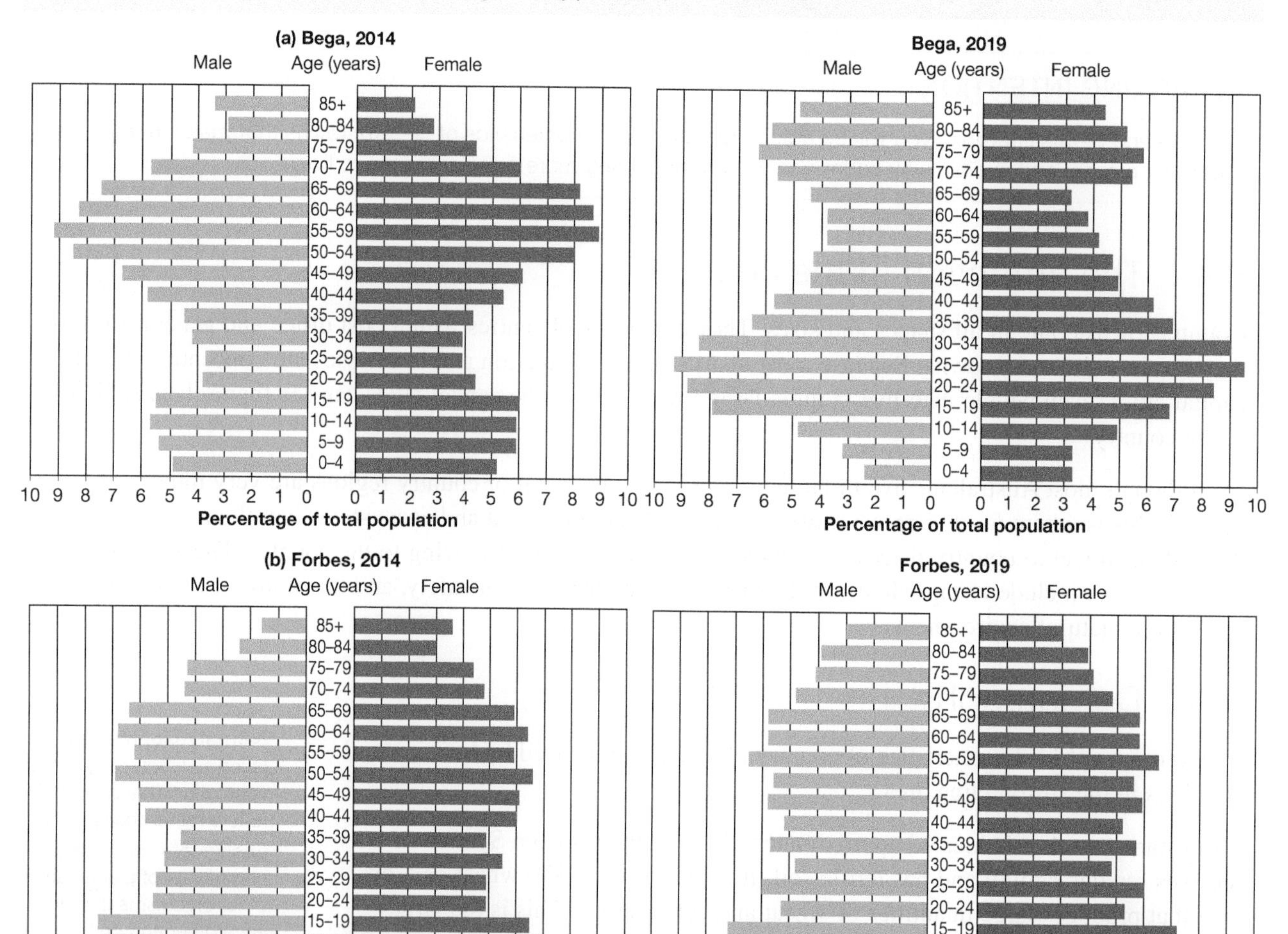

Source: Australian Bureau of Statistics

A sense of belonging

People who live in country centres have a strong sense of identity, connectedness and belonging. Towns like Forbes feature a range of sporting clubs including rugby union, soccer, netball, tennis, basketball, swimming and dragon boating. These clubs cater for all age groups, with a strong focus on junior sport. Cultural activities such as art exhibitions and theatre groups tend to be found in the larger towns. Declining populations have affected the ability of smaller towns to field teams in some major sports, and the range of cultural activities is also shrinking. A recent service now common in country towns of all sizes, however, is the Community Men's Shed. Here, men meet to undertake practical projects with a community focus. Men's Sheds were created to help men make connections within their community as a way to prevent social isolation and encourage communication in order to improve their mental health.

FIGURE 3 Sporting clubs such as lawn bowls help to provide a sense of community in rural places.

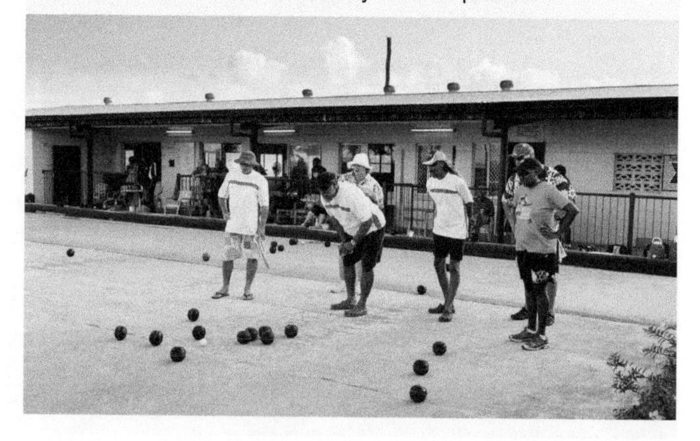

FIGURE 4 What growing up in a country town might mean

A greater understanding of the natural world (e.g. plants, animals, weather) and appreciation of the importance of the quality of the environment

Fewer restrictions on your time

A volunteer fire brigade

A small group of people your age

A sense of safety

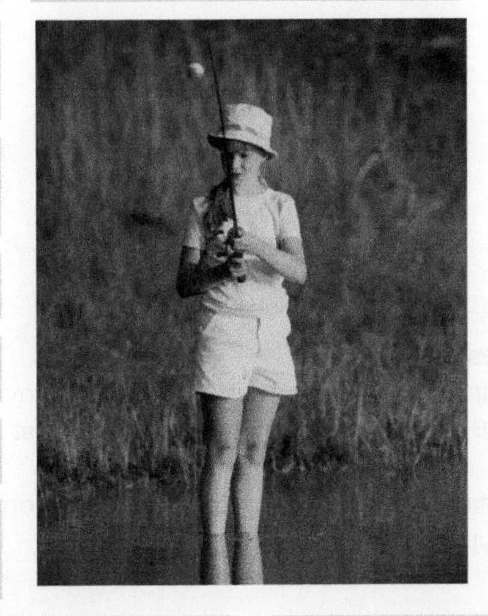

Little public transport; reliance on car travel

Greater likelihood of seeing native animals

An understanding of Aboriginal cultures

Travelling to school by bus

A wider range of pets

Opportunities for physical activity

Leaving home for tertiary study

Plenty of space to move about in

Uncrowded roads

Clean air

A limited range of specialists, shops and employment

Quiet walking and riding trails and fishing spots

Fewer entertainment options and less live music

Community reliance on surrounding resources (e.g. farming, mining, fishing or the environment that attracts tourists)

Knowing many people in the community and many people in the community knowing you

Resources

eWorkbook	Life in a country town (ewbk-9246)
Video eLesson	Life in a country town — Key concepts (eles-5084)
Interactivity	Country town services (int-3092)
Google Earth	Bega, Forbes (gogl-0117)

9.4 ACTIVITIES

1. a. Create a list of Australian country towns that you have heard of.
   b. Collaborate with others to map the location of each of these towns.
   c. For each town, record its latitude, distance and direction from the state capital city.
   d. Complete the following generalisation about the better-known towns: 'Most of the towns we know are located ________________.'
2. How might you feel about living in different places? The following six statements refer to different opinions about living in rural areas or cities:
   - Cities provide more choice in activities and places to live.
   - Pollution and noise in big cities impair living conditions.
   - I don't feel safe in big cities because of crime.
   - Rural areas have great communities with people supporting one another.
   - The natural environment in many rural areas is very peaceful and healthy.
   - Jobs and transport are more accessible in cities.

   Rank these statements from the one you agree with the most, to the one you agree with the least. Explain your ranking to another person. How might these rankings change if they were completed by people who live in places different from where you live? Can you test this hypothesis?

3. As a class or group, discuss the meaning of the term *country life*.
4. Research and compare the regional centre of Wagga Wagga and the rural area of Lockhart. Describe what life might be like in each of these towns and predict the future population structures for each place.

9.4 EXERCISE

Learning pathways

■ LEVEL 1	■ LEVEL 2	■ LEVEL 3
1, 2, 3, 7, 13	4, 5, 8, 10, 14	6, 9, 11, 12, 15

Check your understanding

1. Provide three reasons country places are important.
2. Classify each of the characteristics in **FIGURE 4** as economic, social or environmental.
3. Identify the characteristics in **FIGURE 4** that are attractive to you. Are most of these characteristics social, economic or environmental?
4. Identify and describe the purpose of one community group or organisation in a country area that helps to look after the wellbeing of a specific group in the community.
5. Describe the distribution of ages of:
   a. females in Bega
   b. males in Bega
   c. females in Forbes
   d. males in Forbes.
6. Define the term *demography* in your own words. Identify three factors apart from age and gender that would help to define the demographics of an area.

Apply your understanding

7. Identify three opportunities in a rural community for young people to feel socially connected and explain how they might help people's wellbeing.
8. Considering the main sources of employment in Forbes and Bega, how might a drought affect the number of people moving to or from the town?
9. Consider the ideas presented in **FIGURE 4**. Identify three of the potential features of country life shown that would be subjective (depend on people's judgement of what is positive or negative). Explain why these factors would be subjective.
10. Study the data in **FIGURE 2**. What has changed with the population structure in Bega and Forbes from 2014 to 2019? Explain why this may have occurred in each town.
11. Compare the changes in population structure in Bega and Forbes from 2014 and 2019. Based on this data, which place do you expect to have a higher proportion of people over the age of 60 years in 2030?
12. Explain how Men's Sheds might help to improve the wellbeing of both men and women in country towns.

Challenge your understanding

13. In Australia, the trend is for people to move away from the country to the major cities. Suggest three reasons you think this happens and predict whether it might continue to happen in the future.
14. Country towns are critical to the rest of Australia. Explain what role country towns can play on a national scale.
15. How might governments use demographic information from country towns? Suggest three ways the information might be useful.

To answer questions online and to receive **immediate feedback** and **sample responses** for every question, go to your learnON title at www.jacplus.com.au.

9.5 Lifestyle places

online only

LEARNING INTENTION

By the end of this subtopic, you will be able to explain what a lifestyle place is and identify who might live in a lifestyle place and why.

The content in this subtopic is a summary of what you will find in the online resource.

In the years after 1990, the healthy state of the growing Australian and world economies meant that more people had jobs and were earning higher incomes. This gave them greater choice as to where and how they wanted to live because they had the resources (money) to allow them to choose. For some people, 'lifestyle choice' means escaping the rush of the city by choosing a sea change or tree change. For others, it means living in the inner-city close to shops, cinemas and restaurants.

This, however, is not the case for everyone. Without the money to move or the ability to get work in a new place, the option of choosing where to live is unlikely to be an option. Many people have to live wherever they can afford. What impact do these choices have on people and their lives? Does choice make you happier, more secure or provide more time with your family? Or can living a simpler life with less choice be happier?

FIGURE 3 Contrasting lives in India: (a) The growing middle classes in modern India often purchase modern apartments, like those in this building in New Delhi. (b) For many poor people in urban areas of India, daily activities, such as washing and cooking, take place on the street. Formal housing is well beyond most people's reach, and extended families live together in small, makeshift homes.

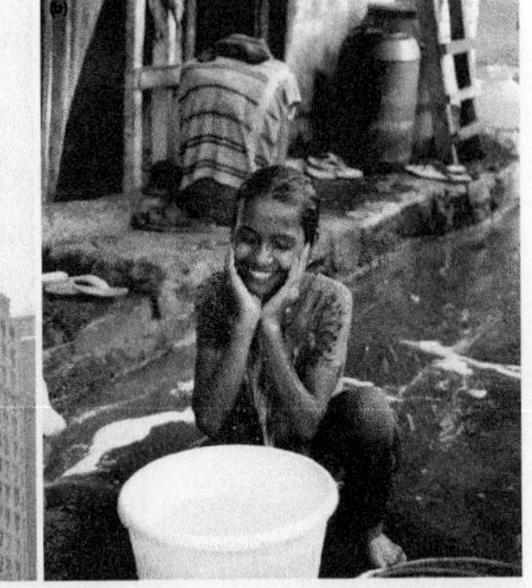

To learn more about lifestyle choices and liveability, go to your learnON resources at www.jacPLUS.com.au.

Contents

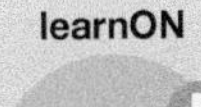

- 9.5.1 Lifestyle
- 9.5.2 Urban renewal projects

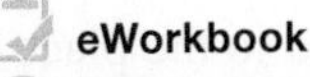

on Resources

eWorkbook	Lifestyle places (ewbk-0329)
Video eLesson	Lifestyle places — Key concepts (eles-5085)
Fieldwork	Sphere of influence survey (fdw-0012)
Google Earth	Canary Wharf, London (gogl-0011) Prague, Czech Republic (gogl-0012) Rio de Janeiro (gogl-0013) New Delhi (gogl-0111) Fishermans Bend (gogl-0112) Venice Beach (gogl-0123)

9.6 Growth cities in Australia

LEARNING INTENTION

By the end of this subtopic, you will be able to explain the interconnection between push and pull factors and the growth of cities, and identify the places in Australia that are growing at a faster rate.

9.6.1 The areas with the greatest growth

Which place in Australia is growing the fastest? If a place is liked by lots of people, does that make it the best? What makes a suburb the most popular? Coastal areas have always been a popular place for Australians to relax and holiday. Is the fastest growing place in Australia near the coast?

People might move to a new place for many different reasons. The attractions that entice people to live somewhere are called its pull factors. Pull factors include cheaper housing, better climate, more job opportunities and improved lifestyle. People can also be forced to leave their home and move to a new place. These reasons are known as push factors. Loss of your job or business, poor school or health facilities, and a natural disaster, such as flood or fire, are examples of push factors. While individual push and pull factors may differ, when there is a lot of overlap in the pull factors, growth in suburban areas occurs.

In the 2018–19 financial year, the Australian Bureau of Statistics recorded the five areas with the highest growth rates. They were mainly located around major cities (see **TABLE 1**). Over recent years, Melbourne has been Australia's fastest growing capital city. It is not surprising that it often tops lists of the world's most liveable cities. In New South Wales, while Sydney experienced a growth rate of 1.7 per cent, the rest of the state only grew by 0.8 per cent. The areas with the greatest growth rates were Riverstone—Marsden Park and Cobbitty—Leppington, both with 20 per cent growth, and Rouse Hill—Beaumont Hills with 14 per cent growth.

TABLE 1 Areas with the highest growth rates

Suburb1	Area	Estimated population 30 June 2019	2018–19 growth rate (%)
Mickleham—Yuroke	Melbourne — North West	11 227	52.5
Moncrieff	Australian Capital Territory	3844	38.1
Rockbank—Mount Cottrell	Melbourne — West	8815	36.6
Ripley	Ipswich	8112	26.6
Pimpama	Gold Coast	19 425	20.5

**Excludes suburbs with fewer than 1000 people at June 2018*

9.6.2 The growth of the Gold Coast

The Gold Coast's warm weather, beach culture and holiday lifestyle have attracted many new residents.

FIGURE 1 shows that most of the new arrivals came from New South Wales. Many were attracted to the place their family visited on holiday, and they later decided to make it their permanent home. The Gold Coast is now the sixth-largest urban area in Australia. It is a major tourist destination, offering a wide range of work opportunities, community facilities, and intercity and interstate transport links by road, rail and air. Many new residents are older Australians who have retired to this place.

The increased population has placed pressure on the coastal environment, as well as on the existing infrastructure of schools, hospitals, roads and housing.

FIGURE 1 Net migration to the Gold Coast between 2011 and 2016

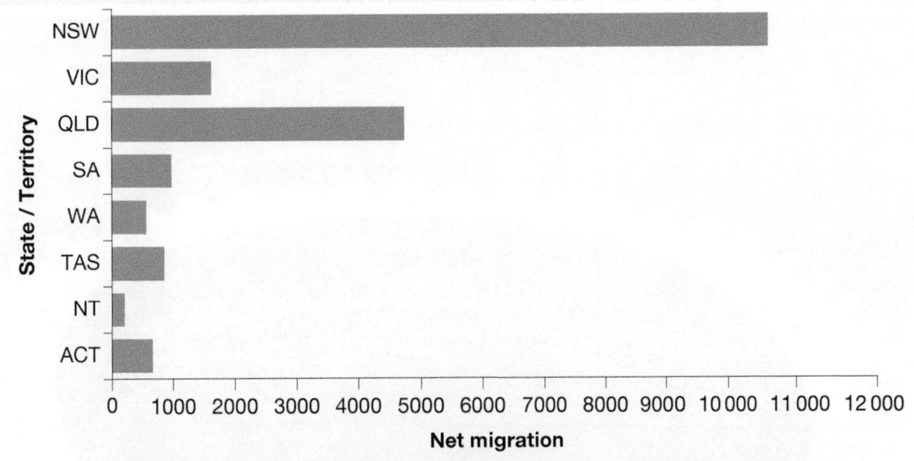

* net migration is the number of people who move to a place minus the number of people who have moved away from the place

FIGURE 2 Topographic map extract of the Gold Coast region in 1967

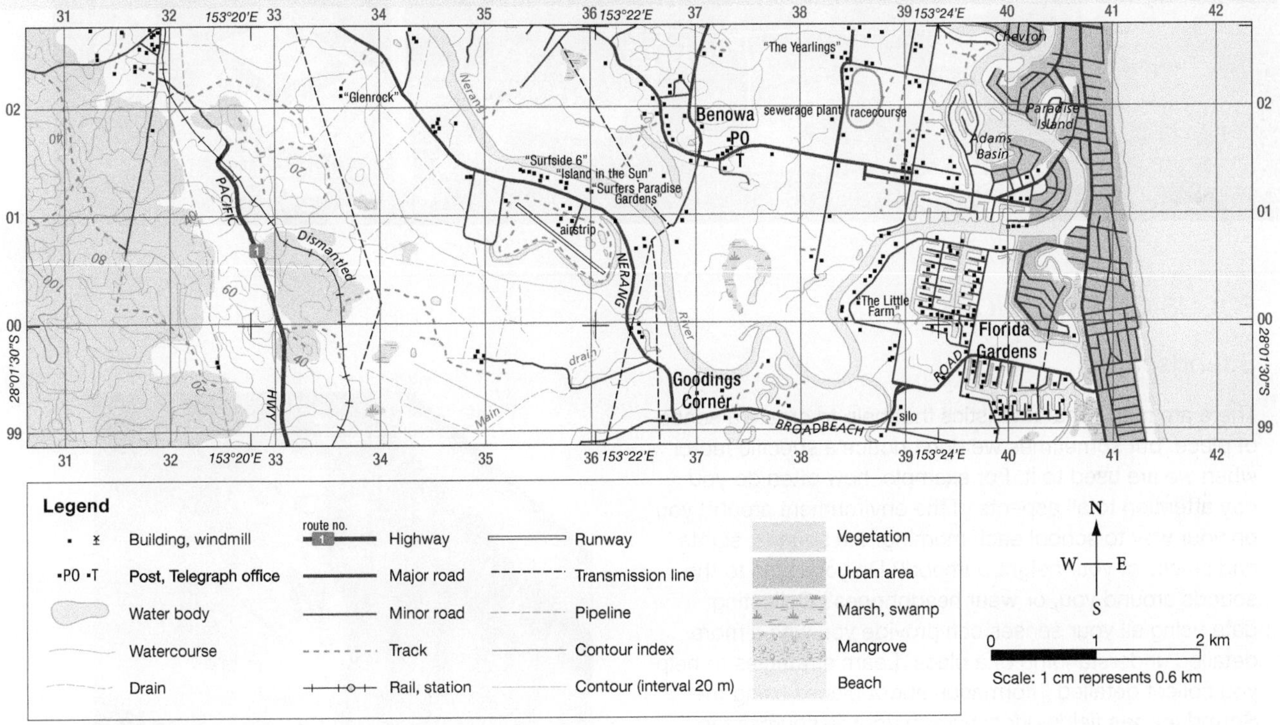

Legend
Building, windmill
Post, Telegraph office
Water body
Watercourse
Drain
Highway
Major road
Minor road
Track
Rail, station
Runway
Transmission line
Pipeline
Contour index
Contour (interval 20 m)
Vegetation
Urban area
Marsh, swamp
Mangrove
Beach

Scale: 1 cm represents 0.6 km

Source: Spatial Vision

AFL expansion team and 2018 Commonwealth Games

The main AFL (Australian Football League) states are Victoria, South Australia, Western Australia and Tasmania: many Aboriginal communities of the Northern Territory also follow AFL. In March 2009, the Gold Coast Football Club, now named the Gold Coast Suns, was established, supported financially by the AFL. The club's establishment on the Gold Coast has seen a rise in youth participation in AFL.

The six largest Australian cities now have at least one AFL club: Sydney (two clubs), Melbourne (nine clubs), Brisbane (one club), Perth (two clubs), Adelaide (two clubs) and Gold Coast (one club).

The Gold Coast was chosen to host the Commonwealth Games in 2018. Metricon Stadium at Carrara, the home of the Gold Coast Suns, was temporarily transformed and increased in capacity to host the athletics events as well as the opening and closing ceremonies for the Commonwealth Games. Australia has hosted five Commonwealth Games, but this was the first time they were not held in a state capital city.

FIGURE 3 The ground developed for the Gold Coast Suns AFL team

fdw-0016

FOCUS ON FIELDWORK

Soundscapes

There are many characteristics that help to create a sense of place, but sometimes we don't notice a specific factor when we are used to it. For example, how often do you pay attention to all aspects of the environment around you on your way to school each morning: the sounds, sights and smells of your neighbourhood? Do you listen to the sounds around you, or wear headphones? Collecting data using all your senses can provide you with a more detailed understanding of a place. Learn strategies to help you collect detailed information about places using the **Soundscapes** fieldwork activity in your Resources tab.

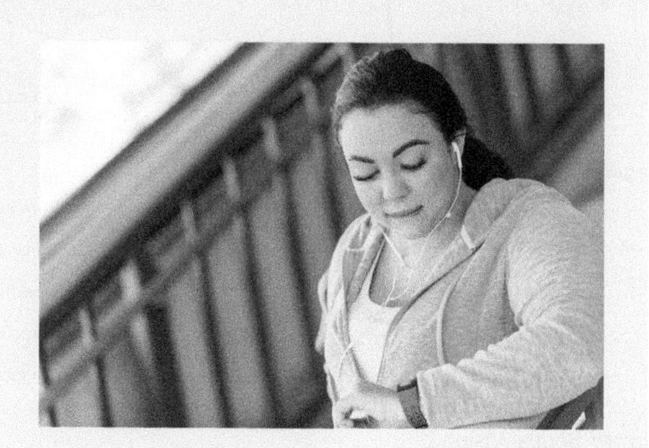

on Resources

eWorkbook	Growth cities in Australia (ewbk-9254)
Digital document	Topographic map extract — Gold Coast region (doc-11400)
Video eLesson	Growth cities in Australia — Key concepts (eles-5086)
Interactivity	Growth cities in Australia (int-8419)
Fieldwork	Soundscapes (fdw-0016)
Google Earth	Gold Coast (gogl-0124)

9.6 ACTIVITIES

1. Compare **FIGURE 2** with an online map or aerial image of the same place today.
   a. Study the map and photo in small groups. Identify the changes to the environment, both built and natural, between the map and the photo.
   b. Collate this information in a table.
   c. Write one sentence to describe the change to the built environment.
   d. Write one sentence to describe the change to the natural or physical environment.
   e. The population of the Gold Coast is predicted to double to 1.2 million people by 2050. Is much space left in this area of the Gold Coast for housing? Suggest where new suburbs could be established.
2. Using your knowledge of the factors that have made the Gold Coast so liveable and grow so quickly, use your atlas to identify another place that could become the 'new' Gold Coast. Identify the pull factors for your location and write a paragraph to sell its advantages to potential residents.

9.6 EXERCISE

Learning pathways

■ LEVEL 1	■ LEVEL 2	■ LEVEL 3
2, 4, 5, 10, 14	1, 3, 7, 9, 13	6, 8, 11, 12, 15

Check your understanding

1. What features of the Gold Coast have made it grow so quickly?
2. Where have most of the new residents of the Gold Coast come from?
3. What services, facilities and environmental attractions does the Gold Coast offer to people wishing to find a more liveable place to retire to?
4. What are *push factors*? Give an example.
5. What are *pull factors*? Give an example.
6. Outline two pull factors that would attract retirees to the Gold Coast.
7. What might be the interconnection between a liveable city and high population growth rates?

Apply your understanding

8. What was significant about the Gold Coast hosting the 2018 Commonwealth Games?
9. Push and pull factors result in the rise or fall of the population of a place. Use examples to explain the differences in environment, services and facilities between a place with push factors and a place with pull factors.
10. Think about the place where you live. Make a list of the pull factors that make your town or suburb more liveable. Then list the possible push factors that might make someone leave your suburb or town to live somewhere else.
11. Imagine you are a town planner. Suggest two new features you can add to your suburb that would make it a more appealing place to live.
12. Do you think building a new sports stadium (including as a home ground for an AFL team) would be a significant pull factor to draw people to where you live? Give reasons for your decision.

Challenge your understanding

13. Many of Australia's towns and cities are growing faster than ever before. What do you believe is behind this growth?
14. Consider the push and pull factors that you've identified and discussed already. Which do you believe is the most important factor people consider when deciding where to live? Justify your response.
15. Can a place grow too quickly? Suggest what factors might influence people's view on this issue. In your answer, discuss push and pull factors and consider long-term and new residents.

To answer questions online and to receive **immediate feedback** and **sample responses** for every question, go to your learnON title at www.jacplus.com.au.

9.7 SkillBuilder — Understanding satellite images

online only

LEARNING INTENTION

By the end of this subtopic, you will be able to identify and describe significant features on a satellite image.

The content in this subtopic is a summary of what you will find in the online resource.

9.7.1 Tell me

eles-1643

What are satellite images?

Satellite images are images that show parts of our planet from space. They are taken from satellites and transmitted to stations on Earth. Satellites can collect a variety of data, including standard photographic imagery, colour, infra-red and radar data.

9.7.2 Show me

How to interpret a satellite image

Step 1

Read the title and check for the date the image was taken. Read any accompanying information.

Step 2

Identify the main features of the image.

Step 3

Look for and label the biophysical features.

Step 4

Look for and label the built features, such as roads, bridges, sports stadiums and residential housing. Again, look at patterns.

FIGURE 1 Currumbin on Australia's Gold Coast, 8 May 2000

Source: Satellite image courtesy of GeoEye. Copyright 2009. All rights reserved.

Step 5

Some colours, patterns and shapes may still be puzzling. Look at Obtain a map of the same area — try an atlas or online mapping tool. Find names of key features to use in your description. If the features you have identified are shown on the map, check whether your analysis so far matches the map. Use the map to investigate the aspects that are still puzzling.

9.7.3 Let me do it

int-3139

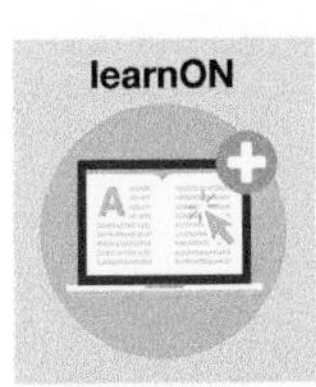

Go to learnON to access the following additional resources to help you build this skill:

- a longer explanation of this skill and its application in Geography (Tell me)
- a video showing the step-by-step process of this skill (Show me)
- an activity and interactivity for you to practise this skill (Let me do it)
- self-marking questions to help you understand and use this skill.

on Resources

eWorkbook	SkillBuilder — Understanding satellite images (ewbk-9258)
Video eLesson	SkillBuilder — Understanding satellite images (eles-1643)
Interactivity	SkillBuilder — Understanding satellite images (int-3139)

9.8 Isolated settlements

online only

LEARNING INTENTION

By the end of this subtopic, you will be able to explain why a settlement might be isolated, and identify the challenges faced by people living in isolated communities.

The content in this subtopic is a summary of what you will find in the online resource.

People choose to live in isolated and very remote places for various reasons. Despite employment being a large pull factor, there are several factors that can make isolated settlements undesirable, including lack of services and limited healthcare. One such place is Dawson City in Yukon, Canada, with a population of 1400. At the latitude of 64°N, it is only about 360 kilometres from the Arctic Circle. What is life in such an isolated place like?

FIGURE 5 Dawson City, Yukon, Canada

To learn more about liveability in very isolated places, go to your learnON resources at www.jacPLUS.com.au.

Contents

- 9.8.1 Living in remote locations

on Resources

eWorkbook	Isolated settlements (ewbk-9262)
Video eLesson	Isolated settlements — Key concepts (eles-5087)
Interactivity	Climate graph for Dawson (int-5300)

9.9 Nomadic ways of life

LEARNING INTENTION

By the end of this subtopic, you will be able to explain why people might choose a nomadic lifestyle, and describe the challenges faced by people living a nomadic lifestyle.

9.9.1 Nomadic cultures

Most people live in the one place, but from time to time they may move to a new location. About 30 million people in the world live a nomadic lifestyle. Nomads do not wander aimlessly. From time to time they pack up all their possessions and move, often returning to a place at some point in the future.

9.9.2 Challenges of living in a harsh environment

The Tuareg people (pronounced *twah*-reg) lead a nomadic way of life mainly in the Sahara Desert. A number of related families live in groups of 30 to 100. These groups usually move to a new site every two or three weeks, because the environment provides very little water and food.

Timbuktu is a town in Mali (see **FIGURE 2**) that has long been famous as a trading town, that played an important part in the nomadic cultures of the region. Here, goods such as salt were brought from the north across the Sahara, and goods such as gold came from the south.

Daytime temperatures are high in Timbuktu and night-time temperatures can be cold. Rainfall is very low and unreliable. There can be long periods without rain and there can be sudden heavy downpours. Strong winds sometimes cause sandstorms that turn the sky yellow or orange, reduce visibility and cover everything with sand.

FIGURE 1 Young Tuareg woman, Ingal, Niger

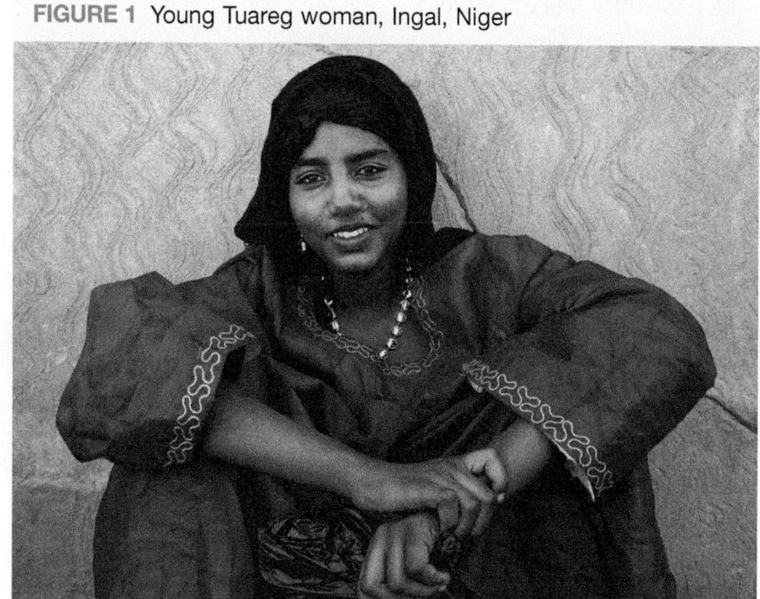

TABLE 1 Average temperature and rainfall, Timbuktu, Mali

	Jan	Feb	Mar	Apr	May	Jun	Jul	Aug	Sep	Oct	Nov	Dec
Average max. temp. (°C)	31	35	38	41	43	42	38	35	38	40	37	31
Average min. temp. (°C)	13	16	18	22	26	27	25	24	24	23	18	14
Average rainfall (mm)	0.5	0.5	0.5	1	4	20	54	93	31	3	0.5	0.5

Source: http://www.timbuktu.climatemps.com/temperatures.php

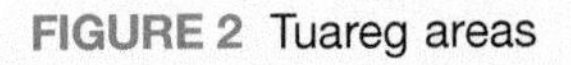

FIGURE 2 Tuareg areas

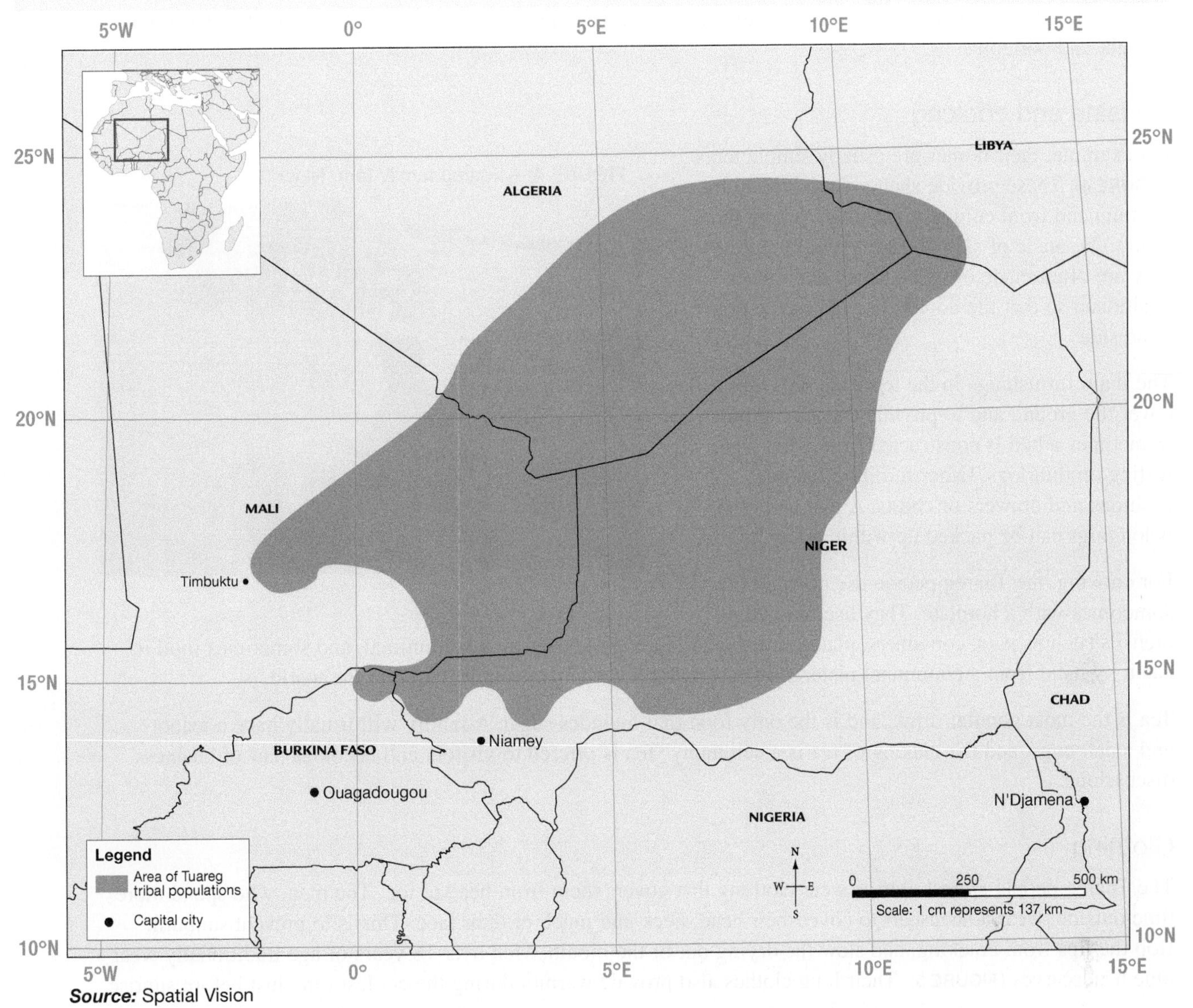

Source: Spatial Vision

The climate means Timbuktu has little vegetation and very little water. Significant plant growth is only found at oases. This means that families have difficulty finding wood for cooking fires. When firewood is unavailable, dried camel dung is used as fuel.

Animals have always been the most important possessions of the Tuareg people, and the need for grass and water for the animals is why the Tuareg people move from place to place. When food for the animals is exhausted, the group must move to a new location. They move across the desert, traditionally finding their way by the stars, the Moon and the landscape.

FIGURE 3 A Tuareg family's camels and goats

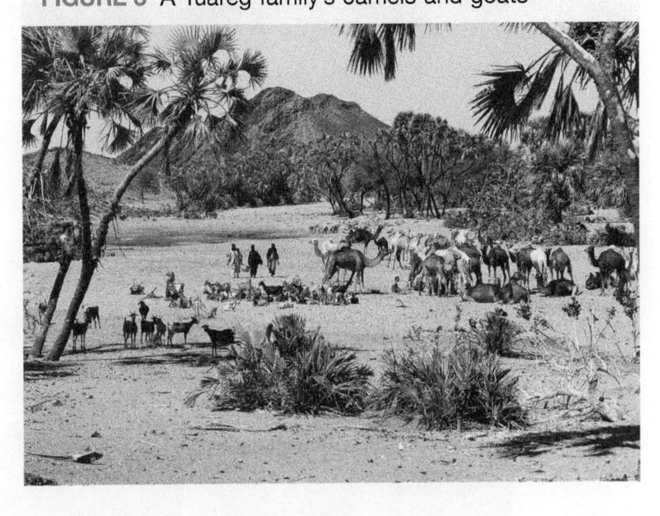

9.9.3 Personal possessions

A nomadic way of life means that Tuareg families do not have many possessions, and the few they have must be light and portable.

Housing and cooking

For example, their homes are usually simple tents (**FIGURE 4**). These provide shelter from heat in the daytime and from cold at night-time. Tuareg tents used to be made of skin and woven cloth but now they are often made of nylon. They are always positioned so that the doorway is facing the non-wind side.

FIGURE 4 A Tuareg family tent, Niger

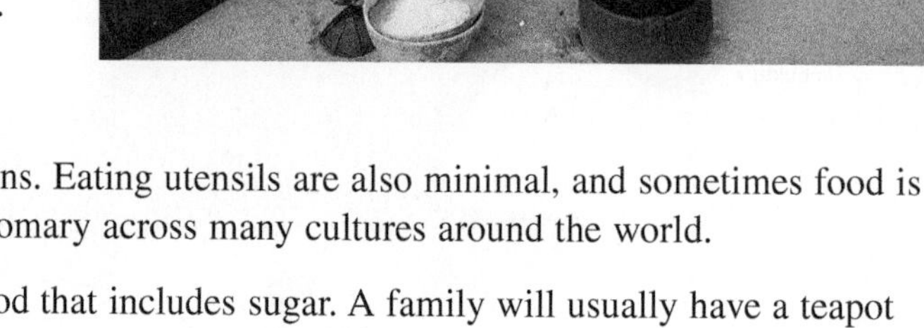

The main furnishings in the tents are rugs to cover the ground and to provide a sleeping mat. Sometimes a bed is constructed from palm slats resting on thin logs. Other furniture includes cushions and drawers or chests. A tent and other belongings can be packed up within a few hours.

For cooking, the Tuareg people use an open fire, sometimes with a hotplate. They use only a few utensils such as pots, containers, plates and spoons. Eating utensils are also minimal, and sometimes food is eaten by hand from a communal plate, as is customary across many cultures around the world.

Tea is the most popular drink, and is the only food that includes sugar. A family will usually have a teapot and small cups, and the making of tea is a ceremony. Tea is offered to visitors and becomes part of business discussions.

Clothing

The Tuareg people almost always wear clothing that covers them from head to toe. The men, who spend more time outside, wear a headscarf to cover their head, neck and much of their face. This is to prevent sunburn, stop the lips from cracking, and slow the drying out of the mouth. Men over 18 years of age traditionally wear blue headscarves (**FIGURE 5**). Their long clothes also provide warmth during the coolest time just before sunrise.

FIGURE 5 Tuareg men traditionally wore blue headscarves.

9.9.4 Challenges to a nomadic lifestyle

In recent times, the Tuareg people have been forced to change their way of life for a number of reasons:

- Drought has reduced the amount of food available for the animals.
- Private ownership of land has reduced the areas in which the Tuareg can move.
- Political unrest has made some areas unsafe.
- Population growth has placed pressure on the available land.

An increasing number of Tuareg people are moving to the south and becoming semi-nomadic. This means they are in one location for a large part of the year. They are building more permanent buildings, such as **adobe** houses, and using some irrigation to help crops grow. Working for money is becoming more common, and children are sometimes able to attend school. Healthcare, such as the provision of vitamins to improve nutrition, is now reaching more people. Solar panels are being used in some areas to produce power to charge mobile phones, run solar cooking ovens and provide lights in schools.

Some traditions remain. Nomadic herding is still valued as the most important activity, and tents are still the main form of housing. Loose clothing is still popular, and the men are still well known for their blue scarves.

adobe bricks made from sand, clay, water and straw and dried by the Sun

FIGURE 6 Technology provides an important means of communication.

FIGURE 7 Classroom in Timia, Aïr, Niger

on Resources

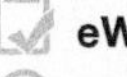 **eWorkbook** Nomadic ways of life (ewbk-9266)

 **Video eLesson** Nomadic ways of life — Key concepts (eles-5088)

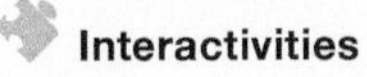 **Interactivities** Tuareg treks (int-3094)
Nomadic ways of life (int-8420)

9.9 ACTIVITIES

1. Refer to an atlas and complete online research to explore where the Tuareg live and how their lives have changed because of modern technology.
2. Select another group of people who lead a nomadic way of life (such as the Awa, Penan, Orang Rimba or Tibetan nomads). Your task is to gather information about your chosen group. To guide your research, develop a key question for each of the following criteria: the environment, the people and the lifestyle. Represent your findings in at least three formats (such as a graph, a map, a satellite image, a photo, text or a diagram). Write three clear sentences to compare your group to the Tuareg people.

9.9 EXERCISE

Learning pathways

■ LEVEL 1	■ LEVEL 2	■ LEVEL 3
1, 2, 7, 8, 14	3, 6, 9, 10, 13	4, 5, 11, 12, 15

Check your understanding

1. Refer to **FIGURE 2**.
   a. Name the five main countries through which the nomadic Tuareg people move.
   b. On which continent do the Tuareg people live?
   c. Do the Tuareg people live in the northern hemisphere or the southern hemisphere?
2. Where is Timbuktu located?
3. Refer to **TABLE 1**.
   a. What is the average yearly total rainfall in Timbuktu?
   b. What is the average maximum temperature in the coolest month?
4. Identify three ways in which the environment has influenced the traditional way of life of the nomadic Tuareg people.
5. Provide four reasons to explain why animals are highly valued by the Tuareg people.
6. Describe what an adobe house looks like.

Apply your understanding

7. Why might having an adobe house make it hard to live a nomadic lifestyle?
8. An increasing number of Tuareg people are becoming semi-nomadic. What does this mean?
9. Explain why Tuareg people might need to use dried camel dung for heating.
10. Look closely at **FIGURE 5**. What purposes might Tuareg men's headscarves serve in the desert?
11. Why might Tuareg people put their tent doors away from the direction of the wind?
12. Explain how portable solar power might make the life of a nomad easier or more comfortable.

Challenge your understanding

13. The Tuareg people may be required to change their way of life in the future. What do you think will happen to the Tuareg's nomadic way of life? Consider the influence of environmental factors (related to the natural world), economic factors (related to businesses and work) and social factors (related to people's welfare, hopes and attitudes).
14. Compare a day in the life of a Tuareg nomad and yourself. Propose how the following parts of your lives might be similar or different:
    a. where and how you both live
    b. diet, clothing, housing type, possessions, settlement size, schooling and travel.
15. Predict whether there will be an increase or decrease in the number of people living a nomadic lifestyle in the next 20 years. Provide reasons for your prediction.

To answer questions online and to receive **immediate feedback** and **sample responses** for every question, go to your learnON title at www.jacplus.com.au.

9.10 Places of change

LEARNING INTENTION

By the end of this subtopic, you will be able to explain why places change over time, describe the changes that might occur in a place over time, and predict how a place might look in the future.

9.10.1 Reasons why places change

A town will change over time if the factors influencing people's decision making about living there also change. Change may be due to government plans, the perception of the natural environment, the economic activities that are carried out in the place and access to resources and other places.

9.10.2 Moving settlements for development

The original town of Tallangatta, in north-east Victoria, about 40 kilometres from Albury—Wodonga, can be seen only when the water level in Lake Hume is very low. The current town was established in 1956. Houses were lifted onto trucks (with parts of the buildings often falling off during the journey) and moved about 8 kilometres (see **FIGURE 1**). The original site, in a valley beside the Mitta Mitta River, was flooded when the size of Lake Hume was increased.

FIGURE 1 Tallangatta house being moved, 1956

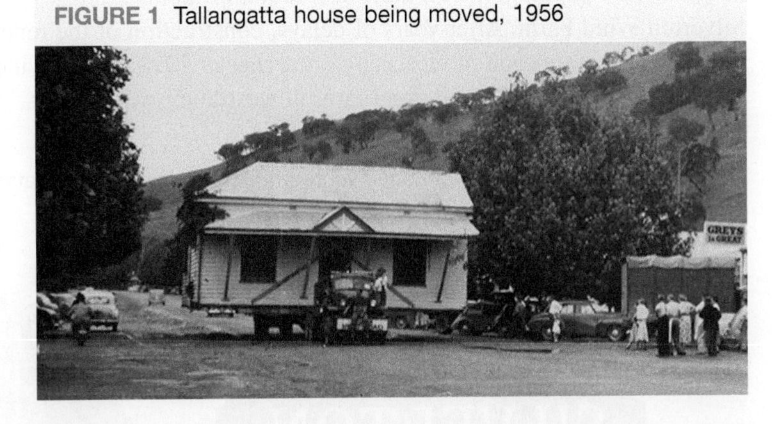

9.10.3 Closing settlements

In 1917, it was decided that a town was needed on the dry and hot Nullarbor Plain to provide services for the Indian Pacific railway (see **FIGURE 2**). With a population of 300, the town of Cook was once big enough to have a school, hospital, shop and accommodation for train drivers. When the railways were privatised in 1997, the town was closed. Cook currently has no known residents and it has effectively become a ghost town.

FIGURE 2 The location of Cook, South Australia, and the Sydney–Perth rail line

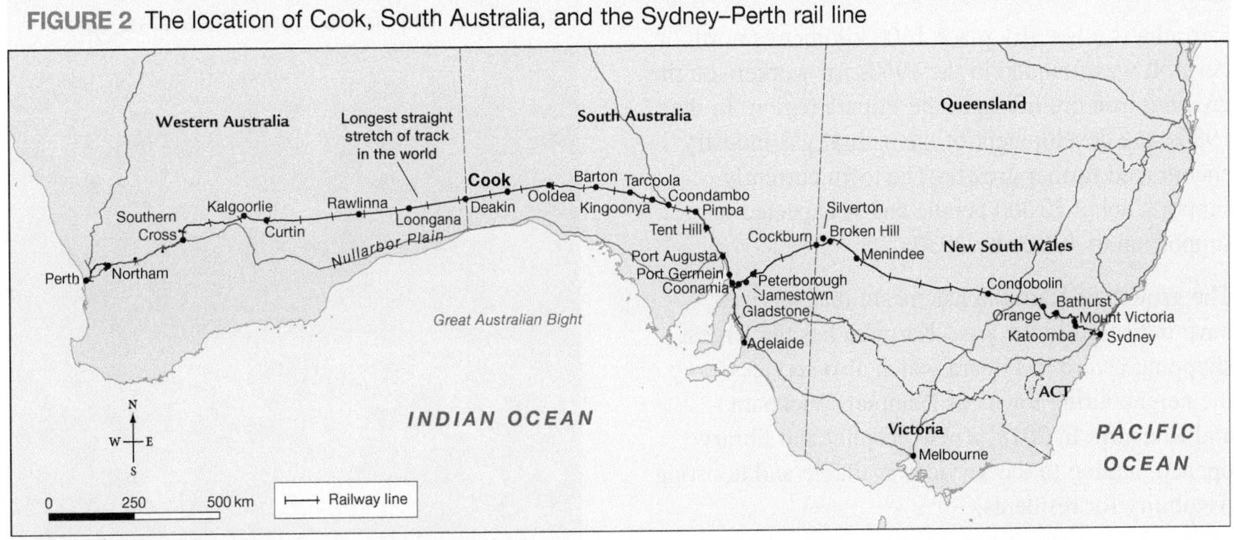

Source: Spatial Vision

9.10.4 Access to resources

Resource depletion

Silverton, 25 kilometres north-west of Broken Hill, was once home to 3000 people who mainly worked in mining. Most people left, often taking their homes with them, when richer mines opened at Broken Hill. According to the 2016 census, the population of Silverton is 50, although the town is now visited by many tourists.

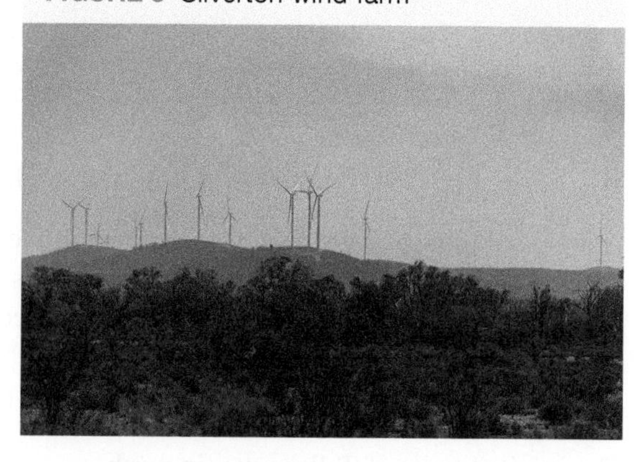

FIGURE 3 Silverton wind farm

Silverton has reinvented itself as a unique location for TV and movie productions. The town and its semi-arid surroundings have been used as the setting in many films, including *Mad Max 2, Dirty Deeds, Strangerland, Mission Impossible 2* and *The Adventures of Priscilla, Queen of the Desert*. In the coming years, we may see another change in Silverton's population with the construction of the Silverton Wind Farm. After years of delays, construction of the renewable energy project commenced in 2017, with the first generation of electricity occurring in 2018. All 58 wind turbines have now been constructed, making it the seventh largest wind farm in Australia.

FIGURE 4 Silverton, NSW, has reinvented itself as a tourist destination.

Resource discovery

Karratha is a hot, dry place 1600 kilometres north of Perth. It was founded in the 1960s for workers on the growing iron ore mines in the Pilbara region. In the 1980s, the development of the natural gas industry encouraged further growth. The town currently supports about 22 000 people and is expected to support up to 40 000 by 2030.

FIGURE 5 The planned town of Karratha, WA

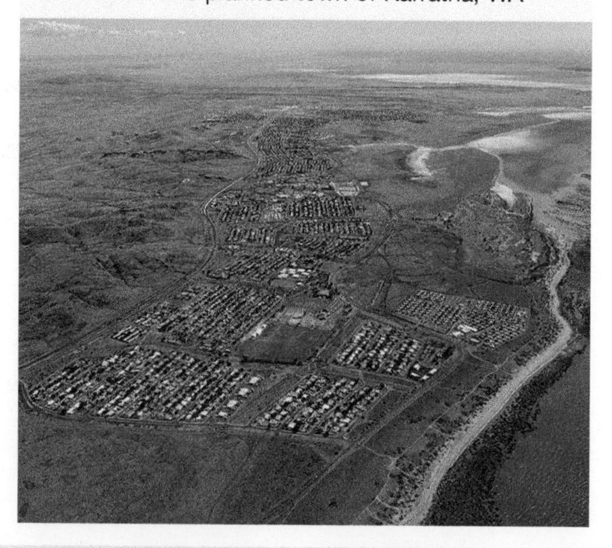

The growth of Karratha has meant that services have increased in the area. Karratha has the largest shopping centre in Pilbara which also services the neighbouring towns of Dampier, Wickham and Roeburn. In 2018, a new hospital and library opened, adding to the services available and boosting liveability for residents.

9.10.5 Sea change and tree change

People who move from the city to the coast are said to have made a sea change. Those who move to an inland location are said to have made a tree change.

Sea change — Yamba

Yamba, located in northern New South Wales at the mouth of the Clarence River, has become a popular coastal town for tourism as well as sea changes.

Change over time

Yamba as a coastal community has changed over time. With a history of fishing and oyster industries in the 1880s and farming from the 1900s, Yamba slowly developed into a tourist destination after a railway line was established in nearby Grafton in 1930. With a current population of more than 6200, Yamba has been known to almost triple its population during peak holiday periods.

Yamba's population is growing at a rate of 0.84 per cent each year with a number of new residents arriving from interstate (see **TABLE 1**).

TABLE 1 Origin of people who moved to Yamba 2011–2016

Previous place of residence	Number of people
New South Wales	704
Victoria	49
Queensland	410
South Australia	16
Western Australia	37
Tasmania	20
Northern Territory	25
Australian Capital Territory	25
Overseas	114

FIGURE 6 The coastal community of Yamba

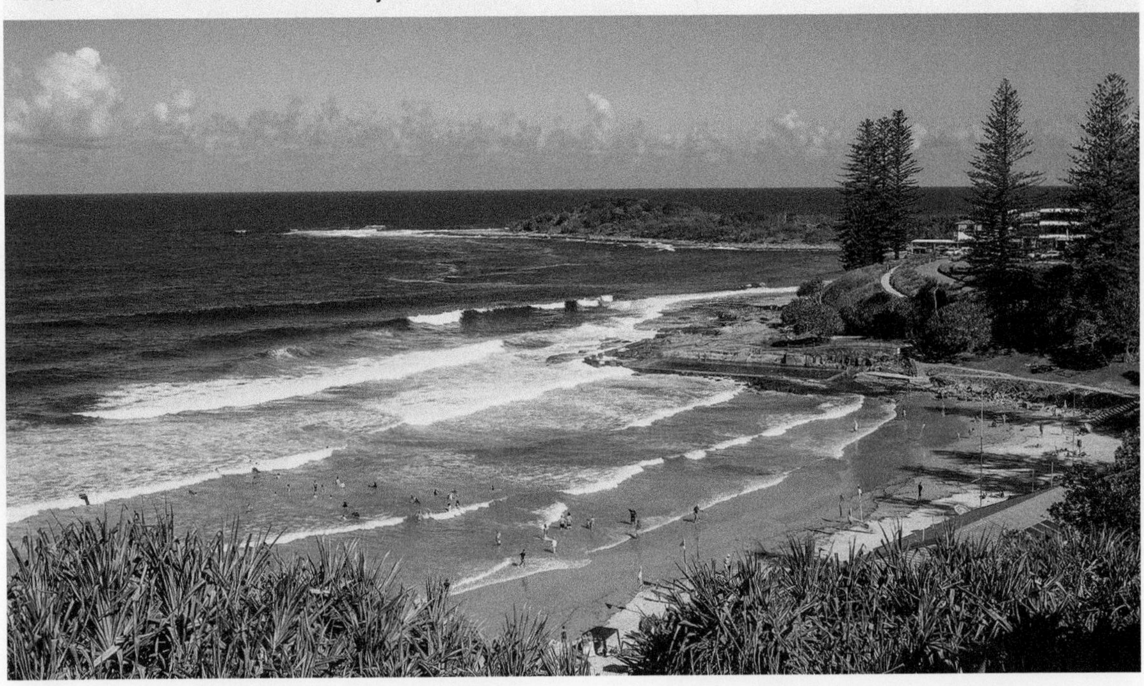

Tree change — Bellingen

Bellingen, located on the mid-north coast of New South Wales, is a small town that has become popular for a tree change.

Change over time

Bellingen has changed over time with its initial beginnings as a cedar cutting location. In the 1870s housing lots were sold and Bellingen soon began to grow in population. As the cedar felling slowed, the cleared land was turned into farmland and the town became a dairying centre. While some farming still occurs in the Bellingen area, the three main employment industries are now hospitality, retail and healthcare. More recently, the town has attracted residents from all over the country, escaping the city for life among the trees.

TABLE 2 Origin of people who moved to Bellingen 2011–2016

Previous place of residence	Number of people
New South Wales	475
Victoria	60
Queensland	77
South Australia	3
Western Australia	20
Tasmania	3
Northern Territory	15
Australian Capital Territory	4
Overseas	43

FIGURE 7 The Gleniffer Valley, or 'Promised Land', near Bellingen, NSW

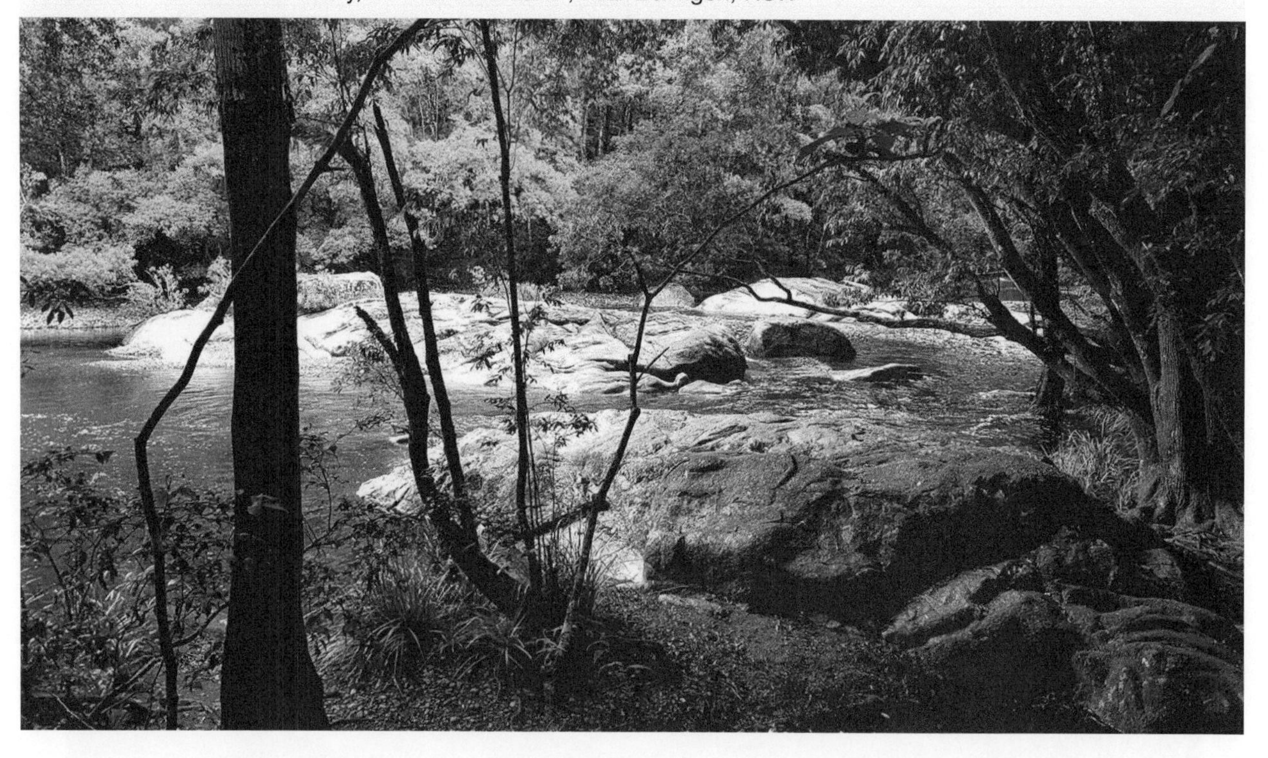

9.10.6 Tourism

Port Douglas, 60 kilometres north of Cairns, was a busy port in the 1870s with a population of more than 10 000. The mining that had attracted people to this hot, wet area did not last. By the 1960s, the population was only 100. In the 1980s, road and air access to the town improved. People were prepared to travel long distances from within Australia and from overseas to enjoy the warm weather, stunning beaches and the World Heritage areas of the nearby Great Barrier Reef and Daintree rainforest. The permanent population is now about 3500. During the peak holiday season (May to November) the population of Port Douglas can increase to more than double its regular size.

FIGURE 8 Port Douglas, 1971

FIGURE 9 Port Douglas, 2009

FIGURE 10 Port Douglas, 2019

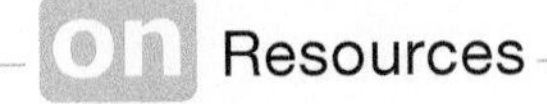

Resources

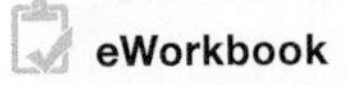

eWorkbook	Places of change (ewbk-9270)
Video eLesson	Places of change — Key concepts (eles-5089)
Interactivity	Places of change (int-8421)
Google Earth	Karratha (gogl-0025) Silverton (gogl-0029) Tallangatta (gogl-0030) Yamba (gogl-0125) Bellingen (gogl-0126) Port Douglas (gogl-0127)

9.10 ACTIVITIES

1. a. Refer to **FIGURES 8, 9** and **10**. Draw a sketch map of natural features at Port Douglas in 1971. Show the ocean, promontory, Dickson Inlet, flat land and hills. Add the settlement features (such as housing, roads and marina).
   b. Using another colour or an overlay, show the settlement features for 2019.
   c. Annotate your map to describe the changes that have occurred and the changes you think will happen in the next ten years.
2. Find maps of Victoria that provide information about landform and climate. Refer to your maps and **FIGURE 11** to complete the following.

FIGURE 11 Projected population change for regional Victoria, 2011–2031

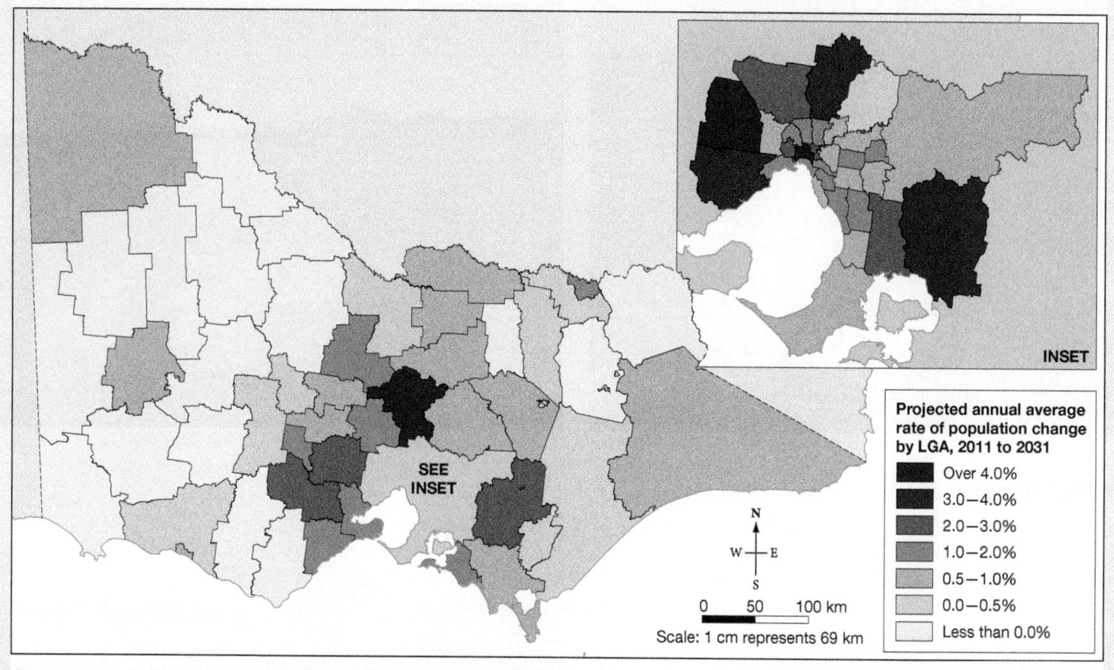

Source: The Department of Environment, Land, Water and Planning.

   a. Think about landform and population change. Are most areas of declining population in places that are not mountainous? Are most areas of increasing population on the coast side of the mountains?
   b. Think about climate and population change. Are most of the highest population growth areas in places where rainfall is over 600 millimetres per year? Are most areas of declining population in places where the rainfall is lower?
   c. What might be reasons for your findings in (a) and (b)?
   d. Repeat the same exercise for NSW to see how Australia's two most populous states compare. You will need to find data about population change in NSW as well as information about landforms and climate.
3. Discuss the following statement with a partner: 'Environmental factors are the main reasons towns change'. Compose a clear paragraph to express your opinion. The first sentence will clearly state your view. The rest of the paragraph should contain at least two pieces of evidence to support your view.
4. Many places change over time. Study the two photographs of Port Douglas in 1971 and in 2019 in **FIGURES 8, 9** and **10**. Discuss how the following people might respond to change that has taken place here:
   - A resident whose family has lived in Port Douglas for three generations
   - A shop owner
   - A travel agent
   - A tourism company owner
   - An angler (person who fishes)
   - A painter/photographer

9.10 EXERCISE

Learning pathways

■ LEVEL 1	■ LEVEL 2	■ LEVEL 3
3, 4, 7, 8, 13	1, 2, 9, 10, 14	5, 6, 11, 12, 15, 16

Check your understanding

1. Outline one reason why a town might be moved.
2. Describe how the discovery of natural resources in an area might change the population over time.
3. What now draws people to Silverton?
4. Outline why the population of Port Douglas increases and decreases significantly throughout the year.
5. Outline the difference between a sea change and a tree change.
6. Outline why Yamba and Bellingen have changed over time.
7. Outline how a place might benefit in the long term from a feature film being made in the local area.

Apply your understanding

8. Refer to **TABLE 1**. What were the three main places people came from to settle in Yamba between 2011 and 2016?
9. Why might May to November be the peak holiday season in Port Douglas?
10. Refer to the map in **FIGURE 11**.
    a. Describe the location of the areas predicted to grow by more than 3 per cent. For example, are they inland or by the coast? Are they in the north, south, east or west of the state? Are they clustered or spread out? Are they close to Melbourne?
    b. What will happen to towns in regional Victoria?
    c. Estimate the proportion of Victoria that is predicted to increase its population and the proportion that is predicted to decrease its population.
11. Factors that cause change can be categorised as social (related to people), economic (related to money) or environmental (related to setting or surroundings). Consider all the reasons for change provided in this subtopic and list each in its correct category.
12. Which of the reasons for population change outlined in this subtopic would create the most significant long-term change to a place? Give reasons to justify your response.
13. Compare and contrast the changes in the populations of Silverton and Karratha over time.

Challenge your understanding

14. What would be the advantages and challenges of living in a town such as Port Douglas, which relies on tourism? Use speech bubbles like the ones provided to create a cartoon to show your ideas.

It would be good because ...

But ...

15. If you were responsible for notifying the people of a town that they were going to be relocated for a development, like Tallangatta was in 1956, how would you justify the decision to move residents to a new place? Write a paragraph defending the choice to move a whole town to build a dam.
16. Predict whether the permanent populations (numbers of people who live in a place all of the time) of four of the towns mentioned in this subtopic will increase or decrease in the next 100 years. Give reasons for your decisions.

To answer questions online and to receive **immediate feedback** and **sample responses** for every question, go to your learnON title at www.jacplus.com.au.

9.11 Investigating topographic maps — Griffith, NSW

LEARNING INTENTION

By the end of this subtopic, you will be able to identify key features of the Griffith area on a topographic map, and discuss how these features affect liveability.

9.11.1 Griffith and surrounding region

Griffith is located within rich farming lands of the Murrumbidgee Irrigation Area in New South Wales. The climate in this area is semi-arid (warm, with unreliable rainfall) and the area supports a variety of agricultural industries such as viticulture, fruit growing and rice production. The picking, processing and distribution of locally grown rural products provides employment and training opportunities for seasonal workers and the local population during harvesting periods.

FIGURE 1 (a) (b) and (c) Griffith is home to Spring Fest, featuring sculptures made from local fruit.

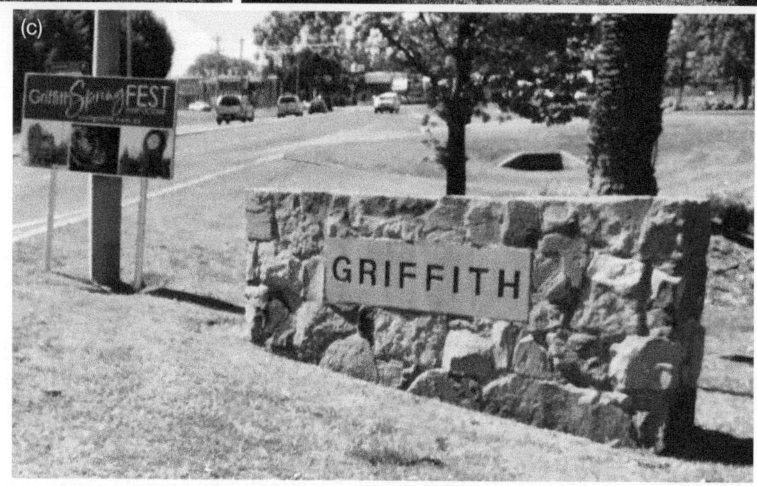

FIGURE 2 Topographic map extract of Griffith

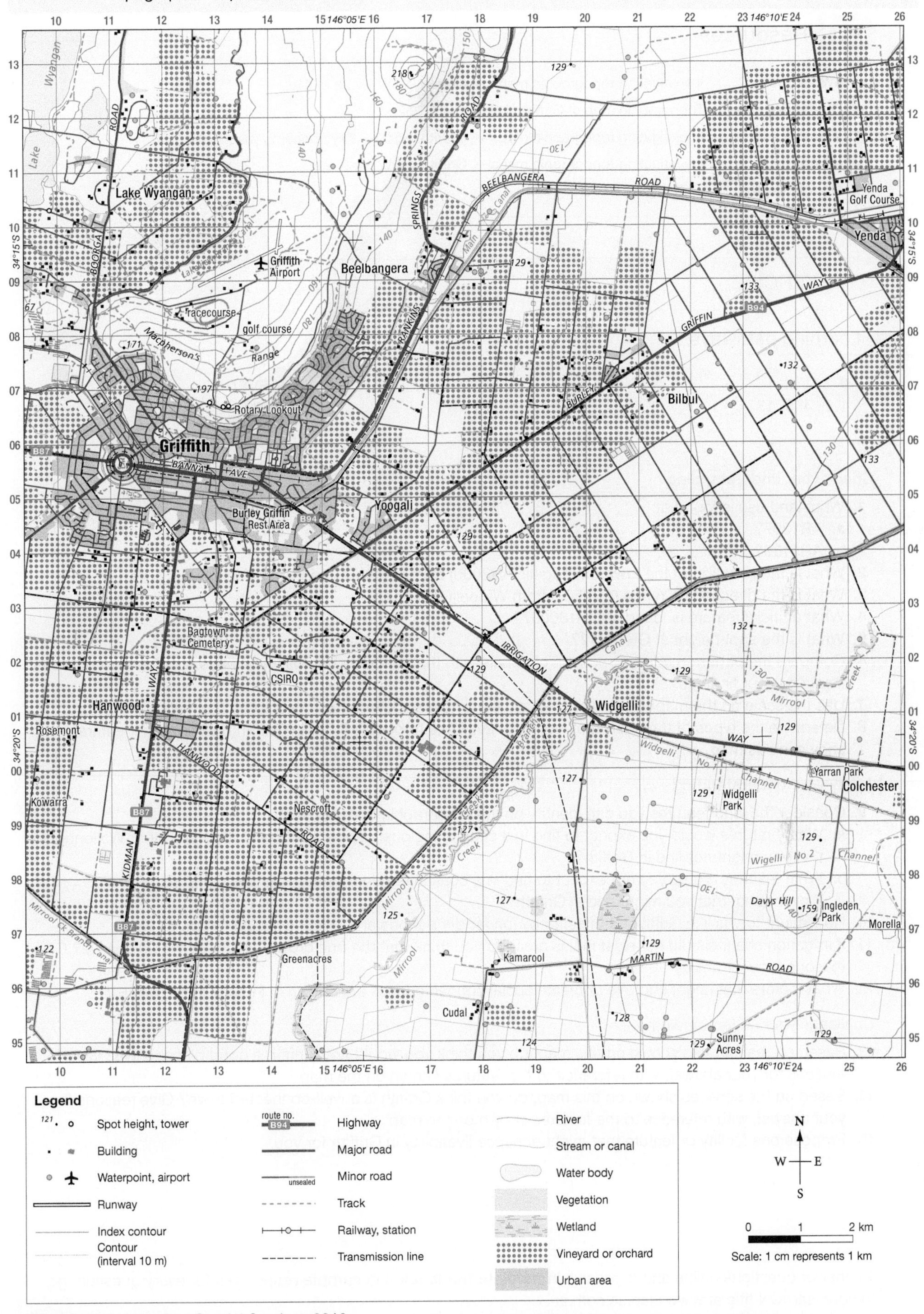

Source: Data based on Spatial Services 2019

on Resources

- **eWorkbook** Investigating topographic maps — Griffith, NSW (ewbk-9274)
- **Digital document** Topographic map of Griffith, NSW (doc-17952)
- **Video eLesson** Investigating topographic maps — Griffith, NSW — Key concepts (eles-5090)
- **Interactivity** Investigating topographic maps — Griffith, NSW (int-8422)
- **Google Earth** Griffith (gogl-0010)

9.11 EXERCISE

Learning pathways

■ LEVEL 1	■ LEVEL 2	■ LEVEL 3
1, 3, 4, 9, 13	2, 5, 6, 10, 14	7, 8, 11, 12, 15

Check your understanding

1. What land use is found at:
   a. AR1297?
   b. AR1105?
2. What is the direction of the town of Widgelli (AR2000) from Griffith Airport (AR1409)?
3. What is the direction of Yenda (AR2509) from Widgelli (AR2000)?
4. What artificial feature is found at GR180097?
5. What is the spot height at GR16712?
6. Compare the pattern made by irrigation channels with that made by natural waterways such as Mirrool Creek. Describe your findings.
7. Use the scale on the map to calculate the number of square kilometres covered by the map.
8. Determine the types of transport that would be available to people living in Griffith based on the features shown in this map.

Apply your understanding

9. Consider the land features you can identify on the topographic map of Griffith.
   a. What features would you expect to find in a coastal regional centre, such as Coffs Harbour or Wollongong, that you might not find in Griffith?
   b. What features would you not expect to find in a coastal regional centre, such as Coffs Harbour or Wollongong, that you might find in Griffith?
10. How might towns like Griffith be affected by longer dry periods and hotter average temperatures?
11. If irrigation channels did not exist in this area, do you think that the fruit growing industry would survive? Give reasons for your answer.
12. What factors that might improve or create social connectedness are shown in **FIGURE 1** or **FIGURE 2**?

Challenge your understanding

13. Based on the features shown on this map, do you think you would find Griffith a good place to live? Give reasons for your answer, with reference to the features shown on the map.
14. Based on the services shown on this map, do you think Griffith is a well-connected town? Give reasons for your answer, with reference to the features shown on the map.
15. Propose one facility or feature that might enhance liveability in Griffith for you.

To answer questions online and to receive **immediate feedback** and **sample responses** for every question, go to your learnON title at www.jacplus.com.au.

9.12 Thinking Big research project — Investigating change over time

online only

The content in this subtopic is a summary of what you will find in the online resource.

Scenario

Have you ever been outside and looked at the landscape in front of you and wondered what this area looked like before humans existed? Have you ever wondered what once stood in the place where your house now stands, or your school? As we've learnt in this topic, it is possible for places to change over time. These changes can be caused by natural phenomena or they can be caused by human activity. You may have even seen some changes in your neighbourhood in your own lifetime. Perhaps new houses have been built in your street or even new shopping strips and parks. Let's have a closer look at these changes and the factors that have caused them. In this task, you will investigate how your local area has changed over time.

Task

In this task, you will investigate how your local area has changed over time. You will make a visual presentation of your findings to the class using either a poster, slides or video. Your presentation should contain a mixture of historical information (using the suggested websites) and demographic data (using the Australian Bureau of Statistics link provided). Demographic data reveals information about the people who live in your local area, including categories such as age, gender and socioeconomic status. You should also attempt to learn what kinds of natural environments used to exist in your suburb.

Go to your Resources tab to access all of the resources you need to complete this research project.

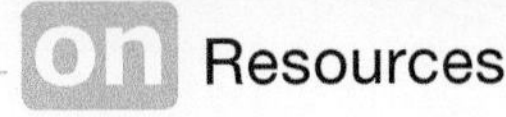
Resources

ProjectsPLUS Thinking Big research project — Investigating change over time (pro-0238)

9.13 Review

9.13.1 Key knowledge summary

9.2 Australians living in remote and rural places

- Mining and farming are the main activities in remote areas in Australia.
- Australians are generally moving towards cities and major towns. In general, many rural communities have been in population decline in Australia.
- A case study of Griffith shows that farming and supporting businesses and social needs are the main attractions to living in the town and surrounding area.

9.4 Life in a country town

- Country communities, though smaller in size, are crucial to Australia's economy because of the agricultural activities that take place in these areas.
- Because country communities are small, they often have an increased sense of belonging.

9.5 Lifestyle places

- Some people have the opportunity to move to new places to improve their lifestyles.
- These changes are sometimes referred to as sea changes (to coastal areas) or tree changes (to rural areas).

9.6 Growth cities in Australia

- The places people choose to live are determined by push and pull factors.
- The number of people living in urban areas continues to rise.

9.8 Isolated settlements

- Life in isolated places can be challenging, especially getting reliable access to resources and medical services. Improvements in technology and transportation have made living in these locations easier.

9.9 Nomadic ways of life

- People living nomadic lifestyles have no permanent settlements and travel from one place to another depending on the availability of resources.
- Currently around 30 million people in the world live nomadic lifestyles.

9.10 Places of change

- Significant changes can occur in towns over time. Often, these changes are related to external factors, such as government decisions and land management issues.
- Changes in tourism patterns can also significantly affect smaller communities.

9.13.2 Key terms

adobe bricks made from sand, clay, water and straw and dried by the Sun
arid lacking moisture; especially having insufficient rainfall to support trees or plants
demographic describes statistical characteristics of a population
ecosystem an interconnected community of plants, animals and other organisms that depend on each other and on the non-living things in their environment
horticulture the growing of garden crops such as fruit, vegetables, herbs and nuts
irrigation water provided to crops and orchards by hoses, channels, sprays or drip systems in order to supplement rainfall
natural resources resources (such as landforms, minerals and vegetation) that are provided by nature rather than people
permafrost permanently frozen ground not far below the surface of the soil
primary industry businesses that mine/farm/harvest natural resources to be made into products
remote a place that is a long way from major towns and cities
rural places that are not part of a major town or city
sea change the act of leaving a fast-paced urban life for a more relaxing lifestyle in a small coastal town
seasonal worker person who moves to a place for a short time to do a specific job (e.g. fruit picking)
tree change the act of leaving a fast-paced urban life for a more relaxing lifestyle in a small country town, in the bush, or on the land as a farmer

9.13.3 Reflection

Complete the following to reflect on your learning.

Revisit the inquiry question posed in the Overview:

Rural, urban, remote or central — the world is full of places. Are any two places the same?

1. Now that you have completed this topic, what is your view on the question? Discuss with a partner. Has your learning in this topic changed your view? If so, how?
2. Write a paragraph in response to the inquiry question outlining your views.

Subtopic	Success criteria			
9.2	I can classify places as remote or rural.			
	I can explain why people live in rural and remote places.			
	I can identify problems faced by people living in rural and remote places and strategies to help the sustainability of rural and remote places.			
9.3	I can identify and describe the age and gender spread shown in a population profile.			
9.4	I can identify the characteristics of country towns that are making them more popular places to live.			
9.5	I can explain what a lifestyle place is.			
	I can describe who might live in a lifestyle place and suggest reasons why they might move there.			
9.6	I can explain the interconnection between push and pull factors and the growth of cities.			
	I can identify the places in Australia that are growing at a faster rate.			
9.7	I can identify and describe significant features on a satellite image.			
9.8	I can identify and describe the challenges faces by people who live in an isolated community.			
9.9	I can explain why people might choose a nomadic lifestyle.			
	I can describe the challenges faced by people living a nomadic lifestyle.			
9.10	I can explain why places change over time.			
	I can describe the changes that might occur in a place over time including how a place might look in the future.			
9.11	I can identify key features of the Griffith area on a topographic map, and discuss how these features affect liveability.			

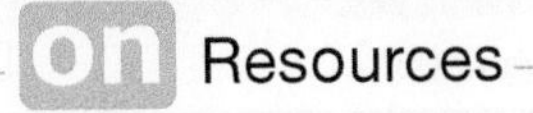

eWorkbook Chapter 9 Student learning matrix (ewbk-8468)
Chapter 9 Reflection (ewbk-8469)
Chapter 9 Extended writing task (ewbk-8470)

Interactivity Chapter 9 Crossword (int-7704)

ONLINE RESOURCES

on Resources

Below is a full list of **rich resources** available online for this topic. These resources are designed to bring ideas to life, to promote deep and lasting learning and to support the different learning needs of each individual.

eWorkbook

- 9.1 Chapter 9 eWorkbook (ewbk-7989) ☐
- 9.2 Australians living in remote and rural places (ewbk-9238) ☐
- 9.3 SkillBuilder — Interpreting population profiles (ewbk-9242) ☐
- 9.4 Life in a country town (ewbk-9246) ☐
- 9.5 Lifestyle places (ewbk-0329) ☐
- 9.6 Growth cities in Australia (ewbk-9254) ☐
- 9.7 SkillBuilder — Understanding satellite images (ewbk-9258) ☐
- 9.8 Isolated settlements (ewbk-9262) ☐
- 9.9 Nomadic ways of life (ewbk-9266) ☐
- 9.10 Places of change (ewbk-9270) ☐
- 9.11 Investigating topographic maps — Griffith, NSW (ewbk-9274) ☐
- 9.13 Chapter 9 Student learning matrix (ewbk-8468) ☐
  - Chapter 9 Reflection (ewbk-8469) ☐
  - Chapter 9 Extended writing task (ewbk-8470) ☐

Sample responses

- 9.1 Chapter 9 Sample responses (sar-0143) ☐

Digital documents

- 9.6 Topographic map extract — Gold Coast region (doc-11400) ☐
- 9.11 Topographic map of Griffith, NSW (doc-17952) ☐

Video eLessons

- 9.1 Why we live where we do (eles-1620) ☐
  - Choosing a place to live — Photo essay (eles-5344) ☐
- 9.2 Australians living in remote and rural places — Key concepts (eles-5083) ☐
- 9.3 SkillBuilder — Interpreting population profiles (eles-1704) ☐
- 9.4 Life in a country town — Key concepts (eles-5084) ☐
- 9.5 Lifestyle places — Key concepts (eles-5085) ☐
- 9.6 Growth cities in Australia — Key concepts (eles-5086) ☐
- 9.7 SkillBuilder — Understanding satellite images (eles-1643) ☐
- 9.8 Isolated settlements — Key concepts (eles-5087) ☐
- 9.9 Nomadic ways of life — Key concepts (eles-5088) ☐
- 9.10 Places of change — Key concepts (eles-5089) ☐
- 9.11 Investigating topographic maps — Griffith, NSW — Key concepts (eles-5090) ☐

Interactivities

- 9.2 Remote living (int-3090) ☐
  - Climate graph for Coober Pedy, South Australia (int-5292) ☐
- 9.3 SkillBuilder — Interpreting population profiles (int-3284) ☐
- 9.4 Country town services (int-3092) ☐
- 9.6 Growth cities in Australia (int-8419) ☐
- 9.7 SkillBuilder — Understanding satellite images (int-3139) ☐
- 9.8 Climate graph for Dawson (int-5300) ☐
- 9.9 Tuareg treks (int-3094) ☐
  - Nomadic ways of life (int-8420) ☐
- 9.10 Places of change (int-8421) ☐
- 9.11 Investigating topographic maps — Griffith, NSW (int-8422) ☐
- 9.13 Chapter 9 Crossword (int-7704) ☐

Fieldwork

- 9.5 Sphere of influence survey (fdw-0012) ☐
- 9.6 Soundscapes (fdw-0016) ☐

ProjectsPLUS

- 9.12 Thinking Big research project — Investigating change over time (pro-0238) ☐

Weblinks

- 9.2 Soldier settlement (web-1103) ☐
  - Roxby Downs (web-6413) ☐
- 9.3 Population pyramid (web-4088) ☐
- 9.5 Fishermans Bend (web-6052) ☐
- 9.8 Bureau of Meteorology (web-0044) ☐

Google Earth

- 9.2 Tom Price township (gogl-0008) ☐
  - Griffith (gogl-0010) ☐
  - Coober Pedy (gogl-0122) ☐
- 9.4 Kellerberrin (gogl-0117) ☐
- 9.5 Canary Wharf, London (gogl-0011) ☐
  - Prague, Czech Republic (gogl-0012) ☐
  - Rio de Janeiro (gogl-0013) ☐
  - New Delhi (gogl-0111) ☐
  - Fishermans Bend (gogl-0112) ☐
  - Venice Beach (gogl-0123) ☐
- 9.6 Gold Coast (gogl-0124) ☐
- 9.10 Karratha (gogl-0025) ☐
  - Silverton (gogl-0029) ☐
  - Tallangatta (gogl-0030) ☐
  - Yamba (gogl-0125) ☐
  - Bellingen (gogl-0126) ☐
  - Port Douglas (gogl-0127) ☐

myWorld Atlas

- 9.2 Population of Australia (mwa-4505) ☐

Teacher resources

There are many resources available exclusively for teachers online.

10 Improving liveability

INQUIRY SEQUENCE

To access a pre-test with **immediate** feedback and **sample responses** to every question in this chapter, select your learnON format at www.jacplus.com.au.

10.1 Overview

Numerous **videos** and **interactivities** are embedded just where you need them, at the point of learning, in your learnON title at www.jacplus.com.au. They will help you to learn the content and concepts covered in this topic.

Why is it important to improve liveability around the world? What are some challenges to improving liveability?

10.1.1 Introduction

Levels of liveability vary around the world and are influenced by many factors such as access to vital services, health and income. It is everyone's job to ensure that we continue to work towards improving liveability for all people around the world. This chapter will explore a range of strategies used to enhance liveability and look at the role that governments, non-government organisations and communities can play in improving liveability.

STARTER QUESTIONS

1. What might life be like in the village shown in **FIGURE 1**?
2. What challenges to liveability might exist?
3. What might improve the liveability of where you live?
4. Compared to other places you have been or seen (online or on TV, for example), do you think where you live compares well?

FIGURE 1 Refilling water bottles at the Dadaab refugee camp, Kenya

Resources

eWorkbook Chapter 10 eWorkbook (ewbk-7990)

Video eLessons Making places liveable for young people (eles-1621)
Improving liveability — Photo essay (eles-5345)

10.2 Liveability and sustainable living

LEARNING INTENTION

By the end of this subtopic, you will be able to define what is meant by the terms *sustainability* and *ecological footprint* and make links between liveability, sustainability and ecological footprints.

10.2.1 Sustainability

Australia's major cities consistently rate among the most liveable. Liveability, however, is not always the same as sustainability. Sustainability considers how well a community is currently meeting the needs and expectations of its population and how well it will be able to continue providing for its population in the future.

Indicators that a place is sustainable include:

- low working hours to meet basic needs
- easy access to education
- satisfactory and affordable housing
- plenty of recycling and composting
- reliable transport
- low emissions and high air quality
- **biodiversity**
- high renewable energy use and low non-renewable energy use
- good water, forest and marine health
- ability to respond to disasters.

FIGURE 1 To achieve sustainability, a city's environmental, economic and social aspects must all be considered.

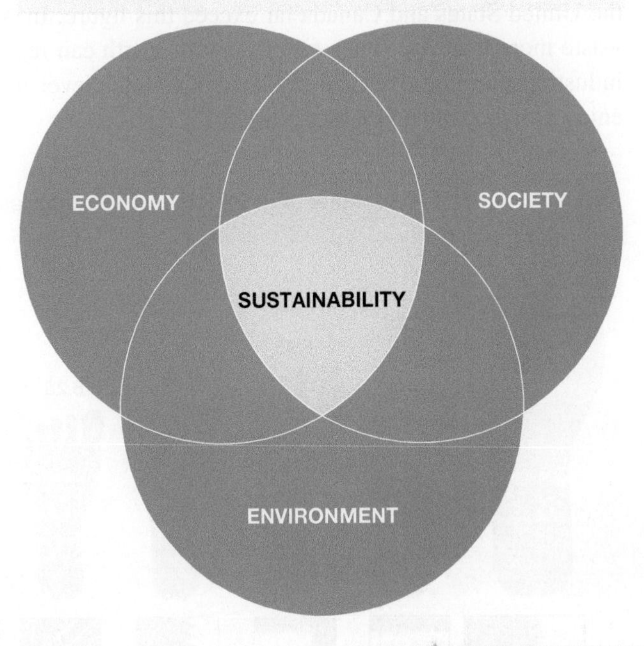

10.2.2 Sustainable cities index

This annual index considers 50 leading cities and ranks them against a range of indicators. The indicators are organised under the three main headings of people (society), planet (environment) and profit (economy).

biodiversity the variety of life in the world or in a particular habitat or ecosystem

TABLE 1 Top ten sustainable cities by main indicator 2018

Ranking	People	Planet	Profit
1	Edinburgh	Stockholm	Singapore
2	London	Frankfurt	London
3	Paris	Zurich	Hong Kong
4	Taipei	Vienna	New York
5	Stockholm	Copenhagen	Munich
6	Prague	Oslo	Edinburgh
7	Seoul	Hamburg	San Francisco
8	Amsterdam	Berlin	Boston
9	San Francisco	Munich	Zurich
10	Madrid	Montreal	Seoul

Source: © Arcadis Sustainable Cities Index 2018, compiled in partnership with Centre for Economics and Business Research

10.2.3 Ecological footprint

Everything we do and consume has an impact on the environment. Land is cleared to grow plants and animals; fish are caught in the sea; water is diverted for homes, businesses and farms; and most transport and other energy-consuming activities such as using an appliance are powered by non-renewable resources. An **ecological footprint** calculates the land area that would be needed to sustain an individual (expressed as hectares per capita). It is used to compare the amount of various resources used per person in countries around the world.

Generally, if you live in a high-income country such as Australia, you are likely to have an ecological footprint that is much larger than a person who lives in a low-income country such as Chad. The average ecological footprint of all people on Earth is 2.84 hectares. The average Australian footprint is about 8.8 hectares. To enjoy a sustainable way of life, the population needs to stay within the Earth's carrying capacity (meaning the maximum number of individuals of a population that the environment can support), and the average footprint should not be more than 1.7 hectares. **FIGURE 2** shows that developed countries such as Qatar, Luxembourg, the United States and Canada far exceed this figure. In Australia, we are using resources and generating waste more than five times faster than the Earth can regenerate and absorb them. As more countries develop industries and improve their standard of living, clever responses will be needed to ensure that everyone can enjoy a high standard of liveability.

FIGURE 2 Top ten countries with the biggest and smallest ecological footprints (hectares per capita), 2017

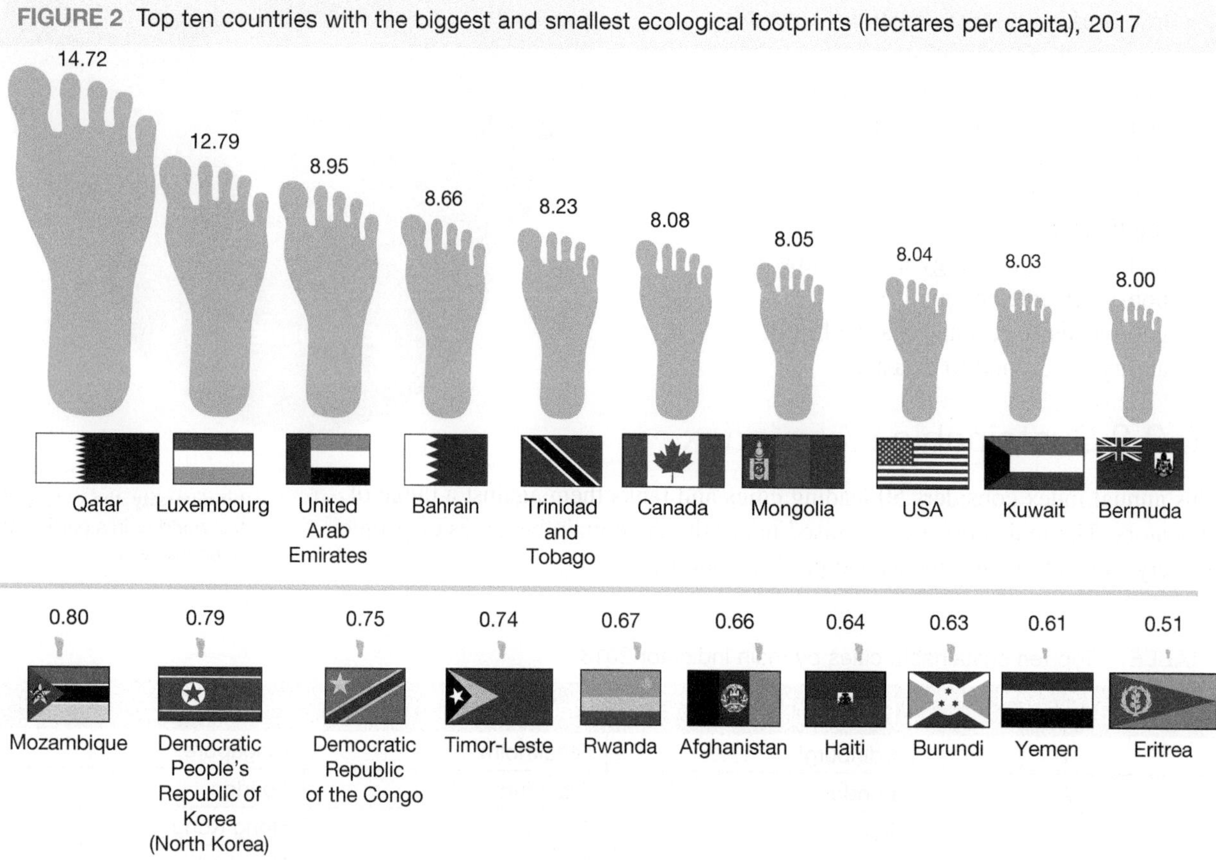

Note: Australia ranks 13th (7.29 ha per capita)

Source: Data based on Global Footprint Network National Footprint and Biocapacity Accounts, 2021 edition, downloaded 21 December 2020 from https://data.footprintnetwork.org.

Government policy can influence the ecological footprint of its population through power generation, transport, water, industry support, rubbish collection and building regulations. Individuals can influence their own ecological footprint through what they eat and buy, how they use water and power, whether they recycle and compost, and how they build their houses and travel.

ecological footprint the total area of land that is used to produce the goods and services consumed by an individual or country

on Resources

eWorkbook	Liveability and sustainable living (ewbk-9314)
Video eLesson	Liveability and sustainable living — Key concepts (eles-5091)
Interactivity	Liveability and sustainable living (int-8423)

10.2 ACTIVITIES

1. Find one image that shows living conditions in a country with an ecological footprint of over 7 hectares per capita and one image that shows living conditions in a country with an ecological footprint of less than 1 hectare per capita. Annotate the images to explain how living conditions affect the ecological footprint of a place.
2. How important and ethical is it to include all aspects of sustainability (environmental, social and economic) when classifying a place and its level of liveability?
3. Refer to **FIGURE 2** and locate and describe the distribution of countries with an ecological footprint of 7 hectares or more per capita. Refer to pattern, directions, continents and latitude.

10.2 EXERCISE

Learning pathways

■ LEVEL 1	■ LEVEL 2	■ LEVEL 3
2, 3, 7, 9, 14	1, 4, 10, 11, 15	5, 6, 8, 12, 13

Check your understanding

1. What are the three aspects that are considered in a definition of sustainability?
2. List three indicators that a place is sustainable.
3. Define the term *biodiversity*.
4. Outline how a country's ecological footprint is calculated.
5. Describe the difference between sustainability and ecological footprint.
6. A sustainable place is not necessarily a good place to live. Outline one reason why this might be the case.

Apply your understanding

7. Explain why high-income countries have a much larger ecological footprint than low-income countries.
8. Compare the Earth's carrying capacity to its current ecological footprint.
9. Choose two indicators that a place is sustainable. Explain why they make a place more sustainable.
10. In your own words, explain and give examples to show how the indicators of sustainability are organised.
11. Compare the data in **TABLE 1**. Which city would you prefer to live in? Give reasons why based on the city's ranking for each of the three criteria and what is important to you.
12. What factors might account for the differences between the countries that rate high on sustainability and those that have a large ecological footprint.

Challenge your understanding

13. What do you think will happen to the global ecological footprint if liveability improves on every continent?
14. Refer to the list of things in section 10.2.1 which indicate that a place is sustainable. Categorise each indicator as applying to society, economy or environment. Suggest one more possible indicator for each category.
15. Consider the ways in which resources have been used to improve the liveability in your area. Which aspects would you be prepared to change a little so that others might improve the liveability of where they live?

To answer questions online and to receive **immediate feedback** and **sample responses** for every question, go to your learnON title at www.jacplus.com.au.

10.3 Improving liveability

LEARNING INTENTION

By the end of this subtopic, you will be able to make links between hunger and liveability, identify reasons for the unequal distribution of food across the world, and explain how help from other countries can improve liveability.

10.3.1 Food security

A basic human requirement is food, and access to enough food is a strong measure of liveability. Even in a world where there is plenty of food and millions of people are overweight, about one person in eight does not have enough to eat. Having access to and being able to afford enough nutritious food to stay healthy is known as **food security**.

There are approximately 815 million **undernourished** people in the world today, which is about one in ten people. Many children in poorer countries are underweight and do not get enough food to be healthy and active.

Three-quarters of all people experiencing hunger live in rural areas, mainly in the villages of Asia and Africa (see **FIGURE 1**). Most of these people depend on **agriculture** for their food. They rarely have other sources of income or employment. As a result, they may be forced to live on one-quarter of the recommended calorie intake and a small amount of water each day.

food security having access to and being able to afford enough nutritious food to stay healthy

undernourished not getting enough food for good health and growth

agriculture the cultivation of land, growing of crops or raising of animals

FIGURE 1 Distribution of hunger, 2018

int-7801

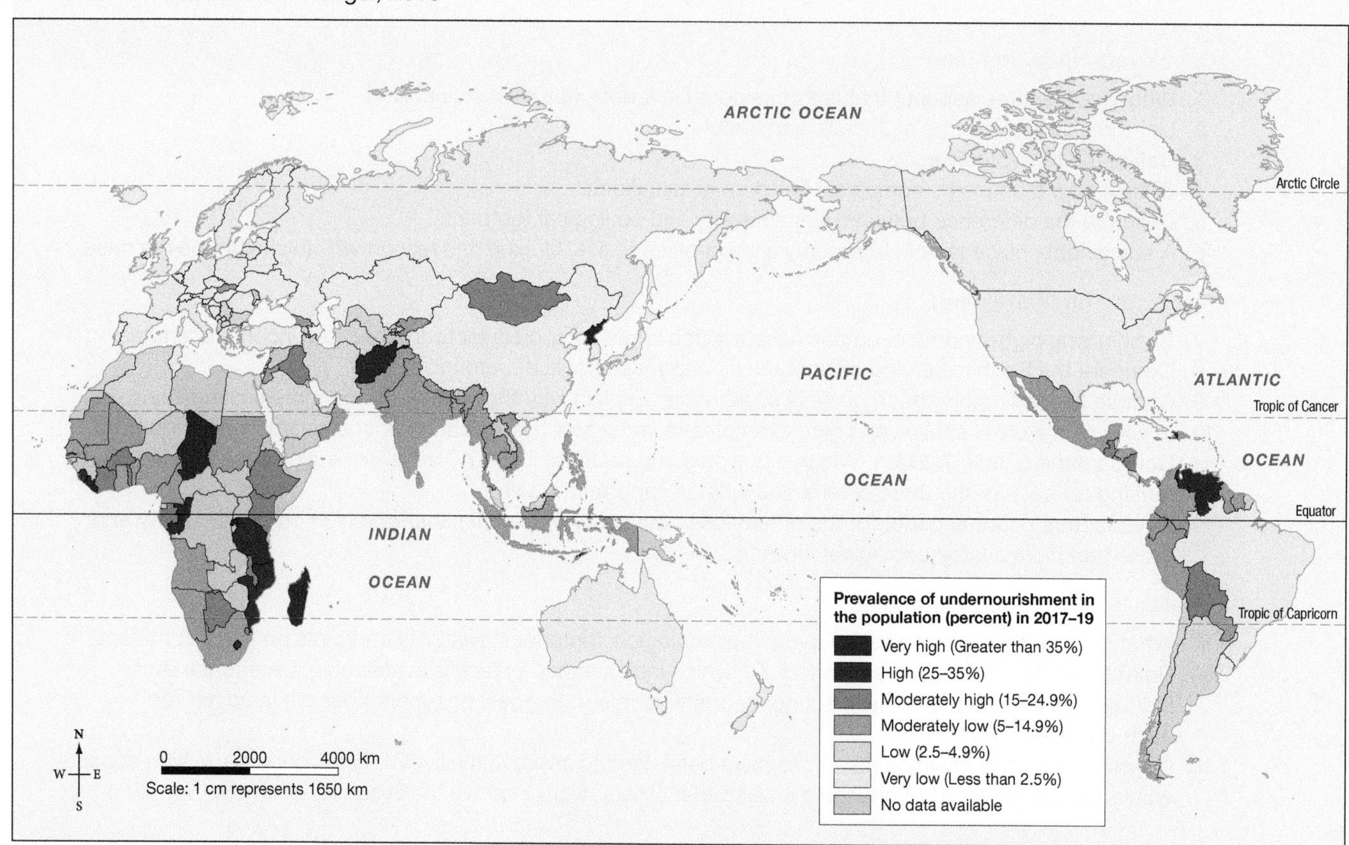

Source: FAO, IFAD, UNICEF, WFP and WHO, (2020). The State of Food Security and Nutrition in the World 2020. Transforming food systems for affordable healthy diets. Rome, FAO. Further information is available at https://www.wfp.org/publications/state-food-security-and-nutrition-world-sofi-report-2020

If enough rain does not fall at the right time of year, crops will not grow well and there will be little grass for **livestock**. However, rainfall is not the only factor contributing to hunger. **FIGURE 2** summarises other causes of hunger.

livestock animals raised for food or other products

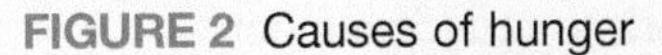

FIGURE 2 Causes of hunger

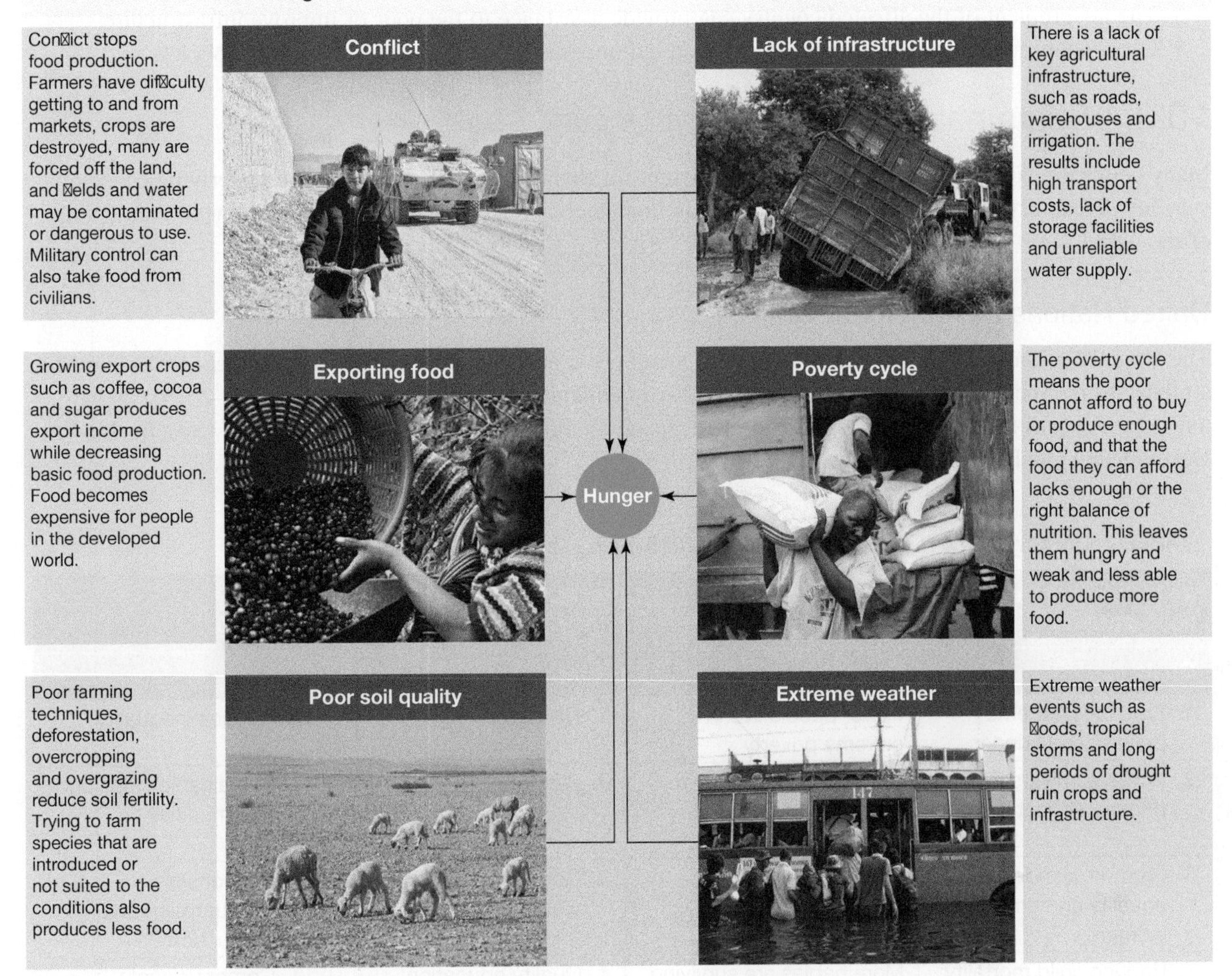

10.3.2 Impact of hunger

TABLE 1 The impact of hunger is felt by individuals, families, communities, regions and whole countries.

Social impacts	Economic impacts	Environmental impacts
People become unwell.	Food production declines.	Soil is overused.
Many people (particularly children) die. (This is called infant mortality.)	The population of cities grows.	Too much land is cleared.
Fathers leave in search of work.	Poverty increases and can also affect the next generations.	Soil fertility and local biodiversity decline.
There is political unrest.	The government cannot afford new infrastructure.	

10.3.3 Ending hunger

There is a range of organisations that focus on reducing hunger. Sometimes they provide food for immediate consumption and sometimes they undertake projects to increase food production in the future. Actions can happen on a range of scales:

- Individuals in any country can join groups or donate to organisations that work to reduce hunger.
- The government of the affected country can provide assistance to the poor or improve infrastructure.
- Other countries can provide financial and food aid or consider the impact of their own policies.

10.3.4 Sustainable Development Goals

Many countries cannot afford to provide infrastructure for their growing population. The underlying cause of very low liveability is poverty. Reducing poverty is fundamental to improving living conditions in many parts of the world.

United Nations Development Goals

The United Nations (UN) is an organisation with members from 193 countries. In 2000, 189 countries signed a pledge to free people from extreme poverty by 2015 (Millennium Development Goals 2000–2015). In 2015, a new pledge was signed with 17 goals, each with specific targets to be reached over 15 years (Sustainable Development Goals 2015–2030).

TABLE 2 UN development goals (a) MDGs 2000-2015 (b) SDGs 2015–2030

(a) Millennium Development Goals 2000–2015		(b) Sustainable Development Goals 2015–2030	
Goals	**Example**	**Goals**	
1. Eradicate extreme poverty and hunger	Fewer people live in extreme poverty	1. No poverty	10. Reduce inequality
2. Achieve universal primary education	Primary school enrolments have increased	2. Zero hunger	11. Sustainable cities and communities
3. Promote gender equality and empower women	Many more girls are attending school	3. Good health and wellbeing	12. Responsible consumption and production
4. Reduce child mortality	More babies are surviving	4. Quality education	13. Combat climate change
5. Improve maternal health	More mothers have access to healthcare when giving birth	5. Gender equality	14. Conserve and use ocean resources sustainably
6. Combat HIV/AIDS, malaria and other diseases	Vaccination has reduced incidence of measles	6. Clean water and sanitation	15. Protect and use Earth's resources sustainably
7. Ensure environmental sustainability	Safe water is available to more people	7. Affordable and clean energy	16. Provide access to justice and promote peaceful societies
8. Develop a global partnership for development	Huge increase in number of people with phone and internet	8. Decent work and economic growth	17. Create stronger global connections and partnerships to improve sustainability
		9. Industry, innovation and infrastructure	

Source: 2021 United Nations Development Programme

Australian government and NGOs

The Australian government recognises that we are **global citizens**, and it supports an overseas aid program through its Department of Foreign Affairs and Trade. Giving or providing overseas aid helps improve outcomes in health, education, economic growth and disaster response in many locations.

The Australian government runs projects to improve living conditions, often working with other countries or with **non-government organisations** (NGOs). NGOs also run programs on their own. Well-known NGOs include World Vision, CARE Australia and Australian Red Cross.

FIGURE 3 Countries receiving assistance from Australia

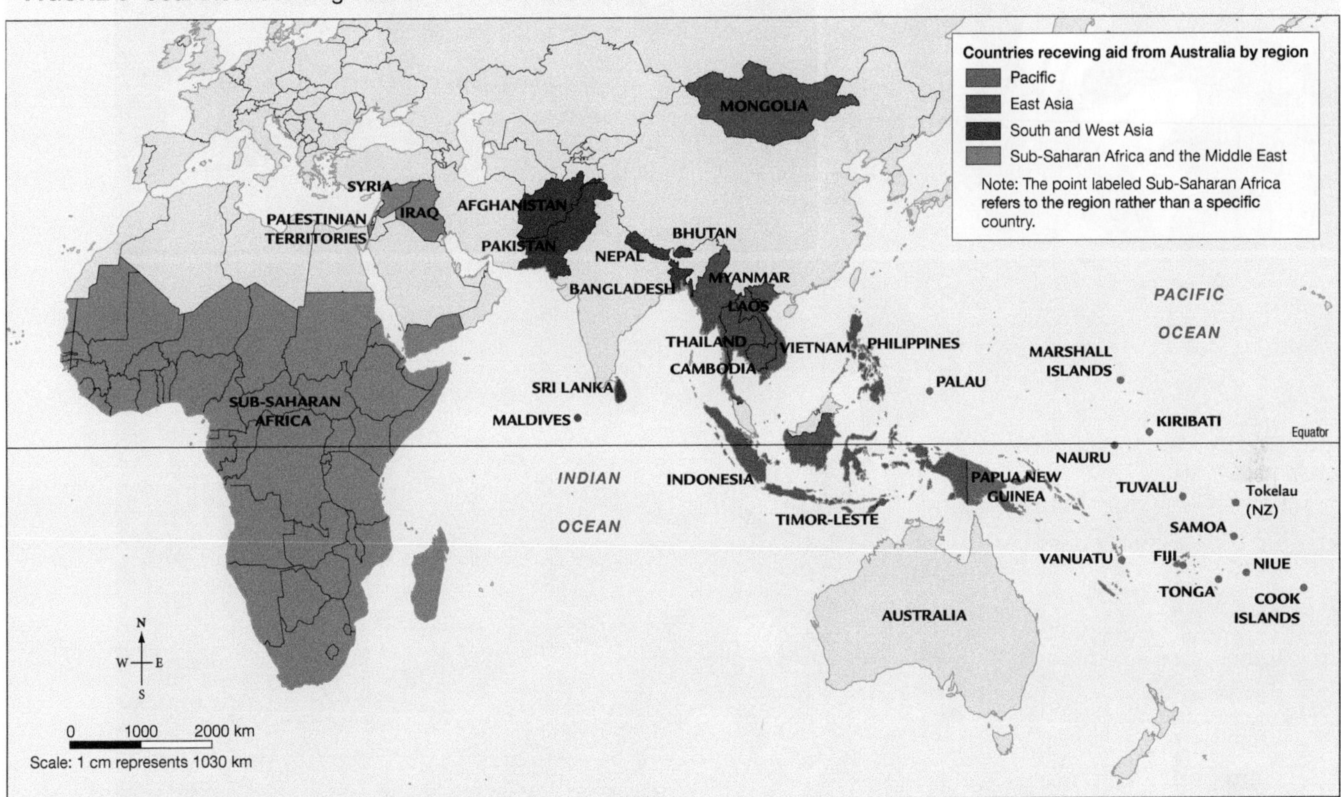

Source: Department of Foreign Affairs and Trade

Small changes, big results

Simple and **appropriate technology** can make an enormous difference to people's lives in developing countries (see **FIGURE 5**).

In addition, a small amount of money can sometimes make a big difference to an individual or community group. Microfinance, or microcredit, is a system of lending small amounts of money, perhaps $150. The money is used to invest in something that can generate income. A person might buy an animal for milking and breeding, equipment for basket-making, stock for a store, or materials for jewellery-making. The loan must be repaid, but at a low interest rate, and further loans can be taken out.

global citizen person who is aware of the wider world, who tries to understand the values of others, and tries to make the world a better place

non-government organisation non-profit group run by people (often volunteers) who have a common interest and perform a variety of humanitarian tasks at a local, national or international level

appropriate technology technology designed specifically for the place and the people who will use it. It is affordable and can be repaired locally.

FIGURE 4 (a) A child immunisation clinic on the Kokoda Track, Papua New Guinea (b) Building schools and improving education in Indonesia (c) Planting grasses in Fiji to stabilise sea banks

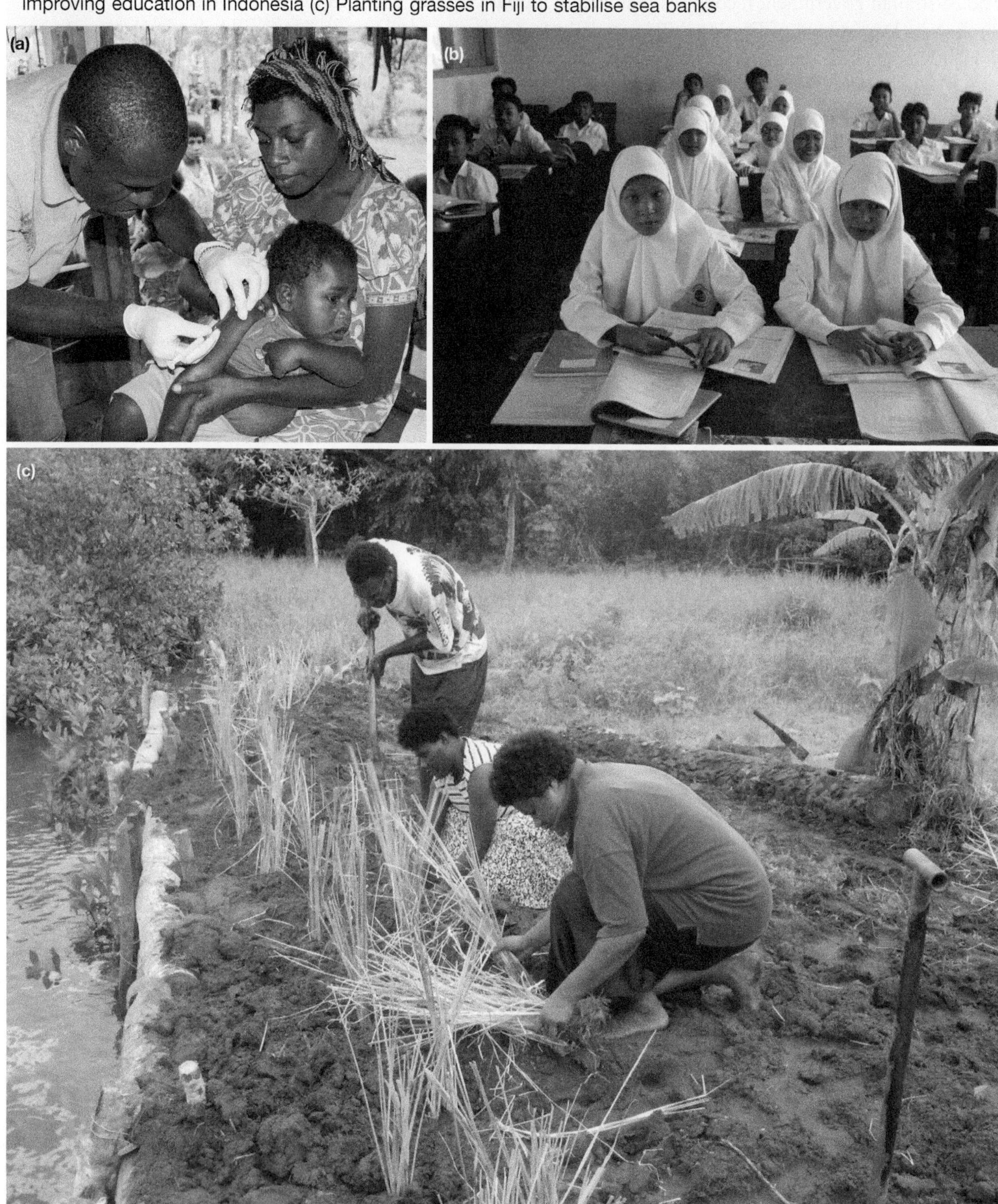

FIGURE 5 (a) Electricity in Nepal is not available to all houses, so a solar lamp increases opportunities to read. (b) In South Africa, people push hippo rollers, which make it easier to collect water from distant wells and bring it home.

on Resources

eWorkbook	Improving liveability (ewbk-9318)
Video eLesson	Improving liveability — Key concepts (eles-5092)
Interactivity	Distribution of hunger, 2018 (int-7801)

10.3 ACTIVITIES

1. Work with a partner to find an example of a project that is trying to solve the immediate issue of hunger and a project that is trying to make food production sustainable. Describe where each project is taking place and which organisation manages the project. Create an outline of each project. Refer to **FIGURE 2** and explain which of the causes of hunger will be reduced by the project.
2. Prepare a report about the work of one NGO involved in programs that aim to improve liveability in an overseas country. Include background information about the location of the project (country and locality); statistical data about living conditions (such as life expectancy, access to safe water, doctors per 100 000 people); and environmental conditions. Describe the NGO and its project and how it is aiming to improve liveability.
3. Choose one of the Sustainable Development Goals. Use a visual organiser to explain how achieving this goal will improve liveability. Consider the flow-on effects and the impact on people, the economy and the environment.

10.3 EXERCISE

Learning pathways

■ LEVEL 1	■ LEVEL 2	■ LEVEL 3
1, 2, 7, 8, 14	3, 6, 9, 12, 15	4, 5, 10, 11, 13

Check your understanding

1. Refer to **FIGURE 1**.
   Which region has the largest number of hungry people? Name three countries in this region.
2. Copy and complete the following sentence to make it accurate. 'Most of the world's hungry people live in ________ villages in __________ and _________.'
3. Refer to **FIGURE 1**.
   a. What percentage of Australia was undernourished?
   b. Which continent has the largest number of countries in which more than 25 per cent of people are undernourished?
4. How might poor roads contribute to hunger?
5. Outline three main differences between the Millennium Development Goals and the Sustainable Development Goals.
6. There are 193 countries that are member states of the UN. What percentage of countries supported both sets of development goals?

Apply your understanding

7. Consider **TABLE 1**. Provide one more example for each category of impact — social, economic and environmental.
8. Study the images in **FIGURES 4** and **5**. Which of the Sustainable Development Goals have been addressed in these projects?
9. Refer to **FIGURE 3**. Describe the distribution of places that receive aid from Australia. Think in terms of region, such as Asia, East Asia, the Middle East, South Asia, West Asia, Pacific, Africa and the Caribbean.
10. Explain how simple technology and microfinance are transforming lives in developing countries.
11. Explain why addressing issues related to hunger is an essential component of improving liveability.
12. Explain how one specific example of providing better access to simple technology can help reduce levels of hunger.

Challenge your understanding

13. Predict what impact climate change might have on hunger.
14. Look closely at the hippo rollers shown in **FIGURE 5**. Draw a plan for another simple invention (or a way to update a wheelbarrow or bicycle) that would help people transport water more easily using simple technology.
15. Suggest why there are still people experiencing hunger in wealthy, developed nations.

To answer questions online and to receive **immediate feedback** and **sample responses** for every question, go to your learnON title at www.jacplus.com.au.

10.4 SkillBuilder — Understanding thematic maps

online only

LEARNING INTENTION

By the end of this subtopic, you will be able to identify key features on a thematic map, and describe patterns in the data shown.

The content in this subtopic is a summary of what you will find in the online resource.

10.4.1 Tell me

What is a thematic map?

A thematic map is a map drawn to show one aspect; that is, one theme. For example, the map may show the location of vegetation types, hazards or weather. Parts of the theme are given different colours or, if only one idea is conveyed, symbols may show location.

FIGURE 1(a) Thematic map of the major landform regions of Australia

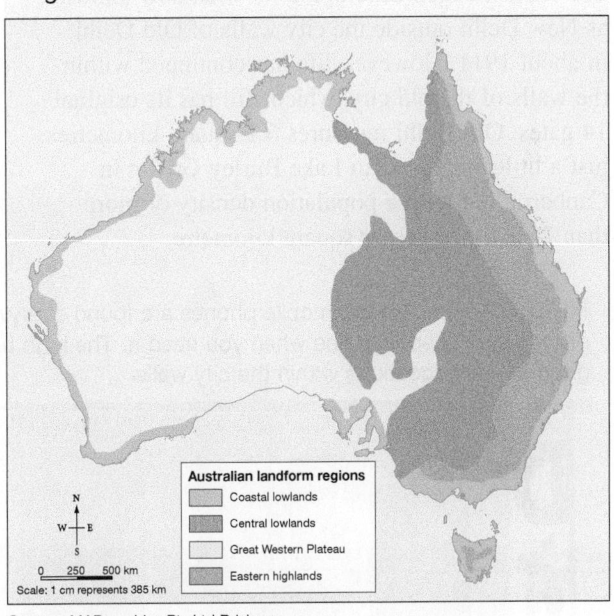

Source: MAPgraphics Pty Ltd Brisbane

10.4.2 Show me

eles-1658

How to understand a thematic map

Step 1

Read the title of the thematic map.

Step 2

Check that the map was put together by a reliable authority.

Step 3

Read the key/legend to understand the colours and/or symbols that are being used. To interpret the colours you need to comment on where the various colours or symbols occur, including the colours or symbols that appear only in small areas of the map.

10.4.3 Let me do it

int-3154

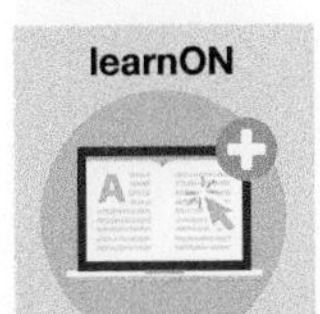

Go to learnON to access the following additional resources to help you build this skill:

- a longer explanation of this skill and its application in Geography (Tell me)
- a video showing the step-by-step process of this skill (Show me)
- an activity and interactivity for you to practise this skill (Let me do it)
- self-marking questions to help you understand and use this skill.

on Resources

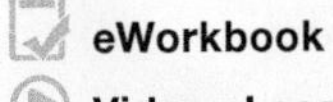

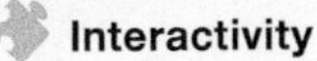

eWorkbook SkillBuilder — Understanding thematic maps (ewbk-9322)

Video eLesson SkillBuilder — Understanding thematic maps (eles-1658)

Interactivity SkillBuilder — Understanding thematic maps (int-3154)

10.5 India, past and present

LEARNING INTENTION

By the end of this subtopic, you will be able to explain the changes that have taken place in India over time and how they have altered liveability.

10.5.1 Life in Old Delhi

Old Delhi is an area within the modern city of New Delhi in India. Old Delhi consists of the original walled city that dates back to 1639. It was founded by the Mughal emperor Shahjahan, and was known then as Shahjahanabad. The local people pride themselves on being a peaceful community in which Muslims, Hindus and Christians have lived together side by side for hundreds of years.

The British began developing the area now known as New Delhi outside the city walls of Old Delhi in about 1911. However, life has continued within the walls of the old city, which still has its original 14 gates. Old Delhi measures 6.1 square kilometres, just a little smaller than Lake Burley Griffin in Canberra, but with a population density of more than 25 000 people per square kilometre.

FIGURE 1 Building a new multi-storey building inside Old Delhi

FIGURE 2 Even though mobile phones are found everywhere in India, there is still room for a business that offers you a telephone line when you need it. The high buildings in Old Delhi mean that mobile reception is not good in many locations within the city walls.

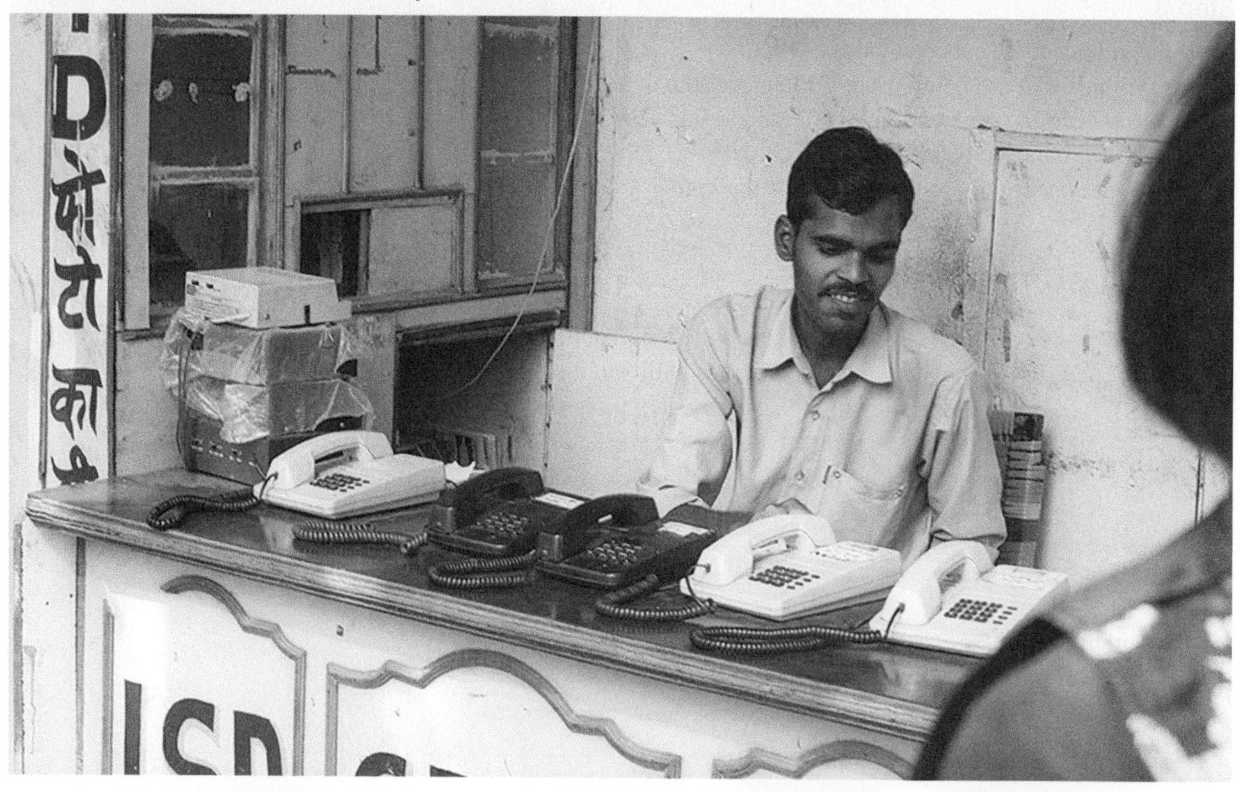

FIGURE 3 People of different religions have lived in Delhi for centuries. About 80 per cent of the inhabitants of Old Delhi are Muslims, whereas in the whole of Delhi, about 80 per cent of the population are Hindu. A number of mosques (place of worship for people who follow Islam), temples and Christian churches are crammed into the old city, including (a) the Jama Masjid (Delhi's largest mosque), (b) small but important mosques such as Kalan Masjid, built around 1387, and (c) the Holy Trinity Church.

FIGURE 4 The bazaars in Old Delhi sell everything from food to bicycles.

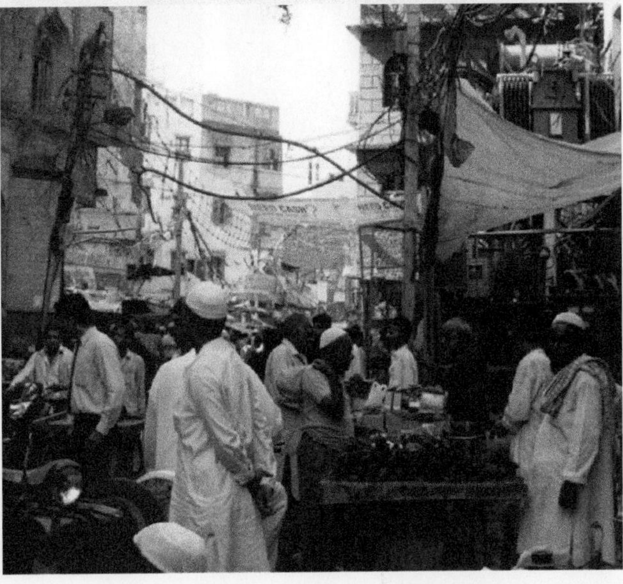

FIGURE 5 Old Delhi is a hub of small industries, crowded together. (a) Power and telephone lines are draped from building to building. (b) Market stall holders (c) Crowded streets (d) A welder works with homemade equipment. (e) A cobbler repairs shoes using his feet as a vice. (f) A tailor uses a foot pedal sewing machine.

FIGURE 6 The narrow streets make rubbish collection impossible, even if local government did offer it. The roofs of lower buildings often become rubbish dumps for those on the higher floors. In Delhi, nearly all rubbish collection is handled by private contractors or by the wastepickers.

FIGURE 7 Getting ready for iftar, the communal Ramadan festival after sunset. The place is full of a range of chaat (tangy and spicy snacks). During Ramadan, communities start to cook a fastbreaking meal in the afternoon so that it is ready for all to eat after sunset. This often takes the form of a neighbourhood banquet, celebrating friendships, family and community.

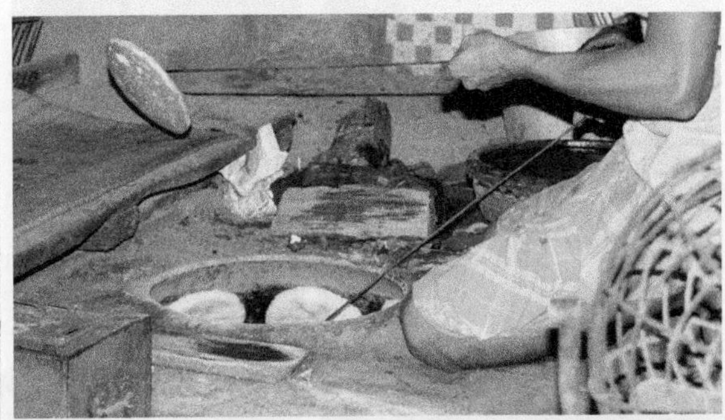

FIGURE 8 Access to clean water is a major problem in Old Delhi. There is no internal plumbing, although some households are able to obtain water through a courtyard distributor system. In general, water supply is communal in Old Delhi, just as it is in much of rural India and older urban areas. Usually, only the middle class and wealthy can afford running water. The courtyards of newer homes sometimes have a water tank on the roof.

FIGURE 9 Schools in Old Delhi exist in small buildings down side alleys. Some schools teach in the English language.

FIGURE 10 With few jobs around, young people learn to be creative in finding a way to make a living in the informal sector. Young boys might sell cool drinks, with cordial made with water from an urn. This is then cooled by ice from a block, which is big enough that it takes all day to melt.

10.5.2 Life in modern India

Modern India is a complex place. Indians themselves often talk about the 'multiple realities' of India. This means that visitors might see many different pictures of life in India. Some of these may contradict each other, yet all of them are true.

A great difference exists between modern India and traditional India. Modern India is in the top ten economies of the world. Since the 1990s, government policies have encouraged industrialisation, which has given jobs to many people and money for them to spend. India has more than 1.3 billion people, with an increasing number of middle-class citizens who are highly educated, earning high incomes and wanting the type of lifestyle that money can bring. In 2020, India had 117 billionaires (in US dollars), and almost 5 per cent of the world's billionaires.

FIGURE 11 In rural areas, many of the poor work in the informal sector. This snake charmer is one of a dying profession, working mainly for tourists and in festivals and markets, despite the fact that snake charming is now illegal. The serpent is sacred in Hinduism, and snake charmers were once thought to have gifts of healing.

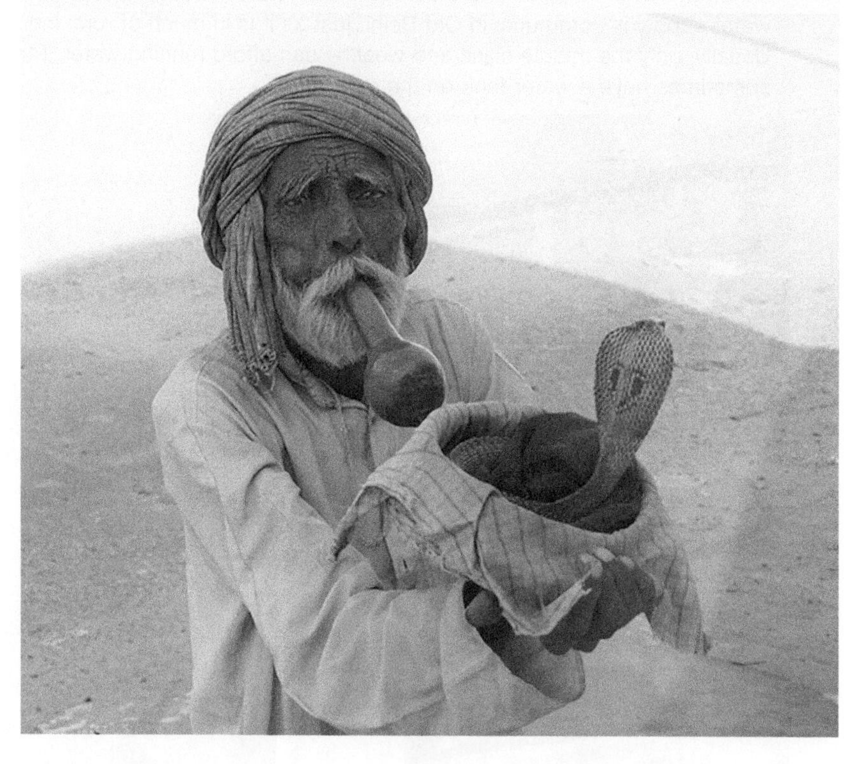

However, the gap between rich and poor is increasing. Economic growth has decreased the number of people living in poverty, but over 40 per cent of the population still lives in very basic conditions. Around 40 per cent of people still work in agriculture, and the rural poor are substantially worse off than other Indians. Incomes for farm workers are well below average.

FIGURE 12 A long drought in the 2000s put great pressure on dwindling water supplies in India, particularly in the countryside. However, urban growth and few environmental controls on local industry have also led to the pollution of small yet important lakes, like this one near Agra.

A large proportion of modern India's economic growth has been in the service sector, where people work in jobs providing skills ranging from gardening and laundry to accountancy. For instance, many businesses in North America, Europe and Australia have 'outsourced' their telecommunications and work to call centres in India. This means that they hire people in India to do jobs that would cost a lot more to pay people to do in their own countries. Many global companies now have bases in India.

FIGURE 13 The contrasts of modern India: (a) When buildings are constructed, bamboo poles are used as scaffolding. (b)–(d) The growing middle class has created a 'westernised' lifestyle for many Indians, where people place great importance on always buying new goods and services. This can be seen in the advertisements within large shopping centres and services such as 'McDelivery from McDonald's'.

FIGURE 14 Three methods of transport in modern India

fdw-0013

FOCUS ON FIELDWORK

Capturing and annotating photographs

You may have heard the saying that a picture is worth a thousand words. Sometimes in your geography work, you will be asked to take photos as part of your fieldwork. These pictures help to show landscapes, landforms or specific parts of an environment and to show your data collection methods. Because these pictures are an important part of a report, it is important to make sure that your photos are clear and show the features that are the most important to your work. Learn how to take useful photographs for your geography projects and how to annotate them well using the **Capturing and annotating photographs** fieldwork activity in your Resources tab.

on Resources

eWorkbook	India, past and present (ewbk-9326)
Video eLesson	India, past and present — Key concepts (eles-5093)
Interactivity	India, past and present (int-8424)
Fieldwork	Capturing and annotating photographs (fdw-0013)
Google Earth	Old Delhi (gogl-0028)

10.5 ACTIVITIES

1. Using the SHEEPT system of classification, construct a table like the following to summarise the major features of Old Delhi and of the lives of the people who live there.

	Social features	Historical features	Economic features	Environmental features	Political features	Technological features
Feature 1						

2. Using the table that you have drawn up, identify five characteristics of Old Delhi that are very different from the place in which you live, and five characteristics that are similar.

3. Use Google Earth to view an image of the space of Old Delhi in the present day. Using the Google Earth options, create an image of Old Delhi showing the distribution of the major human features, such as churches and mosques. Print a copy of the map you have created.
4. One of the areas where the growth of the wealthy class in India can be seen is in sport. Use the **Indian Premier League** weblink in the Resources tab to research the structure of the Indian Premier League (IPL) in cricket. Create a map showing where the different teams are based in India. Construct a table of the names of the owners and how they created their fortunes. What conclusions can you draw about the IPL as a symbol of modern India?

10.5 EXERCISE

Learning pathways

■ LEVEL 1	■ LEVEL 2	■ LEVEL 3
1, 2, 3, 7, 14	4, 6, 9, 10, 13	5, 8, 11, 12, 15

Check your understanding

1. Old Delhi consists of the original walled city that dates back to which year?
2. Which three religions have lived peacefully together for hundreds of years in Old Delhi?
3. Old Delhi has an area of _______ square kilometres and a population density of more than _______ people per square kilometre.
4. What does the phrase 'the multiple realities of India' mean?
5. Based on the size and population density of Old Delhi, calculate about how many people live there.
6. What percentage of the Indian population still lives in very basic conditions?

Apply your understanding

7. Classify each of the following features of Old Delhi as social, historical, economic, environmental, political or technological.
   a. More people, rather than machines, are doing jobs.
   b. There is an emphasis on wasting nothing in production.
   c. Mosques and churches are important.
   d. No motorised transport is used, except for the occasional motorbike.
   e. Most of the buildings are old.
8. Identify the main feature in the environment of Old Delhi that you believe to be the major problem facing its future. Justify your decision, using examples from this subtopic.
9. Compare the living conditions of Old Delhi with those where you live. What similarities and differences do you notice? Explain one difference and one similarity.
10. Compare life in Old Delhi with life in modern parts of India today. Identify and explain one social, one economic and one technological difference, and one social, one economic and one technological similarity.
11. What do you think is the most significant issue facing New Delhi? Explain what impacts this issue has on people and what could be done to solve the problem.
12. Which of the modes of transport shown in **FIGURE 14** do you think is the most environmentally sustainable? Give reasons for your view.

Challenge your understanding

13. Based on the images and descriptions of old and modern India, which lifestyle or way of living do you think is the most environmentally sustainable? Give reasons for your view.
14. There is a large gap between the richest and the poorest people in India. List the problems that might occur in a city with very rich and very poor people living close together. Suggest how one of these problems might be resolved in a way that benefits the whole community.
15. India is described as containing 'multiple realities'. To what extent does this description apply to where you live? Give examples to support your view.

To answer questions online and to receive **immediate feedback** and **sample responses** for every question, go to your learnON title at www.jacplus.com.au.

10.6 Liveable communities and me

LEARNING INTENTION

By the end of this subtopic, you will be able to identify strategies that can be employed at the local level to enhance the liveability of places.

10.6.1 Liveability studies

A study of a region's liveability will reflect its natural and human characteristics. All communities would like a safe, healthy and pleasant place to live, a sustainable environment, the chance to earn a liveable wage, reliable infrastructure and opportunities for social interaction. But not all people enjoy the same standards of living or consider the same things to be important around the world. Some people walk hours for fresh water, some people might see a doctor only once or twice in their lifetime, and some cultures don't consider school to be a necessity for everyone. Because of these different ways of living, liveability *needs* are different to our liveability *desires* (or wishes).

The findings of a liveability survey will be influenced by a range of factors.

- Where a person lives influences their access to services, employment and environmental features, and their address may influence their perception of the quality of the region.
- Different age groups have different views and needs.
- Current economic conditions are an influence; for example, a major employer may have closed or opened.
- Environmental conditions are an influence; for example, a region may be experiencing drought.
- Government policies influence infrastructure, housing assistance and grants to local sports clubs.

TABLE 1 Matching liveability indicators to key themes

Measure	Examples of indicators
Social	• Population characteristics (gender, age) • Education (primary, secondary, tertiary) • Health (life expectancy, health-centre attendance, smoking rates, weight, diseases) • Safety (perception of, crime rates, road deaths and injuries, work safety) • Volunteering • Voting • Aged care accommodation • Access to public transport • Membership of clubs and organisations • Diversity (ethnicity)
Environmental	• Biodiversity • Planning for the future • Water access • Waste management • Ecological footprint • Public spaces • Household recycling • Weather • Land clearing
Economic	• Employment • Variety of businesses • Income • Financial stress • Housing types • House ownership • Infrastructure • Internet access • Car ownership

To find out about the liveability of an area, a number of themes need to be investigated. Some of these can be gained from **census** statistics, while others can be gained only through surveys and fieldwork.

In any community there will usually be agreement about some things that improve liveability. All groups accept that safe water, sealed roads and a reliable power supply are important. If a community wants to obtain certain kinds of items on its liveability 'wish list' (see **TABLE 1**) it sometimes needs help from national, state or local government. Examples of such items include major roads, railways and desalination plants. Sometimes a wish-list item is best obtained by an individual or community. This is the case when setting up sporting clubs, youth groups and local music events.

census a regular survey used to determine the number of people living in Australia. It also has a variety of other statistical purposes

Community wish list

- Playgrounds
- Paths for prams
- Primary schools
- Single-person housing
- Family housing
- Friendly community
- Shopping nearby
- Paths for scooters
- Health services
- Public transport
- Neighbourhood house
- Parks and gardens
- Public seating
- Recognition of those from non-English-speaking backgrounds
- Financial security
- Community groups or classes

FIGURE 1 Community wish list: some aspects of liveability are common to all groups and some are desired by particular groups.

(a)

(b)

10.6.2 Transport strategies

People in towns and cities are always looking for strategies to improve their living conditions. A community is made up of people from a range of age groups, a number of different land uses, a range of needs and a variety of interests. Ideas and plans for improvement may be overarching or targeted.

The movement of people within and between neighbourhoods is an important issue in towns and cities. The bicycle is now seen as a way of increasing mobility, reducing traffic congestion, reducing air pollution and boosting health. Good cycling road infrastructure (where roads are organised and built to suit cyclists) can also help to encourage cycling as transport because it can make cycling safer, quicker and less affected by traffic. Bicycle tracks and shared walking and cycling paths encourage recreational riding for all ages by separating cars and bikes (see **FIGURE 2**) and dedicated bicycle lanes along main roads (see **FIGURE 3**) encourage people to commute by bicycle, rather than car, to work and school.

In 1965, a group in Amsterdam, the Netherlands, introduced the idea of bike sharing — public bicycles that are hired, usually for short trips. This first attempt was not a success but the idea persisted. Modern bike-sharing systems have overcome problems of theft and vandalism by using easily identifiable specialty bicycles, monitoring the bicycles' locations with radio frequency or GPS, and requiring credit-card payment or smart-card-based membership to check out bicycles. In some places, bicycles can be located on your mobile phone, and there are more links between bicycles and existing public transport. Between 2014 and 2018 bike-sharing programs doubled in size. More than 1600 programs are in operation, providing almost 18.2 million bikes to 20 million registered users.

FIGURE 2 Cycling paths, Surrey Hills, NSW

Copenhagen was rated as the world's most bike-friendly city in 2014 and has retained this position every year since. Beijing is the world leader in bike-share programs, with 2.4 million share bikes and 11 million registered users. Bike-sharing programs are an example of a popular strategy that is aimed at improving liveability for a range of ages and locations within a community.

FIGURE 3 Cycling lanes provide commuter routes to train stations in Murrumbeena, a suburb of Melbourne.

An example of a successful bike-sharing scheme is in Paris. The Vélib was introduced in 2007 and quickly doubled in size. By 2012, bicycle trips in the city had grown by 41 per cent. The program continues to grow; an additional 20 000 bikes have been added and the number of share stations expanded to 1400. One-third of the new bikes will have an electric motor with a range of around 50 kilometres when fully charged. These new bikes will also feature a basket with a carrying capacity of 50 kilograms.

It is anticipated that these new bikes will overcome problems associated with maintaining a share-bike program in hilly or uneven terrain, where commuters will ride a bike downhill in the morning, but then elect to return home using alternative transport; leading to a surplus of bikes in one area and a lack of them in others. Bike sharing is part of a plan to reduce car traffic and pollution in Paris. The plan also includes closing streets to cars on weekends, reducing speed limits, encouraging bus travel and extending bicycle lanes.

FIGURE 4 Bike sharing is on the rise globally.

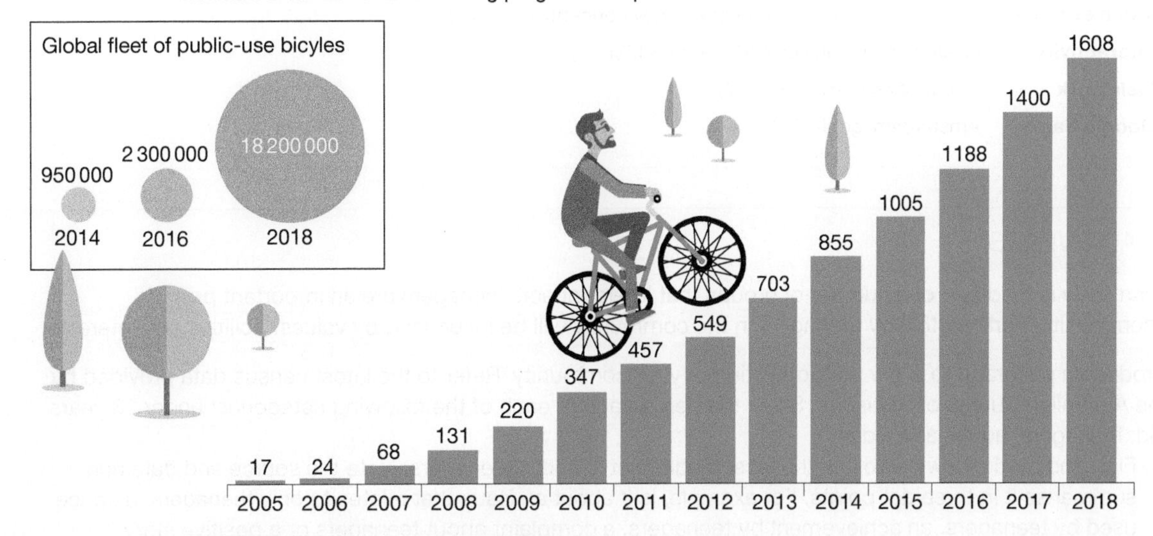

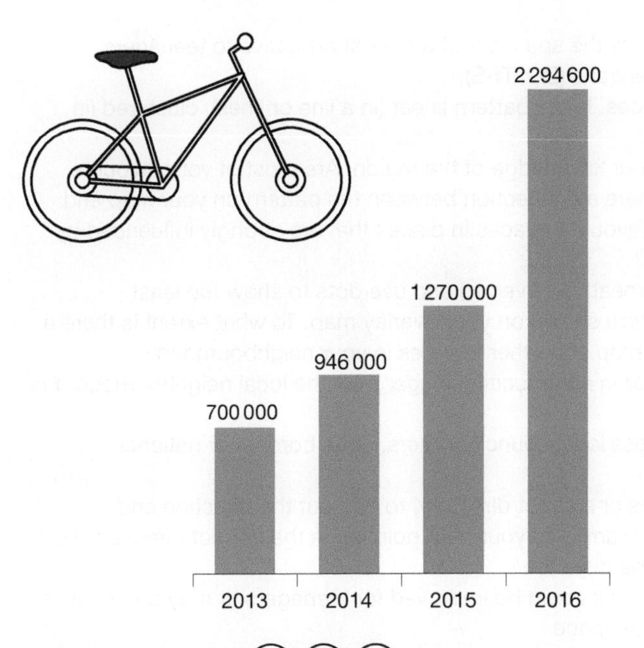

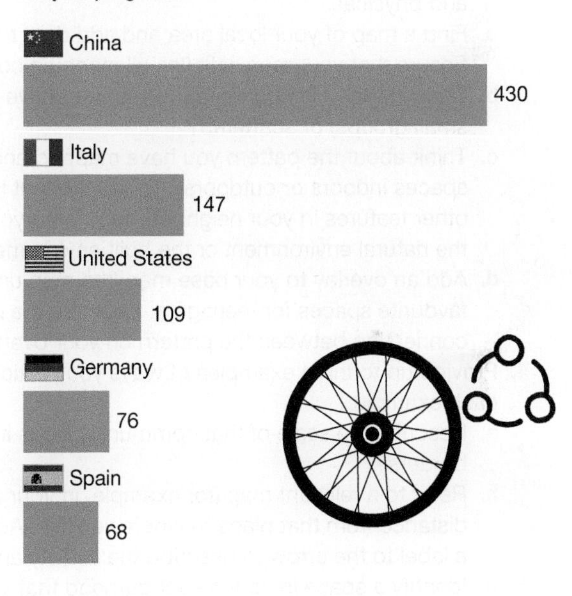

Source: @StatistaCharts (CC) (BY) (=)

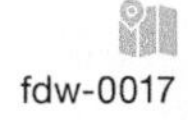

fdw-0017

FOCUS ON FIELDWORK

Liveability survey

You already know that people like different things and make different choices in their lives; this applies to the places they live too. A liveability survey helps you to understand what makes a place more or less liveable for individuals. Learn how to create and complete a liveabilty survey for where you live using the **Liveability survey** fieldwork activity in your Resources tab.

on Resources

eWorkbook	Liveable communities and me (ewbk-9330)
Video eLesson	Liveable communities and me — Key concepts (eles-5094)
Interactivity	Liveable communities and me (int-8425)
Fieldwork	Liveability survey (fdw-0017)
Google Earth	Amsterdam (gogl-0031)

10.6 ACTIVITIES

A community is made up of a number of groups that interconnect. Teenagers are an important part of any community. Participation by teenagers in the community will be influenced by values, abilities and interests.

1. Produce a pie graph to show the population of your community. Refer to the latest census data provided by the Australian Bureau of Statistics. Show the percentage of each of the following categories: under 13 years old, teenagers, adults and elderly.
2. a. Find and read a news article in the local media that is about teenagers. Note the source and date and summarise it in three dot points. For example, is it about an issue that relates to only teenagers, a space used by teenagers, an achievement by teenagers, a complaint about teenagers or a positive story about teenagers?
   b. As a class or in a small group, brainstorm a list of the ways in which teenagers participate in the community. Divide the agreed list into the following categories: informal, formal, social, cultural and physical.
3. a. Find a map of your local area and add dots to show the spaces that are most attractive to teenagers. Ensure that your map satisfies all mapping conventions (BOLTSS).
   b. Describe the distribution pattern of attractive places. Is the pattern linear (in a line or lines), clustered (in small groups) or scattered?
   c. Think about the pattern you have mapped and your knowledge of the region. Are most of your favourite spaces indoors or outdoors? To what extent is there a connection between the pattern on your map and other features in your neighbourhood? Are your favourite spaces in places that are strongly influenced by the natural environment or the built environment?
   d. Add an overlay to your base map (the map underneath an overlay) and use dots to show the least favourite spaces for teenagers. Describe the pattern shown on your overlay map. To what extent is there a connection between the pattern on your overlay map and other features in your neighbourhood?
4. Provide up to three examples of ways you participate in communities bigger than the local neighbourhood. For each example:
   a. Describe the scale of that community. Does it cross local council borders, state borders or national borders?
   b. Refer to a relevant map (for example, in your atlas or a street directory) to find out the direction and distance from that place to where you live. Add an arrow to your map pointing in the correct direction. Add a label to the arrow to describe the activity and the distance.
5. a. Identify a space in your neighbourhood that you think could be improved for teenagers. It may be one that is currently attractive, or it may be a least favourite space.
   b. Provide an image (photograph, diagram or map) of this space. Annotate the image to describe its current characteristics.
   c. Identify the key concerns about this space. You might think about safety, tolerance, sustainability, access, inclusiveness, services, environmental quality, health and respect.
   d. How would you improve this space?
      - To help you think of suggestions, use your research skills to find out about ways in which liveability has been improved for teenagers in other parts of the world. Consider European countries in particular.
      - Discuss the ways in which the European ideas are relevant, or not relevant, to your community.
      - Provide a planning suggestion for each of the concerns you raised in question 5c.
   e. Provide a new image to show the impact of your proposals. This could be a diagram, sketch, annotated photograph, model or whatever helps communicate what the impacts might be.

f. Which are your two most important suggestions? What criteria did you use to choose these suggestions? Which suggestion is most likely to be implemented? Why?
g. Compare your suggestions to the ideas of others in your class. What are the common elements? What would you put in a master plan for teenage spaces in your community?

6. Find a local news story about a change to liveability in your area. Is the change economic, social or environmental? Is the change predicted to be positive or negative? Will the change be permanent?
7. Find out about a bike-sharing scheme in Australia or overseas. Describe its location and the region it covers. Provide three other key facts about the scheme.
8. Some cities provide schemes to encourage people to ride bikes. Find out about the success of bike incentive schemes in European cities. Include the name of the city, the date of the scheme, summary of the scheme and evidence of success or failure.

10.6 EXERCISE

Learning pathways

■ LEVEL 1	■ LEVEL 2	■ LEVEL 3
1, 3, 7, 8, 14	2, 4, 9, 12, 13	5, 6, 10, 11, 15

Check your understanding

1. What are the three themes used when investigating liveability?
2. Identify two transport strategies that might make a place more liveable.
3. Give three examples of features that all communities in Australia would agree make a place liveable.
4. How many more people were estimated to be using public bike share schemes in 2018, compared with the number in 2013?
5. Identify and outline three factors that might affect the outcomes of a liveability survey — one from each theme.
6. Describe two ways that road infrastructure can help to encourage bicycle use.

Apply your understanding

7. What are three advantages of increasing bicycle riding?
8. What problems were faced by the first bike-sharing scheme? Explain why these problems occurred.
9. Refer to **FIGURE 1** and use an organiser like a Venn diagram to compare and contrast the liveability wish lists for young families and older people.
10. How could the improvement in liveability for one age group actually help the liveability of another age group? Provide an example.
11. Refer to **TABLE 1** and identify two aspects that could be placed in a different theme. Justify your suggested change. Suggest one more indicator that should be included. In which theme would it belong?
12. Explain what kinds of information the Australian Census provides about the liveability of specific areas.

Challenge your understanding

13. Refer to **FIGURE 4**.
    a. Suggest a reason for the rapid increase in the number of share bikes in cities around the world.
    b. Why do you think China is the fastest growing market for share bikes?
    c. Copenhagen has been rated the most bike-friendly city in the world but does not appear in the top five countries for bike-share programs. Research the city of Copenhagen and suggest a reason for this.
    d. How might the introduction of electric bikes encourage more people to use share bikes?
    e. What problems might the use of electric bikes cause?
14. Refer to the community wish list in this subtopic. Which three items do you think are most needed in the area you live in? Give reasons for your answer.
15. Would a bike-sharing scheme be a viable option in your area? Give reasons for your answer.

To answer questions online and to receive **immediate feedback** and **sample responses** for every question, go to your learnON title at www.jacplus.com.au.

10.7 SkillBuilder — Drawing a climate graph

online only

LEARNING INTENTION

By the end of this subtopic, you will be able to draw and read a climate graph.

The content in this subtopic is a summary of what you will find in the online resource.

10.7.1 Tell me

What are climate graphs?

Climate graphs, or climographs, are graphs that show climate data for a particular place. They combine a column graph and a line graph. The line graph always shows average monthly temperature, and the column graph always shows average monthly precipitation (rainfall). Temperature can be shown in one line, as mean monthly temperature; or it can be shown in two lines, as maximum and minimum monthly temperatures.

FIGURE 3 Features of a climate graph

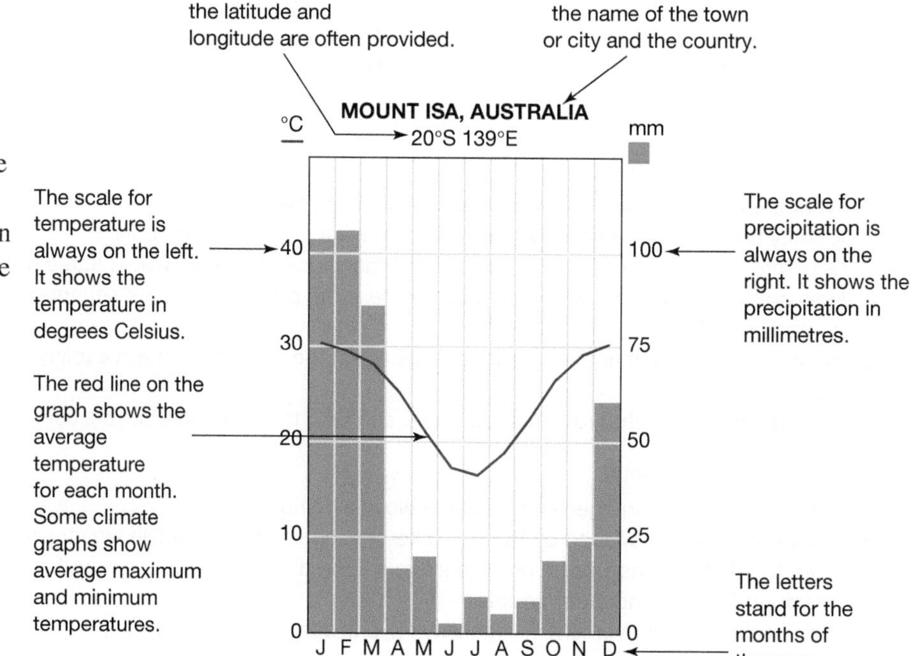

10.7.2 Show me

eles-1644

How to create a climate graph

Step 1

Look at the data in **TABLE 1**. Two sets of data are given: average monthly precipitation and average monthly temperature.

TABLE 1 New Delhi, average monthly precipitation and average monthly temperature

	Jan	Feb	Mar	Apr	May	Jun	Jul	Aug	Sep	Oct	Nov	Dec
°C	14.1	16.9	22.4	28.6	32.8	33.8	31.0	29.8	29.2	26.0	20.3	15.4
mm	22.7	20.1	14.5	10.1	15.0	67.9	200.4	200.3	122.5	18.5	3.0	10.0

Source: www.worldclimate.com

Step 2

Consider the range of the data before you decide what scales will work for the vertical axes. For the right-hand axis, find the wettest month. The precipitation scale begins at 0 and extends far enough to include the wettest month. For the left-hand axis, find the highest and lowest temperatures. A scale of 0 °C to 40 °C will suit most climate graphs.

Step 3

Use a ruler and pencil to draw the axes on graph paper. Label the months, temperatures and precipitation.

Step 4

Construct a column graph showing the average monthly rainfall. Construct a line graph showing the average monthly temperature.

Step 5

Add a title, giving the name of the place, the country and the latitude and longitude. Add the source of your data.

10.7.3 Let me do it

int-3140

Go to learnON to access the following additional resources to help you build this skill:
- a longer explanation of this skill and its application in Geography (Tell me)
- a video showing the step-by-step process of this skill (Show me)
- an activity and interactivity for you to practise this skill (Let me do it)
- self-marking questions to help you understand and use this skill.

on Resources

eWorkbook	SkillBuilder — Drawing a climate graph (ewbk-9334)
Video eLesson	SkillBuilder — Drawing a climate graph (eles-1644)
Interactivity	SkillBuilder — Drawing a climate graph (int-3140)

10.8 Investigating topographic maps — Liveability in Badu and Moa

LEARNING INTENTION

By the end of this subtopic, you will be able to decribe what liveability is like in Badu and Moa.

10.8.1 Badu and Moa islands

Both of these islands are located 40–60 kilometres off the far north Queensland coast in the Torres Strait. Moa Island has a population of approximately 240 people. Badu Island has a slightly larger population of around 850 people.

These small isolated communities rely on the ocean to provide food and as a pathway for trade. Their livelihood is threatened by climate change, particularly rising sea levels. Communities on the islands are heavily reliant on regular rainfall and have reservoirs to ensure a secure clean water supply. More recently, septic tanks and drainage facilities have been installed, improving the environmental health of the islands. With this water security and the development of infrastructure, an increasing number of tourists are travelling to Badu and Moa islands.

FIGURE 1 Moa Island is the second largest island in the Torres Strait.

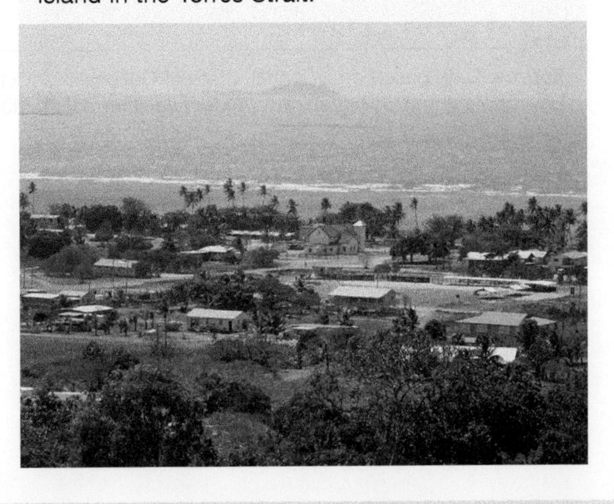

FIGURE 2 Topographic map extract of Badu Island and Moa Island in the Torres Strait

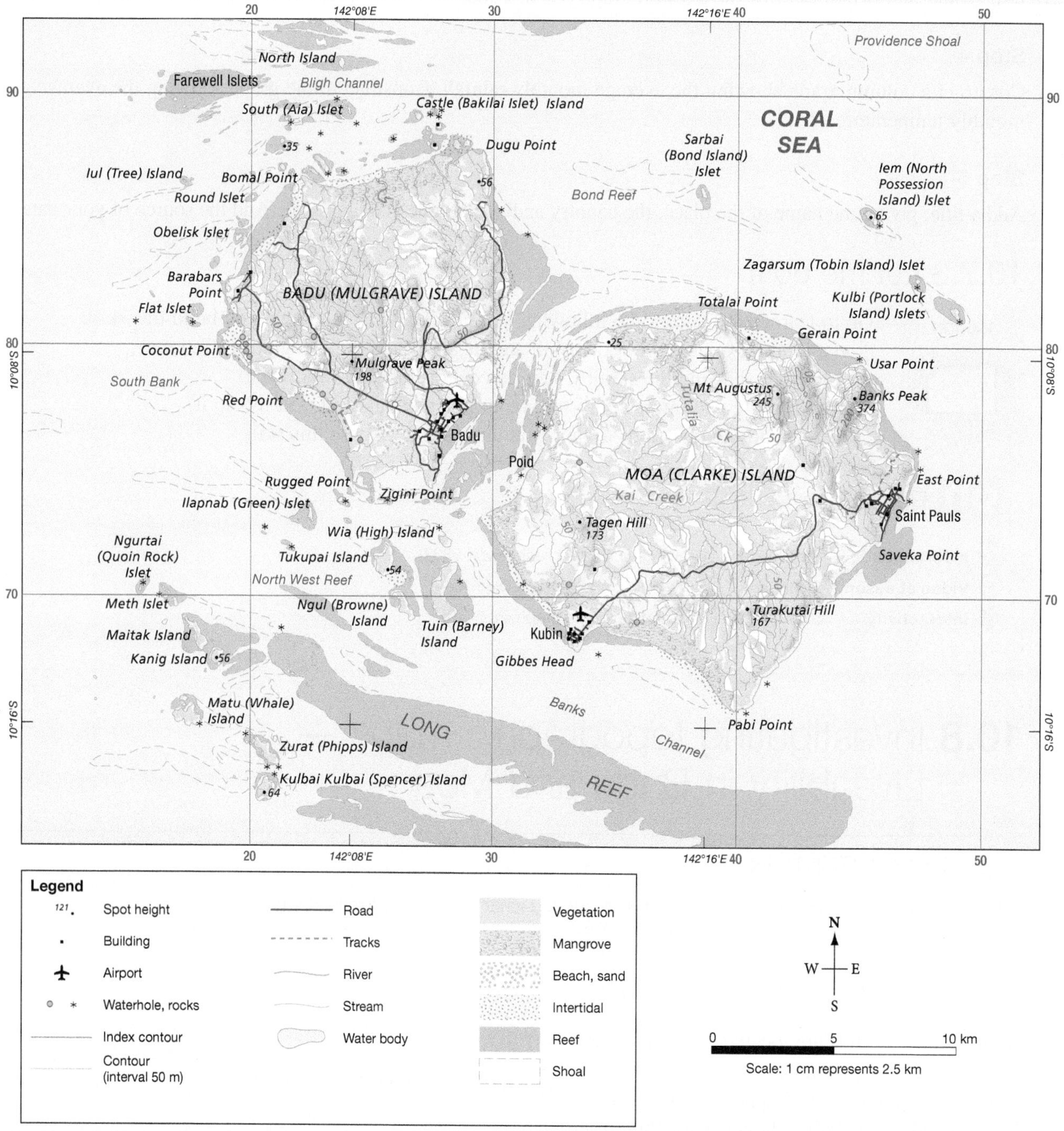

Legend

- 121. Spot height
- Building
- Airport
- Waterhole, rocks
- Index contour
- Contour (interval 50 m)
- Road
- Tracks
- River
- Stream
- Water body
- Vegetation
- Mangrove
- Beach, sand
- Intertidal
- Reef
- Shoal

0 5 10 km

Scale: 1 cm represents 2.5 km

Source: Data based on QSpatial, State of Queensland (Department of Natural Resources, Mines and Energy, Department of Environment and Science), http://qldspatial.information.qld.gov.au/catalogue/; Geoscience Australia.

on Resources

eWorkbook	Investigating topographic maps — Liveability in Badu and Moa (ewbk-9338)
Digital document	Topographic map of Badu and Moa islands (doc-20639)
Video eLesson	Investigating topographic maps — Liveability in Badu and Moa — Key concepts (eles-5095)
Interactivity	Investigating topographic maps — Liveability in Badu and Moa (int-8426)
Google Earth	Badu and Moa (gogl-0091)

10.8 EXERCISE

Learning pathways

■ LEVEL 1	■ LEVEL 2	■ LEVEL 3
1, 2, 3, 9, 13	4, 5, 8, 12, 14	6, 7, 10, 11, 15

Check your understanding

1. Describe where Badu and Moa are located. Use the terms *latitude* and *longitude* in your answer.
2. How many people live on Badu and Moa islands?
3. State the vegetation types on Badu and Moa islands.
4. What is the direction of Saint Pauls from Badu?
5. What is the height and area reference of Mt Augustus?
6. Calculate the approximate population density of each of the islands.
7. Locate and give the area references for three different human features located on these islands.
8. List some of the physical features located on these islands that would appeal to tourists.

Apply your understanding

9. Based on what you can see on the map, how liveable do you think the islands are? Support your answer with evidence from the map.
10. Explain in detail which of the sustainability goals impact the liveability of these islands.
11. How important do you think the airports are to levels of liveability on the islands? Give reasons for your view.
12. What factors might affect how liveable Badu and Moa islands are for different people? Choose one feature of the islands and explain:
    a. Why might someone born and raised on the islands think that feature increases liveability?
    b. Why might someone born and raised on the islands think that feature decreases liveability?
    c. Why might someone born and raised in a big city in Australia think that feature increases liveability?
    d. Why might someone born and raised in a big city in Australia think that feature decreases liveability?

Challenge your understanding

13. Identify a combination of human and natural features that might have influenced people to:
    a. visit the islands as a tourist
    b. permanently move to the islands.
14. a. If you could make these islands more liveable, what would you do?
    b. Create a list of five changes you would make to make them more liveable.
    c. Justify why you would make these changes and explain how they would make it more liveable.
15. What challenges could exist if the island communities are reliant on rainfall for their water source?

To answer questions online and to receive **immediate feedback** and **sample responses** for every question, go to your learnON title at www.jacplus.com.au.

10.9 Thinking Big research project — Liveable cities investigation and oral presentation

online only

The content in this subtopic is a summary of what you will find in the online resource.

Scenario

Every year the Economic Intelligence Unit ranks 140 major cities based on five key indicators:

- Stability — relates to the amount of crime, terror threats and potential military conflict
- Healthcare — relates to the availability and quality of healthcare in both the public and private sectors
- Culture and environment — includes climate, freedom from censorship, sport and corruption
- Education — relates to the availability and quality of education in both the public and private sectors
- Infrastructure — relates to the provision of services and structures in the community and includes road networks, public transport, housing, telecommunications, water quality and energy. Both quality and reliability are considered.

The city that receives the number one ranking is the most liveable city in the world, while the city ranked number 140 is the least liveable of the measured cities.

Task

Your task is to prepare a presentation that examines what it means to be a liveable city and how this is determined. What do you notice about the cities that are considered the most and least liveable?

As part of your investigation you will need to investigate a top ten most liveable city and a top ten least liveable city. You will need to compare them against the key indicators used to rank cities. Finally, you will present your findings to the class.

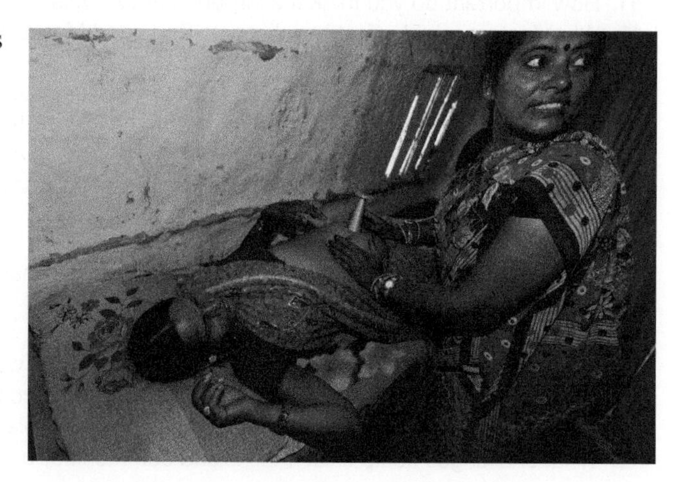

Go to your Resources tab to access all of the resources you need to complete this project.

Resources

ProjectsPLUS Thinking Big research project — Liveable cities investigation and oral presentation (pro-0239)

10.10 Review

10.10.1 Key knowledge summary

10.2 Liveability and sustainable living

- Liveability and sustainability are different.
- High income countries feature strongly in the list of liveable cities; however, these cities also have a high ecological footprint.
- A high ecological footprint means we are using resources at a faster rate than they can be regenerated by the Earth.

10.3 Improving liveability

- For those living below the poverty line, liveability will not improve until they have access to enough food.
- While some people do not have enough to eat, others have too much.
- Hunger has social, environmental and economic impacts.
- The United Nations has identified poverty as the underlying cause of low liveability.
- Australia provides aid to many countries around the world.

10.5 India, past and present

- Modern India is a complex place with huge differences between socioeconomic groups. As the population of cities continues to grow, these gaps are becoming even more significant.

10.6 Liveable communities and me

- People have different perceptions of liveability and this is coloured by their stage in life and where they live.
- Bicycle-sharing schemes are increasing and are a way of reducing traffic congestion, reducing pollution and improving health.
- Beijing in China has more share bikes than any other community.
- Paris is introducing electric bikes to its bike-sharing system.

10.10.2 Key terms

agriculture the cultivation of land, growing of crops or raising of animals

appropriate technology technology designed specifically for the place and the people who will use it. It is affordable and can be repaired locally.

biodiversity the variety of life in the world or in a particular habitat or ecosystem

census a regular survey used to determine the number of people living in Australia. It also has a variety of other statistical purposes

ecological footprint the total area of land that is used to produce the goods and services consumed by an individual or country

food security having access to and being able to afford enough nutritious food to stay healthy

global citizen person who is aware of the wider world, who tries to understand the values of others, and tries to make the world a better place

livestock animals raised for food or other products

non-government organisation non-profit group run by people (often volunteers) who have a common interest and perform a variety of humanitarian tasks at a local, national or international level

undernourished not getting enough food for good health and growth

10.10.3 Reflection

Complete the following to reflect on your learning.

Revisit the inquiry question posed in the Overview:

Why is it important to improve liveability around the world? What are some challenges to improving liveability?

1. Now that you have completed this topic, what is your view on the question? Discuss with a partner. Has your learning in this topic changed your view? If so, how?
2. Write a paragraph in response to the inquiry question, outlining your views.

Subtopic	Success criteria			
10.2	I can define the terms *sustainability* and *ecological footprint*.			
	I can make links between liveability, sustainability and ecological footprints.			
10.3	I can make links between hunger and liveability.			
	I can identify reasons for the unequal distribution of food across the world.			
	I can explain how assistance from other countries can improve liveability.			
10.4	I can identify key features on a thematic map, and describe patterns in the data shown.			
10.5	I can explain the changes that have taken place in India over time and how they have altered liveability.			
10.6	I can identify strategies that can be employed to improve liveability at the local level.			
10.7	I can draw and read a climate graph.			
10.8	I can describe what liveability is like in Badu and Moa.			

on Resources

eWorkbook Chapter 10 Student learning matrix (ewbk-8472)
Chapter 10 Reflection (ewbk-8473)
Chapter 10 Extended writing task (ewbk-8474)

Interactivity Chapter 10 Crossword (int-8427)

ONLINE RESOURCES

on Resources

Below is a full list of **rich resources** available online for this topic. These resources are designed to bring ideas to life, to promote deep and lasting learning and to support the different learning needs of each individual.

eWorkbook

- 10.1 Chapter 10 eWorkbook (ewbk-7990) ☐
- 10.2 Liveability and sustainable living (ewbk-9314) ☐
- 10.3 Improving liveability (ewbk-9318) ☐
- 10.4 SkillBuilder — Understanding thematic maps (ewbk-9322) ☐
- 10.5 India, past and present (ewbk-9326) ☐
- 10.6 Liveable communities and me (ewbk-9330) ☐
- 10.7 SkillBuilder — Drawing a climate graph (ewbk-9334) ☐
- 10.8 Investigating topographic maps — Liveability in Badu and Moa (ewbk-9338) ☐
- 10.10 Chapter 10 Student learning matrix (ewbk-8472) ☐
- Chapter 10 Reflection (ewbk-8473) ☐
- Chapter 10 Extended writing task (ewbk-8474) ☐

Sample responses

- 10.1 Chapter 10 Sample responses (sar-0144) ☐

Digital document

- 10.8 Topographic map of Badu and Moa islands (doc-20639) ☐

Video eLessons

- 10.1 Making places liveable for young people (eles-1621) ☐
- Improving liveability — Photo essay (eles-5345) ☐
- 10.2 Liveability and sustainable living — Key concepts (eles-5091) ☐
- 10.3 Improving liveability — Key concepts (eles-5092) ☐
- 10.4 SkillBuilder — Understanding thematic maps (eles-1658) ☐
- 10.5 India, past and present — Key concepts (eles-5093) ☐
- 10.6 Liveable communities and me — Key concepts (eles-5094) ☐
- 10.7 SkillBuilder — Drawing a climate graph (eles-1644) ☐
- 10.8 Investigating topographic maps — Liveability in Badu and Moa — Key concepts (eles-5095) ☐

Interactivities

- 10.2 Liveability and sustainable living (int-8423) ☐
- 10.3 Distribution of hunger, 2018 (int-7801) ☐
- 10.4 SkillBuilder — Understanding thematic maps (int-3154) ☐
- 10.5 India, past and present (int-8424) ☐
- 10.6 Liveable communities and me (int-8425) ☐
- 10.7 SkillBuilder — Drawing a climate graph (int-3140) ☐
- 10.8 Investigating topographic maps — Liveability in Badu and Moa (int-8426) ☐
- 10.10 Chapter 10 Crossword (int-8427) ☐

Fieldwork

- 10.5 Capturing and annotating photographs (fdw-0013) ☐
- 10.6 Liveability survey (fdw-0017) ☐

ProjectsPLUS

- 10.9 Thinking Big research project — Liveable cities investigation and oral presentation (pro-0239) ☐

Weblink

- 10.5 Indian Premier League (web-0055) ☐

Google Earth

- 10.5 Old Delhi (gogl-0028) ☐
- 10.6 Amsterdam (gogl-0031) ☐
- 10.8 Badu and Moa (gogl-0091) ☐

Teacher resources

There are many resources available exclusively for teachers online.

11 Geographical inquiry — My place

11.1 Overview

Numerous **videos** and **interactivities** are embedded just where you need them, at the point of learning, in your learnON title at www.jacplus.com.au. They will help you to learn the content and concepts covered in this topic.

LEARNING INTENTION

By the end of this subtopic, you will be able to plan a geographical inquiry on the demographic characteristics of a place, collect and record data, and process and communicate information about a local place.

11.1.1 Scenario and task

Task: Create a blog that presents demographic characteristics of your local place.

Every person has their own idea of what their local place is like. For some people, this area can be very large; for others, it can be quite small. It really depends on where you go in your everyday life. For example, homes of relatives or friends, sports clubs, shops and parks. This means that it does not matter if your map representing your place is a different size or shape to those of friends who live in the same area. The differences simply reflect what you do and think as an individual person.

When you draw a mental map of your local place, you identify the features that you think give your neighbourhood a sense of place. All local areas have these special features that create the character or personality of the place. Many of these features can be identified on maps of the area. But there are also characteristics of your local area that you may not know about. How do you find out about these?

The Australian Bureau of Statistics (ABS) is the Commonwealth Government's organisation that has the responsibility to collect, collate and report information about Australia's people. Every five years, the ABS conducts a major survey of all Australians as they are living on one specified day of the year. This is called the census. The ABS compiles this information and releases it for publication, which is when it is used by governments, businesses, companies and individuals to plan for the future.

FIGURE 1 Cooma, NSW

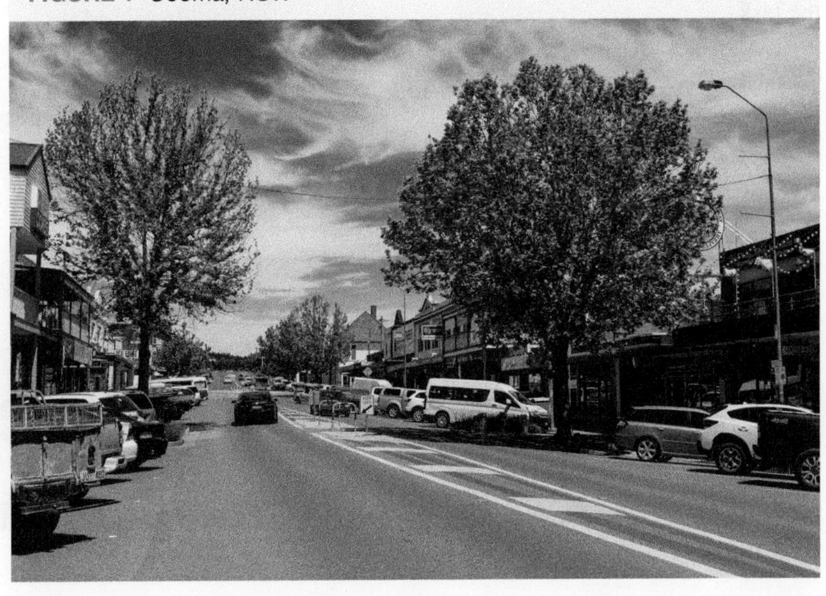

Go to learnON to access the following additional resources to complete this inquiry task:

- a blog model
- a feature article scaffold
- a selection of images and audio and video files to add richness to your blog
- weblinks to research sites and blogging websites
- an assessment rubric.

Your task

Create a blog that presents demographic characteristics of your local place. The Australian Bureau of Statistics website (www.abs.gov.au) provides a pathway for you to find out the demographic characteristics of your chosen postcode area.

FIGURE 2 Coffs Harbour, NSW

11.2 Inquiry process

11.2.1 Process

Open the ProjectsPLUS application for this chapter located in your Resources tab. Watch the introductory video lesson and then click the 'Start Project' button and set up your project group. You will write your blog entries individually, but you could work in groups of three to share your research in the ProjectsPLUS Research Forum and create your blog. Save your settings and the project will be launched.

Planning: You will need to research the demographic characteristics of your chosen local place. Locate and print a map of your chosen place to accompany the data you find. Navigate to your Research Forum. Use the research topics to select the characteristics you are interested in exploring for your place. Note that if you would like to focus on a particular issue that is not on the list, obtain permission from your teacher before starting the task.

FIGURE 3 Lightning Ridge, NSW

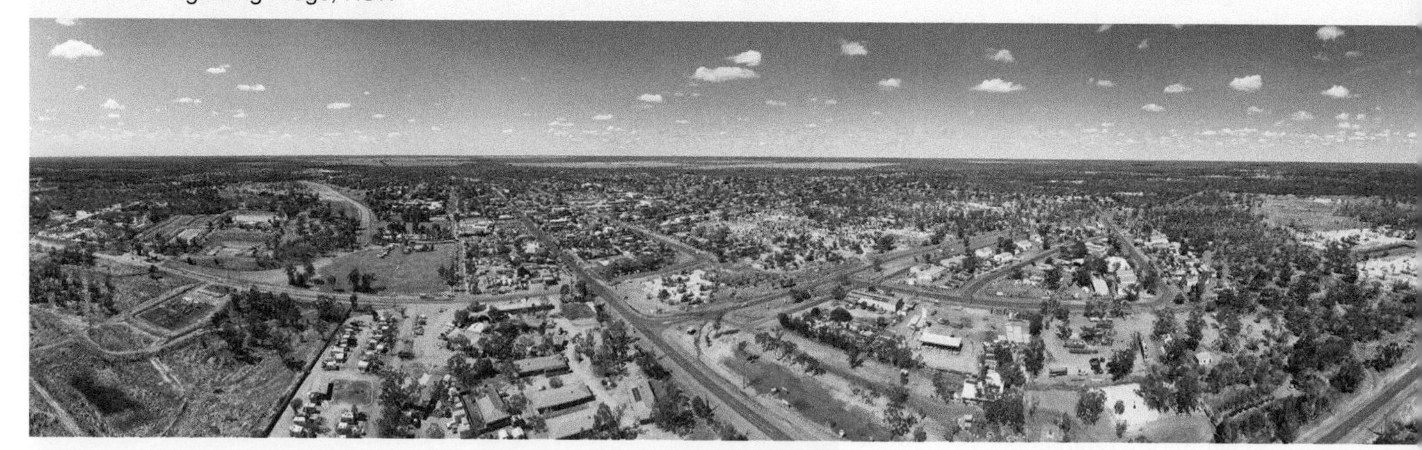

11.2.2 Collecting and recording data

For this inquiry, go to the Australian Bureau of Statistics website (www.abs.gov.au) and select the 'Census' page. Click on 'Data & analysis' in the menu at left and choose 'QuickStats'. Under 'QuickStats Search', select the most recent census year, type your postcode or place name into the search box and select 'Go'. This will bring a map of your local area and related data onto the screen. Choose either People, Families or Dwellings to gather information.

Going further:

- Compare the changes to your place over time. Choose the same topic for different census years and compare the data.
- Compare your local area with another local place in your region or city. The other place could be next to yours or a long distance away.

11.2.3 Processing and analysing your information and data

Having collected the information about your local place, you now need to study the data and describe the patterns you have found.

1. How does your place compare with the neighbouring places?
2. What do the combinations of characteristics you have chosen tell you about the community in your postcode?
3. What do the charts tell you that you did not know about the different places in the region where you live?
4. If you have looked at change over time, compare the two charts and describe how the distribution of that characteristic changed over the years. Suggest reasons to explain these changes.

Visit your Media Centre and download the blog planning template to help you develop your blog. You will also see a sample blog on which you can model your own task. Your Media Centre also includes images, videos and audio files to help bring your blog to life.

11.2.4 Communicating your findings

Use an online blogging site to set up your group's blog and then enter all of the required blog entries. Be sure to create a headline for your article and add relevant tables, graphs, images and videos. Your article should emphasise the important facts, and how and why they have changed over time.

FIGURE 4 The view across Chatswood, NSW

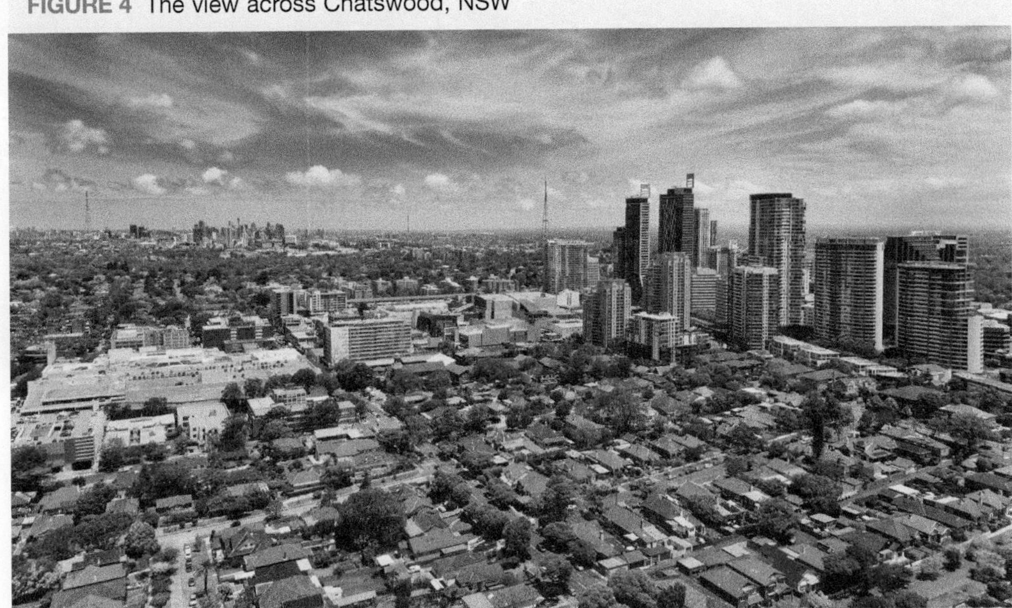

11.3 Review

11.3.1 Reflecting on your work

Think back over how well you worked with your group on the various tasks for this inquiry. Determine strengths and weaknesses and recommend changes if you were to repeat the exercise. Identify one area where you were pleased with your performance, and an area where you would like to improve. Write two sentences outlining how you might be able to do this.

Print your Research Report from ProjectsPLUS and hand it in with your blog and reflection notes.

11.3.2 Reflection

Subtopic	Success criteria			
11.2	I can plan a geographical inquiry on demographic characteristics of a place.			
11.2	I can collect and record data as part of a geographical inquiry into a local place.			
11.2	I can process and communicate information as part of a geographical inquiry into a local place.			

FIGURE 5 Bathurst, NSW

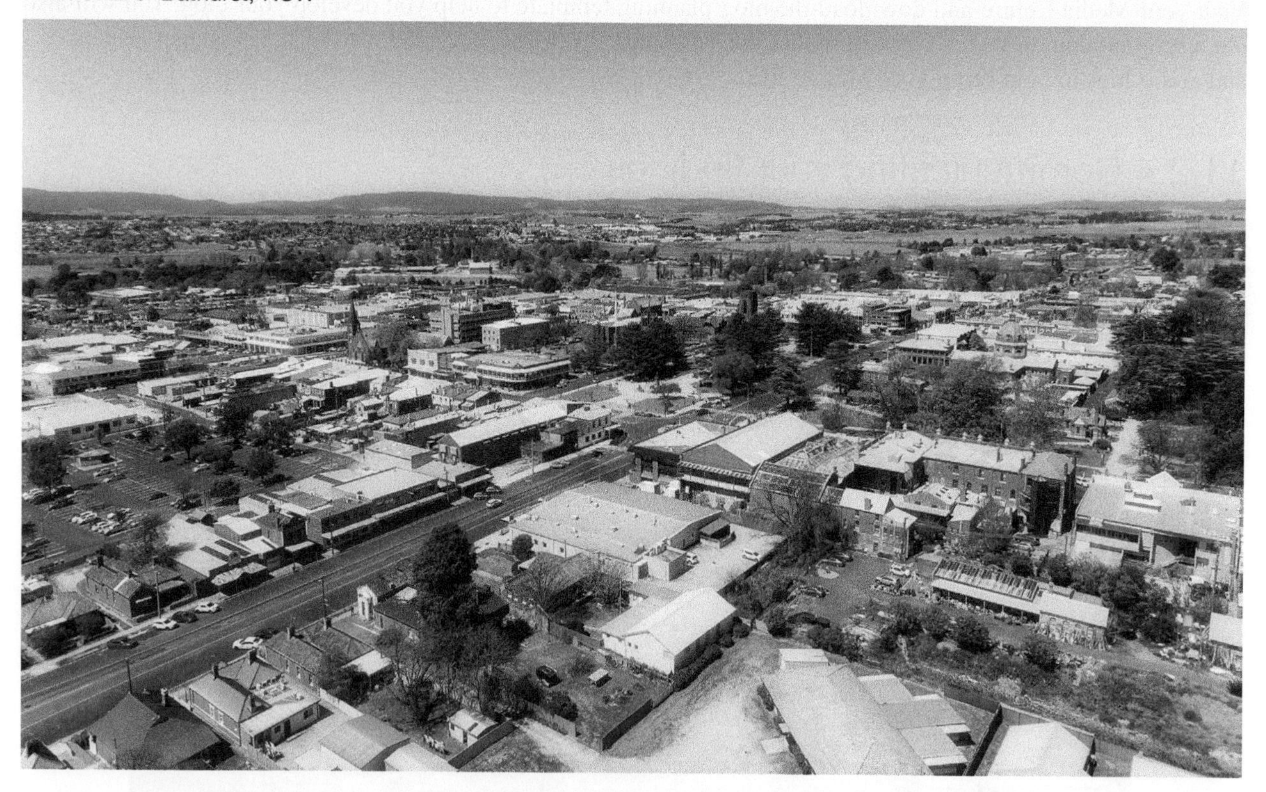

on Resources

ProjectsPLUS Geographical inquiry — My place (pro-0144)

UNIT

3 Water in the world

12 Water in the world

INQUIRY SEQUENCE

To access a pre-test with **immediate feedback** and **sample responses** to every question in this chapter, select your learnON format at www.jacplus.com.au.

12.1 Overview

Numerous **videos** and **interactivities** are embedded just where you need them, at the point of learning, in your learnON title at www.jacplus.com.au. They will help you to learn the content and concepts covered in this topic.

Earth provides us with more than a place to live. How do we access and manage the water resources we need from it to survive?

12.1.1 Introduction

Viewed from space, the Earth is a sphere of blue. Water covers most of our planet. We depend on water for life; in fact, no life is possible without it. Water is a precious and finite resource, yet most of the Earth's water is too salty for humans, animals or plants to use. A number of factors affect where water is available around the globe and as a result, water is not distributed equally around the planet. Weather is one factor that influences the level of precipitation in different places.

STARTER QUESTIONS

1. Describe how the weather affects your everyday life.
2. List some natural resources that are used by your family on a daily basis.
3. List two natural environments that have a lot of water. Name where they are or show their location on a map.
4. Why do you think water is thought of as a precious resource? Justify your answer.
5. Watch the **Water in the world — Photo essay** video eLesson in your Resources tab. How is water used and considered where you live? Is there plenty or never enough? Do you often think about how much water you use?

FIGURE 1 Earth viewed from NASA's Deep Space Climate Observatory (DSCOVR) satellite, 2015

on Resources

eWorkbook Chapter 12 eWorkbook (ewbk-7992)

Video eLessons A world of water (eles-1616)
Water in the world — Photo essay (eles-5346)

12.2 Understanding resources

LEARNING INTENTION

By the end of this subtopic, you will be able to define and classify examples of 'resources' and explain why water is classified as a resource.

12.2.1 Global supply

We depend on natural resources to survive. We need water to drink, soil to produce our food, and forests and mines to supply other materials. Natural resources are raw materials that occur in the environment that are necessary or useful to people. They include soil, water, mineral deposits, fossil fuels, plants and animals.

There are two types of natural resources: non-renewable and renewable.

Renewable resources are those that can be replaced in a short time. For example, solar energy is a renewable resource that can be used for heating water or generating electricity. It is never used up and is constantly being replaced by the Sun.

Non-renewable resources are those that cannot be replaced in a short time. For example, **fossil fuels** such as oil, coal and natural gas are non-renewable because they take thousands of years to be replaced.

We cannot make more non-renewable resources; they are limited and will eventually run out. However, renewable natural resources are things that can grow or be replaced over time if they are carefully managed. Forests, soil and fresh water are renewable.

fossil fuel fuel that comes from the breakdown of living materials, and that is formed in the ground over millions of years. Examples include coal, oil and natural gas.

FIGURE 1 Many resources are used at every meal.

Some renewable resources will always be available for use regardless of human activity, such as energy from the Sun or tides. However, human activities can impact on our ability to use some renewable resources, such as water or soils. We can have a negative impact on the quality of these resources or their ability to regenerate. For example, chemical dumping could make water undrinkable, or overfishing could reduce the ability of a fish species to reproduce at a rate to sustain its population.

The global distribution of natural resources depends on geology (the materials and rocks that make up the Earth) and climate. Some minerals are rare and are found in only a few locations. For example, **uranium** is found mainly in Australia. Several countries in the Middle East, such as Saudi Arabia and Iran, have rich oil resources but are short of water. Many countries in Africa, such as Botswana, have mineral resources but lack the money to mine and process them without investment from companies or other countries.

The human activities of agriculture, fishing, logging and mining all depend directly on natural resources. In developing countries, traditional forms of agriculture such as **subsistence farming** and nomadic herding are still common. These activities are sustainable if farmers move on when an area becomes unproductive, allowing the land to recover. However, poverty and population growth mean that many people now clear forests for farms and overgraze or overcrop small plots of land, resulting in deforestation and land degradation.

FIGURE 2 Natural resources

int-7766

FIGURE 3 The process of shifting cultivation

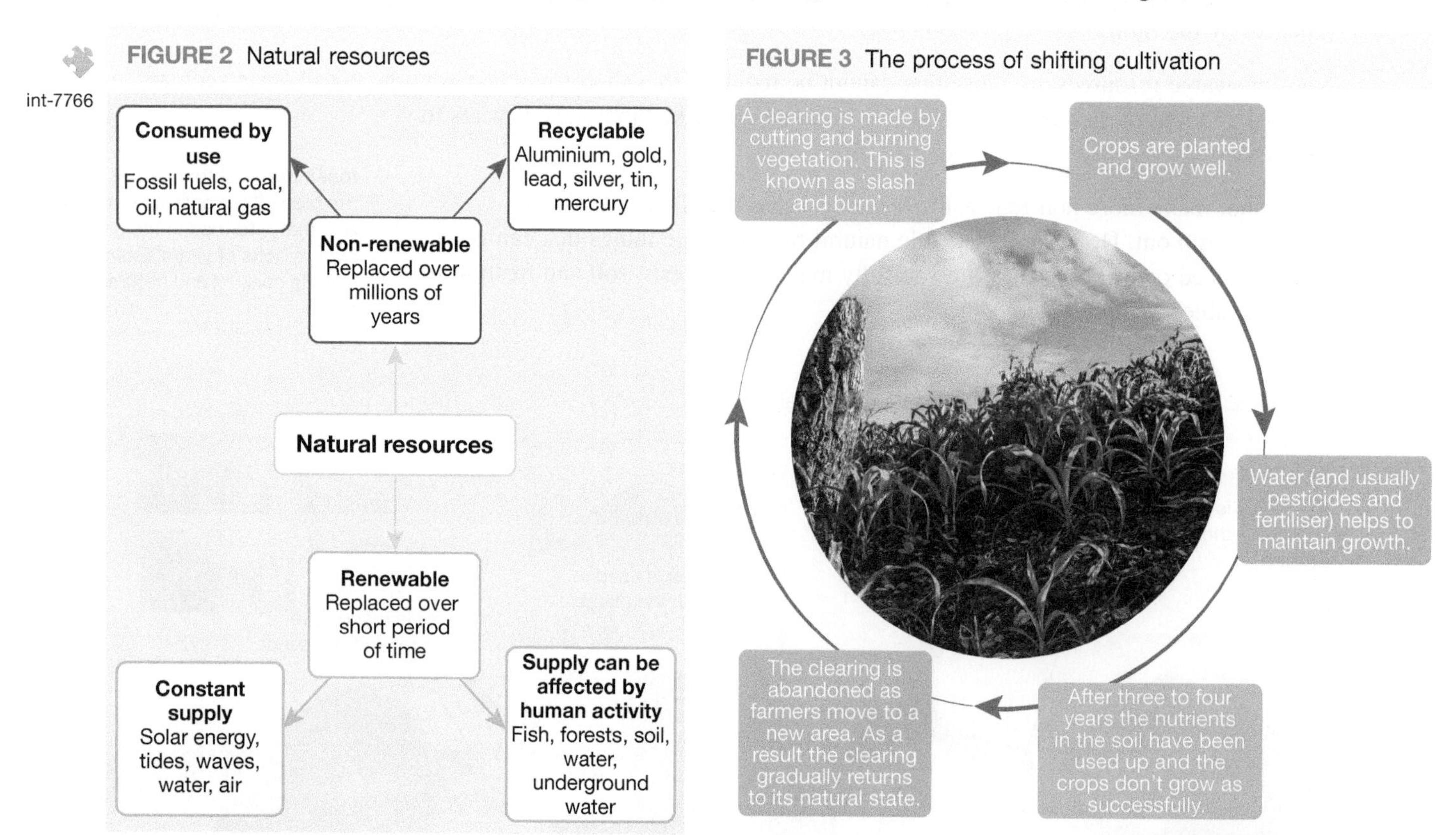

Farms in developed countries are usually much larger. For example, the Anna Creek cattle station in South Australia is the largest working cattle station in the world. It covers 2 400 000 hectares (24 000 square kilometres), roughly the size of Belgium. In contrast, an average intensive rice farm in Bali is only about one hectare (or 0.01 square kilometres). This is about four times the size of an average Australian suburban block of land. Unsustainable agricultural practices in developed countries include the overuse of water, fertilisers and pesticides. For example, fertilisers help crops to grow, but when they end up in rivers and oceans as run-off, they cause algal blooms and damage coral reefs.

uranium radioactive metal used as a fuel in nuclear reactors

subsistence farming a form of agriculture that provides food for the needs of only the farmer's family, leaving little or none to sell

on Resources

eWorkbook Understanding resources (ewbk-9346)
Video eLesson Understanding resources — Key concepts (eles-5096)
Interactivity Understanding resources (int-7766)

12.2 ACTIVITY

Refer to **FIGURE 3** and use the internet to research who uses shifting cultivation around the world. Choose one case study and report back to the class about their way of life. Examples may include tribes from the Amazon, Congo Basin or Papua New Guinea. Compare your chosen tribe's way of life with your own, and explain how they differ when it comes to using resources and accessing food. Upon completion of the presentations, discuss as a class why you think Australian farmers do not use shifting cultivation as their method of agricultural production.

12.2 EXERCISE

Learning pathways

■ LEVEL 1	■ LEVEL 2	■ LEVEL 3
1, 2, 10, 13, 17	3, 4, 6, 9, 11, 15	5, 7, 8, 12, 14, 16

Check your understanding

1. Define the term *natural resource*.
2. Outline the difference between renewable and non-renewable resources.
3. List three examples of non-renewable resources that can be recycled.
4. List three examples of non-renewable resources that are consumed by use.
5. Describe one way that the supply of renewable resources can be affected by human activity.
6. Define the term *subsistence farming*.
7. Describe the process of shifting cultivation.

Apply your understanding

8. Assess which renewable resources are most affected by human activity. Why?
9. When it comes to using natural resources, there are two main problems people face. Explain what these problems are and why they are important.
10. Explain what the phrase 'sustainable use of natural resources' means.
11. Explain why shifting cultivation is a sustainable form of agriculture.
12. Why do farms in developed countries tend to be much larger than in developing countries?
13. List what you had for breakfast today. Explain what resources would have been required to provide it.
14. Provide examples of fossil fuels, and explain how you use them in order to maintain your lifestyle.

Challenge your understanding

15. a. Create a table that lists ten renewable and ten non-renewable resources used by your family. Be specific; for example, list timber used in your furniture.
    b. From your list, note some of the waste and pollution that may be created in the use or creation of these resources.
    c. How could this be reduced to improve environmental sustainability?
16. Do you think the extent of small-scale, traditional farming practices will change in the future? Give reasons to support your answer.
17. Suggest one way that you could reduce your use of non-renewable, consumable resources each day.

To answer questions online and to receive **immediate feedback** and **sample responses** for every question, go to your learnON title at www.jacplus.com.au.

12.3 Water resources

LEARNING INTENTION

By the end of this subtopic, you will be able to identify and explain the processes involved in the water cycle, and describe how they are interconnected.

12.3.1 Water as a resource

A renewable resource is one that can replenish or replace itself in a relatively short period of time. Water is an example of a renewable resource.

The amount of water on Earth is finite, or fixed. The water used by ancient and extinct animals and plants millions of years ago is the same water that today falls as rain. It is cycled and recycled, and constantly changes its state or form from gas, to liquid, to solid, and back again.

Despite there being a finite amount of water on Earth, it is viewed as renewable because it is in a constantly changing state. If water is managed properly it can be renewed or recycled and then reused continuously.

12.3.2 The water cycle

All the water on Earth moves through a cycle that is powered by the Sun. This cycle is called the water cycle, or **hydrologic cycle**. Water is constantly changing its location

hydrologic cycle another term for the water cycle

int-5614

FIGURE 1 The water cycle (hydrologic cycle)

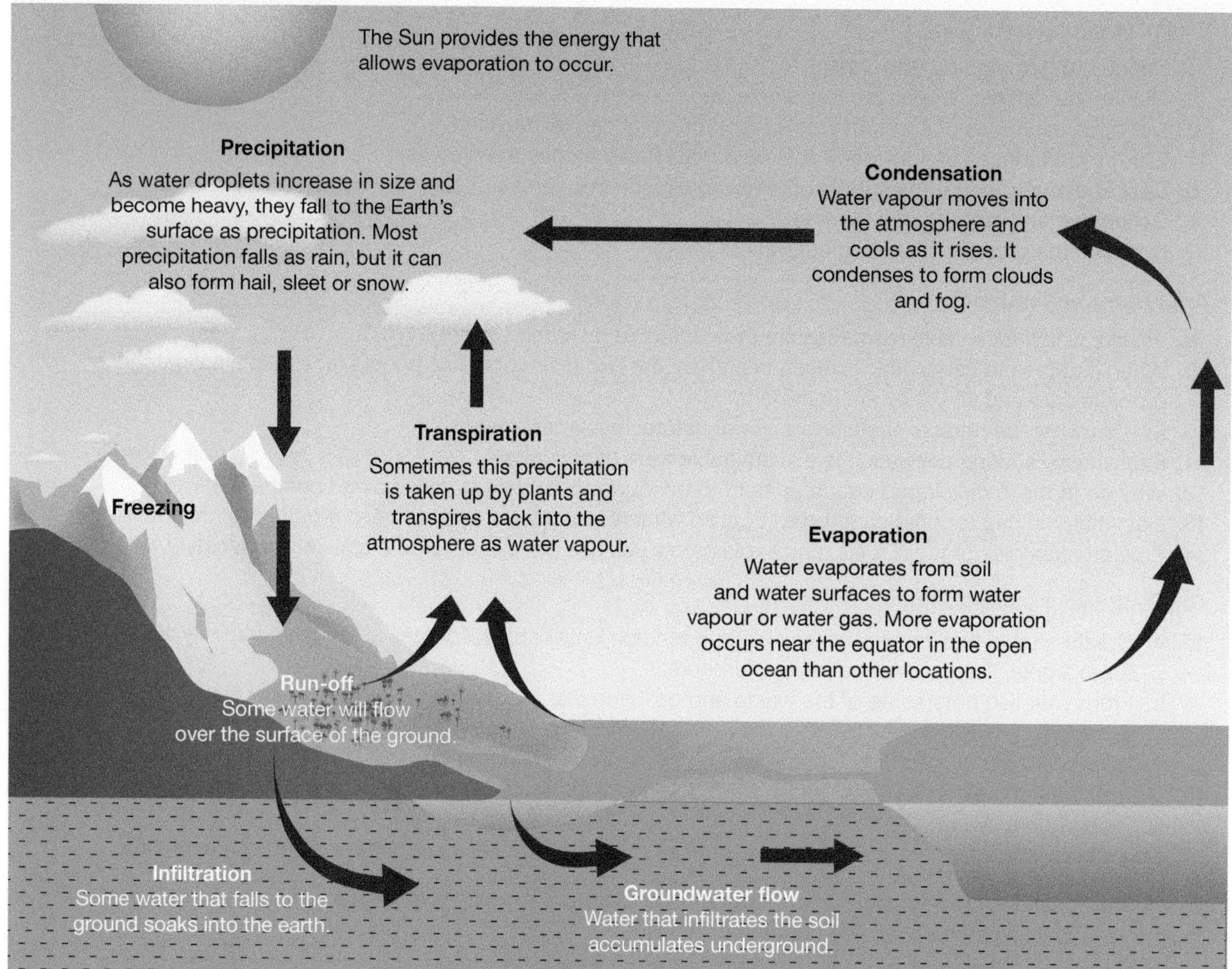

(through constant movement) and its state. Evaporation, condensation and freezing of water occur during the cycle.

Key processes in the water cycle are **precipitation**, **infiltration**, **run-off**, storage, **evaporation**, **evapotranspiration** and **condensation**. In some locations and climates water will also freeze as part of the water cycle.

Water is stored in oceans, lakes and rivers. The Sun's heat changes water from its liquid form into a gas. This is known as evaporation. When the water vapour rises and cools it condenses, and when it becomes heavy enough it will fall back to earth as precipitation. If the land is very steep or the soil is very hard, water is likely to run off to lower ground. If the soil is permeable (having small spaces or gaps) the water may infiltrate underground. Water that has infiltrated will become part of groundwater flow beneath the earth's surface. This water will either accumulate in underground storage or slowly make its way to streams and rivers.

precipitation the form of water falling from the sky, such as rain, snow or hail

infiltration precipitation absorbed into the ground

run-off precipitation not absorbed by soil, which runs over the land and into streams

evaporation the process by which water is converted from a liquid to a gas and thereby moves from land and surface water into the atmosphere

evapotranspiration the process by which plants absorb precipitation and release it back into the air as water vapour

condensation precipitation that collects into droplets of water from humid air

FIGURE 2 Processes in the water cycle

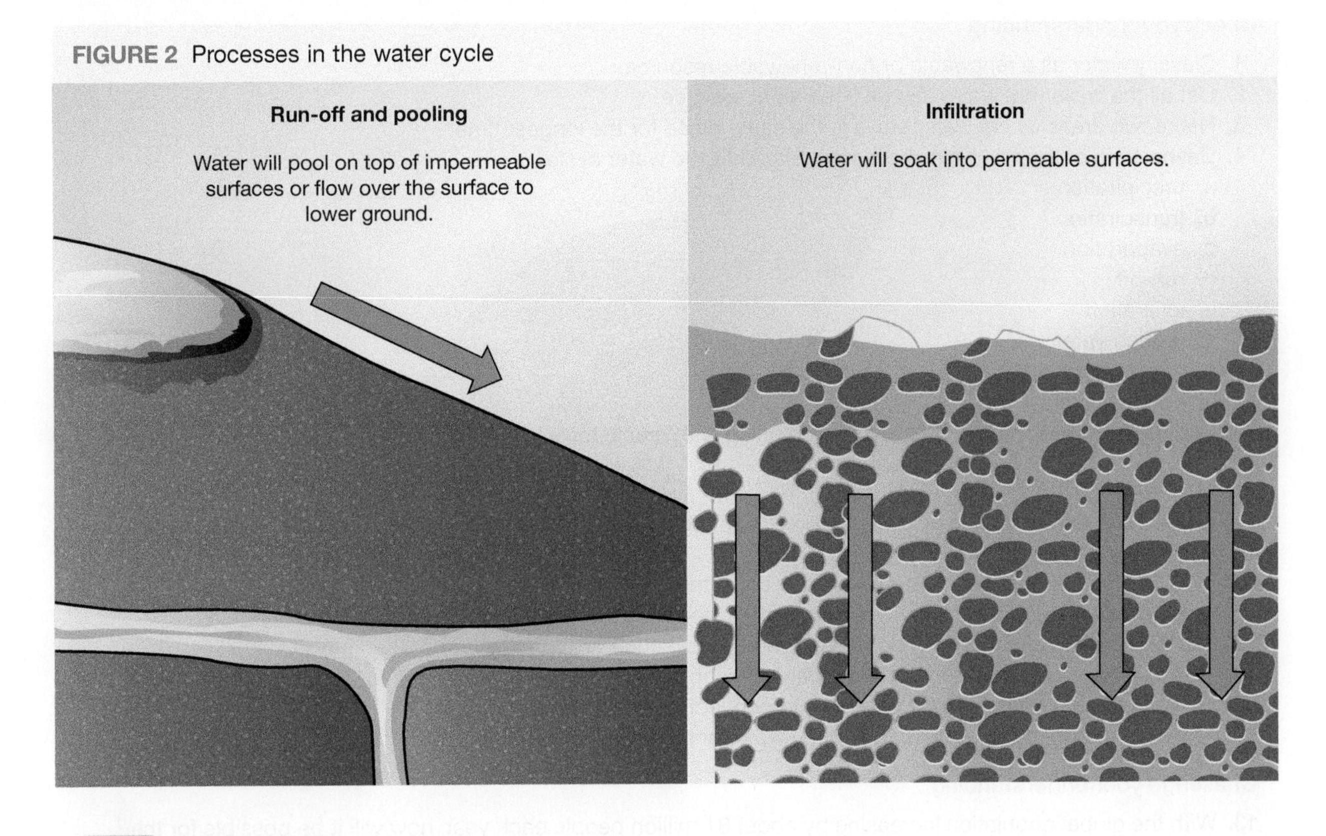

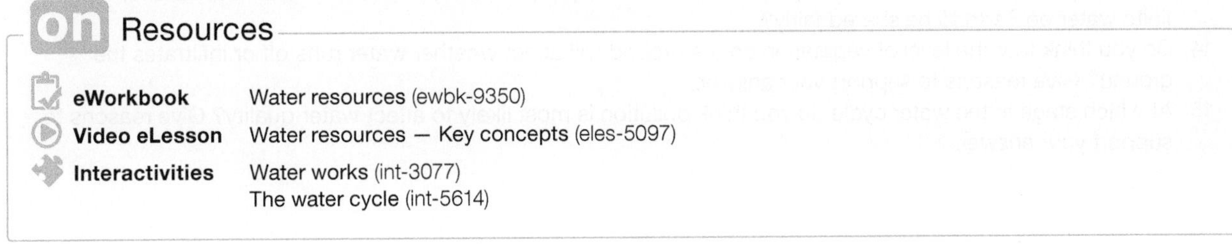

on Resources

eWorkbook Water resources (ewbk-9350)

Video eLesson Water resources — Key concepts (eles-5097)

Interactivities Water works (int-3077)
The water cycle (int-5614)

12.3 ACTIVITIES

1. Describe the water cycle.
2. List one difference and one similarity between transpiration and condensation.
3. Explain how infiltration, groundwater and run-off are related.
4. Using information about the water cycle, write and present 'The incredible journey of water'. Focus on various interconnections between water and the environment. Include diagrams and photographs, and present your story electronically, as a poster, a written piece, a drama piece or a song.

12.3 EXERCISE

Learning pathways

■ LEVEL 1	■ LEVEL 2	■ LEVEL 3
1, 2, 7, 8, 15	3, 4, 9, 12, 14	5, 6, 10, 11, 13

Check your understanding

1. Classify water as a renewable or non-renewable resource.
2. List all the ways that water can be used as a resource.
3. Name two areas where water stays in the same place for the longest time.
4. Define each of the following processes related to the water cycle.
   a. precipitation
   b. transpiration
   c. evaporation
   d. run-off
   e. infiltration
   f. groundwater
   g. evapotranspiration
   h. condensation
5. Outline the stages of the water cycle, beginning with precipitation.

Apply your understanding

6. How does the water cycle prove that we are using the same water that the dinosaurs used — in other words, that water is finite (limited)?
7. Examine **FIGURE 1**. At which stage is water collected for human use?
8. Analyse **FIGURE 2**. Explain why water runs off or pools on some surfaces, but infiltrates others.
9. Explain how the hydrologic cycle moves water across the Earth.
10. How is water vapour related to the process of evaporation?
11. How does the process of precipitation interconnect water in the water cycle?
12. At which stage/s of the water cycle might water enter a river system?

Challenge your understanding

13. With the global population increasing by about 81 million people each year, how will it be possible for the finite water on Earth to be shared fairly?
14. Do you think that the level of vegetation on the ground will affect whether water runs off or infiltrates the ground? Give reasons to support your answer.
15. At which stage in the water cycle do you think pollution is most likely to affect water quality? Give reasons to support your answer.

To answer questions online and to receive **immediate feedback** and **sample responses** for every question, go to your learnON title at www.jacplus.com.au.

12.4 SkillBuilder — Interpreting diagrams

online only

LEARNING INTENTION

By the end of this subtopic, you will be able to identify the features in a diagram and interpret their meaning.

The content in this subtopic is a summary of what you will find in the online resource.

12.4.1 Tell me

What are diagrams?

A diagram is a graphic representation of something. In Geography, it is often a simple way of showing the arrangement of elements in a landscape and the relationships between those elements. Diagrams also have annotations: labels that explain aspects of the illustration. A good diagram is large, clear and easy to interpret, has clear annotations, includes a precise title or caption, and states its source.

12.4.2 Show me

eles-2275

How to interpret a diagram

Step 1

Identify and carefully read the *title* of the diagram, as it helps you to understand its purpose.

FIGURE 1(a) Managing flood risks in new development areas

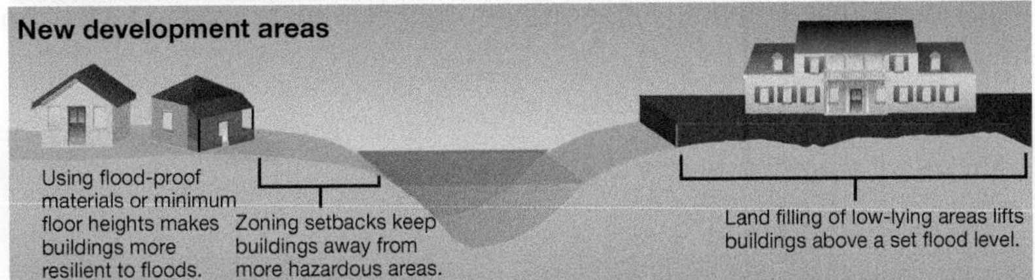

Step 2

Examine the diagram to identify all the features and read the annotations or labels carefully. Look at each *part* of the diagram. In FIGURE 1, you will notice that waterways and development areas are shown, land-filling and levees are mentioned, zoning is illustrated, and ways that developments can be changed or built are shown (e.g. flood-proof materials, raising buildings, removing houses).

Step 3

Now consider the diagram as a *whole* to look for interconnections between each of the factors. For example, if land is filled (or built up) above flood levels, there might be less need for setting houses so far from waterways.

12.4.3 Let me do it

int-3132

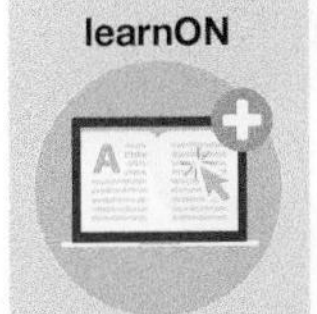

Go to learnON to access the following additional resources to help you build this skill:

- a longer explanation of this skill and its application in Geography (Tell me)
- a video showing the step-by-step process of this skill (Show me)
- an activity and interactivity for you to practise this skill (Let me do it)
- self-marking questions to help you understand and use this skill.

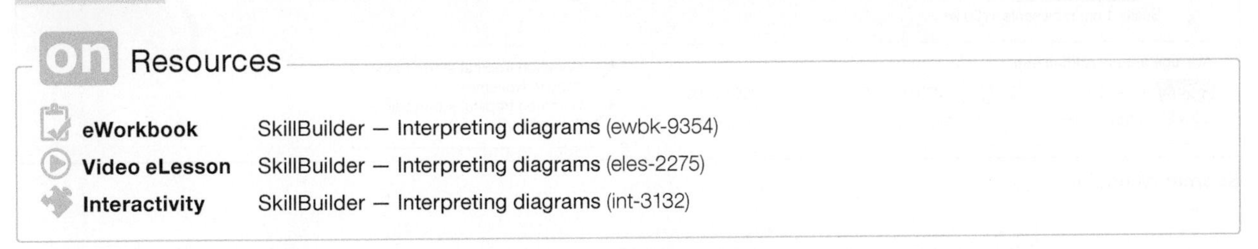

on Resources

eWorkbook SkillBuilder — Interpreting diagrams (ewbk-9354)

Video eLesson SkillBuilder — Interpreting diagrams (eles-2275)

Interactivity SkillBuilder — Interpreting diagrams (int-3132)

12.5 Water in the world

LEARNING INTENTION

By the end of this subtopic, you will be able to describe the spatial distribution of water in the world.

12.5.1 The availability of water

Water is vital to our survival and essential to most human activities. Although Earth seems blue in space, not much of the water we see is available for use. And access to the usable fresh water that can be seen is unequal across the globe.

Water covers about 75 per cent of the Earth's surface. Yet, as **FIGURE 2** shows, almost all this water (97.5 per cent) is salt water and not available for human consumption. Only 2.5 per cent of the world's water is fresh, but most of this is also unavailable for use by people. More than two-thirds (69.5 per cent) of this fresh water is locked up in glaciers, snow, ice and permafrost, and 30.1 per cent of the world's fresh water is found in groundwater. Only 0.4 per cent is left — found in rivers, lakes, wetlands and soil as well as in the bodies of animals and plants.

Global rainfall

The Earth's water is constantly moving. Rainfall patterns show which of the world regions receive more rain than others. The amount of rainfall, or precipitation, is related to the amount of water available for use by people.

FIGURE 1 The distribution of global rainfall

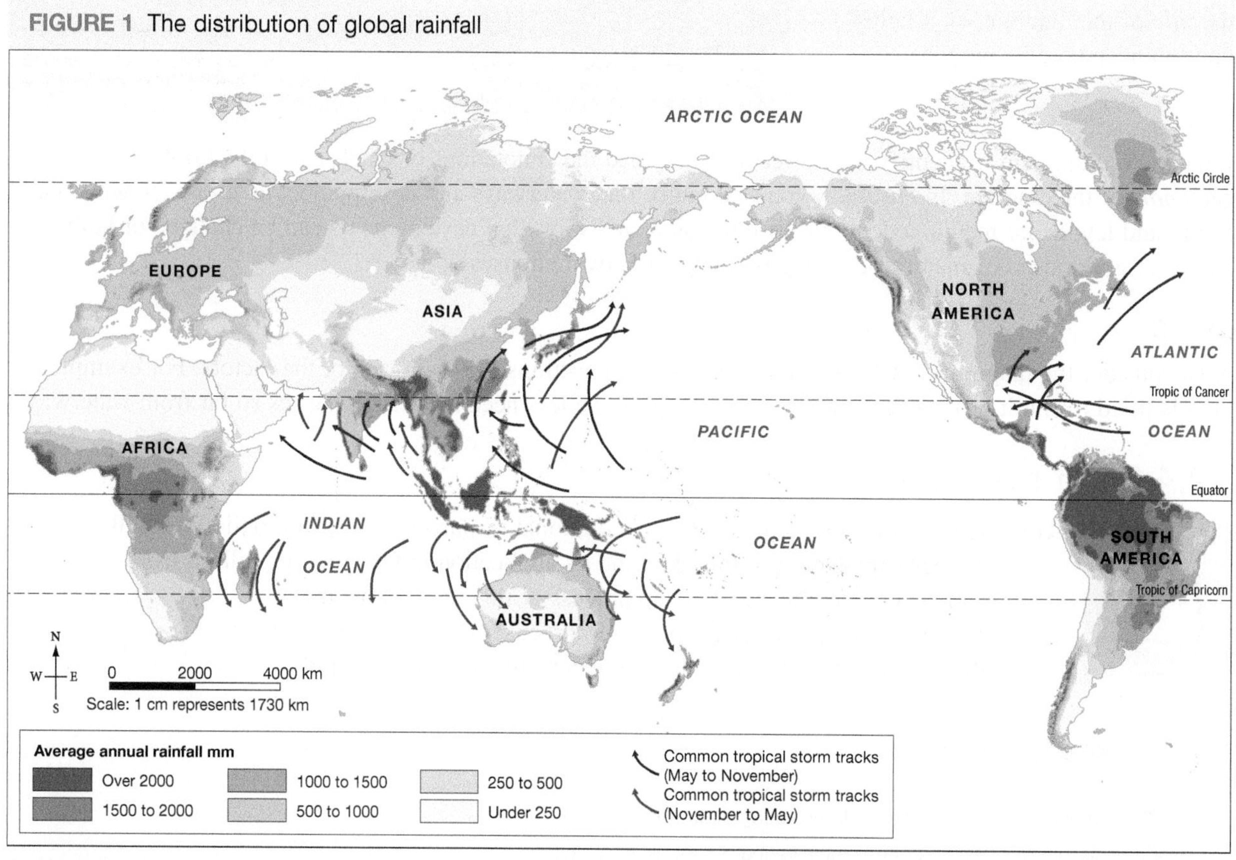

Source: WorldClim

Green and blue water

The key to our survival is being able to use the water that falls on land and into rivers and streams. Water is sometimes categorised as either **blue water** or **green water**.

Green water is the water that does not run into streams or recharge groundwater but is stored in the soil or stays on top of the soil or vegetation. This water eventually evaporates or transpires through plants. Green water is used by crops, forests, grasslands and savannas.

The amount of blue and green water available changes throughout the year, from year to year, and according to changes in the environment.

blue water the water in freshwater lakes, rivers, wetlands and aquifers

green water water that is stored in the soil or that stays on top of the soil or in vegetation

FIGURE 2 The distribution of water on Earth

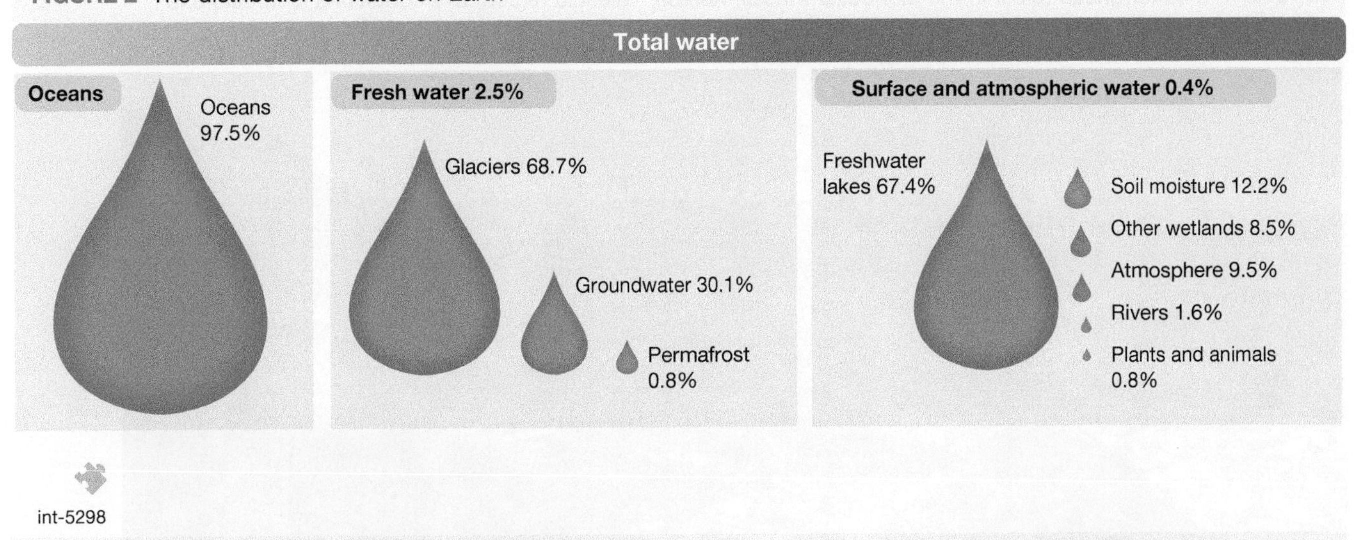

int-5298

FIGURE 3 Predicted change in annual run-off due to climate change, 2084

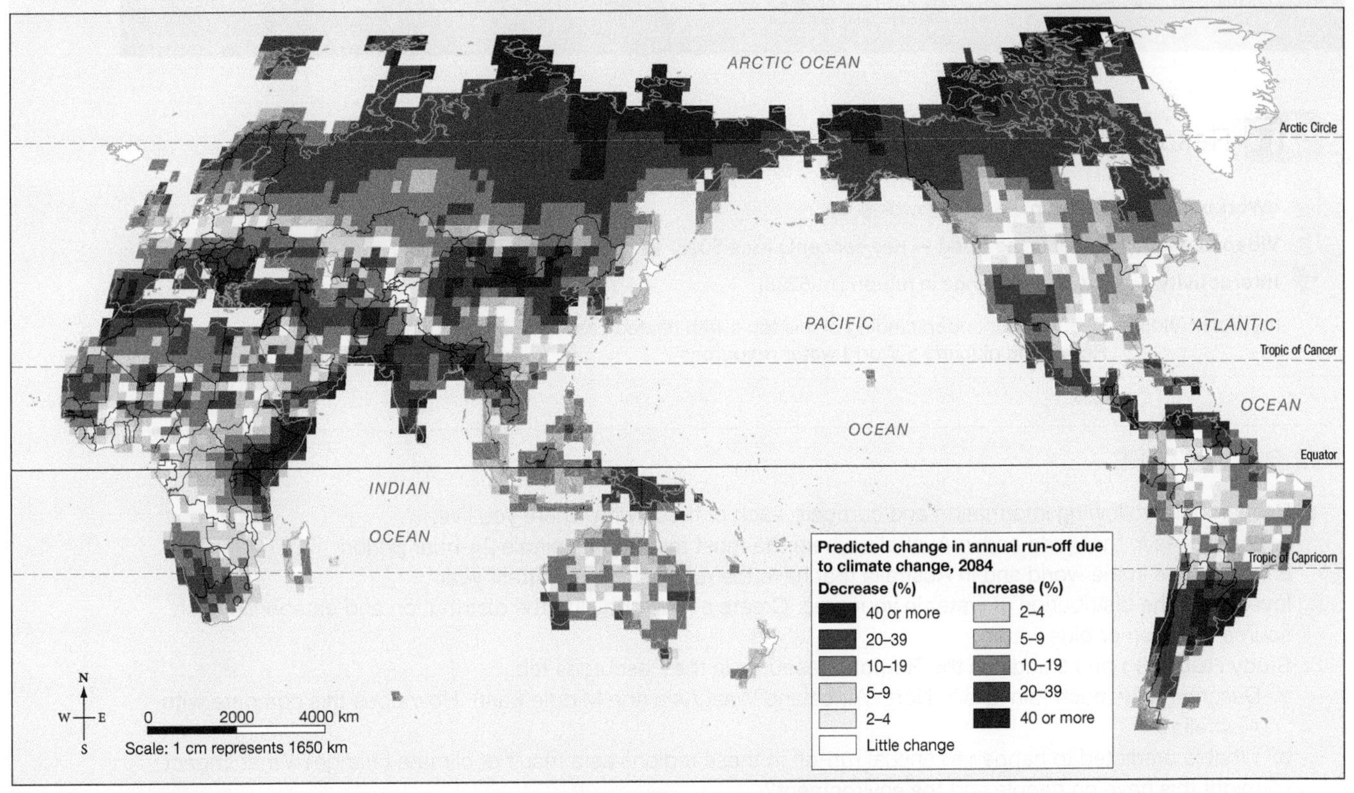

Source: Geophysical Fluid Dynamics Laboratory, National Oceanic and Atmospheric Administration

12.5.2 Climate change and impact on rainfall and run-off

The majority of scientists believe that **climate change** will have an impact on rainfall patterns and run-off. Climate models have shown that areas in the northern latitudes are likely to experience more rain, and areas closer to the equator and mid-latitudes will receive less rain. Some regions will experience droughts, while others will experience high rainfall and even flooding.

Over the past 100 years, global rainfall patterns have changed. In some areas such as North America, South America, northern Europe and northern and central Asia, rainfall has increased significantly. In other areas such as the Sahel, the Mediterranean, southern Africa and parts of Asia, rainfall has decreased.

climate change a change in the world's climate. This can be very long term or short term, and is caused by human activity.

FIGURE 4 Mediterranean forest in the Mondúver massif, Xeresa, Valencia region. Dry conditions in forests and farmlands put towns and cities at greater risk of wildfires.

on Resources

- **eWorkbook** Water in the world (ewbk-9358)
- **Video eLesson** Water in the world — Key concepts (eles-5098)
- **Interactivity** Predicted change in run-off (int-5298)
- **myWorldAtlas** Deepen your understanding of this topic with related case studies and questions > **Retreating glaciers > World water resources**

12.5 ACTIVITIES

1. Research the following information and compare each of these with where you live:
   a. the places in the world and in Australia where the most rain fell in a single 24-hour period
   b. the places in the world and in Australia that have the record for the wettest year.
2. Investigate the distribution of water in your area. Create a map showing the distribution and categorise each source as green or blue water.
3. Study **FIGURES 1** and **3** and use the **Regions** resource in the Resources tab.
   a. Describe how much rain falls in North Africa and West Asia (the Middle East). How does this compare with Australia?
   b. What is predicted to happen to annual run-off in these regions as a result of climate change? What impact might this have on people and the environment?

4. Study **FIGURE 3** (and an atlas or world map if you need to).
   a. Name three places that are predicted to receive more run-off due to climate change.
   b. Name three places that are predicted to receive less run-off due to climate change.
   c. Compare these six places with a global rainfall map. Which of the following statements is true? (Rewrite any false ones and make them true.)
      - Most places with very low rainfall have lower run-off.
      - All places with very high rainfall experience increased run-off.
      - The places with the greatest change in run-off will be northern Russia and northern Canada.

12.5 EXERCISE

Learning pathways

■ LEVEL 1	■ LEVEL 2	■ LEVEL 3
1, 6, 10, 11, 13	2, 3, 4, 12, 14	5, 7, 8, 9, 15

Check your understanding

1. What percentage of the world's water is:
   a. salty?
   b. available for use by people?
2. Outline how climate change will affect rainfall patterns.
3. How much of the Earth's fresh water is contained in glaciers?
4. Based on **FIGURE 1**, identify three continents that have places with over 2200 mm rainfall on average per year.
5. Describe the difference between green and blue water.
6. Generally, is the amount of run-off across NSW expected to increase or decrease because of climate change?
7. Describe the patterns of how run-off is expected to change across Australia due to climate change.

Apply your understanding

8. Explain one negative impact that changes to run-off might have on a farming community.
9. Write a statement about the interconnection between high rainfall and location at the equator and mid-latitudes. Name two places that do not fit this pattern.
10. List two things that might change the amount of blue and green water available.
11. Determine the type of water that farmers would rely on, if:
    a. they use irrigation
    b. they do not have access to irrigation.
12. Study **FIGURE 3**.
    a. Name three places that are predicted to receive more run-off due to climate change.
    b. Name three places that are predicted to receive less run-off due to climate change.
    c. Compare these six places with the global rainfall map in **FIGURE 1**. Which of the following statements is true? (Rewrite any false ones to make them true.)
       i. Most places with very low rainfall have lower run-off.
       ii. All places with very high rainfall experience increased run-off.
       iii. The places with the greatest change in run-off will be northern Russia and northern Canada.

Challenge your understanding

13. Predict what might happen to people and the environment in regions that:
    a. will receive more rainfall than they do now
    b. will receive less rainfall than they do now.
14. Predict what is likely to happen to annual run-off patterns in your area as a result of climate change. Explain the impact this might have on people and the environment where you live.
15. One of the impacts of climate change is that glaciers are melting more rapidly. What effects might this have on the water cycle and the availability of water resources?

To answer questions online and to receive **immediate feedback** and **sample responses** for every question, go to your learnON title at www.jacplus.com.au.

12.6 Water flows through catchments

LEARNING INTENTION

By the end of this subtopic, you will be able to identify and describe the parts of a water catchment area, and explain factors that affect the availability of water in a river system.

12.6.1 Parts of a river system

There are many parts of a river system. Each section of the system has an impact on the health and flow of water in other parts of the system.

FIGURE 1 A river system

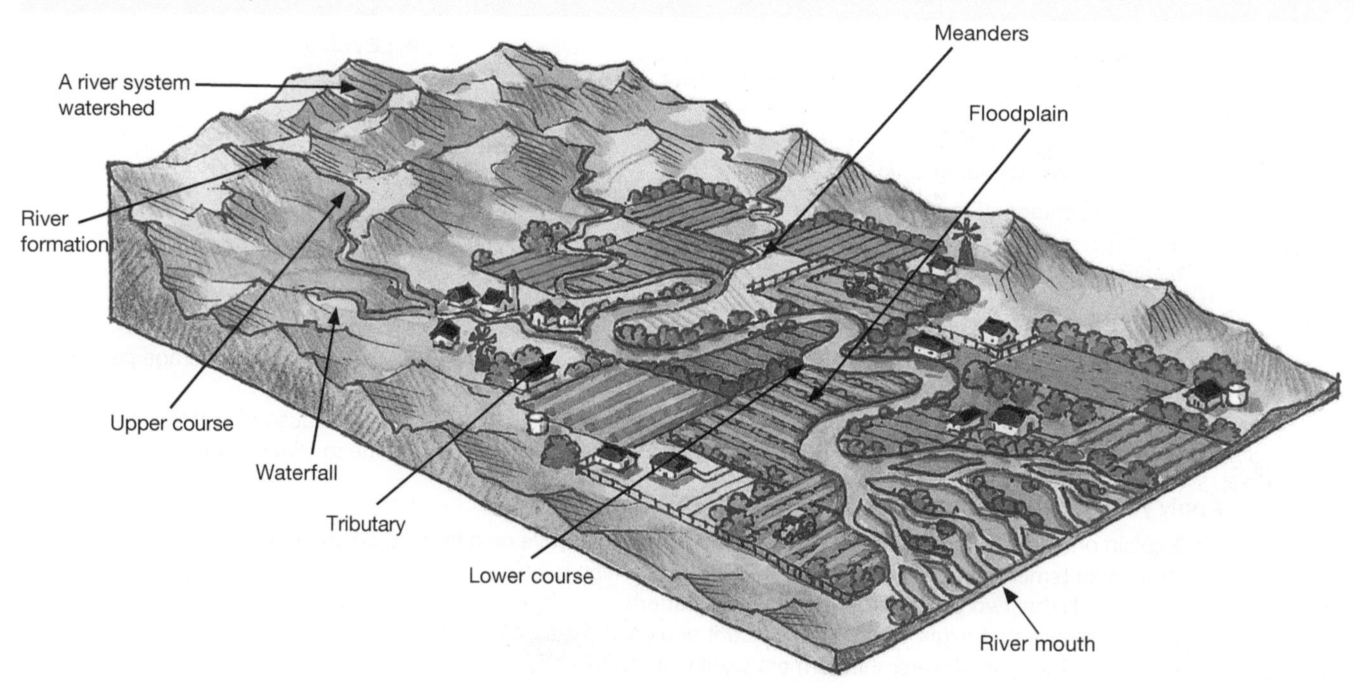

12.6.2 The upper catchment

A catchment is an area of land surrounded by hills or mountains where water collects. The mountains that surround the area separate it from adjoining catchments. These boundary mountains are called watersheds.

Water in the catchment begins in the headwaters. These are steep-sided channels found in the most elevated parts of the catchment. The headwaters often contain fast-flowing water following periods of rainfall. This fast-flowing water will often carry sediment from the upper part of the catchment into lower areas.

FIGURE 2 The watershed of a river system

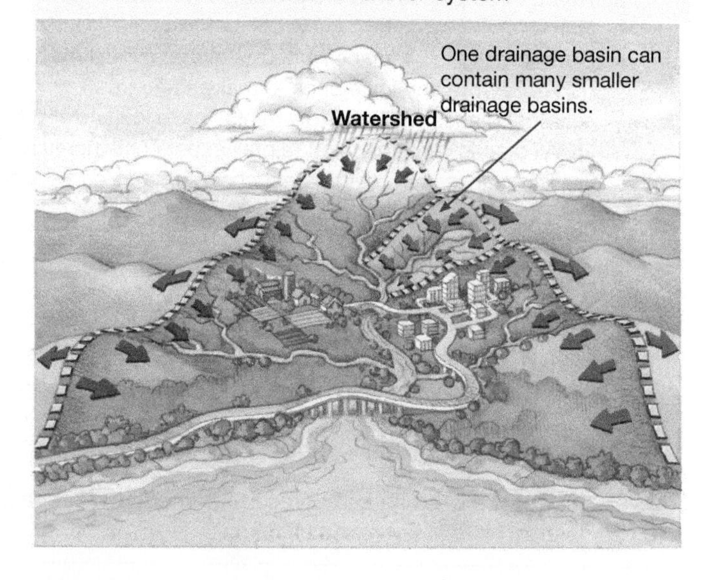

12.6.3 The mid reaches of the catchment

The water flows from the upper part of the catchment into the **floodplain** in the mid reaches of the catchment. Here it begins to slow down. As the water slows it will deposit sediment from the upper catchment. This begins to create landform features such as **meanders**, oxbows and **levees**.

12.6.4 The lower catchment

All rainwater that enters the catchment makes its way to the lowest part of the catchment, eventually flowing into rivers, lakes, creeks or the sea. Where a river meets the sea the water often slows again and deposits more sediment. This creates a **delta** where the river breaks into smaller streams.

floodplain the area of land next to the river, usually reaching to the base of the mountains surrounding the catchment. It experiences flooding during peak rainfall periods.

meander a curve in a river caused by fast-flowing water eroding the bank of one side of the river and slow-flowing water depositing sediment on the other side of the river

levee also known as an embankment, it is a built-up part of the river bank

delta a landform at the mouth of the river where a river splits into smaller streams and sediment is deposited to create an arch of land reaching into the sea

12.6.5 Factors affecting water availability

Latitude

At latitudes near the poles, the Sun's rays hit the Earth at more acute (sharp) angles. This means that the rays are more gentle, so there is less evaporation and less rainfall than close to the equator. Places near the equator tend to be hotter and more humid than places near the poles. As a result, there is generally more precipitation in low latitude areas than in higher latitude areas (see **FIGURE 3**).

FIGURE 3 Temperatures are higher at low latitudes (closer to the equator) because incoming solar radiation has a smaller area of the Earth to heat.

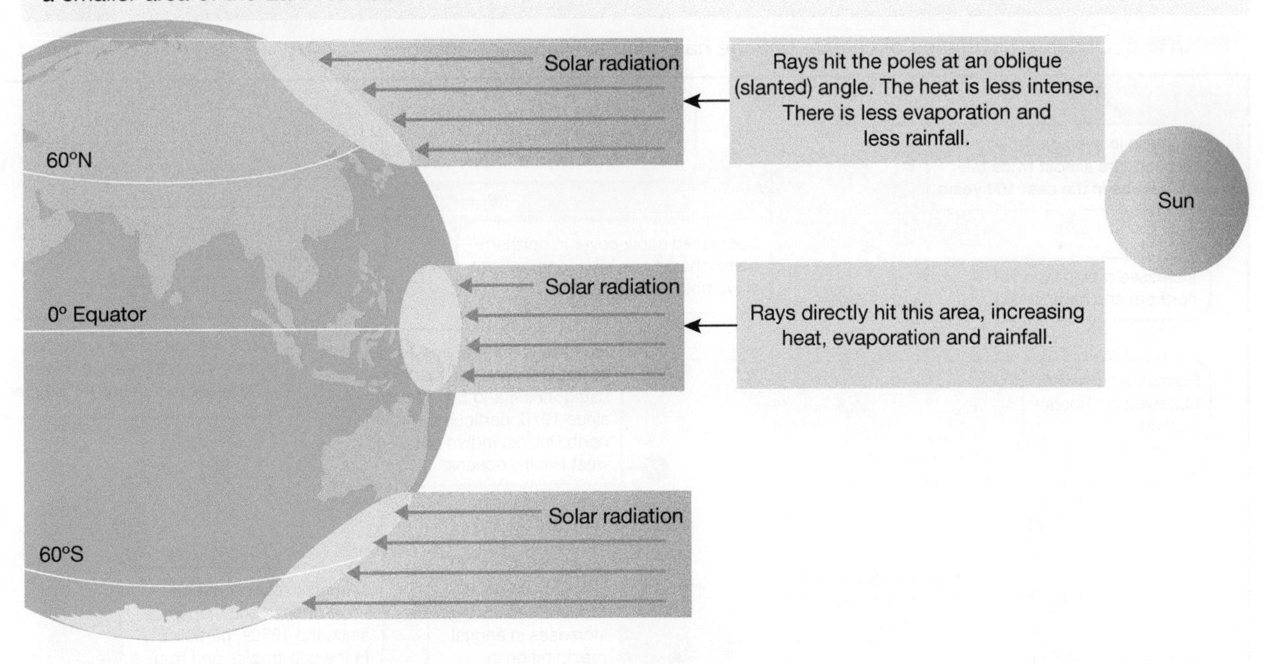

Altitude

As altitude increases, rainfall also increases. Condensation occurs as air cools in higher altitudes. The air can no longer hold moisture and precipitation occurs. This precipitation is often in the form of snow.

Ocean currents

Warm ocean currents generally increase the rate of evaporation and result in higher rainfall. Cold ocean currents reduce the amount of evaporation and as a result less rainfall occurs.

Distance from the sea

Areas close to the sea will receive more rainfall than places further from the sea. This is due to high levels of evaporation over the ocean. As the air mass moves across the land it will drop this moisture.

Geology

Some rocks and soils are more permeable than others. This means they can absorb moisture. Impermeable rocks and soils result in more run-off and higher rates of evaporation. As a result more precipitation occurs.

Topography

As discussed earlier, snow often falls on the tops of mountains. However, topography can also affect precipitation on the lower parts of mountains. On the windward side of the mountain slope, precipitation occurs. On the leeward side of the mountain less precipitation occurs. The leeward side is the side of the mountain that is protected from the wind. This is known as the rain-shadow effect.

12.6.6 The impact of climate change on water availability

Water availability is also affected by changes to our global climate. As shown in **FIGURE 4**, climate change has already had an impact on precipitation levels, the thawing of permafrost and extreme weather events.

FIGURE 4 Observed evidence of climate change has been available for decades.

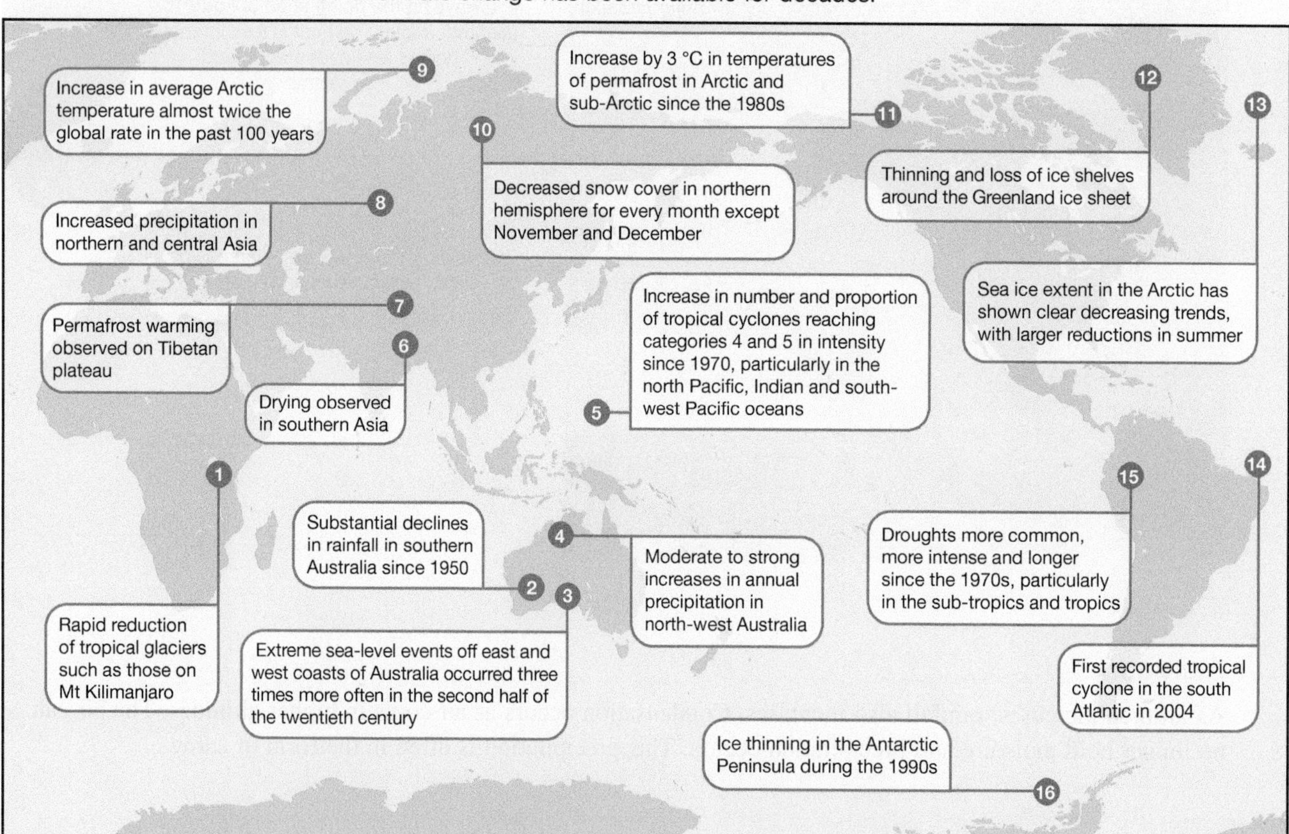

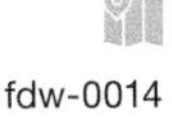
fdw-0014

FOCUS ON FIELDWORK

River study

Have you ever wondered how much water is actually flowing through a river? Or how deep the water is, without wading into the water to find out?

Using some simple measurements and calculations, you can determine the depth and velocity (speed of the water flow) of a river or creek. Learn how to measure the depth and velocity of a river using the **River study** fieldwork activity in your Resources tab.

on Resources

eWorkbook	Water flows through catchments (ewbk-9362)
Video eLesson	Water flows through catchments — Key concepts (eles-5099)
Interactivity	Water flows through catchments — Interactivity (int-8429)
Fieldwork	River study (fdw-0014)

12.6 ACTIVITY

Use the **Bureau of Meteorology** weblink in your Resources tab to investigate rainfall patterns in your local area over time. You can view the data for the weather station nearest to where you live as tables of graphs. Choose the monthly data. Using the information you now have about the factors affecting rainfall, how would you explain these rainfall patterns?

TABLE 1 Monthly and annual rainfall totals, 1950–2020

	Jan	Feb	Mar	Apr	May	Jun	Jul	Aug	Sep	Oct	Nov	Dec	Annual total
1950													
1960													
1970													
1980													
1990													
2000													
2010													
2020													

12.6 EXERCISE

Learning pathways

■ LEVEL 1	■ LEVEL 2	■ LEVEL 3
2, 3, 7, 12, 13	1, 6, 10, 11, 14	4, 5, 8, 9, 15

Check your understanding

1. Define the following terms:
   a. meander
   b. floodplain
   c. levee
   d. watershed
   e. tributary.
2. In which of the following parts of a catchment would you usually expect to find the fastest flowing waters: the river mouth, headwaters, meanders or floodplains?
3. Which part of a water catchment is at the highest altitude?
4. Describe what happens in a river delta.
5. Outline why there is generally more precipitation in areas of low latitude.
6. Categorise the following factors according to whether they will help to produce more or less precipitation:
   a. latitude
   b. altitude
   c. warm ocean currents
   d. distance from the sea
   e. rainshadows
   f. impermeable rocks and soils.
7. List three areas that have already experienced less rainfall because of climate change.

Apply your understanding

8. Explain how topography and altitude can influence precipitation.
9. Create a diagram or flow chart that demonstrates your understanding of how ocean currents have an impact on rainfall.
10. Explain the relationship between the ocean and rainfall. In your answer, use the words *evaporation*, *distance* and *currents*.
11. What is the difference between a watershed and a catchment?
12. Describe how water flows from the upper catchment to the lower catchment.

Challenge your understanding

13. Suggest what might happen to people and the environment in regions that:
    a. will receive more rainfall than they do now
    b. will receive less rainfall than they do now.
14. Complete a consequence chart to show what might happen to places near the equator if precipitation rates decrease quickly because of climate change.
15. Consider the landscape shown in **FIGURE 1**. Imagine you were planning to collect, bottle and sell water from this catchment. Based on the location of the natural and human features of this landscape, propose where you would collect the water and where you would build your bottling plant. In your plans, consider the availability and purity of the water at each point of the river system and the impact your factory might have on the surrounding area and businesses.

To answer questions online and to receive **immediate feedback** and **sample responses** for every question, go to your learnON title at www.jacplus.com.au.

12.7 Weather and climate

LEARNING INTENTION

By the end of this subtopic, you will be able to explain the impacts of climate and weather on water availability.

12.7.1 Understanding weather and climate

Weather is the day-to-day, short-term change in the atmosphere at a particular location. Extreme weather events are often described as unexpected, rare or not fitting the usual pattern experienced at a location.

Climate is the average of weather conditions that are measured over a long time. Places that share the same type of weather are said to lie in the same climatic zone. Because of the size of the Australian continent, its climate varies considerably from one region to another. Synoptic charts provide a visual representation of weather information for a place including rainfall, wind and air pressure. Examine subtopic 12.8 (SkillBuilder — Reading a weather map) to learn more about how to read a synoptic chart, or weather map.

12.7.2 Factors affecting the weather

Earth is surrounded by a band of gases called the atmosphere. The atmosphere protects our planet from the extremes of the Sun's heat and the chill of space, making conditions just right for supporting life. The atmosphere has five layers. The layer that starts at ground level and ends about 16 kilometres above Earth is called the troposphere. Our weather is the result of constant changes to the air in the troposphere. These changes sometimes cause extreme weather events.

Droughts, floods, cyclones, tornadoes, heatwaves and snowfalls — even cloudless days with gentle breezes — all begin with changes to the air in the troposphere. The five main layers in the Earth's atmosphere all differ from one another. For example, the troposphere contains most of the **water vapour** in the atmosphere. As a result, this layer has an important link to precipitation.

All weather conditions result from different combinations of three factors:

- air temperature
- air movement
- the amount of water in the air.

The Sun influences all three.

First, the Sun heats the air. It also heats the Earth's surface, which in turn heats the air even more. How hot the Earth's surface becomes depends on the season and the amount of cloud cover.

Second, the Sun causes air to move. This is because the Sun heats land surfaces more than it heats the oceans. As the warm air over land gets even warmer, it expands and rises. When hot air rises, colder air moves in to take its place.

Third, the Sun creates moisture in the air. The heat of the Sun causes water on the Earth's surface to **evaporate**, forming water vapour. As this water vapour cools, it condenses, forming clouds. It may return to Earth as rain, dew, fog, snow or hail.

At times these three factors — temperature, air movement and water vapour — can create extreme weather events. Very high air temperatures influence heatwaves; rapidly rising air plays a part in the formation of cyclones; and excess rain can create flooding.

weather short-term changes in the atmosphere at a particular location

climate average of weather conditions that are measured over a long time

water vapour water in its gaseous form, formed as a result of evaporation

evaporate to change liquid, such as water, into a vapour (gas) through heat

FIGURE 1 Australia experiences a diversity of weather.

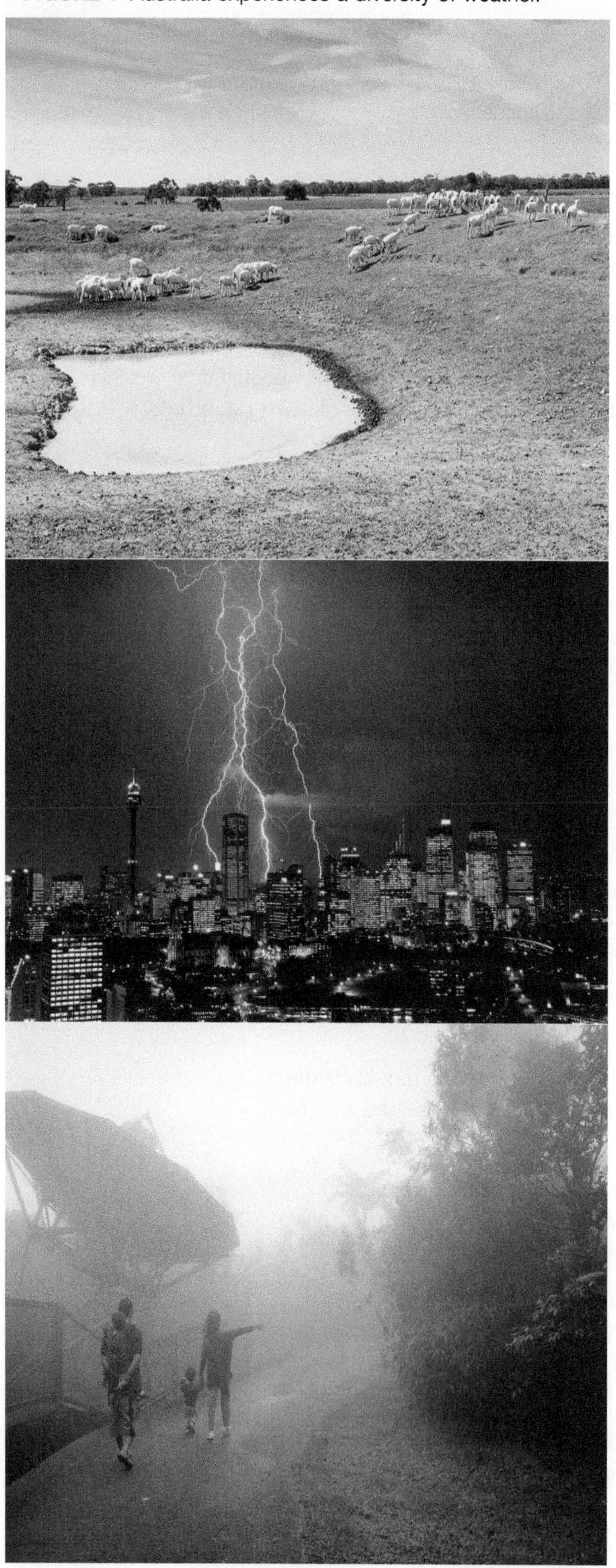

FIGURE 2 Earth's atmosphere (not to scale)

FIGURE 3 Climate zones of Australia

int-8430

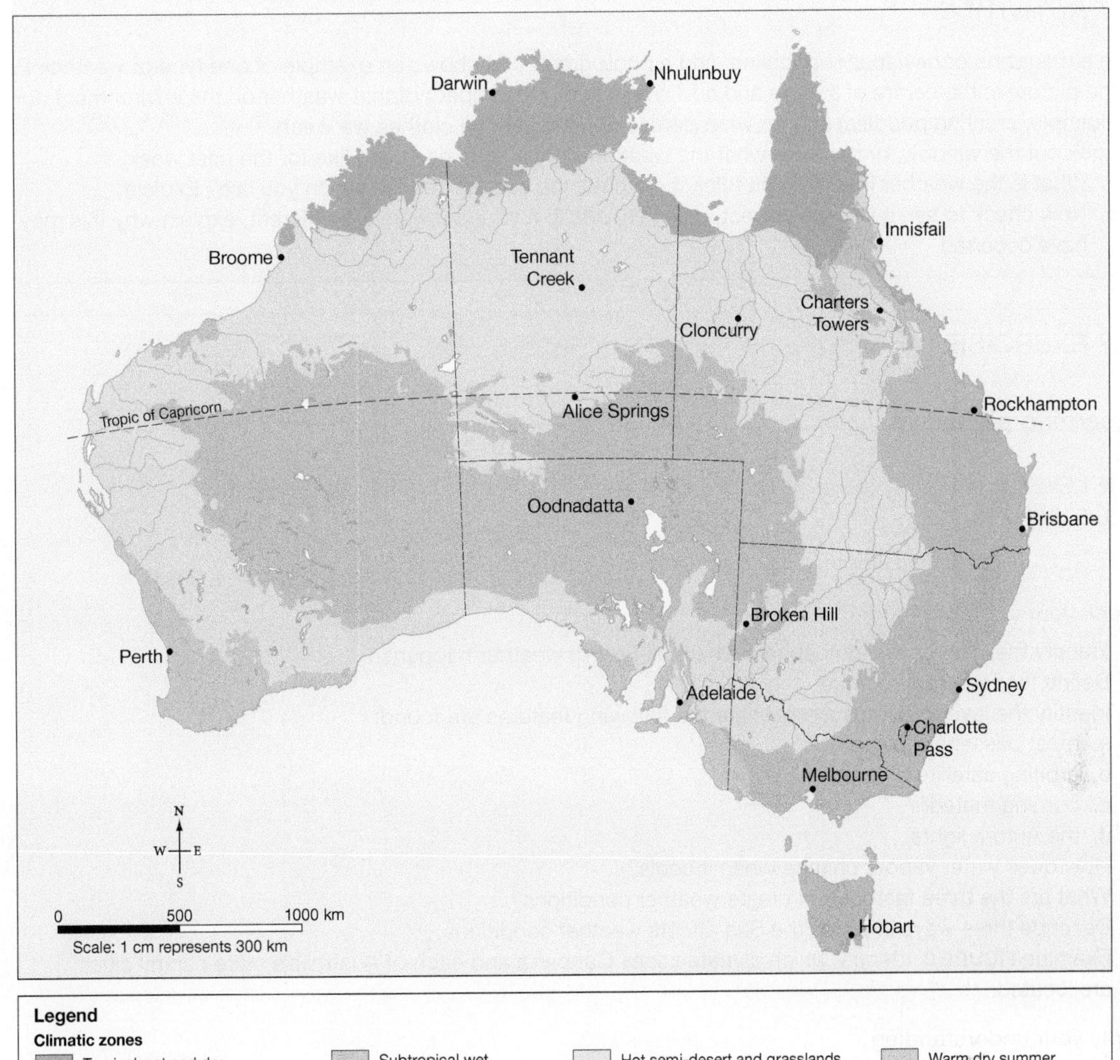

Source: Spatial Vision

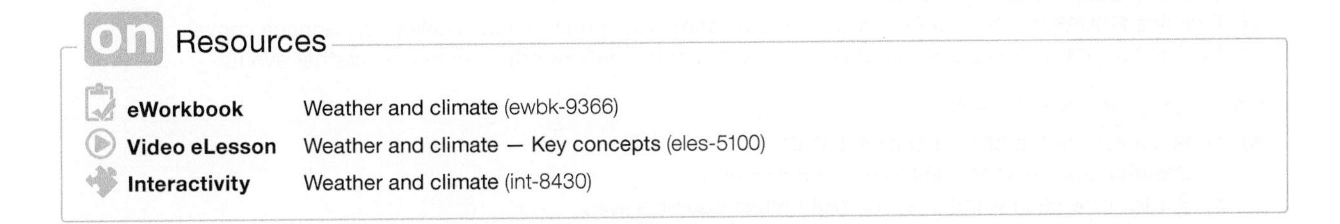

on Resources

eWorkbook Weather and climate (ewbk-9366)

Video eLesson Weather and climate — Key concepts (eles-5100)

Interactivity Weather and climate (int-8430)

12.7 ACTIVITIES

1. In a magazine or newspaper or online, find a photograph that shows an example of one type of weather. Paste the picture in the centre of a page and add labels about the impact of that weather on the environment (for example, creating puddles) and on what people do (such as the clothes we wear).
2. Look out the window, or consider what the weather in your area has been like for the past week.
   a. What is the weather like? Do you think it matches the climate zone in which you live? Explain.
   b. Now check to see if you are correct, using **FIGURE 3**. If your answers are different, explain why this may have occurred.

12.7 EXERCISE

Learning pathways

■ LEVEL 1	■ LEVEL 2	■ LEVEL 3
1, 6, 9, 11, 13	2, 3, 5, 10, 14	4, 7, 8, 12, 15

Check your understanding

1. Identify the layer of the atmosphere where all Earth's weather happens.
2. Define the term *troposphere*.
3. Identify the layer of atmosphere where the following features are found:
   a. most passenger planes
   b. orbiting satellites
   c. burning meteors
   d. the Aurora lights.
4. How does water vapour change when it cools?
5. What are the three factors that create weather conditions?
6. Describe three ways in which the Sun affects weather conditions.
7. Examine **FIGURE 3**. Identify which climate zones Canberra and each of Australia's state capital cities are located.

Apply your understanding

8. In terms of weather conditions, explain the interconnection between air temperature and air movement.
9. Explain the difference between weather and climate.
10. Draw three diagrams to explain the factors that influence the following weather conditions:
    a. air temperature
    b. air movement
    c. the amount of water in the air.
11. Examine **FIGURE 3**. Which of Australia's climate zones would be the most likely to experience snow?
12. Explain how the three factors that produce weather combine to produce extreme weather events.

Challenge your understanding

13. Look carefully at the photographs in **FIGURE 1**.
    a. Describe the weather event in each photograph.
    b. Predict how each weather event might affect people's lives.
14. Look carefully at the map of Australia's climate zones in **FIGURE 3**. Predict which two settlements or places might be at risk of flood. Make sure you explain why you chose them.
15. Predict how climate change will affect the weather where you live.

To answer questions online and to receive **immediate feedback** and **sample responses** for every question, go to your learnON title at www.jacplus.com.au.

12.8 SkillBuilder — Reading a weather map

online only

LEARNING INTENTION

By the end of this subtopic, you will be able to interpret weather maps to read and predict the weather.

The content in this subtopic is a summary of what you will find in the online resource.

12.8.1 Tell me

What are weather maps?

Weather maps (or synoptic charts) appear every day in newspapers and on the news. They may look complex but, once you understand the symbols, you will find these maps easy to use.

12.8.2 Show me

eles-2749

How to interpret a weather map

Step 1

To identify features on a synoptic chart, use this key to interpret the symbols.

TABLE 1 Symbols on synoptic charts for the southern hemisphere

Symbol	Feature	What it means	Impact on the weather
1024 1020 1018	Isobars	Joins places of equal air pressure	The closer together these lines, the stronger the wind.
H 1020	High pressure area	Sinking air	Generally fine weather. Winds rotate around these areas anticlockwise.
L 1008	Low pressure area	Rising air	Generally cooler weather, rain. Winds rotate around these areas clockwise.
T 1004 1008	Tropical cyclone	Rapidly rising air	Strong winds, torrential rain
	Cold front	The 'line' along which an approaching mass of cold air meets warmer air	Fall in temperature, rain. Front moves in direction of arrowheads.
	Warm front	The 'line' along which an approaching mass of warm air meets colder air	Temperature rise, sometimes light rain. Uncommon in Australia.
//////	Rain	Rain in the past 24 hours	Usually associated with low pressure areas and fronts
------------	Trough	A dip in isobars	An area associated with unsettled weather and precipitation

Step 2

Examine the weather map (**FIGURE 2**) to identify its features, based on the key in **TABLE 1**.

- High pressure systems are areas of pressure above 1013 hectopascals. Low pressure systems are areas of pressure below 1013 hectopascals. Pressure systems generally move from west to east as they cross Australia and move around the world.
- Examine any air masses and fronts.
- Precipitation includes snow, hail and dew but the most common form of precipitation is rainfall.
- The closer together the isobars are, the stronger the winds. Some synoptic charts also show wind speed.

Step 3

Look for patterns or changes in the features and consider how they will impact the weather in the places that they pass.

FIGURE 2 Weather map for 10 November

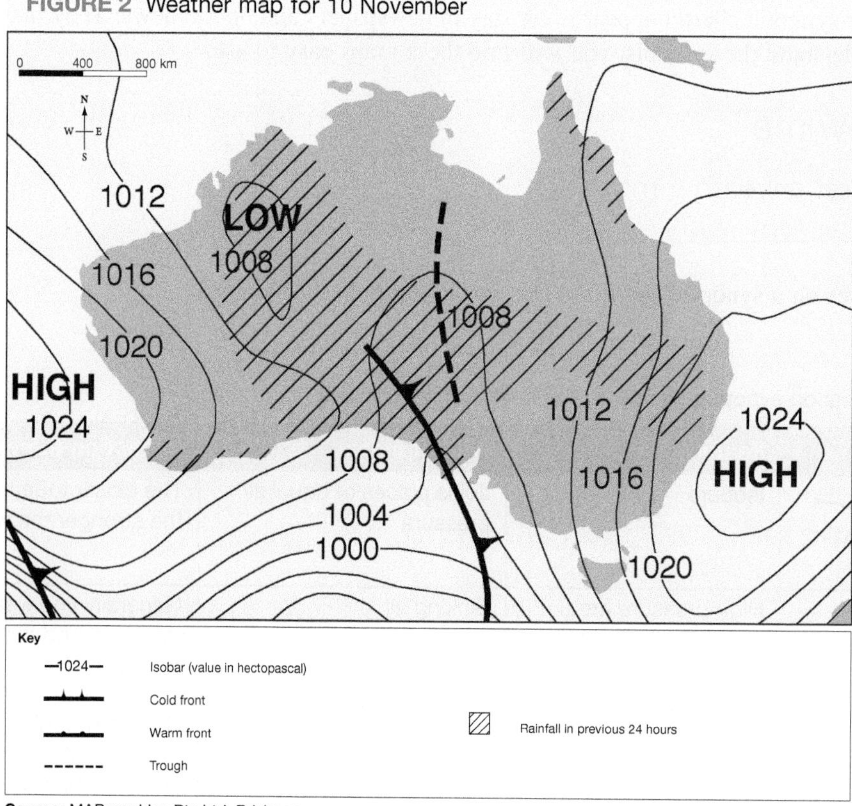

Source: MAPgraphics Pty Ltd, Brisbane

12.8.3 Let me do it

int-6777

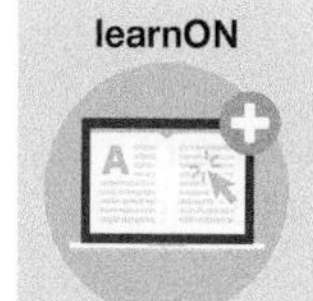

Go to learnON to access the following additional resources to help you build this skill:

- a longer explanation of this skill and its application in Geography (Tell me)
- a video showing the step-by-step process of this skill (Show me)
- an activity and interactivity for you to practise this skill (Let me do it)
- self-marking questions to help you understand and use this skill.

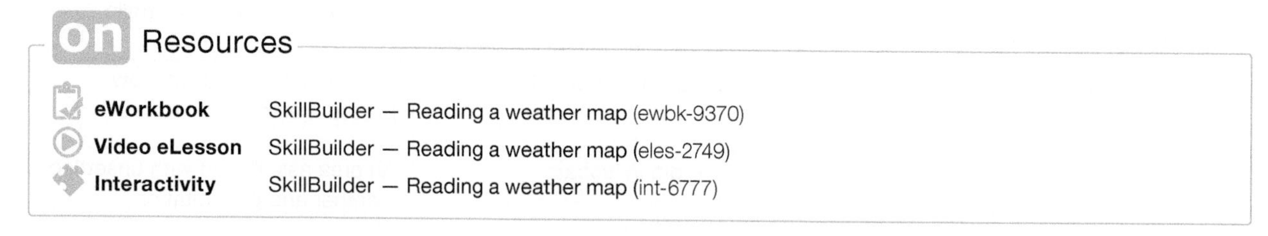

on Resources

eWorkbook SkillBuilder — Reading a weather map (ewbk-9370)

Video eLesson SkillBuilder — Reading a weather map (eles-2749)

Interactivity SkillBuilder — Reading a weather map (int-6777)

12.9 How wind blows

LEARNING INTENTION

By the end of the subtopic, you will be able to explain why wind occurs.

12.9.1 The weight of air

The air around us has weight. The weight of the air above us pushes down, creating pressure. If we could tie a **barometer** to a hot air balloon, we would see the pressure readings fall as the balloon rose in the atmosphere. This is because there is less air higher in the atmosphere. You may have read about mountain climbers and athletes having difficulty breathing when they are at high altitudes.

Air pressure

When a person blows up a balloon, the pressure inside the balloon is higher than the surrounding air. When the neck of the balloon is released, the air rushes out of the balloon. This is the same as wind. If we did not have wind, temperatures would continue to rise over the equator and decrease at the poles as there would be nothing to move the air.

Meteorologists are able to measure air pressure using a unit of measure called a millibar. The average weight of air is about 1013 millibars. Measurements higher than this indicate areas of high pressure; here, the air is sinking. Measurements lower than 1013 millibars indicate areas of low pressure; here, the air is rising. Wind is caused by air moving from areas of high pressure to areas of low pressure.

12.9.2 Variations in air pressure

Variations in air pressure are the result of the heating effect of the Sun and the rotation of the Earth.

Effects of the Sun

The warming influence of the Sun varies with the time of day (see **FIGURE 1**) and latitude (distance from the equator). Temperatures are higher in the middle of the day, and higher at the equator than at the poles. Warm air is also less dense than cold air. This is because as the air heats, it expands, causing it to rise. Air pressure over the equator is less than at the poles. As the warm air over the equator rises and expands, cooler air from near the poles rushes in to replace it. As a result, air is circulated around the Earth, and this movement of air is what we call wind.

barometer an instrument used to measure air pressure

meteorologist a scientist who studies the weather

int-7790

FIGURE 1 The effect of Sun on a sea breeze

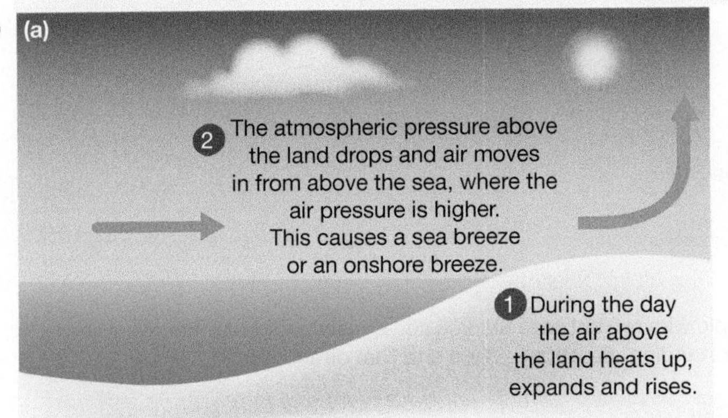

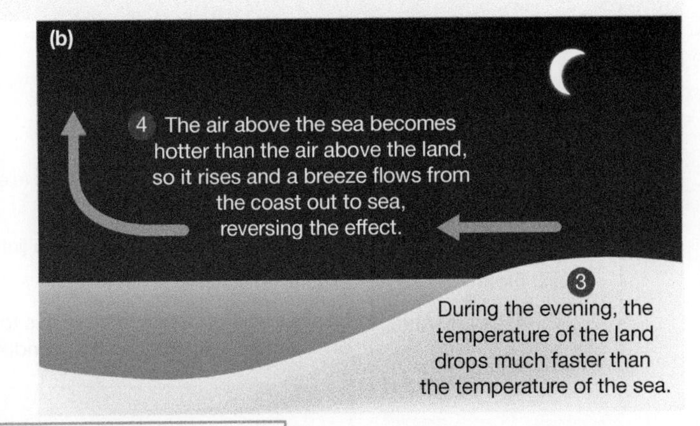

Land heats up and cools down more quickly than the sea.

Effect of the Earth's rotation

Wind is caused by air moving from areas of high pressure to areas of low pressure. Its direction is influenced by the rotation of the Earth.

The rotation of the Earth on its axis causes the air above the surface of the Earth to be deflected (change direction) rather than to travel in a straight line. This causes the wind to circle around high and low pressure systems.

The direction in which winds circle depends on whether you are in the northern or southern hemisphere. As the air moves from an area of high pressure to an area of low pressure, winds circle in the opposite direction in each hemisphere: in an area of high pressure, the winds circle in an anticlockwise direction in the southern hemisphere and a clockwise direction in the northern hemisphere. This deflection of winds is known as the Coriolis effect (see **FIGURE 2**).

FIGURE 2 The movement of air in the southern hemisphere

1004
1008
1012
H
1000
996
992
998
L
984

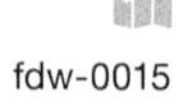
fdw-0015

FOCUS ON FIELDWORK

Clouds

The clouds — especially the kinds and coverage of clouds in the sky — can tell us a lot more about the weather than whether it might be about to rain. Different types of clouds, how they move and how many are in the sky can give us a lot of information about the weather.

Learn how to record cloud coverage and recognise different types of clouds using the **Clouds** fieldwork activity in your Resources tab.

Resources

eWorkbook	How wind blows (ewbk-9374)
Video eLesson	How wind blows — Key concepts (eles-5101)
Interactivities	Highs and lows (int-3086) The effect of Sun on a sea breeze (int-7790)
Fieldwork	Clouds (fdw-0015)
myWorldAtlas	Deepen your understanding of this topic with related case studies and questions > **Australia: weather and climate** > **Building skills** > **Understanding direction** > **Wind and Sun direction**

12.9 ACTIVITIES

1. Working with a partner, try the following activity to illustrate the influence of the Earth's rotation on wind.
   Step 1 Place a pin through the centre of a piece of paper and attach it to a piece of cardboard. Make sure the paper can move freely on the pin.
   Step 2 Have your partner rotate the piece of paper on the pin. At the same time, you should attempt to draw a straight line on the page.
   Step 3 Record your findings.
   Step 4 Compare and discuss your results with the class.
2. Over the course of the coming week, collect weather maps from the daily newspaper and find your location.
   a. Assess whether the weather is being influenced by a high or low pressure system.
   b. Determine whether the wind will be moving in a clockwise or anticlockwise direction. Justify your answer.
3. How easy is it to predict the weather? People often complain that the forecasters don't always get it right.
   a. Using the weather maps you collected earlier, and the observations you made, write a weather forecast for tomorrow. In your forecast, make reference to both wind speed and direction.
   b. Collect tomorrow's weather map and make observations similar to those you made in question 2. Record your findings.
   c. Compare your findings with what you have written for this activity. How accurate were your predictions? Suggest factors that might influent the accuracy of such predictions and the changes that you observed.

12.9 EXERCISE

Learning pathways

■ LEVEL 1	■ LEVEL 2	■ LEVEL 3
3, 5, 6, 11, 13	2, 4, 9, 10, 14	1, 7, 8, 12, 15

Check your understanding

1. What is wind?
2. Identify the two factors that influence wind. Would either of these factors influence the strength of the wind? Outline why.
3. How does the air above the surface of the Earth move: in straight lines around the Earth, in straight lines from pole to pole, in circular patterns around the Earth or not at all?
4. Apart from air pressure, name one other natural factor that influences how air moves around the Earth.
5. Outline what a meteorologist does.
6. What does a barometer measure?
7. Describe how global temperature patterns would change if there was no wind on Earth.

Apply your understanding

8. Explain why you are not affected by the pressure of the atmosphere.
9. Explain the role that the Sun plays in causing wind.
10. What change does a difference in air pressure cause?
11. If you were sitting on a beach, would you be more likely to feel a cool breeze coming from the sea during the day or at night?
12. Explain how the Coriolis effect works.

Challenge your understanding

13. Imagine you are a mountain climber, making your first trip to the top of Mount Everest. Predict how the air pressure might change as you climb from sea level to the top of the mountain. Suggest one way this change might affect you or your expedition.
14. Do you think climate change will affect the way the wind travels around the Earth? Give reasons for your view.
15. Suggest why winds circulate in different ways in the southern and northern hemispheres.

To answer questions online and to receive **immediate feedback** and **sample responses** for every question, go to your learnON title at www.jacplus.com.au.

12.10 Wind strength

LEARNING INTENTION

By the end of the subtopic, you will be able to explain and demonstrate how wind is measured.

12.10.1 How wind is shown on a weather map

Differences in air pressure lead to variations in the strength of the wind. You can work out the strength of the wind by looking at weather maps, the behaviour of objects or by using instruments designed to measure the strength of the wind. Winds are named according to their source. This means that a northerly wind is coming from the north and a southerly from the south.

If you study the **isobars** on a weather map you will notice that they are not evenly spaced. Look closely at the map in **FIGURE 1**. The wind is strongest in the southern regions of this map, where the isobars are close together, and gentler in the northern

isobar line on a map that joins places with the same air pressure

FIGURE 1 A typical weather map

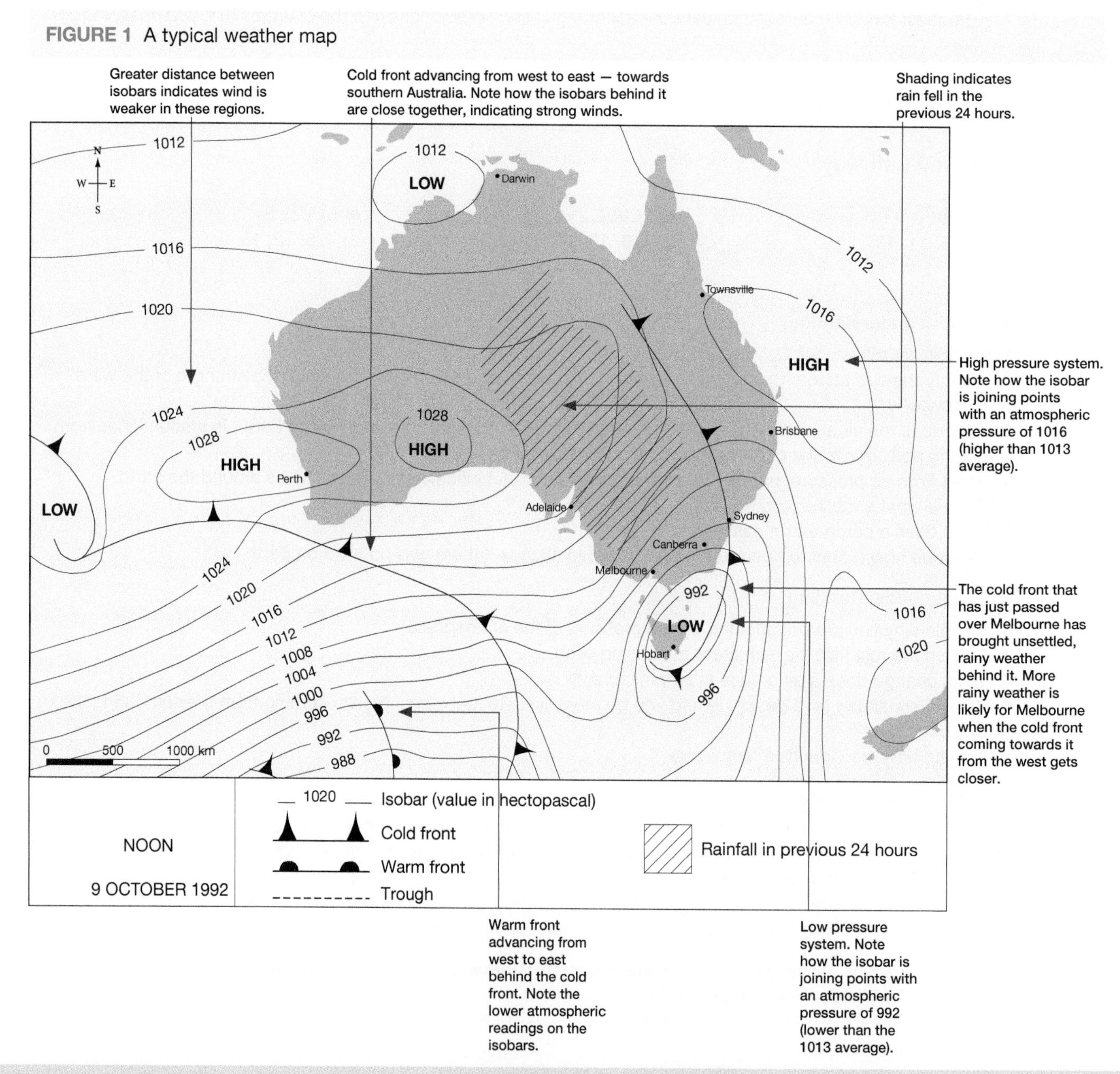

parts of the map, where the spacing between them is much greater. The symbols shown in **FIGURE 2** are also commonly used on weather maps to give a more accurate representation of wind speed and to provide information on the direction of the wind.

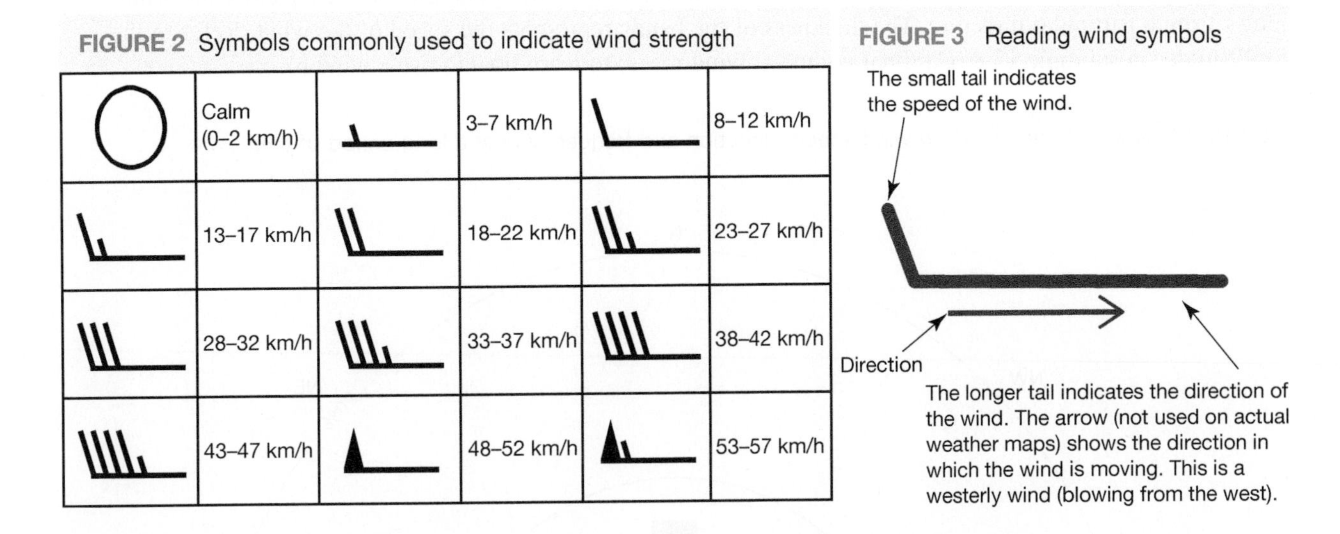

FIGURE 2 Symbols commonly used to indicate wind strength

	Calm (0–2 km/h)		3–7 km/h		8–12 km/h
	13–17 km/h		18–22 km/h		23–27 km/h
	28–32 km/h		33–37 km/h		38–42 km/h
	43–47 km/h		48–52 km/h		53–57 km/h

FIGURE 3 Reading wind symbols

The Beaufort scale (see **FIGURE 4**) relates wind speed to the observable movement of objects.

FIGURE 4 The Beaufort scale is based on the observable impact of winds.

Beaufort number	Description	Wind speed	Observable impact
0	Calm	< 2 km/h	Smoke rises vertically
1	Light air	2–5 km/h	Smoke drift shows wind direction, wind vanes don't move
2	Light breeze	6–12 km/h	Wind felt on face, wind vanes move
3	Gentle breeze	13–20 km/h	Leaves and small twigs in motion, hair disturbed, clothing flaps
4	Moderate breeze	21–30 km/h	Dust and loose paper move, small branches move
5	Fresh breeze	31–40 km/h	Small trees with leaves begin to sway, wind force felt on body
6	Strong breeze	41–51 km/h	Large branches move, umbrellas difficult to use, difficult to walk steadily
7	Moderate gale	52–63 km/h	Whole trees in motion, inconvenience felt when walking
8	Gale	64–77 km/h	Twigs broken off trees, difficult to walk
9	Strong gale	78–86 km/h	People blown over, slight structural damage, including tiles blown off houses
10	Whole gale	88–101 km/h	Trees uprooted, considerable structural damage
11	Storm	102–120 km/h	Widespread damage
12	Hurricane/cyclone	> 120 km/h	Widespread devastation

12.10.2 Using a wind rose

A wind rose such as that shown in **FIGURE 5** uses data collected over long periods of time to visually represent wind information. The spokes represent wind direction; the longer the spoke the more frequently the wind blows from a particular direction. The thickness of the bands represents the speed of the wind. Refer to the SkillBuilder in subtopic 15.8 (Cardinal points — wind roses) to learn how to use a wind rose.

FIGURE 5 A wind rose can show wind speed, direction and frequency over a long period of time.

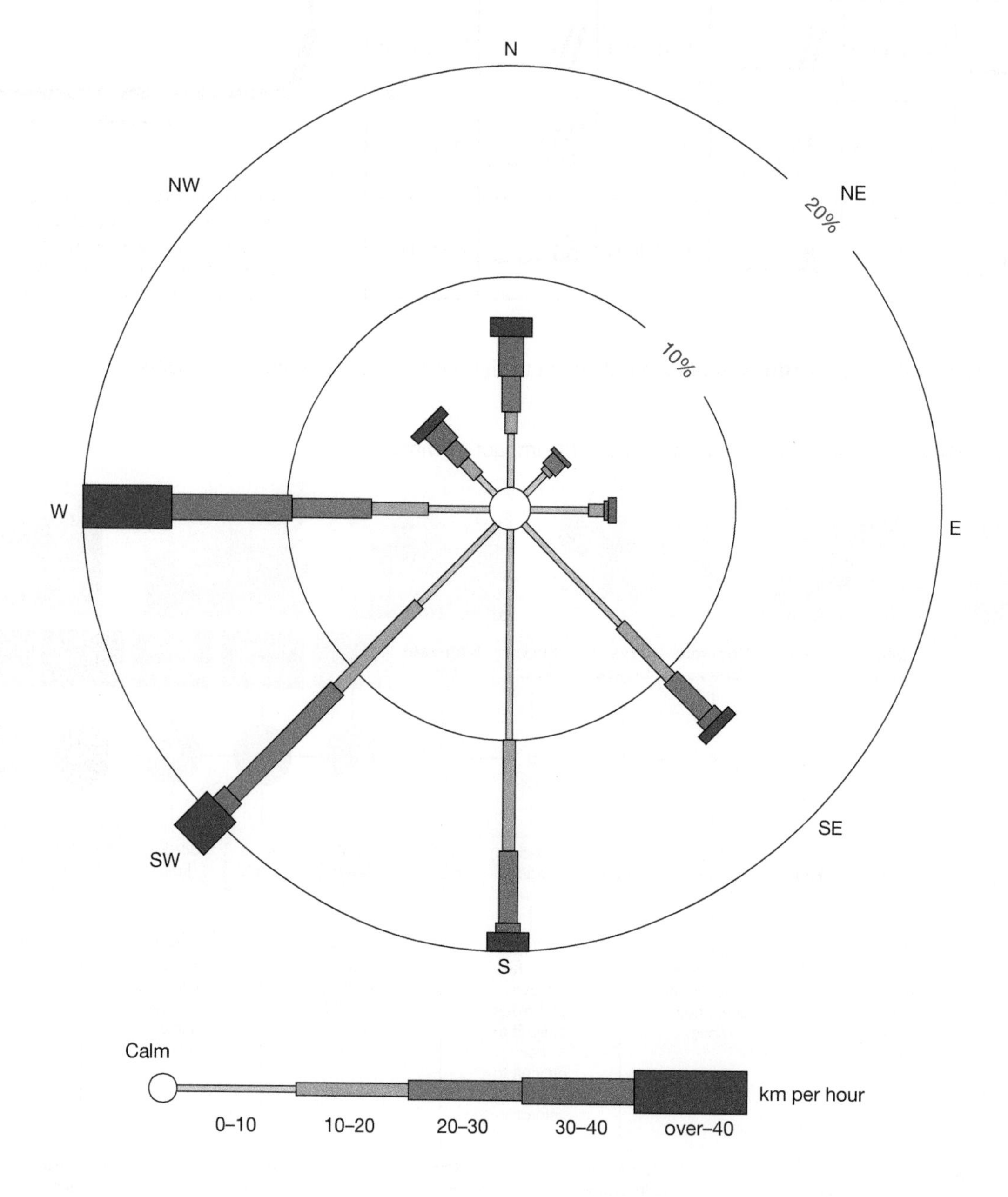

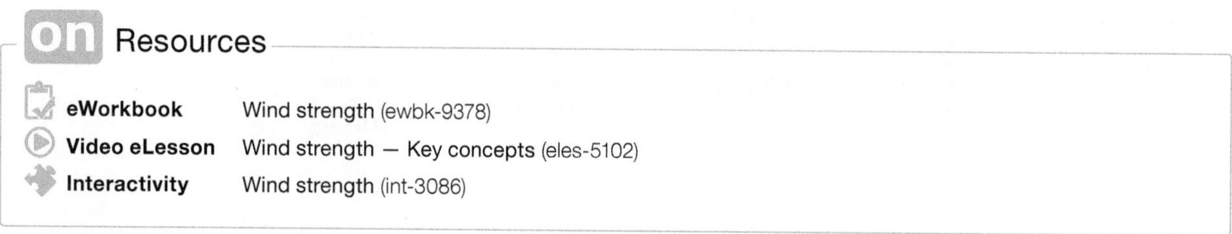

12.10 ACTIVITIES

1. Collect a weather map from a newspaper or online. Predict what the wind conditions will be like in each of the Australian capital cities over the next two days.
   Use the **Wind rose maps** weblink in the Resources tab to compare your predictions with what is shown on these maps. Make sure you select the current month. Note any similarities and differences. Why might differences occur?
2. Devise your own symbols similar to those shown in **FIGURE 2**. Obtain a current weather map from a newspaper or online. Paste it onto a sheet of paper and annotate your map with your symbols for describing wind speed. Swap maps with a partner and further annotate each other's maps with written descriptions of the symbols.

12.10 EXERCISE

Learning pathways

■ LEVEL 1	■ LEVEL 2	■ LEVEL 3
2, 4, 10, 12, 13	1, 3, 7, 9, 15	5, 6, 8, 11, 14

Check your understanding

1. What does the Beaufort scale measure?
2. What does a wind rose show?
3. What symbol would be used on a weather map to show the following:
   a. calm weather
   b. 38-42 km/h westerly wind
   c. a cold front
   d. a warm front
   e. a trough
   f. a pressure system
4. If the isobars on a weather map are very close together, what does this indicate?
5. What change does a difference in air pressure cause?
6. Describe two methods you could use to determine wind speed.

Apply your understanding

7. Explain how differences in air pressure can cause changes to the weather.
8. In your opinion, which method of determining wind speed gives the most useful information about wind speed? Give reasons for your answer.
9. Using **FIGURE 1** describe the wind speeds and directions in Western Australia and along the east coast of Australia on that day.
10. At what level on the Beaufort scale do you think it would be dangerous to go outside?
11. Would you suggest to someone living in Adelaide on the day shown in **FIGURE 1** that they take an umbrella or raincoat if they go outside? Justify your advice.
12. Discuss how understanding weather maps might be helpful for you if you rely on public transport to get to school.

Challenge your understanding

13. What do weather maps not show us about the weather? Suggest one thing you might not learn about the day's weather by looking at a weather map.
14. Suggest how examining weather maps over time might be an important tool for meteorologists to predict the paths of future storms.
15. Based on the Beaufort scale, create a safety plan for your school that gives advice about when specific outdoor activities become unsafe and when equipment and outdoor furniture might need to be secured or brought inside.

To answer questions online and to receive **immediate feedback** and **sample responses** for every question, go to your learnON title at www.jacplus.com.au.

12.11 Groundwater resources

online only

LEARNING INTENTION

By the end of this subtopic, you will be able to identify the locations of major groundwater basins in Australia and explain why the quality of and access to groundwater can vary.

The content in this subtopic is a summary of what you will find in the online resource.

An important part of the water cycle, groundwater is the water that is found under the Earth's surface. Many settlements — especially those in arid and semi-arid areas — rely on groundwater for their water supply.

When rain falls to the ground, some flows over the surface into waterways and some seeps into the ground. Any seeping water moves down through soil and rocks. Groundwater is water held within water-bearing rocks, or aquifers, in the ground. These work like sponges, holding water in the tiny holes between the rock particles.

FIGURE 1 An artesian aquifer

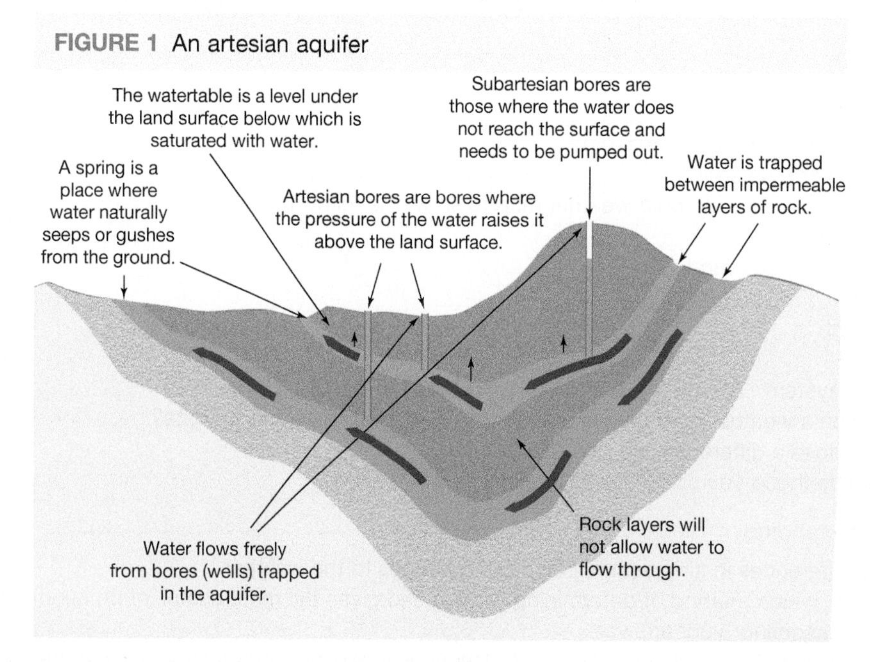

To learn more about groundwater resources, go to your learnON resources at www.jacPLUS.com.au.

Contents

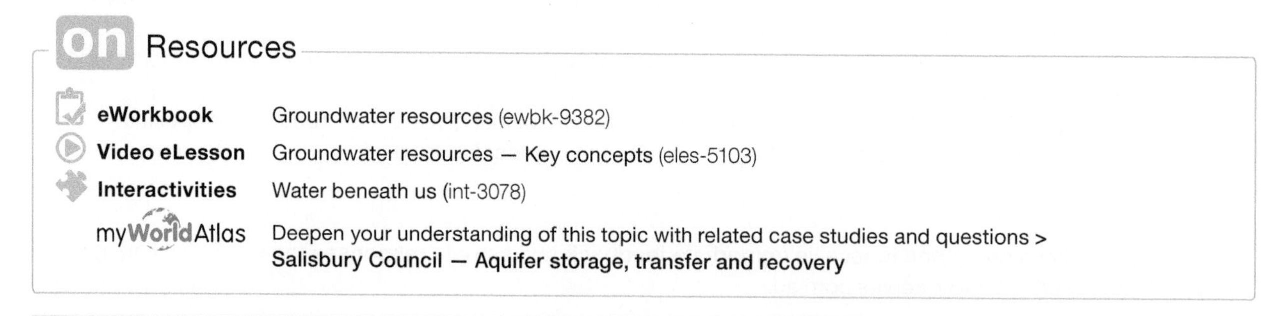

12.12 Investigating topographic maps — Water flows in the Haast River

LEARNING INTENTION

By the end of this subtopic, you will be able to describe how water flows through a catchment by referencing a topographic map.

12.12.1 New Zealand's Haast region

Haast is a World Heritage site located on the west coast of the South Island of New Zealand. The region is untamed, with majestic mountains, pristine lakes and rugged coastlines.

The Haast River discharges as much as 6 cubic kilometres of water annually, and the catchment extends into the Southern Alps. There are some areas of seasonal and permanent snows and very high rates of run-off in the region due to the steep mountainous terrain.

FIGURE 1 Annual rainfall, New Zealand

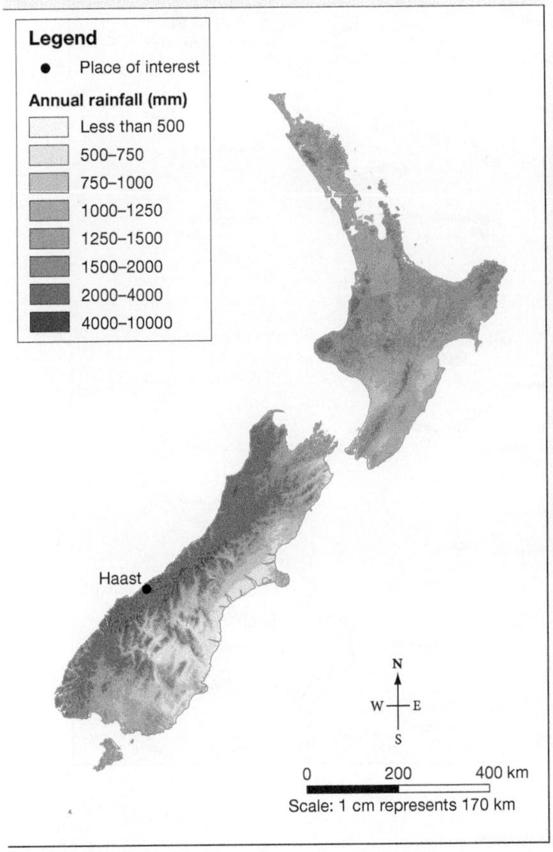

Legend
● Place of interest
Annual rainfall (mm)
Less than 500
500–750
750–1000
1000–1250
1250–1500
1500–2000
2000–4000
4000–10000

Haast

N
W E
S

0 200 400 km
Scale: 1 cm represents 170 km

Source: WorldClim

FIGURE 2 The Gates of Haast

int-8433

FIGURE 3 Topographic map extract of Haast, New Zealand

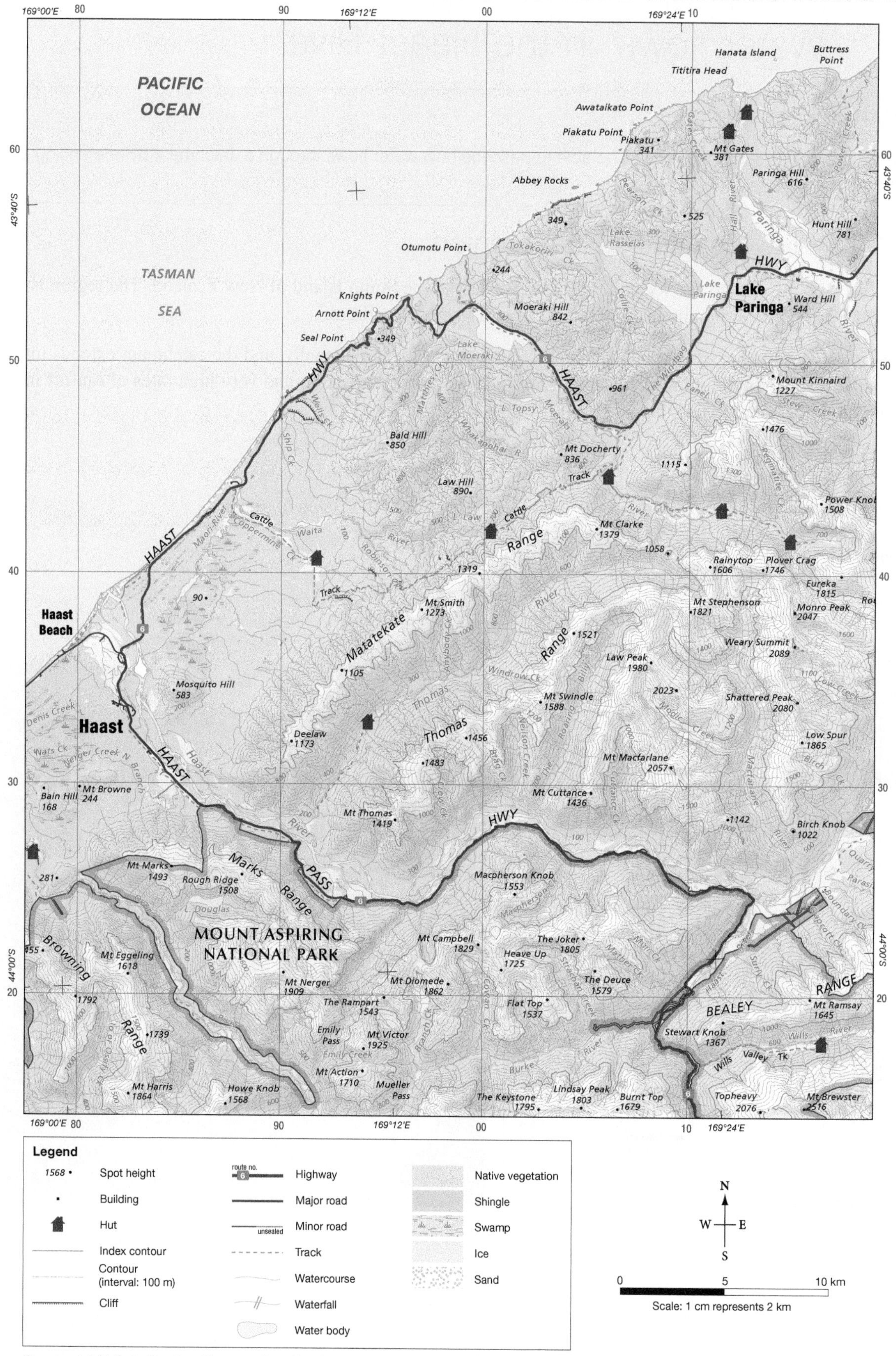

Source: LINZ Data Service. License: Creative Commons Attribution 3.0 New Zealand https://data.linz.govt.nz/license/attribution-3-0-new-zealand/

Resources

eWorkbook	Investigating topographic maps — Water flows in the Haast River (ewbk-9386)
Digital document	Topographic map of Haast, New Zealand (doc-20640)
Video eLesson	Investigating topographic maps — Water flows in the Haast River — Key concepts (eles-5104)
Interactivity	Investigating topographic maps — Water flows in the Haast River (int-8433)
Google Earth	Haast, New Zealand (gogl-0136)

12.12 EXERCISE

Learning pathways

■ LEVEL 1	■ LEVEL 2	■ LEVEL 3
1, 5, 6	2, 4, 7, 8	3, 9, 10

Check your understanding

1. Examine **FIGURE 3**. On a hike, what direction would you have to travel to get from:
   a. Mosquito Hill to the Haast Pass Highway
   b. the hut in AR9030 to the hut on Cattle Track near Coppermine Creek
   c. Buttress Point to the Haast Highway at Lake Paringa
   d. Mt Browne to Haast Beach
2. Examine **FIGURE 3**. State the spot height and area reference of:
   a. Mt Swindle
   b. Plover Crag
   c. Mosquito Hill.
3. What is the local relief in AR8030?
4. Describe the location of swamp areas found in the Haast region.
5. Examine **FIGURE 1**. Would you describe Haast as being in a high, medium or low rainfall area?

Apply your understanding

6. A tributary is a stream that flows into a larger river. Explain why there are so many tributaries that flow into the Haast River.
7. Examine **FIGURE 2**. Explain the reasons for a high volume of river flow in this area.
8. Explain how each of the following factors could affect water availability in the Haast region:
   a. latitude
   b. altitude
   c. topography
   d. climate change.
9. With reference to locations shown in **FIGURE 3**, describe how the water cycle might function in this region.

Challenge your understanding

10. Possible impacts of climate change are increased rainfall in north-eastern New Zealand and reduced snow cover. Consider how this may affect river flow in the Haast region. Predict the possible impacts of this for local communities and environments.

To answer questions online and to receive **immediate feedback** and **sample responses** for every question, go to your learnON title at www.jacplus.com.au.

12.13 Thinking Big research project — Water access comparison

online only

The content in this subtopic is a summary of what you will find in the online resource.

Scenario

The world's natural resources are not distributed equally. Some countries have an abundance of natural resources and fertile land, while other countries are entirely composed of desert biomes. The same can be said for the distribution of water resources. Countries that have rivers that are fed by melting glaciers (such as India and Bangladesh) have a steady supply of water. Yet countries that rely on rainfall to fill rivers, streams and lakes are often at the mercy of prevailing weather patterns. And then there are countries that receive plenty of rain, but whose people have poor access to water resources. In these situations, there are both natural and human factors causing poor water access.

Task

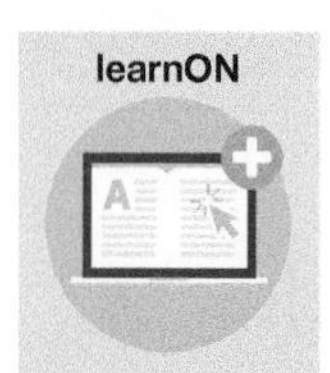

In this task, you will compare two countries that have poor access to fresh water. Go to learnON to access the resources you need to complete this research project.

on Resources

ProjectsPLUS Thinking Big research project — Water access comparison (pro-0235)

12.14 Review

12.14.1 Key knowledge summary

12.2 Understanding resources

- People depend on the Earth's environmental resources for survival.
- There are renewable and non-renewable environmental resources.
- Careful use of environmental resources will ensure their sustainability.
- Australia has an abundance of environmental resources, both renewable and non-renewable. These include minerals, soils, forests and natural scenery.

12.3 Water resources

- Water is a renewable resource in most forms.
- The water cycle is a process that cycles water in different forms across places.
- Water connects places as it moves through the environment.
- Groundwater is water stored in aquifers under the ground and is an important source of water for people living in arid and semi-arid regions.
- If more groundwater is used than is being recharged, aquifers may dry up; groundwater can therefore be regarded as a finite and non-renewable resource.

12.5 Water in the world

- Of all the water in the world, only a small fraction is available to people to use.
- Rainfall varies widely across the world.
- Climate change is affecting the amount of water available in many locations around the world.

12.6 Water flows through catchments

- Catchments are areas of land where water collects.
- The water moves quickly from the headwaters after rain, slows over the floodplains and collects into rivers/streams towards lakes or the sea.
- Low latitudes generally have more precipitation.
- High altitude places tend to have higher rainfall and snow.
- Warm ocean currents create higher rainfall; cold ocean currents create less rainfall.
- Evaporation from the ocean generally leads to higher rainfall in areas close to the coast.
- Areas with a lot of impermeable rocks tend to have higher evaporation of run-off and more precipitation.
- More precipitation occurs on the windward slope of a mountain than leeward (the rain-shadow effect).
- Climate change is affecting rainfall and precipitation patterns.

12.7 Weather and climate

- Weather is the result of constant changes to the air in the troposphere. All weather conditions are caused by three factors: air temperature, air movement and the amount of water in the air.
- Climate is the average of weather conditions measured over a period of time.

12.9 How wind blows

- Weather is the result of changes in the lowest levels of our atmosphere.
- Wind is caused by air moving from high pressure to low pressure areas in the atmosphere.
- Changes and differences in air pressure in the atmosphere occur because of the Sun's heat and the rotation of the Earth.

12.10 Wind strength

- Wind is named after the direction it comes from (a north wind comes from the north).
- Isobars on a weather map join places with the same air pressure; the closer together the isobars are, the stronger the wind.
- Symbols are used on weather maps to provide more specific details about conditions.
- Wind roses show wind speed, direction and frequency over a long period of time.
- Wind is measured using the Beaufort scale.

12.11 Groundwater resources

- Groundwater is water stored in aquifers under the ground.
- It is an important source of water for people living in arid and semi-arid regions.
- If more groundwater is used than is being recharged, aquifers may dry up; groundwater can therefore be regarded as a finite and non-renewable resource.
- Mound springs are mounds of built-up minerals and sediments brought up by water discharging from a groundwater aquifer.

12.14.2 Key terms

aquifer a body of permeable rock below the Earth's surface that contains water, known as groundwater. Water can move along an aquifer.

artesian aquifer an aquifer confined between impermeable layers of rock. The water in it is under pressure and will flow upward through a well or bore.

barometer an instrument used to measure air pressure

blue water the water in fresh-water lakes, rivers, wetlands and aquifers

climate average of weather conditions that are measured over a long time

climate change a change in the world's climate. This can be very long term or short term, and is caused by human activity.

condensation precipitation that collects into droplets of water from humid air

delta a landform at the mouth of the river where a river splits into smaller streams and sediment is deposited to create an arch of land reaching into the sea

evaporation the process by which water is converted from a liquid to a gas and thereby moves from land and surface water into the atmosphere

evaporate to change liquid, such as water, into a vapour (gas) through heat

evapotranspiration the process by which plants absorb precipitation and release it back into the air as water vapour

floodplain the area of land next to the river, usually reaching to the base of the mountains surrounding the catchment. It experiences flooding during peak rainfall periods.

fossil fuel fuel that comes from the breakdown of living materials, and that is formed in the ground over millions of years. Examples include coal, oil and natural gas.

green water water that is stored in the soil or that stays on top of the soil or in vegetation

groundwater recharge a process in which water moves down from the Earth's surface into the groundwater

hydrologic cycle another term for the water cycle

infiltration precipitation absorbed into the ground

isobar line on a map that joins places with the same air pressure

levee also known as an embankment, it is a built-up part of the river bank

meander a curve in a river caused by fast-flowing water eroding the bank of one side of the river and slow-flowing water depositing sediment on the other side of the river

meteorologist a scientist who studies the weather

precipitation the form of water falling from the sky, such as rain, snow or hail

run-off precipitation not absorbed by soil, which runs over the land and into streams

subsistence farming a form of agriculture that provides food for the needs of only the farmer's family, leaving little or none to sell

troposphere the layer of the atmosphere closest to the Earth. It extends about 16 kilometres above the Earth's surface, but is thicker at the tropics and thinner at the poles, and is where weather occurs

uranium radioactive metal used as a fuel in nuclear reactors

water vapour water in its gaseous form, formed as a result of evaporation

weather short-term changes in the atmosphere at a particular location

12.14.3 Reflection

Complete the following to reflect on your learning.

Revisit the inquiry question posed in the Overview:

Earth provides us with more than a place to live. How do we access and manage the water resources we need from it to survive?

1. Now that you have completed this topic, what is your view on the question? Discuss with a partner. Has your learning in this topic changed your view? If so, how?
2. Write a paragraph in response to the inquiry question, outlining your views.

Subtopic	Success criteria			
12.2	I can identify resources and classify them as either renewable or non-renewable.			
	I can explain why water is classified as a resource.			
12.3	I can identify and explain the processes involved in the water cycle, and describe how they are interconnected.			
12.4	I can interpret diagrams.			
12.5	I can describe the spatial distribution of water in the world.			
12.6	I can identify and describe the parts of a water catchment area.			
	I can explain factors that affect the availability of water in a river system.			
12.7	I can explain the impacts of climate and weather in water availability.			
12.8	I can read synoptic charts to predict the weather.			
12.9	I can explain why wind occurs.			
12.10	I can explain and demonstrate how wind is measured.			
12.11	I can identify the locations of major groundwater basins in Australia.			
	I can explain why the quality of and access to groundwater can vary.			
12.12	I can describe how water flows through a catchment by referencing a topographic map.			

on Resources

eWorkbook Chapter 12 Student learning matrix (ewbk-8480)
Chapter 12 Reflection (ewbk-8481)
Chapter 12 Extended writing task (ewbk-8482)

Interactivity Chapter 12 Crossword (int-8434)

ONLINE RESOURCES

on Resources

Below is a full list of **rich resources** available online for this topic. These resources are designed to bring ideas to life, to promote deep and lasting learning and to support the different learning needs of each individual.

eWorkbook

- 12.1 Chapter 12 eWorkbook (ewbk-7992) ☐
- 12.2 Understanding resources (ewbk-9346) ☐
- 12.3 Water resources (ewbk-9350) ☐
- 12.4 SkillBuilder — Interpreting diagrams (ewbk-9354) ☐
- 12.5 Water in the world (ewbk-9358) ☐
- 12.6 Water flows through catchments (ewbk-9362) ☐
- 12.7 Weather and climate (ewbk-9366) ☐
- 12.8 SkillBuilder — Reading a weather map (ewbk-9370) ☐
- 12.9 How wind blows (ewbk-9374) ☐
- 12.10 Wind strength (ewbk-9378) ☐
- 12.11 Groundwater resources (ewbk-9382) ☐
- 12.12 Investigating topographic maps — Water flows in the Haast River (ewbk-9386) ☐
- 12.14 Chapter 12 Student learning matrix (ewbk-8480) ☐
- Chapter 12 Reflection (ewbk-8481) ☐
- Chapter 12 Extended writing task (ewbk-8482) ☐

Sample responses

- 12.1 Chapter 12 Sample responses (sar-0145) ☐

Digital document

- 12.12 Topographic map of Haast, New Zealand (doc-20640) ☐

Video eLessons

- 12.1 A world of water (eles-1616) ☐
- Water in the world — Photo essay (eles-5346) ☐
- 12.2 Understanding resources — Key concepts (eles-5096) ☐
- 12.3 Water resources — Key concepts (eles-5097) ☐
- 12.4 SkillBuilder — Interpreting diagrams (eles-2275) ☐
- 12.5 Water in the world — Key concepts (eles-5098) ☐
- 12.6 Water flows through catchments — Key concepts (eles-5099) ☐
- 12.7 Weather and climate — Key concepts (eles-5100) ☐
- 12.8 SkillBuilder — Reading a weather map (eles-2749) ☐
- 12.9 How wind blows — Key concepts (eles-5101) ☐
- 12.10 Wind strength — Key concepts (eles-5102) ☐
- 12.11 Groundwater resources — Key concepts (eles-5103) ☐
- 12.12 Investigating topographic maps — Water flows in the Haast River— Key concepts (eles-5104) ☐

Interactivities

- 12.2 Understanding resources (int-7766) ☐
- 12.3 Water works (int-3077) ☐
- The water cycle (int-5614) ☐
- 12.4 SkillBuilder — Interpreting diagrams (int-3132) ☐
- 12.5 Predicted change in run-off (int-5298) ☐
- 12.6 Water flows through catchments (int-8429) ☐
- 12.7 Weather and climate (int-8430) ☐
- 12.8 SkillBuilder — Reading a weather map (int-6777) ☐
- 12.9 Highs and lows (int-3068) ☐
- The effect of Sun on a sea breeze (int-7790) ☐
- 12.10 Wind strength (int-3086) ☐
- 12.11 Water beneath us (int-3078) ☐
- 12.12 Investigating topographic maps — Water flows in the Haast River (int-8433) ☐
- 12.14 Chapter 12 Crossword (int-8434) ☐

Fieldwork

- 12.6 River study (fdw-0014) ☐
- 12.9 Clouds (fdw-0015) ☐

ProjectsPLUS

- 12.13 Thinking Big research project — Water access comparison (pro-0235) ☐

Weblinks

- 12.6 Bureau of Meteorology (web-0044) ☐
- 12.10 Wind rose maps (web-0048) ☐

Google Earth

- 12.12 Haast, New Zealand (gogl-0136) ☐

myWorld Atlas

- 12.5 Retreating glaciers > World water resources (mwa-7328) ☐
- 12.9 Wind and Sun direction (mwa-4374) ☐
- 12.11 Salisbury Council — Aquifer storage, transfer and recovery (mwa-7295) ☐

Teacher resources

There are many resources available exclusively for teachers online.

13 The value of water

INQUIRY SEQUENCE

To access a pre-test with **immediate feedback** and **sample responses** to every question in this chapter, select your learnON format at www.jacplus.com.au.

13.1 Overview

Numerous **videos** and **interactivities** are embedded just where you need them, at the point of learning, in your learnON title at www.jacplus.com.au. They will help you to learn the content and concepts covered in this topic.

We can't survive without water, but does it mean more than just survival — do we value water in other ways?

13.1.1 Introduction

Water is an important resource used by people. It has great economic, cultural, spiritual and aesthetic value. The health and wellbeing of a community can be greatly impacted by its access to water, and water can be central to the beliefs and practices of indigenous cultures. Access to water is a basic human right. The amount of available fresh water on Earth needs to be shared among an ever-growing global population. It is a resource that must be used carefully so that current and future populations can have adequate supplies.

STARTER QUESTIONS

1. List some of the different ways water is used in our lives.
2. How would your life be different if you didn't have easy access to water?
3. Do you value water? What are the different ways in which water can have a value to you?
4. Watch the video using the **Introduction to water** weblink in your Resources tab. List all the ways that water is used by people, animals, plants and the environment. How much of the Earth is covered by water? How much of this water can be used by people?
5. Watch the **Value of water — Photo essay** Video eLesson in your Resources tab. Consider some of the different ways that people in your community might value or use water.

FIGURE 1 Traditional methods are used to fish sustainably.

on Resources

eWorkbook	Chapter 13 eWorkbook (ewbk-7993)
Video eLessons	Water — A vital resource (eles-1615) Value of water — Photo essay (eles-5106)

13.2 Water use

LEARNING INTENTION

By the end of this subtopic, you will be able to describe the many ways water is used by people around the world and give examples of agricultural, commercial, industrial and recreational uses.

13.2.1 Global water use

There are three main uses made of water by all people: growing food, producing goods and electricity, and using it in the home. The amount of water consumed for each of these uses differs from one place to another. The problem remains that while the total amount of fresh water is fixed, the amount used on average per person is increasing.

It is interesting to look at water consumption on a global scale. The global average is 1240 cubic metres per person per year but some countries consume more water than others. Examples of countries that consume nearly twice as much as the global average are the United States and Thailand. Some countries that consume the least amount of water per person are Peru, Somalia and China.

FIGURE 1 World water use

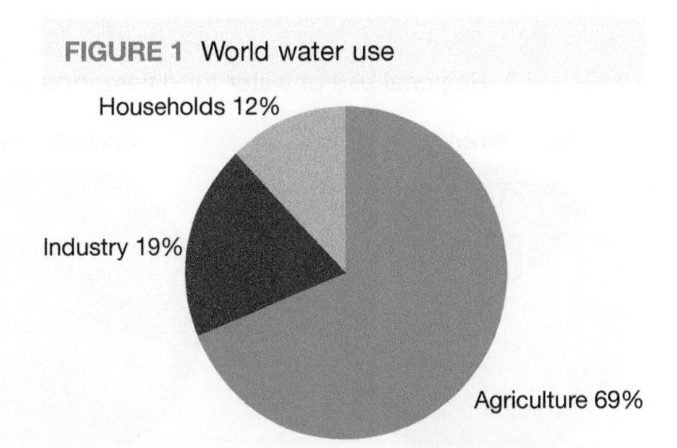

Source: United Nations World Water Development Report, UNESCO, 2019

FIGURE 2 Countries in the world differ in their use of water.

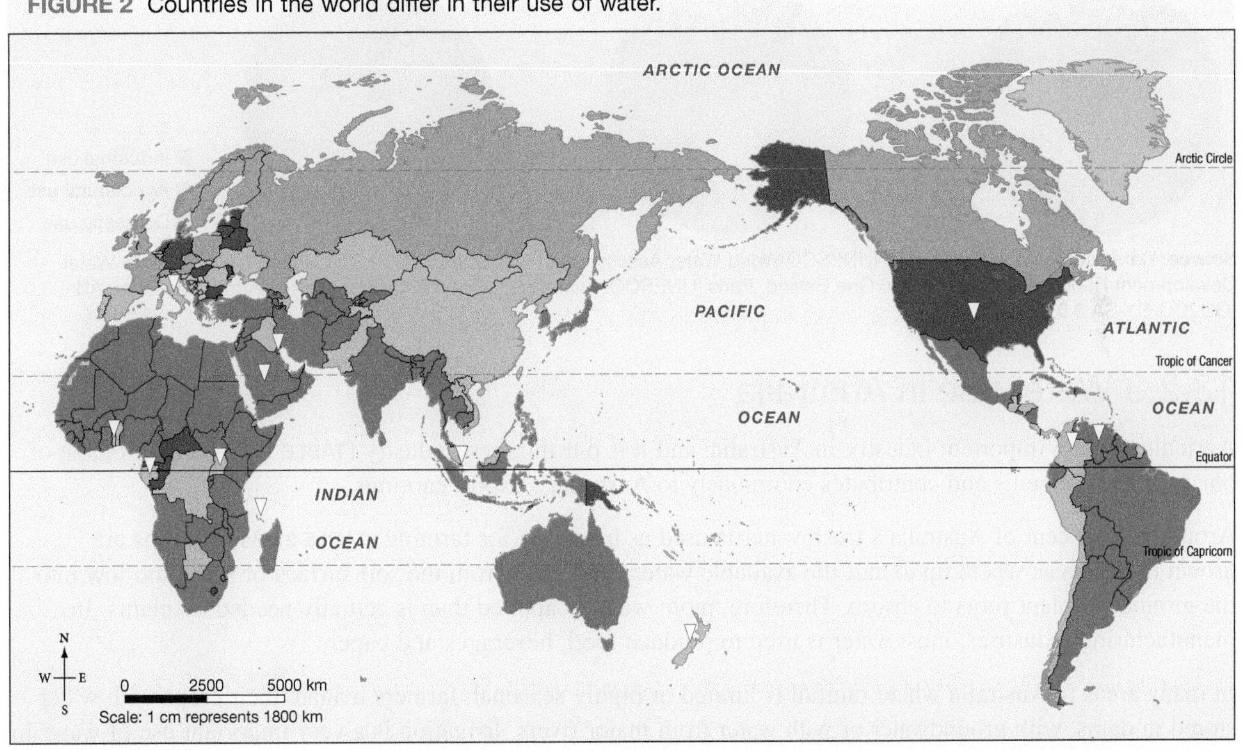

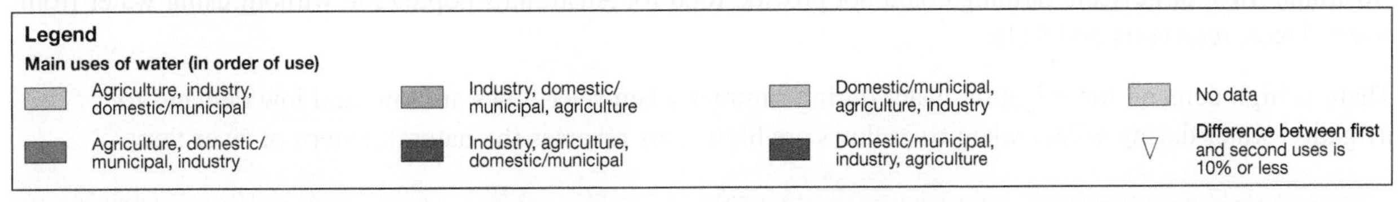

Source: Mekonnen, M.M. and Hoekstra, A.Y., 2011. National water footprint accounts: The green, blue and grey water footprint of production and consumption', *Value of Water Research Report Series No. 50,* UNESCO-IHE, Delft, The Netherlands

13.2.2 How we use water

FIGURE 2 shows that most of the world's water is used in agriculture, to grow food for the world's increasing population. This is especially the case in the drier parts of the world where there is not enough rainfall to grow crops or grass for animals. There is a strong interconnection between the amount of rainfall in a region and the amount of water used in agriculture.

It is interesting to see how this pattern varies among countries (**FIGURE 3**). In some countries, the water used in agriculture and industry is greater than the amount of water used in homes for domestic use. In other places, people consume more water for domestic use than for either agriculture or industry.

FIGURE 3 Regional use of water for different purposes

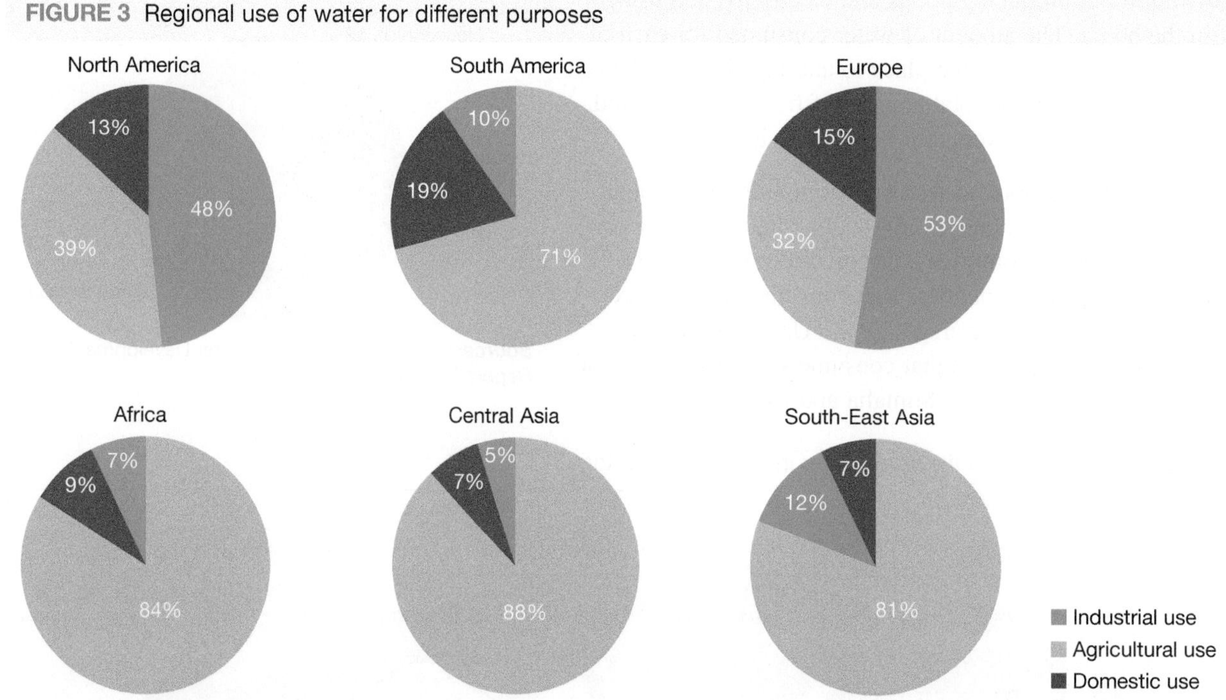

Source: Data source based on WWAP (UNESCO World Water Assessment Programme). 2019. The United Nations World Water Development Report 2019: Leaving No One Behind. Paris, UNESCO. Available in Open Access under the Attribution-ShareAlike 3.0 IGO (CC-BY-SA 3.0 IGO)

13.2.3 Water use in Australia

Agriculture is an important industry in Australia, and it is our thirstiest industry (**TABLE 1**). It produces most of our food requirements and contributes enormously to Australia's export earnings.

Around 70 per cent of Australia's fresh water is used as irrigation for farming (**TABLE 2**). Many crops are grown in dry areas where up to half the available water evaporates from the soil surface or seeps too low into the ground for plant roots to absorb. Therefore, more water is applied than is actually needed by plants. In manufacturing industries, most water is used to produce food, beverages and paper.

In many areas in Australia where rainfall is limited or highly seasonal, farmers irrigate their crops with water stored in dams, with groundwater or with water from major rivers. Irrigation is a very important use of water in Australia. Most large-scale farming could not provide food for Australia's population without using water from rivers, lakes, reservoirs and wells.

There is high demand for irrigation water during summer when river flows are low, and low demand for irrigation water during winter when river flows are high. This reverses the natural pattern of river flow.

TABLE 1 Fresh water use in Australia

Types of use of fresh water	%
Agriculture • pasture 35% • crops 27% • rural and domestic stock 8%	70
Urban	12
Horticulture	10
Industry	3
Mining	2
Services	2
Hydro-electricity	1

FIGURE 4 The Hunter River in NSW provides water for agricultural irrigation.

TABLE 2 Fresh water used to irrigate different crops in Australia

Crop type	Water (gigalitres)	%
Livestock, pasture, grains and other agriculture	8795	56
Cotton	1841	12
Rice	1643	11
Sugar	1236	8
Fruit	704	5
Grapes	649	4
Vegetables	635	4

Note: One gigalitre = 1 000 000 000 litres or one thousand million litres or 400 Olympic-size swimming pools

on Resources

eWorkbook	Water use (ewbk-9394)
Video eLesson	Water use — Key concepts (eles-5107)
Interactivity	Water use (int-7780)
myWorldAtlas	Deepen your understanding of this topic with related case studies and questions > **Three rivers in Africa** Deepen your understanding of this topic with related case studies and questions > **How is water used in Australia?**

13.2 ACTIVITIES

1. Watch the video **Three rivers in Africa** in your Resources tab and look at the information on these rivers in myWorld Atlas. Use the atlas map to trace the flow of the river and upload images to the map showing all the different water uses in the video.
2. Use your atlas to look at patterns in **FIGURE 2** and compare them to a map that shows global wealth.
   a. Identify a country with low wealth and high water use in industry.
   b. Identify two wealthy countries that do not have high water use in industry.
   c. Write a general statement about wealth and water use. Add two exceptions to your general statement.

13.2 EXERCISE

Learning pathways

■ LEVEL 1	■ LEVEL 2	■ LEVEL 3
1, 2, 3, 11, 13	4, 6, 8, 9, 15	5, 7, 10, 12, 14

Check your understanding

1. What is most of the world's water used for?
2. Which regions of the world use the majority of their water in agriculture?
3. Which countries use water mainly for industrial purposes?
4. How much of Australia's water is used for agriculture?
5. How many Olympic-size swimming pools are used for irrigating the following in Australia each year:
   a. livestock
   b. cotton
   c. rice
   d. sugar
   e. fruit?
6. Which region uses the highest percentage of its water for industry:
   a. agriculture
   b. domestic use?
7. Why might some farmers need to put more water on their crops than the plants need? Give two reasons why crops need to be watered more frequently.
8. Refer to **FIGURE 1**. Describe the patterns you notice over space of countries with:
   a. very high and high water footprints
   b. very low and low water footprints.

Apply your understanding

9. Explain why some countries might use more water in industry than in agriculture or domestic use.
10. Discuss how water is used in Australia, including which crops use the least and which use the most water.
11. Study **FIGURE 2** and decide which of the following statements are true and which are false. Provide evidence to support your decision.
    a. Australia uses most water for agriculture, then industry, then domestic/municipal.
    b. Countries in North Africa use most water for industry, then domestic/municipal, then agriculture.
    c. Belarus uses most water for industry, then agriculture, then domestic/municipal.
    d. Colombia uses most water for agriculture, then domestic/municipal, then industry.
    e. Belize uses most water for industry, then agriculture, then domestic/municipal.
    f. Malaysia uses most water for industry, then agriculture, then domestic/municipal.
12. Explain how water can have a social value.

Challenge your understanding

13. Write three summary statements that describe water use in Africa. How do these patterns compare with Australia?
14. Suggest what can be done on a global scale to protect water resources.
15. Predict how the water use shown for Australia in **FIGURE 2** will change in the next 20 years if Australia's population continues to grow.

To answer questions online and to receive **immediate feedback** and **sample responses** for every question, go to your learnON title at www.jacplus.com.au.

13.3 SkillBuilder — Drawing a line graph

online only

LEARNING INTENTION

By the end of this subtopic, you will be able to represent information using line graphs.

The content in this subtopic is a summary of what you will find in the online resource.

13.3.1 Tell me

What is a line graph?

A line graph displays information as a series of points on a graph that are joined to form a line. A line graph can be drawn by hand or by using a spreadsheet program such as Excel.

FIGURE 2 Use of rainwater tanks by household, 2001–2010

Use of rainwater tanks by household, 2001–2010

Source: © Australian Bureau of Statistics

13.3.2 Show me

eles-1635

How to complete a line graph

Step 1

Select the data you wish to compare or interpret. Draw a horizontal and vertical axis using a ruler.

Evenly space and then label the years along the horizontal axis. Look carefully at your range of data and work out appropriate increments for the vertical axis, then evenly space and label this information on the axis. Start at zero where the axes join. Label the x and y axes.

Step 2

Plot the statistics. Draw a dot at the point where the position on the horizontal axis meets the relevant position on the vertical axis. Once you have plotted all the statistics, join the dots. This can be done freehand or using a ruler. Add a title and a source to the graph.

13.3.3 Let me do it

int-3131

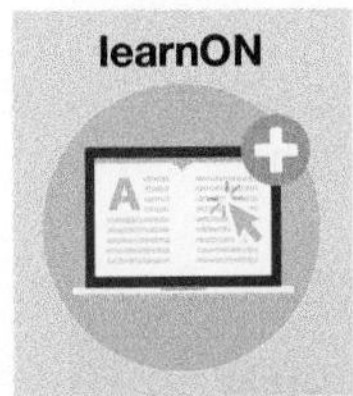

Go to learnON at www.jacplus.com.au to access the following additional resources to help you build this skill:

- a longer explanation of this skill and its application in Geography (Tell me)
- a video showing the step-by-step process of this skill (Show me)
- an activity and interactivity for you to practise this skill (Let me do it)
- self-marking questions to help you understand and use this skill.

on Resources

eWorkbook	SkillBuilder — Drawing a line graph (ewbk-9398)
Video eLesson	SkillBuilder — Drawing a line graph (eles-1635)
Interactivity	SkillBuilder — Drawing a line graph (int-3131)

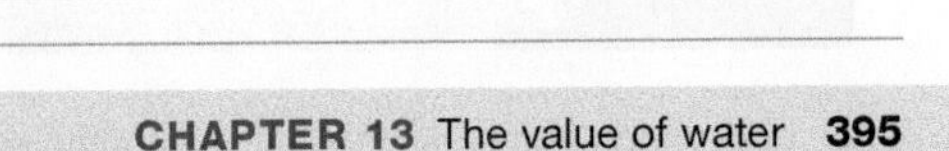

13.4 The value of clean water

LEARNING INTENTION

By the end of this subtopic, you will be able to explain how water quality may affect perceptions of the value of water, and explain some of the ways that water quality is measured.

13.4.1 Water quality

Value refers to how important something is or what it is considered to be worth. The value people place on water will vary according to their standard of living, income, culture, location and the way they use water. In this way the value of water is subjective. This means that people have different opinions or perspectives. People who live in cities in developed countries with easy access to piped water may not value water as much as farmers in rural areas in the same country. Those living in developing countries with no or limited access to water are likely to feel that water has a much higher value.

FIGURE 1 The different values of water

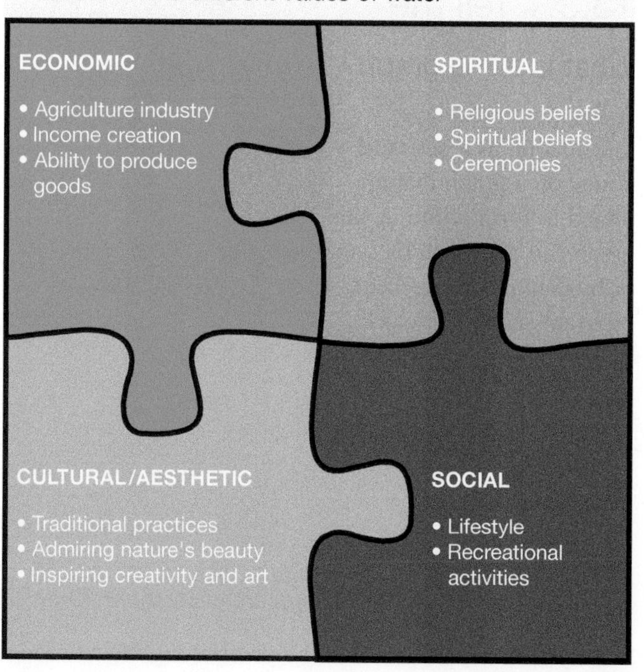

The value of water and the way that it can be used is determined by the quality of the water. Water that is not polluted is better for humans to drink and use, and it is also thought of as more beautiful (valuable aesthetically).

Different pollutants — faeces (human and animal), food wastes, pesticides, chemicals and heavy metals — can come from industrial wastewater, domestic sewage, cars, gardens, farmland, mining sites and roads, and flow into waterways.

aquifer a body of permeable rock below the Earth's surface which contains water, known as groundwater. Water can move along an aquifer.

Some countries, cities and local areas are better than others at providing services and enforcing laws to prevent pollutants from entering water. Some of the worst polluted rivers and lakes in the world include rivers and **aquifers** in China (such as the Songhua River and the Yellow River), the Citarum River in Indonesia (**FIGURE 2**), the Yamuna and Ganges rivers in India, the Buriganga River in Bangladesh and the Marilao River in the Philippines. These water sources provide drinking water to millions of people, so levels of pollution can have a significant negative impact on the health and wellbeing of many people.

FIGURE 2 (a) and (b) The Citarum River, Bandung, Indonesia, is one of the most polluted rivers in the world, but local people still rely on it for fishing.

13.4.2 Measuring water quality

Investigating the physical properties of a river or creek at various points along its length is one way to measure water quality and any evidence of human impact. Does the quality of the water change between upstream and downstream sites? Are there human factors that can account for these changes? Differences may be more obvious if the waterway passes through a built-up area or a farm.

Temperature

Aquatic plants and animals have a particular temperature range in which they can survive. High water temperatures can result in reduced oxygen available for plants and animals. Comparing temperature readings with biodiversity counts provides data to help investigate this relationship.

FIGURE 3 A Secchi disc is used to measure turbidity.

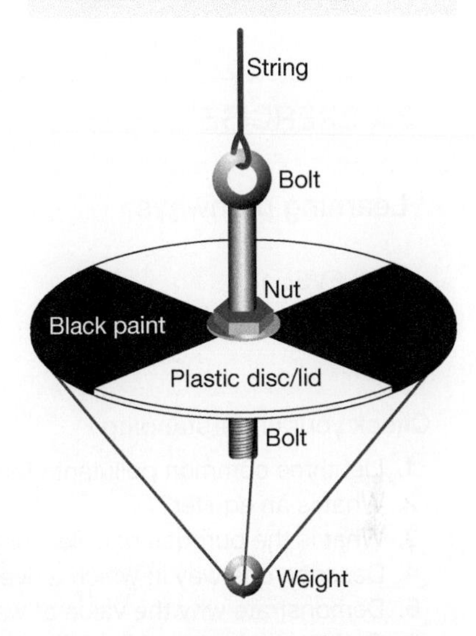

pH

A pH value is used to measure the acidity or alkalinity of water on a scale of 1 to 10. Drinking water should have a pH reading of around 6. A reading either side of this may indicate that water is polluted. You can test pH by taking a sample of water from different parts of a waterway and using pH paper or chemical reagents.

Turbidity

Water that lets little sunlight through is said to be turbid. **Turbidity** is the amount of suspended sediment in water — sediments such as clay, silt, industrial waste or sewage. A Secchi disc (see **FIGURE 3**) is used to measure turbidity.

Salinity

Salinity measures the amount of salt in the water. It is measured with an electrical conductivity meter, or EC meter.

turbidity the amount of sediment suspended in water

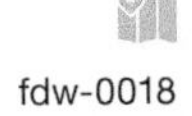
fdw-0018

FOCUS ON FIELDWORK

Investigating waterways

Some schools are located close to a waterway, even if it is a highly modified one like a concrete drain. Do you know what the water in that source is like? There are many simple tests that you can conduct on water to measure how clean it is.

Learn how to measure the quality of the water in a waterway near you with the **Investigating waterways** fieldwork activity in your Resources tab.

Resources

eWorkbook	The value of clean water (ewbk-9402)
Video eLesson	The value of clean water — Key concepts (eles-5108)
Interactivity	The value of clean water (int-8535)
Fieldwork	Investigating waterways (fdw-0018)

13.4 ACTIVITIES

1. As a class, discuss whether people always see natural, clean water sources as being more valuable and beautiful than waterways that have been changed by people.
   a. Are there examples of waterways that are not clean or natural that people still appreciate as beautiful?
   b. Investigate the importance of the Ganges River to the Hindu religion, and pilgrimages involving ceremonial bathing in the river. Does the cleanliness of water affect its spiritual value? How might people who see the river as sacred feel when they see it being polluted?
2. Work in groups or pairs to investigate one of the polluted rivers mentioned in the text. Show where it is located on a map. What does it look like? What has caused the pollution? Is anything being done to improve conditions? What strategies would you suggest for improving the water quality?

13.4 EXERCISE

Learning pathways

■ LEVEL 1	■ LEVEL 2	■ LEVEL 3
1, 4, 7, 8, 13	2, 3, 10, 11, 15	5, 6, 9, 12, 14

Check your understanding

1. List three common pollutants found in waterways.
2. What is an aquifer?
3. What is the purpose of a Secchi disc?
4. Describe one way in which a river might become polluted.
5. Demonstrate why the value of water is subjective, using three examples.
6. Outline what each of the following measures can reveal about the quality of water:
   a. temperature
   b. pH
   c. turbidity
   d. salinity.

Apply your understanding

7. Explain why someone living on a farm during a drought might value water differently to someone living in Sydney or another major city in Australia.
8. Do you value water? Support your answer with at least two reasons.
9. Explain why pollutants released onto the ground can find their way into aquifers.
10. If you wanted to test a water source to see if it is polluted, what tools would you use and what would you use them for?
11. Examine the image of the Citarum River in Indonesia (**FIGURE 2**). Compare this with a river you have visited in Australia.
12. Should governments enforce laws to prevent pollutants from entering waterways? Provide three reasons to justify your answer.

Challenge your understanding

13. Based on your own water use, do you think your habits reflect the importance of water in the world?
14. If water quality in a river decreases to the point at which it is dangerous for people to come into contact with it, is that waterway still valuable?
15. Predict whether the way people in your community value water will change in the next 50 years, providing at least two reasons for your opinion.

To answer questions online and to receive **immediate feedback** and **sample responses** for every question, go to your learnON title at www.jacplus.com.au.

13.5 SkillBuilder — Annotating a photograph

online only

LEARNING INTENTION

By the end of this subtopic, you will be able to identify and annotate important geographical information in a photograph.

The content in this subtopic is a summary of what you will find in the online resource.

13.5.1 Tell me

Why annotate photographs?

Photographs are used to show aspects of a place, but often people will notice different elements in the same photograph. Annotations are added to photographs to draw the reader's attention to what can be seen and concluded from the details.

FIGURE 1 Campaspe River near Axedale

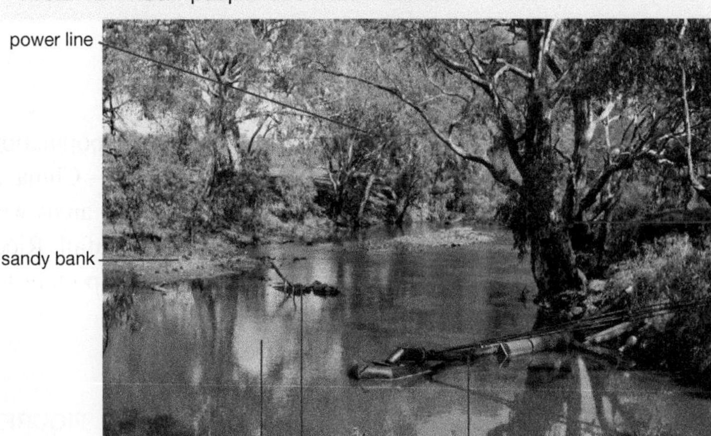

Source: Taken by Mattinbgn, 17 March 2012. © Creative Commons

13.5.2 Show me

eles-1633

How to annotate a photograph

Step 1

Examine the photograph carefully for key information — use SPICESS to remind you what might be important.

Step 2

Identify the source, place and date of the photograph. Provide the source underneath as this enables the reader to understand its author and the time of year.

Step 3

Add labels to the image, focusing on the key aspects you wish the viewer to notice. This might include natural or human features. It is often much easier to add labels if you took the photograph and made field notes while you were there. Labels can be placed outside the photograph with fine lines connecting them to the feature.

13.5.3 Let me do it

int-3129

learnON

Go to learnON to access the following additional resources to help you build this skill:

- a longer explanation of this skill and its application in Geography (Tell me)
- a video showing the step-by-step process of this skill (Show me)
- an activity and interactivity for you to practise this skill (Let me do it)
- self-marking questions to help you understand and use this skill.

on Resources

eWorkbook SkillBuilder — Annotating a photograph (ewbk-9406)

Video eLesson SkillBuilder — Annotating a photograph (eles-1633)

Interactivity SkillBuilder — Annotating a photograph (int-3129)

13.6 The economic value of water

LEARNING INTENTION

By the end of this subtopic, you will be able to explain why water has economic value, and demonstrate its value with examples from the rice farming and aquaculture industries in Asia.

13.6.1 Economic value

The **economic** value of water must include all the different ways that we use water to produce a good or service. This includes drinking water, water used for irrigation, water used for **industry** and in recreation, and power generation. It can also relate to the importance of water as a breeding ground and nursery for fish stocks. In Australia and many other countries, the basic economic value of water is recognised in the charges we pay to access piped water. Water plays a significant role in the production of food through the agricultural industry. It can contribute to domestic food security and the reduction of poverty.

13.6.2 Rice farming in Asia

In Asia, rice forms the staple food for the majority of the population. The world's top five rice producers in 2018–2019 were all found in Asia — China, India, Indonesia, Bangladesh and Vietnam. Rice cultivation is best suited to areas with a tropical climate, with warm temperatures and high amounts of rainfall. Rice production requires a high amount of water availability. Rice is an important form of foreign income in Thailand, India and Vietnam.

economic a system made up of producing, buying and selling goods and services
industry the production of goods and services

FIGURE 1 Terraced rice fields

FIGURE 2 A Vietnamese farmer in a rice field

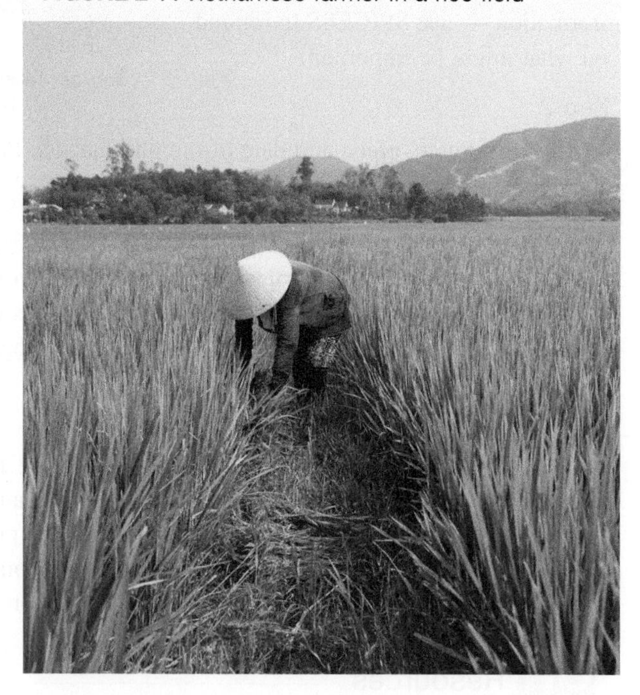

13.6.3 Aquaculture in Asia

Aquaculture, also known as aqua farming, involves breeding and harvesting plants and fish in water environments. It can include farms in both salt water and fresh water. The aqua farms are set up in ponds (**FIGURE 3**), in cages and on rafts and long lines (for shellfish).

Overfishing, destructive fishing practices and pollution have taken a toll on fish populations in the wild. As wild fish populations have plunged, fish catches by traditional means have declined, and aquaculture is a way of addressing this shortage. Aquaculture contributes to world fish supplies and food security in Asia.

Aquaculture is heavily concentrated in Asia, in areas where there are tropical climates, vast coastlines and large inland water bodies. These features can make many Asian countries suitable for aquaculture industries.

Aquaculture is an important source of foreign income and an important factor in encouraging economic development. Aquaculture can support local communities by generating employment and income, and provides a means to significantly improve standards of living. Traditional aquaculture operations tend to be small-scale and farmer owned and managed, providing autonomy to the owner. Aquaculture can often complement other farming practices such as rice farming, and a rice–fish culture has developed in some regions of China and India. Aquaculture provides opportunities for participation by both the men and women of the community. Much of the daily management and often the sale of the products is carried out by women.

Despite the economic benefits of aquaculture, it has a number of environmental drawbacks. These include natural habitat destruction, release of chemical and organic effluent, and the depletion of wild fish stocks.

FIGURE 3 An aquaculture worker feeding fish

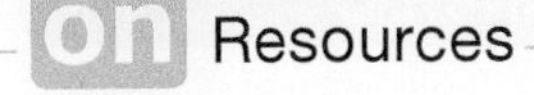

Resources

eWorkbook	The economic value of water (ewbk-9410)
Video eLesson	The economic value of water — Key concepts (eles-5109)
Interactivity	The economic value of water (int-8536)

13.6 ACTIVITY

Research the environmental impacts of aquaculture such as natural habitat destruction, release of chemical and organic effluent, and the depletion of wild fish stocks. Share your findings with your class.

13.6 EXERCISE

Learning pathways

■ LEVEL 1	■ LEVEL 2	■ LEVEL 3
2, 3, 4, 10, 14	1, 6, 8, 9, 15	5, 7, 11, 12, 13

Check your understanding

1. Define the term *economic resource*.
2. What does it mean for something to have economic value?
3. What is the staple food of many Asian countries?
4. List the top five rice-producing countries in the world.
5. In what kind of climate does rice grow well?
6. Define the term *aquaculture*.
7. What does it mean to 'overfish' an area of the ocean?

Apply your understanding

8. What aspects of the climate in tropical areas makes it good for rice growing?
9. Do parts of Australia have a suitable climate for aquaculture? Give reasons for your decision.
10. How does aquaculture help to support local communities?
11. Explain one of the negative environmental impacts of aquaculture.
12. Why might rice- and fish-farming practices work well together within a community?

Challenge your understanding

13. Might aquaculture help to reduce extinctions of species in the ocean? Give reasons for your answer.
14. Predict whether most of the fish we eat in the future will come from fish farms or the wild. Give reasons for your answer.
15. Is Australia well suited to rice farming? Consider the images of Australian rice farming in **FIGURE 4**, and refer to the processes involved in farming rice in your answer.

FIGURE 4 (a) Aerial view of a rice crop, NSW, and (b) Rice harvesting, Murrami, NSW

(a)

(b)

To answer questions online and to receive **immediate feedback** and **sample responses** for every question, go to your learnON title at www.jacplus.com.au.

13.7 The value of water to health

LEARNING INTENTION

By the end of this subtopic, you will be able to explain the value of clean water to human health, and evaluate strategies to improve access to clean water around the world.

13.7.1 The impact of dirty water on health

Water has great value in improving the health and living standards of people around the world.

Nearly 780 million people in the world have no access to clean water, and 2.4 billion people have no safe way of disposing of human waste. Lack of toilets means many people defecate in open spaces or near the same rivers from which they drink. It is estimated that 90 per cent of sewage in poor countries ends up flowing straight into rivers and creeks.

This is an unacceptable situation. Dirty water and lack of proper hygiene kill 3.4 million people around the world every year, most of them children younger than five. People who are sick are often unable to work properly, to look after their families or to attend school, adding to the poverty cycle they may already be in. The diseases that can be passed on to people as a result of contaminated water include diarrhoeal diseases such as cholera, typhoid and dysentery. Malaria, a disease transmitted by mosquitoes that breed in still water, kills about a million people every year.

FIGURE 1 Women collect water for cooking and washing, Abeche, Chad

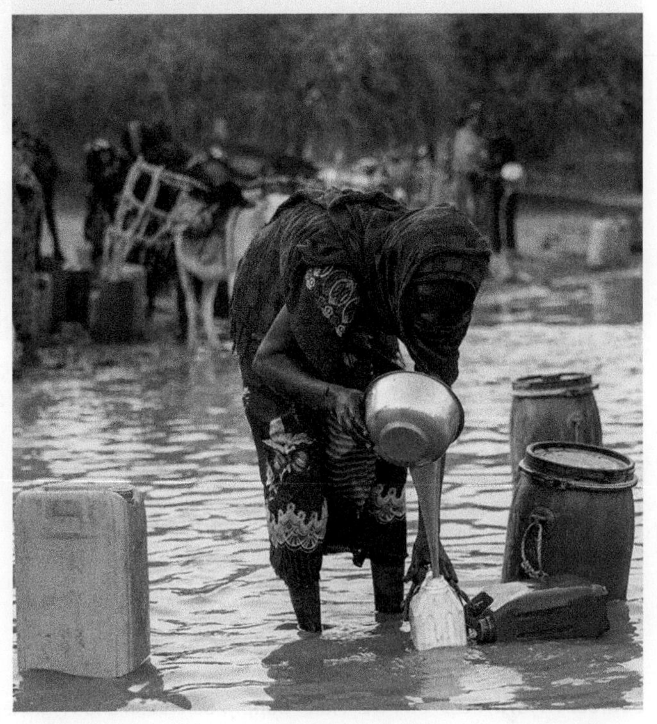

13.7.2 COVID-19

Access to water, hand washing facilities and hygiene services are vital in reducing the spread of COVID-19 and other viruses. Communities with limited access to water, facilities and services are likely to be disproportionately impacted by these viruses. These includes the homeless, rural communities, people living in informal settlements, migrants and refugees. As well, women, the very young, very old and people with disabilities are often disproportionately impacted by rapidly spreading disease.

13.7.3 Reducing water-borne diseases

People use different methods to treat the water they have collected. They can let it stand and settle, strain it through a cloth, filter it, add bleach or chlorine, or boil it. Some people do not treat their water at all.

When there is barely enough water to drink or to cook with, it is difficult for people to set aside water for washing hands and cleaning clothes. However, hygiene and sanitation are very important for health.

A number of aid groups, such as WaterAid, Water.org, CARE and A Glimmer of Hope, work on projects to improve sanitation and access to clean water. Washing hands, building cheap and effective toilets, and teaching the community about good hygiene all help to reduce disease.

13.7.4 Sustainable Development Goals

In 2015, the United Nations agreed on 17 Sustainable Development Goals to achieve a better and more sustainable world for all. Goal 6 of the Sustainable Development Goals was 'Ensure access to water and sanitation for all', and in 2016 the United Nations General Assembly adopted the 'International Decade for Action – Water for Sustainable Development' (2018–2028). As **FIGURE 2** shows, some improvements have already been made.

FIGURE 2 People with access to water services, 2000 and 2017

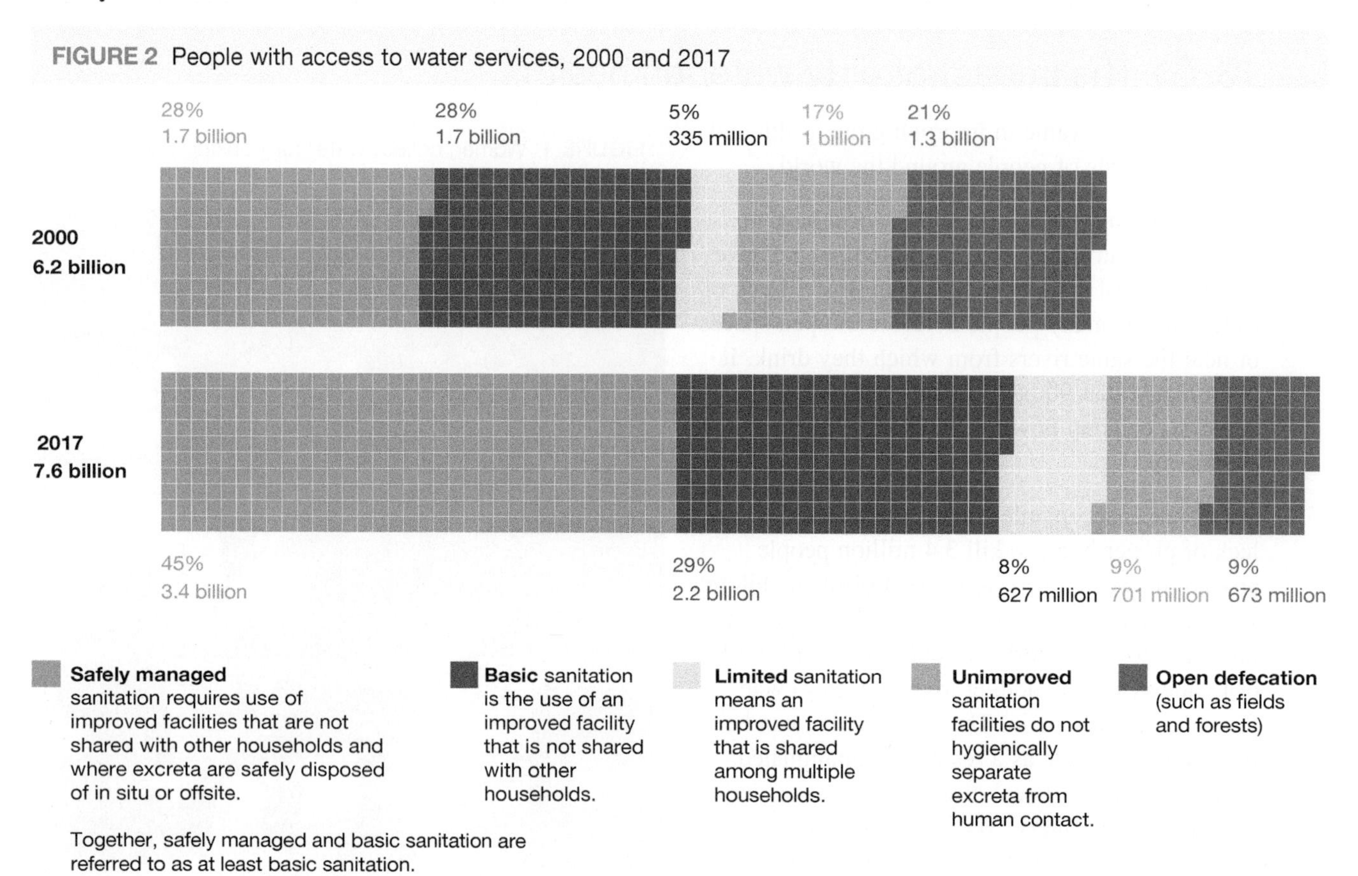

Source: WHO/UNICEF JMP for Water Supply, Sanitation and Hygiene, https://washdata.org. WDI (SH.STA.SMSS.ZS; SH.STA.BASS.ZS).

CASE STUDY

Community-led total sanitation (CLTS), Nigeria

WaterAid went to a village called Olorioko in the state of Ekiti to see if they could improve sanitation. When they arrived, there were very high rates of illness and death due to water-related diseases. The people in the village used the bush near their houses as their toilet.

CLTS leaders developed a relationship with the villagers and taught them how faeces can enter their food and make them sick. Once they understood this, the people wanted to change their practices so this would no longer happen. Action plans were drawn up and eventually the villagers created clean water points, built simple but effective toilets, and were given lessons in sanitation.

The health of the entire village has improved and they are also increasing their wealth. The CLTS project, which started in Bangladesh some years ago, has been a success and is spreading throughout Nigeria.

Use the **Health is wealth** weblink in your Resources tab to learn about a community-led sanitation projects around the world.

FIGURE 3 Location of Nigeria in Africa

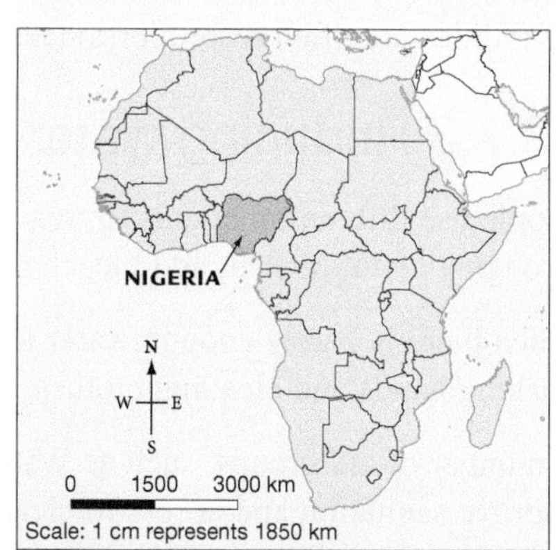

Source: Spatial Vision

on Resources

eWorkbook — The value of water to health (ewbk-9414)

Video eLesson — The value of water to health — Key concepts (eles-5110)

Interactivity — The value of water to health (int-8537)

myWorldAtlas — Deepen your understanding of this topic with related case studies and questions > **World health**

13.7 ACTIVITIES

1. Work in groups of three or four. Use the information about the importance of safe water and sanitation to create a video promoting understanding of this issue in your school. Make particular reference to North and sub-Saharan Africa, and find out what is being done by aid organisations to improve the situation in these regions.
2. In developed countries that generally have good access to clean water and sanitation, are there individual communities that struggle to access clean water? Investigate this question and report back to your class. In your investigation, consider the reasons why clean water scarcity might occur in developed nations (both natural processes such as drought, and human processes such as pollution), and how these might impact on the availability of water in the future.

13.7 EXERCISE

Learning pathways

■ LEVEL 1	■ LEVEL 2	■ LEVEL 3
1, 7, 8, 9, 13	2, 3, 4, 11, 14	5, 6, 10, 12, 15

Check your understanding

1. How many people in the world do not have access to clean water?
2. Where is Nigeria located within Africa?
3. What does CLTS stand for?
4. Identify one way (other than by drinking or cooking with dirty water) that not having access to clean water is dangerous to human health.
5. Outline the differences between safely managed, basic and unimproved sanitation.
6. What percentage of people around the world are estimated to be using sanitation facilities that are not safely managed?
7. What does SDG stand for?

Apply your understanding

8. Explain why so much human waste ends up in rivers in some countries.
9. Why are clean rivers important for human health?
10. Explain why many people living in poverty who fall sick also suffer economically.
11. Explain what a water-borne disease is, and give two examples of this kind of disease.
12. Which do you think is more important for human health: access to water or good sanitation? Or are they both equally important? Explain your view.

Challenge your understanding

13. Predict what might happen to people's health in North and sub-Saharan Africa if access to water and sanitation is not improved.
14. Suggest one change that could help a community improve their sanitation from limited to basic.
15. Access to clean water is an important tool in fighting viruses such as COVID-19. Predict whether COVID-19 spread more quickly through areas with limited or no access to clean water, and provide a justification for your prediction.

To answer questions online and to receive **immediate feedback** and **sample responses** for every question, go to your learnON title at www.jacplus.com.au.

13.8 The value of water for Aboriginal Peoples

LEARNING INTENTION

By the end of this subtopic, you will be able to describe some of the ways that Aboriginal Peoples find, use and manage water resources, and outline how Culture and water resource management are interconnected.

13.8.1 The importance of water management

Water is vital for the survival of human beings. Aboriginal Peoples' knowledge of the land and how to survive in it has been passed from generation to generation through **Dreaming** Stories. Understanding water sources plays an important role in Culture as well as in daily life and survival.

Evidence from the passing down of Culture through stories, song, dance and art, from physical changes made to the land and artefacts, and from historical records left by early European settlers and explorers shows that Aboriginal Peoples managed their water very carefully. They channelled and filtered water, covering it to keep it clean and to stop it from evaporating. They also created wells and tunnel reservoirs, and managed and cultivated stocks of food sources in lakes, rivers and oceans.

Dreaming in Aboriginal cultures, the time when the Earth took on its present form, and cycles of life and nature began; also known as the Dreamtime. Dreaming stories pass on important knowledge laws and beliefs.

FIGURE 1 Location map of Lake Condah (Tae Rak) and the Budj Bim World Heritage site

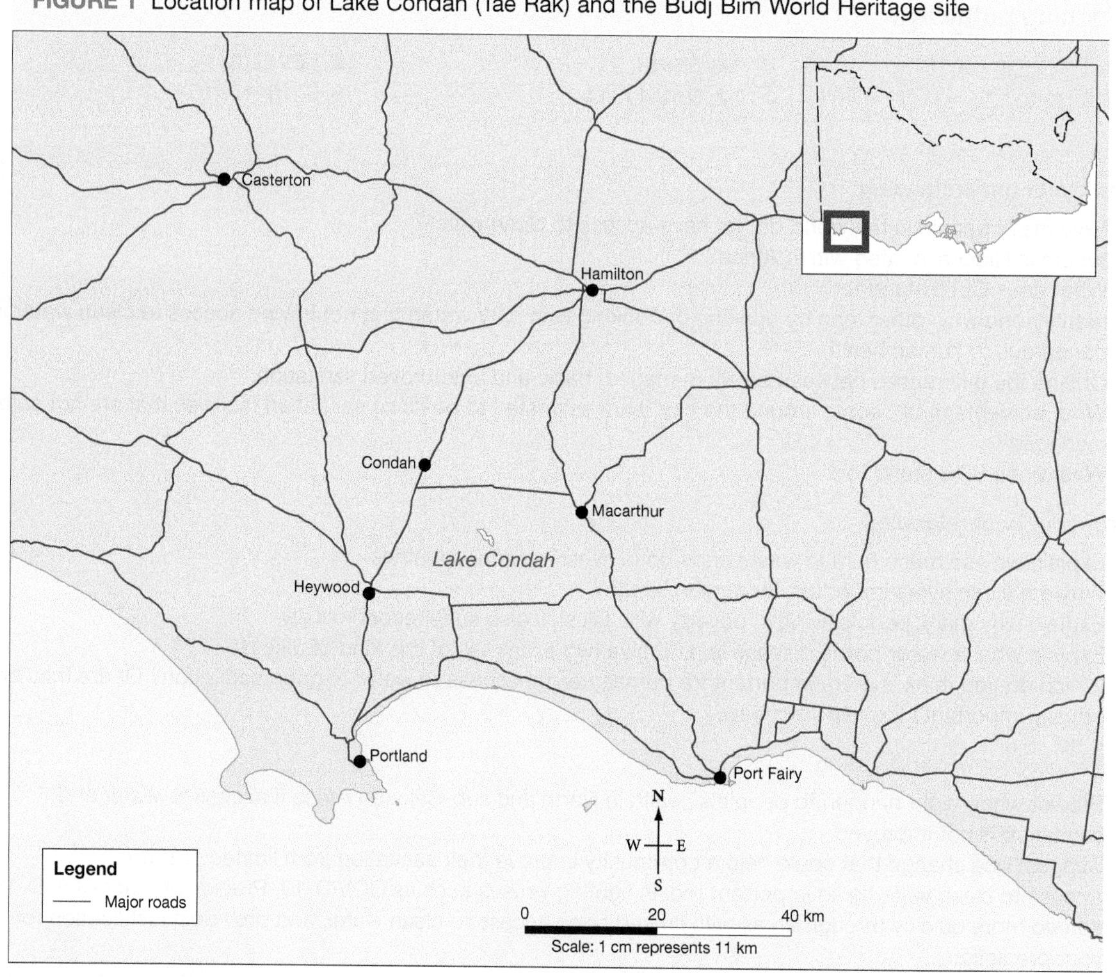

Source: Spatial Vision

FIGURE 2 Tae Rak channel and holding pond, Budj Bim, Victoria

Source: © Gunditj Mirring Traditional Owners Aboriginal Corporation (taken by Tyson Lovett-Murray)

On 30 March 2008, the Victorian government returned the World Heritage-listed Budj Bim Cultural landscape at Lake Condah (Tae Rak) in Victoria to the Gunditjmara traditional owners. Budj Bim is considered one of Australia's earliest and largest aquaculture ventures. The Gunditjmara People are preserving their culture by engaging in tourism, water restoration and sustainability projects. One example is the plan to restore the ancient stone aquaculture system at the lake for eel farming (**FIGURE 2**).

FIGURE 3 Brewarrina Fish Traps on the Barwon River

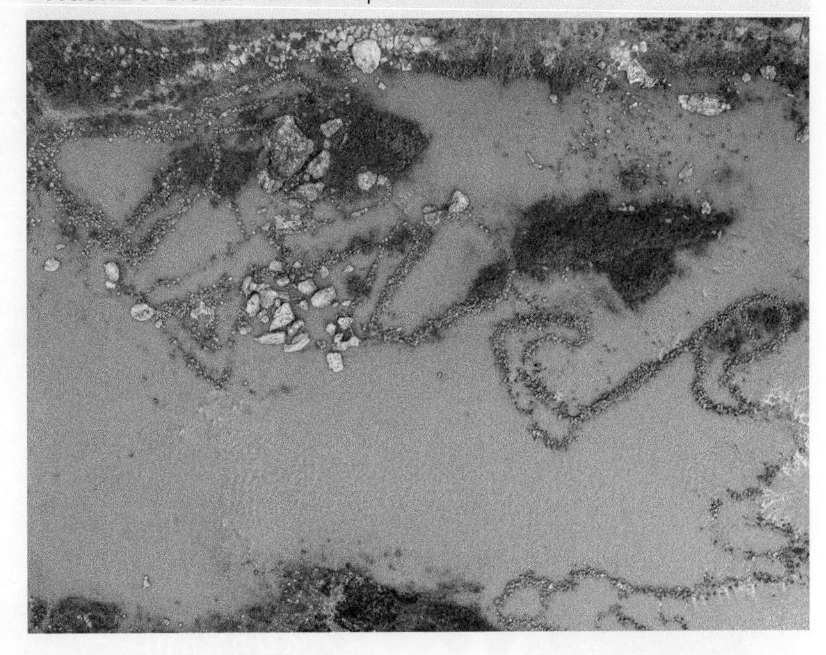

In western New South Wales, the Ngemba Peoples have a similarly sophisticated aquaculture system on the Barwon River at Brewarrina. This is commonly known as the Brewarrina Fish Traps (**FIGURE 3**).

The Ngemba Peoples used this system to manage and maintain a source of fresh water and further a plentiful and sustainable supply of food resources. They maintain its cultural and spiritual importance by carefully caring for Country.

13.8.2 Finding water

Aboriginal Peoples sourced water to live on Country through many different methods depending on where they lived, the seasons and the environment. These methods included collecting surface water from creeks, rivers and waterholes; accessing underground water supplies such as soaks and springs; and obtaining it directly from plants, including tree roots.

An understanding of the seasons also helps manage the water resources and provides a key to knowing where water is likely to be found. During the dry seasons and periods of drought, mound springs are incredibly important. These springs are linked by **songlines** and stories, and are often connected to rain-making rituals.

The rainbow serpent is a key symbol of creation for Aboriginal Peoples, but its journey from underground to the surface also represents groundwater rising to the top via springs. The creation of water sources and where to find them was often told in stories or through artwork. Use the **How the water got to the plains** weblink in your Resources tab to hear one story that describes how billabongs appeared in the dry inland plains, told by Butchulla Elder Olga Miller.

The importance of the seasons to water management is clear in the Miriwoong seasonal calendar (**FIGURE 4**), which maps the changing seasons of Miriwoong Country in northern Western Australia. Use the **Miriwoong seasonal calendar** weblink in your Resources tab to view this calendar and learn more about Miriwoong Language and Culture.

songline directly linked to Dreamings and Creation Stories, which depict the paths travelled by ancient creator spirits. They incorporate knowledge, spirituality and beliefs, which are embedded in Country and demonstrate the interconnectedness between the spiritual, physical and living worlds.

FIGURE 4 Miriwoong seasonal calendar

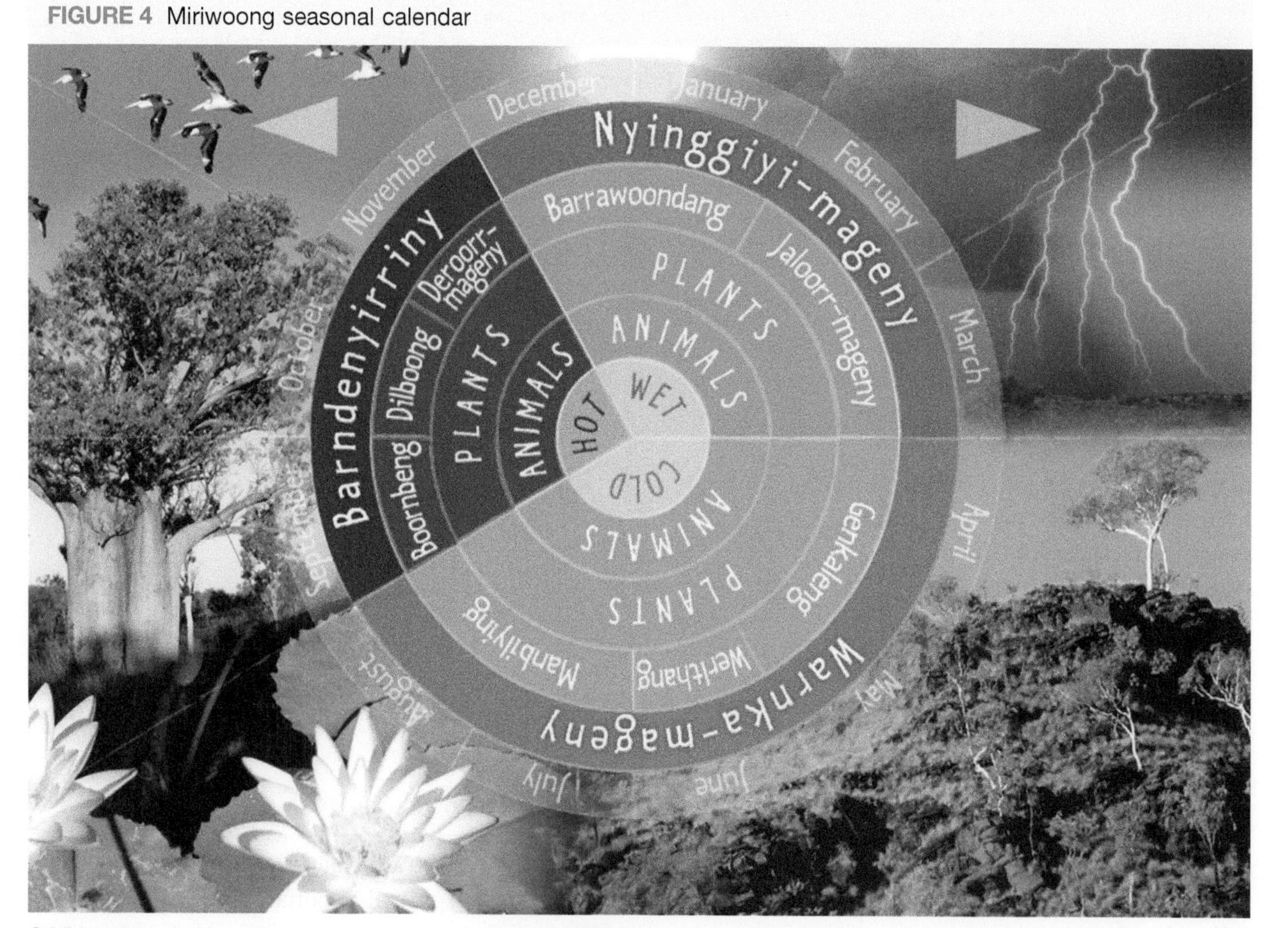

13.8.3 Groundwater

Particularly in Australia's desert regions, groundwater was vital for Aboriginal Peoples. There are many groundwater sources throughout Australia. One of these sources is **soaks**: groundwater that comes to the surface, often near rivers and dry creek beds, and which can be identified by certain types of vegetation. Another source is **mound springs**: mounds of built-up minerals and sediments brought up by water discharging from an aquifer (**FIGURE 5**).

FIGURE 5 A cross-section of a typical mound spring

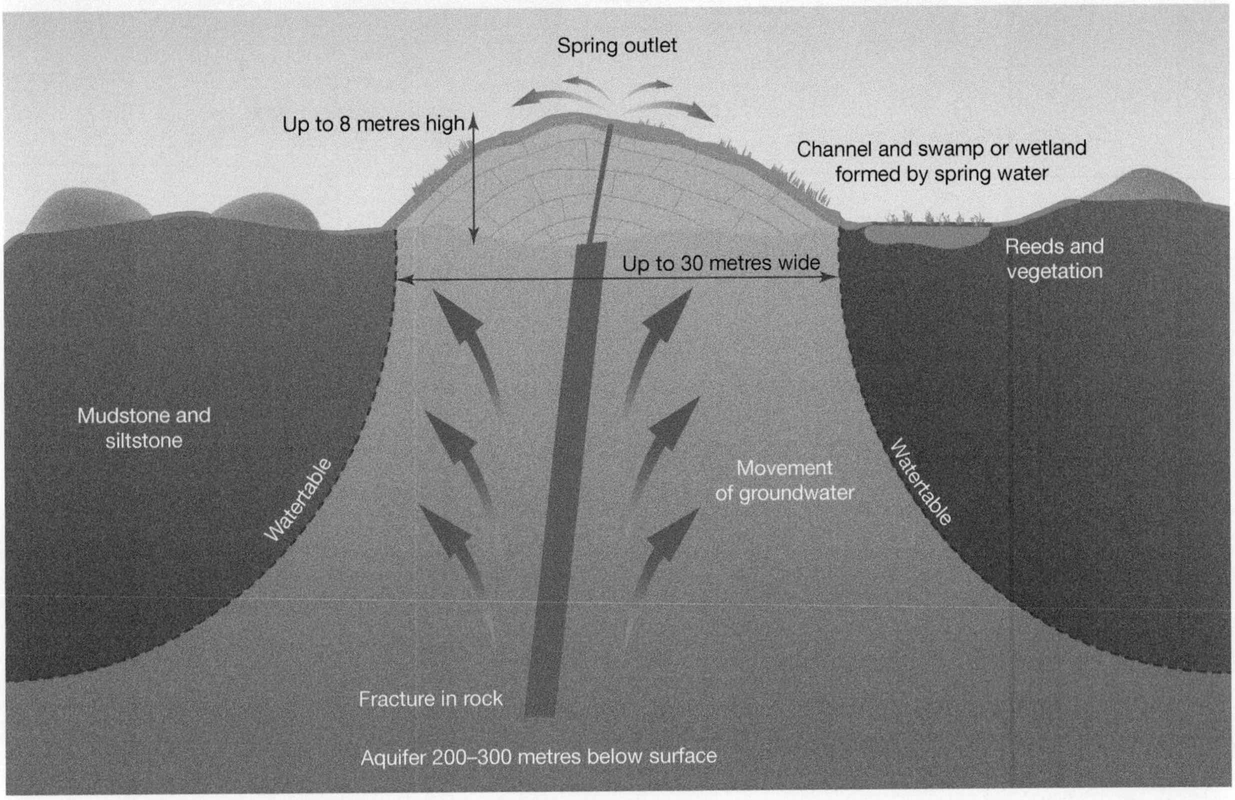

Mound springs of the Oodnadatta Track

The Oodnadatta Track is located in the north-east of South Australia (**FIGURE 6** and **7**). The track follows the edge of the Great Artesian Basin and the south-western edge of Kati Thanda–Lake Eyre and, along its route, groundwater makes its way to the surface in several locations.

The Oodnadatta Track crosses the traditional lands of three Aboriginal Peoples. In the south, between Lake Torrens and Kati Thanda–Lake Eyre, is the Country of the Kuyani Peoples; most of the west of Kati Thanda–Lake Eyre is the Country of the Arabana Peoples; and to the north is the Country of the Arrernte Peoples.

soak place where groundwater moves up to the surface

mound spring mound formation with water at its centre, which is formed by minerals and sediments brought up by water from artesian basins

FIGURE 6 The Oodnadatta Track area, South Australia

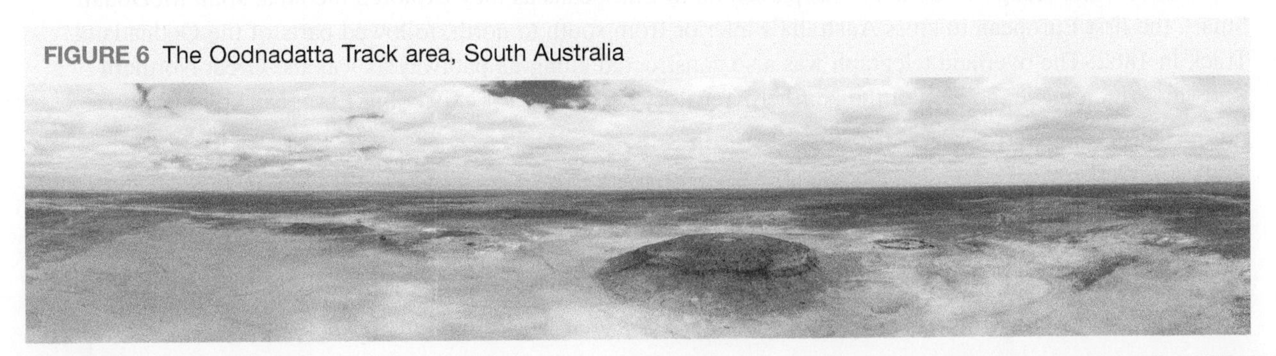

FIGURE 7 The Oodnadatta Track and Great Artesian Basin

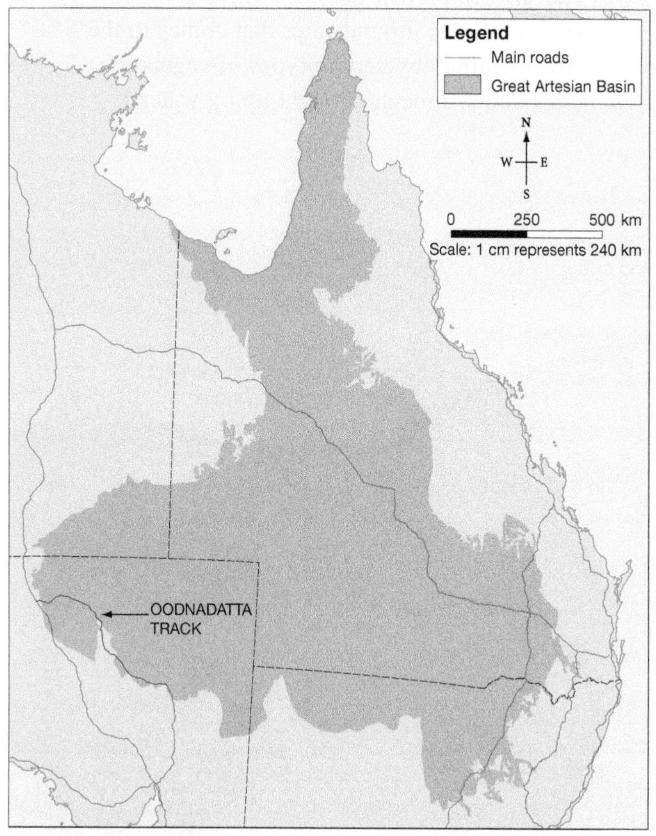

Source: Spatial Vision

FIGURE 8 The Oodnadatta Track passes close to the Old Ghan, the Great Northern Railway

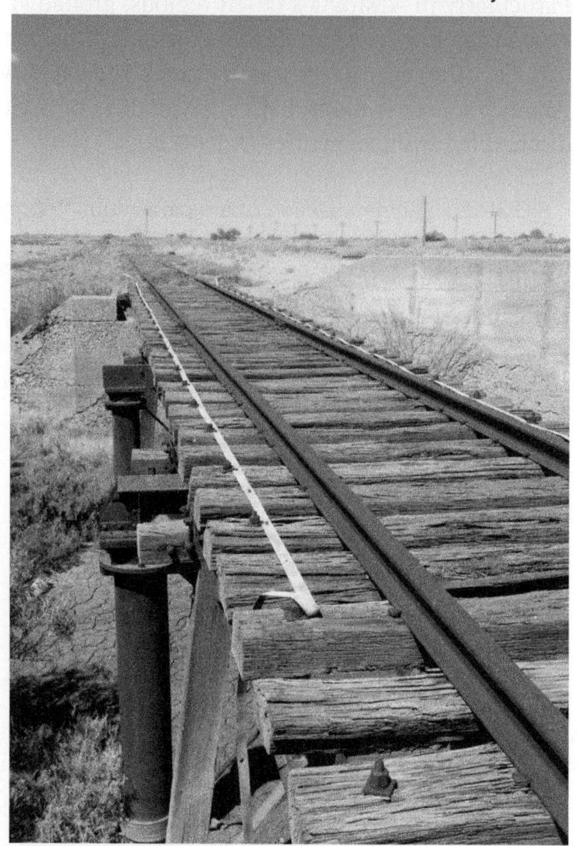

The Ancestors established the springs as water sources and as sites for important ceremonies and events. Knowledge of these water sources and the cultural practices associated with them have been passed down through the generations. Mound springs were particularly important for Aboriginal Peoples. They could rely upon springs as reliable sources of water in a very harsh, dry environment.

Old campsites provide archaeological evidence that people remained at such places for varying time periods. However, because the plant and animal life around these regions is quite sparse during some seasons, people also moved away from the springs when rainfall was plentiful in other parts of their Country.

As the springs were spread over hundreds of kilometres, they were also part of an important network of trading and communication routes across Australia. As Aboriginal Peoples moved around the region, they traded goods and shared information with other groups. This interconnection between Peoples allowed for a valuable trade in resources such as ochre, stone and wooden tools, bailer shells, and pituri (a spindly shrub used during ceremonies and to spike waterholes to catch animals for food).

Some of this knowledge of Country was passed on to Europeans as they explored the area. John McDouall Stuart, the first European to cross Australia's interior from south to north, followed parts of the Oodnadatta Track in 1862. The overland telegraph was also constructed along its pathway as was the Great Northern Railway, which made the land of the Northern Territory accessible for expanding European colonisation.

CASE STUDIES

Dreaming Stories that discuss the creation and location of water

1. **Thutirla Pula (Two Boys Dreaming)**
   This is one of the most important stories of the Wangkangurru and other peoples of Central Australia. Thutirla Pula is how the spirits of the Dreaming first crossed the desert called Munga-Thirri (land of sandhills). One of the most important songlines is the story of the Two Boys Dreaming.

   The story tells of two boys crossing the Simpson Desert, through Queensland and back to just north of Witjira (Dalhousie) in the Finke River area. The songline contains information on every waterhole or soak that was known in the Simpson Desert. Following this songline meant you could cross the Simpson Desert using available groundwater along the way, taking 600 kilometres off the usual journey south of the Simpson Desert to Kati Thanda–Lake Eyre, then back north along the Diamantina River.

2. **Bidalinha (or the Bubbler)**
   The Kuyani ancestor Kakakutanha followed the trail of the rainbow serpent Kanmari to Bidalinha (the Bubbler **FIGURE 9**) where he killed it. He then threw away the snake's head, which is represented by Hamilton Hill, and cooked the body in a *dirga* (oven), which is now Blanche Cup.

   Kakakutanha's wife, angry at missing out on the best meat from the snake, cursed her husband, and he went on to meet a gruesome death at Kudna-ngampa (Curdimurka). The bubbling water represents the movements of the dying serpent.

FIGURE 9 The 'Old Bubbler' on the Oodnadatta Track

3. **Thinti-Thintinha Spring (Fred Springs)**
   The *thunti-thuntinha* (willy wagtail) danced his circular dance to create this spring and the surrounding soils, which are easily airborne in windy conditions. The moral to the story is that, while it is easy to catch the skilful *thunti-thuntinha*, you must never do so because of the terrible dust storms that may follow.
4. **Camp of the Mankarra-kari (Kewson Hill: The Seven Sisters)**
   The Seven Sisters came down here to dig for *yalkapakanha* (bush onions). As they peeled the onions, they tossed the skins to one side, creating the dark-coloured extinct mound spring on the south-west side of the track. The peeled bulbs created the light-coloured yalka-parlumarna hill to the north-east, which is also an extinct mound spring.

FIGURE 10 Groundwater springs along the Oodnadatta Track

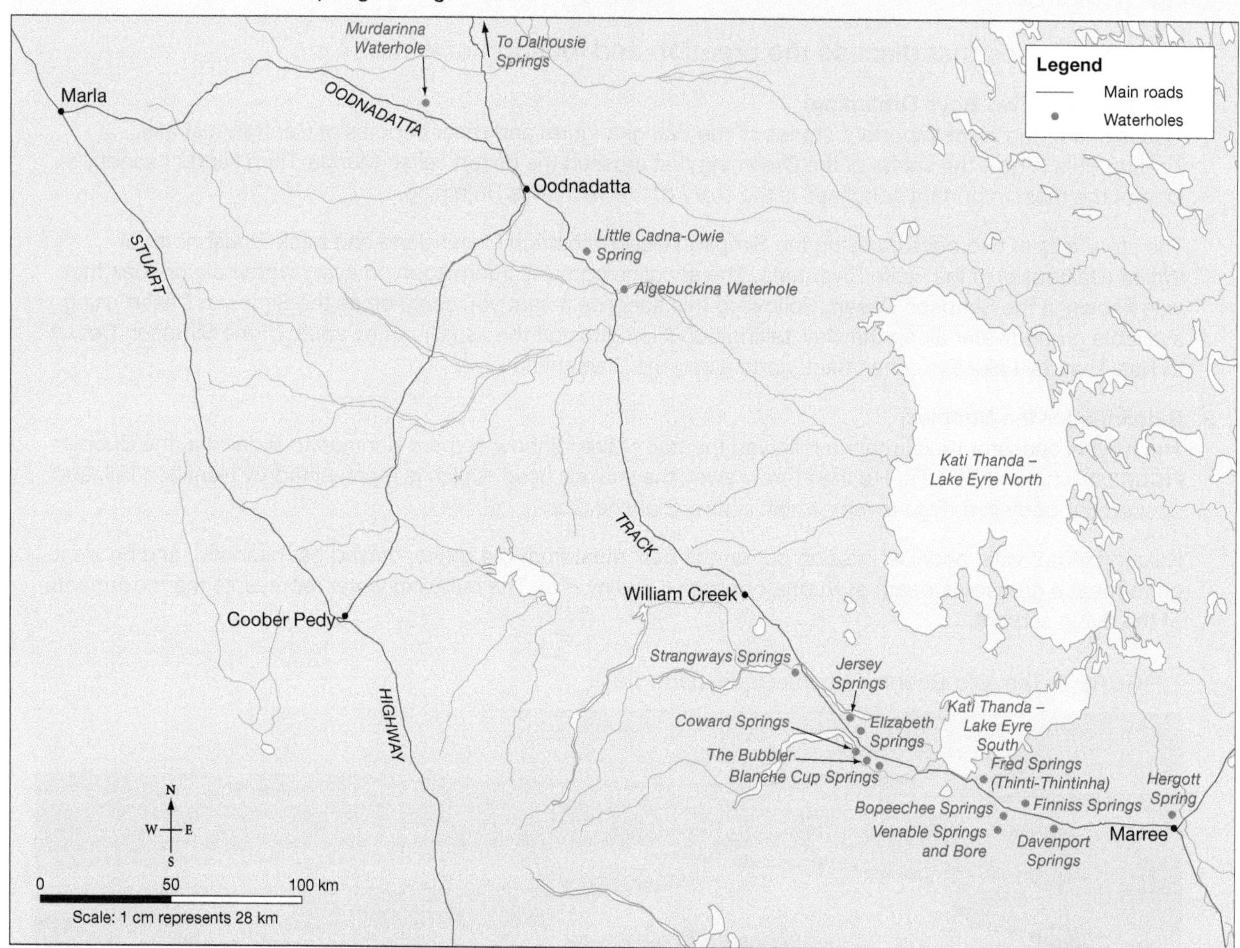

Source: Spatial Vision

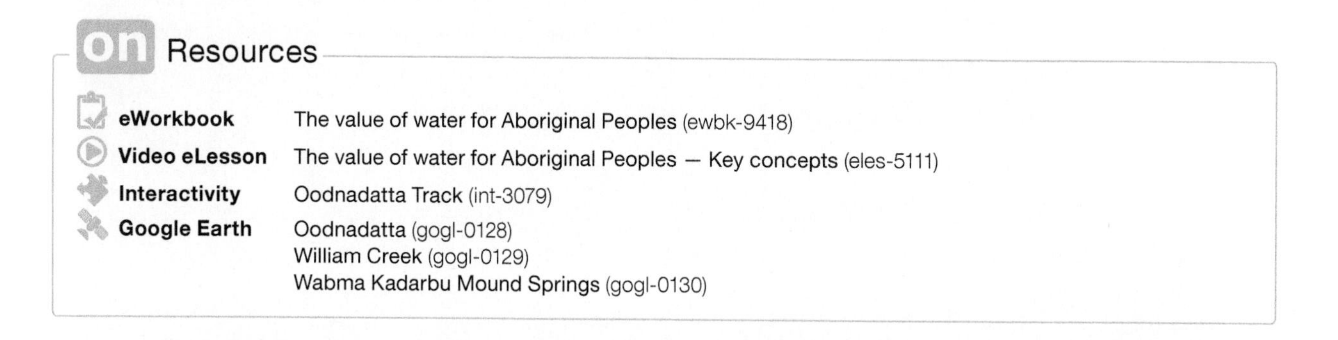

Resources

eWorkbook The value of water for Aboriginal Peoples (ewbk-9418)

Video eLesson The value of water for Aboriginal Peoples — Key concepts (eles-5111)

Interactivity Oodnadatta Track (int-3079)

Google Earth Oodnadatta (gogl-0128)
William Creek (gogl-0129)
Wabma Kadarbu Mound Springs (gogl-0130)

13.8 ACTIVITIES

1. Use Google Earth and enter the search terms Oodnadatta or William Creek to locate the Oodnadatta Track. Describe the landscape you see. Why is finding groundwater so important in this environment?
2. Use an atlas or Google Earth to map the Two Boys Dreaming story from the case study and describe how following the story would have saved a lot of time travelling across the desert.
3. Look at the **Miriwoong seasonal calendar** in your Resources tab to answer the following questions.
   a. Which Australian region does this calendar represent?
   b. How are the seasons divided? How does this compare with a European calendar?
   c. How are seasons and water linked in this calendar?

4. Use Google Earth to locate the Wabma Kadarbu Mound Springs Conservation Park. Place a pin on this location. Now zoom in and out to help you complete the following tasks to annotate the map.
   a. Where is this park located within South Australia? Where is this place in relation to where you live? Use distance and direction in your answer.
   b. What is the name of the nearest road?
   c. Describe the surrounding area.
   d. Why would these springs be so important to Aboriginal Peoples and European explorers?
   e. Do some research to find out why these springs are protected today.

13.8 EXERCISE

Learning pathways

■ LEVEL 1	■ LEVEL 2	■ LEVEL 3
1, 3, 7, 10, 14	2, 4, 8, 11, 13	5, 6, 9, 12, 15

Check your understanding

1. Define the following terms in your own words:
   a. soak
   b. mound spring
   c. groundwater
   d. waterhole.
2. Outline one reason why songlines and Dreaming stories helped Aboriginal Peoples survive in Australia's desert environments.
3. Describe one way that Aboriginal Peoples managed water sources to ensure their food sources were protected.
4. What does evidence show about traditional Aboriginal Peoples managing water supplies?
5. Outline why Dreaming Stories were important to Aboriginal Peoples living on Country in the Oodnadatta region.
6. Explain why water sources were important to the interconnection between Aboriginal Peoples.

Apply your understanding

7. Why might groundwater be important to the communities in the Oodnadatta region today?
8. In what ways have aquaculture practices provided vital resources for Aboriginal Peoples?
9. Assess the importance of Dreaming stories in the creation of water sources.
10. Explain two ways that access to water sources may be difficult today.
11. Analyse two of the case studies in this subtopic and summarise the similarities and differences between them.
12. Using **FIGURE 5** as a reference, explain how people obtain water from a mound spring.

Challenge your understanding

13. What difficulties might Aboriginal Peoples face today in accessing sacred or significant water sites on songlines?
14. Suggest reasons why water sources became important meeting and trading places for Aboriginal Peoples.
15. Provide three recommendations for how Aboriginal knowledge of water can benefit Australian society today.

To answer questions online and to receive **immediate feedback** and **sample responses** for every question, go to your learnON title at www.jacplus.com.au.

13.9 The cultural and spiritual value of water

LEARNING INTENTION

By the end of this subtopic, you will be able to outline why water has cultural value for many First Nations peoples, and provide specific examples of its cultural value for Aboriginal Peoples in Australia.

13.9.1 Water in culture and religion

Water features in the cultures of most peoples around the world. Many famous artworks include water; for example, the famous woodblock print by Japanese artist Katsushika Hokusai shown in **FIGURE 1**. Water also plays an important role in religious traditions around the world. Water is used in practices such as ritual washing or cleansing, and many religious traditions include a belief in water gods or deities, or include stories about floods.

FIGURE 1 'The great wave off Kanagawa', by Katsushika Hokusai (1830), one of 36 images of waves and Mount Fuji created by the artist

13.9.2 First Nations peoples' connections to water

For many First Nations peoples, water has a strong spiritual importance that connects their cultures to the natural world. Water is central to both maintaining life and the cultural practices of many peoples around the world. Water is integral to some sacred landscapes in Aboriginal cultures and for the peoples of the Torres Strait Islands, giving these places significant spiritual value.

The Indigenous Peoples' Water Declaration demonstrates this cultural importance of water. The declaration came from a meeting in 2008 in north-east Arnhem Land, Northern Territory, where First Nations peoples from around the world met to discuss water management issues in their communities.

FIGURE 2 An extract from the Indigenous Peoples' Water Declaration, 2008

We, the indigenous peoples from all parts of the world assembled here, reaffirm our relationship to Mother Earth and responsibility to future generations to raise our voices in solidarity to speak for the protection of water. We were placed in a sacred manner on this earth, each in our own sacred and traditional lands and territories, to care for all of creation and to care for water.

Source: www.indigenouswater.org

13.9.3 The cultural value of water for Aboriginal Peoples

Water has significant cultural value for Australian Aboriginal Peoples in both historical and contemporary contexts. Waterways including rivers, oceans, lakes, lagoons, waterholes and underground water sources are directly linked to Dreaming stories and Creation stories. Explanations of how the world and its landscapes came to be are derived from these stories which are locational and may vary depending on the area.

Water as a common factor of identity usually relates to either of the two distinct water areas: Freshwater Peoples and Saltwater Peoples. Depending on geographical location, cultural practices that incorporate water may vary. For example, Saltwater Peoples will have specific stories and practices that relate to marine life in the ocean, such as whales and sharks. Ceremony, dance, art, hunting practices and totem systems will vary when caring for water sources.

Water has enabled Aboriginal Peoples to sustain societies in one of the driest countries on Earth. Knowledge embedded in Country and maintained through songlines that interconnect the Freshwater Peoples and Saltwater Peoples have also ensured cultural interconnection through water.

FIGURE 3 The cultural value of water

In a report on the cultural values of water for the Anmatyerr Peoples in the Northern Territory it was stated:

> Our cultural values of water are part of our law, our traditional owner responsibilities, our history and our everyday lives. Everyone and everything is related. Our law has always provided for the values we place on water. It is the rules for men, women and country. Anmatyerr Law is strong today, but it is invisible to other people. Australian law should respect Anmatyerr Law so we can share responsibility for looking after water.

Source: Dr N Rea & Anmatyerr Water Project Team, *Provision for Cultural Values in Water Management: The Anmatyerr Story*, Land & Water Australia Final Report (2008)

Water in art

It is highly common, particularly for Desert Peoples, to represent water in paintings of Dreaming stories. This is possibly due to water being scarce in their areas and is a vital way to ensure that knowledge and lore for particular stories are protected, cared for and maintained. The representations of water differ due to different symbols and art styles used across Australia.

13.9.4 Ongoing connections to water

Water continues to be of spiritual and cultural importance for the many Aboriginal Peoples and Torres Strait Islander Peoples across Australia. Due to water being a core aspect of Country (and intrinsically linked to heritage, identity and a sense of belonging) contemporary Aboriginal Peoples and Torres Strait Islander Peoples are still connected with the significant waters within their Country. These connections also help to ensure that ancient beliefs and stories are maintained.

CASE STUDY

Gamay Botany Bay Rangers

In New South Wales, an Aboriginal organisation called the Gamay Botany Bay Rangers is providing care for the waterways of Botany Bay and its surrounding areas. Using Aboriginal cultural knowledge and understanding they are ensuring conservation of Country. In particular, Botany Bay is protected and cared for through educational and cultural strategies.

Key to the delivery of their programs and initiatives is that Aboriginal Peoples are at the forefront, leading and providing awareness for everyone who accesses Botany Bay and its surrounding areas.

FIGURE 4 The Gamay Botany Bay Rangers

Source: ABC Radio Sydney

on Resources

eWorkbook The cultural and spiritual value of water (ewbk-9422)

Video eLesson The cultural and spiritual value of water — Key concepts (eles-5112)

Interactivity The cultural and spiritual value of water (int-8539)

13.9 ACTIVITIES

1. Research a community in Asia and explain the cultural and spiritual significance of water for that community.
2. Using the case study and the **Gamay Botany Bay Rangers** and **Caring for Country** weblinks, assess the ways in which the Gamay Botany Bay Rangers care for water. As a class, list the reasons why it is important for Aboriginal Peoples to be involved in the conservation and management of their local water resources.
3. Use the **Water Dreaming** weblink in your Resources tab to hear an interview with artist Malcolm Maloney Jagamara about water imagery in his art. Research and locate one Aboriginal artwork that tells the story of water and write a short paragraph explaining its meaning and symbolism.
4. Identify any waterways in your local area and region.
   a. Construct a map of these identified waterways.
   b. Research your local area and identify the traditional owners of the land in your area. Identify their significant waters and how they are protected and cared for.

13.9 EXERCISE

Learning pathways

■ LEVEL 1	■ LEVEL 2	■ LEVEL 3
1, 4, 9, 10, 15	2, 3, 7, 8, 13	5, 6, 11, 12, 14

Check your understanding

1. List three reasons why water is important for Aboriginal Peoples.
2. How is water linked to Aboriginal Culture?
3. Describe the connection between water and landscapes.
4. Name the two distinct water areas that Aboriginal Peoples identify with.
5. Identify similarities that First Nations peoples from different places may have in relation to water sources.
6. Describe what it means for a place to have spiritual significance.

Apply your understanding

7. Explain the relationship between water and Dreaming stories.
8. How does the work of groups such as Gamay Botany Bay Rangers benefit all Australians' access to water?
9. Explain and give two examples of the cultural value of waterways for Aboriginal Peoples.
10. Provide two reasons why cultural practices related to water will vary depending on location.
11. Explain how Aboriginal law is connected with water.
12. How is Aboriginal art used to tell the story of water?

Challenge your understanding

13. Propose a strategy that includes Aboriginal Peoples and peoples not from an Aboriginal background working together to conserve and protect their water sources.
14. Why do First Nations people care so deeply about water? Give examples from the Indigenous Peoples' Water Declaration and the cultural values of water for the Anmatyerr people in your answer.
15. Suggest one way that you could learn more about the cultural and spiritual importance of water to the Aboriginal Peoples of your area.

To answer questions online and to receive **immediate feedback** and **sample responses** for every question, go to your learnON title at www.jacplus.com.au.

13.10 The aesthetic and social value of water

LEARNING INTENTION

By the end of this subtopic, you will be able to explain some of the ways in which water has aesthetic and social value.

13.10.1 Aesthetic value

Water can have an aesthetic value. **Aesthetics** is about the nature and appreciation of beauty, and can be used to describe the way people respond to the environment. It includes the way people interact with a place through their senses and the emotions they experience.

Water contributes to the visual quality of an environment. Houses and units near rivers, lakes and the ocean tend to be much more expensive than properties some distance from the water. In part, this takes into consideration the aesthetic value of a water view. People enjoy being able to watch the waves breaking, or to watch sailboats on a lake from their homes.

Together with vegetation, water is one of the most important features of an environment in providing emotional and **psychological benefits** to people. Water is required for basic human survival, and environments that contain a large amount of water make people feel safe and nurtured. Being close to water and appreciating its beauty can help people feel relaxed and calm.

FIGURE 1 Water can help people feel calm and relaxed.

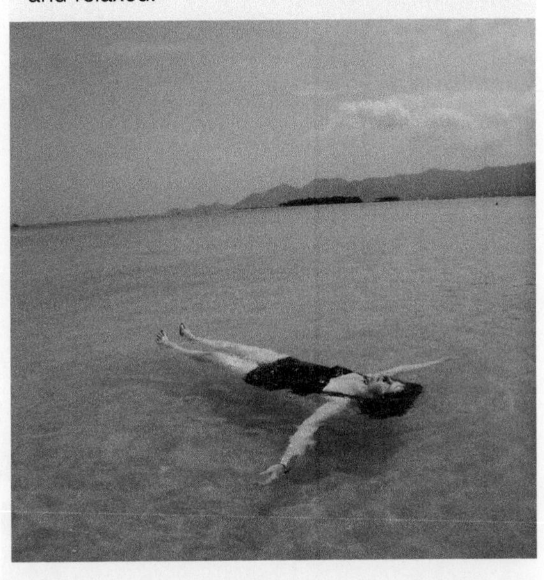

FIGURE 2 The aesthetic characteristics of water

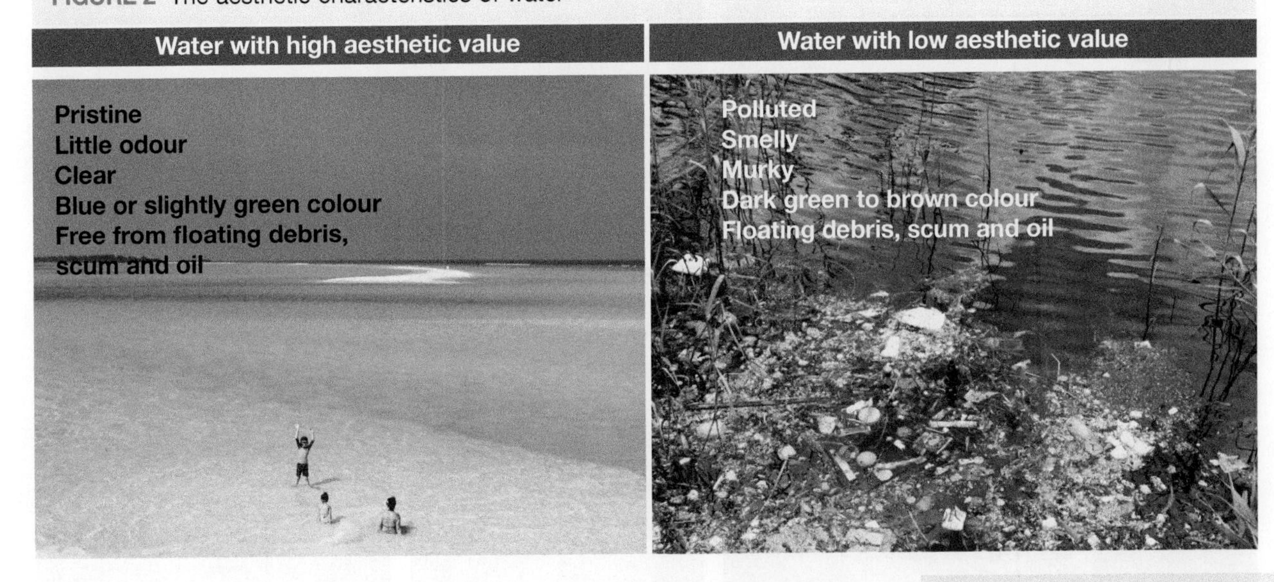

aesthetics (aesthetic qualities) the beautiful or attractive features of something

psychological benefits positive effects on emotional or mental wellbeing

CASE STUDY

The aesthetic value of the Great Barrier Reef

When people visit the Great Barrier Reef they feel a sense of beauty, naturalness and remoteness. The amazing contrasting colours of the water, from deep, dark blues to **iridescent** aqua, contribute to the natural beauty of the reef. The bright white sand, the green islands and the unusual shapes of the reefs and cays create beautiful mosaic patterns in the water.

iridescent bright and colourful

FIGURE 3 The beauty of living creatures, and vegetation above and below the waters surrounding Fitzroy Island, on the Great Barrier Reef, Queensland

13.10.2 Social value

Water has a social value as it helps people undertake activities that maintain a healthy lifestyle. People are able to undertake many activities such as swimming, kayaking, sailing, and a range of other water sports and social interactions.

FIGURE 4 Surfing can provide a physical activity, connection with the water and a social outlet.

FOCUS ON FIELDWORK

fdw-0019

Perceptions of water's value

There are many factors that contribute to the way we value water. Conducting your own survey of a local waterway, or examining images or footage of waterways, can help you collect qualitative data about the value of a waterway from your perspective.

Qualitative data is subjective (based on opinion). This means that it cannot easily be calculated or measured, but there are ways to record your views and attitudes to a place. For example, you might create a rating system or use descriptions based on your senses (what you saw, heard, smelled, felt and, perhaps, tasted during your visit).

Examine your own perceptions of a waterway using the **Perceptions of water's value** fieldwork activity in your Resources tab.

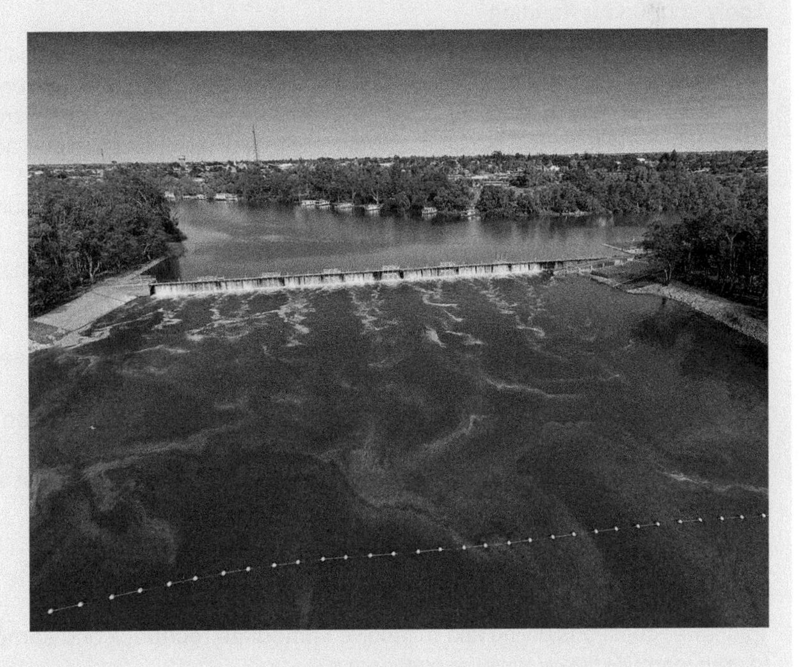

Resources

eWorkbook	The aesthetic and social value of water (ewbk-9426)
Video eLesson	The aesthetic and social value of water — Key concepts (eles-5113)
Interactivity	The aesthetic and social value of water (int-8540)
Fieldwork	Perceptions of water's value (fdw-0019)

13.10 ACTIVITY

As a class or in small groups, brainstorm the kinds of water environments you have visited.

- What makes them beautiful?
- Are they also useful to humans or used by us in other ways?
- Are they protected because of their beauty? Or are there other reasons?

13.10 EXERCISE

Learning pathways

■ LEVEL 1	■ LEVEL 2	■ LEVEL 3
2, 3, 8, 9, 14	1, 5, 6, 11, 13	4, 7, 10, 12, 15

Check your understanding

1. In your own words, define the term *aesthetics*.
2. List three features of a beach that has high aesthetic value.
3. List three features of a river that gives it social value.
4. Give one example of how water adds to the aesthetics of an inland environment.
5. Environments with lots of water generally make people feel _______________.
6. Identify two psychological benefits of being near or watching the ocean.
7. Apart from water, what other natural feature provides important emotional and psychological benefits to people?

Apply your understanding

8. Would you expect a home in a coastal town to be more expensive than the same kind of home in a remote inland town of the same size with the same facilities? Give reasons for your decision.
9. Explain why it would be important to have clean water in an area where there are swimming activities.
10. Discuss the following statement: Water has many psychological and aesthetic benefits but does not have any social value.
11. Coral bleaching is a process from which coral loses its beautiful colours. How would this change the aesthetic value of the Great Barrier Reef.
12. Explain the social value of water. Use one specific example to demonstrate your explanation.

Challenge your understanding

13. Suggest whether people might still want to visit Fitzroy Island for its aesthetic value if it wasn't close to the Great Barrier Reef.
14. Rubbish on beaches decreases the area's aesthetic value, and yet some people still leave rubbish on beaches. Propose one strategy to stop people littering beaches.
15. Should places be protected if they are considered beautiful? Suggest why deciding which places should be protected solely for their beauty might be complicated.

To answer questions online and to receive **immediate feedback** and **sample responses** for every question, go to your learnON title at www.jacplus.com.au.

13.11 Virtual water

LEARNING INTENTION

By the end of this subtopic, you will be able to define the term *virtual water* and explain the value of water to food production.

13.11.1 Virtual water

The water we consume is not just what we use in cooking, drinking, washing, flushing or in the garden. Water is used to manufacture everything we use, including mobile phones, toys, cars and newspapers. This **virtual water** needs to be accounted for in our water footprint.

Virtual water is also known as embedded water, embodied water or hidden water. It includes all the water used to produce goods and services. Food production uses more water than any other type of production.

Hidden in a cup of coffee are 140 litres of water used to grow, produce, package and ship the coffee beans. That is roughly the same amount of water used by an average person per day in Australia for drinking and household needs. There is a lot of water hidden in a hamburger too: 2400 litres. This includes the water needed to grow the feed for the cattle over a number of years, to grow wheat for the bread roll, to grow all the other ingredients in the hamburger, and to process all the food.

Virtual water varies from food to food. For example, it takes about 3400 litres of water to grow 1 kilogram of rice, whereas it takes 200 litres to grow 1 kilogram of cabbage. Regions that are water stressed and that export food and other products (such as Australia and some countries in Africa and Asia) are effectively exporting their precious water in these goods.

FIGURE 1 Virtual water is the water used to produce our goods and services, such as the water used to irrigate food crops.

A country that imports rice, rather than growing it locally, therefore saves 3400 litres of water for every kilogram it imports. Some countries (such as Japan) have very little land on which to grow food; other countries have very few cubic metres of renewable water per person. Singapore, for example, has only about 130 cubic metres per person. Both types of countries survive by virtual water imports: they import food rather than attempt to grow and produce all their food themselves. This means that small, wealthy countries can import food that needs a lot of water to produce, and export products that need little water to produce. This makes water available for domestic purposes such as drinking and cooking.

The major exporters of virtual water are found in North and South America (the United States, Canada, Brazil and Argentina), South Asia (India, Pakistan, Indonesia, Thailand) and Australia. The major virtual water importers are North Africa and the Middle East, Mexico, Europe, Japan and South Korea.

virtual water all the water used to produce goods and services. Food production uses more water than any other production.

13.11.2 Water footprint

The **water footprint** of an individual or country is the total volume of fresh water that is used to produce the goods and services consumed by the individual or country. It includes the use of:

- blue water (rivers, lakes, aquifers)
- green water (rainfall used for crop growth)
- grey water (water polluted after agricultural, industrial and household use).

Not all goods consumed in one particular country are produced in that country — some foods and products are imported. Therefore the water footprint consists of two parts: use of domestic water resources and use of water outside the borders of the country.

In the United States, the average water footprint per person per year is 2480 cubic metres, which is enough to fill an Olympic-size swimming pool. In China, the average water footprint per person is 1071 cubic metres per year. The figure for Australia is 1393 cubic metres per person per year. Japan, with a footprint of 1200 cubic metres per person per year, has about 65 per cent of its total water footprint outside its borders, meaning a lot of its water is imported in the form of consumer goods and food. On the other hand, only about 7 per cent of the Chinese water footprint falls outside China.

water footprint the total volume of fresh water that is used to produce the goods and services consumed by an individual or country

The Middle East and Northern Africa rely heavily on imports for agricultural products. This means that many items, such as food, are bought from other countries. As a result the region also imports large amounts of virtual water.

FIGURE 2 Average water footprints

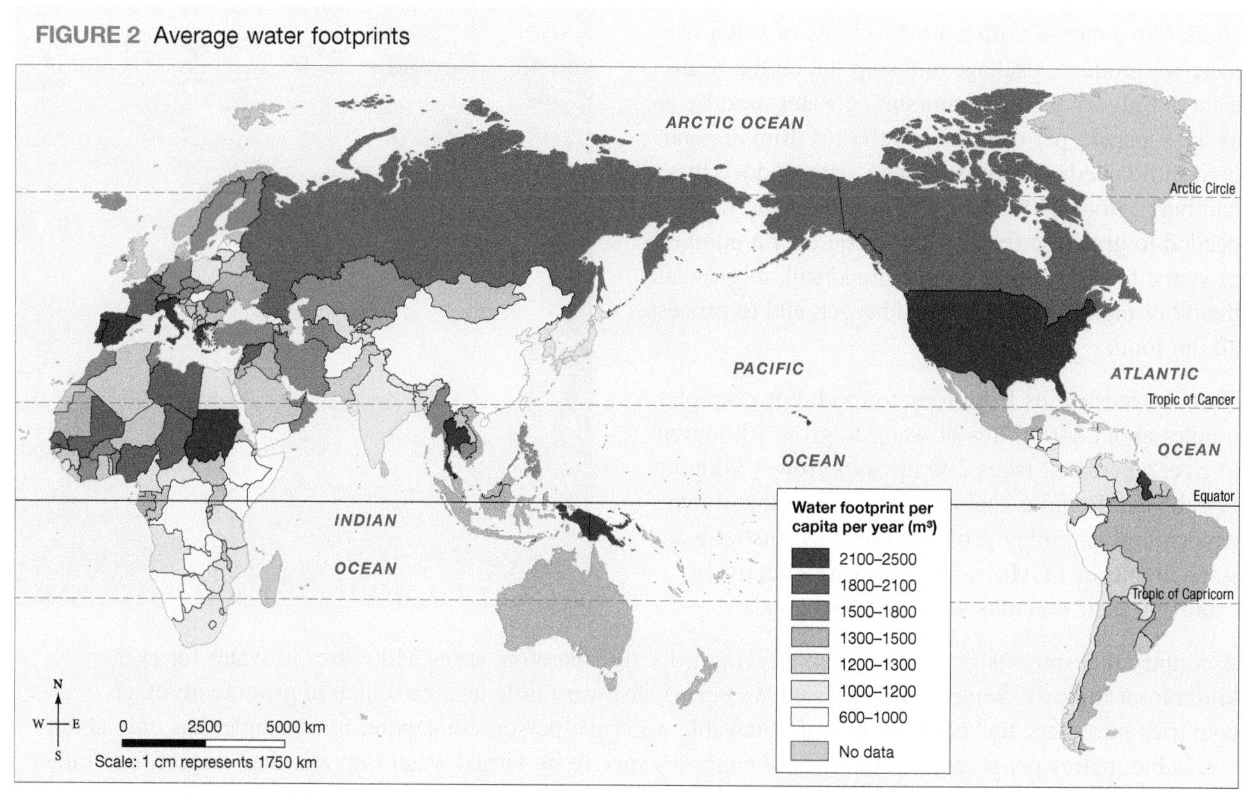

Source: World Water Exchange (2016)

TABLE 1 The water used to grow food and other household items varies from product to product.

Item	Unit	Global average water (litres)
Apple or pear	1 kg	700
Barley	1 kg	1300
Banana	1 kg	860
Beef	1 kg	15500
Beer (from barley)	250 mL	75
Bread (from wheat)	1 kg	1300
Cabbage	1 kg	200
Cheese	1 kg	5000
Chicken	1 kg	3900
Chocolate	1 kg	24000
Coconut	1 kg	2500
Coffee (roasted)	1 kg	21000
Cotton shirt	1	2700
Cucumber or pumpkin	1 kg	240
Dates	1 kg	3000
Eggs	1	200
Goat meat	1 kg	4000
Hamburger	1	2400
Lamb	1 kg	6100
Leather	1 kg	16600
Lettuce	1 kg	130
Maize	1 kg	900
Mango	1 kg	1600
Millet	1 kg	5000
Milk	250 mL	250
Olives	1 kg	4400
Orange	1	50
Paper	A4 sheet	10
Peach or nectarine	1 kg	1200
Peanuts (in shell)	1 kg	3100
Pork	1 kg	4800
Potato	1 kg	4800
Rice	1 kg	3400
Soya beans	1 kg	1800
Sugar (sugar cane)	1 kg	1500
Tea	250 mL	30
Tomato	1 kg	180
Wheat	1 kg	1300
Wine	125 mL	120

Source: John Wiley & Sons Australia

on Resources

eWorkbook Virtual water (ewbk-9430)

Video eLesson Virtual water — Key concepts (eles-5114)

Interactivity Unreal (int-3080)

myWorldAtlas Deepen your understanding of this topic with related case studies and questions > **Our water footprint**

13.11 ACTIVITIES

1. Study **TABLE 1**. Choose three meat, five grain, two dairy, two non-food, four fruit, four vegetable and two processed products from the list. Create a bar graph to show how much water is used to produce a vegetarian diet and a meat-based diet. Which diet uses more water?
2. Use the **Just add water** weblink in the Resources tab to listen to an audio program about the water footprint in food production.
   a. What is the relationship between water-stressed countries and food production?
   b. Give an example where the water footprint figure is in conflict with the opinion of farmers.
   c. Find as much information as you can about a product produced in Wodonga that includes both virtual and domestic water.
3. Conduct a debate on the following statement: That people should eat less meat in order to consume less water.

13.11 EXERCISE

Learning pathways

■ LEVEL 1	■ LEVEL 2	■ LEVEL 3
2, 3, 5, 9, 14	1, 7, 8, 10, 13	4, 6, 11, 12, 15

Check your understanding

1. Define the terms *virtual water* and *water footprint* in your own words.
2. Refer to **FIGURE 2**. Describe the patterns you notice between countries with (i) very high and high water footprints and (ii) very low and low water footprints.
3. Describe the difference between virtual water and a water footprint.
4. Describe how the scale of footprints is different for countries with (i) very high and high water footprints and (ii) very low and low water footprints.
5. Outline the differences between blue, green and grey water.
6. Approximately how many litres of water is used to make:
   a. a hamburger
   b. a cup of tea
   c. a sheet of A4 paper
   d. an egg?
7. Write three summary statements that describe the water footprints of two regions.

Apply your understanding

8. Explain the difference between virtual water and a water footprint.
9. Explain how water is embedded (or hidden) in a cup of coffee.
10. Create a bar graph to show how much water is used to produce a vegetarian diet and a meat-based diet.
11. How do water footprints connect places that are far from each other?
12. 'In a dry country such as Australia, it makes more sense to import food and resources that require a lot of water to produce.' Drawing on your understanding of virtual water, provide an argument for and an argument against this viewpoint. Ensure that your arguments are supported with evidence and are logical.

Challenge your understanding

13. Should people eat less meat in order to consume less water? Give reasons, based on data, for your opinion.
14. Predict whether the proportion of Australia's water footprint outside its borders will increase or decrease in the future. Give reasons for your view.
15. Propose one way that more grey water could be used to reduce Australia's green and blue water footprints.

To answer questions online and to receive **immediate feedback** and **sample responses** for every question, go to your learnON title at www.jacplus.com.au.

13.12 Investigating topographic maps — The value of water in Noosa

LEARNING INTENTION

By the end of this subtopic, you will be able to describe perceptions of the social, cultural and economic value of water at Noosa.

13.12.1 Noosa's beaches and river

Noosa is about 140 kilometres north of Brisbane, Queensland, between the mouth of the Noosa River and the headland of Noosa National Park. Noosa demonstrates how water can have value in different ways.

Noosa is a popular tourist destination and, as such, the water at nearby beaches and in local rivers and lakes contributes economic value to the area. Hotels with water views will charge tourists higher prices than those without water views. Similarly, Noosa's water has an aesthetic value — it is visually appealing and is part of the attraction of Noosa. People enjoy looking out to the ocean or over the river. The aesthetic and social value of water helps visitors relax and unwind, and the many water-based activities in Noosa such as cruises, canoeing and kite-surfing contribute to the culture of place.

FIGURE 1 Visitors enjoy the spectacular views from Noosa National Park.

FIGURE 2 An oblique aerial photograph of Noosa, 2009

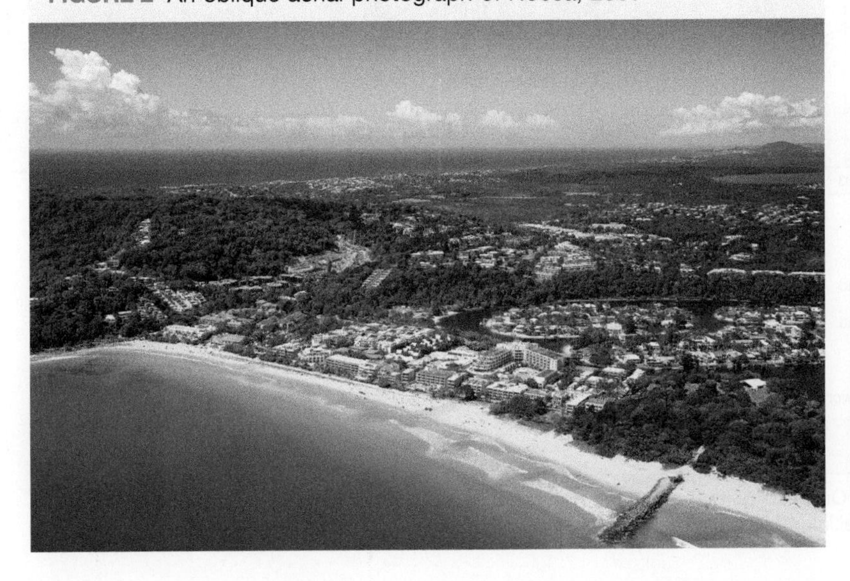

FIGURE 3 Many activities for holiday makers at Noosa are water-based.

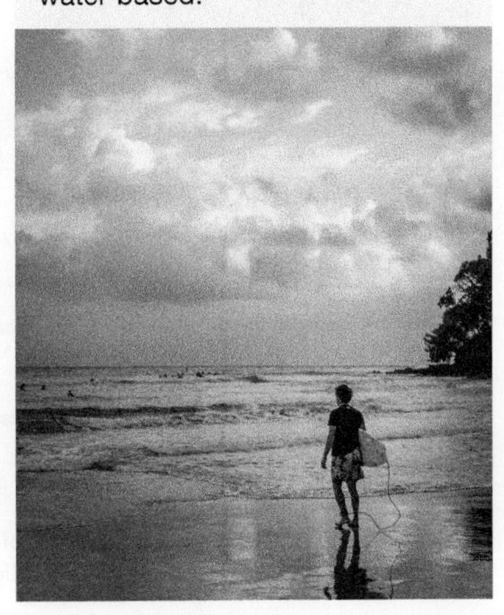

int-8541

FIGURE 4 Topographic map extract of Noosa

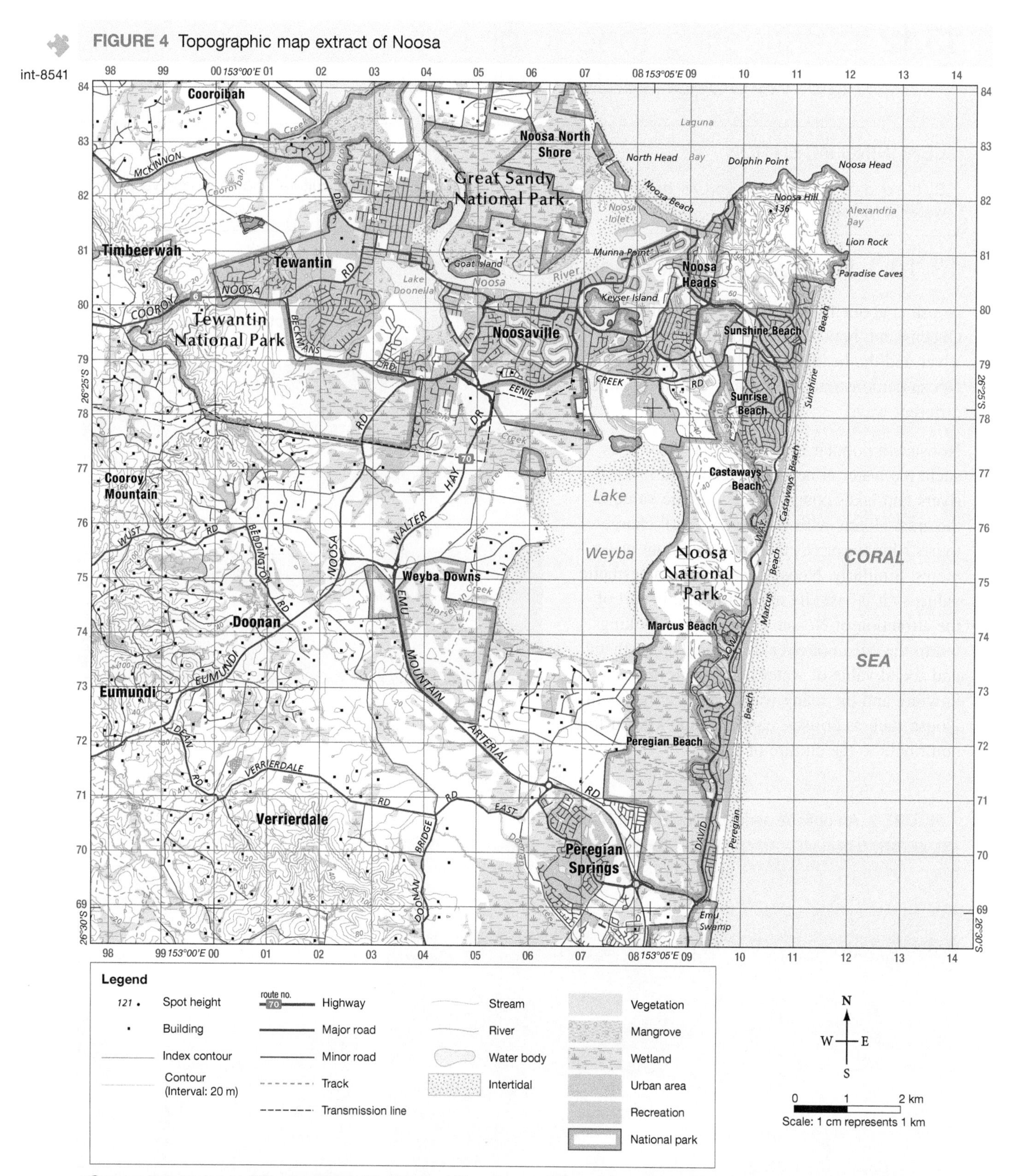

Source: Data based on QSpatial, State of Queensland (Department of Natural Resources, Mines and Energy, Department of Environment and Science), http://qldspatial.information.qld.gov.au/catalogue/

Resources

eWorkbook	Investigating topographic maps — The value of water in Noosa (ewbk-9434)
Digital document	Topographic map of Noosa (doc-36257)
Video eLesson	Investigating topographic maps — The value of water in Noosa — Key concepts (eles-5115)
Interactivity	Investigating topographic maps — The value of water in Noosa (int-8541)
Google Earth	Noosa Heads (gogl-0092)

13.12 EXERCISE

Learning pathways

■ LEVEL 1	■ LEVEL 2	■ LEVEL 3
1, 2, 8	3, 4, 5, 7	6, 9, 10

Check your understanding

Examine **FIGURE 4**.

1. State the scale of the map in one sentence.
2. State the contour interval of the map.
3. State the feature at the following area references:
   a. AR0376
   b. AR0470
   c. AR0583.
4. State the distance from the breakwater at AR0781 to the causeway at AR0380.
5. State the altitude of:
   a. the spot height at AR1081
   b. the sewage treatment works at AR0878
   c. the caravan park at AR0780.
6. List different ways in which people use the water around Noosa. What evidence is there on the map for these uses?

Apply your understanding

7. Explain why you think the settlements around Noosa have developed in their current locations. Support your answer with observations from **FIGURE 4**.
8. Explain why the settlements located in AR1079 and AR0880 are separated.
9. Throughout this chapter we have examined the economic, aesthetic, cultural and spiritual value of water. How do you think the settlement patterns around Noosa have been influenced by:
   a. the economic value of water
   b. the aesthetic value of water
   c. the cultural and spiritual value of water?

Challenge your understanding

10. Predict how increased tourism and development around Noosa may have an impact on the value of water in the region in the future. In your answer, refer to the economic, aesthetic and cultural/spiritual value of water.

To answer questions online and to receive **immediate feedback** and **sample responses** for every question, go to your learnON title at www.jacplus.com.au.

13.13 Thinking Big research project — The Great Artesian Basin

online only

The content in this subtopic is a summary of what you will find in the online resource.

Scenario

It's the year 2100, and Australia is relying more and more on groundwater for its water supply. The federal government wants to supply Victoria and Western Australia with groundwater from the Great Artesian Basin (GAB), but the Queensland, New South Wales, South Australian and Northern Territory governments are against this proposal as they think the water will be depleted if it is not used sustainably.

Task

Your team of engineers has been commissioned by the Queensland, New South Wales, South Australian and Northern Territory governments to present a poster display to the federal Minister for Water. You need to convince her that using the GAB to supply water to distant states is not a solution to the water crisis.

Go to your Resources tab to access the resources you need to complete this research project.

Resources

ProjectsPLUS Thinking Big research project — The Great Artesian Basin (pro-0233)

13.14 Review

13.14.1 Key knowledge summary

13.2 Water use

- The global average water use is 1240 cubic metres per person per year.
- Most of the world's water is used for agriculture to grow food for the world's population.
- Around 70 per cent of Australia's fresh water is used as irrigation for farming.
- Rainfall in Australia is limited and highly seasonal in some areas.

13.4 The value of clean water

- Water has economic, cultural, spiritual and social value.
- The use of water in agriculture, income creation and to produce goods means that water has economic value.
- Water can have a spiritual value because it can be used in religious and spiritual beliefs and ceremonies.
- Water can have cultural and social value because it is used in recreational activities, traditional practices and lifestyle choices.

13.6 The economic value of water

- Water is used in a variety of ways to produce goods and services, including drinking water, irrigation, industry, recreation and power generation.
- The world's top five rice producers are found in Asia. Rice requires a high amount of water availability.
- Aquaculture involves breeding and harvesting plants and fish in water environments such as ponds, cages, rafts or long lines. It can support local communities, improve standards of living and generate employment and income.

13.7 The value of water to health

- Nearly 780 million people in the world have no access to clean water and 2.4 billion people have no safe way of disposing of human waste.
- Dirty water and lack of hygiene kill 3.4 million people around the world every year.
- Goal 6 of the Sustainable Development Goals of 2015 was 'Ensure access to water and sanitation for all', and in 2016 the United Nations General Assembly adopted the resolution 'International Decade for Action — Water for Sustainable Development' (2018–2028).

13.8 The value of water for Aboriginal Peoples

- Before European colonisation, Aboriginal Peoples managed Australia's water resources very carefully because of its importance to maintaining life (through food and drink) and Culture.
- Knowledge of where to find water is passed down from generation to generation; for example, through Dreaming stories.
- Groundwater and mound springs are very important water sources in dry environments.

13.9 The cultural and spiritual value of water

- Water is very important to the cultures of many First Nations peoples around the world.
- Aboriginal Peoples' sense of identity and belonging is tied to their connection to the water of their Country.
- Symbols of water used in Aboriginal Cultures' art differ between different Peoples.
- Water still connects Aboriginal Peoples to their Country today, and so local Aboriginal Peoples are often asked to help conserve special places using Cultural knowledge.

13.10 The aesthetic and social value of water

- Water contributes to the visual quality of the environment, and improves the value of property and people's enjoyment.
- Water provides emotional and psychological benefits to people.
- Water has a social value as it helps people undertake activities to maintain a healthy lifestyle and social interactions.

13.11 Virtual water

- Virtual water is embedded water and includes all the water used to produce goods and services.
- Food production uses more water than any other type of production.
- The water footprint of a country is the total volume of fresh water that is used to produce the goods and services consumed by that country and includes blue water and green water.

13.14.2 Key terms

aesthetics (aesthetic qualities) the beautiful or attractive features of something

aquifer a body of permeable rock below the Earth's surface which contains water, known as groundwater. Water can move along an aquifer.

Dreaming in Aboriginal cultures, the time when the Earth took on its present form, and cycles of life and nature began; also known as the Dreamtime. Dreaming stories pass on important knowledge laws and beliefs.

economic a system made up of producing, buying and selling goods and services

industry the production of goods and services

iridescent bright and colourful

mound spring mound formation with water at its centre, which is formed by minerals and sediments brought up by water from artesian basins

psychological benefits positive effects on emotional or mental wellbeing

soak place where groundwater moves up to the surface

songline directly linked to Dreamings and Creation Stories, which depict the paths travelled by ancient creator spirits. They incorporate knowledge, spirituality and beliefs, which are embedded in Country and demonstrate the interconnectedness between the spiritual, physical and living worlds.

turbidity the amount of sediment suspended in water

virtual water all the water used to produce goods and services. Food production uses more water than any other production.

water footprint the total volume of fresh water that is used to produce the goods and services consumed by an individual or country

13.14.3 Reflection

Complete the following to reflect on your learning.

Revisit the inquiry question posed in the Overview:

We can't survive without water, but does it mean more than just survival — do we value water in other ways?

1. Now that you have completed this topic, what is your view on the question? Discuss with a partner. Has your learning in this topic changed your view? If so, how?
2. Write a paragraph in response to the inquiry question outlining your views.

Subtopic	Success criteria	●	●	●
13.2	I can describe the ways water is used by people, and give examples of agricultural, commercial, industrial and recreational uses.			
13.3	I can represent information using line graphs.			
13.4	I can explain how water quality may affect perceptions of the value of water.			
	I can explain some of the ways that water quality is measured.			
13.5	I can identify and annotate important geographical information in a photograph.			
13.6	I can explain why water has economic value.			
	I can demonstrate the economic value of water with examples from the rice farming and aquaculture industries in Asia.			
13.7	I can explain the value of clean water for people's health.			
	I can evaluate strategies to improve access to clean water around the world.			
13.8	I can describe some of the ways that Aboriginal Peoples find, use and manage water resources.			
	I can outline how Aboriginal Culture and water resource management are interconnected.			
13.9	I can outline why water has cultural value for many First Nations peoples.			
	I can provide specific examples of water's cultural value for Aboriginal Peoples in Australia.			
13.10	I can analyse and describe the aesthetic and social value of water.			
13.11	I can define the term *virtual water*.			
	I can explain the value of water to food production.			
13.12	I can describe perceptions of the social, cultural and economic value of water at Noosa.			

on Resources

eWorkbook Chapter 13 Student learning matrix (ewbk-8484)
Chapter 13 Reflection (ewbk-8485)
Chapter 13 Extended writing task (ewbk-8486)

Interactivity Chapter 13 Crossword (int-8542)

ONLINE RESOURCES

on Resources

Below is a full list of **rich resources** available online for this topic. These resources are designed to bring ideas to life, to promote deep and lasting learning and to support the different learning needs of each individual.

eWorkbook

- 13.1 Chapter 13 eWorkbook (ewbk-7993) ☐
- 13.2 Water use (ewbk-9394) ☐
- 13.3 SkillBuilder — Drawing a line graph (ewbk-9398) ☐
- 13.4 The value of clean water (ewbk-9402) ☐
- 13.5 SkillBuilder — Annotating a photograph (ewbk-9406) ☐
- 13.6 The economic value of water (ewbk-9410) ☐
- 13.7 The value of water to health (ewbk-9414) ☐
- 13.8 The value of water for Aboriginal Peoples (ewbk-9418) ☐
- 13.9 The cultural and spiritual value of water (ewbk-9422) ☐
- 13.10 The aesthetic and social value of water (ewbk-9426) ☐
- 13.11 Virtual water (ewbk-9430) ☐
- 13.12 Investigating topographic maps — The value of water in Noosa (ewbk-9434) ☐
- 13.14 Chapter 13 Student learning matrix (ewbk-8484) ☐
  Chapter 13 Reflection (ewbk-8485) ☐
  Chapter 13 Extended writing task (ewbk-8486) ☐

Sample responses

- 13.1 Chapter 13 Sample responses (sar-0146) ☐

Digital documents

- 13.2 Regions (doc-17950) ☐
- 13.12 Topographic map of Noosa (doc-36257) ☐

Video eLessons

- 13.1 Water — A vital resource (eles-1615) ☐
  Value of water — Photo essay (eles-5106) ☐
- 13.2 Water use — Key concepts (eles-5107) ☐
- 13.3 SkillBuilder — Drawing a line graph (eles-1635) ☐
- 13.4 The value of clean water — Key concepts (eles-5108) ☐
- 13.5 SkillBuilder — Annotating a photograph (eles-1633) ☐
- 13.6 The economic value of water — Key concepts (eles-5109) ☐
- 13.7 The value of water to health — Key concepts (eles-5110) ☐
- 13.8 The value of water for Aboriginal Peoples — Key concepts (eles-5111) ☐
- 13.9 The cultural and spiritual value of water — Key concepts (eles-5112) ☐
- 13.10 The aesthetic and social value of water — Key concepts (eles-5113) ☐
- 13.11 Virtual water — Key concepts (eles-5114) ☐
- 13.12 Investigating topographic maps — The value of water in Noosa — Key concepts (eles-5115) ☐

Interactivities

- 13.2 Water use (int-7780) ☐
- 13.3 SkillBuilder — Drawing a line graph (int-3131) ☐
- 13.4 The value of clean water (int-8535) ☐
- 13.5 SkillBuilder — Annotating a photograph (int-3129) ☐
- 13.6 The economic value of water (int-8536) ☐
- 13.7 The value of water to health (int-8537) ☐
- 13.8 Oodnadatta Track (int-3079) ☐
- 13.9 The cultural and spiritual value of water (int-8539) ☐
- 13.10 The aesthetic and social value of water (int-8540) ☐
- 13.11 Unreal (int-3080) ☐
- 13.12 Investigating topographic maps — The value of water in Noosa (int-8541) ☐
- 13.14 Chapter 13 Crossword (int-8542) ☐

Fieldwork

- 13.4 Investigating waterways (fdw-0018) ☐
- 13.10 Perceptions of water's value (fdw-0019) ☐

ProjectsPLUS

- 13.13 Thinking Big research project — The Great Artesian Basin (pro-0233) ☐

Weblinks

- 13.7 Health is wealth (web-6053) ☐
- 13.8 How the water got to the plains (web-4029) ☐
  Miriwoong seasonal calendar (web-4033)
- 13.9 Water Dreaming (web-2860) ☐
  Gamay Botany Bay Rangers (web-6054) ☐
  Caring for Country (web-6055) ☐
- 13.11 Just add water (web-4032) ☐

Google Earth

- 13.8 Oodnadatta (gogl-0128) ☐
  William Creek (gogl-0129) ☐
  Wabma Kadarbu Mound Springs (gogl-0130) ☐
- 13.12 Noosa Heads (gogl-0092) ☐

myWorld Atlas

- 13.2 How is water used in Australia? (mwa-4537) ☐
  Three rivers in Africa (mwa-4541) ☐
- 13.7 World health (mwa-4430) ☐
- 13.11 Our water footprint (mwa-4540) ☐

Teacher resources

There are many resources available exclusively for teachers online.

14 Water scarcity and management

INQUIRY SEQUENCE

To access a pre-test with **immediate feedback** and **sample responses** to every question in this chapter, select your learnON format at www.jacplus.com.au.

14.1 Overview

Numerous **videos** and **interactivities** are embedded just where you need them, at the point of learning, in your learnON title at www.jacplus.com.au. They will help you to learn the content and concepts covered in this topic.

Our planet is covered with water, so why do we have to be so careful with it?

14.1.1 Introduction

Every person on the planet interacts with the weather on a daily basis. Sometimes we feel hot, sometimes we feel cold. Climate patterns affect water availability in Australia and around the world. Change in the weather and climate patterns also affects the environment. If there is too little rain, drought can develop, sometimes producing heatwaves — days of dry, hot weather. If there is too much rain, flooding will occur. Changes to climate patterns can make water availability very unpredictable, and places that once had water readily available may now suffer from water scarcity. Water management can help communities and individuals gain more predictable access to water.

STARTER QUESTIONS

1. List three things you would expect to happen if an area did not get its usual rainfall for a long time.
2. How would water scarcity impact on the way you live your life?
3. List two natural environments that have very little water. Name where they are or show their location on a map.
4. What are ways you could manage water use in your own home?
5. Watch the **Water scarcity — Photo essay** Video eLesson in your Resources tab. Consider some of the different ways that people in your community would be affected by a water shortage.

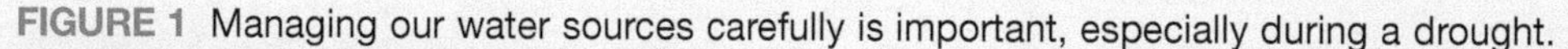

FIGURE 1 Managing our water sources carefully is important, especially during a drought.

Resources

eWorkbook	Chapter 14 eWorkbook (ewbk-7994)
Video eLessons	Of droughts and flooding rains (eles-1617)
	Water scarcity – Photo essay (eles-5347)

14.2 Australia's water resources

LEARNING INTENTION

By the end of this subtopic, you will be able to identify the types and locations of Australia's water resources.

14.2.1 Australia's rivers and lakes

Like any other resource, water has great value when a use is found for it. Environments where water is found are also a resource. A river can be the site of a settlement that provides transport as well as food for the people who live there. A riverine environment that includes fish, birds, wildlife, wetlands, plants and micro-organisms is also valuable as a living system and can therefore be regarded as a resource.

Many parts of Australia have very limited freshwater resources.

FIGURE 1 The Daintree River is a perennial river, meaning that it flows permanently.

Australia's rivers

Rivers can be either **perennial** (permanent flow) or non-perennial (only flow at certain times, usually seasonally). Many rivers in Australia are non-perennial, particularly in drier parts of the continent. Only a small proportion of rainfall received makes its way into river systems, and as a result the flow of rivers in Australia is extremely variable.

The area of land that contributes water to a river and its tributaries is called a **drainage basin**. Major drainage basins (large catchments) in Australia are shown in **FIGURE 2**. The major rivers in Australia are shown in **TABLE 1**.

Australia's lakes

Australia's largest lakes are Kati Thanda–Lake Eyre and Lake Torrens. Generally, Australia's lakes do not hold much water compared to lakes found elsewhere in the world because our climate is hot and dry and our landscape is relatively flat. Kati Thanda–Lake Eyre is often dry.

TABLE 1 Major rivers and river systems in Australia

Major rivers	States
Murray–Darling system	New South Wales, South Australia, Victoria
Murrumbidgee	New South Wales, Australian Capital Territory
Lachlan	New South Wales
Cooper	Queensland, South Australia
Flinders	Queensland
Diamantina	Queensland, South Australia
Gascoyne	Western Australia
Goulburn	Victoria

perennial permanent or continually available

drainage basin the entire area of land that contributes water to a river and its tributaries

FIGURE 2 Australia's drainage basins

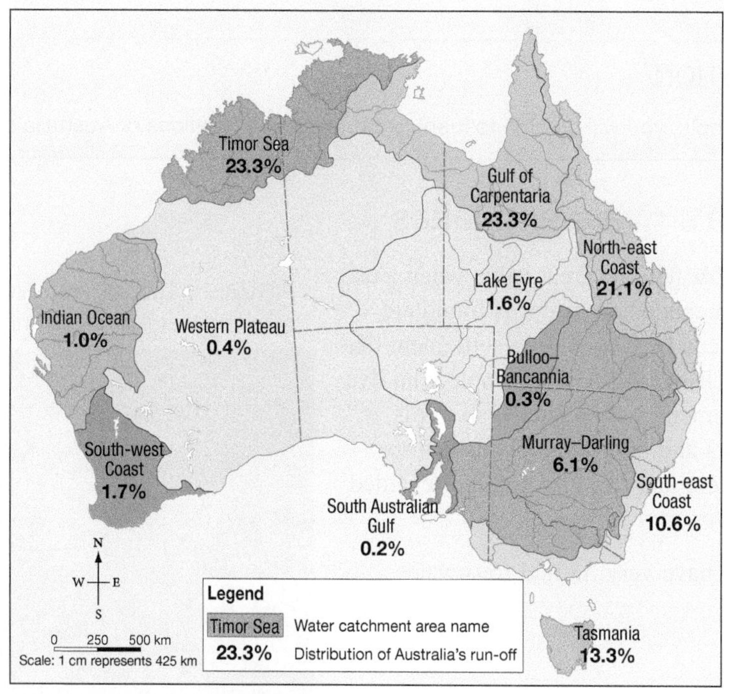

Source: MAPgraphics Pty Ltd Brisbane

fdw-0020

FOCUS ON FIELDWORK

Water infiltration

Something that influences water availability is the way that water moves over the landscape or is absorbed into the ground. Water will move very differently through hilly or steep environments compared with how it moves over flat areas. The extent and type of vegetation covering the location will also have an impact on the way water flows. These factors affect how much water runs off the soil it lands on (run-off) and how much soaks in to the soil (infiltration).

Learn how to observe the movement of water and its impact with the **Water infiltration** fieldwork activity in your Resources tab.

on Resources

eWorkbook Australia's water resources (ewbk-9442)

Video eLesson Australia's water resources — Key concepts (eles-5116)

Interactivity Australia's water resources (int-8543)

Fieldwork Water infiltration (fdw-0020)

14.2 ACTIVITIES

1. Locate each of the following significant rivers in Australia:
   Burdekin, Hunter, Daly, Victoria, Gascoyne, Ord, Fitzroy, Leichhardt, Tamar and Derwent. Choose one to create a multimedia presentation that follows the flow of the river from its source to the mouth of the river. Describe the size of the river, environments it passes through, settlements and important features along the river.
2. Visit the **Australian groundwater explorer** weblink in your Resources tab. You can manoeuvre around the map and zoom in and out using buttons on the map. In the left-hand column you can select the information you would like displayed on the map as different map layers. Click on the boxes to choose the layer you want to display. Explore the location of river regions, groundwater management areas and sedimentary basins. Along the top of the map you can see a range of icons that give you extra functionality. Write a summary of the location of Australia's groundwater resources.

14.2 EXERCISE

Learning pathways

■ LEVEL 1	■ LEVEL 2	■ LEVEL 3
1, 2, 4, 7, 14	3, 5, 8, 11, 13	6, 9, 10, 12, 15

Check your understanding

1. Define the term *drainage basin*.
2. Outline the difference between perennial and non-perennial rivers.
3. Define the terms *run-off* and *infiltration* as they relate to the way water moves across an environment.
4. Identify two landforms that would affect the way water moves across a landscape.
5. Refer to **FIGURE 2** to answer the following. Which of Australia's drainage basins:
   a. is the largest in area
   b. receives the highest percentage of run-off
   c. covers the most of NSW
   d. crosses the most state boundaries.
6. Name the largest lake in Australia and outline why it doesn't hold as much water as other large lakes around the world.

Apply your understanding

7. Explain how rainfall and run-off contribute to the amount of water Australia has to use.
8. Would you expect the quality of water to be the same at all points in a river system?
9. Would you expect larger population centres (such as major cities) to be located near perennial or non-perennial rivers?
10. Assess the factors that determine whether rain infiltrates or runs off the ground when it lands.
11. Contrast how water moves over hilly and flat landscapes.
12. Do you think the Murray–Darling Basin could be considered a significant drainage basin? Give reasons to support your view.

Challenge your understanding

13. Do you think that it is possible for a perennial river to become non-perennial? Give reasons for your decision.
14. Suggest a reason why Australian lakes hold less water, generally, than lakes in other parts of the world.
15. Predict what would happen to the quality of water in a river if the land in its drainage basin was stripped of vegetation completely.

To answer questions online and to receive **immediate feedback** and **sample responses** for every question, go to your learnON title at www.jacplus.com.au.

14.3 Australia's climate and water availability

LEARNING INTENTION

By the end of this subtopic, you will be able to evaluate how climate affects Australia's water availability.

14.3.1 Australia's climate

Australia's **climate** is the most important factor affecting how much water is available for drinking, for use by humans, for plant growth, in river flows and for a range of other purposes. Australia is the driest inhabited continent (only Antarctica is drier), and there is very little fresh water available for our use. Rain falls unevenly across the country and from season to season.

The driest part of Australia is around the Kati Thanda–Lake Eyre Basin, and the wettest locations are in north-east Queensland and western Tasmania. Rainfall is also highly variable, meaning that the amount of rainfall can vary or change from year to year. For example, one year a location might have very high rainfall, the next year it might have very low rainfall, and the following might be an average year.

int-5284

FIGURE 1 The amount of rain that falls in Australia varies from place to place, as this rainfall map shows.

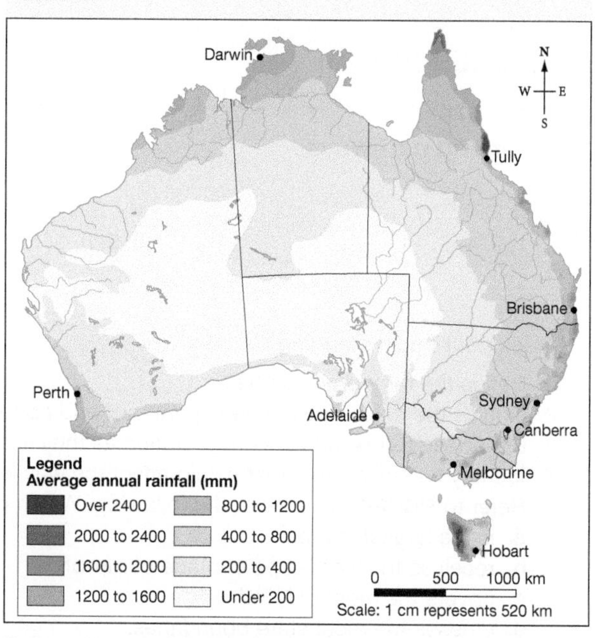

Source: Bureau of Meteorology

14.3.2 Rainfall variability

Rainfall variability is the way rainfall totals in a given area vary from year to year. For example, if an area has low rainfall variability, it means rainfall will tend to be fairly consistent from one year to the next. Many coastal areas show this kind of rainfall pattern. In contrast, high rainfall variability means rainfall is likely to be irregular from one year to the next; there may be heavy rainfall in some years and little or no rainfall in others. Desert areas in central Australia tend to have low rainfall but high rainfall variability.

Australia's seasons result in variations in rainfall. In northern Australia a wet season is experienced from November to April, and a dry season from May to October. Some parts of southern and south-western Australia experience winter rain peaks. Rainfall is more evenly distributed throughout the year in most parts of Australia.

climate long-term weather patterns in an area

rainfall variability the change from year to year in the amount of rainfall in a given location

14.3.3 Evaporation

Another problem for Australia is that most of its rainfall does not end up in rivers; much of it evaporates. Of all the water carried by the world's rivers, Australian rivers contain only 1 per cent of that total — even though Australia has 5 per cent of the world's land area. On average, only 10 per cent of our rainfall runs off into rivers and streams or is stored as groundwater. This figure drops to 3 per cent in dry areas and rises to 24 per cent in wetter places. The rest evaporates, is used by plants, or is stored in lakes, wetlands or underground storages. Areas in central Australia are very dry and, as a result, have high **evaporation** rates. Coastal areas have lower evaporation rates because they are close to the sea.

Relative humidity is a measure of the air's moisture content expressed as a percentage of the maximum moisture the air can contain at a certain temperature. Keep in mind that warm air can contain more moisture than cool air. Relative humidity does not measure the exact amount of moisture in the air because that depends on air temperature. For example, if Brisbane has a day of 30 °C and Melbourne has a day of 15 °C, and the relative humidity in both places is 60 per cent, there will be much more moisture in the air in Brisbane than in Melbourne.

Relative humidity tends to be higher in coastal regions, as does rainfall. The parts of Australia with higher relative humidity are north Queensland and western Tasmania.

evaporation the process by which water is converted from a liquid to a gas and thereby moves from land and surface water into the atmosphere

relative humidity the amount of moisture in the air

FIGURE 2 Average annual evaporation, Australia

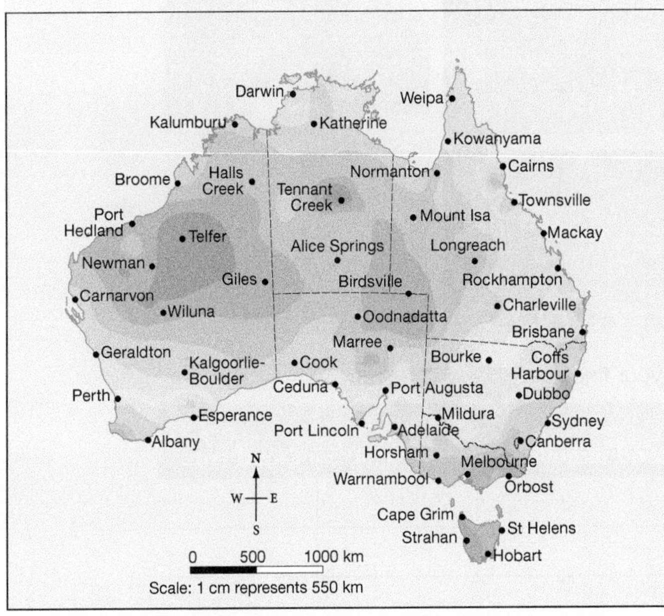

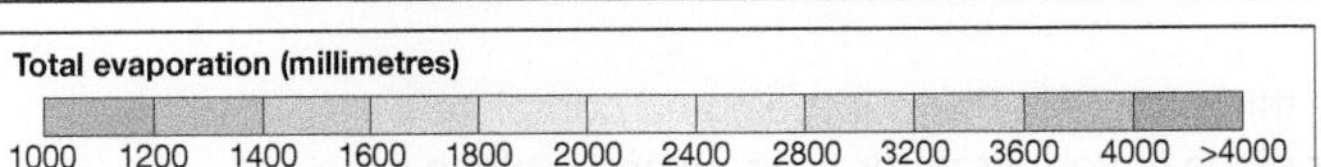

Source: Bureau of Meteorology

FIGURE 3 Average relative humidity across Australia

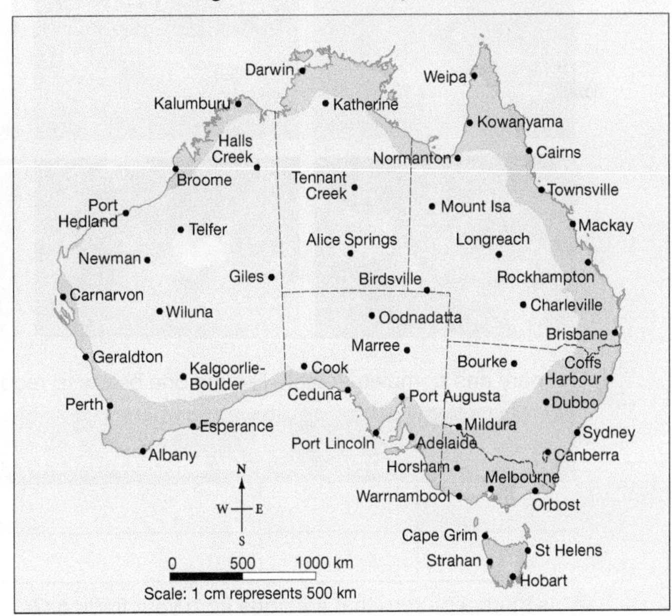

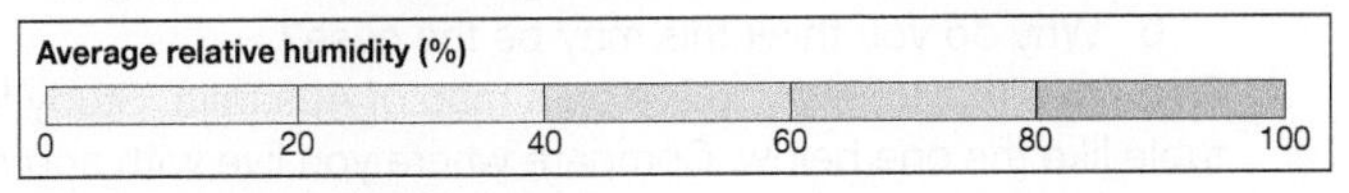

Source: Bureau of Meteorology

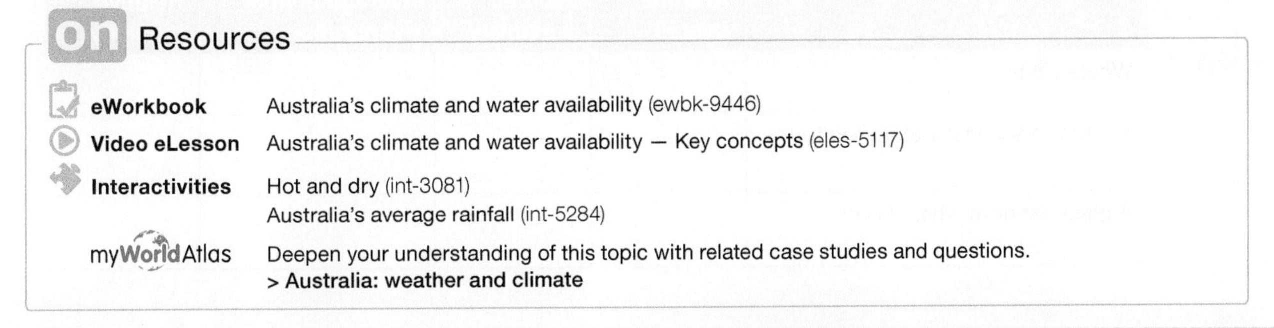

on Resources

eWorkbook Australia's climate and water availability (ewbk-9446)

Video eLesson Australia's climate and water availability — Key concepts (eles-5117)

Interactivities Hot and dry (int-3081)
Australia's average rainfall (int-5284)

myWorldAtlas Deepen your understanding of this topic with related case studies and questions.
> Australia: weather and climate

14.3 ACTIVITIES

1. In small groups, without looking directly at the Sun, observe the atmosphere above your school over a period of five days.
   - Measure and record the temperature of the air using a thermometer.
   - Measure and record wind direction.
   - Estimate the percentage of the sky covered by cloud.
   - Create a cloud identification chart or use the **Cloud identification** weblink in your Resources tab to help you identify the cloud type.

   Prepare and complete a table like the one below to record your findings.

Date	Time	Air temperature (°C)	Wind direction	Cloud cover (%)	Cloud type

   a. Is there a pattern that emerges from your findings?
   b. Why do you think this may be the case?
2. Find the place where you live on a map of Australia. Study the three maps in this subtopic and complete a table like the one below. Compare where you live with another place in your state or territory and a place a long way from where you live.

	Average rainfall	Rainfall variability	Average annual evaporation	Relative humidity
Where I live:				
Another place in my state/territory:				
A place far from where I live:				

3. Find out the average rainfall for Kati Thanda–Lake Eyre and the wettest locations and heaviest rainfalls in north-east Queensland and western Tasmania. Record the rainfall variability, evaporation and relative humidity. Outline the differences between these locations.

14.3 EXERCISE

Learning pathways

■ LEVEL 1	■ LEVEL 2	■ LEVEL 3
1, 2, 3, 7, 13	4, 6, 8, 9, 14	5, 10, 11, 12, 15

Check your understanding

1. Identify the two regions that receive the most rainfall in Australia.
2. Identify the two regions that receive the least rainfall in Australia.
3. Identify the region that has the most variable rainfall.
4. Define rainfall variability and outline how it is related to the seasons in Australia. Support your answer with at least two examples from **FIGURES 1–3**.
5. Summarise why rainfall variability, evaporation rates and relative humidity present problems in Australia. Use data from **FIGURES 1–3** to support your answer. Use **FIGURE 3** to find the relative humidity of the place where you live.

Apply your understanding

6. Explain why Australia has high evaporation rates.
7. Study the rainfall, humidity and evaporation maps. Fill in the missing word in each of the following statements in order to describe the interconnections between these features of our climate.
   a. Areas with low rainfall and low humidity tend to have a _______ evaporation rate.
   b. Areas with high rainfall and high humidity tend to have a ________ evaporation rate.
8. Refer to **FIGURES 1–3** to compare data for Dubbo in NSW and Telfer in WA.
   a. Write a sentence comparing the relative humidity of the two places.
   b. Write a sentence comparing the average evaporation rates in the two places.
   c. Locate Dubbo and Telfer on **FIGURE 1**. What is the range of annual rainfall in each place?
   d. Based on this data, which place do you think has the greater difficulty accessing large volumes of water?
9. Using the data in **FIGURES 1–3**, account for Australia being considered a dry continent.
10. Compare the evaporation, relative humidity and rainfall averages for Darwin and Sydney.
11. If Sydney and Bourke are both experiencing days of 30 °C, which is likely to have the more humid conditions? Explain why.
12. Coastal areas tend to have lower evaporation rates than inland areas. What factors might contribute to the high evaporation rates in some coastal places, such as Port Hedland in WA?

Challenge your understanding

13. Australia has high evaporation rates and high rainfall variability. List all the ways that this environment makes water delivery to people a challenge.
14. Large water storage facilities in Australia tend to be on open lakes or dams. How might water loss from evaporation in these storage facilities be reduced?
15. Do you think that access to water will change in Australia if more land is cleared for farming? Provide at least two reasons for your answer.

To answer questions online and to receive **immediate feedback** and **sample responses** for every question, go to your learnON title at www.jacplus.com.au.

14.4 The right to water

online only

LEARNING INTENTION

By the end of this subtopic, you will be able to discuss people's rights to water and explain some of the impacts of water stress.

The content in this subtopic is a summary of what you will find in the online resource.

Access to water is a human right that is protected by many international agreements, yet not everyone has access to this life-giving resource. Everyone has the right to enough safe, accessible and affordable water for all their needs. Water is more important to survival than food is.

FIGURE 3 Women carrying water near Ambalavao, Madagascar

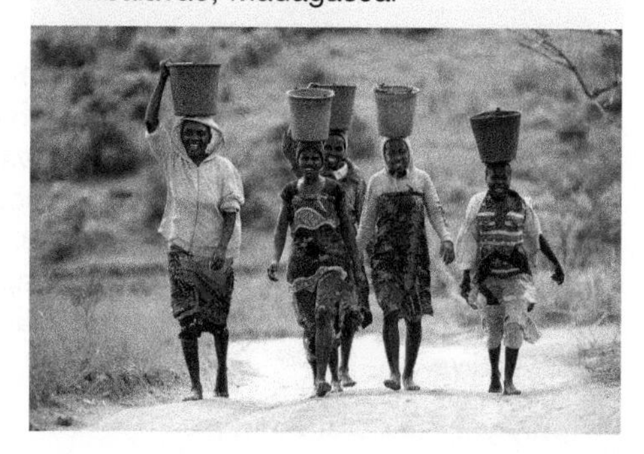

To learn more about the right to water, go to your learnON resources at www.jacPLUS.com.au.

Contents

learnON

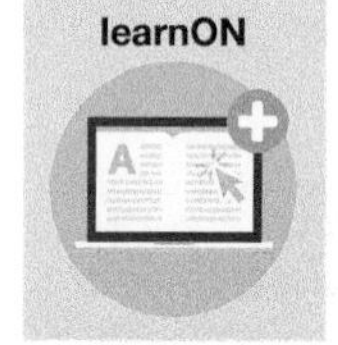

- 14.4.1 The human right to water
- 14.4.2 The water carriers

on Resources

 **eWorkbook** The right to water (ewbk-9450)

 **Video eLesson** The right to water — Key concepts (eles-5118)

Interactivity The right to water (int-8544)

myWorldAtlas Deepen your understanding of this topic with related case studies and questions.
> The Dead Sea — overcoming water scarcity > Russia and Eurasia

14.5 SkillBuilder — Constructing a pie graph

online only

LEARNING INTENTION

By the end of this subtopic, you will be able to represent geographical information in a pie graph.

The content in this subtopic is a summary of what you will find in the online resource.

14.5.1 Tell me

What is a pie graph?

A pie graph, or pie chart, is a graph in which slices or segments represent the size of different parts that make up the whole. The circle of 360 degrees represents the total, or 100 per cent of whatever is being looked at. The size of the segments is easily seen. By presenting the parts in order, from largest to smallest, it is easier to interpret. A pie graph can be drawn by hand or by using a spreadsheet program.

14.5.2 Show me

eles-1632

How to create a pie graph

Step 1

Order the statistics from largest to smallest. If there is an 'other' category, put it last.

Step 2

If there are raw figures, convert them to percentages: divide each category by the total figure and multiply by 100.

Step 3

Convert the percentages to degrees of a circle by multiplying by 3.6. (100 per cent of the circle = 360 degrees, so 1 per cent of the circle = 3.6 degrees.)

Step 4

Draw a straight line from the centre of the circle to 12 o'clock.

Step 5

Use the protractor to mark the first and largest segment, working clockwise. To do this, place the 0 degrees line on the protractor along the line you have just drawn. Now mark in the second largest group. Use the protractor to mark each of the other segments in descending size, marking the 'other' category last.

Step 6

Label and colour each segment, making sure you include the percentage. Provide a clear title and source.

FIGURE 1 Percentage of electricity generated from renewables in Australia by energy source (2020)

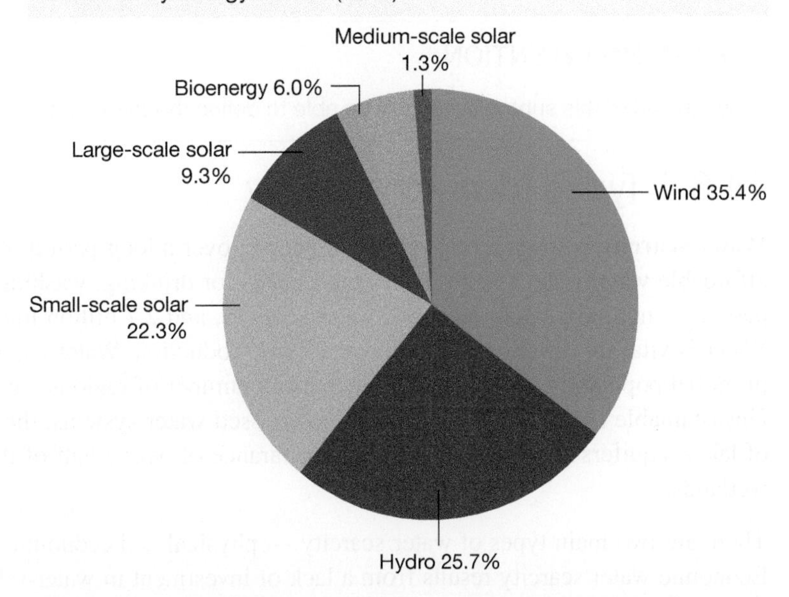

TABLE 1 Percentage of electricity generated from renewables in Australia by energy source (2020)

Energy source	Percentage (%)
Wind	35.4
Hydro	25.7
Small-scale solar	22.3
Large-scale solar	9.3
Bioenergy	6.0
Medium-scale solar	1.3

Source: Clean Energy Council Fact Sheet, 2020

14.5.3 Let me do it

int-3128

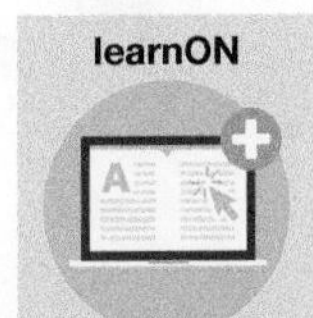

Go to learnON to access the following additional resources to help you build this skill:

- a longer explanation of this skill and its application in Geography (Tell me)
- a video showing the step-by-step process of this skill (Show me)
- an activity and interactivity for you to practise this skill (Let me do it)
- self-marking questions to help you understand and use this skill.

on Resources

eWorkbook SkillBuilder — Constructing a pie graph (ewbk-9454)

Video eLesson SkillBuilder — Constructing a pie graph (eles-1632)

Interactivity SkillBuilder — Constructing a pie graph (int-3128)

14.6 Water scarcity

LEARNING INTENTION

By the end of this subtopic, you will be able to define the term *water scarcity* and explain the factors that affect it.

14.6.1 Types of water scarcity

Water scarcity is when a large group of people, over a long period of time, do not have access to safe, affordable water to be able to satisfy their needs for drinking, washing or their livelihoods. Globally, 2.2 billion people do not have safely managed water services, and 2.7 billion find water scarce for at least a month a year. Water is vital for drinking, sanitation and food production. Water use is increasing at more than twice the rate of global population growth, and an increasing number of regions are experiencing chronic water shortages. Unsustainable use of water has resulted in stressed water systems, the drying up of lakes, aquifers and rivers, and the disappearance of over a half of the world's wetlands.

water scarcity when annual water supplies drop below 1000 cubic metres per person. Absolute water scarcity describes the situation when annual water supplies drop below 500 cubic metres per person.

There are two main types of water scarcity — physical and economic (**FIGURE 1**). Economic water scarcity results from a lack of investment in water-related infrastructure. Physical water scarcity refers to water not being abundant enough to meet all demands.

FIGURE 1 Global physical and economic water scarcity

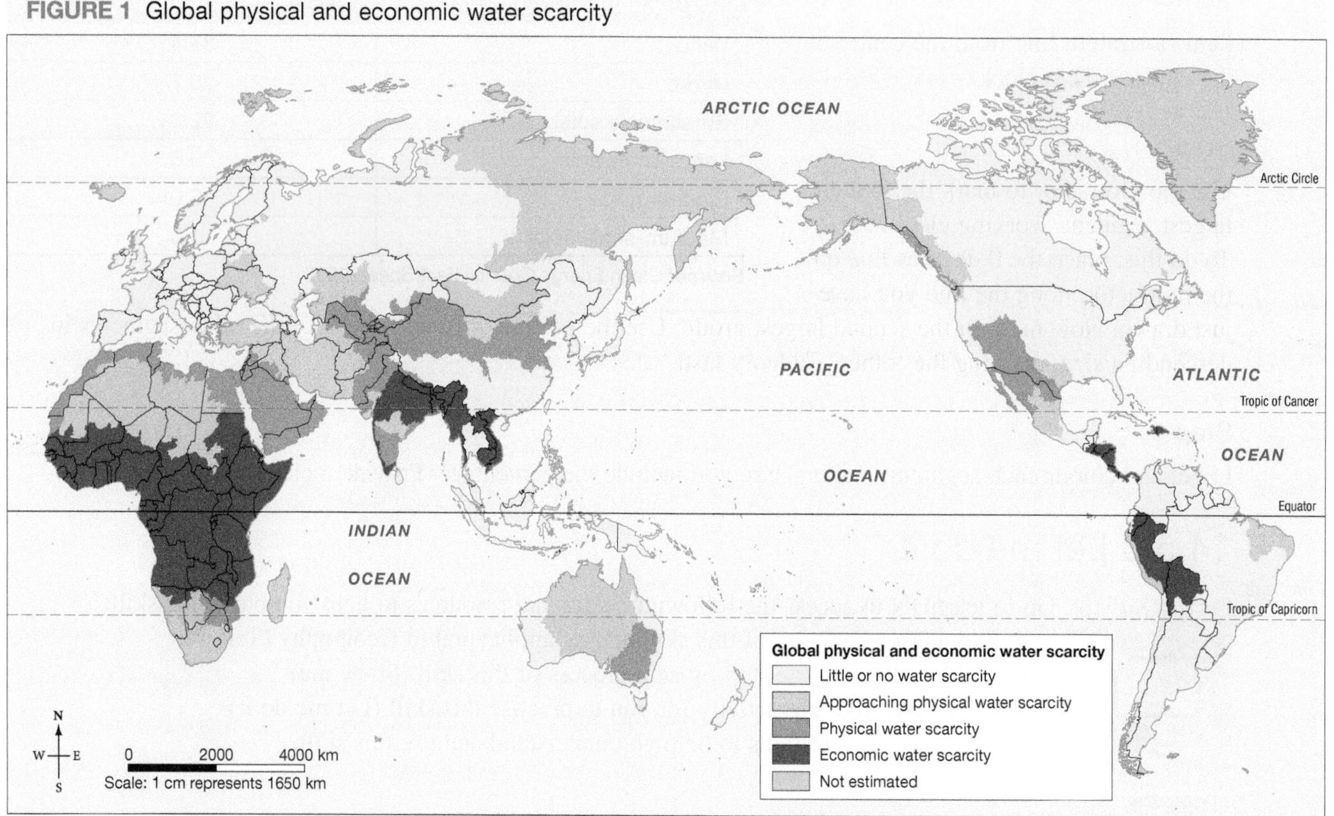

Source: International Water Management Institute (IWMI)

14.6.2 Distribution of water scarcity

Uneven distribution of water resources occurs naturally. As discussed in Chapter 12, factors such as latitude, altitude, ocean currents, distance from the sea, geology and topography naturally have an impact on the amount of water available in certain places. Places that receive less rainfall are more likely to suffer from water vulnerability or water scarcity.

14.6.3 Human causes of water scarcity

There is a range of human causes that contribute to water vulnerability and water scarcity.

Population

More than 1.7 billion people are currently living in river basins where water use exceeds recharge, and this is likely to increase as the world's population increases. Further pressure on local water supplies will result as people withdraw more water and pollution increases.

Agriculture

Agriculture consumes around 70 per cent of the world's accessible fresh water. Due to inefficiencies and leaky irrigation systems much of this water is wasted. Many irrigation systems draw on groundwater to supplement supplies and, as a result, many of these groundwater sources are being used at unsustainable rates and are being depleted. Farmers also use fertiliser and pesticides to help their crops grow and to reduce pests; however, when these chemicals make their way into local waterways, this can further reduce the amount of potable fresh water available.

FIGURE 2 Water scarcity affects landscapes and environments, as seen in this mangrove area.

FIGURE 3 Global areas of groundwater stress

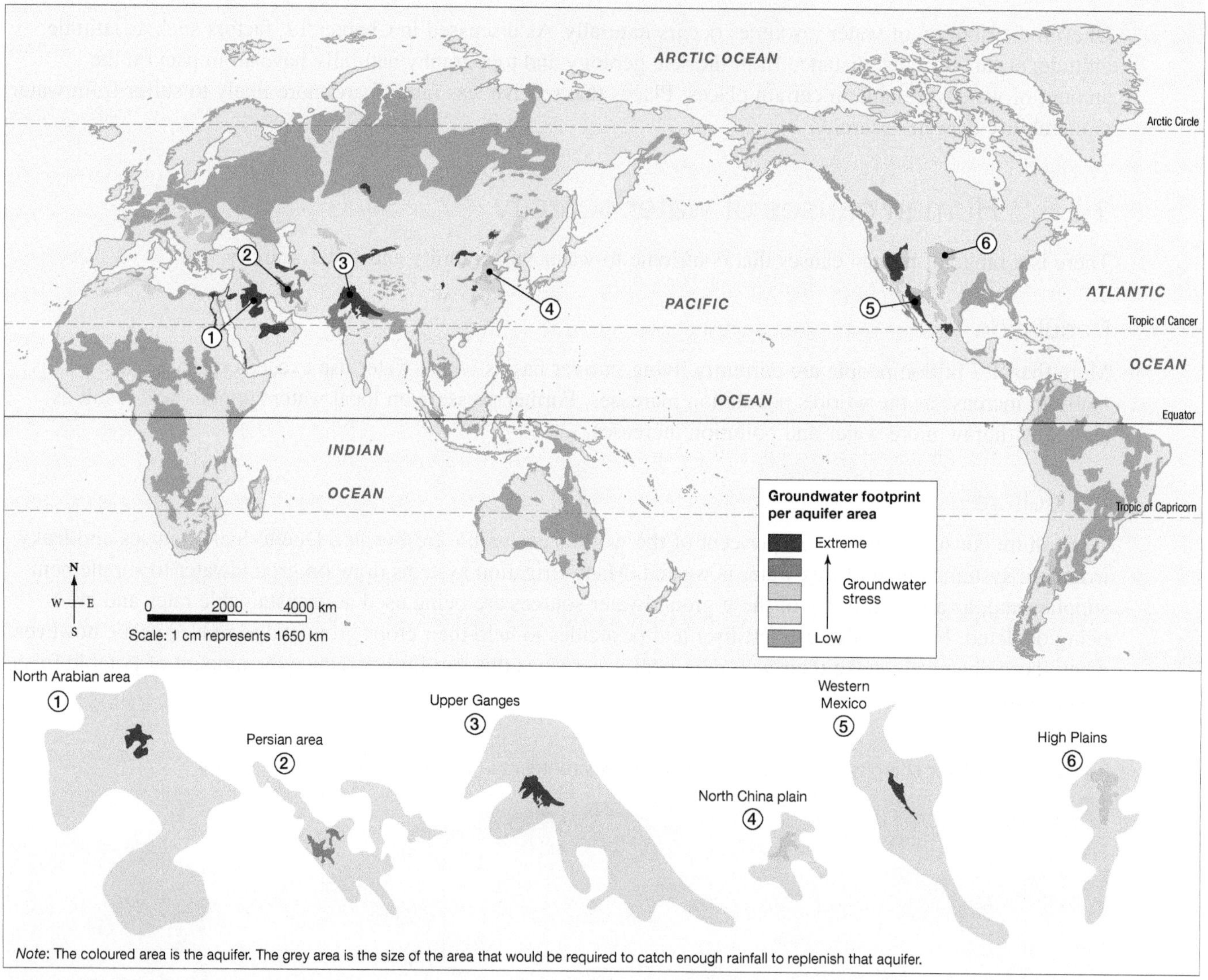

Note: The coloured area is the aquifer. The grey area is the size of the area that would be required to catch enough rainfall to replenish that aquifer.

Source: BGR & UNESCO 2008: Groundwater Resources of the World 1 : 25 000 000. Hannover, Paris.

Energy production

Every year 580 billion cubic metres of fresh water globally is used in energy production. This equates to about 15 per cent of the world's fresh water. Water is used to cool thermal power plants, in drilling and refining processes, and rivers can be diverted for hydro-electric schemes. This use can threaten both quantity and quality of water supplies. If power plants are built in areas with limited water resources, this can have direct impact on water availability for local communities. If water is used in processing, it can result in the creation of polluted water known as slurry.

Climate change

Climate change is expected to drastically modify weather patterns, and result in an increasing number of droughts and floods in particular areas. This will make seasonal predictions and crop planning more difficult. The disappearance of glaciers and pack ice will restrict water availability in downstream communities. Based on the existing climate change scenario, it is anticipated that by 2030 half of the world's population will live in areas of high water stress. The impact of climate change on water availability will disproportionately impact equatorial regions that rely heavily on water received from rainfall rather than irrigation for agriculture.

FIGURE 4 Global projected water scarcity

ARCTIC OCEAN
PACIFIC OCEAN
ATLANTIC OCEAN
INDIAN OCEAN
Arctic Circle
Tropic of Cancer
Equator
Tropic of Capricorn

Projected 2025 water scarcity
- Little or no water scarcity
- Economic water scarcity
- Physical water scarcity
- Not estimated
- Countries that will import more than 10% of their cereal consumption

0 2000 4000 km
Scale: 1 cm represents 1650 km

Source: International Water Management Institute.

Ownership

There is concern regarding the sale of long-term water rights to investors along with the sale of land in some African countries such as Mali and Sudan. This jeopardises water access for local farmers, traditional water users and downstream users.

FIGURE 5 Individuals and communities are affected by water scarcity in Kathmandu.

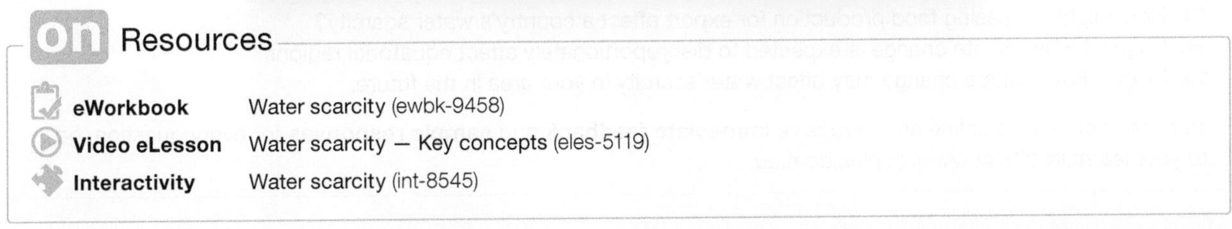

on Resources

eWorkbook	Water scarcity (ewbk-9458)
Video eLesson	Water scarcity — Key concepts (eles-5119)
Interactivity	Water scarcity (int-8545)

14.6 ACTIVITIES

1. Draw your own cartoon or illustration showing the impact of water scarcity on communities or individuals.
   a. Use **FIGURE 6** as a starting point to brainstorm some of the issues regarding water access that you might want to highlight.
   b. Identify one aspect of inequality of access to water that is unequal around the world.
   c. Consider how you might show this inequality visually, with only a few words.
   d. When you create your drawings, be careful to keep the focus of the cartoon on the issue.
2. Create a flow chart that shows how water scarcity impacts on the quality of life of people.

FIGURE 6 Access to water globally

14.6 EXERCISE

Learning pathways

■ LEVEL 1	■ LEVEL 2	■ LEVEL 3
1, 2, 7, 8, 13	3, 4, 10, 11, 14	5, 6, 9, 12, 15

Check your understanding

1. Describe the difference between water scarcity and water stress.
2. Describe the difference between physical water scarcity and economic water scarcity.
3. Locate the following countries on **FIGURE 1** and rank them in terms of their levels of physical water scarcity (from highest levels to lowest): Australia, New Zealand, Canada, England, Saudi Arabia.
4. Describe two ways in which agriculture might affect levels of potable water available to a community.
5. List three areas in the world that experience extreme groundwater stress.
6. Describe the link between water scarcity and population growth.

Apply your understanding

7. Explain why there are differences in water availability naturally. How do these factors result in water scarcity?
8. Refer to **FIGURE 1**.
   a. Describe the areas affected by little or no water scarcity.
   b. Explain why these areas might have little or no water scarcity.
   c. Describe the regions most affected by economic water scarcity.
   d. Why do you think these regions are likely to be affected by this type of water scarcity?
9. Examine **FIGURE 6**, a cartoon on water scarcity. Explain how this demonstrates the unequal impacts of water scarcity on different regions and communities.
10. Identify a region or country that is likely to experience water scarcity as a result of climate change. Discuss the impact this will have on the quality of life of people, the economy of the region or country, and the environment.
11. Explain the connection between water scarcity and the need to import food, such as cereals and grains.
12. Refer to **FIGURE 3**. What factors are likely to contribute to the areas in Australia that experience low levels of groundwater stress?

Challenge your understanding

13. How might increasing food production for export affect a country's water scarcity?
14. Suggest why climate change is expected to disproportionately affect equatorial regions.
15. Predict how climate change may affect water scarcity in your area in the future.

To answer questions online and to receive **immediate feedback** and **sample responses** for every question, go to your learnON title at www.jacplus.com.au.

14.7 SkillBuilder — Reading and describing basic choropleth maps

online only

LEARNING INTENTION

By the end of this subtopic, you will be able to analyse geographical information presented in choropleth maps.

The content in this subtopic is a summary of what you will find in the online resource.

14.7.1 Tell me

What is a basic choropleth map?

A basic choropleth map is a shaded or coloured map that shows the density or concentration of a particular aspect of an area. The key/legend shows the value of each shading or colouring. The darkest colours show the highest concentration, and the lightest colours show the lowest concentration.

FIGURE 1 Population density in Brazil

Source: MAPgraphics Pty Ltd, Brisbane

14.7.2 Show me

eles-1706

How to read and describe a choropleth map

Step 1

Read the title of the map to get an impression of what the map is about. Check that the source of the information is reliable.

Step 2

Read the key/legend next. Check the units of measurement that are used. Think about the divisions that are used for colours. Cast your eye over the map, taking in the colours and trying to work out any general patterns that emerge.

Step 3

Comment on where the darkest colours or the more intense/higher values occur; for example, a specific region, a country, a continent, coastal areas, near rivers or in cities. Can you discuss the map by continent, or by region? In **FIGURE 1**, the highest density of people in Brazil occurs in the cities, such as São Paulo and Fortaleza, on the Atlantic Ocean coastline.

Step 4

Comment on where the lightest colours or the least intense/lower values occur. For example, the lowest density of people in Brazil occurs in the large inland region, especially along and around the Amazon River and its tributaries.

Step 5

Identify and comment on any places that stand out from their surrounding areas or the general patterns overall. Is there a patch of colour that isn't expected based on the general pattern? This is referred to as an anomaly. For example, the population densities around Brasilia and Goiania are unusual as these appear to be isolated clusters of higher population, whereas most of the area contains fewer than 10 people per square kilometre.

14.7.3 Let me do it

int-3286

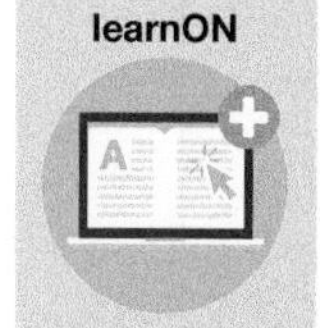

Go to learnON to access the following additional resources to help you build this skill:

- a longer explanation of this skill and its application in Geography (Tell me)
- a video showing the step-by-step process of this skill (Show me)
- an activity and interactivity for you to practise this skill (Let me do it)
- self-marking questions to help you understand and use this skill.

on Resources

eWorkbook	SkillBuilder — Reading and describing basic choropleth maps (ewbk-9462)
Video eLesson	SkillBuilder — Reading and describing basic choropleth maps (eles-1706)
Interactivity	SkillBuilder — Reading and describing basic choropleth maps (int-3286)

14.8 Strategies to address water scarcity

LEARNING INTENTION

By the end of this subtopic, you will be able to describe the roles of governments, non-government organisations, communities and individuals in water management.

14.8.1 United Nations

Globally, 3 in 10 people lack access to safely managed drinking water, 1 in 4 health facilities lack basic water facilities, 2.4 billion people lack access to basic sanitation services, and 40 per cent of the world's population are affected by water scarcity. Goal 6 of the UN's Sustainable Development Goals was 'Ensure access to water and sanitation for all'. This goal is one of the 17 Sustainable Development Goals agreed to by the United Nations in 2015. The years 2018–2028 is the 'International Decade for Action — Water for Sustainable Development', or the Water Action Decade.

14.8.2 Role of governments

Governments are responsible for some of the practical management of water sources to reduce the severity of water scarcity. To achieve this, they need to create and enforce laws and fund institutions to regulate water use. Governments need to listen to the views and take into account the needs of different stakeholders such as local residents and businesses, and monitor the needs of the environment. They have to manage conflict between different groups over water use; coordinate water management across boundaries; and ensure that management zones take into account natural boundaries, such as catchments, to ensure effective management.

Governments may fund innovations in water management such as recycling, purification, 3D modelling and simulations, screens to separate salt water and fresh water, and desalination.

14.8.3 Role of non-government organisations

Non-government organisations (NGOs), such as the Water Project, provide training and finance for construction of water projects for communities in sub-Saharan Africa. This involves activities such as digging wells, constructing dams, protecting springs and water filtration.

Environmental NGOs such as the World Wide Fund for Nature (WWF) undertake small-scale local projects to measure water use and the impacts of this use on river basins. They provide funding to advance the science of water conservation and promote methods of sustainable water use.

14.8.4 Role of communities

The ability of communities to address water scarcity is limited by access to funding and the much larger government structures in place. Communities can ensure that they have effective local government representation so that they are able to voice their concerns. They can collaborate with competing stakeholders to try to bring about equitable access to existing water sources.

FIGURE 1 Young children in Zambia travel many kilometres to fetch water for their families.

FIGURE 2 Water is transported long distances, taking up much of the day.

14.8.5 Role of individuals

It is generally very difficult for people in water-scarce areas to address water scarcity and options are often limited. They may be forced to drink contaminated water, or use water from the same sources that animals use to drink and bathe. Individuals and communities may become involved in conflict in their fight to maintain access to water sources. Individuals can improve their access to water by fixing leaks in pipes and taps, and reducing pollution of local water sources.

Individuals in more developed countries with abundant water sources are much more able to reduce their water consumption. They can do this by using water-efficient appliances and taps, using water-efficient methods of cooking, and reducing the use of running water in cleaning food, brushing teeth and rinsing dishes.

fdw-0021

FOCUS ON FIELDWORK

Monitoring water use

The water use of homes and businesses in Australia is measured by water meters. This allows for water supply companies to charge homes and businesses for the water they use.

Learn how to monitor how much water is being used at your home or at your school using the **Monitoring water use** fieldwork activity in your Resources tab.

on Resources

- **eWorkbook** Strategies to address water scarcity (ewbk-9466)
- **Video eLesson** Strategies to address water scarcity — Key concepts (eles-5120)
- **Interactivity** Strategies to address water scarcity (int-8546)
- **Fieldwork** Monitoring water use (fdw-0021)

14.8 ACTIVITY

Research one example of how an NGO has assisted a community in improving their access to fresh water. Describe the role that the NGO played, providing statistics that show how access to water was improved, and describe the impact of these changes on quality of life within the community. Include a map showing the location of the community and the project.

14.8 EXERCISE

Learning pathways

■ LEVEL 1	■ LEVEL 2	■ LEVEL 3
1, 2, 3, 12, 13	4, 5, 7, 11, 14	6, 8, 9, 10, 15

Check your understanding

1. Identify the proportion of the world's population that is still without access to safe drinking water.
2. Outline the aim of Goal 6 of the United Nations Sustainable Development Goals.
3. Complete the following sentence: The United Nations has named the years ________ as the 'International Decade for Action — ________ for Sustainable Development'.
4. List three ways that governments might act to protect water sources.
5. How does the WWF help to protect water sources?
6. How can members of communities act to protect their water sources?
7. List three ways that you can limit your water use at home.
8. Outline whether people in developed communities are more or less able to employ strategies to conserve water than people living in developing countries. Support your answer with examples.

Apply your understanding

9. What factors affect the impact an individual can have on water scarcity in the community? Determine which three factors have the greatest impact.
10. To what extent do you think governments should be responsible for ensuring their people have access to clean drinking water? Provide reasons for your view, taking into consideration the variation in resources available to governments.
11. If you were going to make a donation to an NGO for a water project, would you choose to donate to a project that provided water filtering equipment, constructed dams or dug wells? Provide reasons for your decision that show how you weighed up the benefits and costs of each type of project.
12. How might water scarcity create conflict in a community?

Challenge your understanding

13. Do you think that building more desalination plants in Australia (to remove the salt from sea water to make it potable) would help to improve our water security? Provide reasons for your answer.
14. Do you think that more or less money will be needed to fund water security projects in the next 50 years? Give reasons for your opinion.
15. Propose one strategy that the United Nations might use to achieve Goal 6 of its Sustainable Development Goals.

To answer questions online and to receive **immediate feedback** and **sample responses** for every question, go to your learnON title at www.jacplus.com.au.

14.9 Managing water

LEARNING INTENTION

By the end of this subtopic, you will be able to give examples of and assess the likely impact of water management strategies.

14.9.1 Managing our water supply

More water cannot be created, but it can be managed better. With a growing global population, and the predicted changes due to **climate change**, the pressure on this finite resource requires a number of solutions.

Introducing effective water management can be a challenge at any scale, whether local, national or global. It needs the cooperation of all users, including farmers, industry, individuals, and upstream and downstream people in different countries or different states. With all the competing demands on water, management is often easier to approach at a local scale.

climate change a change in the world's climate. This can be very long term or short term, and is caused by human activity.

FIGURE 1 Kurnell desalination plant in southern Sydney provides water to the city's population. It is powered by wind energy produced in Canberra.

Because agriculture uses the greatest amount of water, it makes sense to make irrigation systems more efficient. The aim is to get more production for every drop of water used. Some irrigation systems waste up to 70 per cent of their water through leaks and evaporation, so changing the irrigation method can save water. Other management practices include recycling, using desalinated water, and using stormwater.

Managing water across borders

About 260 drainage basins across the world are shared by two or more countries. Thirteen river basins are shared by five or more countries. Depending on their location in the catchment, some countries can suffer reduced access to water because of other countries' usage. This shows the interconnection between places — what happens in one place affects another. Diverting rivers, building dams, taking large amounts of water out for irrigation, and creating pollution can all lead to conflict between countries, states and political groups.

Country disputes have occurred in the Nile Basin in North Africa, along the Mekong River in Asia, the Jordan River Basin in West Asia (the Middle East) and along the Silala River in South America. This can also happen within a country, such as with the Murray–Darling Basin in Australia, which sits across four states and one territory.

FIGURE 2 The Jordan River passes through three countries, and has tributaries in the north that flow in from another two countries.

Source: Humanitarian Information Unit

Some countries sign international agreements or treaties to try to share water between nations. These include the Rhine and Danube rivers in Europe, the Nile River in North Africa, the Ganges and Brahmaputra rivers in Asia and the Parana River in South America.

Other examples of water basins that are shared by states within a country are the Colorado River and Ogallala Aquifer in the United States, and the Kaveri River in India. The Great Artesian Basin in Australia shares groundwater between Queensland, New South Wales, South Australia and the Northern Territory.

Other water management solutions for Australia

Over recent years, and especially after prolonged droughts, many solutions have been suggested to solve our water problems. Some of these seem impractical, such as towing icebergs from Antarctica; others have generated much discussion, such as fixing all the leaking pipes in towns, cities and outback bores.

Each of the suggestions has to be considered in light of various factors: cost, impact on people, impact on the environment, technology and politics.

FIGURE 3 Possible water management solutions

Possible water management solutions

- Fix all the leaking water pipes.
- Every house should have a water tank.
- Divert rivers back inland to stop water reaching the sea.
- Move water from wet areas such as Tasmania and northern Australia to dry areas using ship tankers.
- Use more groundwater.
- Build more desalination plants.
- Relocate most of Australia's farms to the north where most of the rain is.
- Build more dams.
- Use cloudseeding to make it rain more.
- Use pipes to move water from areas of high rainfall to areas of low rainfall.
- Use drip irrigation on all crops to save water.
- Move water from wet areas such as Tasmania and northern Australia to dry areas using ship tankers.
- Cover reservoirs to stop evaporation.
- Increase the price of water to homes.

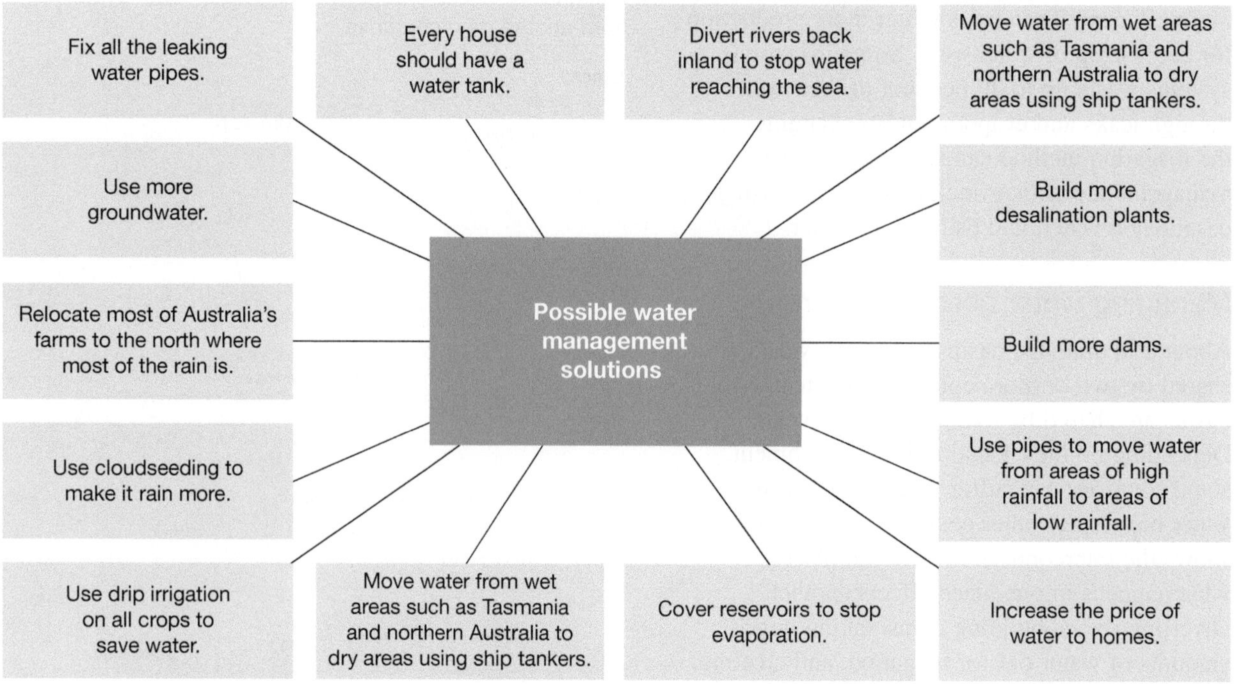

CASE STUDY

Managing water within a country — The Murray–Darling Basin

The Murray–Darling Basin (MDB) is Australia's largest catchment area, covering 14 per cent of the country's total landmass. It stretches across four Australian states (Queensland, New South Wales, Victoria and South Australia) and the Australian Capital Territory, and includes 23 separate river catchments. This is a changed system, in which water is taken out all along its length for storage in dams and use in irrigation. During long periods of drought, little water reaches the Murray mouth, often causing it to close. The Queensland floods in the summers of 2011 and 2012 provided the largest river flows for the MDB in many years.

FIGURE 4 Mass fish kill at the Menindee Lakes in 2019

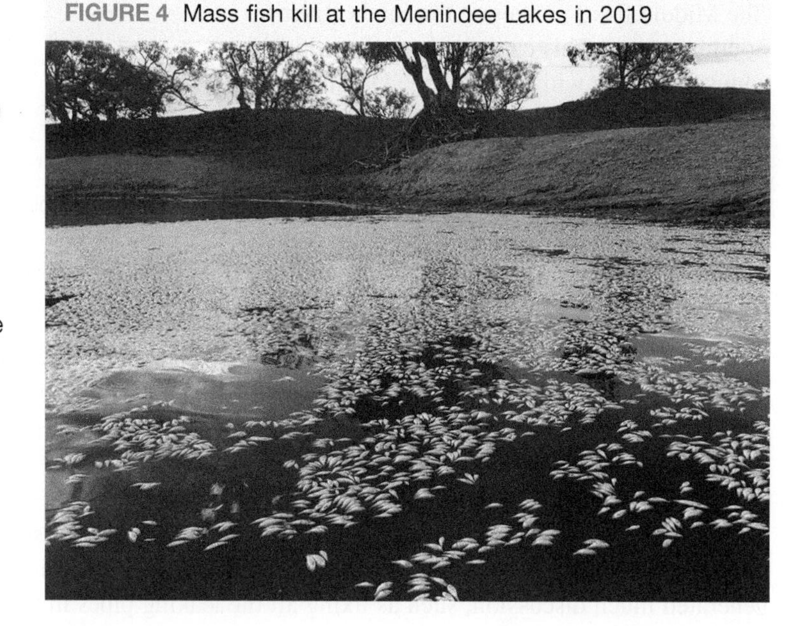

Over the years the management of this river system was the responsibility of each of the four states and the territory, resulting in a lot of conflict. In 2008, the Commonwealth Government took control of the MDB and prepared plans for how it should be managed.

Despite the federal government taking over management of the MDB, there has continued to be controversy over decisions made about sharing the water. In 2018–19, very low rainfall and inequitable water use in the MDB resulted in high fish kills in the Menindee Lakes in NSW. Other impacts include reduced water allocations to farmers, who rely on water from the system for crops and stock.

FIGURE 5 The Australian Government manages the basin through the Murray–Darling Basin Authority.

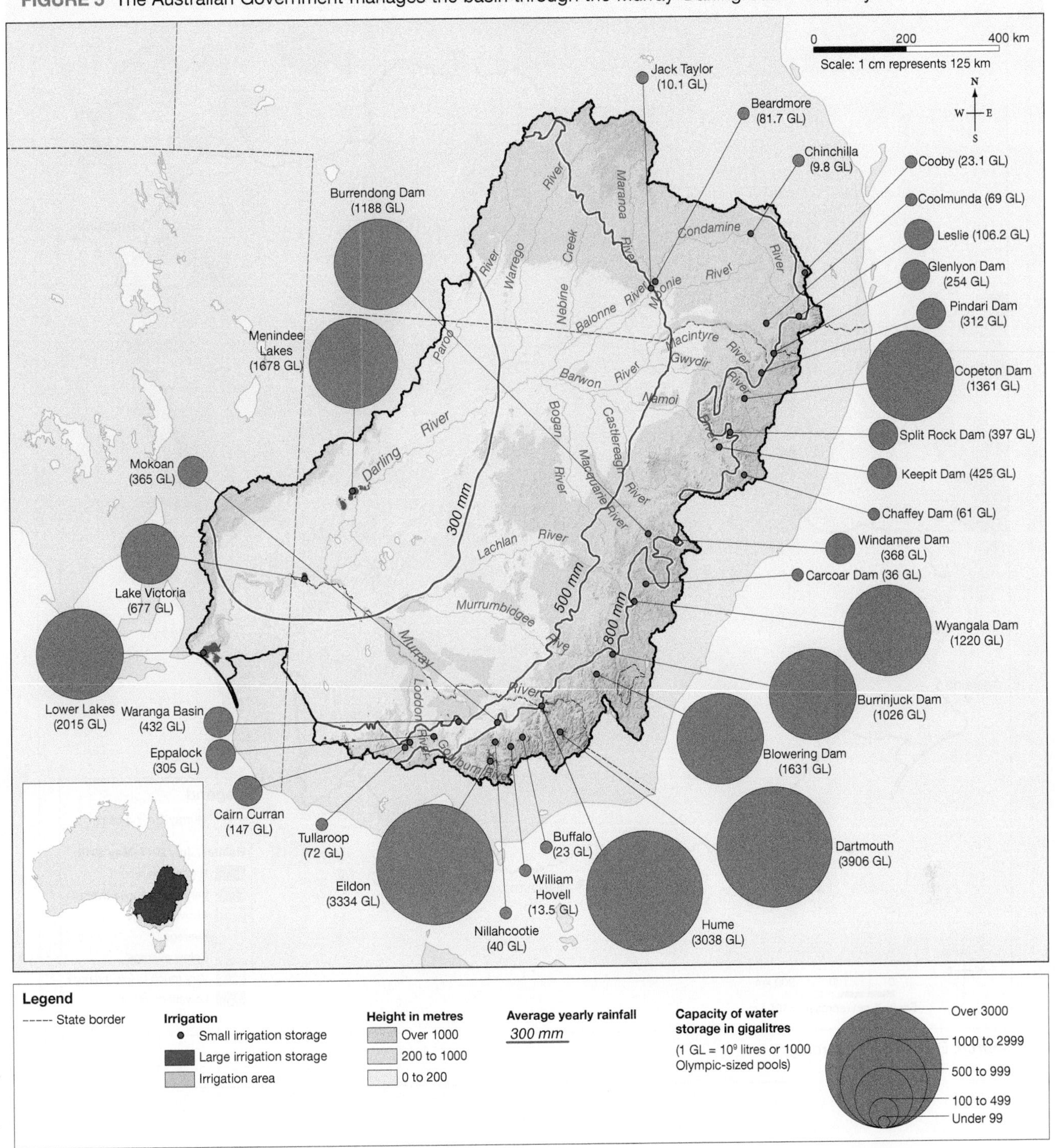

Source: Geoscience Australia, Murray Darling Basin Authority, Spatial Vision.

FIGURE 6 Rainfall in the Murray–Darling Basin from July 2017 to May 2018

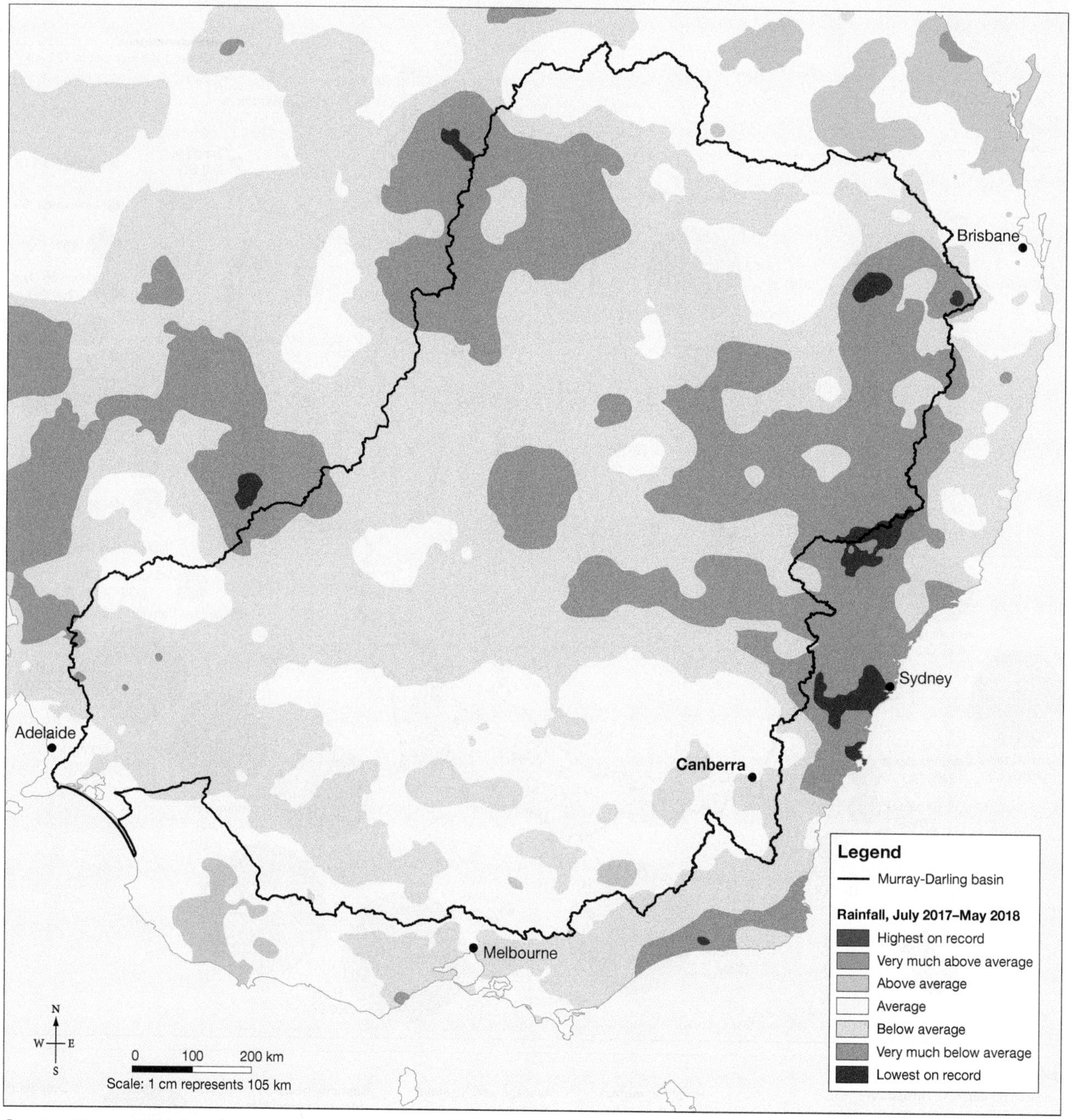

Brisbane
Sydney
Adelaide
Canberra
Melbourne

Legend
— Murray-Darling basin

Rainfall, July 2017–May 2018
- Highest on record
- Very much above average
- Above average
- Average
- Below average
- Very much below average
- Lowest on record

N W E S

0 100 200 km
Scale: 1 cm represents 105 km

Source: Bureau of Meteorology

14.9.2 Managing water use at home

About 50 per cent of household water in Australia is used in the garden. Inside the house, approximately 80 per cent of water used is in the shower, toilet and laundry. It is predicted that our growing Australian population will, by 2051, need nearly twice as much water as we do now.

Policy changes such as water restrictions can reduce water usage. Some restrictions include bans on hosing down driveways, washing cars with hoses and watering private lawns, and cutting back sprinkler use during the day.

The long drought experienced by Australia in the early twenty-first century forced many governments to offer rebates on purchases of water-saving products. Items and services included buying and installing water tanks and grey water systems; dual flush toilets; water-saving shower heads and water-efficient washing machines. People were encouraged to spend money on these in order to save water.

Tips to save water in the house

- Take shorter showers.
- Turn off taps firmly and fix any leaks.
- Use water-efficient shower heads.
- Install a dual-flush toilet or adaptor.
- Use water-efficient appliances and use them only when they are full.
- Keep a jug of cold water in the fridge so you don't need to run the tap until the water is cold enough to drink.
- When replacing appliances, choose water-efficient models (AAA or AAAA rating).

Tips to save water in the garden

- Plant local native plants that need only rainfall.
- Group plants with similar watering needs and water them together.
- Use mulch on the garden beds to stop soil drying out — evaporation can be reduced by up to 70 per cent this way.
- Use a trigger nozzle on your hose.
- Install a timer on your outdoor taps.
- Use trickle irrigation systems rather than sprinklers for garden beds. They direct water where it is needed, and less water is lost to evaporation and wind drift.
- Use a pool cover when the pool is not being used.

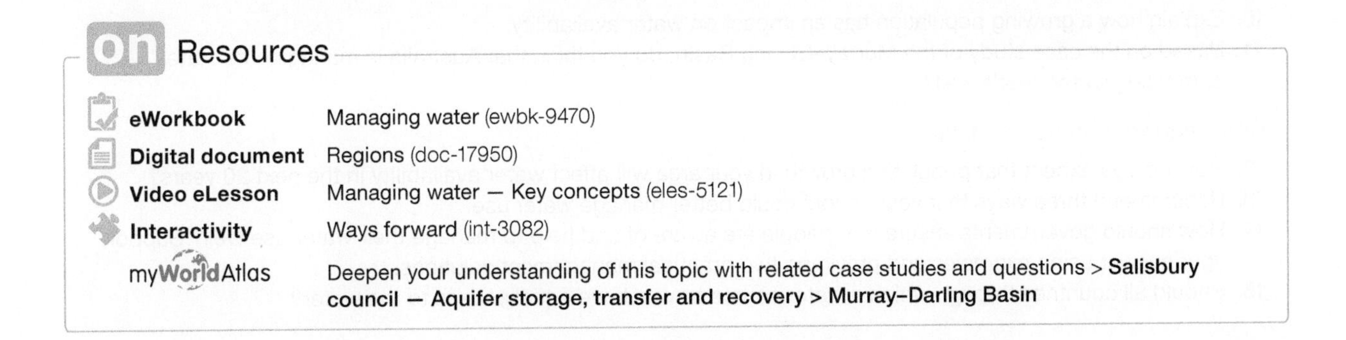

14.9 ACTIVITIES

1. Conduct some further research to find out how water is being managed along the Jordan River in West Asia.
2. Investigate the use of desalination in countries in West Asia (the Middle East). Map the locations of desalination plants using the **Regions** resource in the Digital documents of your Resources tab.
3. Watch the video on the Murray–Darling Basin and look at the information on this region in *myWorld Atlas*. Find out about the MDB plan and how it will help manage water. Conduct further research online to see how healthy the basin is today, and whether the plan is working.
4. If water were oil, leaking pipes would be fixed immediately. However, there is still a perception that water is not as valuable as oil, so the same investment is not made in it. Imagine you work for an advertising company and have to convince your audience that water is more valuable than oil. Film your advertisement and use a video editing tool to create music and voiceovers.

14.9 EXERCISE

Learning pathways

■ LEVEL 1	■ LEVEL 2	■ LEVEL 3
1, 3, 5, 11, 13	2, 4, 9, 12, 15	6, 7, 8, 10, 14

Check your understanding

1. Describe why it is difficult to manage water when the water supply crosses country or state borders.
2. What similarities and differences might exist between the way water is managed in the Murray–Darling Basin and the Jordan River.
3. Define the term *climate change*.
4. How might relying on a drainage basin that spans multiple countries lead to water shortages for some countries but not others?
5. Might disputes over water access occur within countries, or do they only occur between countries?
6. List three possible water management solutions that could be implemented on an individual level in Australia.
7. Might different water management strategies be more effective in different parts of Australia? Support your answer with examples.

Apply your understanding

8. Refer to **FIGURE 3**. Categorise each of the possible solutions as being more likely to be put in place on a local, national or international scale.
9. Choose two of the possible management solutions described in **FIGURE 3**. Compare their benefits and potential complications.
10. Explain how a growing population has an impact on water availability.
11. Based on the case study of the Murray–Darling Basin, do you think that Australia is managing its states' competing water needs well?

Challenge your understanding

12. How do you expect that population growth in your area will affect water availability in the next 20 years?
13. Recommend three ways that your school could better manage water use.
14. How should governments ensure that people are aware of and help to manage their water use well? Support your answer using examples and providing two practical management solutions.
15. Should all countries through which a water source travels have equal access to the water?

To answer questions online and to receive **immediate feedback** and **sample responses** for every question, go to your learnON title at www.jacplus.com.au.

14.10 Investigating topographic maps — Sources of water at Whataroa

LEARNING INTENTION

By the end of this subtopic, you will be able to use topographic maps to draw conclusions about water management issues.

14.10.1 Whataroa region

Whataroa is a town on the northern edge of Te Wāhipounamu, a World Heritage site in the south-west of New Zealand. It is an area of abundant water and incorporates Fiordland, glaciers and spectacular lakes.

Visitors travel from the town down the Waitangitaona River to view the White Heron Sanctuary, the only breeding area of the white heron (kotuku). Glaciers and snow feed the Whataroa River and its tributaries. South Westland Maori people used the area to hunt birdlife and for fresh-water fishing.

FIGURE 1 Kotuku (white heron) nesting site in Waitangi Roto Nature Reserve

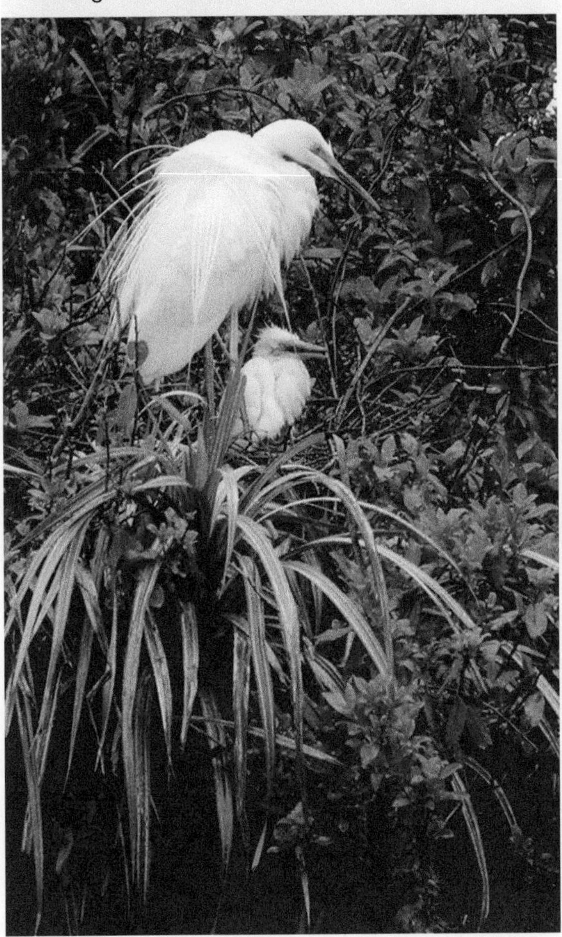

FIGURE 2 View of Franz Josef Glacier, New Zealand

FIGURE 3 Topographic map of the Whataroa area

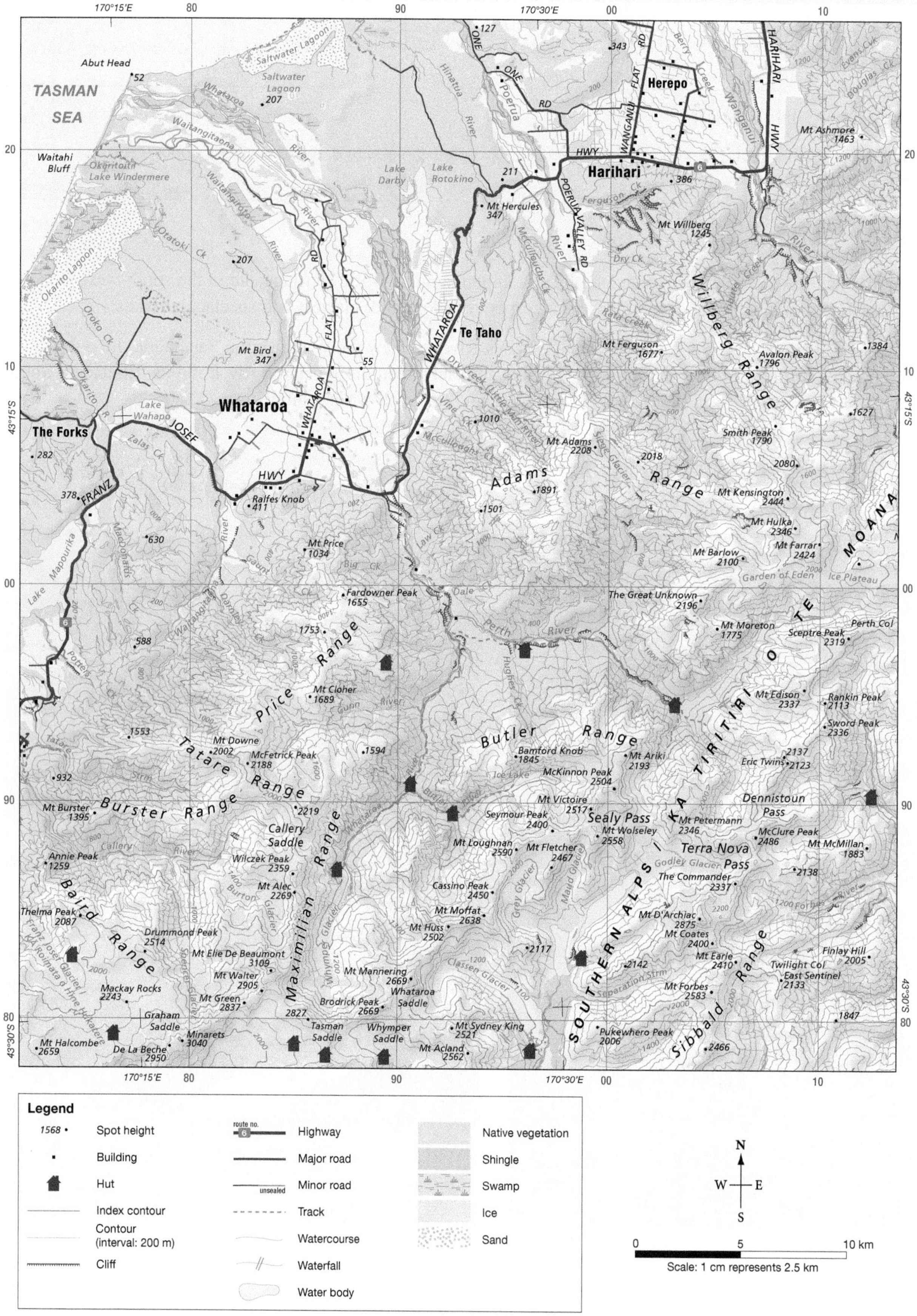

Legend

1568 •	Spot height	route no.	Highway		Native vegetation
▪	Building		Major road		Shingle
	Hut	unsealed	Minor road		Swamp
	Index contour		Track		Ice
	Contour (interval: 200 m)		Watercourse		Sand
	Cliff		Waterfall		
			Water body		

N, S, E, W

0 5 10 km

Scale: 1 cm represents 2.5 km

Source: LINZ Data Service. License: Creative Commons Attribution 3.0 New Zealand https://data.linz.govt.nz/license/attribution-3-0-new-zealand/

Resources

eWorkbook	Investigating topographic maps — Sources of water at Whataroa (ewbk-9474)
Digital document	Topographic map of the Whataroa area (doc-36258)
Video eLesson	Investigating topographic maps — Sources of water at Whataroa — Key concepts (eles-5122)
Interactivity	Investigating topographic maps — Sources of water at Whataroa (int-8547)
Google Earth	Whataroa, New Zealand (gogl-0093)

14.10 EXERCISE

Learning pathways

■ LEVEL 1	■ LEVEL 2	■ LEVEL 3
1, 4, 7	3, 5, 8	2, 6, 9, 10

Check your understanding

1. What is the height of:
   a. Mount Hercules (AR3921)
   b. Mount Cloher (AR3819)
   c. Drummond Peak (AR3718)
   d. Bamford Knob (AR3919)
   e. Mount Price (AR3820)?
2. What is the direction of:
   a. The Forks (AR3720) from Harihari (AR4021)
   b. Abut Head (AR 3722) from Whataroa (AR3820)
   c. Franz Josef Glacier (AR3718) from Te Taho (AR3921)?
3. Examine the north-west quadrant of **FIGURE 3**. What are the grid references of the following places?
   a. Waitahi Bluff
   b. Abut Head
   c. Lake Rotokino
4. In what direction are the following creeks flowing?
   a. Potters Creek (AR3719)
   b. Guant Creek (AR3820)
   c. Vine Creek (AR3920)
   d. Oroko Creek (AR3721)
   e. Hughes Creek (AR3919)

Apply your understanding

5. Describe the distribution of swamps in the vicinity of Whataroa. Explain why they may be located where they are.
6. Find Franz Josef Glacier in AR3718 and examine the photo in **FIGURE 2**. Provide a grid reference of where the photographer could have been standing when taking the photograph. Justify your answer.
7. Why might this area have been declared a World Heritage site? Use evidence from the map to support your reasoning.
8. Might water scarcity be an issue in this area? Give reasons why or why not.
9. Identify one challenge that the settlement of Whataroa may face in terms of its water management. Explain why this issue might occur, using the details shown in **FIGURE 3** to support your observations.

Challenge your understanding

10. How might increased glacier melting from climate change affect this area?

To answer questions online and to receive **immediate feedback** and **sample responses** for every question, go to your learnON title at www.jacplus.com.au.

14.11 Thinking Big research project — Desalination plant advertising

online only

The content in this subtopic is a summary of what you will find in the online resource.

Scenario

It is 2025 and the drought that began in 2019 has continued without foreseeable sufficient rain. The New South Wales Government has recognised that there is a need for a new desalination plant to create fresh water to meet the state's needs and, in particular, the growing population of Sydney and the metropolitan regions on the coast.

Task

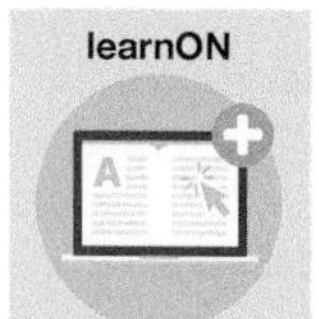

Your media company has been asked to develop an advertisement to explain the need to spend government money on this new plant. The government plans to reclaim part of the Belmont Wetlands State Park to build the plant, which will supply water to communities in Newcastle, Central Coast and Northern Beaches regions.

Go to your Research tab to access the resources you need to complete this research project.

Resources

ProjectsPLUS Thinking Big research project — Desalination plant advertising (pro-0234)

14.12 Review

14.12.1 Key knowledge summary

14.2 Australia's water resources

- Australia has limited freshwater resources.
- Many rivers in Australia are non-perennial (only flow at certain times). The flow of rivers in Australia is highly variable due to the small proportion of rainfall that makes its way into river systems.
- Major rivers in Australia are the Murray–Darling, Murrumbidgee, Lachlan, Cooper, Flinders, Diamantina, Gascoyne and Goulburn rivers.

14.3 Australia's climate and water availability

- Australia's climate is the most important factor affecting water availability.
- Australia's coastal areas have low rainfall variability. Desert areas tend to have low rainfall but high rainfall variability.
- Much of the rainfall received in Australia evaporates. Central Australia has high evaporation rates while coastal areas have low evaporation rates.

14.4 The right to water

- Everyone has the right to enough safe, accessible and affordable water for all their needs.
- Water scarcity occurs when the demand for water is greater than the available supply.
- Water stress occurs when a country has annual water supplies less than 1700 cubic metres per capita.

14.6 Water scarcity

- Water scarcity is when annual water supplies drop below 1000 cubic metres per person.
- Absolute water scarcity is when annual water supplies drop below 500 metres per person.
- Factors that affect water scarcity are the naturally occurring, uneven distribution of water resources; population; agriculture; energy production; climate change and water ownership.

14.8 Strategies to address water scarcity

- Globally, 3 in 10 people lack access to safely managed drinking water, and 1 in 4 health facilities lack basic water facilities.
- Goal 6 of the UN's Sustainable Development Goals was 'Ensure access to water and sanitation for all'.
- Governments, non-government organisations, communities and individuals play an important role in managing water.

14.9 Managing water

- Water management involves cooperation of different users such as farmers, industry and individuals.
- Water can be managed at a local, regional, state, national and global scale.
- Water management strategies include building dams and desalination plants, fixing leaking pipes, increasing the price of water, cloudseeding, drip irrigation, reusing and recycling water, installing water tanks.

14.12.2 Key terms

climate long-term weather patterns in an area

climate change a change in the world's climate. This can be very long term or short term, and is caused by human activity.

drainage basin the entire area of land that contributes water to a river and its tributaries

evaporation the process by which water is converted from a liquid to a gas and thereby moves from land and surface water into the atmosphere

improved drinking water water that is safe for human consumption

rainfall variability the change from year to year in the amount of rainfall in a given location

relative humidity the amount of moisture in the air

perennial permanent or continually available

water scarcity when annual water supplies drop below 1000 cubic metres per person. Absolute water scarcity describes the situation when annual water supplies drop below 500 cubic metres per person.

water stress a situation that occurs in a country with less than 1000 cubic metres of renewable fresh water per person

14.12.3 Reflection

Complete the following to reflect on your learning.

Revisit the inquiry question posed in the Overview:

Our planet is covered with water, so why do we have to be so careful with it?

1. Now that you have completed this topic, what is your view on the question? Discuss with a partner. Has your learning in this topic changed your view? If so, how?
2. Write a paragraph in response to the inquiry question outlining your views.

Subtopic	Success criteria			
14.2	I can identify and locate Australia's water resources.			
14.3	I can evaluate how climate affects Australia's water availability.			
14.4	I can discuss people's rights to water.			
	I can explain some impacts of water stress.			
14.5	I can represent geographical information is a pie graph.			
14.6	I can define 'water scarcity'.			
	I can explain the factors that impact on water scarcity.			
14.7	I can analyse geographical information presented in a choropleth map.			
14.8	I can identify and describe the roles of governments, non-government organisations, communities and individuals in water management.			
14.9	I can give examples of water management strategies at local, country and global scales.			
	I can assess the likely impact of a water management strategy.			
14.10	I can interpret a topographic map to draw conclusions about water management issues.			

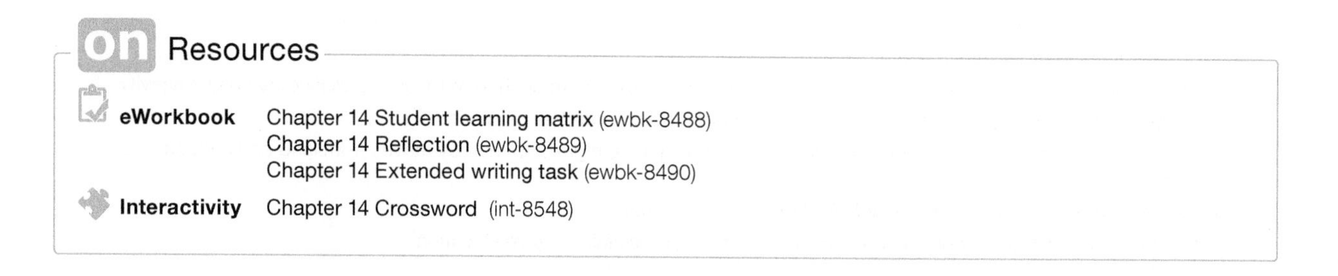

ONLINE RESOURCES

on Resources

Below is a full list of **rich resources** available online for this topic. These resources are designed to bring ideas to life, to promote deep and lasting learning and to support the different learning needs of each individual.

eWorkbook

- 14.1 Chapter 14 eWorkbook (ewbk-7994) ☐
- 14.2 Australia's water resources (ewbk-9442) ☐
- 14.3 Australia's climate and water availability (ewbk-9446) ☐
- 14.4 The right to water (ewbk-9450) ☐
- 14.5 SkillBuilder — Constructing a pie graph (ewbk-9454) ☐
- 14.6 Water scarcity (ewbk-9458) ☐
- 14.7 SkillBuilder — Reading and describing basic choropleth maps (ewbk-9462) ☐
- 14.8 Strategies to address water scarcity (ewbk-9466) ☐
- 14.9 Managing water (ewbk-9470) ☐
- 14.10 Investigating topographic maps — Sources of water at Whataroa (ewbk-9474) ☐
- 14.12 Chapter 14 Student learning matrix (ewbk-8488) ☐
  - Chapter 14 Reflection (ewbk-8489) ☐
  - Chapter 14 Extended writing task (ewbk-8490) ☐

Sample responses

- 14.1 Chapter 14 Sample responses (sar-0147) ☐

Digital documents

- 14.9 Regions (doc-17950) ☐
- 14.10 Topographic map of the Whataroa area (doc-36258) ☐

Video eLessons

- 14.1 Of droughts and flooding rains (eles-1617) ☐
  - Water scarcity – Photo essay (eles-5347) ☐
- 14.2 Australia's water resources — Key concepts (eles-5116) ☐
- 14.3 Australia's climate and water availability — Key concepts (eles-5117) ☐
- 14.4 The right to water — Key concepts (eles-5118) ☐
- 14.5 SkillBuilder — Constructing a pie graph (eles-1632) ☐
- 14.6 Water scarcity — Key concepts (eles-5119) ☐
- 14.7 SkillBuilder — Reading and describing basic choropleth maps (eles-1706) ☐
- 14.8 Strategies to address water scarcity — Key concepts (eles-5120) ☐
- 14.9 Managing water — Key concepts (eles-5121) ☐
- 14.10 Investigating topographic maps — Sources of water at Whataroa — Key concepts (eles-5122) ☐

Interactivities

- 14.2 Australia's water resources (int-8543) ☐
- 14.3 Australia's average rainfall (int-5284) ☐
  - Hot and dry (int-3081) ☐
- 14.4 The right to water (int-8544) ☐
- 14.5 SkillBuilder — Constructing a pie graph (int-3128) ☐
- 14.6 Water scarcity (int-8545) ☐
- 14.7 SkillBuilder — Reading and describing basic choropleth maps (int-3286) ☐
- 14.8 Strategies to address water scarcity (int-8546) ☐
- 14.9 Ways forward (int-3082) ☐
- 14.10 Investigating topographic maps — Sources of water at Whataroa (int-8547) ☐
- 14.12 Chapter 14 Crossword (int-8548) ☐

Fieldwork

- 14.2 Water infiltration (fdw-0020) ☐
- 14.8 Monitoring water use (fdw-0021) ☐

ProjectsPLUS

- 14.11 Thinking Big research project — Desalination plant advertising (pro-0234) ☐

Weblinks

- 14.2 Australian groundwater explorer (web-2864) ☐
- 14.3 Cloud identification (web-2865) ☐

Google Earth

- 14.10 Whataroa, New Zealand (gogl-0093) ☐

myWorld Atlas

- 14.3 Australia: weather and climate (mwa-4409) ☐
- 14.4 Russia and Eurasia (mwa-4452) ☐
- 14.9 Salisbury council — Aquifer storage, transfer and recovery > Murray–Darling Basin (mwa-7295) ☐

Teacher resources

There are many resources available exclusively for teachers online.

15 Natural hazards — Water and wind

INQUIRY SEQUENCE

To access a pre-test with **immediate feedback** and **sample responses** to every question in this chapter, select your learnON format at www.jacplus.com.au.

15.1 Overview

Numerous **videos** and **interactivities** are embedded just where you need them, at the point of learning, in your learnON title at www.jacplus.com.au. They will help you to learn the content and concepts covered in this topic.

What are natural hazards? Can we predict when they will occur and prepare ourselves for them?

15.1.1 Introduction

People have harnessed wind and energy sources throughout history to generate power and to supply human settlements with supplies of fresh water. Intense weather events generate strong winds and destructive water flows when accompanied by heavy rainfall. Severe storms can tear roofs from houses and pull trees from the ground.

STARTER QUESTIONS

1. As a class, identify locations at your school that are more exposed to wind than sheltered spots.
2. Create a list of the ways the wind can influence your personal interests, activities and hobbies.
3. Brainstorm a list of extreme weather events related to the wind and water.
4. Have you ever experienced an extreme weather event? Describe how it made you feel.
5. Watch the **Wind and water hazard — Photo essay** Video eLesson in your Resources tab. What wind and water hazards impact your community?

FIGURE 1 Dust storm driven by strong wind, central west New South Wales

on Resources

eWorkbook Chapter 15 eWorkbook (ewbk-7995)

Video eLessons Wind hazards (eles-1618)
Wind and water hazard — Photo essay (eles-5123)

15.2 Features of natural hazards

LEARNING INTENTION

By the end of this subtopic, you will be able to define the terms *natural hazard* and *natural disaster*, and locate places where atmospheric and hydrologic hazards occur in Australia.

15.2.1 Natural hazards defined

Australia is prone to a wide variety of natural hazards, which range from drought and bushfire to flooding. Many of these events are part of the weather's natural cycle. However, human actions such as overgrazing, deforestation and the alteration of natural waterways have sometimes increased the impact of these hazards.

FIGURE 1 Flooding of the Nepean River and Sydney International Regatta Centre in Penrith, New South Wales, 2021

A **natural hazard** is an event that is a *potential* source of harm to a community. It is the result of natural processes and may cause serious damage to buildings, roads, communication and transport, and loss of life. In contrast, a **natural disaster** occurs because of a hazardous event that dramatically affects a community. According to the World Meteorological Organization, natural hazards become disasters when people's lives and livelihoods are destroyed.

Atmospheric and hydrologic hazards

Atmospheric hazards are extreme natural events that originate in the Earth's atmosphere; they include tropical cyclones, storms, blizzards and tornadoes. **Hydrologic hazards** originate in the Earth's hydrosphere from changes to the water cycle and include flooding and drought. (See section 12.2 Resources). They are closely linked to atmospheric processes. For example, a storm surge following the passage of a tropical cyclone (atmosphere) can lead to flooding (hydrosphere) in low-lying areas of coastline.

natural hazard an extreme event that is the result of natural processes and has the potential to cause serious material damage and loss of life

natural disaster an extreme event that is the result of natural processes and causes serious material damage or loss of life

atmospheric hazard extreme natural event that begins in the atmosphere (in the air)

hydrologic hazard extreme natural event that begins in the hydrosphere (in the water)

15.2.2 Spatial distribution of natural hazards

A wide variety of atmospheric and hydrologic hazards occur around the world. Some places experience a particular type of hazard more frequently than other locations. For example, **FIGURE 2** shows that coastal areas of northern Australia are more vulnerable to the threat of tropical cyclones and flash flooding is common along Australia's inland river systems.

The ability of an area to recover from the impacts of a natural hazard may be hampered by a lack of resources. People may lose their personal belongings, homes and livestock, which are often linked to their incomes. However, regions that are adequately prepared, and where support services are available, can rebuild their lives and recover more quickly. Communities living in places more susceptible to natural hazards have decided that the benefits of living there outweigh the risks.

int-8609

FIGURE 2 Some of Australia's natural hazards and disasters

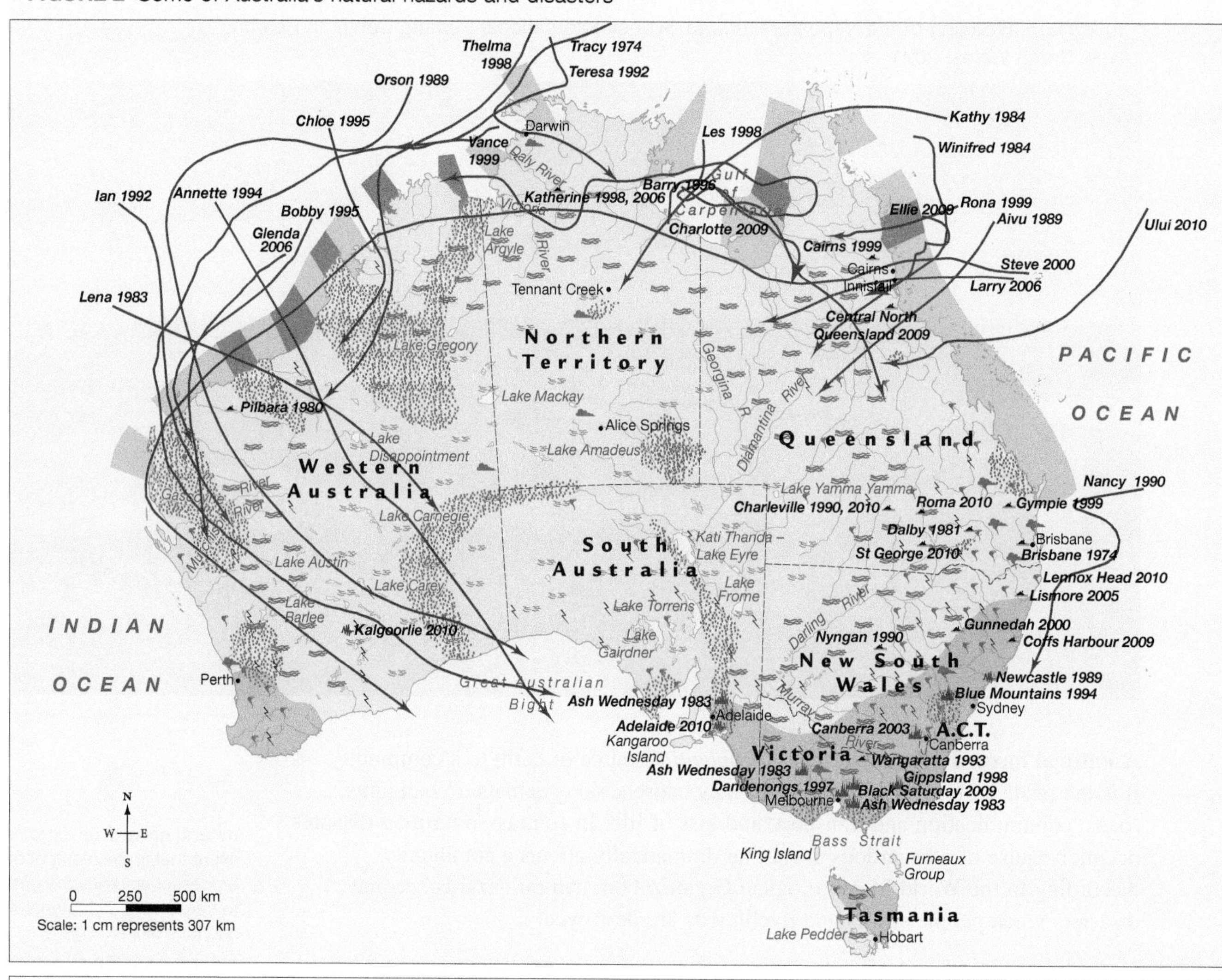

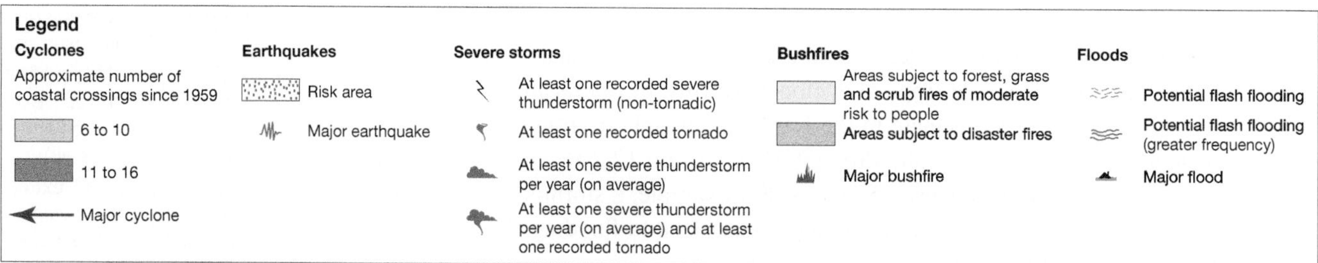

Source: MAPgraphics Pty Ltd, Brisbane

15.2.3 Managing impacts of natural hazards

The impacts of some natural hazards can be influenced by the actions of people. For example, humans can influence the severity of a flood through land clearing and construction within a river catchment. After a period of sustained rainfall, a flood may be expected to occur in low-lying areas, and a greater impact will be felt in areas where effective planning measures have not been taken.

When hazardous events take place in populated areas, a disaster can occur more easily. Action that is designed to prepare for such a disaster, prevent it, and respond quickly, and that involves individuals, groups and the government, allows recovery to occur more effectively (see **FIGURE 3**).

FIGURE 3 Some strategies to manage the impacts of natural hazards

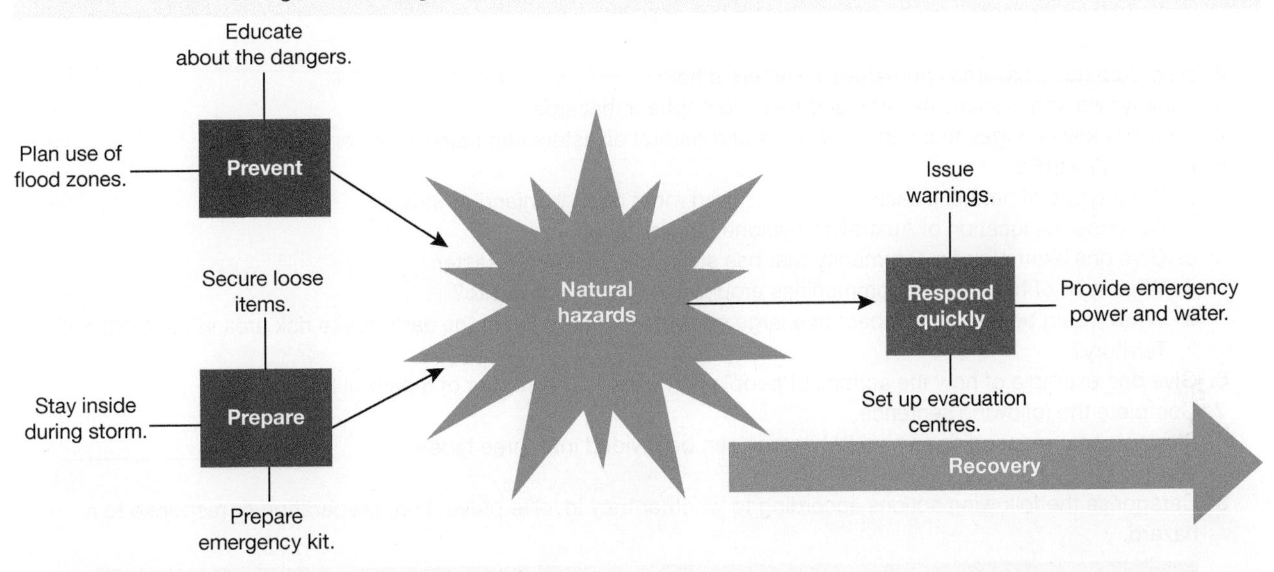

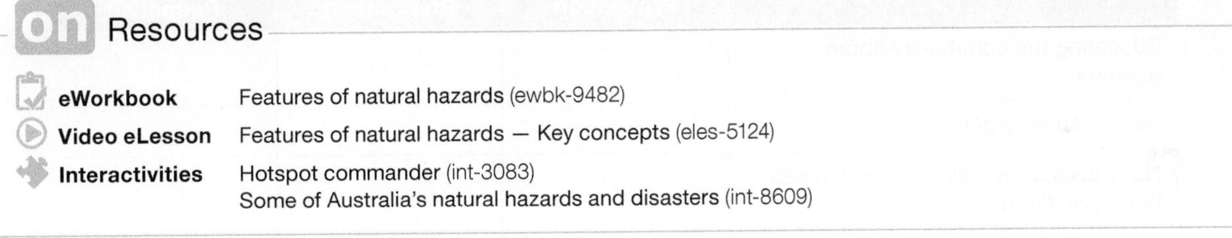

on Resources

eWorkbook	Features of natural hazards (ewbk-9482)
Video eLesson	Features of natural hazards — Key concepts (eles-5124)
Interactivities	Hotspot commander (int-3083) Some of Australia's natural hazards and disasters (int-8609)

15.2 ACTIVITIES

1. Investigate the ways a natural hazard has created challenges for heavily populated areas. Find examples of specific hazards and the challenges they presented. Discuss as a class how these issues might have been prevented, how people might have prepared, and what responses were required.
2. Research natural disasters in Australia in the past ten years. Use the **Disaster resilience** weblink in your Resources panel as a starting point. Add recent disasters to a map of Australia (you could do this on a printed map or digitally).
   a. Which types of hazards and disasters have become more frequent or more severe?
   b. Have the patterns of where and when hazards occur changed? For example, has a specific type of hazard started to occur in different areas?
   c. Investigate the potential reasons for these patterns.
3. Research two ways that 'the impacts of some natural hazards can be influenced by the actions of people'. Present your findings in a one-page A4 document and include all source details.

FIGURE 4 Cyclone warning sign, Townsville, Queensland

15.2 EXERCISE

Learning pathways

■ LEVEL 1	■ LEVEL 2	■ LEVEL 3
1, 2, 7, 9, 13	3, 5, 6, 10, 14	4, 8, 11, 12, 15

Check your understanding

1. Define the following terms in your own words:
   a. atmospheric hazard
   b. hydrologic hazard
   c. natural disaster.
2. How do natural hazards and natural disasters differ?
3. Identify two atmospheric hazards and two hydrospheric hazards.
4. Describe key changes that natural hazards and natural disasters can cause to an environment.
5. Refer to **FIGURE 2**.
   a. What types of natural disasters have occurred most often in inland Australia?
   b. Describe the location of Australia's cyclone hazard zone.
   c. Give one example of a community that has suffered a bushfire disaster.
   d. What type of hazards are communities around Newcastle subject to?
   e. What would be the likely impact of a large earthquake occurring in the earthquake risk area in the Northern Territory?
6. Give one example of how the actions of people can worsen the impact of a natural hazard.
7. Complete the following sentence.
   Strategies for responding to natural hazards can be divided into three types: ____________, ____________ and ____________
8. Categorise the following actions according to whether they involve prevention, preparation or response to a hazard.

	Prevention	Preparation	Response
Educating the community about dangers			
Having an emergency kit at home			
Not allowing people to build on areas that often flood			
Downloading an app to receive emergency warnings from authorities			
Fixing fallen power lines			

Apply your understanding

9. Explain how a flood can be both a natural and human hazard.
10. Explain three strategies that might reduce property damage and injury in the event of a natural hazard in your local area.
11. Explain why the risk of experiencing a natural disaster depends on the geographical location of a community.
12. How can human actions have an impact on some natural hazards? Give an example to support your answer.

Challenge your understanding

13. Predict what kind of natural hazard is most likely to affect where you live. Give reasons for your decision.
14. Suggest three actions that might help your school prepare for severe storms in your area.
15. Are some natural hazards more difficult to prevent or prepare for? How might this impact on the way communities respond?

To answer questions online and to receive **immediate feedback** and **sample responses** for every question, go to your learnON title at www.jacplus.com.au.

15.3 SkillBuilder — Creating a simple column or bar graph

online only

LEARNING INTENTION

By the end of this subtopic, you will be able to construct and interpret a column or bar graph.

The content in this subtopic is a summary of what you will find in the online resource.

15.3.1 Tell me

What are column or bar graphs?

Column graphs show information or data in vertical columns. In a bar graph, the data is shown in bars that are drawn horizontally. Column and bar graphs can be drawn by hand or constructed using a computer spreadsheet. Column graphs are useful for comparing quantities. They can help us understand and visualise data, see patterns and gain information.

FIGURE 1 A labelled column graph

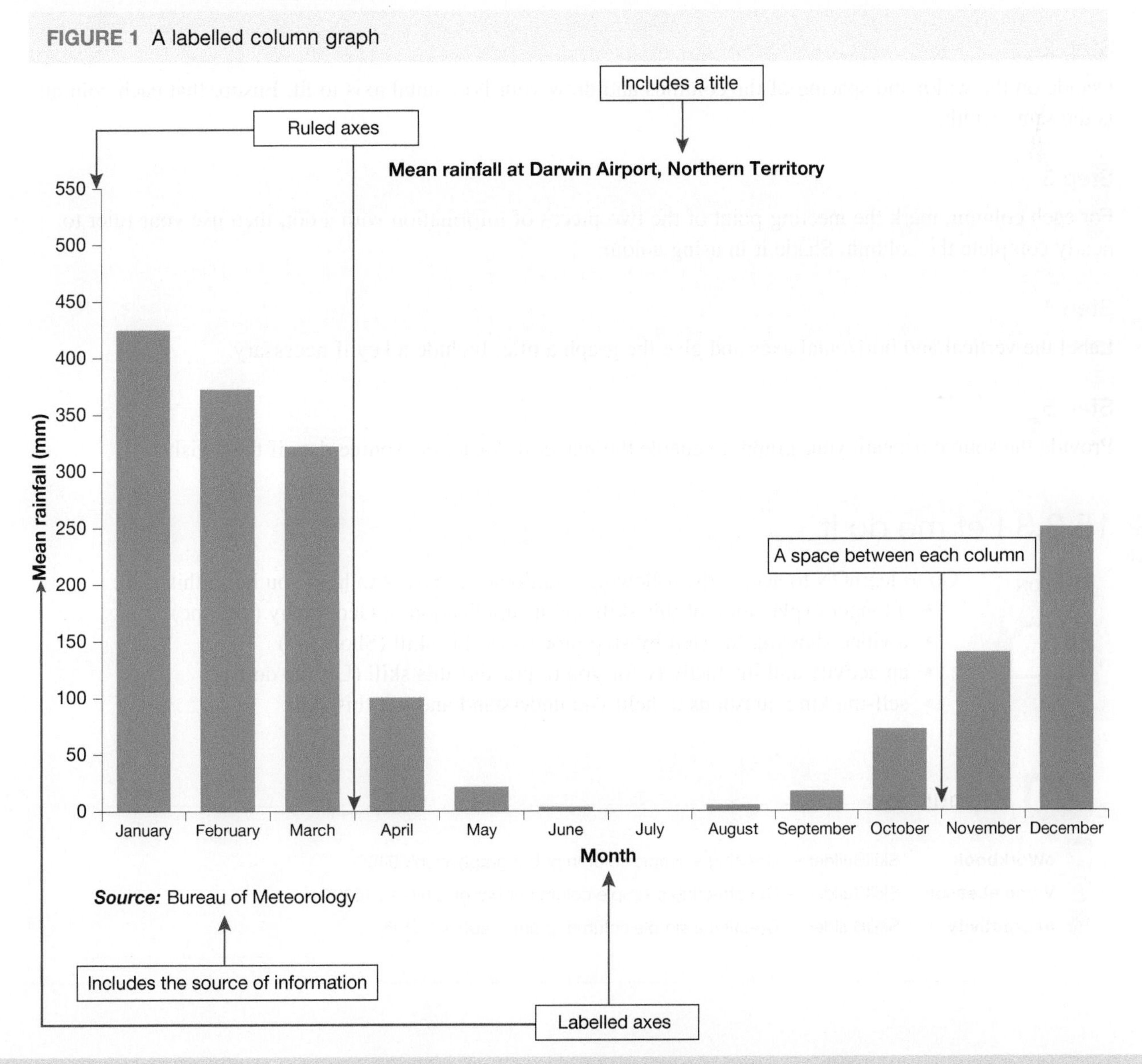

15.3.2 Show me

eles-1639

How to complete a column graph

Step 1

Examine the data. Decide on the scale to use for your vertical axis. For this example the vertical axis should start at zero and increase at intervals to suit the data. As the highest rainfall for any month for Cardwell is 465.9 millimetres, intervals of 50 would be suitable. For this exercise you could use 1 centimetre to represent 50 millimetres of rainfall. Draw your vertical axis according to the scale you have devised.

TABLE 1 Mean monthly rainfall for the years 1871 to 2016, Cardwell, Queensland

Statistics	Jan	Feb	Mar	Apr	May	Jun	Jul	Aug	Sep	Oct	Nov	Dec
Mean rainfall (mm) for years 1871 to 2016	438.5	465.9	400	208.6	94.7	47	32.4	29.2	38.5	54.4	115.2	193.5

Source: © Bureau of Meteorology

Step 2

Decide on the width and spacing of the columns and draw your horizontal axis to fit. Ensure that each column is the same width.

Step 3

For each column, mark the meeting point of the two pieces of information with a dot, then use your ruler to neatly complete the column. Shade it in using colour.

Step 4

Label the vertical and horizontal axes and give the graph a title. Include a key if necessary.

Step 5

Provide the source beneath your graph, to enable the reader to locate the source data if they wish.

15.3.3 Let me do it

int-3135

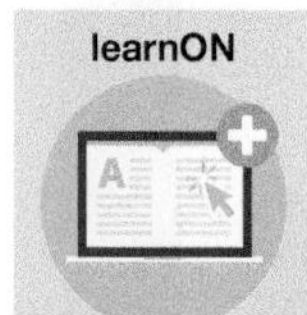

Go to learnON to access the following additional resources to help you build this skill:

- a longer explanation of this skill and its application in Geography (Tell me)
- a video showing the step-by-step process of this skill (Show me)
- an activity and interactivity for you to practise this skill (Let me do it)
- self-marking questions to help you understand and use this skill.

on Resources

eWorkbook SkillBuilder — Creating a simple column or bar graph (ewbk-9486)

Video eLesson SkillBuilder — Constructing a simple column or bar graph (eles-1639)

Interactivity SkillBuilder — Creating a simple column or bar graph (int-3135)

15.4 Thunderstorms

LEARNING INTENTION

By the end of this subtopic, you will be able to identify the causes of thunderstorms and suggest ways to minimise the damage they cause.

15.4.1 Types of storms

A storm is any violent disturbance of the atmosphere and the effects associated with it. The main types of storms are thunderstorms, tropical cyclones, cold fronts and tornadoes. In this section we look at thunderstorms and hailstorms. Hailstorms are a specific type of thunderstorm.

15.4.2 Causes of thunderstorms

Thunderstorms, also referred to as electrical storms, form in unstable, moist atmospheres where powerful updrafts occur. Air may be forced upwards when a cold front approaches, when air is pushed upwards to pass over higher land or in places where warm temperatures create convectional currents. It is estimated that, around the Earth, there are 1800 thunderstorms each day. Between 2000 and 2018, an average of around 100 severe thunderstorms were reported in Australia each year.

Some 1000 or so years ago, the Vikings thought thunder was the rumble of Thor's chariot. (He was their god of thunder and lightning.) Lightning marked the path of his mighty hammer Mjöllnir when he threw it across the sky at his enemies.

Today we know that thunderstorms occur when large **cumulonimbus clouds** build up enough static electricity to produce lightning, as shown in **FIGURE 2**. Lightning instantly heats the air through which it travels to about 20 000 °C — more than three times as hot as the surface of the Sun. This causes the air to expand so quickly that it produces an explosion (thunder). The time between a lightning flash and the crash of thunder tells you how far away the lightning is (5 seconds means that the lightning is 1.6 kilometres away).

cumulonimbus clouds huge, thick clouds that produce electrical storms, heavy rain, strong winds and sometimes tornadoes. They often appear to have an anvil-shaped flat top and can stretch from near the ground to 16 kilometres above the ground.

FIGURE 1 A severe thunderstorm developing over the NSW Central Coast

int-7794

FIGURE 2 How a thunderstorm works

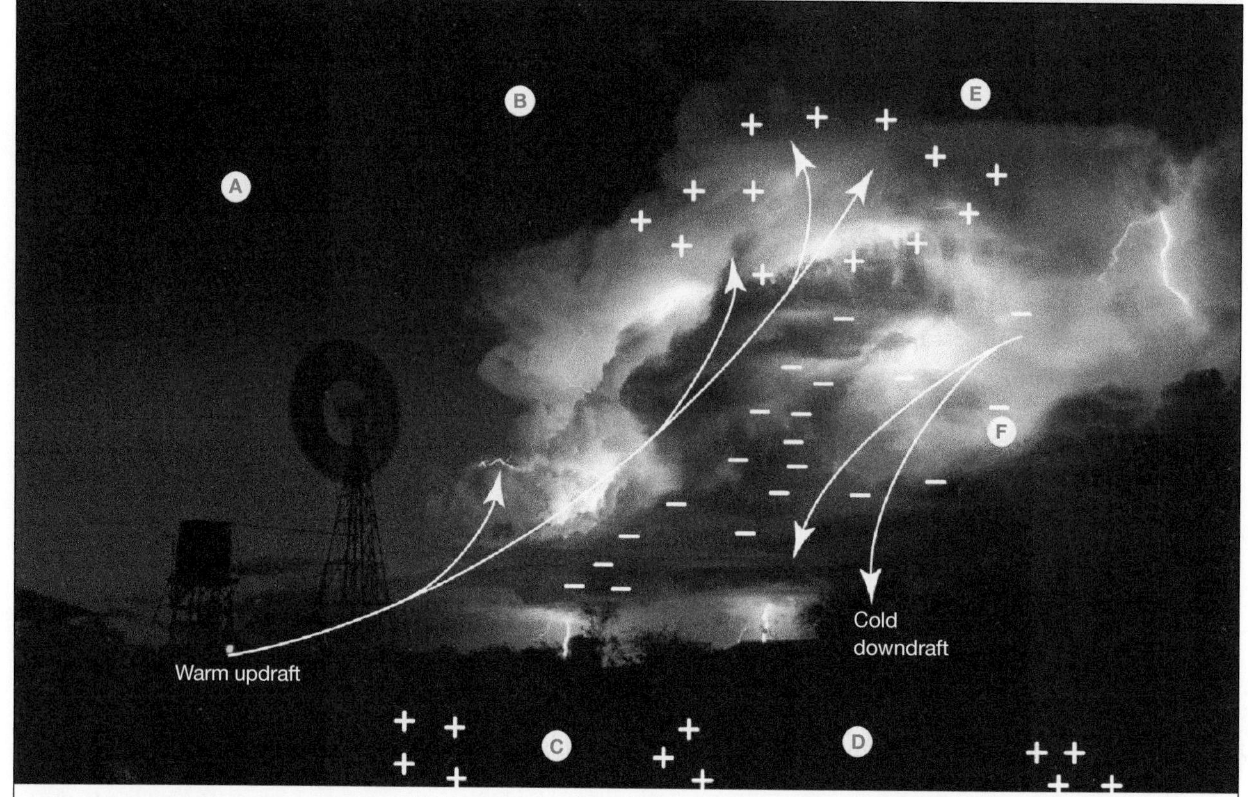

A As air currents in a cumulonimbus cloud become more violent, they fling ice crystals and water droplets around faster. The more these crystals and droplets smash into one another, the more friction builds up. This creates huge energy stores of static electricity in the cloud.

B Lighter particles with a positive electric charge drift upwards. Heavier particles with a negative charge sink.

C The ground below the cloud has a positive charge.

D Lightning travels to the ground via the shortest route. This is why it sometimes strikes buildings or tall trees.

E A bolt of lightning actually consists of a number of flashes that travel up and down between the cloud and the ground. This happens so quickly we can't see it.

F The difference in energy between the positive charge on the ground and the massive negative charge at the bottom of the cloud becomes huge. A lightning bolt corrects some of this difference.

+ positive charge

– negative charge

15.4.3 Features of severe thunderstorms

According to the Bureau of Meteorology, a thunderstorm can be classified as severe if it has one or more of the following features.

- Flash flooding. Thunderstorms often move slowly, dropping a lot of precipitation in one area. The rain or hail may consequently be too heavy and long-lasting for the ground to absorb the moisture. The water then runs off the surface, quickly flooding local areas.
- **Hailstones** that are 2 centimetres or more in diameter. The largest recorded hailstone had a circumference of 47 centimetres.
- Wind gusts of 90 kilometres per hour or more. Cold blasts of wind hurtle out of thunderclouds, dragged down by falling rain or hail. When the drafts hit the ground, they gust outwards in all directions.

hailstone an irregularly shaped ball of frozen precipitation

In the right conditions, rapidly spinning updrafts of air can develop from thunderstorm activity. Although severe **tornadoes** are not common in Australia, around 400 tornadoes have been recorded here. Social media warnings, as shown in **FIGURE 3**, are issued by the Bureau of Meteorology to warn the community.

FIGURE 3 Bureau of Meteorology warnings are issued via social media.

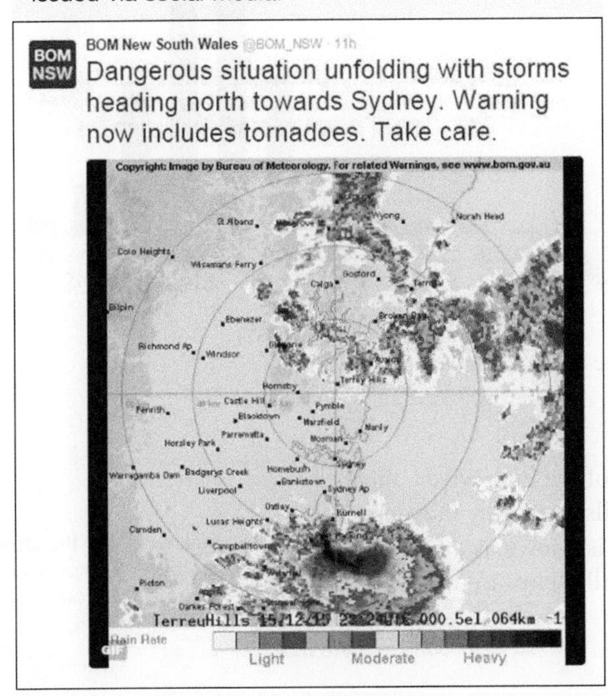

FIGURE 4 Flash flooding may occur after a thunderstorm.

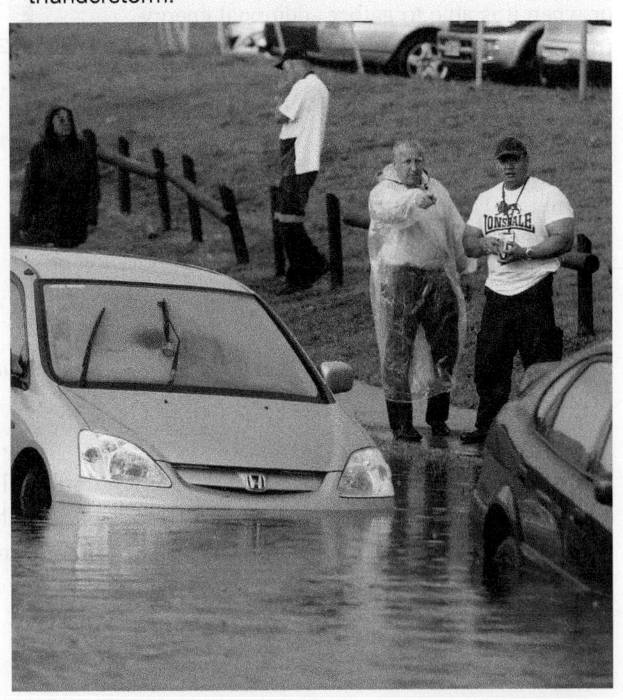

15.4.4 Thunderstorm activity

Thunderstorms can occur at any time of the year, but they are more likely to occur during spring and summer, as shown in **FIGURES 5** and **6**. This is due mainly to the warming effects of the Sun and the fact that warm air can hold more moisture than cold air.

tornado violent, wildly spinning column of air that forms over land surfaces. It drops from under a cumulonimbus cloud and contacts the ground.

FIGURE 5 Average monthly distribution of thunderstorms in Darwin

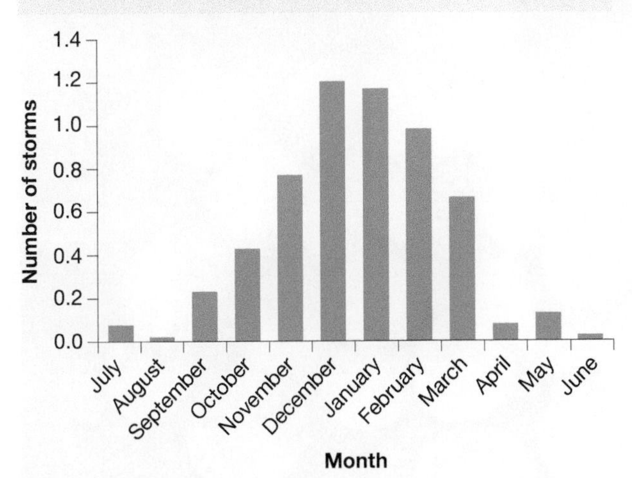

Source: Bureau of Meteorology

FIGURE 6 Average monthly distribution of thunderstorms in Melbourne

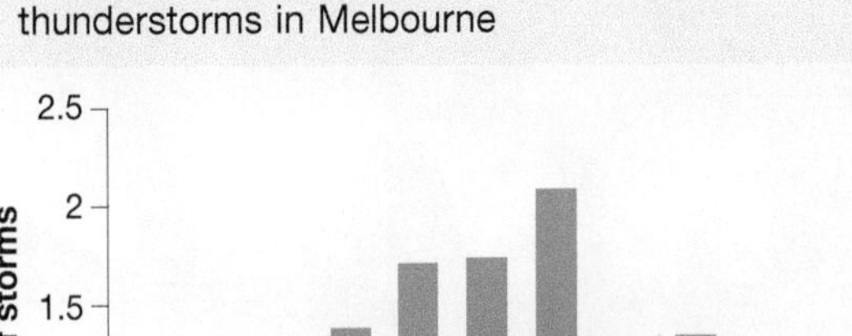

Source: Bureau of Meteorology

Thunderstorms are created when cooler air begins to push warmer, humid air upwards. As the warm air continues to rise rapidly in an unstable atmosphere, the cloud builds up higher and begins to spread. Thunderstorms can quickly develop when the atmosphere remains unstable or when it is able to gather additional energy from surrounding winds.

The time of day when thunderstorms are more likely is shown in **FIGURE 7**. You will notice that thunderstorm activity is greater in the afternoon. This is linked to the daily heating of the Earth by the Sun, which peaks in the afternoon.

FIGURE 7 Hourly distribution of thunderstorms in New South Wales and the Australian Capital Territory

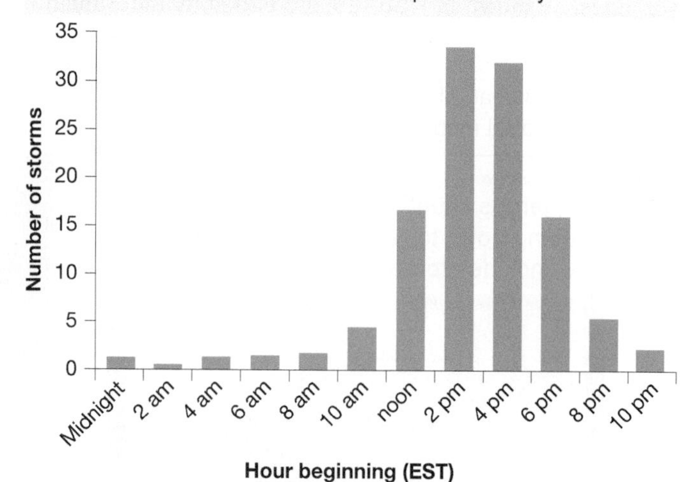

Source: Bureau of Meteorology

15.4.5 Features of a hailstorm

When we think about thunderstorms, we often think only of the high winds, thunder and lightning, but significant damage is also caused by hailstones. Any thunderstorm that produces hailstones large enough to reach the ground is known as a **hailstorm**. Hailstones in Australia tend to range in size from a few millimetres to the size of a tennis ball (see **FIGURE 8**)

hailstorm any thunderstorm that produces hailstones large enough to reach the ground

FIGURE 8 Hailstones, some of which are almost the size of tennis balls, fell in Leppington, NSW, on 20 December 2016.

int-8550

FIGURE 9 How a supercell hailstorm forms

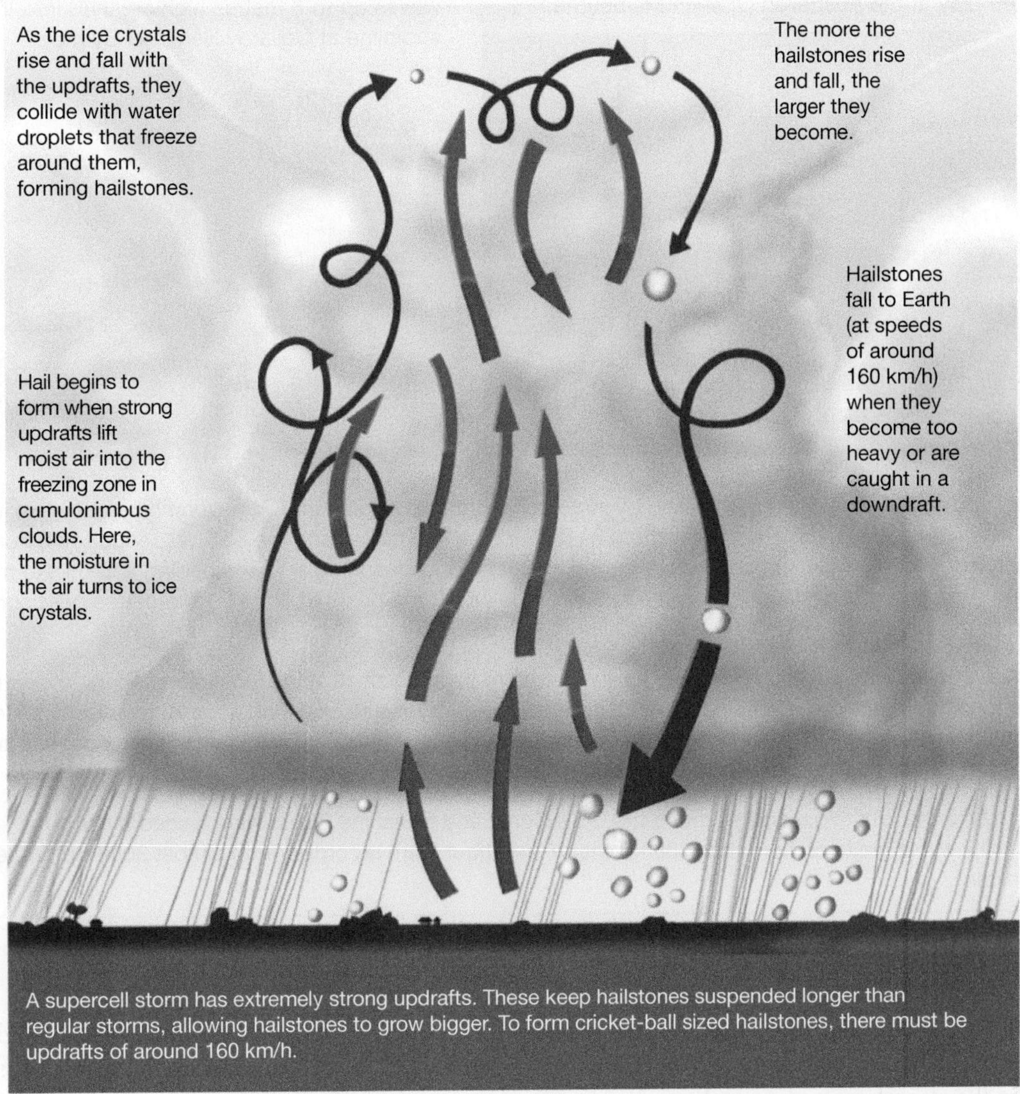

CASE STUDY

Timeline of supercell storm event across South East Australia, 2016

On 13 January 2016, Melbourne sweltered through temperatures of about 43 °C. Intense thunderstorm activity with wind gusts up to 100 kilometres per hour swept through in the early evening, causing the city to be blanketed by a cloud of dust. Up to 1000 homes were left without power.

The following day a severe storm struck Sydney, with winds gusting up to 98 kilometres per hour bringing down power lines, damaging buildings and cars and causing flash flooding. More than 40 000 homes and businesses reported power outages. The temperature plummeted by more than 10 °C in five minutes. Emergency services responded to 145 storm-related incidents, including a gas leak.

On 16 January, Townsville recorded 91 millimetres of rain in 30 minutes, resulting in flash flooding that left many motorists stranded. The rain continued to fall, with 181 millimetres recorded in two hours. Wind gusts of more than 100 kilometres accompanied the massive storm that has been described as a once-in-a-100-year event. Unfortunately, while large areas were inundated, the rain had little impact on the region's water storages.

FIGURE 10 In January 2016, a supercell storm interrupted play at the Australian Open in Melbourne.

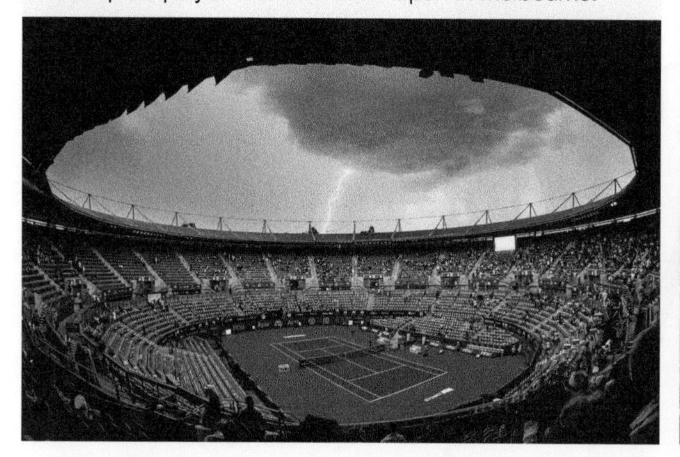

FIGURE 11 Later in 2016, in a supercell storm, waves up to 8 metres high crashed into the shoreline at Collaroy Beach in Sydney.

Both Adelaide and Sydney were pummelled by supercell storms on 22 January. The worst hit areas were in the Adelaide Hills and Fleurieu Peninsula where 20 000 homes lost power and the SES responded to 61 calls for help. Thirty-five millimetres of rain was recorded in half an hour, resulting in flash flooding and hailstorms measuring 2 centimetres in diameter carpeted parts of the city. Wind gusts of up to 90 kilometres per hour were recorded at the airport.

Meanwhile, Sydney was warned to prepare for the worst, to secure vehicles and loose items, unplug electronic equipment and to stay indoors. The city braced itself for more intense storms, following from those experienced in previous days that were the result of a large number of hot days. Flash flooding, damaging winds, hail and lightning were set to continue.

On 29 January, the tourist hot spots around the Gold Coast and Sunshine Coast were lashed by severe storm activity. Wind gusts of more than 100 kilometres per hour were recorded, with almost 9000 properties losing power.

15.4.6 Steps to take during a thunderstorm

During storms, damage and injury are often caused by loose objects blown around by the wind, by lightning strikes, and by people being caught in flash floods. To protect yourself, take the following precautions:

- Before the storm approaches, make sure loose objects outside your home are secure.
- Stay inside during the storm.
- Unplug electrical equipment such as computers, televisions and gaming consoles.
- Avoid using the phone until the storm has passed.
- Use torches rather than candles as a source of light.
- Stay indoors, and stay away from windows.
- If caught in a storm, try to find shelter.
- If caught in the open, move away from objects that could fall, such as trees.
- Crouch down; don't huddle in a group.
- Never try to walk or drive through floodwater.
- Do not touch or approach fallen power lines.

FIGURE 12 The roof of a house sits in the middle of the road at The Gap in Brisbane's north-west. The roof is from a home 50 metres away.

Resources

eWorkbook Thunderstorms (ewbk-9490)

Video eLesson Thunderstorms — Key concepts (eles-5125)

Interactivities How a thunderstorm works (int-7794)
How a supercell hailstorm forms (int-8550)

15.4 ACTIVITIES

1. Use the diagrams in this subtopic to make your own sketch of a supercell storm. Using words such as *evaporation*, *condensation* and *precipitation*, annotate your diagram to explain how storms develop.
2. Use the information in this subtopic to annotate a map of Australia to show the dates when thunderstorms were recorded around Australia and the damage they caused. (Find information online to annotate places that are not mentioned in this subtopic.)

15.4 EXERCISE

Learning pathways

■ LEVEL 1	■ LEVEL 2	■ LEVEL 3
2, 3, 9, 12, 14	1, 4, 7, 8, 15	5, 6, 10, 11, 13

Check your understanding

1. What is a thunderstorm?
2. Identify three features associated with thunderstorms.
3. List the two changes to the natural environment that take place during thunderstorm activity.
4. a. List three ways in which the natural environment may be damaged during thunderstorm activity.
   b. Next to each type of damage indicate:
      - whether the damage is caused predominantly by wind or water
      - whether the damage tends to occur to the natural or built environment.
5. Study **FIGURE 9** and outline how hailstones are formed.
6. In which season of the year are thunderstorms most likely to occur?

Apply your understanding

7. Explain why thunderstorms can cause so much damage to human environments.
8. Study **FIGURE 9**, which shows a supercell storm. Write a paragraph explaining why hailstones can vary so much in size.
9. During which seasons of the year are thunderstorms more likely? Give reasons for your answer.
10. Study **FIGURE 7**. During which hours of the day do most severe thunderstorms occur? Why?
11. Explain why storms with stronger updrafts produce larger hailstones.
12. Consider the steps recommended for preparing for and managing the impact of a thunderstorm. Which of the steps would be the most important to follow if you were caught outside in a severe hailstorm?

Challenge your understanding

13. Why might people in earlier civilisations have assumed weather events were the work of the gods? Draw on your understanding of the features of severe storms and how they impact the environment to support your answer.
14. Suggest a further three ways a person can protect themselves from harm during a thunderstorm.
15. Predict whether climate change will increase or decrease the frequency and severity of storms on Australia's east coast. Use your knowledge of the factors affecting storm severity to support your answer.

To answer questions online and to receive **immediate feedback** and **sample responses** for every question, go to your learnON title at www.jacplus.com.au.

15.5 Floods

LEARNING INTENTION

By the end of this subtopic, you will be able to identify the causes of floods and understand where flooding is more likely to take place.

15.5.1 Causes of flooding

Even though Australia is the driest of all the world's inhabited continents, there are periods of very heavy rainfall and **flood**. Flood disasters in Australia damage property, kill livestock, and cause the loss of human life. Since 1788 more than 2000 people have died in floods, equalling the number of deaths from cyclones. In some cases, entire sections of a town have been washed away, as in 1852, when one-third of the town of Gundagai disappeared.

FIGURE 1 Flash flood in Toowoomba, Queensland, 2011

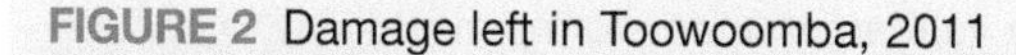

Floods are an unusual accumulation of water that overflows from rivers, lakes or the ocean onto land that is not normally covered by water. The flooding can be caused by a number of factors including storms, and low-pressure systems such as cyclones, typhoons and hurricanes. Heavy rainfall during storms can lead to a rapid rise in water level over a very short period of time. This is known as flash flooding. The water is unable to soak into the ground quickly, and so it runs down to the river and rushes towards the sea. Fast-moving floodwater (**FIGURE 1**) causes great risk to property and human life because there is less time to prepare.

FIGURE 2 Damage left in Toowoomba, 2011

flood an unusual accumulation of water onto land that is not normally covered by water

Riverine flooding occurs when rivers burst their banks and cover the surrounding land. It is a relatively slower process and may isolate communities for many weeks. There are other forms of flooding including those caused by storm surges and tsunamis that lead to sea water flooding.

15.5.2 Floods and floodplains

Floods are a natural occurrence, but they are a natural hazard to humans, who tend to build farms, towns and transport routes in areas such as floodplains. A **floodplain** is an area of relatively flat land that borders a river and is covered by water during a flood. Floodplains are formed when the water in a river slows down in flat areas. The river begins to meander and gradually deposits **alluvium**, which builds up the floodplain and other landforms such as deltas.

These fertile, flat areas are used for farming and settlement around the world. In Australia, many of our richest farmlands are on floodplains, and towns are often built on them because they are close to rivers. Such towns are subject to flooding. The possibility of flood is also increased when vegetation in **catchment areas** has been cleared or modified. Native vegetation can slow down run-off and reduce the chance of flooding.

riverine relating to rivers

floodplain relatively flat land by a river

alluvium the loose material brought down by a river and deposited on its bed, or on the floodplain or delta

catchment area the area of land that contributes water to a river and its tributaries

La Niña a pattern in the climate that includes warmer ocean waters off Australia's east coast

15.5.3 La Niña and floods

A **La Niña** event in Australia is often associated with floods. Very cold waters dominate the eastern Pacific near South America, and the oceans off the coast of Australia become warmer than normal. As a result, large areas of low-pressure extend over much of Australia; warm, moist air moves in; and above-average rainfall occurs. There can also be torrential rain, widespread floods and tropical cyclones.

FIGURE 3 A La Niña event leads to more rain in Australia.

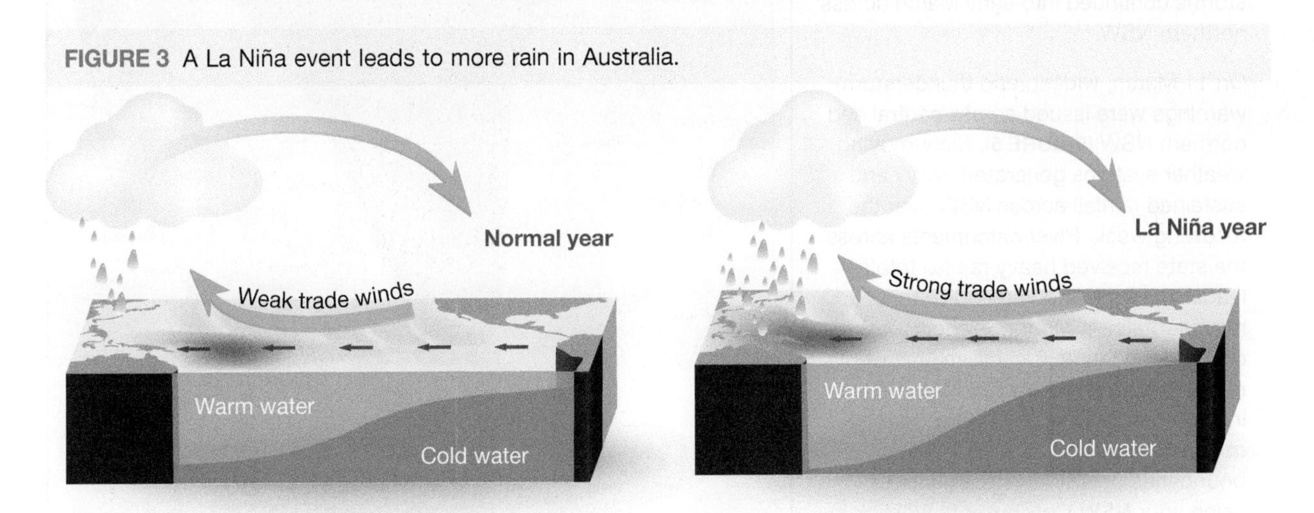

During 2010–2011, many parts of Queensland, New South Wales and Victoria were flooded (**FIGURE 4**) during a La Niña event. Further severe flooding took place during March 2021. Historical flood levels were recorded across south-eastern Australia during this flooding event, which was influenced by a La Niña weather phase.

int-5291

FIGURE 4 Rockhampton, Queensland (a) before and (b) during the 2011 flood

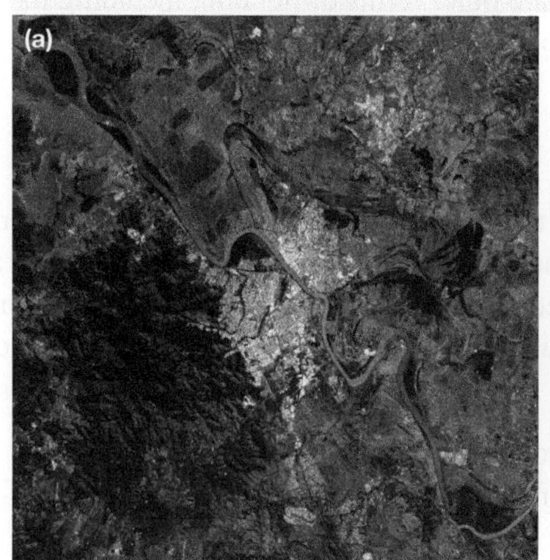

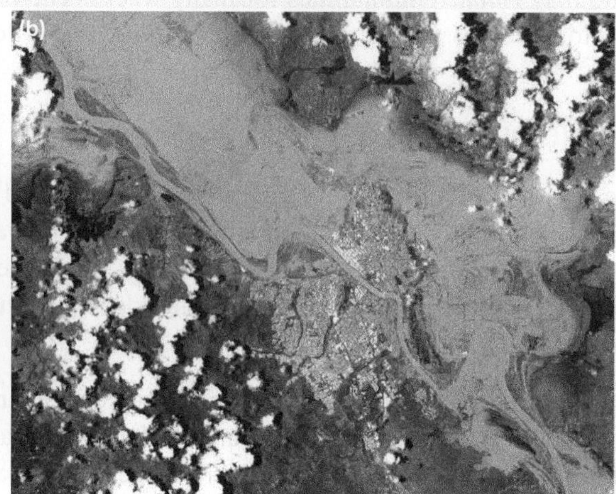

CASE STUDY

NSW floods March 2021

Many parts of Australia experienced intense weather patterns during February and March 2021, which resulted in historical flooding and rainfall records. In late February, severe storms in south-east Queensland generated large hailstones, flash flooding and widespread power outages. A series of storms continued into early March across northern NSW.

On 11 March, widespread thunderstorm warnings were issued across central and northern NSW (**FIGURE 5**). Slow-moving weather systems generated heavy and sustained rainfall across NSW over the following week. River catchments across the state received heavy rainfall totals leading to flash flooding and historic rainfall totals. At Mount Seaview west of Port Macquarie, 815 millimetres of rain fell between 19 and 23 March, and inland at Moree the Mehi River peaked at 10.38 metres on 25 March. (View the boundaries of NSW river catchments using your **NSW Catchments** weblink in your online Resources.)

As heavy rainfall continued to fall within the Hawkesbury-Nepean River catchment, western and north-western suburbs of Sydney faced widespread flooding and evacuations. The Warragamba Dam, upstream of the suburbs near Penrith, Richmond and Windsor, reached full capacity on 21 March. As further rain fell within the catchment, the water of Warragamba Dam started to spill over the dam wall, contributing to higher river levels downstream. Major road networks were cut off by the flood waters and property was destroyed. The 'bathtub effect' of flooding within the Hawkesbury-Nepean catchment is demonstrated in **FIGURE 10**.

FIGURE 5 An alert issued on 11 March 2021 for widespread thunderstorm activity.

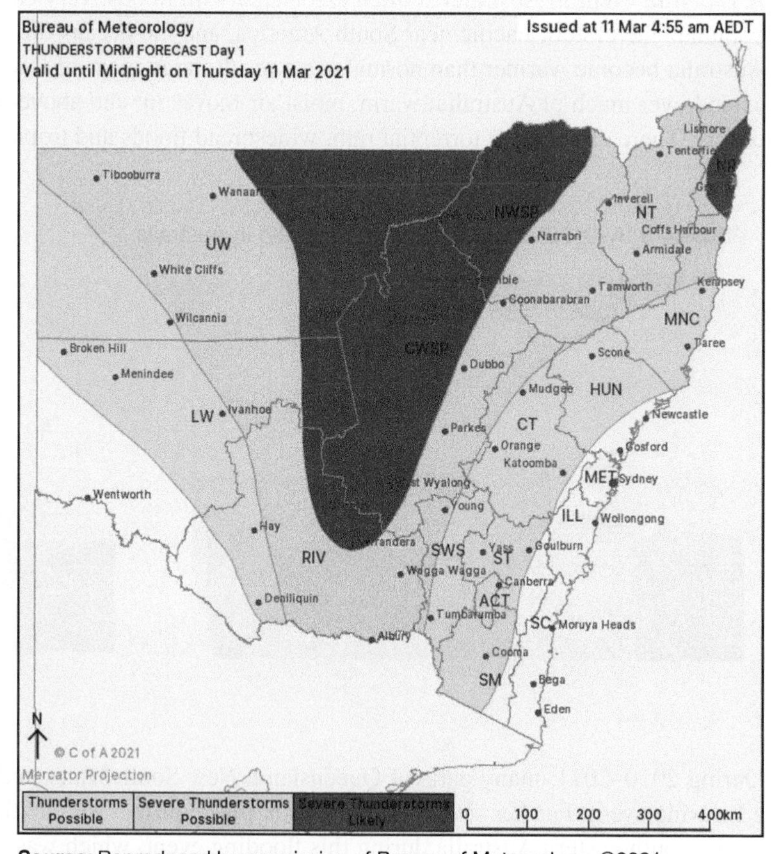

Source: Reproduced by permission of Bureau of Meteorology, ©2021 Commonwealth of Australia.

FIGURE 6 Flood waters cut off access for motorists and residents living in Windsor.

FIGURE 7 Properties along the Hawkesbury-Nepean River were damaged by rising floodwaters.

FIGURE 8 The flood indicator at Windsor indicating 13 metre flood levels

FIGURE 9 Windsor, a semi rural suburb of Sydney, (a) before and (b) during the 2021 flood

Source: https://earthobservatory.nasa.gov/images/148093/historic-floods-in-new-south-wales

FIGURE 10 The 'bathtub' effect of flooding within the Hawkesbury-Nepean River catchment

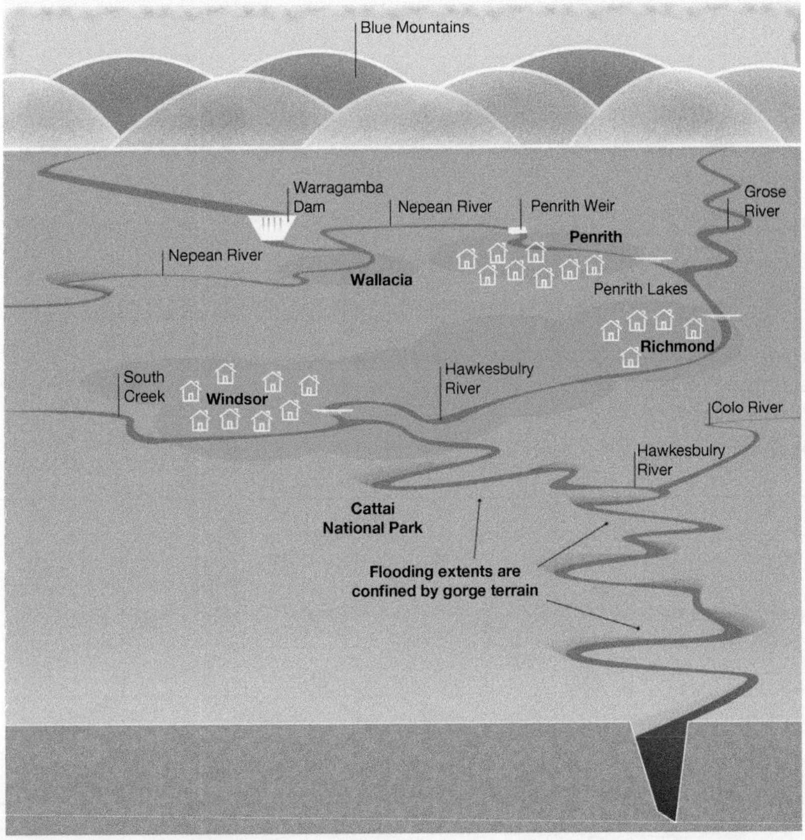

Source: © NSW Government

eWorkbook	Floods (ewbk-9494)
Video eLesson	Floods — Key concepts (eles-5126)
Interactivities	Floods (int-8551) 2011 floods in Rockhampton (int-5291)

15.5 ACTIVITIES

1. Make a sketch of a local river catchment near you. Using words such as *floodplain*, *tributary* and *watershed*, annotate your diagram to explain how rivers make their way to the sea.
2. Research and write a news-style report on a recent flood in Australia. (You could use the case study of the NSW 2021 floods as a starting point for the kinds of information you might collect.) Using the **El Niño–Southern Oscillation (ENSO)**, **El Niño — What is it?** and **The impacts of La Niña in Australia** weblinks in your Resources panel, include information about the impact of El Niño and La Niña on the frequency and severity of Australian flood events.
3. Create an overlay map to show the distribution of water in the aerial photos of Rockhampton or Windsor before and after the 2011 or 2021 floods. Estimate the percentage of the area covered in both maps. Describe the changes to the environment.
4. Collect a media source that describes the impact of the high rainfall received in March 2021 on one location within NSW that experienced flooding.
   a. Identify the catchment that the area is located within.
   b. Outline the impact of the high rainfall on this location.
   c. Suggest two reasons why this location flooded.

15.5 EXERCISE

Learning pathways

■ LEVEL 1	■ LEVEL 2	■ LEVEL 3
1, 2, 8, 12, 15	4, 6, 7, 9, 13	3, 5, 10, 11, 14

Check your understanding

1. What is a flood?
2. What is a floodplain?
3. Why is alluvium important to agriculture?
4. Construct a simple flow diagram to show the process of a La Niña event.
5. What does the term *La Niña* mean?
6. Describe how floodplains form.
7. Label the missing elements on the following diagram.

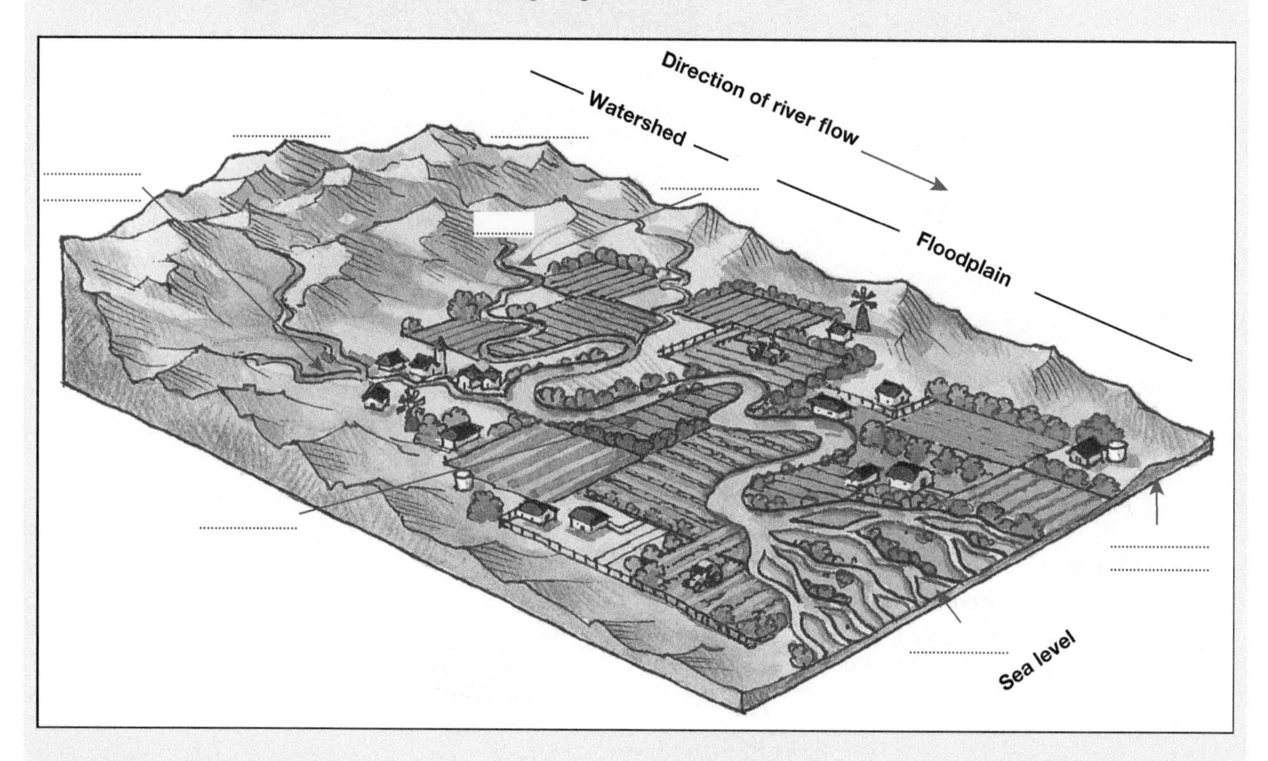

Apply your understanding

8. Create an infographic warning potential homeowners about the consequences of buying a house in floodplain regions.
9. Why do floods occur on floodplains and in deltas?
10. How might the effects of floods in urban spaces differ from effects in rural spaces?
11. Explain how native vegetation might help to reduce flooding.
12. Explain one benefit and one possible danger of farming on a floodplain.

Challenge your understanding

13. Should people continue to build on floodplains? Why or why not?
14. Climate change is warming overall ocean temperatures. Predict how this will affect how often Australia experiences La Niña conditions.
15. Suggest one strategy that farmers might use to protect their riverside properties from flooding.

To answer questions online and to receive **immediate feedback** and **sample responses** for every question, go to your learnON title at www.jacplus.com.au.

15.6 Impacts of a flooding disaster

LEARNING INTENTION

By the end of this subtopic, you will be able to identify the impacts of a flooding disaster on communities and the environment.

15.6.1 Impacts of flooding

The impact of flooding will depend on a number of factors including steps taken to prepare for, prevent and respond quickly to the event. The disruption flooding causes to an environment and space can be extensive and recovery can take many years.

When the Brisbane River broke its banks on 11 January 2011, many residents were shocked by the devastation left in its wake. Across the world in Brazil, devastating floodwaters were also having a large impact on the people and the land.

CASE STUDY 1

Queensland, Australia, 2010–2011

Country background

Australia is considered a developed nation with a strong economy. Australians earn on average $38 200 per person. Approximately 25.5 million people reside in Australia, with 4.6 million of those living in Queensland. About 84 per cent of all Australians are located within 50 kilometres of the coast.

Why did the flooding occur?

The flooding that affected this region was due to a strong La Niña event. Long periods of heavy rain over Queensland catchments caused rivers to burst their banks.

The sources of the flooding:

- Floodwaters from Lockyer Creek, which flows into Brisbane River — the Lockyer Valley was hit by more than 200 mm of rain.
- More than 490 000 million litres were released from Wivenhoe Dam into Brisbane River.
- Floodwaters from the Bremer River, which is also fed by the Lockyer Valley. After passing Ipswich, where it burst its banks, the Bremer flows into the Brisbane River.

Town heights above sea level:

- Toowoomba 700 metres
- Murphys Creek 704 metres
- Withcott 262 metres
- Helidon 143 metres
- Grantham 110 metres
- Gatton 111 metres
- Forest Hill 95 metres
- Laidley 135 metres
- Ipswich 54.8 metres
- Brisbane 28.4 metres.

Impacts of the flooding

- Three-quarters of the state was declared a disaster zone.
- At least 70 towns and more than 200 000 people were affected.
- There were 35 deaths.
- The cost to the Australian economy was estimated to be at least $10 billion.
- Up to 300 roads were closed, including nine major highways.
- More than 20 000 homes were flooded in Brisbane alone.
- There was massive damage and loss of property.

FIGURE 1 Rainfall in the days before the flood

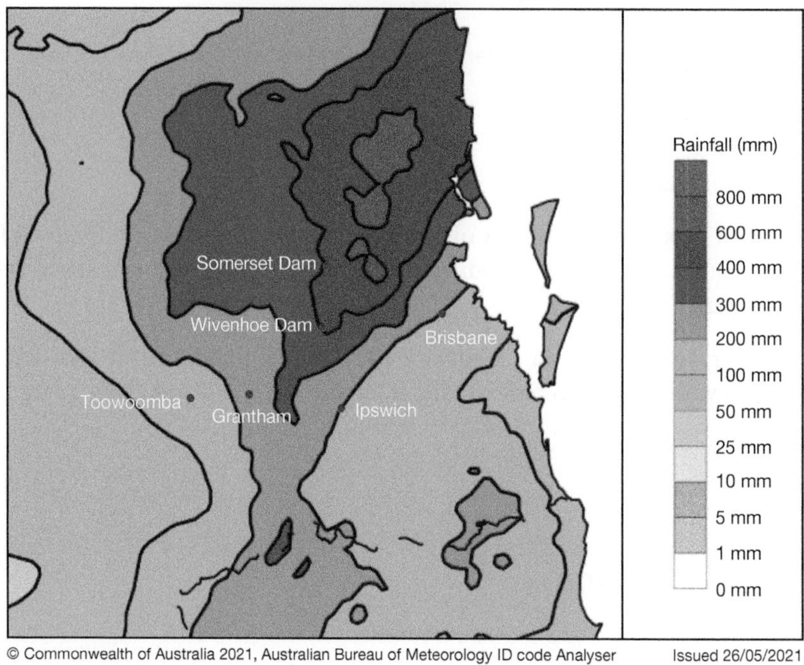

Assistance and recovery strategies

- $1.725 billion is being raised by the Federal Government via a flood levy in the tax system.
- $281.5 million is in the Disaster Relief Appeal set up by the then Premier Anna Bligh.
- Over $20 million was donated to aid agencies such as the St Vincent de Paul Society to help those suffering.
- About $1.2 million was raised through charity sporting events such as Rally for Relief, Legends of Origin and Twenty 20 cricket.
- The Australian Defence Force was mobilised to help with the clean-up.
- The Mud Army was formed: 55 000 volunteers registered to help clean up the streets, and thousands more unregistered people joined them.
- Improvements will be made to dam manuals to help manage the release of water from dams during floods.

FIGURE 2 The Wivenhoe Dam is the primary drinking source for south-east Queensland.

FIGURE 3 Brisbane floods were declared a natural disaster on 12 January 2011.

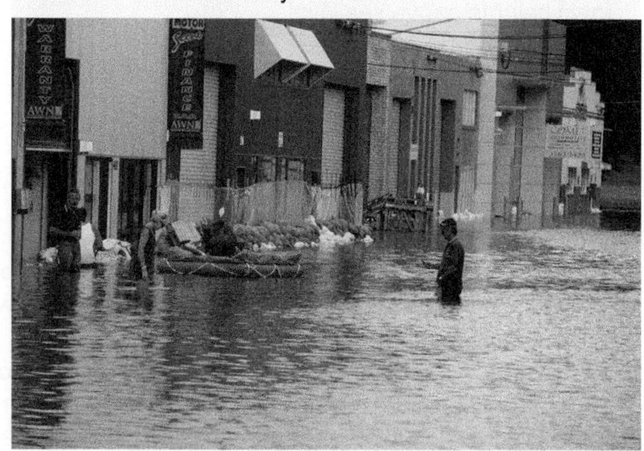

CASE STUDY 2

State of Rio de Janeiro, Brazil, 2011

Country background

Brazil is considered a developing nation. Brazilians earn on average $10 200 per person per year. Approximately 198 million people reside in Brazil, with 650 000 living in the three towns worst affected by the flooding.

Why did the flooding occur?

Due to the equivalent of a month's rain falling in 24 hours, flash flooding occurred in a mountainous region in Rio de Janeiro State and São Paulo State. Hillsides and riverbanks collapsed due to landslides. It is believed that illegal construction and deforestation may have contributed to the instability of the land.

Impacts of the flooding

- Approximately 900 people died — most of them in poverty-stricken areas with poor housing conditions and no building policies.
- Forty per cent of the vegetable supply for the city of Rio de Janeiro was destroyed.
- Around 17 000 people were left homeless.
- There was widespread property damage, most of it to homes built riskily at the base of steep hills.

Assistance and recovery strategies

- $460 million was set aside by the president for emergency aid and reconstruction.
- Troops were deployed to help.
- There were donations of clothes and food to the area from other Brazilians.
- About $450 million was lent by the World Bank.
- Support was given by domestic and international charities.

FIGURE 4 Areas affected by the floods in Brazil, 13 January 2011

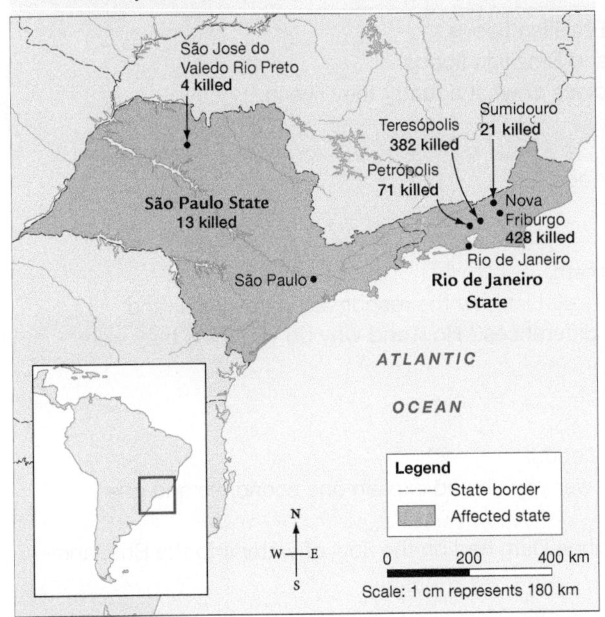

Source: Spatial Vision

FIGURE 5 Hills collapsed after the heavy rains, destroying homes.

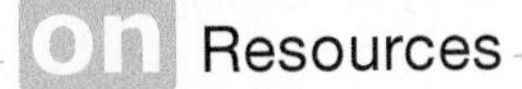

Resources

eWorkbook	Impacts of a flooding disaster (ewbk-9498)
Video eLesson	Impacts of a flooding disaster — Key concepts (eles-5127)
Interactivity	Impacts of a flooding disaster (int-8552)

15.6 ACTIVITIES

1. Imagine you had to evacuate your home because it was under threat from a flood.
   a. What five things would you take with you and why?
   b. Compare the list with a partner, then negotiate the five things you would both agree to include on your emergency list.
2. In a table, classify (in point form) the impacts on people, the economy and the environment of the Brisbane and Brazilian floods.

15.6 EXERCISE

Learning pathways

■ LEVEL 1	■ LEVEL 2	■ LEVEL 3
1, 4, 8, 9, 13	2, 5, 6, 10, 15	3, 7, 11, 12, 14

Check your understanding

1. When did the floods in Queensland and Brazil occur?
2. Identify one similarity between the areas affected by the Brazilian and Queensland flooding.
3. What climate pattern was affecting Queensland at the time of the floods?
4. Complete the table to compare Brazil and Australia.

	Brazil	Australia
Population		
Average earnings		

5. Identify two economic impacts of the Queensland and Brazilian floods.
6. Identify two environmental impacts of the Queensland and Brazilian floods.
7. Outline two ways that flooding can affect people, regardless of what country they live in.

Apply your understanding

8. Compare the causes of the Brazilian and Queensland floods.
   a. Explain the causes of the Brazilian floods.
   b. Explain the three causes of the Queensland floods.
   c. Summarise the similarities and differences between them.
9. Compare the responses to the floods in Brisbane and Brazil. List how the responses were similar and how they were different. What factors might explain the differences? How and why do you think they were different?
10. Compare the scale of the floods in Queensland and Brazil.
    a. Explain the scale of each flood.
    b. Give reasons for the differences in the scale of these floods.
11. Explain why the impacts of floods can differ. (In your answer you should explain one economic and one environmental reason.)
12. Explain what impact the release of water from the Wivenhoe Dam had on the flow of water into the Brisbane River.

Challenge your understanding

13. Suggest three strategies that could be implemented to lessen the impact of floods if they occurred in these regions again.
14. Predict how warmer ocean temperatures would affect the spatial distribution and frequency of flooding on the east coast of Australia.
15. Suggest why more people might have died in the Brazilian floods than in Queensland's flood event. (Consider the location, economy and population.)

To answer questions online and to receive **immediate feedback** and **sample responses** for every question, go to your learnON title at www.jacplus.com.au.

15.7 Flood management around the world

LEARNING INTENTION

By the end of this subtopic, you will be able to list different strategies used to manage floods around the world.

15.7.1 Managing floods

Floods occur in many countries around the world. It is important for these countries to learn how to live with this natural hazard. Managing the effects of floods is important if the amount of damage caused is to be minimised. Unfortunately, not all countries have the same resources to tackle this problem. Those countries that are able to invest in flood-prevention **infrastructure** have a greater chance of reducing the risk of flood.

FIGURE 1 Lake Wivenhoe, Queensland, at 190 per cent capacity, January 2011

Weirs, floodgates and dams

The most common form of flood management is to build a barrier that prevents excess water from reaching areas that would suffer major damage. Levees (see FIGURE 2), **weirs** and dams are a few examples of structures that are built to contain floodwaters. Dams that are used to stop flooding need to be kept below a certain level to allow space for floodwater to fill. Wivenhoe Dam in south-east Queensland was built in response to the floods in 1974. However, there was some debate about whether this dam could have been used more effectively during the 2010–2011 floods.

To prevent London being flooded during unusually high tides and storm surges, the city constructed the Thames Barrier, a system of **floodgates** that stretch across the width of the Thames (see FIGURE 3). The barrier is triggered if predicted water levels are above a certain height. If this happens, the gates rise to stop the incoming water. Once the water recedes, the danger has passed and the gates are lowered.

infrastructure the facilities, services and installations needed for a society to function, such as transportation and communications systems, water and power lines

weir a barrier across a river, similar to a dam, which causes water to pool behind it. Water is still able to flow over the top of the weir.

floodgate a barrier across a river that is raised when water levels are predicted to be dangerously high

FIGURE 2 An artificial levee

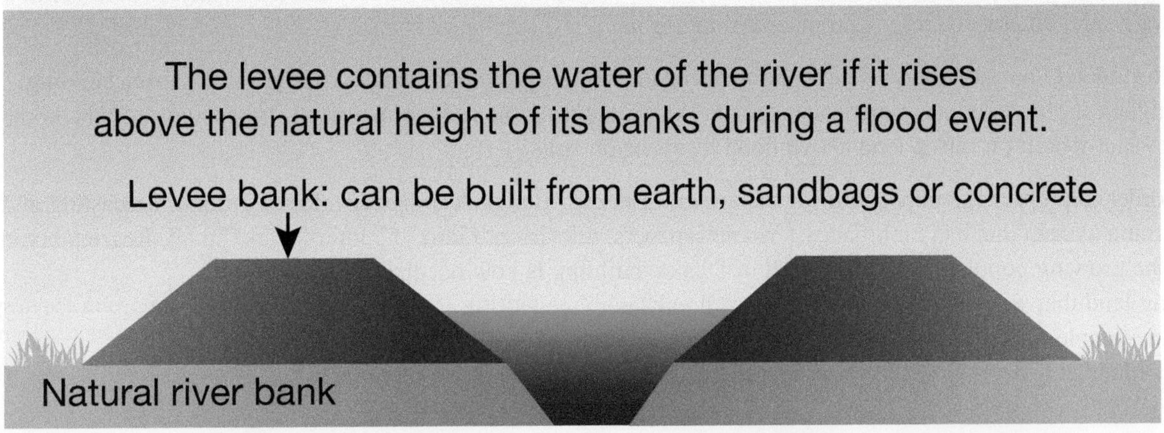

FIGURE 3 The Thames flood barrier with its gates up, near Woolwich, England

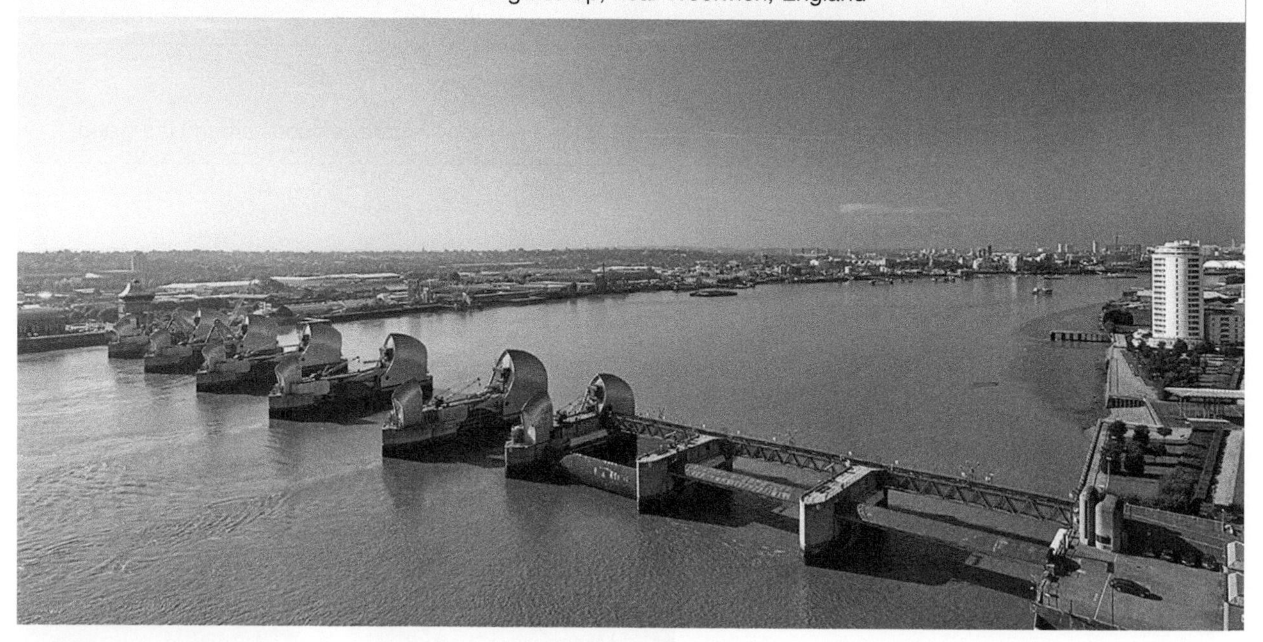

Building management

Another way to manage the risk of damage from floods is to stop building on low-lying land that is guaranteed to flood. Unfortunately, in many urban areas this land has been developed, which increases the chance of property damage in a flood.

FIGURE 4 In Dhaka, Bangladesh, homes are built on stilts to avoid the floodwaters.

Since 2006, the Brisbane City Council has offered a residential property buy-back scheme to homeowners living in flood-prone areas. This scheme gives people the opportunity to sell their property to the council if they live in a low-lying area that has a 50 per cent chance of flooding every year. People will not be allowed to build on this land again. For this initiative to be successful, it is essential that the price offered by the council is similar to what the owners would get in a private sale; otherwise there is no incentive to use it.

Unfortunately not all countries have the finances to fund property buy-backs or large-scale barrier building. Bangladesh, for example, experiences annual flooding during the **monsoon** season. In response to this, homes are usually built on raised land above flood levels or on stilts.

In order to prepare the population for the arrival of floods, Bangladesh has developed a flood forecasting and warning system that can be broadcast via newspapers, television, radio, the internet and email. Regrettably, due to the growing population in the capital of Dhaka, building is now occurring on low-lying land that was previously used to store floodwater (see **FIGURE 4**). As a result, many people are still being affected by flooding. In 1998, 65 per cent of Bangladesh was inundated. Twenty million people needed shelter and food aid for two months.

monsoon the rainy season in the Indian subcontinent and South-East Asia

Community responses

There is an increased awareness of the valuable role mangrove species play in reducing the level of flooding experienced along flood-prone areas of land. Human activities such as land clearing for agriculture, pollution and urbanisation have led to the widespread destruction of mangrove forests. These wetland environments play an important role in our ecosystems by water-quality filtering, carbon storage and promoting biodiversity.

FIGURE 5 Mangrove replanting can reduce the effects of flooding.

Conservation and replanting of mangrove areas helps communities take action to reduce coastal flooding from storm-generated waves and intercepts sediment-filled runoff from the land. Coastlines and riverbanks can be rebuilt, and local communities can take ownership of an ongoing threat to their local environment and sources of employment such as the fishing industry.

FIGURE 6 The role of mangroves in reducing coastal flooding

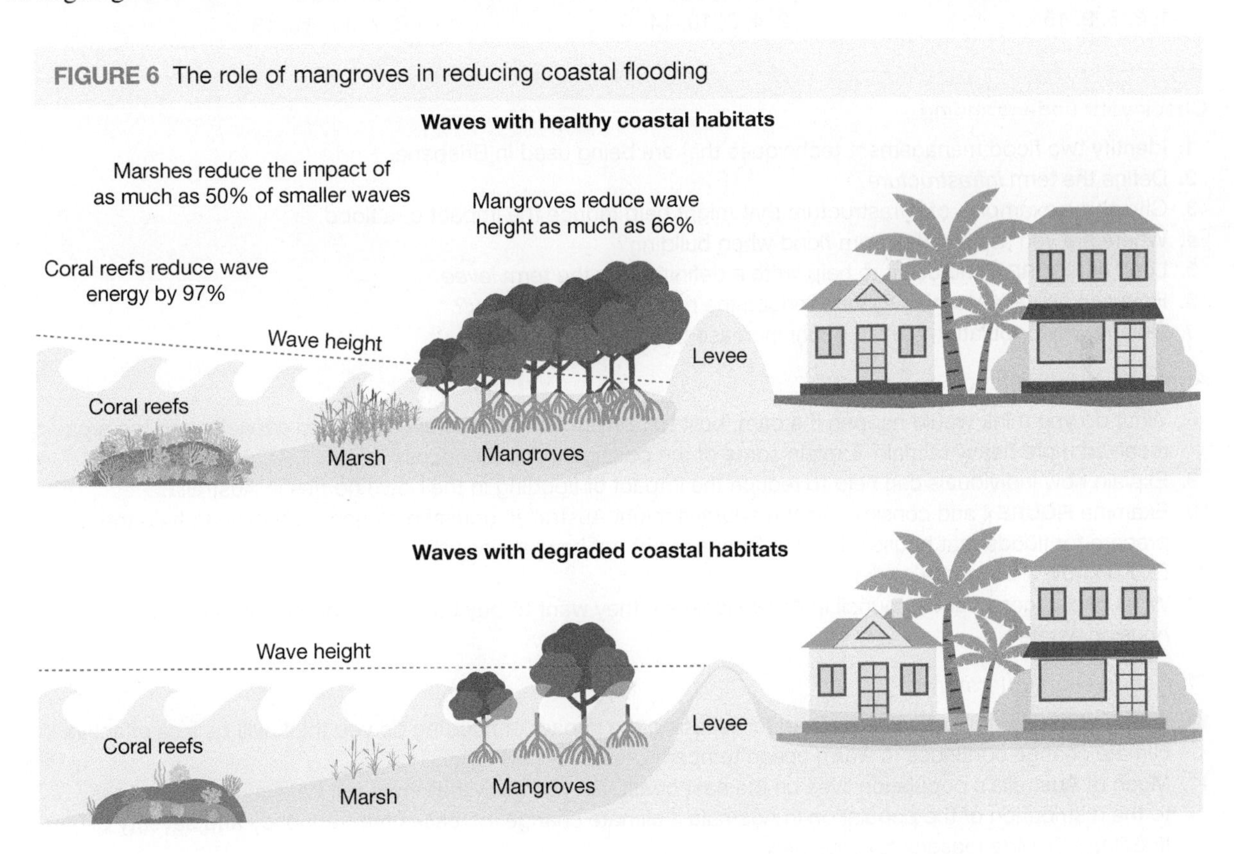

on Resources

eWorkbook Flood management around the world (ewbk-9502)

Video eLesson Flood management around the world — Key concepts (eles-5128)

Interactivity Responding to floods (int-3085)

15.7 ACTIVITIES

1. Use the **Bureau of Meteorology** weblink in your Resources tab to find out more about flood warnings. Prepare an information sheet that could be released to a rural community about to be affected by a major flood event. It should include tips on what to do before, during and after the event.
2. Research a community that has a mangrove rehabilitation program. Assess the effectiveness of this effort to reduce flooding in the local area.
3. Around the world, floods are becoming more frequent and their impact on people and the environment more damaging and costly. Write an argument that supports and an argument that challenges the following viewpoint: People should not be allowed to live in areas that are prone to flooding.

15.7 EXERCISE

Learning pathways

■ LEVEL 1	■ LEVEL 2	■ LEVEL 3
1, 3, 5, 9, 15	2, 4, 8, 10, 14	6, 7, 11, 12, 13

Check your understanding

1. Identify two flood management techniques that are being used in Brisbane.
2. Define the term *infrastructure*.
3. Give three examples of infrastructure that might help reduce the impact of a flood.
4. Where are you at most risk from flood when building?
5. Look at **FIGURE 3** and use it to help write a definition for the term *levee*.
6. How can an early warning system reduce the risk of a flood disaster?
7. Outline why population growth might increase the risks from floods.

Apply your understanding

8. What do you think would happen if a dam, built to prevent floods, was already full to capacity and the area received more heavy rainfall? Explain some of the possible consequences.
9. Explain how individuals can help to reduce the impact of flooding in their own homes in Australia.
10. Examine **FIGURE 4** and consider what resources might Australian households generally have to help them prepare for floods that householders in Dhaka might not have access to.
11. Explain how a floodgate works.
12. What challenges might a council in Australia face if they want to buy back flood-prone land in a housing estate?

Challenge your understanding

13. Which of the common strategies used for reducing the impact of flooding do you think will be less effective if climate change continues to warm ocean temperatures?
14. Much of Australia's population lives on the east coast. Do you think that there will be significant changes to the distribution of the population in Australia if climate change increases the frequency and severity of flooding? Provide reasons for your view.
15. What responsibilities do you think the government should have in reducing the possible impact of floods on peoples' homes?

To answer questions online and to receive **immediate feedback** and **sample responses** for every question, go to your learnON title at www.jacplus.com.au.

15.8 SkillBuilder — Cardinal points — Wind roses

online only

LEARNING INTENTION

By the end of this subtopic, you will be able to read a wind rose to determine direction and strength of wind patterns.

The content in this subtopic is a summary of what you will find in the online resource.

15.8.1 Tell me

What is a wind rose?

A wind rose is a diagram that shows the direction, speed and frequency of wind. We name wind direction according to the direction from which the wind is blowing. A wind rose such as that shown in **FIGURE 2** uses data collected over long periods of time to visually represent wind information. The spokes represent wind direction; the longer the spoke the more frequently the wind blows from a particular direction. The thickness of the bands represents the speed of the wind.

FIGURE 2 A wind rose

15.8.2 Show me

How to read a wind rose

eles-1638

Step 1

Determine the direction of wind with the greatest frequency by finding the longest ray.

Step 2

Determine the direction of wind with the highest speed by finding the widest ray.

Step 3

Work out the general pattern and main features of wind direction and strength.

15.8.3 Let me do it

int-3134

Go to learnON to access the following additional resources to help you build this skill:

- a longer explanation of this skill and its application in Geography (Tell me)
- a video showing the step-by-step process of this skill (Show me)
- an activity and interactivity for you to practise this skill (Let me do it)
- self-marking questions to help you understand and use this skill.

on Resources

eWorkbook	SkillBuilder — Cardinal points — Wind roses (ewbk-9506)
Video eLesson	SkillBuilder — How to interpret wind roses (eles-1638)
Interactivity	SkillBuilder — How to interpret wind roses (int-3134)

15.9 Cyclones, typhoons and hurricanes

LEARNING INTENTION

By the end of this subtopic, you will be able to identify the causes of cyclones and understand where cyclones are more likely to take place.

15.9.1 The formation of cyclones

Tropical **cyclones** (called hurricanes in the Americas and typhoons in Asia) can cause great damage to property and significant loss of life. Some 80 to 100 tropical cyclones occur around the world every year in tropical coastal areas located north and south of the equator. Australia experiences, on average, about 13 cyclones per year.

Cyclones form when a cold air mass meets a warm, moist air mass lying over a tropical ocean with a surface temperature greater than 27° C. Cold air currents race in to replace rapidly rising, warm, moist air currents, creating an intense low-pressure system. Winds with speeds over 119 kilometres per hour can be generated. Cyclones are classified using the scale in **TABLE 1**.

int-7791

FIGURE 1 World distribution of tropical cyclones by names used in different regions

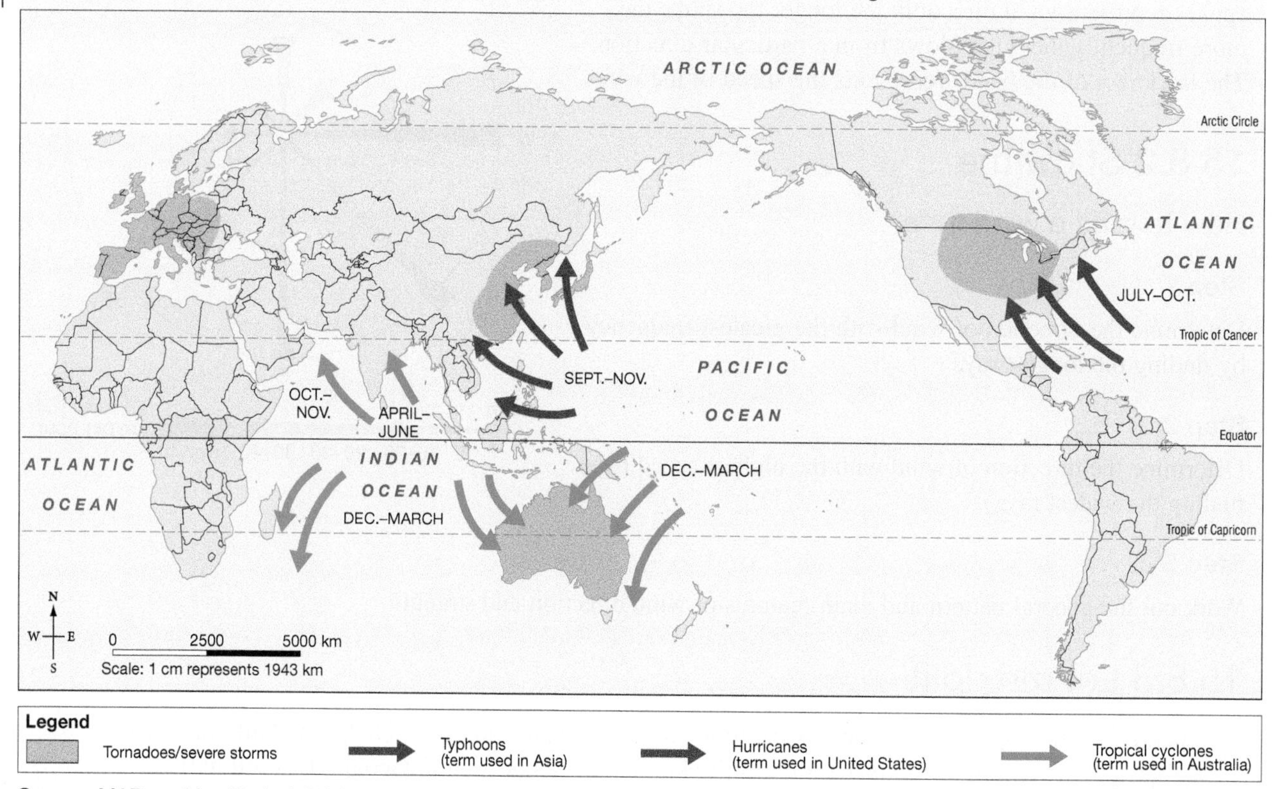

Source: MAPgraphics Pty Ltd, Brisbane.

FIGURE 2 shows the continuous cycle of evaporation, condensation and precipitation associated with cyclones. At first the winds spin around an area about 200 to 300 kilometres wide. As the winds gather energy by sucking in more warm moist air, they get faster. In severe cyclones, winds may reach speeds of 295 kilometres per hour. The faster the winds blow, the smaller the area around which they spin; this is called the eye. It might end up being only about 30 kilometres wide. Around the edge of the eye, winds and rain are at their fiercest. However, in the eye itself, the air is relatively still, and the sky above it may be cloudless.

cyclone intense low-pressure system producing sustained wind speeds in excess of 65 km/h. It develops over tropical waters where surface water temperature is at least 27 °C.

TABLE 1 Classification of cyclones using the Saffir–Simpson scale

Category	Wind gust speed	Ocean swell	Damage
1	Less than 125 km/h	1.2–1.6 m	Mild damage
2	126–169 km/h	1.7–2.5 m	Significant damage to trees
3	170–224 km/h	2.6–3.7 m	Structural damage, power failures likely
4	225–279 km/h	3.8–5.4 m	Most roofing lost
5	More than 280 km/h	More than 5.4 m	Almost total destruction

int-7792

FIGURE 2 How a cyclone forms. The winds within a cyclone spin because of the rotation of the Earth. In the southern hemisphere, they rotate in a clockwise direction. In the northern hemisphere, they rotate in an anticlockwise direction.

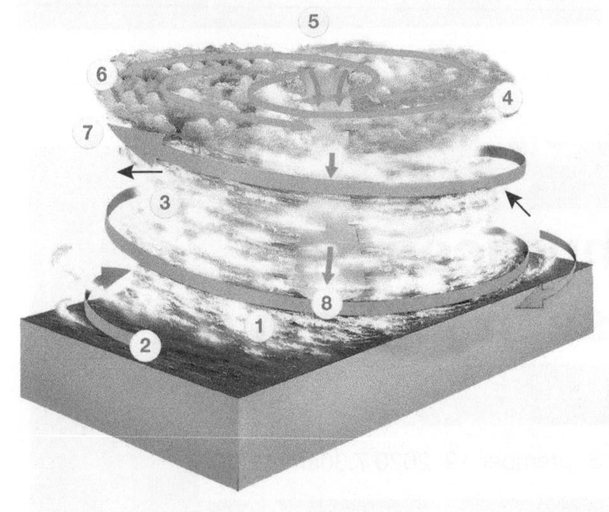

1. Warm sea water evaporates and rises.
2. Low-pressure centre creates converging winds, which replace rising air.
3. Warm air spirals up quickly.
4. Warm moist air is drawn in, providing additional energy.
5. Water vapour fuels cumulus clouds.
6. In the upper atmosphere, the air moves away from the eye.
7. Storm moves in direction of prevailing wind.
8. Descending air in the eye of cyclone

15.9.2 The impact of tropical cyclones

Tropical cyclones can cause extensive damage if they cross land. **Gale force winds** can tear roofs off buildings and uproot trees. **Torrential rain** can often cause flooding, as can **storm surges.**

When a tropical cyclone approaches or crosses a coastline, the very low atmospheric pressure and impact of strong winds on the sea surface combine to produce a rise in sea level, as shown in FIGURE 3.

FIGURE 3 Flooding caused by storm surges

gale force wind wind with speeds of over 62 km/h

torrential rain heavy rain often associated with storms, which can result in flash flooding

storm surge a sudden increase in sea level as a result of storm activity and strong winds. Low-lying land may be flooded.

FIGURE 4 Satellite image of Typhoon Maysak, 31 March 2015

FIGURE 5 Damage to homes caused by Cyclone Fani, Puri, Odisha, India, 2019

CASE STUDY

The 2020 Atlantic hurricane season is so intense, it just ran out of storm names

September 19, 2020 3.29am AEST Updated September 19, 2020 7.30am AEST

This year, sea surface temperatures have been above average across much of the Atlantic Ocean and wind shear has been below average. That means it's been more conducive than usual to the formation of tropical cyclones.

... La Niña is El Niño's opposite – it happens when sea surface temperatures in the eastern and central Pacific are below average. That cooling affects weather patterns across the U.S. and elsewhere, including weakening wind shear in the Atlantic basin. NOAA determined in early September that we had entered a La Niña climate pattern. That pattern has been building up for weeks, so these trending conditions could have contributed to how favourable the Atlantic has been to tropical cyclones this year.

Source: The Conversation, 19 September 2020, article by Kimberly Wood

fdw-0022

FOCUS ON FIELDWORK

Tracking a cyclone

There are many names for tropical revolving storms: cyclones, typhoons and hurricanes. Tropical revolving storms that reach Australia are called cyclones. One important tool used by meteorologists to help inform people about the path a cyclone will take is a cyclone tracking map. This shows the predicted path of the cyclone and its intensity.

Learn how to plot the path of a cyclone with the **Tracking a cyclone** fieldwork activity in your Resources tab.

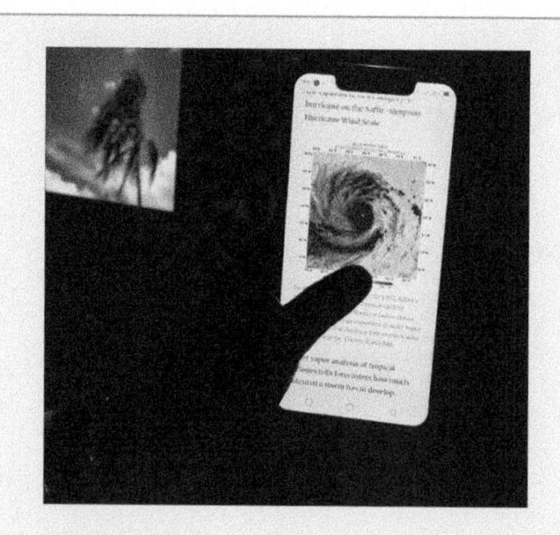

on Resources

eWorkbook	Cyclones, typhoons and hurricanes (ewbk-9510)
Video eLesson	Cyclones, typhoons and hurricanes — Key concepts (eles-5129)
Interactivities	Spiralling sea storm (int-3087) World distribution of tropical cyclones by names used in different regions (int-7791) How a cyclone forms (int-7792)
Fieldwork	Tracking a cyclone (fdw-0022)

15.9 ACTIVITY

Refer to **FIGURE 1**, showing the world pattern of tropical cyclones.

a. When do most cyclones occur north of the equator? When do most cyclones occur south of the equator? Suggest a reason for this difference.
b. Name the parts of Australia most at risk from cyclone activity.

15.9 EXERCISE

Learning pathways

■ LEVEL 1	■ LEVEL 2	■ LEVEL 3
2, 4, 10, 11, 13	1, 3, 8, 9, 14	5, 6, 7, 12, 15

Check your understanding

1. What conditions do tropical cyclones need in order to develop?
2. What are tropical cyclones called in other places around the world?
3. Why do tropical cyclones die out if they move inland?
4. Outline the changes that a storm surge can cause to a coastal area.
5. How might the scale of a cyclone vary?
6. Describe how the damage would differ between a category 1 and category 5 cyclone.

Apply your understanding

7. What is the interconnection between the warmth of sea water and cyclones?
8. If the water source for cyclones is the ocean over which they form, explain why strong winds and flooding occur in places located inland from the coast.
9. Technology allows us to view cyclones on a global scale. Refer to **FIGURE 4** and **FIGURE 5**. What are the benefits of using these images to understand the power of cyclones?
10. Explain how a storm surge occurs.
11. If you live in a coastal town, are you necessarily safe after a cyclone has passed?
12. Why was 2020 such a bad year for hurricanes?

Challenge your understanding

13. Why do you think many people are killed after the eye of the storm has passed over an area?
14. Cyclones are associated with destructive winds and the displacement of large volumes of water. What emergency strategies should be introduced in coastal areas to manage the risk of cyclones?
15. a. Predict how the frequency and severity of cyclones might change because of climate change.
    b. Suggest ways that individuals, communities and governments might reduce the impact of these changes.

To answer questions online and to receive **immediate feedback** and **sample responses** for every question, go to your learnON title at www.jacplus.com.au.

15.10 Case studies — Impacts of Typhoon Haiyan and Cyclone Veronica

online only

LEARNING INTENTION

By the end of this subtopic, you will be able to describe the events leading up to one severe cyclone event in Asia and one severe cyclone event in Australia.

The content in this subtopic is a summary of what you will find in the online resource.

FIGURE 3 The aftermath of Typhoon Haiyan in the Philippines

Typhoon Haiyan (locally known as Yolanda) developed into a category 5 storm over the Pacific Ocean and directly affected countries such as the Philippines, Palau, Vietnam and China. Strong wind gusts of up to 314 kilometres per hour, heavy rain, and waves taller than palm trees were experienced in parts of the Philippines on 8 November 2013.

On 18 March 2019, a tropical low-pressure system formed about 550 kilometres north of Broome in the waters off the West Australian coast. The system drifted south-west before intensifying into a category 1 cyclone at 2 am on 20 March. Named Veronica by the Australian Bureau of Meteorology, this cyclone quickly intensified into a category 5 storm by 8 am the following day.

Contents

learnON

on Resources

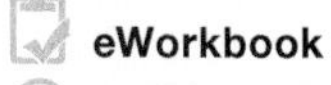

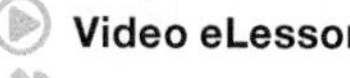

eWorkbook	Case studies — Impacts of Typhoon Haiyan and Cyclone Veronica (ewbk-9514)
Video eLesson	Case studies — Impacts of Typhoon Haiyan and Cyclone Veronica — Key concepts (eles-5130)
Interactivity	Case studies — Impacts of Typhoon Haiyan and Cyclone Veronica (int-8553)

15.11 Responding to extreme weather events

LEARNING INTENTION

By the end of this subtopic, you will be able to describe the efforts made by one country to help another during a time of crisis caused by a natural disaster.

15.11.1 Features of Cyclone Gita, Tonga

On 9 February 2018, a tropical low-pressure system intensified in the warm South Pacific waters near American Samoa and was named Cyclone Gita. The category 1 tropical cyclone strengthened as it moved westwards and on 14 February Cyclone Gita passed directly over the Kingdom of Tonga as a category 4 storm. Most of the impact was felt by the largest of Tonga's 169 islands, Tongatapu, which was home to 24000 residents living in the nation's capital, Nuku'alofa.

FIGURE 1 Cyclone Gita over the South Pacific in 2018

Cyclone Gita continued to intensify into a category 5 storm generating strong wind gusts of 205 kilometres per hour and rainfall totals up to 296 millimetres. For the following week it continued to move westwards and then southwards, influencing weather patterns around the nations of Fiji, New Caledonia and New Zealand. The most significant damage took place in Tonga where 4000 homes were damaged, and 800 homes were destroyed. Eighty per cent of homes in Tonga lost access to power, more than 4500 people were evacuated and the majority of schools were destroyed.

15.11.2 Humanitarian aid efforts

The Kingdom of Tonga declared a State of Emergency on 12 February 2018 to ensure emergency resources and trained responders could be accessed by the government. Emergency relief partners responded quickly to the need for short- and long-term humanitarian support and recovery.

Non-government organisations, such as Red Cross and Care Australia, and international organisations such as UNICEF and the World Bank partnered with the Tongan government and local Tongan relief agencies to deliver essential items and support to affected people.

FIGURE 2 The path taken by Cyclone Gita across the Kingdom of Tonga

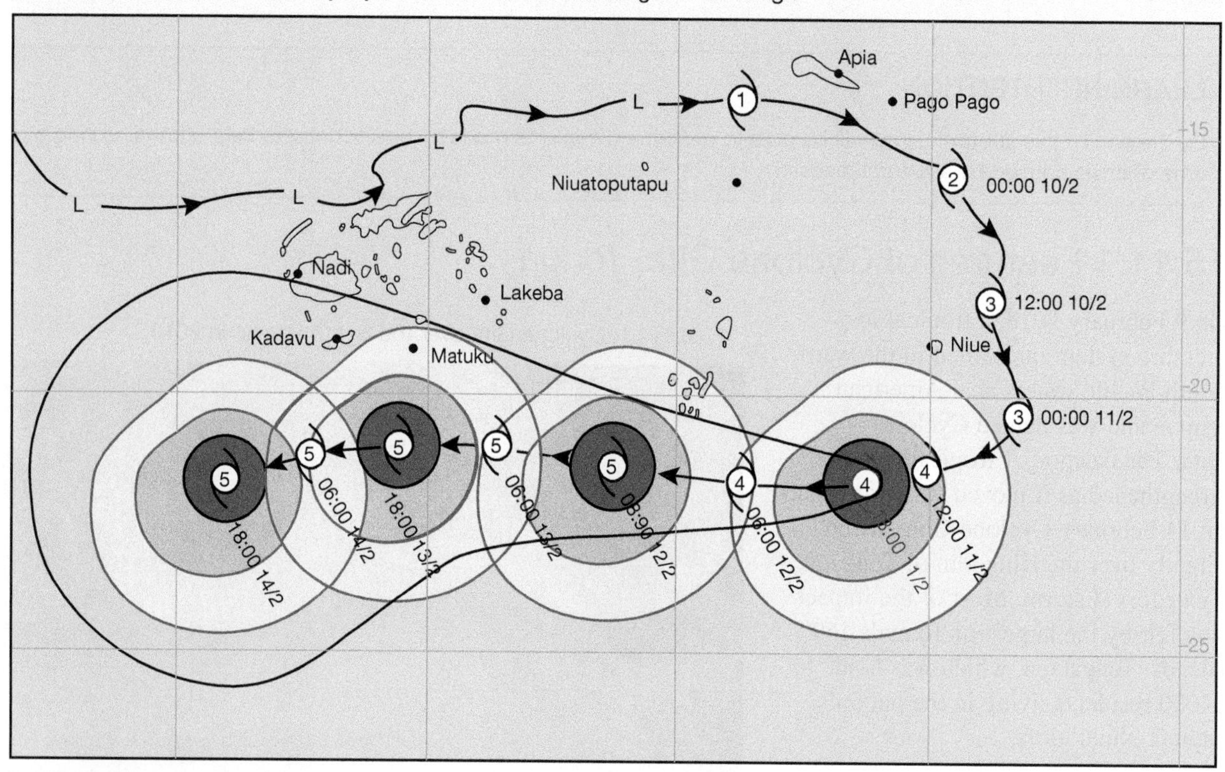

Source: © Fiji Meteorological Service

FIGURE 3 The response by Australia's government to Cyclone Gita

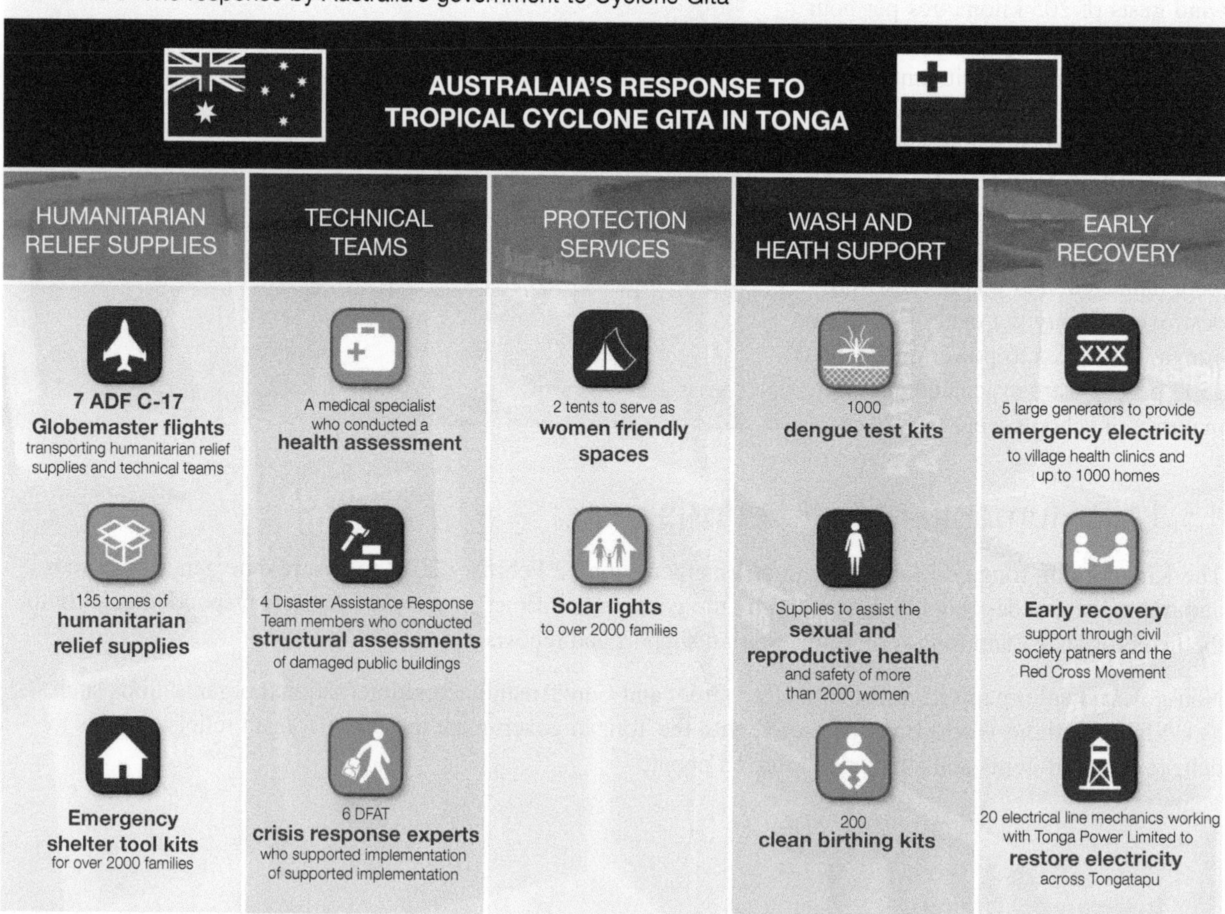

Source: © 2018 Commonwealth of Australia

The Australian and New Zealand governments, along with others around the world, joined to support many of the Pacific nations impacted by Cyclone Gita. Australia's Disaster Assisting Response Teams (DART) worked with New Zealand's Rescue Teams to provide general emergency management assistance and assess the structural integrity of buildings impacted by the storm.

FIGURE 4 Tonga required ongoing aid and support following Cyclone Gita.

Source: Dr Yutaro SETOYA/WHO Tonga Office

15.11.3 Individual responses to a severe storm

With today's modern technology we have access to a wealth of information that enables individuals and communities to prepare themselves for the wild winds over which they have no control. In many cases, the winds also bring vast amounts of rainfall and the land is often inundated. While we can prepare for such events in some ways, it is inevitable that both the natural and built environments will be impacted.

People who live in disaster-prone areas should know the risks associated with the potential hazards they face and the time of the year when they are at greatest risk. In Queensland, for example, where tropical cyclones bring flooding rains, houses are often built on stilts.

The key to survival is to be prepared. Securing your home and having an emergency kit are two important things that can be done on a regular basis.

FIGURE 5 An unprepared home **FIGURE 6** A well-prepared home

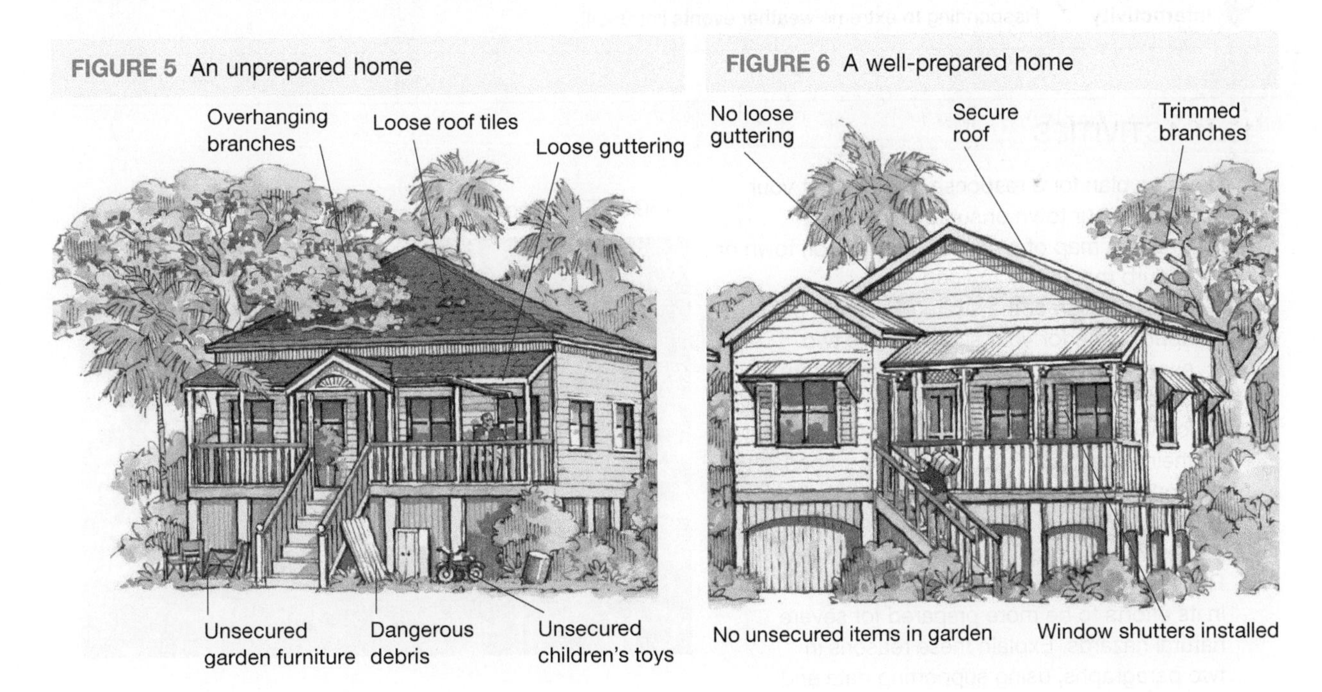

FIGURE 7 An emergency kit

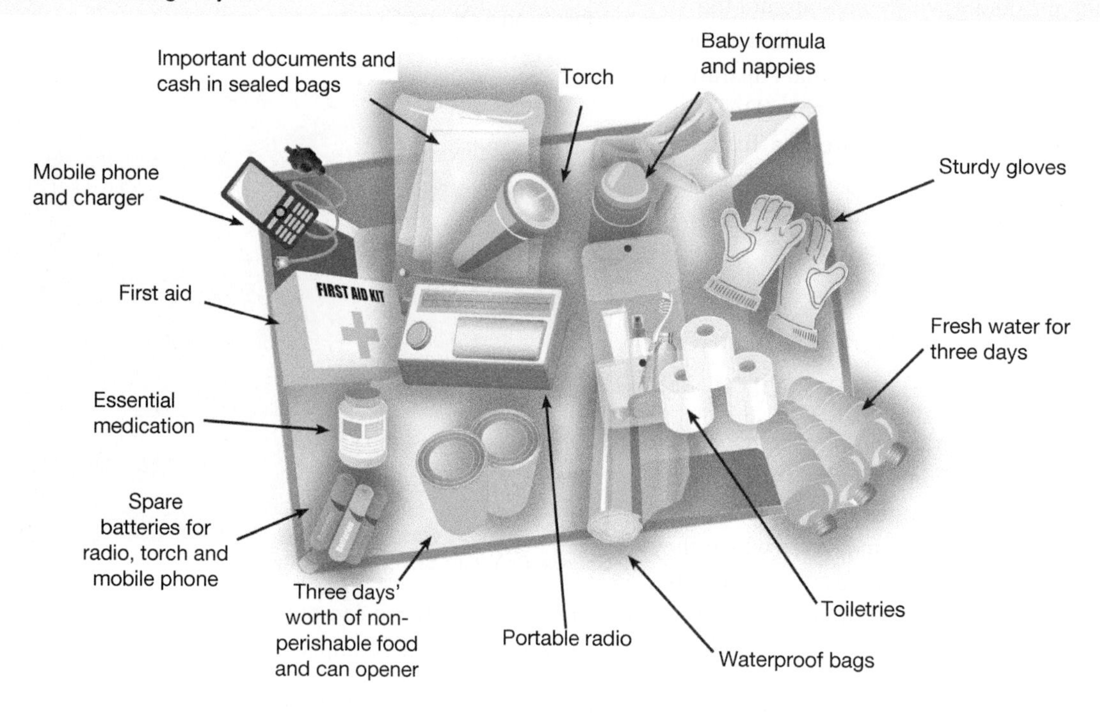

Resources

eWorkbook	Responding to extreme weather events (ewbk-9518)
Video eLesson	Responding to extreme weather events — Key concepts (eles-5131)
Interactivity	Responding to extreme weather events (int-8554)

15.11 ACTIVITIES

1. Devise a plan for a response to a flood at your school, in your town or suburb.
   a. Create a map of a part of your school, town or suburb *most likely* to flood.
   b. Decide which of the strategies would be appropriate for your site. Propose two evacuation points and photograph them.
   c. Create a warning sign to display at the location to notify people of the dangers and their expected responses during a flood.
   d. Consider what makes a good evacuation point. Which of your evacuation points is best? Justify your choices.
2. Research two challenges that Tonga faces in its efforts to be more prepared for severe natural hazards. Explain these reasons in two paragraphs, using supporting data and examples.

FIGURE 8 Flood hazard warning sign

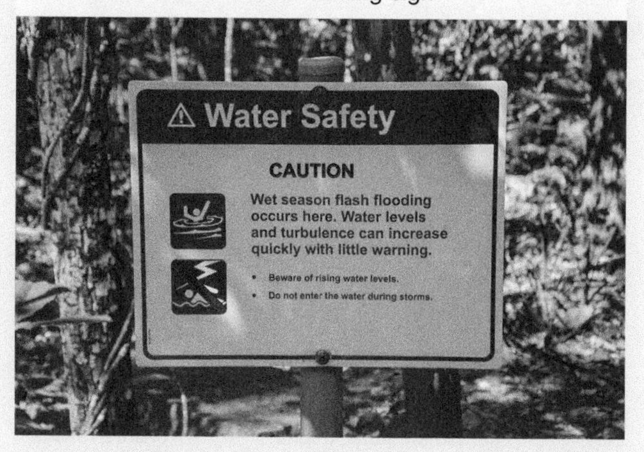

15.11 EXERCISE

Learning pathways

■ LEVEL 1	■ LEVEL 2	■ LEVEL 3
1, 5, 8, 13, 15	3, 6, 7, 9, 12	2, 4, 10, 11, 14

Check your understanding

1. What category storm was Cyclone Gita when it passed over the Kingdom of Tonga?
2. Why was a State of Emergency declared before Cyclone Gita had reached Tonga?
3. What was the wind strength of Cyclone Gita when it passed over Tonga?
4. What were the immediate needs of the population impacted by Cyclone Gita?
5. List five things you should include in a home emergency kit.
6. List three things that you might normally keep in your yard that could become dangerous during a cyclone.

Apply your understanding

7. Select one area of support that Australian responders delivered to the relief effort. Explain how this effort would have supported the people of Tonga.
8. What can Australian citizens do to support an overseas emergency such as the Cyclone Gita natural disaster?
9. How could the residents of Tonga have been better prepared for Cyclone Gita?
10. What steps could be taken to ensure that the residents of Tonga are well prepared for the next natural hazard?
11. Are cyclones the strength of Cyclone Gita likely to make landfall in NSW? Give reasons for your view based on the spatial distribution and extent of cyclones in the region.

Challenge your understanding

12. Australia and New Zealand helped Tonga to rebuild after Cyclone Gita. What responsibilities should developed nations have to less developed nations in their region to ensure that everyone is well prepared for natural disasters? Propose a series of recommendations for being good regional neighbours.
13. Suggest two ways that you could prevent being injured from flying glass from windows in your home during a cyclone. Propose one temporary and one permanent change to the building.
14. Should governments enforce special building codes in areas prone to tornadoes and cyclones? Give reasons for your answer.
15. Suggest why it is recommended that when cyclones and tornadoes are approaching, you should turn off all power. Consider **FIGURE 9** as a starting point for your thinking.

FIGURE 9 Damage caused by a tornado, Greensboro, North Carolina, 2018

To answer questions online and to receive **immediate feedback** and **sample responses** for every question, go to your learnON title at www.jacplus.com.au.

15.12 Management of future extreme weather events

online only

LEARNING INTENTION

By the end of this subtopic, you will be able to describe the strategies being used to predict future extreme weather events and to minimise loss and damage.

The content in this subtopic is a summary of what you will find in the online resource.

Climate change is any change in climate over time, whether due to natural or human processes. Evidence shows that extreme weather events are occurring more frequently in a range of places. Around the globe, temperatures are increasing, polar ice is melting and sea levels are rising. Studies of this change to the environment help inform the development of management strategies to help vulnerable communities.

To learn more about how we might manage extreme weather events in the future, go to your learnON resources at www.jacPLUS.com.au.

FIGURE 7 Restoring and managing law and order is important after a disaster.

Contents

Resources

eWorkbook	Management of future extreme weather events (ewbk-9522)
Video eLesson	Management of future extreme weather events — Key concepts (eles-5132)
Interactivity	Management of future extreme weather events (int-8555)

15.13 Investigating topographic maps — Potential flooding on Sydney's northern beaches

LEARNING INTENTION

By the end of this subtopic, you will be able to identify on a topographic map some places that are likely to be affected by flooding.

15.13.1 Narrabeen Lagoon

Narrabeen Lagoon is 55 square kilometres in size and is the largest coastal lagoon in the Sydney region. The suburbs around the lagoon support many residents who are attracted to the relaxed coastal lifestyle of the northern beaches. Tourists are attracted to the area for the range of outdoor recreational opportunities such as fishing, kayaking, cycling, sailing, walking and picnics.

The main body of Narrabeen Lagoon is fed by stream flow that drains off surrounding elevated suburbs. The lagoon is connected to the sea by a narrow channel that is often blocked by sand movement along North Narrabeen beach. Severe storms can result in flooding of the low-lying suburbs and roads. Rainfall totals and river levels are monitored throughout the catchment, and warnings are issued to motorists and residents when flooding is predicted.

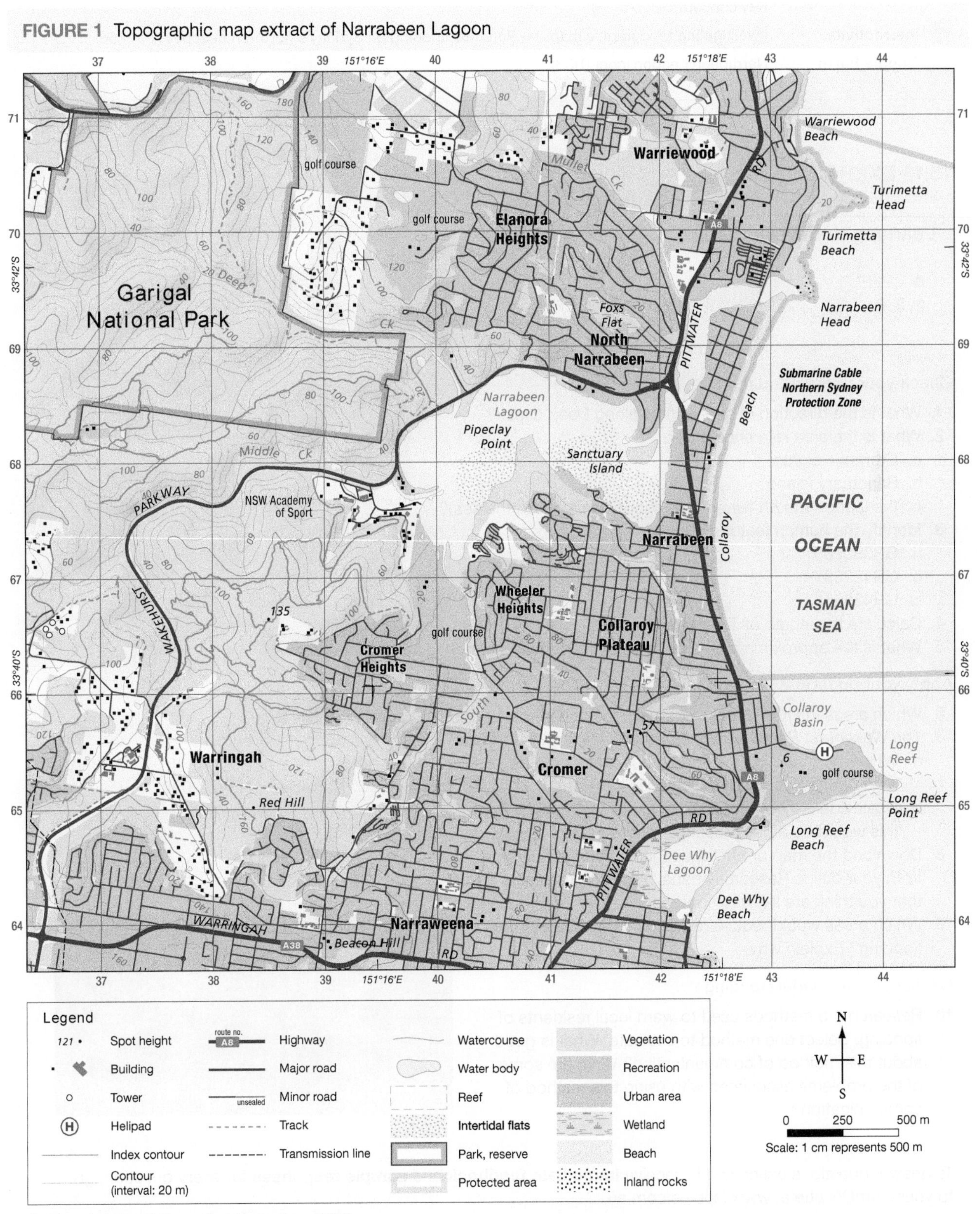

FIGURE 1 Topographic map extract of Narrabeen Lagoon

Source: Land and Property Information, NSW.

Resources

eWorkbook	Investigating topographic maps — Potential flooding on Sydney's northern beaches (ewbk-9526)
Digital document	Topographic map of Narrabeen Lagoon (doc-36259)
Video eLesson	Investigating topographic maps — Potential flooding on Sydney's northern beaches — Key concepts (eles-5133)
Interactivity	Investigating topographic maps — Potential flooding on Sydney's northern beaches (int-8556)
Google Earth	Narrabeen Lagoon (gogl-0138)

15.13 EXERCISE

Learning pathways

■ LEVEL 1	■ LEVEL 2	■ LEVEL 3
2, 3, 6	1, 4, 7	5, 8, 9, 10

Check your understanding

1. What is the direction of stream flow along Deep Creek?
2. What is the area reference of:
   a. Cromer Heights
   b. Sanctuary Island
   c. the lagoon mouth (where the lagoon flows out to the sea)?
3. Identify the human feature found at:
   a. GR390705
   b. GR422687
   c. GR435655
4. Calculate the length of Turimetta Beach.
5. What is the approximate area covered by Narrabeen Lagoon in square kilometres?

Apply your understanding

6. Which areas in **FIGURE 1** are *less likely* to flood in the Narrabeen Lagoon region? Why?
7. The Wakehurst Parkway is an important link between places in Sydney's north.
   a. Where do you think the flood gates that block the road during flooding are located?
   b. Use evidence from the map to explain why you think this would be a good location for the gates.
8. Download the map or print the digital document version from your online Resources panel. Highlight the areas that you think are likely to flood.
9. Which areas would require residents to evacuate during flooding? Explain why.

Challenge your understanding

10. Research the methods used to warn local residents of flooding. Select one method to evaluate. What is good about this method of communication? What are some of the problems associated with using this method of communication?

FIGURE 2 Narrabeen, NSW, 2020

To answer questions online and to receive **immediate feedback** and **sample responses** for every question, go to your learnON title at www.jacplus.com.au.

15.14 Thinking Big research project — Weather hazard documentary

online only

The content in this subtopic is a summary of what you will find in the online resource.

Scenario

The Bureau of Meteorology has issued a severe cyclone warning for northern Australia. Tropical Cyclone Trevor is bearing down on the far north Queensland coast and Tropical Cyclone Veronica is heading for Western Australia's northern coast.

Task

You have been engaged by an independent media company to follow one of these cyclones and prepare a documentary that will air later this year. The company is keen for viewers to witness the birth of a cyclone, its destructive power and its dissipation. You have also been asked to highlight human responses to and preparation for cyclones.

Go to your Resources tab to access the resources you need to complete this research project.

on Resources

ProjectsPLUS Thinking Big research project — Weather hazard documentary (pro-0236)

15.15 Review

15.15.1 Key knowledge summary

15.2 Features of natural hazards

- Natural hazards are a potential source of harm to a community.
- Natural disasters are extreme events that cause serious damage to property and/or loss of life.
- Some places are more prone to natural hazards than other places.
- Some atmospheric and hydrologic natural hazards occur seasonally.

15.4 Thunderstorms

- Thunderstorms, also referred to as electrical storms, form in unstable atmospheric conditions where warm and cold air collide causing powerful updrafts.
- The lightning generated in a thunderstorm is three times as hot as the surface of the Sun.
- Severe thunderstorms are associated with flash flooding, hailstones and wind gusts over 90 kilometres per hour.
- Thunderstorms are more common in spring and summer and in the afternoon, and are linked to the heating effects of the Sun.

15.5 Floods

- Floods can be formed by factors such as storms and low-pressure systems.
- Floods occur when large amounts of water cover places that are not normally covered by water.
- Flash flooding takes place when water levels rise rapidly.
- Riverine flooding takes place when river flow bursts the riverbanks.
- Storm surges and tsunamis cause sea water flooding.

15.6 Impacts of a flooding disaster

- Communities may be isolated from services and support as a result of flooding.
- Repairs to damaged infrastructure may take many months to repair causing disruption to recovery efforts.
- There are positive and negative elements of the social, economic and environmental impacts of flooding.
- Loss of property and life are devastating for people and their communities.
- A renewed sense of community, economic stimulus and flood-proof rebuilding programs can improve liveability in flood-prone places.

15.7 Flood management around the world

- Engineered structures constructed to manage flooding include levee banks, weirs and flood barriers.
- Flood-prone communities build homes high above the river level.
- Replanting mangrove forests is an effective way to minimise flood damage in coastal areas.

15.9 Cyclones, typhoons and hurricanes

- Tropical cyclones are intense low-pressure systems that form in the South Pacific and Indian Oceans.
- Cyclones spin in a clockwise direction in the southern hemisphere.
- Hurricanes form over the North Atlantic Ocean.
- Northern hemisphere intense low-pressure systems spin in an anti-clockwise direction.

15.10 Case studies — Impacts of Typhoon Haiyan and Cyclone Veronica

- Typhoon is the name given to a tropical cyclone in Asia.
- As there are several islands in the Pacific, typhoons can strengthen again once they pass over land and head back out to sea, resulting in multiple land masses being affected.

15.11 Responding to extreme weather events

- Homes and other buildings need to be specially prepared to cope with extreme weather events.
- Emergency kits need to include medical supplies, batteries, food and water.
- Places in tornado-prone areas often have storm shelters and early warning systems.
- Often there is not enough time to fully evacuate an area ahead of an extreme weather event.

15.12 Management of future extreme weather events

- Climate change is predicted to cause more intense storm activity into the future.
- Technology is being widely used and shared between nations to ensure communities can be well prepared for predicted weather patterns and the flood risks they bring.
- Predicting natural hazards allows relief agencies to prepare for humanitarian and emergency support responses.

15.15.2 Key terms

alluvium the loose material brought down by a river and deposited on its bed, or on the floodplain or delta

atmospheric hazard extreme natural event that begins in the atmosphere (in the air)

catchment area the area of land that contributes water to a river and its tributaries

cumulonimbus clouds huge, thick clouds that produce electrical storms, heavy rain, strong winds and sometimes tornadoes. They often appear to have an anvil-shaped flat top and can stretch from near the ground to 16 kilometres above the ground.

cyclone intense low-pressure system producing sustained wind speeds in excess of 65 km/h. It develops over tropical waters where surface water temperature is at least 27 °C.

flood an unusual accumulation of water onto land that is not normally covered by water

floodgate a barrier across a river that is raised when water levels are predicted to be dangerously high

floodplain relatively flat land by a river

gale force wind wind with speeds of over 62 km/h

hailstone an irregularly shaped ball of frozen precipitation

hailstorm any thunderstorm that produces hailstones large enough to reach the ground

hydrologic hazard extreme natural event that begins in the hydrosphere (in the water)

infrastructure the facilities, services and installations needed for a society to function, such as transportation and communications systems, water and power lines

intensify to become stronger

La Niña a pattern in the climate that includes warmer ocean waters off Australia's east coast

monsoon the rainy season in the Indian subcontinent and South-East Asia, followed by a dry season

natural disaster an extreme event that is the result of natural processes and causes serious material damage or loss of life

natural hazard an extreme event that is the result of natural processes and has the potential to cause serious material damage and loss of life

riverine relating to rivers

shanty town slum settlement housing the poor on the outskirts of major cities in developing nations. Lacks basic services such as clean water, sanitation and electricity.

storm surge a sudden increase in sea level as a result of storm activity and strong winds. Low-lying land may be flooded.

tornado violent, wildly spinning column of air that forms over land surfaces. It drops from under a cumulonimbus cloud and contacts the ground.

torrential rain heavy rain often associated with storms, which can result in flash flooding

typhoon the name given to cyclones in the Asian region

weir a barrier across a river, similar to a dam, which causes water to pool behind it. Water is still able to flow over the top of the weir.

15.15.3 Reflection

Complete the following to reflect on your learning.

Revisit the inquiry question posed in the Overview:

What are natural hazards? Can we predict when they will occur and prepare ourselves for them?

1. Now that you have completed this topic, what is your view on the question? Discuss with a partner. Has your learning in this topic changed your view? If so, how?
2. Write a paragraph in response to the inquiry question, outlining your views.

Subtopic	Success criteria			
15.2	I can define the terms *natural hazard* and *natural disaster*.			
	I can locate places where atmospheric and hydrologic hazards occur in Australia.			
15.3	I can construct and interpret a column or bar graph.			
15.4	I can identify the causes of thunderstorms.			
	I can describe ways to minimise the damage they cause.			
15.5	I can identify the causes of floods.			
	I can identify locations where flooding is more likely to take place based on their characteristics.			
15.6	I can identify the impacts of a flooding disaster on communities and the environment.			
15.7	I can list different strategies used to manage floods around the world.			
15.8	I can read a wind rose to determine direction and strength of wind patterns.			
15.9	I can identify the causes of cyclones.			
	I can explain where cyclones are more likely to take place.			
15.10	I can describe and compare the events leading up to one severe cyclone event in Asia and one severe cyclone event in Australia.			
15.11	I can give examples of ways that countries help one another during a time of crisis caused by a natural disaster.			
15.12	I can describe strategies being used to predict future extreme weather events and to minimise loss and damage.			
15.13	I can identify on a topographic map some places that are likely to be affected by flooding.			

on Resources

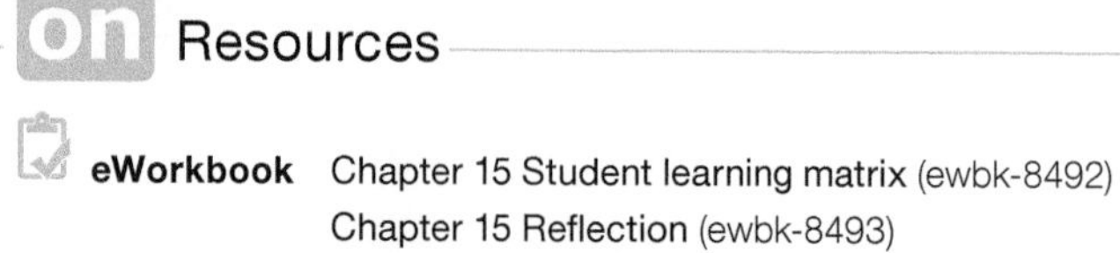

eWorkbook Chapter 15 Student learning matrix (ewbk-8492)
Chapter 15 Reflection (ewbk-8493)
Chapter 15 Extended writing task (ewbk-8494)

Interactivity Chapter 15 Crossword (int-7702)

ONLINE RESOURCES

on Resources

Below is a full list of **rich resources** available online for this topic. These resources are designed to bring ideas to life, to promote deep and lasting learning and to support the different learning needs of each individual.

eWorkbook

- 15.1 Chapter 15 eWorkbook (ewbk-7995) ☐
- 15.2 Features of natural hazards (ewbk-9482) ☐
- 15.3 SkillBuilder — Creating a simple column or bar graph (ewbk-9486) ☐
- 15.4 Thunderstorms (ewbk-9490) ☐
- 15.5 Floods (ewbk-9494) ☐
- 15.6 Impacts of a flooding disaster (ewbk-9498) ☐
- 15.7 Flood management around the world (ewbk-9502) ☐
- 15.8 SkillBuilder — Cardinal points — Wind roses (ewbk-9506) ☐
- 15.9 Cyclones, typhoons and hurricanes (ewbk-9510) ☐
- 15.10 Case studies — Impacts of Typhoon Haiyan and Cyclone Veronica (ewbk-9514) ☐
- 15.11 Responding to extreme weather events (ewbk-9518) ☐
- 15.12 Management of future extreme weather events (ewbk-9522) ☐
- 15.13 Investigating topographic maps — Potential flooding on Sydney's northern beaches (ewbk-9526) ☐
- 15.15 Chapter 15 Student learning matrix (ewbk-8492) ☐
- Chapter 15 Reflection (ewbk-8493) ☐
- Chapter 15 Extended writing task (ewbk-8494) ☐

Sample responses

- 15.1 Chapter 15 Sample responses (sar-0148) ☐

Digital documents

- 15.10 Blank world map template (doc-36260) ☐
- 15.13 Topographic map of Narrabeen Lagoon (doc-36259) ☐

Video eLessons

- 15.1 Wind hazards (eles-1618) ☐
- Wind and water hazard — Photo essay (eles-5123) ☐
- 15.2 Features of natural hazards — Key concepts (eles-5124) ☐
- 15.3 SkillBuilder — Constructing a simple column or bar graph (eles-1639) ☐
- 15.4 Thunderstorms — Key concepts (eles-5125) ☐
- 15.5 Floods — Key concepts (eles-5126) ☐
- 15.6 Impacts of a flooding disaster — Key concepts (eles-5127) ☐
- 15.7 Flood management around the world — Key concepts (eles-5128) ☐
- 15.8 SkillBuilder — How to interpret wind roses (eles-1638) ☐
- 15.9 Cyclones, typhoons and hurricanes — Key concepts (eles-5129) ☐
- 15.10 Case studies — Impacts of Typhoon Haiyan and Cyclone Veronica — Key concepts (eles-5130) ☐
- 15.11 Responding to extreme weather events — Key concepts (eles-5131) ☐
- 15.12 Management of future extreme weather events — Key concepts (eles-5132) ☐
- 15.13 Investigating topographic maps — Potential flooding on Sydney's northern beaches — Key concepts (eles-5133) ☐

Interactivities

- 15.2 Hotspot commander (int-3083) ☐
- Some of Australia's natural hazards and disasters (int-8609) ☐
- 15.3 SkillBuilder — Creating a simple column or bar graph (int-3135) ☐
- 15.4 How a thunderstorm works (int-7794) ☐
- How a supercell hailstorm forms (int-8550) ☐
- 15.5 Floods (int-8551) ☐
- 15.6 Impacts of a flooding disaster (int-8552) ☐
- Floodplains (int-7788) ☐
- 2011 floods in Rockhampton (int-5291) ☐
- 15.7 Responding to floods (int-3085) ☐
- 15.8 SkillBuilder – How to interpret wind roses (int-3134) ☐
- 15.9 Spiralling sea storm (int-3087) ☐
- World distribution of tropical cyclones by names used in different regions (int-7791) ☐
- How a cyclone forms (int-7792) ☐
- 15.10 Case studies — Impacts of Typhoon Haiyan and Cyclone Veronica (int-8553) ☐
- 15.11 Responding to extreme weather events (int-8554) ☐
- 15.12 Management of future extreme weather events (int-8555) ☐
- 15.13 Investigating topographic maps — Potential flooding on Sydney's northern beaches (int-8556) ☐
- 15.15 Chapter 15 Crossword (int-7702) ☐

ProjectsPLUS

- 15.14 Thinking Big research project — Weather hazard documentary (pro-0236) ☐

Weblinks

- 15.2 Disaster resilience (web-6056) ☐
- 15.5 El Niño — What is it? (web-6057) ☐
- The impacts of La Niña in Australia (web-6058) ☐
- El Niño–Southern Oscillation (ENSO) (web-6059) ☐
- NSW catchments (web-6396) ☐
- 15.7 Bureau of Meteorology (web-1583) ☐
- 15.10 Get ready Queensland (web-6060) ☐
- 15.12 World Meteorological Organization (WMO) (web-6061) ☐

Google Earth

- 15.13 Narrabeen Lagoon (gogl-0138) ☐

Fieldwork

- 15.9 Tracking a cyclone (fdw-0022) ☐

Teacher resources

There are many resources available exclusively for teachers online.

16 Fieldwork inquiry — Water quality in my local catchment

16.1 Overview

Numerous **videos** and **interactivities** are embedded just where you need them, at the point of learning, in your learnON title at www.jacplus.com.au. They will help you to learn the content and concepts covered in this topic.

LEARNING INTENTION

By the end of this subtopic, you will be able to plan fieldwork to investigate the water quality in a local catchment, collect and record data, and process and communicate information about your findings.

16.1.1 Scenario and task

Task: Produce a report and presentation about the water quality of a local catchment or waterway.

Water is our most valuable resource, and the management of this vital resource should be a priority at a local, regional and global scale. Everybody lives near a catchment — there is usually a river, creek, drain or other waterway close to your home, school or neighbourhood. If you live in an urban area, the creek may have been highly modified and may look like a concrete drain. Water quality in these waterways will vary from place to place and is influenced by many factors.

Water quality can affect health in many ways. Rivers and streams act as drainage systems. When it rains, water transports rubbish, chemicals and other waste into drains and, eventually, rivers.

Go to learnON to access the following additional resources to complete this fieldwork task:

- fieldwork data recording templates
- a report template
- a presentation template
- a selection of images and audio and video files to add richness to your presentation
- weblinks to sites to assist in your catchment research and to provide sample presentations
- an assessment rubric.

FIGURE 1 A student conducting research in the field

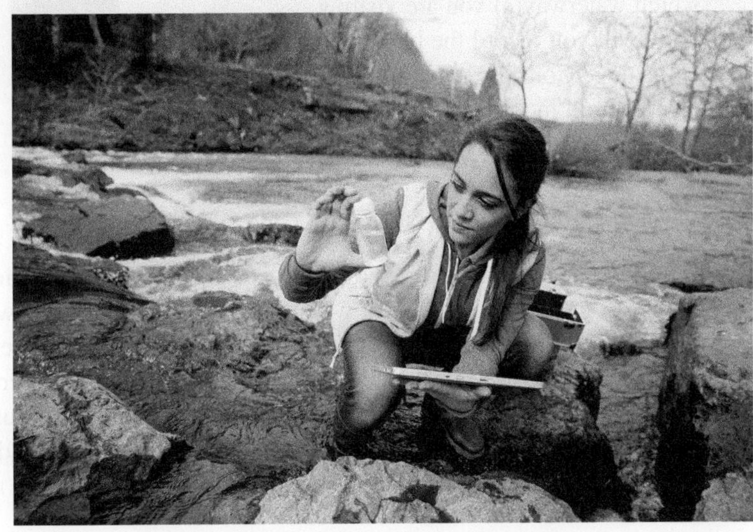

Your task

Your team has been selected to research the water quality of a local catchment or waterway and produce a report and presentation on your findings. Be sure to measure water quality at different locations along the river, creek or stream, and try to determine the causes of different water quality.

16.2 Inquiry process

16.2.1 Process

Open the ProjectsPLUS application for this chapter located in your Resources tab. Watch the introductory video lesson and then click the 'Start Project' button and set up your project group. You can complete this project individually or invite members of your class to form a group. Save your settings and the project will be launched.

FIGURE 2 It is important that you learn about the area before visiting your fieldwork site.

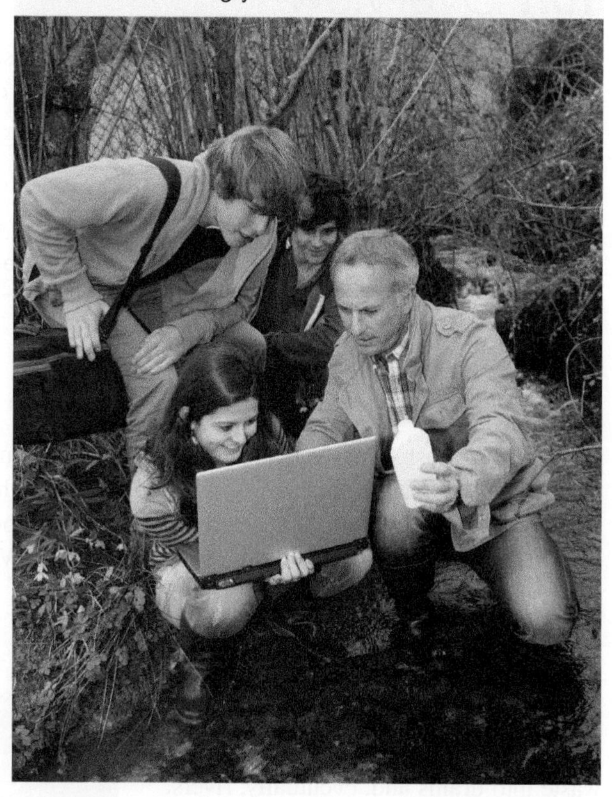

Planning: Navigate to your Research Forum. You will need to research the characteristics of your local catchment area. In order to complete sufficient research, you will need to visit a number of sites within the catchment, comparing different locations upstream and downstream of one creek or river. Research topics have been loaded in the system to provide a framework for your research:

- **What** sort of data and information will you need to study water quality at your fieldwork sites?
- **How** will you collect and record this information?
- **Where** would be the best locations to obtain data? You can determine this once you know which waterway(s) you are visiting.
- **How** will you record the information you are collecting? Consider using GPS, video recorders, cameras and mobile devices (laptop computer, tablet, mobile phone).

16.2.2 Collecting and recording data

It is important that you have some knowledge of the fieldwork location before you visit the site. Access to topographic maps and Google Earth will help you become familiar with the location. Using these tools, complete a sketch map of the waterway(s) and label the sites you are going to visit. You can then scan your sketch map and have it available electronically on the field trip. Alternatively, use Google Maps to record all the sites you visit. Ensure that you bring all the equipment and resources you will need to collect the data with you to each site. It would be useful to work in groups to collect the data, with each group collecting different data at each site. Use the supplied data collection templates on your mobile device, or print copies.

16.2.3 Processing and analysing your information and data

Once you have collected, collated and shared your data, you will need to decide what information to include in your report and the most appropriate way to show your findings. If using spreadsheet data, make total and percentage calculations. Some measurements are best presented in a table; others in graphs or on maps. If you have used a spreadsheet, you may like to produce your graphs electronically. Use photographs as map annotations (either scanned and attached to your electronic map or attached to your hand-drawn map) to show features recorded at each site. You may also like to annotate each photograph to show the geographical features you observed. Describing and interpreting your data is important. There are broad descriptions that can be made of your findings, which might include the following.

FIGURE 3 The floods in Windsor, NSW, in 2021 affected water quality in the local area.

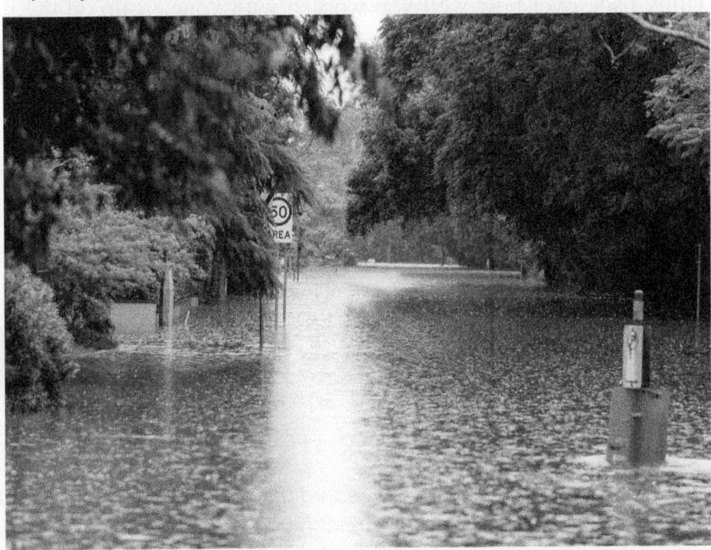

- Where is water quality highest (best) in the waterway studied?
- Is water quality better in the upper reaches of the river or creek?
- Does an urban waterway have better water quality than a rural waterway?
- Does surrounding land use have an impact on water quality?
- Do large waterways have better water quality than smaller waterways?
- What were the main contributors to poor and good water quality?
- How does surrounding vegetation impact on water quality?

Visit your Media Centre and download the report template and the presentation planning template to help you complete this project. Your Media Centre also includes images, videos and audio files to help bring your presentation to life. Use the report template to create your report, and use the presentation template to create an engaging presentation that showcases all of your important findings.

16.2.4 Communicating your findings

You will now produce a fieldwork report and presentation to present your findings. Your report and presentation should include all of the research that you completed and all evidence to support your findings. Ensure that your report includes a title, an aim, a hypothesis (what you think you will find, which is written before you go into the field), your findings and a conclusion. You will also need to recommend some type of action that needs to be taken to improve water quality in the creek or river you visited.

FIGURE 4 Cooks River, Croydon Park

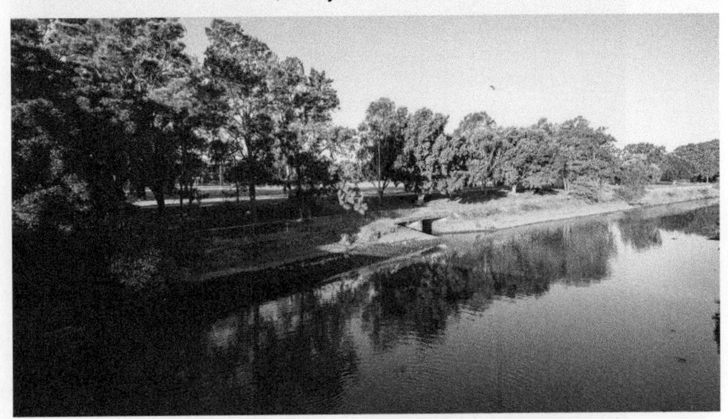

16.3 Review

16.3.1 Reflecting on your work

Think back over how well you worked with your group on the various tasks for this inquiry. Determine strengths and weaknesses and recommend changes if you were to repeat the exercise. Identify one area where you were pleased with your performance, and an area where you would like to improve. Write two sentences outlining how you might be able to do this.

Print your Research Report from ProjectsPLUS and hand it in with your fieldwork report and presentation, and reflection notes.

16.3.2 Reflection

Subtopic	Success criteria			
16.2	I can plan fieldwork to investigate the water quality in a local catchment.			
	I can collect and record data as part of a fieldwork inquiry into the water quality in a local catchment.			
	I can process and communicate information about the water quality in a local catchment.			

on Resources

 **ProjectsPLUS** Fieldwork inquiry — Water quality in my local catchment (pro-0143)

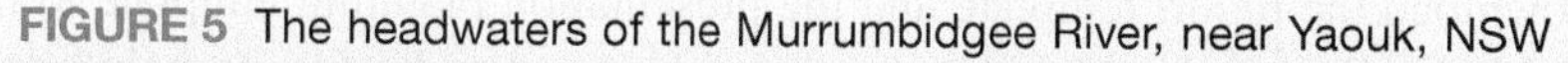

FIGURE 5 The headwaters of the Murrumbidgee River, near Yaouk, NSW

UNIT

4 Interconnections

17 Personal connections

INQUIRY SEQUENCE

To access a pre-test with **immediate feedback** and **sample responses** to every question in this chapter, select your learnON format at www.jacplus.com.au.

17.1 Overview

Numerous **videos** and **interactivities** are embedded just where you need them, at the point of learning, in your learnON title at www.jacplus.com.au. They will help you to learn the content and concepts covered in this topic.

How do the interconnections formed by tourism affect people, places and environments?

17.1.1 Introduction

The World Tourism Organization estimates that by 2030, 5 million people will move each day. Where will these people go and what will influence their choices? What impact will these choices have on the places they visit?

STARTER QUESTIONS

1. Explain how tourism interconnects people and places.
2. When you visit other places, do you meet and get to know the locals, or spend most of your time with other tourists?
3. How might your hobbies or sport activities affect the environment?
4. Are some types of travel or destinations more sustainable than others?
5. Watch the **Personal connections — Photo essay** video in your Resources tab. Where in the world would you like to travel? What might someone learn from travelling to different places — about the world, the place they visit or themselves? What impact might your visit have?

FIGURE 1 Popular destinations such as Venice, Italy, have been impacted significantly by rising numbers of tourists.

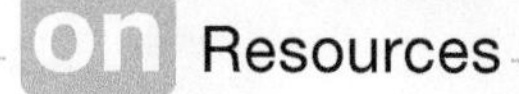

Resources

eWorkbook Chapter 17 eWorkbook (ewbk-7997)

Video eLessons Moving around (eles-1723)
Personal connections — Photo essay (eles-5134)

17.2 Global tourism

LEARNING INTENTION

By the end of this subtopic, you will be able to describe the influences on people's travel connections with different places.

17.2.1 Tourism defined

Travel is the movement between places, often a journey between distant geographical locations. The World Tourism Organization defines **tourism** as the temporary movement of people away from the places where they normally work and live. This movement can be for business, leisure or cultural purposes (see **FIGURE 1**), and it involves a stay of more than 24 hours but less than one year.

Prior to the COVID-19 pandemic, global tourism was expected to increase by 3.3 per cent per year between 2010 and 2030, to reach 1.8 billion tourists traveling annually by 2030. These projections took into account global economic issues, political change in the Middle East and Africa and a range of natural disasters. However, the unprecedented impact of the global COVID-19 pandemic has had a profound impact on tourism operations and projections for the future of tourism. Prior to COVID-19, globally, 308 million jobs existed because of tourism. Around 6 per cent of global **gross domestic product (GDP)** was directly linked to the tourism industry. For many developing countries, tourism is the primary source of income. For these countries, the pandemic's economic impacts are particularly significant.

FIGURE 1 Purpose of people's travel, 2017

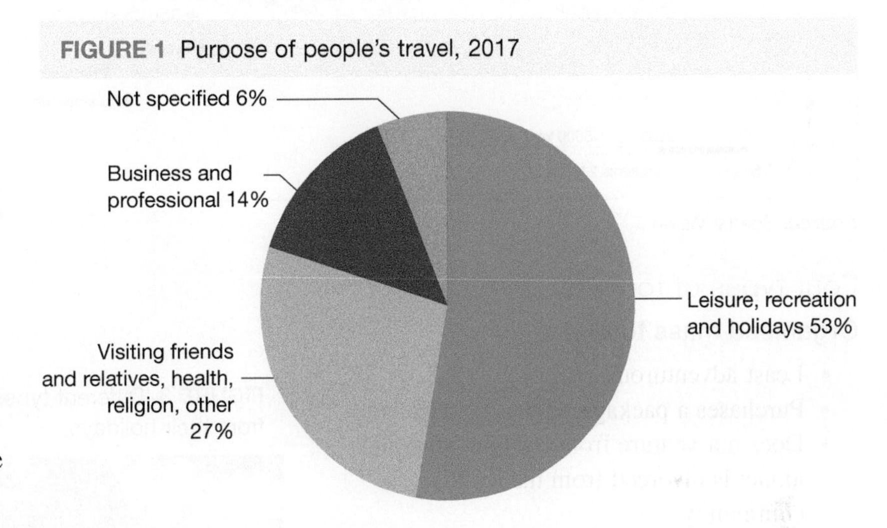

17.2.2 Changes in tourist numbers over time

Tourism has increased dramatically over the past 50 years. Advances in transport technology have reduced not only travel times but also cost. **FIGURE 2** illustrates why particular destinations are popular.

- Today, you can fly from Australia to Europe in around 20 hours, whereas 50 years ago the same journey took six weeks by boat.
- Today, airline and tour companies offer a range of cut-price deals, and the increased number of competitors for the tourist dollar means that travel is more affordable.
- Increased awareness and knowledge of the world has sparked people's desire to see new places and experience different cultures.
- In general, the travelling public has more leisure time and more disposable income, making both domestic and international travel viable.

tourism moving away from where you normally live/work temporarily (for between 24 hours and 1 year)

gross domestic product (GDP) the value of all the goods and services produced within a country in a given period of time. It is often used as an indicator of a country's wealth.

FIGURE 2 Types of tourist destinations

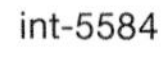

int-5584

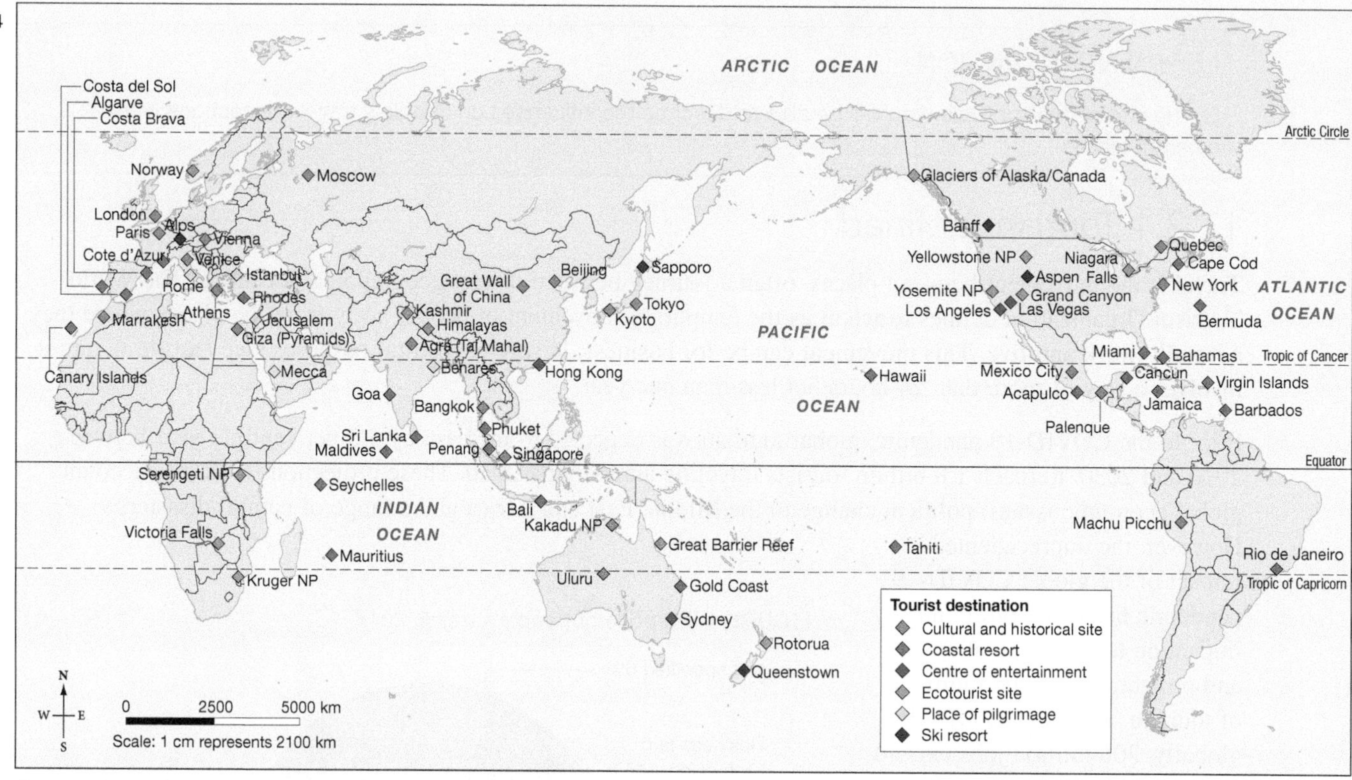

Source: Spatial Vision

Four types of tourists

Organised mass tourist

- Least adventurous
- Purchases a package with a fixed itinerary
- Does not venture from the hotel complex alone; is divorced from the local community
- Makes few decisions about the holiday

Individual mass tourist

- Similar to the organised mass tourist, generally purchases a package
- Maintains some control over their itinerary
- Uses accommodation as a base and may take side tours or hire a car

FIGURE 3 Different types of tourist want different things from their holidays.

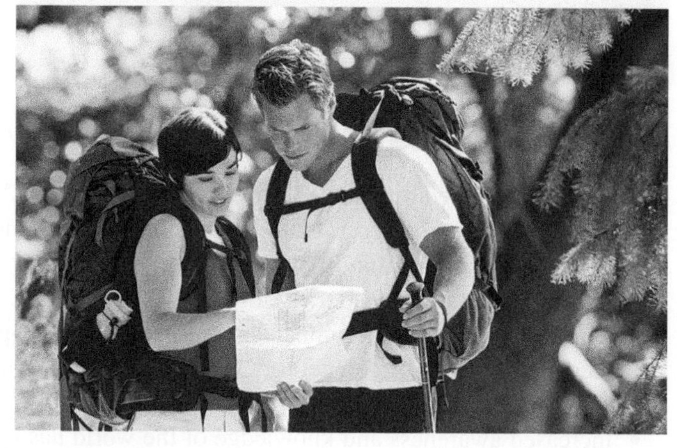

The drifter

- Identifies with local community and may live and work within it
- Shuns contact with tourists and tourist hotspots
- Takes risks in seeking new experiences, cultures and places

The explorer

- Arranges their own trip
- May go off the beaten track but still wants comfortable accommodation
- Motivated to associate with local communities and may try to speak the local language

FIGURE 4 Contribution of tourism to the world economy

Source: World Tourism Organization (2015), *UNWTO Tourism Highlights, 2015 Edition,* UNWTO, Madrid, pp. 2, 13 and 14, DOI: https://doi.org/10.18111/9789284416899

17.2.3 Trends in tourism

Tourism is an important component in world economies. One in 11 jobs worldwide is linked either directly or indirectly to the tourism industry. This industry grew from 674 million international travellers annually in 2000 to 1.46 billion in 2019.

It was predicted that the number of international tourist arrivals worldwide would increase to 1.6 billion by 2020, but COVID-19 severely impacted global tourism operations.

17.2.4 The impact of COVID-19

As a result of the COVID-19 global pandemic, Australia's international borders were closed to all travellers except Australian citizens and residents returning from overseas, and immediate family members. Limited flights were available to and from Australia and **biosecurity** measures such as health screening and travel restrictions were put in place. All travellers arriving in Australia, including Australian citizens, were subject to **mandatory quarantining** for 14 days at a designated facility. Arrangements were made for transport from the flight to the quarantine site, and travellers were required to undertake COVID-19 testing in the first 48 hours after their return and again between days 10 and 12 of quarantine. Source countries were categorised as either red zones or green zones. Green-zone countries were low-risk countries such as New Zealand, while red-zone countries were those with high numbers of COVID-19 infections, such as the United States, United Kingdom and India.

Domestic tourism in Australia has been impacted by lockdowns, border closures and changing rules regarding movements into and out of particular local government areas. Australia's state and territory governments took different approaches to managing the COVID-19 pandemic. The first interstate border closure was implemented in March 2020 by Tasmania. This was followed by a similar approach in Western Australia, South Australia and the Northern Territory later in March, and Queensland in April. In July, Victoria and New South Wales closed their common border. The states and territories required a 14-day quarantine

biosecurity actions taken to stop specific organisms, animals or plants entering the country

mandatory quarantine enforced restriction of movement and separation from the rest of the community, monitored by officials period for returning residents or visitors entering states. Some states also imposed restrictions on movement between different regions or places. For example, for more than two months, residents of the greater Melbourne area were prevented from moving further than 5 kilometres from their home address.

Certain locations such as Lord Howe Island and remote communities had additional restrictions in place. In July a decision was made by the New South Wales Government for the $3000 cost of mandatory quarantine to be charged to the returning traveller.

FIGURE 5 Within Australia, during the COVID-19 pandemic, border closures prevented movement between states. In some places, localised travel was also not allowed.

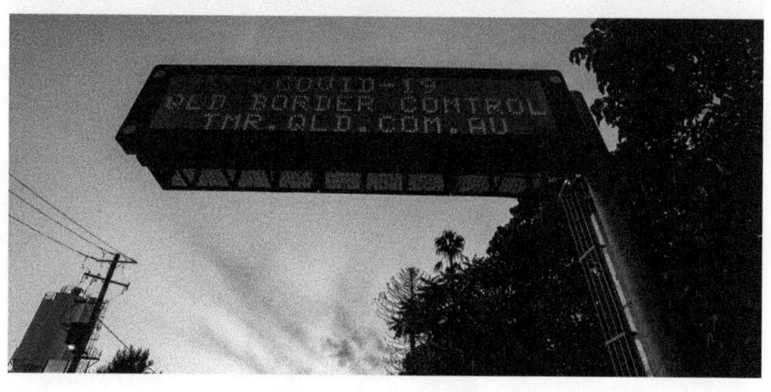

While international arrivals remained limited, the states and territories opened domestic borders in December, prior to the Christmas holiday travel period. Some of these borders closed again in January 2021, as each state government began restricting people's movements for varying lengths of time, depending on where outbreaks occurred.

on Resources

eWorkbook	Global tourism (ewbk-9534)
Video eLesson	Global tourism — Key concepts (eles-5135)
Interactivities	Types of tourist destinations (int-5584) Global tourism (int-8557)
myWorldAtlas	Deepen your understanding of this topic with related case studies and questions > Investigate additional topics > Tourism > **World tourism** Investigate additional topics > Tourism > **Victoria Falls**

17.2 ACTIVITIES

1. Using your atlas as a primary source of information, select three places from different categories shown in **FIGURE 2** that you might like to visit.
   a. Calculate the distance between them.
   b. Explain how you would travel to each place.
   c. Explain what you might expect to see and do in each place.
   d. Work out how long it might take to visit each place.
   e. Describe each location using geographical concepts such as latitude and longitude, direction and scale.
   f. Explain why you have chosen each place.
2. What type of tourist are you? Make a sketch of yourself, similar to the one shown at right.
   Annotate your cartoon to describe yourself as a tourist, using information in this section to help you. Include information about your ideal holiday and explain why you appear as you do in your cartoon.

3. Use the **ABS COVID-19** weblink in your online Resources panel to investigate the impact of COVID-19 on travel within and to Australia. Compare the 2020 statistics with 2019 and 2021. Based on this data, discuss the challenges of rebuilding the Australian tourism industry. What solutions can you think of that might help businesses while international travel is restricted?

17.2 EXERCISE

Learning pathways

■ LEVEL 1	■ LEVEL 2	■ LEVEL 3
1, 3, 8, 9, 14	2, 4, 6, 10, 13	5, 7, 11, 12, 15

Check your understanding

1. Define the term *tourist*.
2. Describe why tourism has been one of the fastest growing industries.
3. What are the two most common purposes for people to travel?
4. Describe each of the following types of tourist:
   a. organised mass tourist
   b. individual mass tourist
   c. explorer
   d. drifter.
5. Define the term *gross domestic product* (GDP).
6. Identify one example of each of the following types of tourist destinations shown in **FIGURE 2**:
   a. coastal resort
   b. place of pilgrimage
   c. ecotourist site
   d. cultural and historical site.
7. Before the COVID-19 pandemic, how much was global tourism predicted to increase each year to 2030?

Apply your understanding

8. What type of tourist do you like to be? Choose the type of travel you prefer from the list of tourist types and explain why you prefer this type of travel.
9. Is a person who flies from Sydney to Melbourne for work, and returns home later that day, considered a tourist? Explain why or why not.
10. Why is tourism an important component of GDP?
11. Consider and outline the possible advantages and disadvantages of placing returned travellers in mandatory quarantine (apart from reducing the spread of illness).
12. Many states made people pay for their mandatory hotel quarantine. Do you agree with this approach? Explain why or why not.

Challenge your understanding

13. Create a mind-map demonstrating the impact of the COVID-19 global pandemic on tourism within Australia.
14. If you had one week and unlimited finances, where would you travel and why?
15. What reasons for travel might be included in the 6 per cent of people in **FIGURE 1** that were 'not specified'?

FIGURE 6 International and interstate travel were severely impacted by the COVID-19 epidemic.

To answer questions online and to receive **immediate feedback** and **sample responses** for every question, go to your learnON title at www.jacplus.com.au.

17.3 Travel patterns and trends

LEARNING INTENTION

By the end of this subtopic, you will be able to analyse patterns and trends in people's travel choices.

17.3.1 Increases in global tourism

Tourism is one of the world's largest industries. Even when global economies are experiencing a downturn, people still travel. After natural disasters, countries rely on the return of the tourist dollar to help stimulate their economies. However, the spread of tourism is not uniform across the globe.

FIGURE 1 Backpackers often travel further and stay longer than other tourists.

Over time, travel has become faster, easier, cheaper and safer. Economic growth in many parts of the world has helped many people to have more money to spend and they can afford to travel. Increases in annual leave also provide people with time to travel. For example, many Australians receive long service leave when they stay in a job for a long time — this is in addition to the four weeks' annual leave given to full-time employees in Australia. It has also become common for young people to spend time travelling during a 'gap year' after finishing secondary school or **tertiary study**.

Many young travellers see backpacking as the best way to travel. Generally this group:
- is on a tight budget
- wants to mix with other young travellers and local communities
- has a flexible itinerary
- seeks adventure
- is prepared to work while on holiday to extend their stay.

At the other end of the scale, there has also been a dramatic growth in **mature-aged** tourist movements. The number of older people visiting developing countries is growing. Older people might choose to travel when their children are older and are no longer dependent on their parents for support. Some of these travellers have savings, access to superannuation funds, and the opportunity to retire early; thus, they have both the time and the money to travel.

17.3.2 Popular destinations

As each tourist enters or leaves a country, they are counted by that country's customs and immigration officials. This data is collected by the World Tourism Organization, and the results can be shown spatially (see **FIGURE 2**).

tertiary study the educational level following high school, e.g. university

mature-aged describes individuals aged over 55

FIGURE 2 World's top ten tourist destinations, 2018

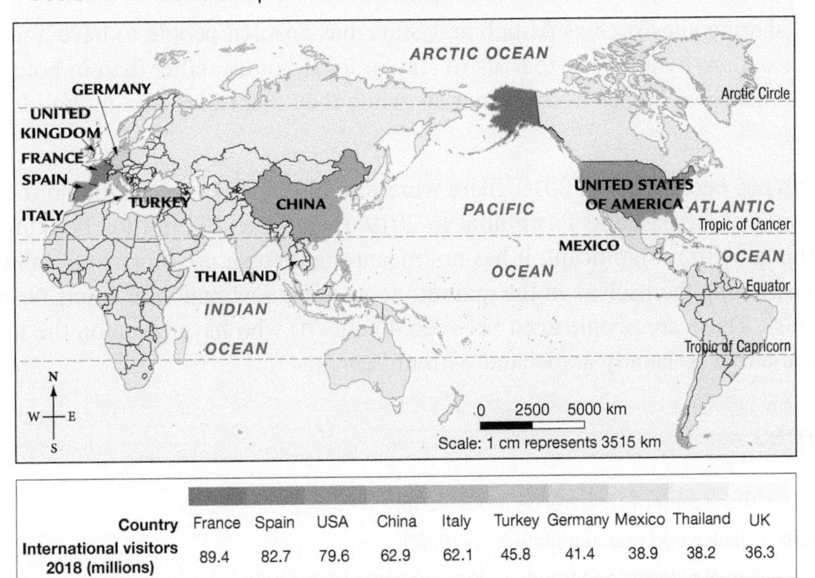

Country	France	Spain	USA	China	Italy	Turkey	Germany	Mexico	Thailand	UK
International visitors 2018 (millions)	89.4	82.7	79.6	62.9	62.1	45.8	41.4	38.9	38.2	36.3

Source: World Tourism Organization (2020), UNWTO World Tourism Barometer, volume 18, issue 1, UNWTO, Madrid, p. 28, DOI https://doi.org/10.18111/wtobarometereng

17.3.3 Spending on travel

FIGURE 2 shows the countries that attract the most tourists, but which countries do these tourists come from, and what do they spend? **FIGURE 3** shows the top ten countries in terms of the money they spend on international tourism. **FIGURE 4** shows the per capita expenditure of the world's top ten tourist spenders.

FIGURE 3 Top ten tourist spenders by expenditure

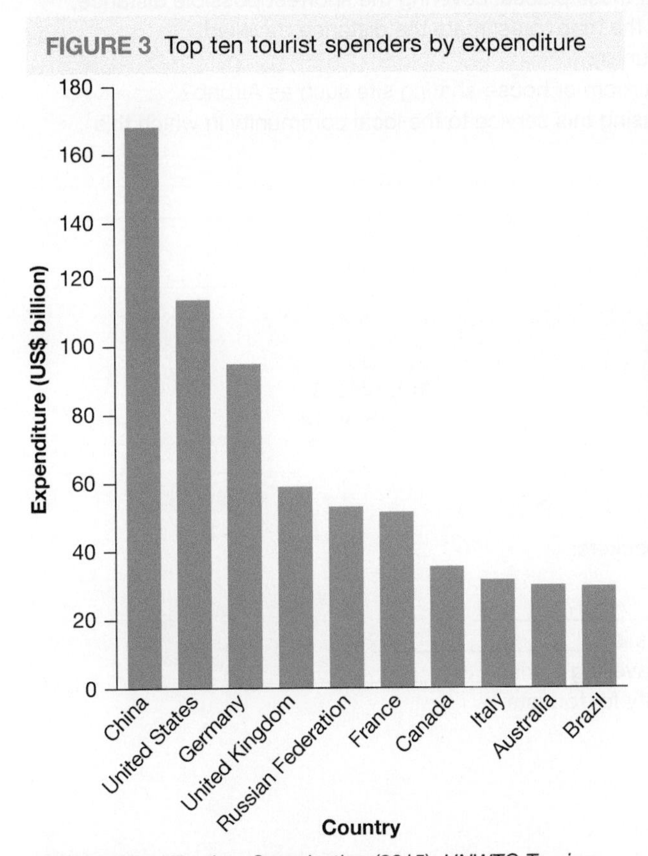

Source: World Tourism Organization (2015), *UNWTO Tourism Highlights, 2015 Edition,* UNWTO, Madrid, pp. 2, 13 and 14, DOI: https://doi.org/10.18111/9789284416899

FIGURE 4 Top ten tourist spenders per capita

Source: World Tourism Organization (2015), *UNWTO Tourism Highlights, 2015 Edition,* UNWTO, Madrid, pp. 2, 13 and 14, DOI: https://doi.org/10.18111/9789284416899

17.3.4 Accommodation trends

The growth of home sharing sites such as Airbnb and Stayz has enabled people to have a more authentic experience of another culture by allowing tourists to stay in local homes rather than in hotels. Many people use these sites to rent out their own homes to other tourists while they themselves are on holiday, in a paid form of house swapping.

The growth of Airbnb has been huge. In 2010, there were roughly 50000 stays in the northern hemisphere summer. By 2015, this had increased to 17 million. In 2019, there were 272 million bookings made worldwide. While this growth is significant, it has not meant that people no longer stay in hotels. Many hotel groups are now repositioning themselves in the market, developing new boutique hotels designed to appeal to 'ageless millennials'. These are people aged between 40 and 70 who have taken on the love of technology, trends and social media that is usually associated with millennials.

on Resources

eWorkbook	Travel patterns and trends (ewbk-9538)
Digital document	Blank world map template (doc-36260)
Video eLesson	Travel patterns and trends — Key concepts (eles-5136)
Interactivity	Travel patterns and trends (int-8558)

17.3 ACTIVITIES

1. What is the difference between a mature-aged traveller and a backpacker? Create a Venn diagram to show the similar and different needs and wants of someone typical of each group.
2. a. On a blank map of the world, locate and label the capital cities of each of the top ten tourist destinations.
   b. Plot a trip from your nearest capital city to all ten of these places, covering the shortest possible distance, and returning to your capital city. Use the scale on the map to estimate the distance travelled.
   c. Calculate the time it might take to complete this journey.
3. a. What are the advantages/disadvantages of using a room or house-sharing site such as Airbnb?
   b. What would be the advantages/disadvantages of using this service to the local community in which this accommodation is situated?

17.3 EXERCISE

Learning pathways

■ LEVEL 1	■ LEVEL 2	■ LEVEL 3
1, 2, 3, 11, 15	4, 6, 10, 12, 13	5, 7, 8, 9, 14

Check your understanding

1. Complete the following statements describing backpackers:
   a. A backpacker's budget is usually ______________.
   b. The itinerary followed by a backpacker is usually ______________.
   c. Backpackers' attitude to visiting local communities is ________________.
   d. Backpackers' attitude to working while they are travelling is that ________________.
2. Rank the following countries in order of their popularity for tourists:
   a. United States
   b. Thailand
   c. France
   d. Germany
   e. Mexico
   f. Turkey

3. What defines a mature-aged tourist?
4. On which three continents are the top ten tourist destinations located?
5. Describe what distinguishes an 'ageless millennial' from other travellers of the same age group.
6. Which continent:
   a. is generating the most tourist spending?
   b. is home to the most 'top ten' tourist destinations?
7. Outline three reasons people might choose to rent out their home or spare room to tourists.

Apply your understanding

8. Why are more mature-aged people choosing to travel than in the past?
9. Explain the interconnection between destinations and tourism spending.
10. Discuss how home sharing sites have changed tourism.
11. Why is tourism more accessible to the broader community than it was 100 years ago?
12. Examine **FIGURE 5**, which shows how tourist numbers were expected to grow before the COVID-19 pandemic.
    a. How many international tourists were predicted to be travelling in 2020?
    b. If the predicted number of tourists for 2020 decreased by 80 per cent because of COVID-19, approximately how many tourists travelled internationally in 2020?
    c. Do you think the predicted number of tourists will reach 3.4 billion by 2050? Give reasons for your answer.

FIGURE 5 Projected future growth in world tourism

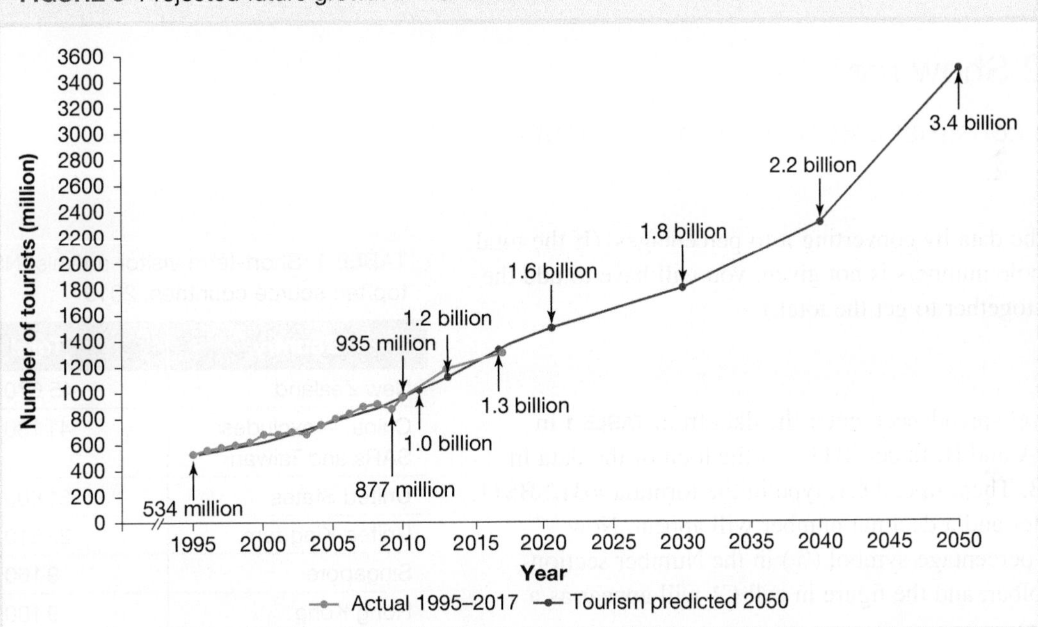

Challenge your understanding

13. Suggest and explain a strategy that would have helped the Australian tourist industry recover from the COVID-19 downturn.
14. Do you think the trend in tourism growth (pre-COVID-19) was sustainable? Give reasons to support your answer.
15. Predict whether home-sharing services will continue to grow in popularity in the future.

To answer questions online and to receive **immediate feedback** and **sample responses** for every question, go to your learnON title at www.jacplus.com.au.

17.4 SkillBuilder — Constructing and describing a doughnut chart

online only

LEARNING INTENTION

By the end of this subtopic, you will be able to represent geographical data in graphs such as a doughnut chart.

The content in this subtopic is a summary of what you will find in the online resource.

17.4.1 Tell me

What is a doughnut chart?

A doughnut chart is a circular chart with a hole in the middle. Each part of the doughnut is divided as if it were a pie chart with a cut-out. The circle represents the total, or 100 per cent, of whatever is being looked at. The size of the segments is easily seen. Presenting the parts in order from largest to smallest makes it easy to interpret.

17.4.2 Show me

How to construct and describe a doughnut chart

eles-1738

Step 1

Prepare the data by converting it to percentages. (If the total of the whole numbers is not given, you will have to add the numbers together to get the total.)

Step 2

In an Excel spreadsheet, enter the data from **TABLE 1** in columns A and B. In cell B11, put the total of the data in column B. Then, in cell C1, type in the formula =B1/B11. Press Enter and a decimal number will appear. Now click the percentage symbol (%) in the Number section of the toolbar, and the figure in cell C1 will appear as a percentage.

Step 3

Click on cell C1, and you will see a small black square in the bottom right corner. Using the cursor, drag this down over the remaining rows to be calculated. Do not include the row in which the total is shown (row 11 in this example). The percentages for all remaining cells will appear.

TABLE 1 Short-term visitor arrivals, NSW — top ten source countries, 2019

Country	Total
New Zealand	45 670
China — excludes SARs and Taiwan	41 130
United States	31 600
United Kingdom	22 510
Singapore	9 160
Hong Kong	9 100
India	9 000
Philippines	6 820
Tonga	530
Vanuatu	350

Source: Australian Bureau of Statistics, Overseas Arrivals and Departures, Australia

Step 4

To generate the doughnut chart, select all columns and rows in which your data appears, excluding the total figure (columns A, B and C, and rows 1 to 10 in this example), and then click the Insert tab on your toolbar. In the Charts section, select Other Charts and choose the simple doughnut chart.

Step 5

Click on the doughnut and select Chart Layouts in the Design tab on the toolbar. Try different chart layouts to see which works best.

Click on the outer rectangular border around your doughnut chart and drag it. As you do this, the chart will enlarge.

You can also decrease the type size of the labels by clicking on one, going to the Home tab and reducing the font size.

Step 6

To add the title, click on the heading Chart Title, type your title and then press Enter.

FIGURE 1 Short-term visitor arrivals, NSW — top ten source countries, 2019

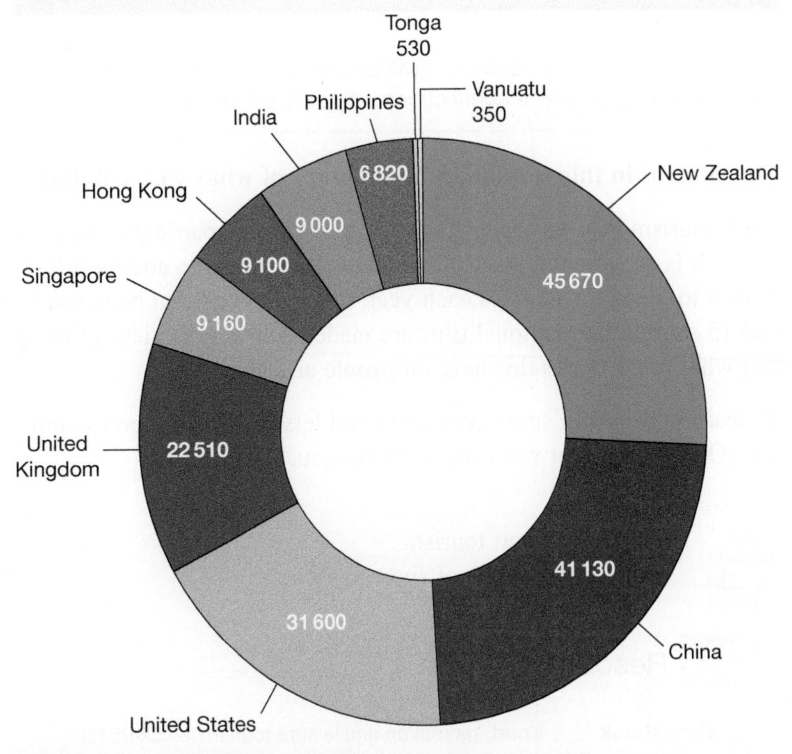

Note: China excludes SARs and Taiwan.

Source: Australian Bureau of Statistics, Overseas Arrivals and Departures, Australia https://www.abs.gov.au/statistics/industry/tourism-and-transport/overseas-arrivals-and-departures-australia/latest-release

17.4.3 Let me do it

int-3356

Go to learnON to access the following additional resources to help you build this skill:

- a longer explanation of this skill and its application in Geography (Tell me)
- a video showing the step-by-step process of this skill (Show me)
- an activity and interactivity for you to practise this skill (Let me do it)
- self-marking questions to help you understand and use this skill.

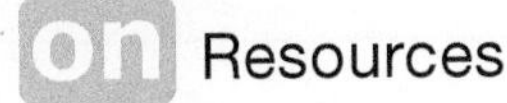

Resources

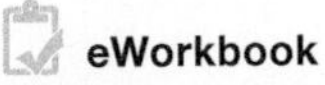

eWorkbook SkillBuilder — Constructing and describing a doughnut chart (ewbk-9542)

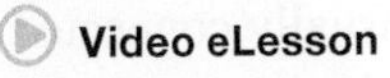

Video eLesson SkillBuilder — Constructing and describing a doughnut chart (eles-1738)

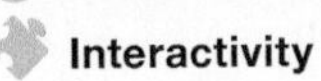

Interactivity SkillBuilder — Constructing and describing a doughnut chart (int-3356)

17.5 Sport, recreation and leisure tourism

online only

LEARNING INTENTION

By the end of this subtopic, you will be able to identify influences on and analyse the effect of people's recreational and leisure connections with different places.

The content in this subtopic is a summary of what you will find in the online resource.

Sport tourism involves people travelling to view or participate in a sporting event. It is an expanding sector of the tourism industry, adding billions of dollars to global economies each year. It is estimated that between 12 million and 15 million international trips are made each year to view sporting events. But what impact does this have on people and places?

To learn more about sport, recreation and leisure tourism, go to your learnON resources at www.jacPLUS.com.au.

FIGURE 4 The Barmy Army comprises thousands of fans who come to Australia to cheer on the English cricket team.

Contents

- 17.5.1 Sport tourism
- 17.5.2 Impacts of sport tourism

on Resources

eWorkbook	Sport, recreation and leisure tourism (ewbk-9546)
Video eLesson	Sport, recreation and leisure tourism — Key concepts (eles-5137)
Interactivity	Are the Olympics worth gold? (int-3337)
Fieldwork	Coding an article (fdw-0023)
myWorldAtlas	Deepen your understanding of this topic with related case studies and questions Investigate additional topics > Australia's links with the world > **Sport** Investigating Australian Curriculum topics > Year 9: Geographies of interconnections > **The FIFA World Cup**

17.6 Cultural tourism

LEARNING INTENTION

By the end of this subtopic, you will be able to describe cultural influences on people's travel connections with different places.

17.6.1 Defining cultural tourism

Cultural tourism is concerned with the way of life of people in a geographical region. It is usually connected to elements that have shaped their values or culture, such as a shared history, traditions or religion.

According to the UN World Tourism Organization:

> Cultural tourism is a type of tourism activity in which the visitor's essential motivation is to learn, discover, experience and consume the tangible and intangible cultural attractions/products in a tourism destination. These attractions/products relate to a set of distinctive material, intellectual, spiritual and emotional features of a society that encompasses arts and architecture, historical and cultural heritage, culinary heritage, literature, music, creative industries and the living cultures with their lifestyles, value systems, beliefs and traditions.

Whatever the reason, the mass movement of people associated with these events has a significant impact on both people and places.

FIGURE 1 The Louvre Museum in Paris is a cultural tourism destination.

Examples of cultural tourism include the following.

- The Day of the Dead (FIGURE 2) is a festival originating in Mexico that celebrates the dead. The festivities span three days, and the public holiday encourages people to remember and pray for family and friends who have passed away.
- Holi is a religious festival marking the arrival of spring, predominantly celebrated in the Hindu nations of India and Nepal. It celebrates the start of a plentiful spring harvest.
- The Hajj pilgrimage (FIGURE 3) is expected to be made by each Muslim person at least once during their life. Every year, over five or six days in the last month of the Islamic calendar, 3 to 5 million people travel to the sacred city of Mecca, Saudi Arabia.
- In many cultures where Christianity is the predominant religion, people come together to celebrate Christmas, commemorating the birth of Christ; and Easter, to remember his resurrection.

FIGURE 2 Thousands gather to attend the Day of the Dead parade in Mexico City each year.

FIGURE 3 Muslims from all over the world make the Hajj pilgrimage to Mecca, Saudi Arabia.

17.6.2 Thanksgiving in the United States

Thanksgiving is held in the United States on the fourth Thursday in November. It dates back to the seventeenth-century celebration of the harvest. Today it is a time for families to get together and give thanks for what they have.

The Thanksgiving holiday runs from Wednesday to Sunday. In 2019, around 55.3 million people travelled to celebrate with family and friends during Thanksgiving, with about 108 million travelling throughout the country during the extended holiday period that also includes Christmas and New Year (see **TABLE 1**). As millions of people travel across the country, transport systems are stretched to their limits, creating delays and traffic congestion. Because the holiday season is so close to the start of winter, the weather can further complicate people's travel plans, especially those who live in the northern states. Early winter storms can bring ice and snow, resulting in airport closures and impassable roads.

There were still 50.6 million Thanksgiving travellers in the United States in 2020, despite warnings to reduce travel and to have a socially distanced Thanksgiving due to the COVID-19 pandemic. A greater proportion of travellers used private vehicles and fewer relied on planes, trains and other means of travel than in previous years (see **TABLE 1**). Travellers also tended to travel shorter distances than in previous years.

TABLE 1 Modes of transport use, end-of-year holiday period (Thanksgiving to New Year), United States

Transport type	Estimated travellers (millions)		Change
	2019	2020	
Air	7.33	2.94	–59.9%
Personal vehicle (driving)	108	81.1	–24.9%
Other transport	3.89	0.48	–87.8%

Source: https://newsroom.aaa.com/2020/12/at-least-34-million-fewer-americans-to-travel-this-holiday-season

17.6.3 Chinese New Year

Chinese New Year is the longest and most important of traditional Chinese holidays. Dating back centuries, it is steeped in ancient myths and traditions. The festivities begin on the first day of the first month in the traditional Chinese calendar, and last for 15 days. They conclude with the lantern festival on Chinese New Year's Eve, a day when families gather for their annual reunion dinner. It is considered a major holiday, and it influences not only China's geographical neighbours but also the nations with whom China has economic ties.

Of special significance is the fact that the date on which Chinese New Year occurs varies from year to year. This date coincides with the second **new Moon** after the Chinese **winter solstice**, which can occur any time between 21 January and 20 February.

Chinese New Year, or Lunar New Year, is celebrated as a public holiday in many countries with large Chinese populations or with calendars based on the Chinese lunar calendar. The changing nature of this holiday has meant that many governments have to shift working days to accommodate this event.

In China itself, many manufacturing centres close down for the 15-day period, allowing tens of millions of people to travel from the industrial cities where they work to their hometowns and rural communities. This means that retailers and manufacturers in overseas countries such as the United States and Australia have to adjust their production and shipping schedules to ensure they have enough stock on hand to deal with the closure of factories in China. For those shopping online, delays in delivery are to be expected during this period.

The logistics of movement in China

Chinese New Year has been described as the world's biggest annual movement of people. During the Chinese New Year period (known as the Spring Festival) people

new Moon the phase of the Moon when it is closest to the Sun and is not normally visible

winter solstice the shortest day of the year, when the Sun reaches its lowest point in relation to the equator travel to see family and friends. Many travel to other Asian countries, with Thailand, Japan and Indonesia being the most popular destinations in 2019. However, holidays to Europe are becoming increasingly popular during this period.

In 2018, there were 6.5 million Chinese outbound tourists during the Chinese New Year period. In 2019, both the number of travellers and the length of travel increased. COVID-related travel bans and disruptions encouraged people to travel domestically in 2020 rather than undertake overseas trips, with around 600 million people travelling around the country during the period. The pandemic has also encouraged personal car use over other forms of transportation.

FIGURE 4 To ensure prosperity and good fortune in the year ahead, parades, dragons and lion dances feature in Chinese New Year celebrations.

on Resources

eWorkbook	Cultural tourism (ewbk-9550)
Video eLesson	Cultural tourism — Key concepts (eles-5138)
Interactivity	Cultural tourism (int-8560)

17.6 ACTIVITIES

1. As a class, brainstorm a list of cultural or celebratory events that occur in Australia.
2. Use the internet to find out more about either Chinese New Year or Thanksgiving. Investigate the history, myths and traditions associated with your chosen event. Prepare an annotated visual display comparing your findings with a cultural or celebratory event in Australia. Make sure you include references to the scale of your chosen event and the place in which it occurs.
3. Copy the table below into your workbook and fill it in. Use the **Thanksgiving** weblink in your Resources tab to find out more and help you complete your table.

	Thanksgiving	Chinese New Year
Number of trips		
Most common form of transport		
Length of holiday period		
Purpose of trip/activities		

   a. What is the preferred mode of transport for Thanksgiving and for Chinese New Year? Suggest reasons for differences in travel arrangements. In your response, include reference to the scale of movement.
   b. Make a list of problems associated with the mass movement of people.
   c. Select one of the problems you have identified and explain the impact it might have on people, places and the environment. Suggest a strategy for the sustainable management of this problem in order to reduce its impact.

17.6 EXERCISE

Learning pathways

■ LEVEL 1	■ LEVEL 2	■ LEVEL 3
2, 4, 9, 11, 15	1, 3, 6, 8, 13	5, 7, 10, 12, 14

Check your understanding

1. Define *cultural tourism*.
2. Identify three cultural or celebratory events that occur in Australia that might attract tourists.
3. What is a pilgrimage?
4. Complete the following sentences, comparing Thanksgiving and Chinese New Year travel.
   a. In 2020, approximately ______________ people travelled in the US during the winter holiday period by air. This was a decrease of ______________ on 2019.
   b. In 2020, approximately ______________ people travelled domestically in China during the New Year holiday period.
5. List three problems that might be faced by people who choose each of the following forms of transport to travel during busy holiday periods.
   a. Air
   b. Their own car
   c. Train or bus

Apply your understanding

6. Explain why Thanksgiving and Chinese New Year are regarded as cultural events.
7. Explain why Chinese New Year leads to industries shutting down for 15 days.
8. Write a paragraph explaining how cultural events can change people, places and the environment.
9. Write a paragraph describing a traditional cultural event that you and your family celebrate. Is it an example of cultural tourism? Give reasons for your answer.
10. What impacts might the increase in travel during cultural holiday periods have on the natural environment?
11. What impacts would tourism have on your life if you lived near a place that hosts a significant religious event, such as Mecca? Consider the possible positive and negative impacts.
12. Some cultural events, such as Thanksgiving, occur at approximately the same time each year, whereas others such as Chinese New Year vary more in their timeframe. Explain why this is so.

FIGURE 5 Chinese New Year decorations at the Kek Lok Si Buddhist temple, Air Itam, Penang, Malaysia

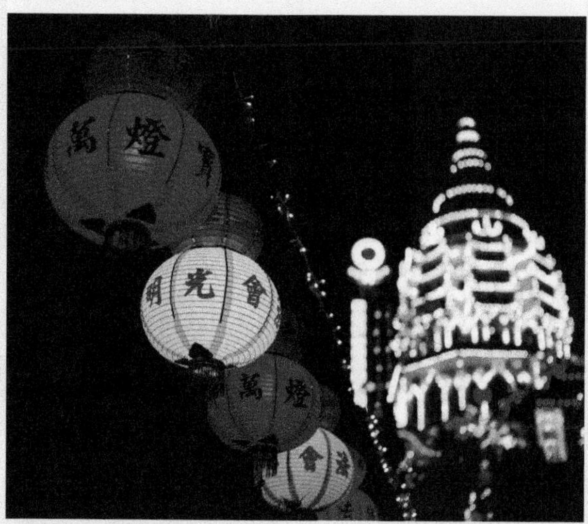

Challenge your understanding

13. Suggest how a place might be changed over time by hosting a large cultural event, such as a festival. Consider how the place might need to change in the first few years, and how it might have changed after 50 years.
14. Predict the impact Chinese New Year might have on a clothing import business in Australia. In your answer, explain what business owners might need to do to ensure their business is not affected by this event.
15. Driving might mean being caught in a traffic jam and not getting home. Flying, trains and buses might mean delays or catching COVID-19. Weigh up all of the factors that impact on travel plans, and recommend the best way to travel during busy times when people are eager to get home.

To answer questions online and to receive **immediate feedback** and **sample responses** for every question, go to your learnON title at www.jacplus.com.au.

17.7 Impacts of tourism

LEARNING INTENTION

By the end of this subtopic, you will be able to assess the impact of people's travel on the future of places.

17.7.1 Costs and benefits

In some ways, tourism seems like the perfect industry. It can encourage greater understanding between people and bring prosperity to communities. However, tourism development can also destroy people's cultures and the places in which they live. There is sometimes a fine line between exploitation and sustainable tourism.

Many travellers don't consider or are unaware of the consequences their choices have on both the environment and people of the destination they are visiting. Conversely, many tourists leave their holiday destinations feeling uplifted by their positive experiences and seek to return the favour. The mind-maps in **FIGURE 1** and **FIGURE 2** outline just some of the impacts that tourism has on people and places.

FIGURE 1 The positive impacts of tourism

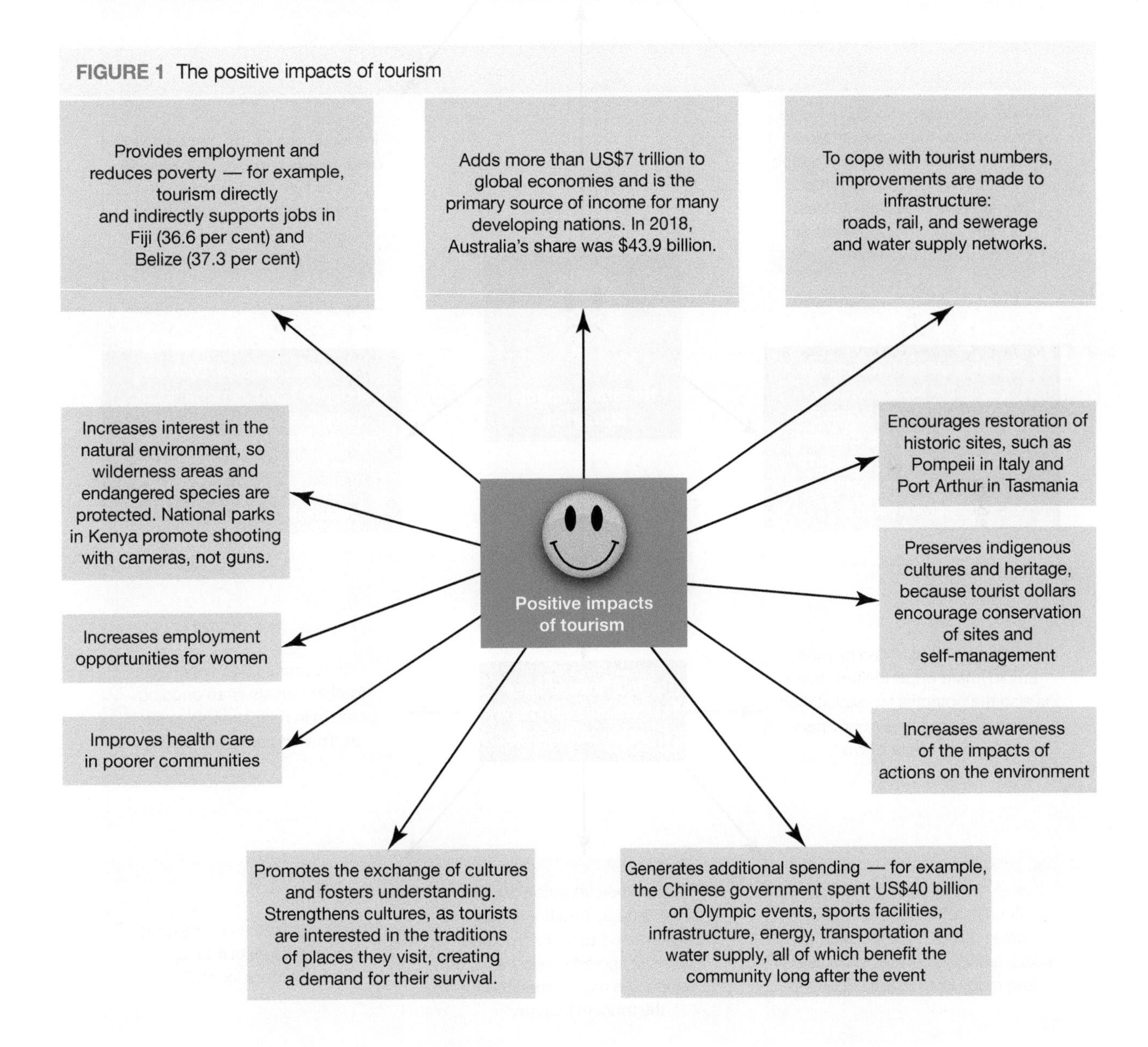

FIGURE 2 The negative impacts of tourism

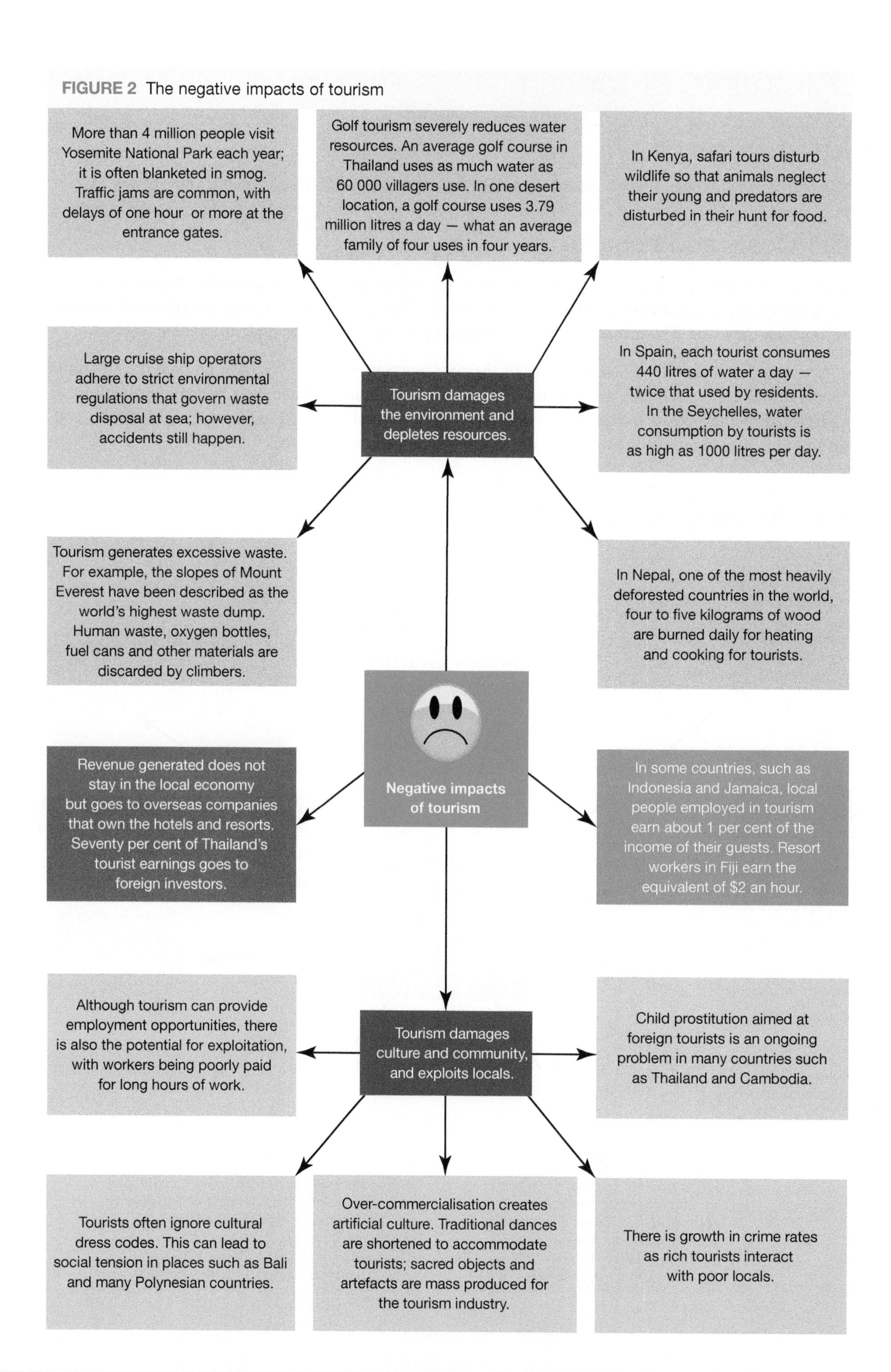

FIGURE 3 Promoting cultural understanding or commercialising traditional culture? Protecting or exploiting the natural environment? Tourism can have both positive and negative impacts on people and the environment.

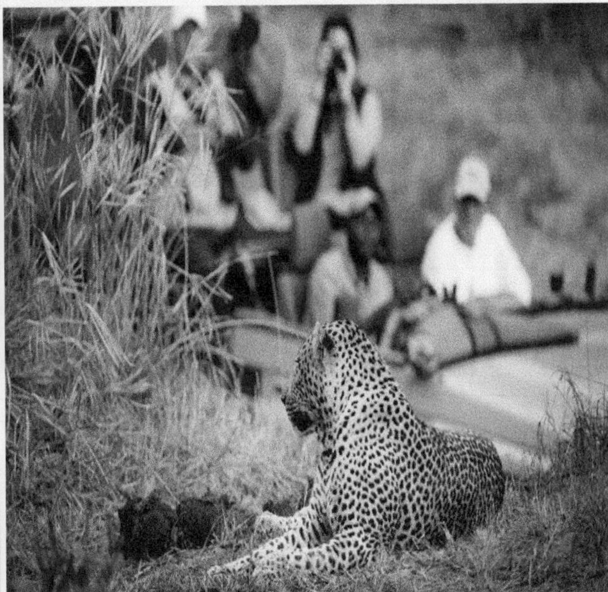

on Resources

eWorkbook	Impacts of tourism (ewbk-9554)
Video eLesson	Impacts of tourism — Key concepts (eles-5139)
Interactivity	Tourism trauma (int-3335)

17.7 ACTIVITIES

1. The impacts of tourism can be classified as environmental, cultural and economic. Study **FIGURE 2**, showing the negative impacts of tourism. Working in groups of three, select an impact from each group. Explain the scale of each impact and devise a strategy for sustainable tourism.
2. Many Australians who were travelling, studying or living overseas when the COVID-19 pandemic struck found themselves stranded without the money or opportunity to fly home. Similarly, many international tourists were stranded in Australia. This raised questions about whether government resources should be used to help get people home or support them, and the impact of stranded tourists on local communities and government resources. Discuss this issue as a class, considering:
   - If flights are grounded because of a natural disaster or global crisis, should a country be responsible for getting their citizens home?
   - If international citizens are stranded in Australia because of a natural disaster or global crisis, should our government provide them with financial assistance or support until they can get home?
   - If Australian citizens are stranded overseas because of a natural disaster or global crisis, who — if anyone — should provide them with support and assistance?
3. a. **FIGURE 4** shows how the tourist dollar (money spent by tourists) can flow from one job to the next. Those jobs in the centre of the diagram interact directly with the tourist, while those on the outside do not. Copy the diagram into your workbook at an enlarged size. Complete it by adding other jobs. Add to it if you can. Study your completed diagram and write a paragraph explaining the interconnection between tourism and the economy.
   b. Repeat this exercise looking at either the social or environmental impacts of tourism.

FIGURE 4 One view on tourism

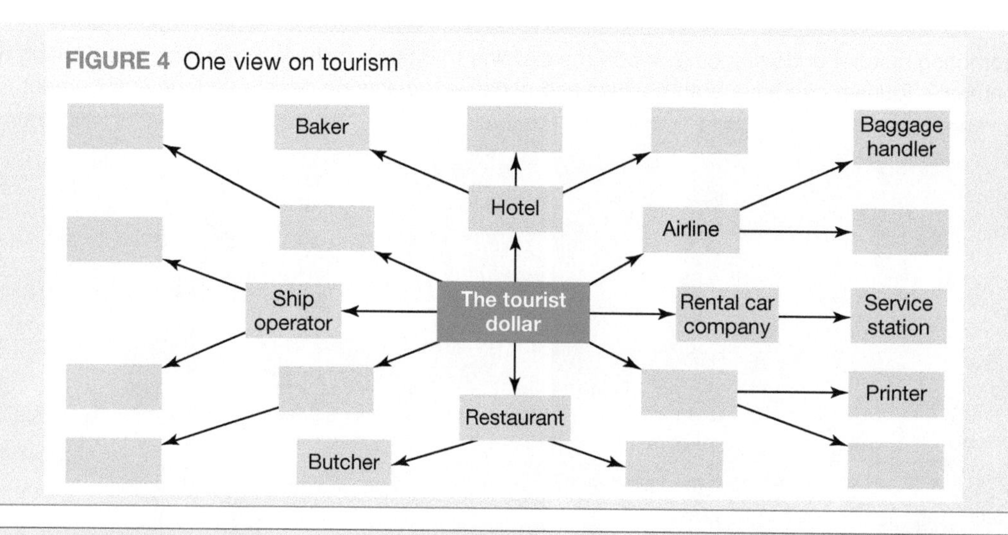

17.7 EXERCISE

Learning pathways

■ LEVEL 1	■ LEVEL 2	■ LEVEL 3
1, 6, 9, 10, 14	2, 3, 8, 11, 15	4, 5, 7, 12, 13

Check your understanding

1. Identify one example of each of the following:
   a. the positive social impact of tourism
   b. the negative social impact of tourism
   c. the positive economic impact of tourism
   d. the negative economic impact of tourism.
2. List three examples of direct employment provided by tourism.
3. Define the term *indirect employment*.
4. Identify one of the possible risks for the local people when tourism creates more jobs in their community.
5. Identify one of the possible benefits for local people when tourism creates more jobs in their community.
6. Describe one situation in which tourists can have a lasting negative impact on the natural environment; for example, affecting rare or endangered animals.

Apply your understanding

7. The type of interconnection shown between industries in **FIGURE 4** is sometimes called the multiplier effect. Explain what you think this means.
8. One criticism of tourism is that it is 'over-commercialised'. Explain your understanding of this term.
9. Explain how tourism can improve the living conditions for individuals living in developing nations.
10. Explain how tourism may lead to an increase in the crime rate in a popular tourist destination.
11. How might tourism lead to the preservation and conservation of ancient ruins and the creation of nature reserves?
12. Analyse the following view: 'Tourism only has positive effects on culture, community and the environment'. Write a paragraph in response.

Challenge your understanding

13. One popular tourist destination is the Inca trail in Peru. Suggest what the negative impacts of tourism in this hard-to-reach region might be and suggest possible strategies for sustainable tourism in remote mountain regions.
14. Which of the following would be the best to develop as a tourist resource in your region: an art gallery, a museum, a cinema complex or a sports stadium? Why?
15. Recommend three strategies that governments could put in place to protect vulnerable places and cultures from the negative impacts of tourism.

To answer questions online and to receive **immediate feedback** and **sample responses** for every question, go to your learnON title at www.jacplus.com.au.

17.8 SkillBuilder — Creating a survey

online only

LEARNING INTENTION

By the end of this subtopic, you will be able to develop and conduct a survey.

The content in this subtopic is a summary of what you will find in the online resource.

17.8.1 Tell me

What is a survey?

Surveys collect primary data. A survey involves asking questions, recording and collecting responses, and collating and interpreting the number of responses. Because your survey is taken from a relatively small number of people in a population, it is called a sample.

17.8.2 Show me

How to create a survey

eles-1764

Step 1

Determine the topic that you want to gather data about, and consider why you want this data.

Step 2

Write a series of questions that you think will generate the data you need. These should be closed questions, with a series of four to six answers to choose from.

Step 3

Test your questions to see whether they are clear and whether they prompt quick, short responses. If necessary, reword your questions.

Step 4

Discuss the order of questions with a classmate. Review and reorganise the order of the questions if necessary.

Step 5

Write one or two open-ended questions. These are questions the respondent can answer in their own words and give more detail.

FIGURE 1 A survey of shoppers

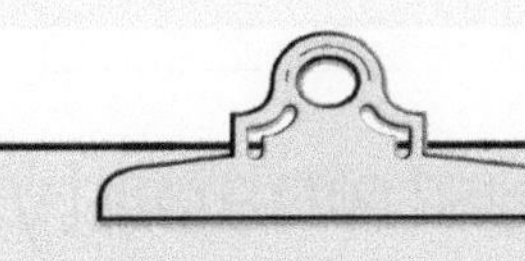

QUESTIONNAIRE FOR SHOPPERS

1. What suburb do you live in? ______
2. How did you get to the centre?
   Taxi Bus Bicycle
   Train Car or motorcycle Walk
3. Did you use the car park provided by the centre?
   Yes No
4. How often do you shop at the centre?
   This is the first time Once a fortnight
   Several times a week Once a month
   Once a week Only very occasionally
5. What types of goods and services will you buy today?
   Clothes Groceries
   Household/electrical goods Fresh fruit and vegetables
   Financial/banking services Light meal/refreshments
6. Do you often shop at any other major shopping centre?
   Yes No If yes, which one? ______
7. What attracts you to this centre? ______
8. Apart from shopping, are there any other reasons for you coming to the centre?
   Work Post office Bank
   Hairdresser Doctor Dentist
   Solicitor Restaurants Entertainment
   Other ______

Step 6

Check that you have not worded your questions in an offensive way or asked anything embarrassing. Reword or delete questions as needed.

Step 7

Check your survey for bias. Bias is when you have unfairly influenced the respondent to your survey. You do not want to lead your respondent in a particular direction and thus skew your research.

Step 8

Review your work before asking people to complete your survey.

Step 9

When you go out into the field, introduce yourself, take care for your own safety and the comfort of others, and always thank people for their time.

17.8.3 Let me do it

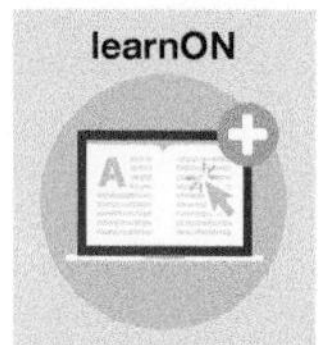

Go to learnON to access the following additional resources to help you build this skill:

- a longer explanation of this skill and its application in Geography (Tell me)
- a video showing the step-by-step process of this skill (Show me)
- an activity and interactivity for you to practise this skill (Let me do it)
- self-marking questions to help you understand and use this skill.

Resources

eWorkbook	SkillBuilder — Creating a survey (ewbk-9558)
Video eLesson	SkillBuilder — Creating a survey (eles-1764)
Interactivity	SkillBuilder — Creating a survey (int-3382)

17.9 Sustainability and ecotourism

online only

LEARNING INTENTION

By the end of this subtopic, you will be able to identify and evaluate strategies to make tourism more environmentally and culturally sustainable.

The content in this subtopic is a summary of what you will find in the online resource.

Tourism has the capacity to benefit environments and cultures or destroy them. Ecotourism has developed in response to this issue. Its aim is to manage tourism in a sustainable way. This might be through educational programs related to the environment or cultural heritage, or through controlling the type and location of tourist activities or the number of tourists visiting an area.

FIGURE 3 Ningaloo Reef, Western Australia

To learn more about ecotourism and sustainability, go to your learnON resources at www.jacPLUS.com.au.

Contents

- 17.9.1 Ecotourism and the environment

Resources

eWorkbook	Sustainability and ecotourism (ewbk-9566)
Video eLesson	Sustainability and ecotourism — Key concepts (eles-5140)
Interactivities	Sustainable sightseeing (int-3336) Anatomy of an ideal ecotourism resort (int-5574)
 myWorldAtlas	Deepen your understanding of this topic with related case studies and questions Investigate additional topics > Tourism > **Kakadu National Park** Investigate additional topics > Managing environments > **Wilsons Promontory** Investigate additional topics > Tourism > **Ningaloo Reef and ecotourism**

17.10 Future trends and sustainability

LEARNING INTENTION

By the end of this subtopic, you will be able to make informed predictions about how places and people might be affected by tourism in the future.

17.10.1 Factors shaping the future of tourism

Tourism is no longer just for the elite. Improvements in transport and technology have increased our awareness of the world around us. Improved living standards, increased leisure and greater disposable incomes have all created opportunities for people to travel and to experience new places and cultures. These factors are also shaping the tourist of the future (see **FIGURE 1**). A particular growth area is the 18–35 market — young people travelling while studying, taking a break from study, or seeing the world before they settle down. The other major area of growth is the over-60 age bracket.

Factors that are shaping the future of tourism include:
- COVID-19
- sustainability
- technology
- medical tourism
- targeted tourism (for example, women-only, retiree and solo travel).

17.10.2 The COVID-19 global pandemic

The COVID-19 global pandemic is shaping the future of tourism. Increased concerns for safety, demand for higher hygiene standards and new government restrictions are encouraging tourism operators to implement more thorough cleaning, social distancing (for example extra spaces between seats) and contactless payments. Many locations are still grappling with getting the virus under control and trying to reopen the economy, including tourism, at the same time.

An important future direction for tourism will be restoring traveller confidence and designing tourism recovery plans. For example, when the ban on cruise ships in Australia is lifted, cruise ships will likely impose mandatory COVID-19 tests and temperature checks for guests and crew, and limit passenger numbers. There will be an increase in domestic tourism while some international borders remain closed.

FIGURE 1 The tourist of the future

FIGURE 2 Domestic and international travel is being changed by the COVID-19 global pandemic.

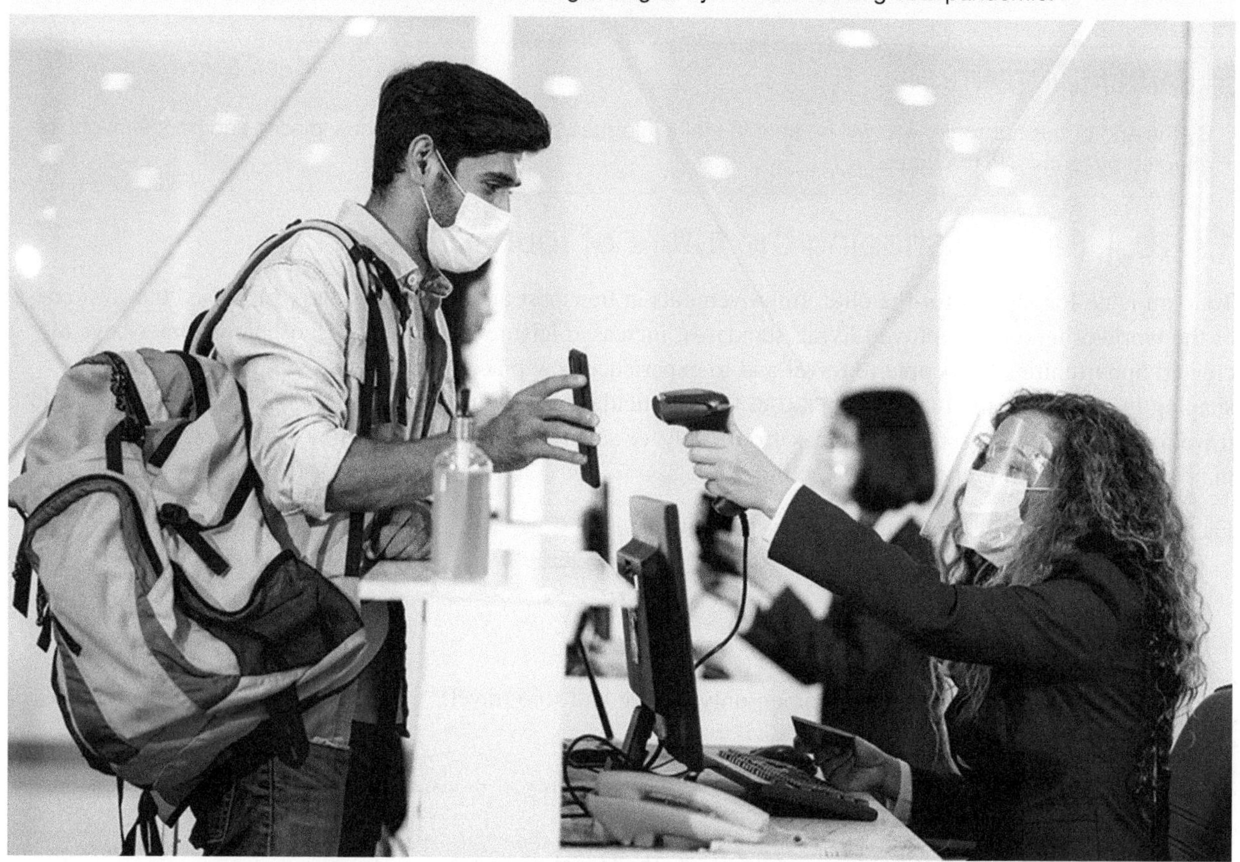

17.10.3 Sustainability

People recognise the benefits that responsible and sustainable tourism can bring to destinations, in terms of employment and the preservation of cultural and natural heritage. However, we are also aware of the growing contribution that flying makes to global warming.

FIGURE 3 Nature-based wellness retreats and hotels target their services to tourists looking to travel sustainably and experience the natural environment.

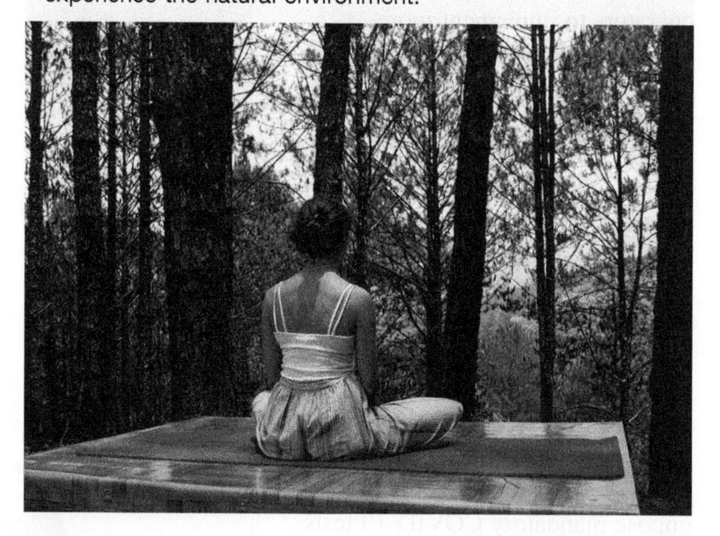

Currently aviation accounts for 2 per cent of all CO_2 emissions and 12 per cent of CO_2 from the transport industry. **Carbon offsetting** is an initiative to manage carbon emissions as a result of tourism. It involves calculating the amount of carbon emitted through flights and transport and putting in place programs that compensate for the carbon produced, using strategies such as reforestation. Sustainability is driving the popularity of eco-travel, green hotels, eco-lodges, nature-based wellness vacations and volunteering. Many hotels are also phasing out single-use plastics and miniature toiletries and moving toward larger pump packs of shampoo and liquid soap to reduce environmental impact.

carbon offsetting calculating the amount of carbon released during an activity and investing in strategies to pay off or cancel out the carbon

17.10.4 Technology

Mobile technology and travel-related apps have revolutionised tourism planning over the past decade. Apps are used for accessing maps, booking accommodation and tours, and restaurant and campsite recommendations. Tourism promotions are targeted through social media sites, particularly Instagram. Augmented Reality and Virtual Reality are increasingly being used to allow travellers to preview cruise ship cabins or hotel rooms, and experience tourist sites before visiting in person. Virtual assistants such as Siri and Alexa help tourists learn about the city they are visiting. Automation, voice search and voice control are increasingly playing a role in travel, including translation apps that break down the language barriers that once made travelling in some countries more difficult. In contrast, some travellers are keen to visit locations where they can take part in digital detoxing, away from their emails and social media.

Travel to space is an emerging industry. In 2004, the first non-government organisation launched a crewed craft into space twice in two weeks. Now there are already people booking to stay in hotels on the Moon. In 2020, seven private citizens spent time on the International Space Station. In 2021, SpaceX plans to send three tourists to the International Space Station on a 10-day trip, for a price of around US$55 million. SpaceX has announced that it is working on a space tourism company, Space Adventures. Virgin Galactic is another spaceflight company, and is developing spacecraft to undertake suborbital spaceflights.

FIGURE 4 SpaceX Crew Dragon spacecraft docking to the International Space Station

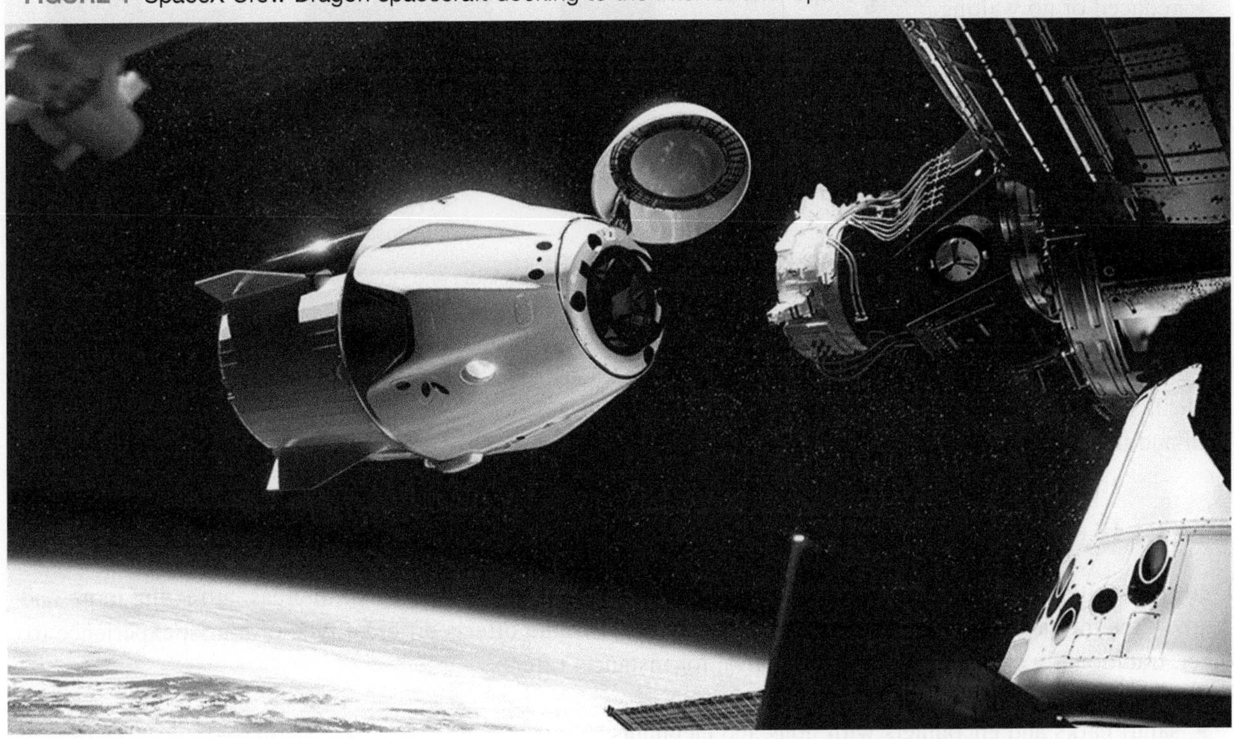

17.10.5 Medical tourism

Medical tourism involves people travelling to overseas destinations for medical care and procedures. The low cost of travel, advances in technology and lengthy waiting lists caused by increased demand for elective surgery are turning medical tourism into a multi-billion dollar industry. In 2017, the global medical tourism market was valued at almost $54 billion; it was expected at the time that the figure would rise to more than $140 billion by 2025.

Medical tourism only used to be for cosmetic procedures, such as face-lifts and tummy tucks, but the range of services offered has expanded dramatically over recent years. It now includes:

- knee replacements
- hip replacements
- fertility treatments (IVF)
- surrogacy services
- complex heart surgery.

Medical tourism added more than US$8.5 billion to Asian economies in 2019; however, countries all over the world are attracting patients for a variety of reasons:

- high standard of medical care
- outstanding reputation of particular facilities
- cost savings
- reduced or no waiting time
- opportunity to include a holiday and luxury accommodation as part of the package.

FIGURE 5 Cost savings that can be made by having medical treatment in Asia rather than Australia

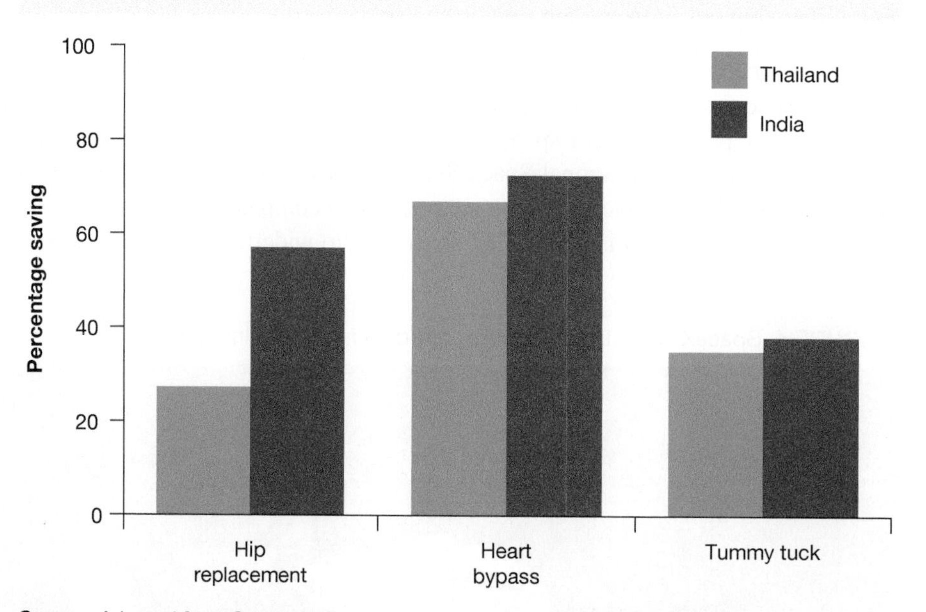

Source: Adapted from Cosmetic Surgeon India and Rowena Ryan/News.com.au

Asia is the market leader in the medical tourism industry, with Thailand and India vying for the top spot. Thailand is slightly more expensive, but offers a better tourist experience and has greater services available. India is less expensive and boasts state-of-the-art facilities staffed by western medical staff, predominantly from the United States. **FIGURE 5** shows the relative costs of surgery in India and Thailand in comparison to Australia.

17.10.6 Growth areas

Predictions suggest that Africa and the Asia–Pacific region will be particular growth areas, attracting more and more of the tourist dollar. Countries such as Kenya and Tanzania offer a different type of tourist experience to other, traditionally popular destinations. Kenya, for instance, offers:

- beaches and a tropical climate
- safari parks and encounters with lions and elephants
- a unique cultural experience with the Maasai people.

The resulting influx of tourists to Kenya has led to the establishment of national parks to protect endangered wildlife and promote this aspect of the tourism experience. Money flowing into the region helps improve water quality and infrastructure such as water pipes, roads and airports. In 2017, tourism accounted for about 9.7 per cent of the Kenyan GDP. Kenya had hoped to increase their tourist visitors to around 3 million by 2017, but the figure was just under 1.5 million. Part of this lower figure was due to terrorist activity in Kenya, some of which specifically targeted western tourists. This demonstrates that even before the worldwide tourism decline of 2020, the opportunities offered to developing countries from tourism depended on balancing the economic, political, social and natural environments.

FIGURE 6 Trends and forecasts in tourist arrival by region as predicted in 2010. Forecasts predicted steady growth after the 2008 economic crisis based on a return to economic growth — was this a strong prediction?

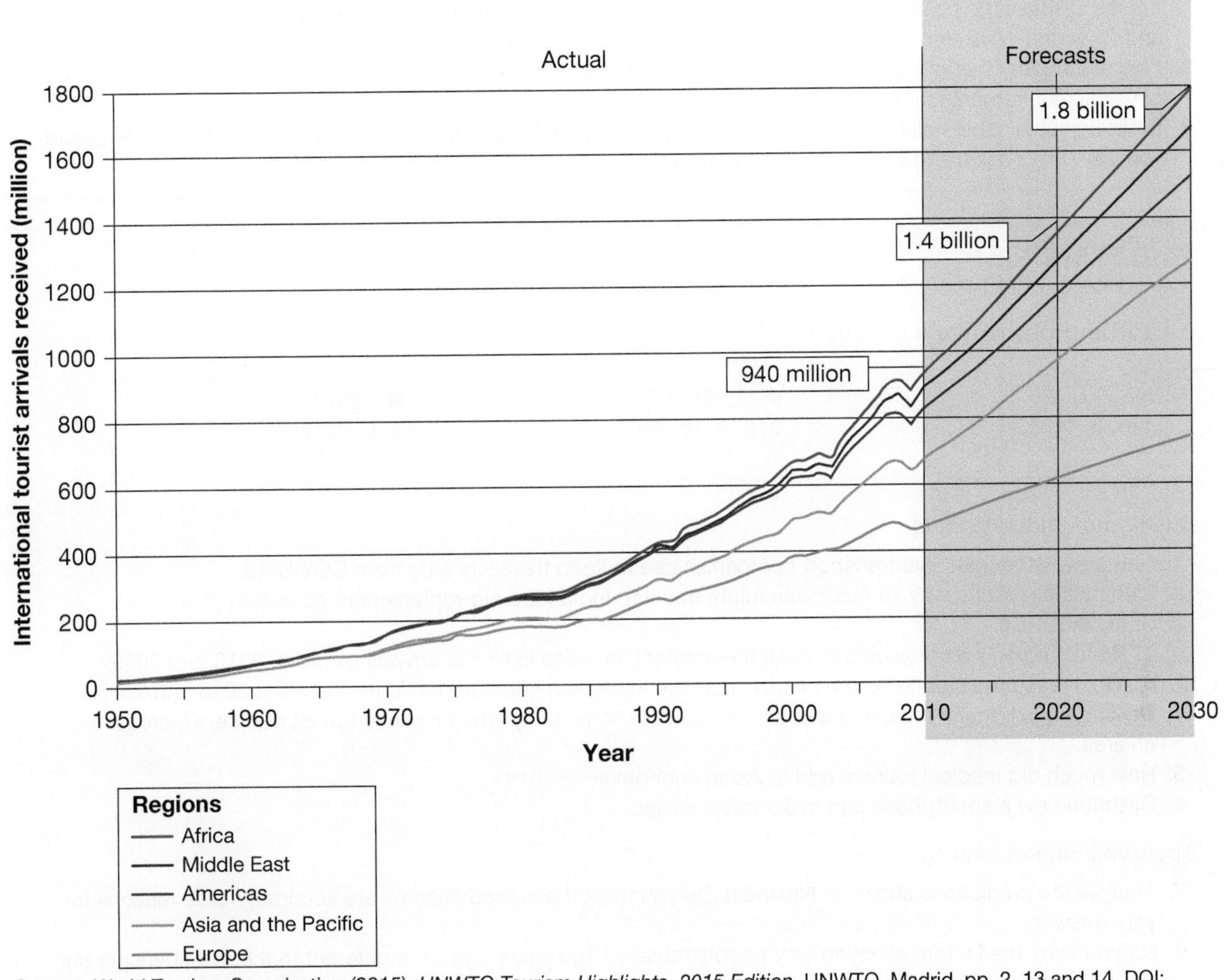

Source: World Tourism Organization (2015), *UNWTO Tourism Highlights, 2015 Edition,* UNWTO, Madrid, pp. 2, 13 and 14, DOI: https://doi.org/10.18111/9789284416899

The true challenge for tourism in the future is to ensure that:

- money remains in the local economy rather than in the hands of developers, and is used to improve local services, not just tourist services
- the need of First Nations peoples to live on the land is balanced with tourist development
- tourist numbers are controlled, to ensure that the environment is not damaged.

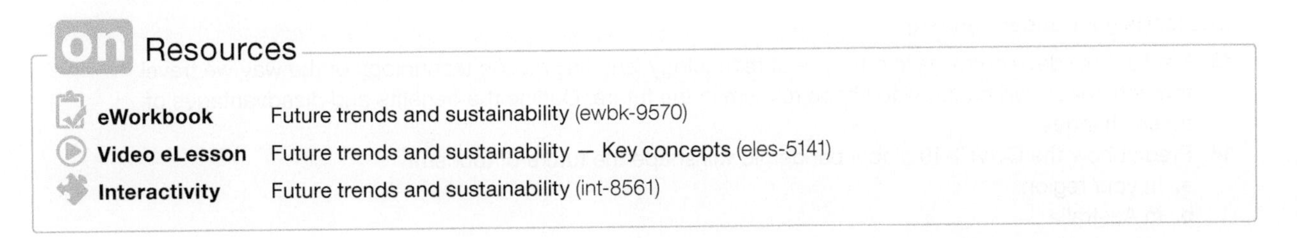

17.10 ACTIVITIES

1. Do you think the COVID-19 pandemic will have a significant impact on how and where tourists travel in the future? Discuss the possibilities for the industry as a class; consider what might happen in 2 years, 10 years and 20 years. What trends do you predict? How will tourist behaviour change, if at all?
2. Find a company that offers space flight holidays or 'out of atmosphere' experiences. What are the costs and physical requirements for potential passengers?
3. Devise a sustainable holiday that has no negative impact on the environment or the communities in the places you visit. Is it possible to travel with 'zero impact'?

17.10 EXERCISE

Learning pathways

■ LEVEL 1	■ LEVEL 2	■ LEVEL 3
1, 5, 8, 11, 13	2, 3, 9, 10, 14	4, 6, 7, 12, 15

Check your understanding

1. Name two strategies that transport companies use to keep travellers safe from COVID-19.
2. Identify the reasons why an Australian might travel to India for a hip-replacement operation.
3. Study **FIGURE 6**.
   a. Which region was projected to have the greatest increase in tourist arrivals between 2010 and 2030?
   b. What was predicted to be the relative increase in tourism numbers for Africa between 2010 and 2030?
4. Describe one benefit and one drawback to the local community when wealthy tourist numbers increase in an area.
5. How much did medical tourism add to Asian economies in 2019?
6. Describe how a smartphone can make travel easier.

Apply your understanding

7. Analyse the predictions shown in **FIGURE 6**. Do you think these predictions were accurate? Give reasons for your answer.
8. Do you think the factors affecting why people travelled 100 years ago were different to those influencing our travel decisions today? Explain your answer giving examples to support your view.
9. Explain how environmental awareness and sustainability are shaping the future of tourism.
10. Analyse how an unexpected decline in the tourism industry in Kenya might affect the people working in the at industry.
11. Evaluate whether space tourism is an environmentally sustainable activity.
12. Consider the process of carbon offsetting.
    a. Explain how the process works.
    b. Evaluate the benefits of and problems associated with carbon offsetting; is this a good long-term strategy for creating sustainable travel? Explain.

Challenge your understanding

13. Predict how developments in one type of technology (such as mobile technology or the way we travel through space) will continue to shape tourism in the future. Outline the benefits and disadvantages of these changes.
14. Predict how the COVID-19 global pandemic will shape the future of tourism:
    a. in your region
    b. in Australia
    c. globally.
15. Predict whether space travel will become a common and affordable tourism option in the future.

To answer questions online and to receive **immediate feedback** and **sample responses** for every question, go to your learnON title at www.jacplus.com.au.

17.11 Investigating topographic maps — Nature-driven tourism at Victoria Falls

LEARNING INTENTION

By the end of this subtopic, you will be able to identify features on a topographic map and examine their impacts on tourism at a specific location.

17.11.1 Tourist mecca

Victoria Falls, or Mosi-oa-Tunya, has been recognised as a World Heritage site due to its spectacular waterfalls. The falls attract several hundred thousand visitors from around the world each year, which has led to the development of numerous tourist businesses to cater to them.

FIGURE 1 Oblique aerial photo of Victoria Falls

UNESCO has deemed Victoria Falls to be a site of exceptional natural beauty and geological importance. The main falls, located on the Zimbabwe side of the border with Zambia, drop in excess of 100 metres. Due to the outstanding universal value posed by this collection of waterfalls, a significant tourism industry has sprung up on both sides of the border. Victoria Falls cover an area of 6860 hectares. The region is used for a variety of tourist activities including rafting, helicopter flights, walking with lions, abseiling and bungee jumping, bush walks, mountain biking, jet boats, horseback and elephant-back safaris, as well as fly fishing.

FIGURE 2 Topographic map extract of Victoria Falls, Zambia and Zimbabwe, 2021

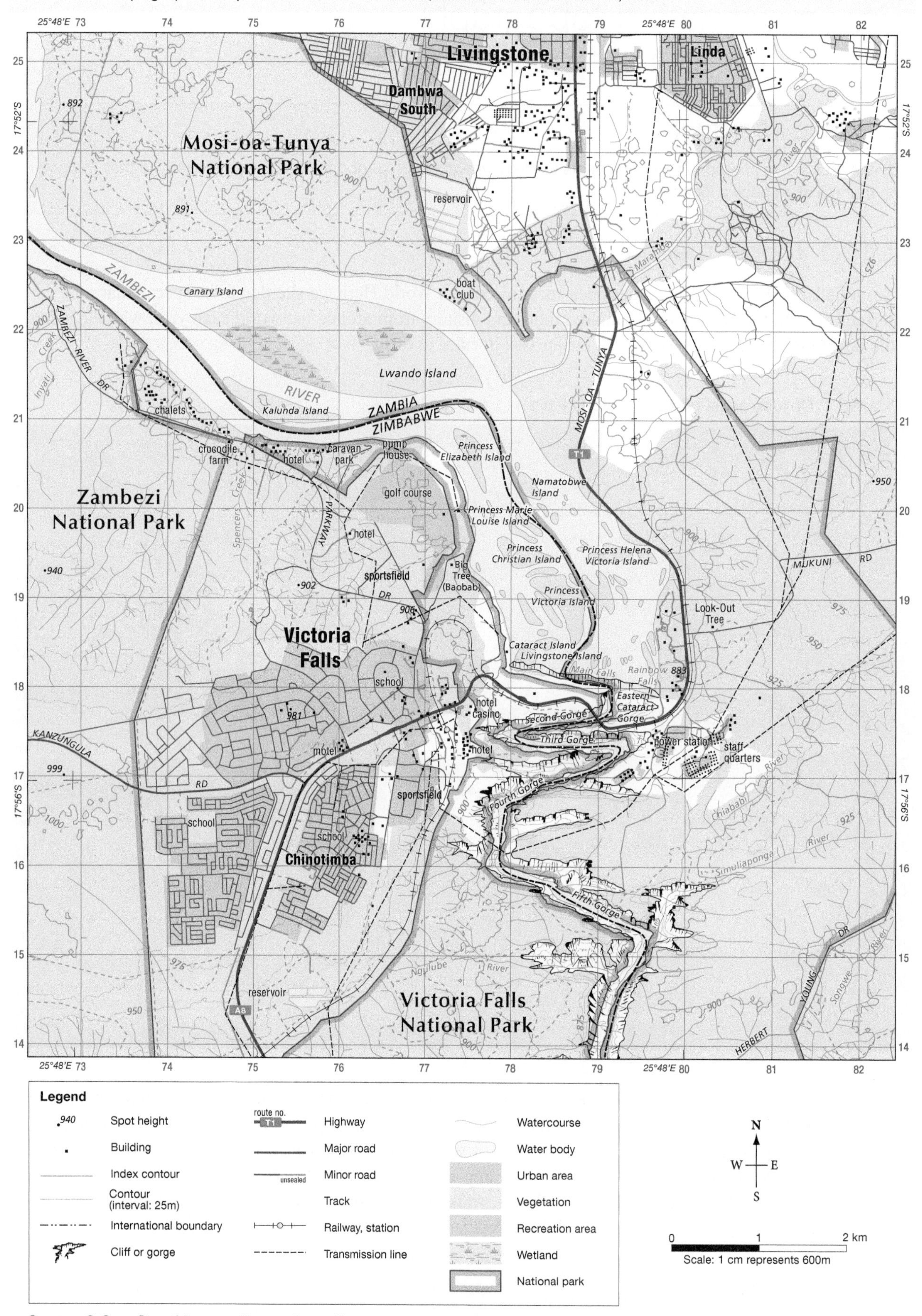

Source: © OpenStreetMap contributors, https://openstreetmap.org, Data is available under the Open Database Licence, https://opendatacommons.org/licenses/odbl/; Spatial Vision.

on Resources

eWorkbook	Investigating topographic maps — Nature-driven tourism at Victoria Falls (ewbk-9574)
Digital document	Topographic map of Victoria Falls, Zambia and Zimbabwe (doc-36261)
Video eLesson	Investigating topographic maps — Nature-driven tourism at Victoria Falls — Key concepts (eles-5142)
Interactivity	Investigating topographic maps — Nature-driven tourism at Victoria Falls (int-8562)
Google Earth	Victoria Falls (gogl-0143)

17.11 EXERCISE

Learning pathways

■ LEVEL 1	■ LEVEL 2	■ LEVEL 3
1, 3, 5, 7, 11	2, 4, 6, 12, 14	8, 9, 10, 13, 15

Check your understanding

1. Describe the main vegetation types on the map in **FIGURE 2**.
2. Which of the following physical features forms a natural boundary between Zambia and Zimbabwe?
   a. A gorge
   b. A perennial lake
   c. A dam
3. What physical feature can be found at AR7415?
4. Give the area reference for Canary Island.
5. What is the contour interval of the map?
6. In what direction is Main Falls from Lwando Island?
7. What is found at GR757183?
8. Calculate the distance by road from the boat club to Big Tree.
9. How do people staying in Livingstone, Zambia, get to Victoria Falls?
10. Refer to **FIGURE 1**.
    a. In what direction would the photographer have been facing when they took the photo?
    b. What natural features would attract visitors to Victoria Falls?
11. List as many human features on the map as you can that have been built because Victoria Falls is a tourist destination.

Apply your understanding

12. What would be some of the main impacts of the tourist facilities, established in and around Victoria Falls, on the environment and on the people who live in this region?
13. Evaluate the impacts (positive and negative) of tourism to this area.
14. How might Victoria Falls allow for the interconnection of people from different cultures?
15. Suggest how the COVID-19 pandemic might have affected tourism in this region.

FIGURE 3 Victoria Falls

To answer questions online and to receive **immediate feedback** and **sample responses** for every question, go to your learnON title at www.jacplus.com.au.

17.12 Thinking Big research project — Design a seven-day cruise adventure

online only

The content in this subtopic is a summary of what you will find in the online resource.

Scenario

Mystic Cruises is about to add a new cruise ship to its fleet. As a member of the company's cruise-development team, you need to design a seven-day cruise, including exotic ports of call and shore excursions that allow cruise guests to take in the sites and cultures of the places they visit. Remember, though, your tour must be COVID-safe. Consider where to go, but also the size of the ship, the number of passengers and the strategies you could put in place on board to keep everyone well.

If your plan is accepted, you will also have the honour of naming the ship!

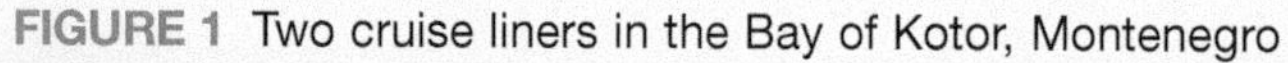

FIGURE 1 Two cruise liners in the Bay of Kotor, Montenegro

Task

Design a seven-day cruise. You can choose where in the world the cruise ship will operate, and at which ports it will dock (choose at least four). Go to your Resources tab to access the resources you need to complete this research project.

Resources

ProjectsPLUS Thinking Big research project — Design a seven-day cruise adventure (pro-0194)

17.13 Review

17.13.1 Key knowledge summary

17.2 Global tourism

- Tourism is the temporary movement of people away from the places where they normally live and work for business, leisure or cultural purposes.
- Prior to the COVID-19 pandemic, global tourism was expected to increase by 3.3 per cent per year between 2010 and 2030, to reach 1.8 billion tourists by 2030.
- The unprecedented impact of the global COVID-19 pandemic has had a profound impact on tourism operations and projections for the future of tourism.

17.3 Travel patterns and trends

- Travel has become faster, easier, cheaper and safer, and economic growth has meant that many people now have more money to spend and can afford travel.
- Top destination countries include the United States, France, Spain, China, the United Kingdom and Germany.
- The world's highest tourist spenders are those from China, the United States, Germany, the United Kingdom and the Russian Federation.

17.5 Sport, recreation and leisure tourism

- Sport tourism triggers the construction of new stadiums, expansion of transport networks, improvements to airport facilities and the clean-up of cities.
- Hosting the Olympic Games can generate tourism and increase international exposure.
- The COVID-19 global pandemic has resulted in the delay of the 2020 Olympic Games in Tokyo. This will result in considerable additional costs for the hosts.

17.6 Cultural tourism

- Cultural events such as Thanksgiving and Chinese New Year are associated with an increase in tourism, particularly domestic tourism within the home country.
- In 2019, 55.3 million people travelled within the United States for Thanksgiving.
- Chinese New Year festivities begin on the first day of the first month in the traditional Chinese calendar and last for 15 days. In 2020, around 600 million people travelled around China during the holiday period, with more travelling internationally.

17.7 Impacts of tourism

- Negative impacts of tourism include the destruction of communities and natural environments to make way for tourist developments; increased crime rates; increased rubbish; damage to sacred sites; pressure on resources such as water; and commercialisation of culture.
- Positive impacts of tourism include promoting cultural awareness; improving access to health care; increased employment opportunities; increased interest in the natural environment; improvements to infrastructure (roads, water supply, sewage); and preserving indigenous cultures.

17.9 Sustainability and ecotourism

- Ecotourism manages tourism in a sustainable way through strategies such as education programs, limiting numbers of tourists and controlling the type and location of activities.
- Ecotourism Australia (EA) aims to promote ecotourism throughout Australia.
- Ecotourism is the fastest growing sector in the tourism industry.

17.10 Future trends and sustainability

- Factors shaping the future of tourism include sustainability, technology, demand for medical tourism and COVID-19.
- Increased concerns for safety, demand for higher hygiene standards and new government restrictions as a result of the COVID-19 global pandemic are encouraging tourism operators to implement more thorough cleaning, social distancing (for example, extra spaces between seats) and contactless payments.
- People recognise the benefits that responsible and sustainable tourism can bring to destinations, in terms of employment and the preservation of cultural and natural heritage. Sustainability is driving the popularity of eco-travel, green hotels, eco-lodges, nature-based wellness vacations, volunteering and the phasing out of single-use plastics.

17.13.2 Key terms

biosecurity actions taken to stop specific organisms, animals or plants entering the country

carbon offsetting calculating the amount of carbon released during an activity and investing in strategies to pay off or cancel out the carbon

ecotourism tourism that interprets the natural and cultural environment for visitors, and manages the environment in a way that is ecologically sustainable

gross domestic product (GDP) the value of all the goods and services produced within a country in a given period of time. It is often used as an indicator of a country's wealth.

hard sport tourism tourism in which someone travels to either actively participate in or watch a competitive sport as the main reason for their travel

mandatory quarantine enforced restriction of movement and separation from the rest of the community, monitored by officials

mature-aged describes individuals aged over 55

new Moon the phase of the Moon when it is closest to the Sun and is not normally visible

soft sport tourism tourism in which someone participates in recreational and leisure activities, such as skiing, fishing and hiking, as an incidental part of their travel

tourism moving away from where you normally live/work temporarily (for between 24 hours and 1 year)

tertiary study the educational level following high school, e.g. university

winter solstice the shortest day of the year, when the Sun reaches its lowest point in relation to the equator

17.13.3 Reflection

Complete the following to reflect on your learning.

Revisit the inquiry question posed in the Overview:

How do the interconnections formed by tourism affect people, places and environments?

1. Now that you have completed this topic, what is your view of the question? Discuss with a partner. Has your learning in this topic changed your view? If so, how?
2. Write a paragraph in response to the inquiry question outlining your views.

FIGURE 1 What impact will your travels have on the places you visit?

Subtopic	Success criteria			
17.2	I can describe what influences people's travel connections with different places.			
17.3	I can analyse patterns and trends in people's travel, recreational, cultural and leisure activities.			
17.4	I can represent geographical data in a doughnut chart.			
17.5	I can identify influences on and analyse the effect of people's recreational and leisure connections with different places.			
17.6	I can describe cultural influences on people's travel connections with different places.			
17.7	I can assess the impact of people's travel and leisure activities on the future of places.			
17.8	I can develop and conduct a survey.			
17.9	I can identify and evaluate strategies to make tourism more environmentally and culturally sustainable.			
17.10	I can make informed predictions about how places and people might be affected by tourism in the future.			
17.11	I can identify features on a topographic map and examine their impacts on tourism at a specific location.			

on Resources

eWorkbook Chapter 17 Student learning matrix (ewbk-8500)
Chapter 17 Reflection (ewbk-8501)
Chapter 17 Extended writing task (ewbk-8502)

Interactivity Chapter 17 Crossword (int-8563)

FIGURE 2 The Rialto Bridge, Venice, Italy

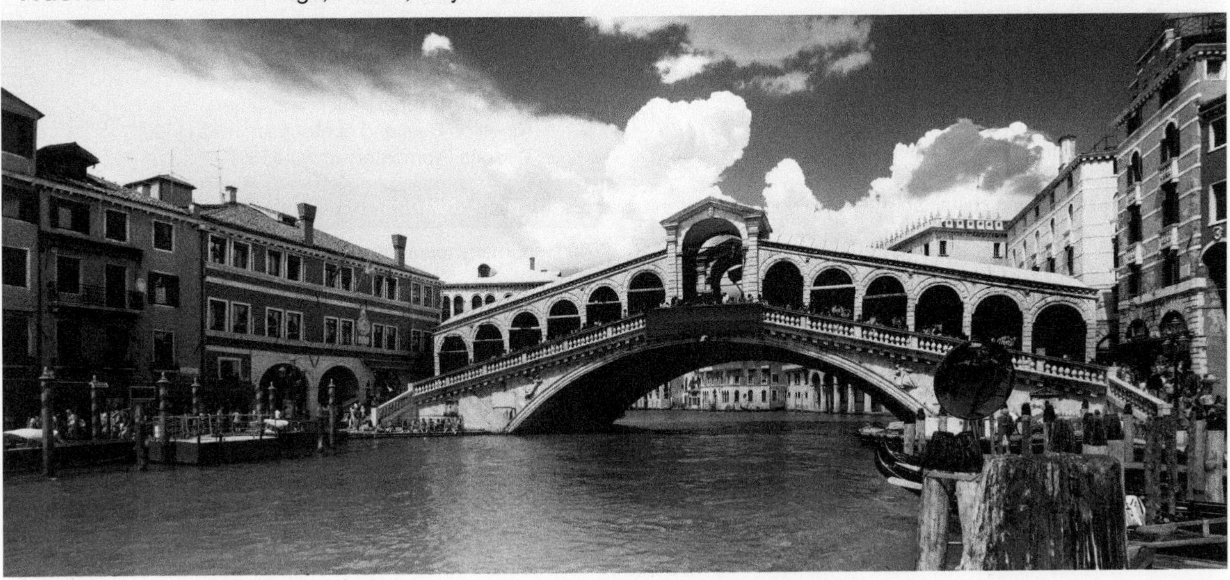

ONLINE RESOURCES

on Resources

Below is a full list of **rich resources** available online for this topic. These resources are designed to bring ideas to life, to promote deep and lasting learning and to support the different learning needs of each individual.

eWorkbook

- 17.1 Chapter 17 eWorkbook (ewbk-7997) ☐
- 17.2 Global tourism (ewbk-9534) ☐
- 17.3 Travel patterns and trends (ewbk-9538) ☐
- 17.4 SkillBuilder — Constructing and describing a doughnut chart (ewbk-9542) ☐
- 17.5 Sport, recreation and leisure tourism (ewbk-9546) ☐
- 17.6 Cultural tourism (ewbk-9550) ☐
- 17.7 Impacts of tourism (ewbk-9554) ☐
- 17.8 SkillBuilder — Creating a survey (ewbk-9558) ☐
- 17.9 Sustainability and ecotourism (ewbk-9566) ☐
- 17.10 Future trends and sustainability (ewbk-9570) ☐
- 17.11 Investigating topographic maps — Nature-driven tourism at Victoria Falls (ewbk-9574) ☐
- 17.13 Chapter 17 Student learning matrix (ewbk-8500) ☐
  - Chapter 17 Reflection (ewbk-8501) ☐
  - Chapter 17 Extended writing task (ewbk-8502) ☐

Sample responses

- 17.1 Chapter 17 Sample responses (sar-0149) ☐

Digital documents

- 17.3 Blank world map template (doc-36260) ☐
- 17.11 Topographic map of Victoria Falls, Zambia and Zimbabwe (doc-36261) ☐

Video eLessons

- 17.1 Moving around (eles-1723) ☐
  - Personal connections — Photo essay (eles-5134) ☐
- 17.2 Global tourism — Key concepts (eles-5135) ☐
- 17.3 Travel patterns and trends — Key concepts (eles-5136) ☐
- 17.4 SkillBuilder — Constructing and describing a doughnut chart (eles-1738) ☐
- 17.5 Sport, recreation and leisure tourism — Key concepts (eles-5137) ☐
- 17.6 Cultural tourism — Key concepts (eles-5138) ☐
- 17.7 Impacts of tourism — Key concepts (eles-5139) ☐
- 17.8 SkillBuilder — Creating a survey (eles-1764) ☐
- 17.9 Sustainability and ecotourism — Key concepts (eles-5140) ☐
- 17.10 Future trends and sustainability — Key concepts (eles-5141) ☐
- 17.11 Investigating topographic maps — Nature-driven tourism at Victoria Falls — Key concepts (eles-5142) ☐

Interactivities

- 17.2 Types of tourist destinations (int-5584) ☐
  - Global tourism (int-8557) ☐
- 17.3 Travel patterns and trends (int-8558) ☐
- 17.4 SkillBuilder — Constructing and describing a doughnut chart (int-3356) ☐
- 17.5 Are the Olympics worth gold? (int-3337) ☐
- 17.6 Cultural tourism (int-8560) ☐
- 17.7 Tourism trauma (int-3335) ☐
- 17.8 SkillBuilder — Creating a survey (int-3382) ☐
- 17.9 Sustainable sightseeing (int-3336) ☐
  - Anatomy of an ideal ecotourism resort (int-5574) ☐
- 17.10 Future trends and sustainability (int-8561) ☐
- 17.11 Investigating topographic maps — Nature-driven tourism at Victoria Falls (int-8562) ☐
- 17.13 Chapter 17 Crossword (int-8563) ☐

Fieldwork

- 17.5 Coding an article (fdw-0023) ☐

ProjectsPLUS

- 17.12 Thinking Big research project — Design a seven-day cruise adventure (pro-0194) ☐

Weblinks

- 17.2 ABS COVID-19 (web-6062) ☐
- 17.6 Thanksgiving (web-3324) ☐

Google Earth

- 17.11 Victoria Falls (gogl-0143) ☐

myWorld Atlas

- 17.2 World tourism (mwa-4495) ☐
  - Victoria Falls (mwa-4497) ☐
- 17.5 Australia's links with the world > Sport (mwa-4545) ☐
  - FIFA World Cup (mwa-7347) ☐
- 17.9 Kakadu National Park (mwa-4498) ☐
  - Ningaloo Reef and ecotourism (mwa-4499) ☐
  - Wilsons Promontory (mwa-4530) ☐

Teacher resources

There are many resources available exclusively for teachers online.

18 Transport and technology

INQUIRY SEQUENCE

To access a pre-test with **immediate feedback** and **sample responses** to every question in this chapter, select your learnON format at www.jacplus.com.au.

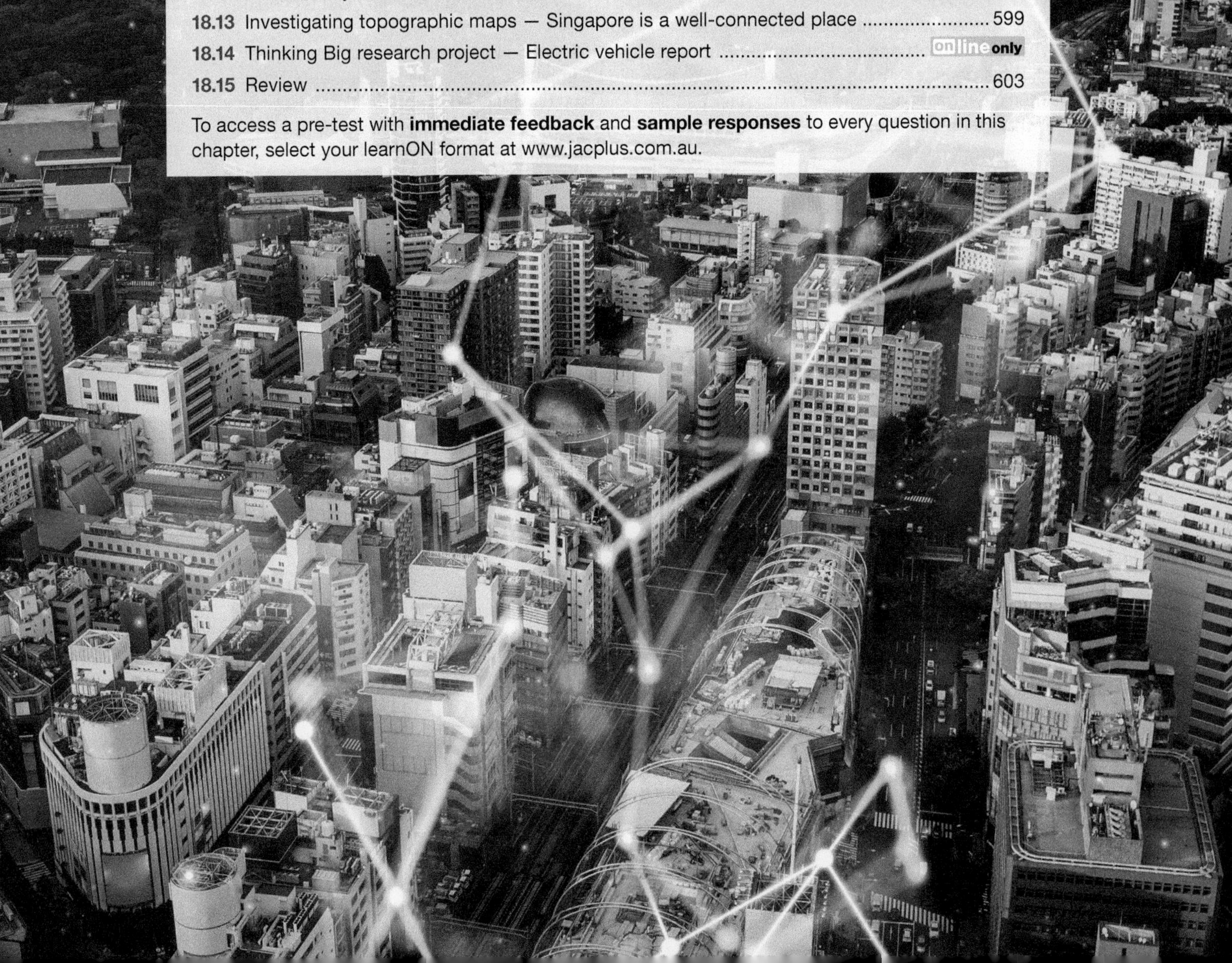

18.1 Overview

Numerous **videos** and **interactivities** are embedded just where you need them, at the point of learning, in your learnON title at www.jacplus.com.au. They will help you to learn the content and concepts covered in this topic.

What are the different ways that technology connects people and places?

18.1.1 Introduction

Developments in transport and information and communication technologies have allowed people and places to become more interconnected. Advancements in transport technologies have resulted in a shrinking world through vastly reduced travel times and costs. Our ability to communicate with each other has been transformed by innovations in information and communication technology. These developments have allowed people, goods and services to move and communicate easily across great distances. While the benefits of these developments are enormous, these benefits are not equally shared.

STARTER QUESTIONS

1. List the different ways you travel to a variety of places both near and far. What transport do you use for the different distances?
2. Identify and describe the different ways you interact and communicate with different people in your life.
3. How did the COVID-19 pandemic affect the way people communicated and travelled? Did technology help you stay connected with friends and family around Australia or in other countries?
4. Watch the **Transport and technology — Photo essay** in the Video eLesson tab in your Resources panel. What does technology allow you to do that would be impossible otherwise? Think about how you communicate, travel and connect with other people. How has this changed in your lifetime? How do you hope it will change for you in the future; for example, might it change the kinds of jobs you could do?

FIGURE 1 The COVID-19 pandemic changed the way people in some places travelled and connected with other people.

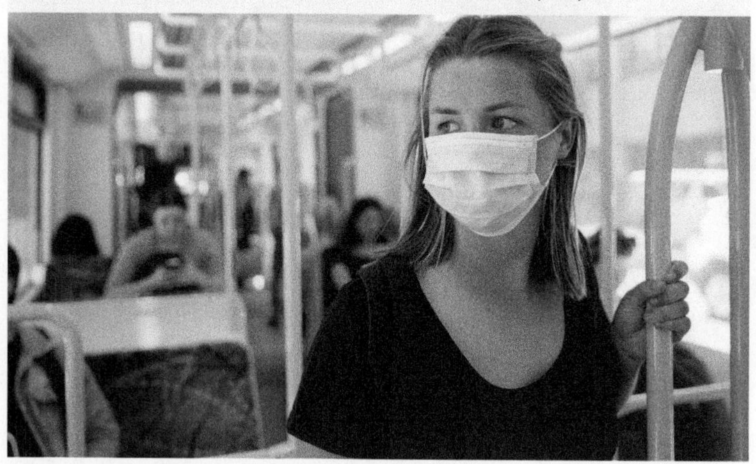

on Resources

eWorkbook	Chapter 18 eWorkbook (ewbk-7998)
Video eLessons	Making connections (eles-1722) Transport and technology — Photo essay (eles-5143)

18.2 Connecting people with places

LEARNING INTENTION

By the end of this subtopic, you will be able to identify the major transport technologies that helped connect people and places, describe how transport technologies developed, and outline future innovations in transport technologies.

18.2.1 Transport changes

Improvements in transport technologies and the **infrastructure** built for them over the past 200 years have helped people and places connect. The inventions of the steam engine and the internal combustion engine allowed ships to travel faster and led to the development of trains, planes and automobiles.

When Europeans on board the ships of the First Fleet came to Australia in the late eighteenth century, the trip took over eight months (252 days). Today, ships can leave England and arrive in Sydney in 40 days, and a plane trip can take just under 24 hours.

18.2.2 Trains

Trains were developed in England in the early 1800s by using the steam engine invented by Bolton and Watt. Railways not only changed travel, but also changed the speed at which goods, people and information were transported from one end of a country to the other — this could now take hours rather than days.

These new railway networks connected people to distant places within their own country. Cities and their outer areas became more connected, as people could live out in the countryside but travel into the city each day to work.

Trains moved from steam power to diesel and electric power, and are still an important transport technology. High-speed trains, such as the Japanese Shinkansen, travel at speeds of up to 320 km/h. In 2020, there were 32 868 km of railways in Australia, with 3446 km of electric railways for public transport. In Australian capital cities with rail networks, a combined 11.46 billion passenger kilometres were travelled in 2019–20.

18.2.3 Planes

Airliners began to be developed in the decade after the Wright brothers made the world's first flight in 1903. But air travel was, for some time, a luxury afforded only by the wealthy; a person travelling on one of the earliest commercial flights on a return journey between London and Sydney would need to spend an amount equivalent to two and a half years' average income.

FIGURE 1 The Wright Brothers' plane, in which they made the first powered controlled flight, 17 December 1903

In 2020, based on average weekly income, it took just 11 days' salary for the same return journey. The drop in price is due to the creation of wide-body planes (planes with two or more aisles), such as Boeing's 747 and Douglas' DC-10 in the late 1960s. The ability to carry more people, along with improvements in other technologies, allowed long-distance flights to become more affordable. Thus, more people are able to travel to a variety of places around the globe.

infrastructure the features that are built by humans to allow other activities to occur; for example, roads, electricity networks and water supply pipes

airliner an aircraft that transports people and goods

TABLE 1 Number of flights and passenger travellers, domestic and international, in Australia

Financial year	Flights available		Passengers travelled	
	Domestic	International	Domestic	International
1977–78	374 866	24 082	11 958 560	3 036 960
1978–79	397 242	20 764	12 587 854	3 506 753
1979–80	415 879	20 478	13 540 872	4 019 316
2017–18	634 994	201 374	60 779 500	40 619 162
2018–19	634 061	205 814	60 981 780	42 121 004
2019–20	491 897	159 671	45 241 761	30 732 112

Source: Australian Infrastructure Statistics—Yearbook 2020 © Commonwealth of Australia

FIGURE 2 Wide-body planes made flying cheaper and more accessible.

18.2.4 Automobiles

Automobiles have allowed many people to connect to other places within a country. Trucks can transport goods door-to-door, and service providers and people can travel to multiple locations easily within a day. Automobiles are used in public transport, moving many people through buses, taxis and ride-sharing services to places that trains cannot access.

Cars were made more affordable through **mass production**. Henry Ford's Model T was the first mass-produced car, in 1908 (see **FIGURE 3**), but it was not until the 1950s that cars truly became accessible to the average person. Mass production made cars cheaper, but the strong economies of many countries after World War I (including Australia) also meant that more people could afford a family car.

mass production producing a large volume of one product by automating or standardising the process

Improvements in car safety technology through seatbelts, airbags and anti-lock braking systems (ABS) have also made cars an appealing form of transport. In 2020, for every 1000 people in Australia there were 778 vehicles registered, compared to the United Kingdom's rate of 580 per 1000 people. This demonstrates the importance of automobiles as a form of transport technology in Australia. Extensive road networks were developed to accommodate the increase in car ownership and movement of goods by trucks – especially the shipping of goods to and from remote and rural areas Australia has more than 1 027 545 kilometres of roads across the country connecting people and places (**FIGURE 4**). These road networks and developments in road safety and efficiency helped to improve bus services, which provide key public transport links in both urban and rural parts of Australia.

FIGURE 3 An advertisement for a Model T Ford. In order to speed up the production process, Henry Ford started to produce only black cars, famously saying, 'Any customer can have a car painted any colour that he wants so long as it is black'.

18.2.5 The future

Emissions created by the burning of fossil fuels in engines, safety concerns, and the high cost of congested roads and infrastructure have led to new innovations in transport technologies. Many car manufacturers are now producing electric or hybrid cars as a way of reducing engine emissions, but battery technologies and the infrastructure for charging batteries are not fully developed; however, hybrids are the fastest growing segment of the car industry at present. Despite imperfections or inefficiencies, many customers are choosing to buy hybrids and electric vehicles for their lower environmental impact.

Driverless transport, which uses artificial intelligence (AI) technology, is already being used in train and bus networks around the world. It is estimated that 96 per cent of accidents are caused by human error, and it is hoped that using AI in driverless transport will reduce accidents, ease congestion and emissions, and provide more mobility options for elderly and young people, and people living with a disability.

Another emerging transport technology. is the Hyperloop system (see **FIGURE 5**). In this system, which uses magnetic levitation, a pressurised electric carriage 'floats' and is propelled at great speed inside a low-pressure tube (which reduces air friction). Because of its fast travel times and potential low cost to passengers, it is believed that it would be a better alternative to domestic air and rail travel. If Hyperloop systems are successful, they may help to alleviate housing problems in cities by allowing people to live further from where they work or go to school.

FIGURE 4 A topological map showing Australia's highway system

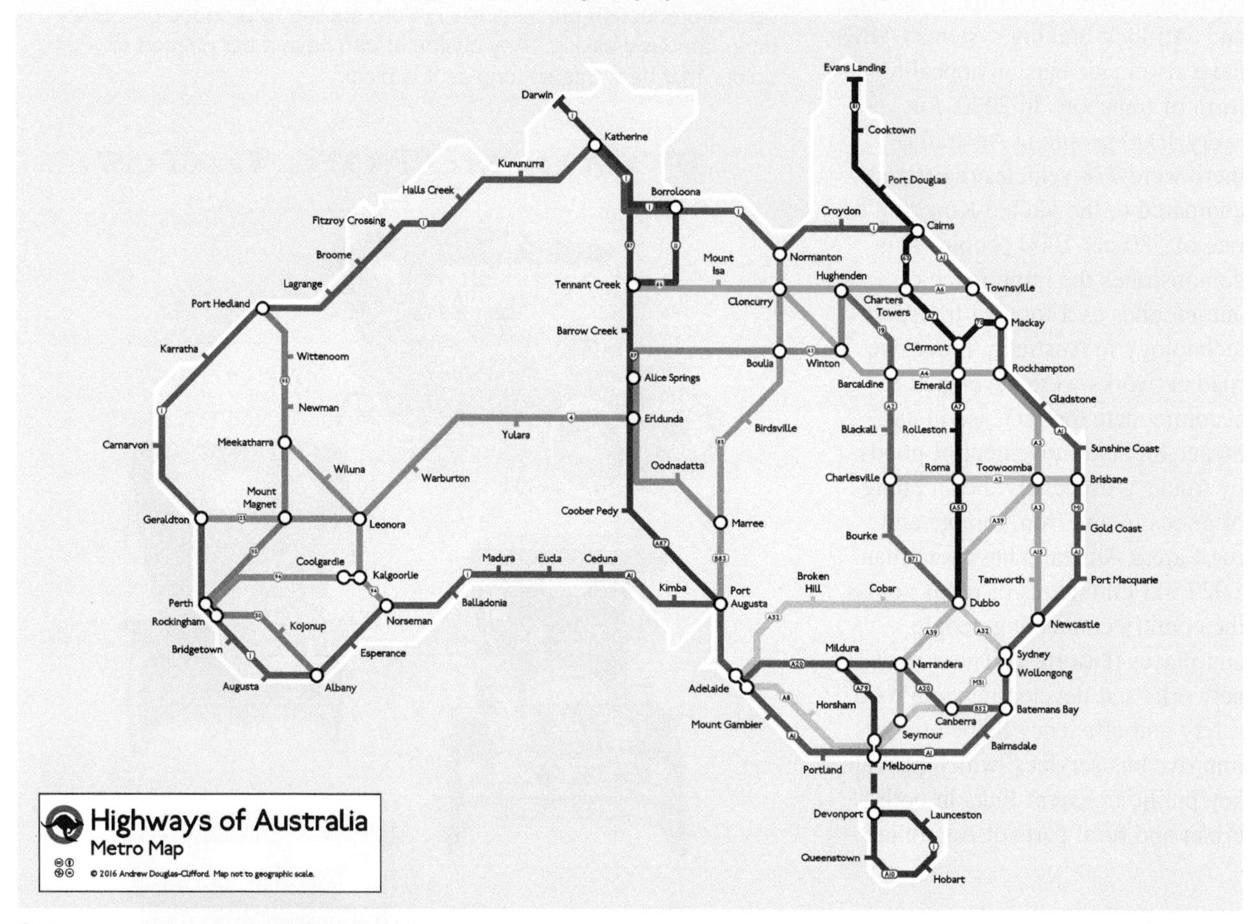

Source: www.andrewdc.co.nz/?s=australia

FIGURE 5 The Virgin Hyperloop One prototype, Dubai Motor Show, 2019

Resources

- **eWorkbook** Connecting people with places (ewbk-9582)
- **Video eLesson** Connecting people with places — Key concepts (eles-5144)
- **Interactivity** Connecting people with places (int-8564)

18.2 ACTIVITIES

1. Investigate the history of trains, planes or automobiles. Create a 1–2 minute multimedia presentation (with visuals and audio), such as a slide show or video, which briefly outlines how your chosen transport technology developed over time.
2. Research Hyperloop technology and write a mini-report that describes how this technology works and where it currently is being developed around the world. Using a map of Australia, predict travel times between Brisbane, Sydney, Canberra, Melbourne and Adelaide if a Hyperloop network was built connecting these cities.

18.2 EXERCISE

Learning pathways

■ LEVEL 1	■ LEVEL 2	■ LEVEL 3
1, 2, 6, 7, 13	3, 5, 8, 9, 14	4, 10, 11, 12, 15

Check your understanding

1. What two inventions led to the development of trains, planes and automobiles?
2. How did the development of trains affect where people worked and lived?
3. What is an airliner?
4. What type of plane made airline travel affordable?
5. List two reasons why cars became more affordable and an attractive form of transport.

Apply your understanding

6. Why does Australia have a large amount of railway lines?
7. Consider the data in **TABLE 1**. What global event affected the flight availability and amount of passengers travelling in 2019–2020?
8. Australia's car registration per 1000 people is higher than the United Kingdom's, even though the UK has a higher population. What geographical spatial factors can you think of that explain this?

Challenge your understanding

9. Evaluate the changes that have been made to automobiles since they were invented. What do you consider the most important change? Give reasons for your decision.
10. Explain why more families in Australia were able to afford cars in the second half of the twentieth century.
11. Explain how a Hyperloop would improve transport in Australia.
12. Wide-bodied planes are able to carry more people, but what disadvantages might they have for passengers? Explain one way that a smaller aircraft might be better for travellers.
13. Suggest two safety improvements that could be made to cars. Explain why your suggestions would make drivers, passengers and/or pedestrians safer.
14. If the Hyperloop was built between Brisbane, Sydney, Canberra, Melbourne and Adelaide, what impact might it have on where people live and go to work and school?
15. Suggest one way that the transport infrastructure where you live might be improved. Explain why your suggestion would improve the ease, cost or efficiency of travel.

To answer questions online and to receive **immediate feedback** and **sample responses** for every question, go to your learnON title at www.jacplus.com.au.

18.3 Public transport

LEARNING INTENTION

By the end of this subtopic, you will be able to identify places that have the best public transport systems, outline who uses public transport in Australia and name technologies that make public transport more accessible.

18.3.1 Public transport

Public transport provides a relatively low-cost way for people to interconnect with places, and can reduce traffic congestion and pollution. It comes in many forms, with the most popular types being trains, buses and light rail (trams). Other forms of transport are also available to the public, such as taxis, ride share services, limousines and motorised rickshaws to name a few. Public transport availability and use is considerably higher in large cities, partly because cities have relatively large populations and better public transport infrastructure.

public transport transport of passengers by group travel systems available for use by the general public, typically managed on a schedule, operated on established routes, and that charges a posted fee for each trip. Public transport modes include city buses, trolleybuses, trams (or light rail), passenger trains, rapid transit (e.g. metro/subway/underground) and ferries.

patron a customer, especially a regular one

FIGURE 2 shows the ten cities with the best public transport systems in the world. Singapore's public transport system is considered the best due to its affordability, convenient electronic contactless ticketing systems, and availability of a variety of public transport options (see **FIGURE 3**). The move to contactless electronic payments (see **FIGURE 4**), using prepaid cards such as the Opal Card in New South Wales, smartphones and smartwatches, adds to the convenience of using public transport. **Patrons** no longer have to wait in queues and risk missing their bus or train.

FIGURE 1 Familiar public transport options

FIGURE 2 Top ten cities with the best public transport systems, 2018

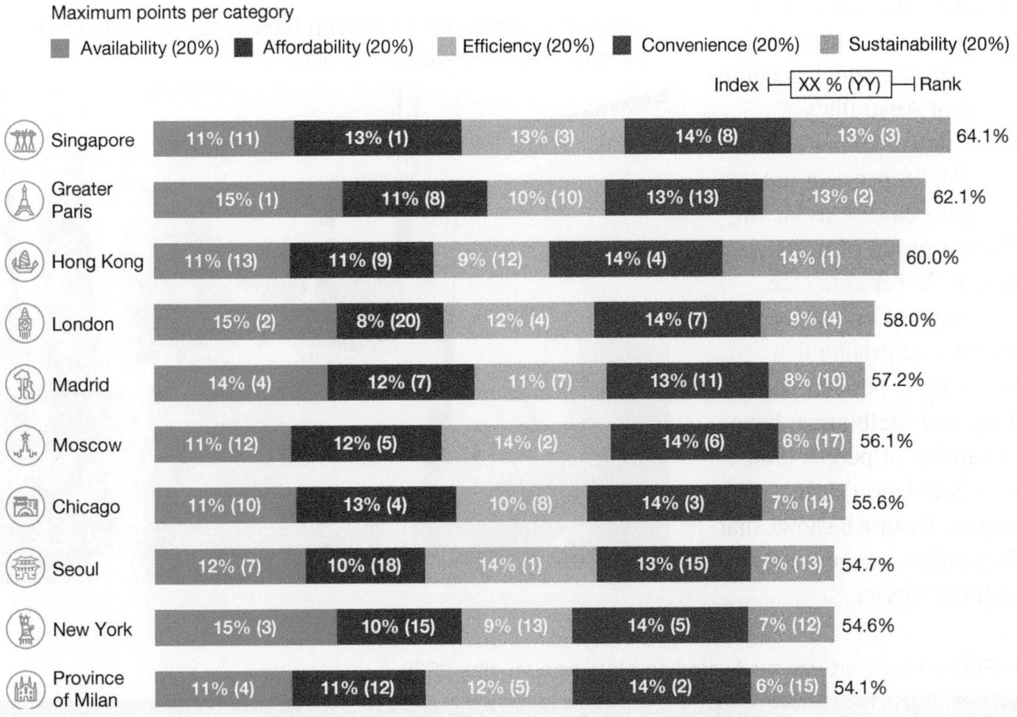

Source: Exhibit from 'Elements of success: Urban transportation systems of 24 global cities', June 2018, McKinsey & Company, www.mckinsey.com.

FIGURE 3 Radar graph showing Singapore's rankings in five domains of effective transport systems

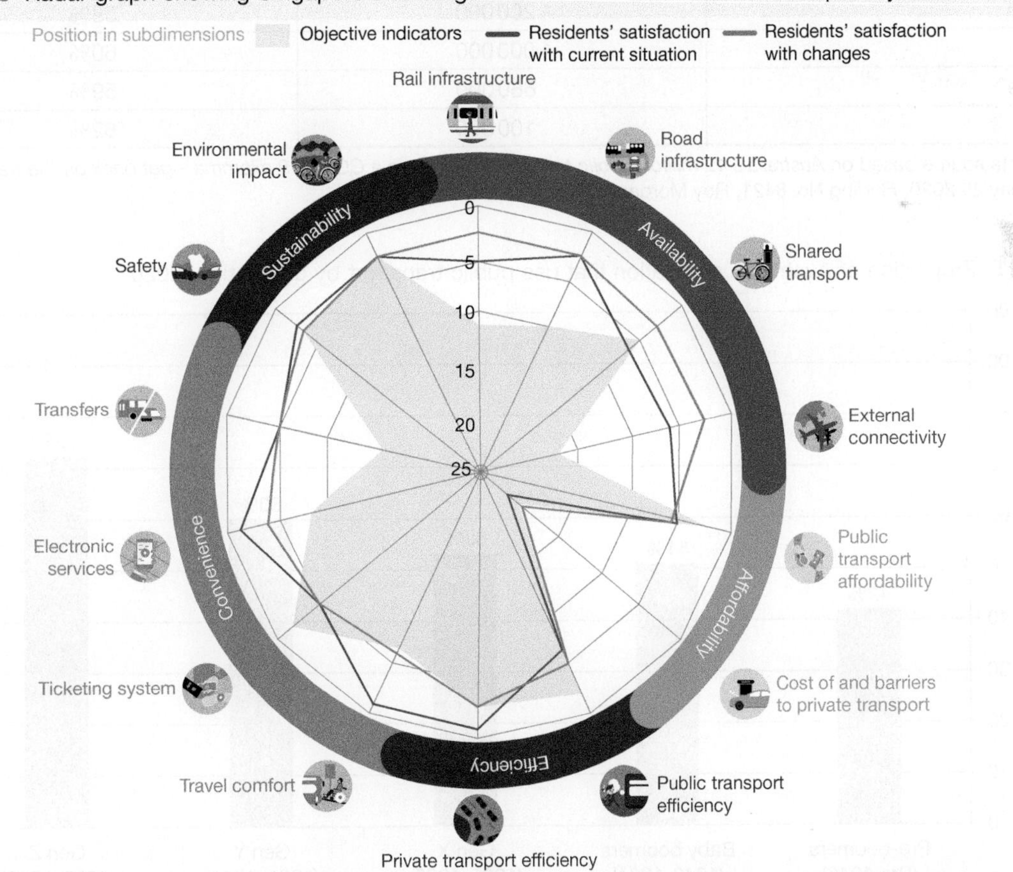

Source: Exhibit from 'Elements of success: Urban transportation systems of 24 global cities', June 2018, McKinsey & Company, www.mckinsey.com.

18.3.2 People who use public transport

On average, more than 12 million people aged 14 and over use public transport in Australia, which accounts for 58 per cent of Australia's population. Of those, 16 per cent use trams (light rail), 38 per cent use buses and 44 per cent use trains. In 2020, the largest group of users was Generation Z at 72 per cent (see **FIGURE 5**). In terms of cities (see **TABLE 1**), it is not surprising that Australia's two most populated cities, Sydney and Melbourne, have the highest number of people (and proportion of population) using public transport. **FIGURE 6** shows that many Sydney-siders are generally satisfied with the service.

FIGURE 4 Contactless ticketing technology is making public transport more convenient to use.

TABLE 1 Public transport use by Australian state capital city, 2020

City	Number of users	Proportion of population
Sydney	3 400 000	76%
Melbourne	3 000 000	71%
Brisbane	1 200 000	59%
Perth	1 000 000	60%
Adelaide	660 000	59%
Hobart	100 000	52%

Source: Data source based on *Australia's 12 million public transport users face a COVID19 dilemma – get back on the train or drive to work?* May 28 2020. Finding No. 8421, Roy Morgan.

FIGURE 5 Proportion of Australian population that use public transport by generation, 2020

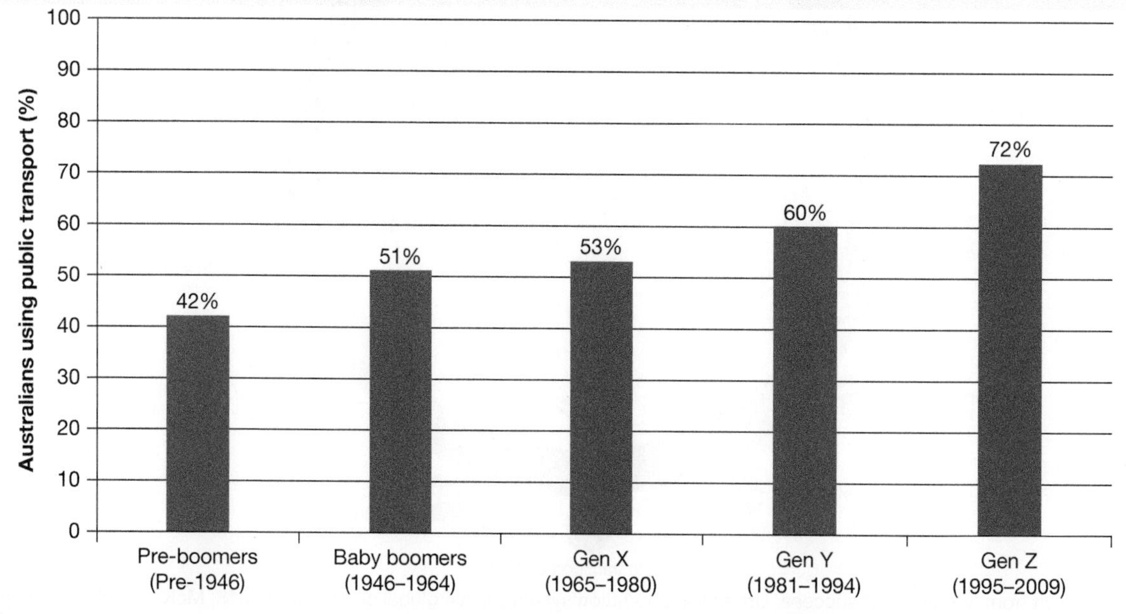

Source: Roy Morgan Single Source, April 2019–March 2020, sample n = 13,208. Base: Australians aged 14+.

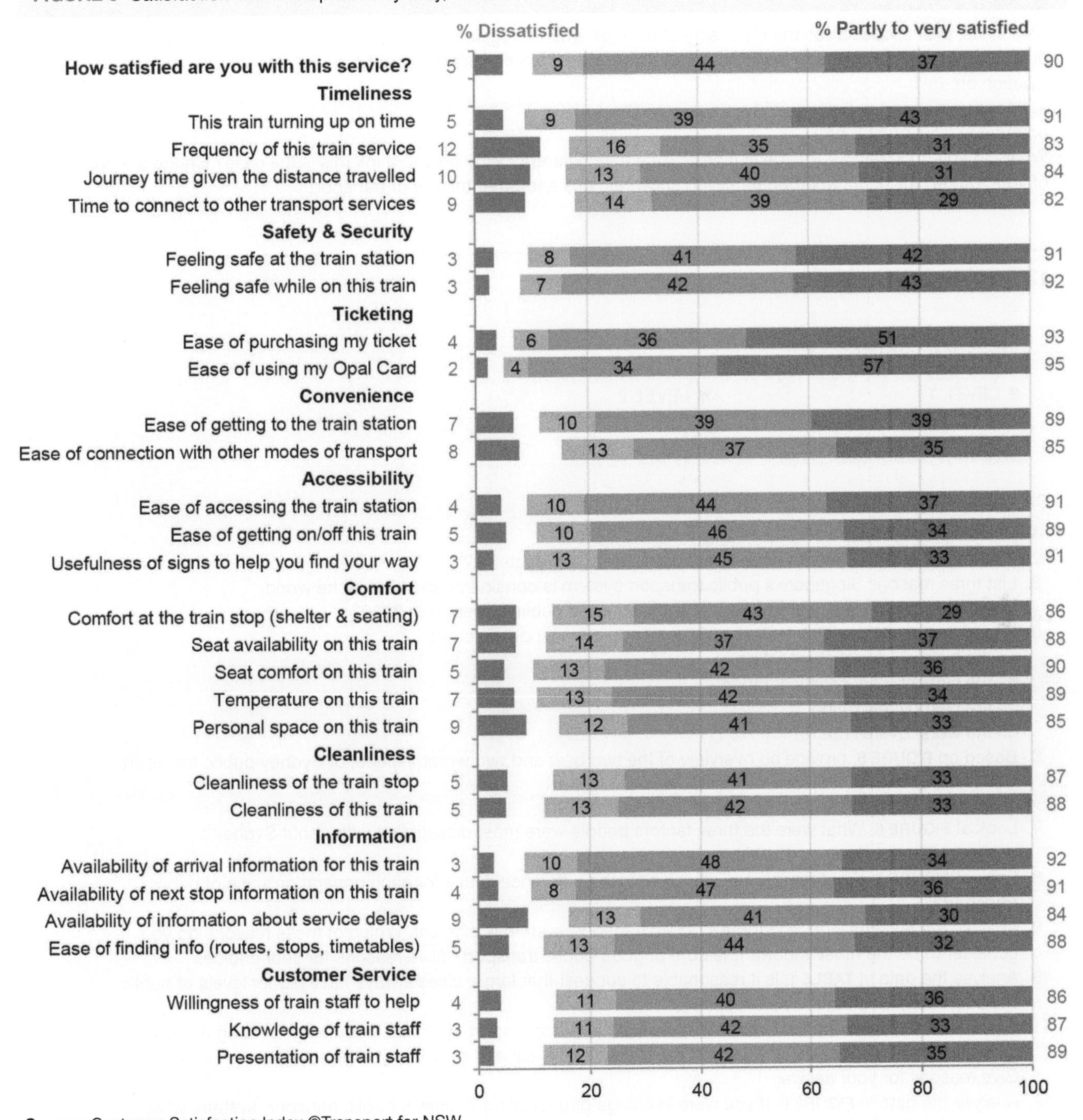

FIGURE 6 Satisfaction with transport in Sydney, 2019

Source: Customer Satisfaction Index ©Transport for NSW

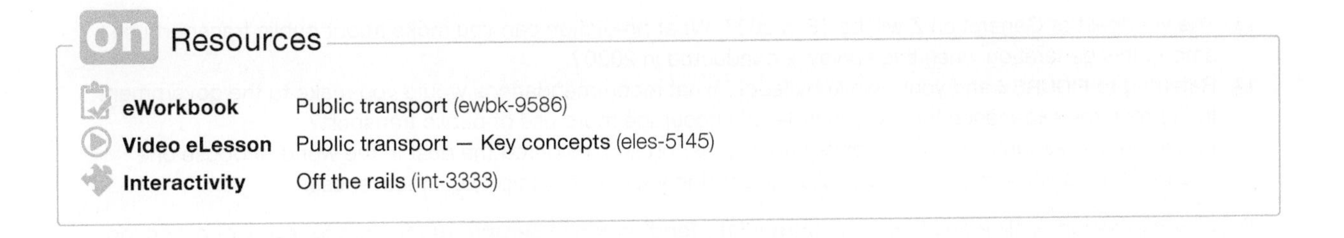

18.3 ACTIVITY

As a class, pick a location on the other side of town or another regional place near you. Using your rail, bus or other public transport provider website, find out how long it would take to travel from your school or home to this location on:

a. Monday morning at 9 am
b. Sunday evening at 6 pm.

What did you notice about the travel times? Were they different? Why do you think this is the case? Create a map of your journey, using an appropriate key to show rail, bus and other modes of transport.

18.3 EXERCISE

Learning pathways

LEVEL 1	LEVEL 2	LEVEL 3
1, 3, 7, 12, 15	4, 5, 8, 9, 13	2, 6, 10, 11, 14

Check your understanding

1. What is public transport?
2. Why do major cities tend to have better public transport systems?
3. List three reasons Singapore's public transport system is considered the best in the world.
4. How many people in Australia aged 14 and over used public transport in 2020?
5. Based on **FIGURE 6**, which feature of Sydney public transport received:
   a. the highest 'dissatisfied' rating
   b. the highest 'very satisfied' rating
   c. the best overall rating
   d. the worst overall rating?
6. Based on **FIGURE 6**, provide an overview of the two best and two worst aspects of Sydney public transport.

Apply your understanding

7. Look at **FIGURE 6**. What were the three factors people were most dissatisfied with about Sydney's public transport system in 2019?
8. Look at **FIGURE 5**. What reason can you give for the difference in use for each generation's use of public transport?
9. Analyse the reasons why Singapore's public transport system is efficient. Which of these reasons do you consider to be the most important feature of good public transport? Give reasons for your choice.
10. Analyse the data in **TABLE 1**. Is it reasonable to suggest that larger cities always have higher levels of public transport use? Use data in your response.
11. Consider the criteria used in **FIGURE 2** to evaluate public transport systems. If you were creating a similar ranking system, would you rate each of the five elements equally, or would you give some more weighting? Give reasons for your answer.
12. Analyse the data in **FIGURE 6**. If you were in charge of upgrading Sydney's public transport system, what feature would you focus on as your highest priority? Give reasons for your answer, based on the data.

Challenge your understanding

13. The youngest of Generation Z will be 18 in 2027. What prediction can you make about public transport use among this generation when this survey is conducted in 2030?
14. Referring to **FIGURE 6** and your own knowledge, what recommendations would you make to the government to improve the experience for users in order to encourage more use of public transport?
15. Based on the reasons why Singapore's public transport is considered the best in the world, propose one change that could be made to the public transport in your area to improve effectiveness.

To answer questions online and to receive **immediate feedback** and **sample responses** for every question, go to your learnON title at www.jacplus.com.au.

18.4 Transport accessibility

LEARNING INTENTION

By the end of this subtopic, you will be able to identify groups of people and places that have limited access to public transport and describe some ways that technology is helping people living with a disability to access public transport.

18.4.1 Public transport and liveability

Public transport is one of the key factors used when compiling **liveable city** scores. Many, often newer, suburbs that score poorly overall or that have low rankings receive these scores because they do not possess the same levels of infrastructure and public transport as other suburbs. This is usually because older, more established suburbs have existing networks and tend to be located closer to major centres of trade and entertainment.

Despite the best efforts of governments and service providers, there is no doubt that some people are better connected by public transport than others. 'Black spots' for public transport leave many without access to essential services, and can force people into driving cars because there simply are no alternatives. These low levels of access to public transport generally occur on the edges of cities and in rural areas (see **FIGURE 1**).

liveable city an informal name given to cities that rank highly on a reputable annual survey of living conditions

FIGURE 1 Sydney's public transport accessibility, 2016

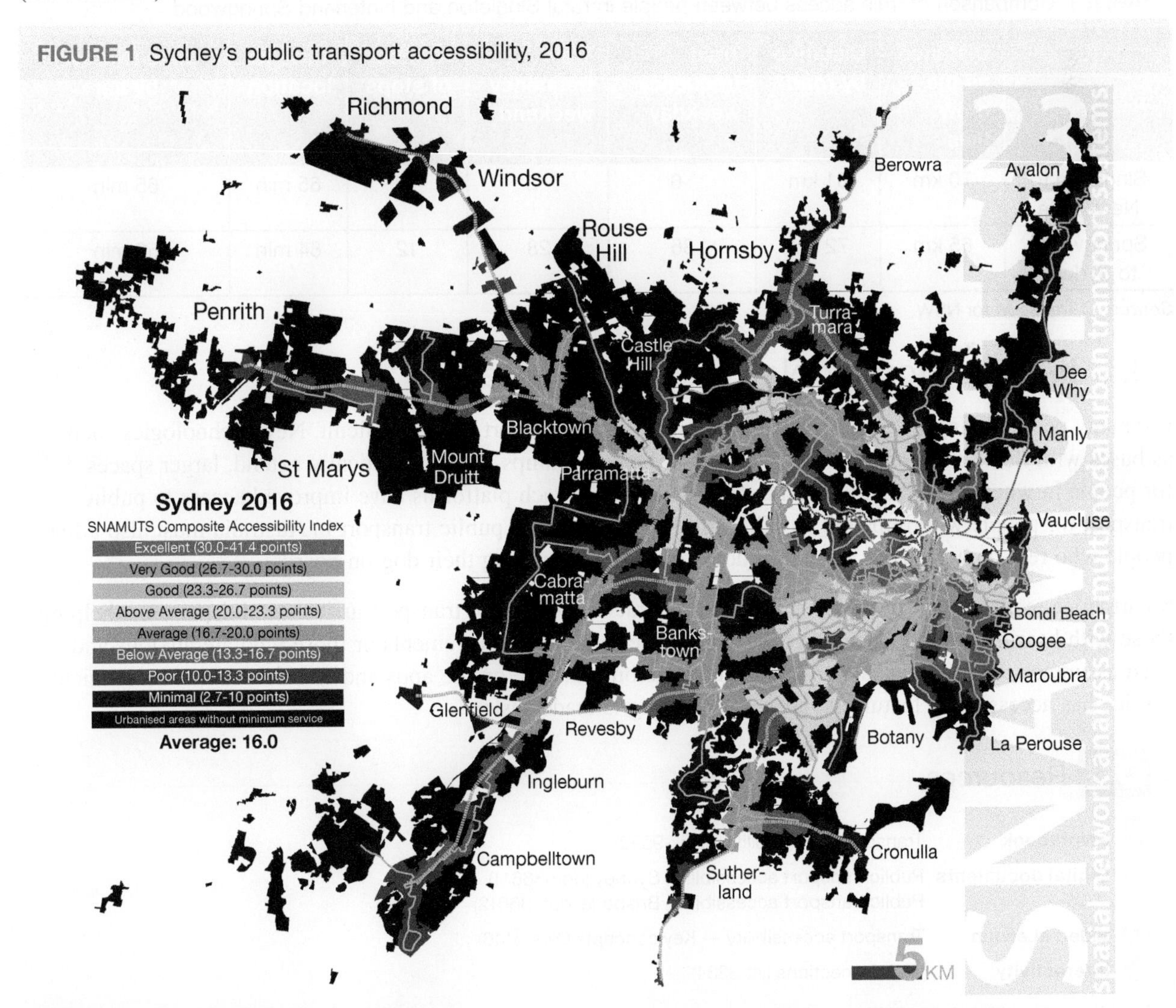

Source: © SNAMUTS 2016, www.snamuts.com

Public transport also needs to be affordable because owning a car can be too expensive for many people. In 2018, Sydney's public transport was ranked the fourth most affordable transport system in the world based on an average resident's income and the cost of available public transport.

FIGURE 2 Location of Singleton and Springwood, NSW

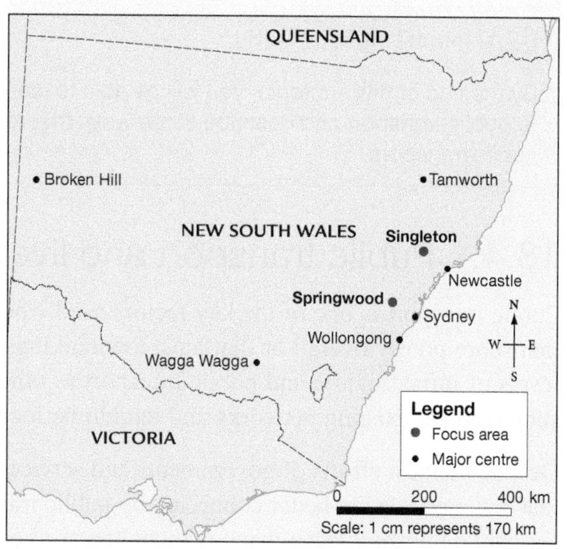

Source: Spatial Vision

18.4.2 Access in rural areas

One of the largest issues for people in rural areas is the lack of public transport. If public transport exists, it is often limited to local private providers, such as taxis, ride share services and private bus operators, often with few scheduled services. Even if there is a rail service, the trains often run on a low number of scheduled trips that leave early and return early (see **TABLE 1**). This means that people in rural areas are highly dependent on automobiles to connect to other people and places, or on information communication technologies to access essential services, such as Telehealth.

TABLE 1 Comparison of train access between people in rural Singleton and hinterland Springwood

	Distance as the crow flies	Distance by road	Train frequency weekday	Train frequency weekend	Number of stops	Time to location	Time to travel by car
Singleton to Newcastle	70 km	81 km	6	2	13	65 min	65 min
Springwood to Sydney	65 km	72 km	36	28	12	84 min	59 min

Source: © Transport for NSW

18.4.3 Access for people living with a disability

For many people living with a disability, accessing public transport can be difficult. New technologies such as buses with floors that can be lowered closer to the curb, ramps that automatically extend, larger spaces for people in wheelchairs, and the addition of elevators to reach platforms have improved access to public transport for people with impaired mobility. By law, all forms of public transport in Australia must also allow people who rely on accredited assistance dogs for mobility to bring their dog on board.

Smartphone technologies and apps show real-time tracking of public transport arrivals and departures, helping those with hearing impairments who may not be able to hear announcements or those who are anxious and need assurance for their mental wellbeing. Information available on the apps and websites also shows which service has accessibility features so that trip planning is made easier.

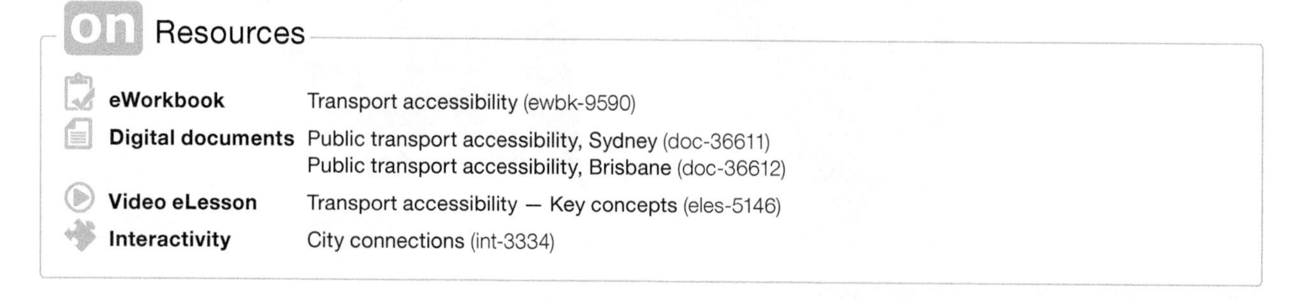

18.4 ACTIVITIES

1. As a class, develop a liveability score, based on access to public transport, for the places where everyone in the class lives. You can develop a scorecard based on how close the nearest public transport is to each person's neighbourhood the number and frequency of services, and the directness of travel to the nearest major town centre, cinema or shopping complex etc. Map the location of each classmate's block and use a colour code to indicate the score.
2. **TABLE 1** and **FIGURE 2** compare public transport options for a town on the fringes of major city (Springwood) and a rural town (Singleton). If the latest blockbuster movie was on at 1 pm on a Saturday in the nearest major city (Newcastle for Singleton and Sydney for Springwood), plan a return public transport trip for a group of 14-year-old friends from these two towns. Discuss the differences between these two trips.
3. Using the **public transport accessibility** maps of Sydney and Brisbane in the Digital Documents in your Resources panel, compare public transport access in the two cities. What patterns are evident in the locations of areas with each level of public transport access? Are there any anomalies?

18.4 EXERCISE

Learning pathways

■ LEVEL 1	■ LEVEL 2	■ LEVEL 3
1, 2, 3, 8, 14	4, 7, 9, 11, 15	5, 6, 10, 12, 13

Check your understanding

1. Identify three places in Sydney that do not have access to minimum public transport services.
2. Why do older suburbs generally have better public transport than others?
3. Why are people from outer suburbs of cities and rural areas sometimes forced into using cars rather than public transport?
4. What type of public transport is generally available to those who live in regional and rural areas?
5. How is public transport improving its services for people living with a disability?
6. Define the term *liveable city* and outline the role that transport plays in liveability.
7. Referring to **FIGURE 1**, describe the spatial pattern of urban areas that have poor access to public transport in Sydney.

Apply your understanding

8. Explain why it might be difficult for people living with a disability to access public transport.
9. Why might transport companies in rural areas provide fewer services than companies in urban areas?
10. Discuss the benefits and costs of having to rely on taxis, ride share or your own vehicle rather than having access to public transport.
11. Explain the term 'public transport black spot'.
12. Based on **FIGURE 1**, evaluate which area of Sydney has the most accessible public transport network. Use data to support your answer.

Challenge your understanding

13. Suggest why there are ribbons or threads throughout the transport maps of Sydney and Brisbane that seem to have slightly better access to transport than the surrounding areas.
14. Using the example of Singleton from **FIGURE 2** and **TABLE 1**, create a proposal to improve Singleton's public transport options, from a young person's perspective.
15. Referring to **FIGURE 1**, deduce what you think housing prices would be like in the dark green-shaded areas compared to in the black and deep red areas.

To answer questions online and to receive **immediate feedback** and **sample responses** for every question, go to your learnON title at www.jacplus.com.au.

18.5 SkillBuilder — Interpreting topological maps

online only

LEARNING INTENTION

By the end of this subtopic, you will be able to read and interpret a topological map.

The content in this subtopic is a summary of what you will find in the online resource.

18.5.1 Tell me

What is a topological map?

Topological maps are very simple maps, with only the most vital information included. These maps generally are not drawn to scale and give only general directions. However, everything is correct in its interconnection to other points. Topological maps are useful as a mental map to help you locate important features. Large areas can be drawn to show the viewer the important points.

18.5.2 Show me

eles-1736

How to interpret a topological map

Step 1

Examine the topological map and identify the key features being shown.

Step 2

Think about the interconnection between features.

Step 3

Since there is no scale, consider possible distances. Interpreting a topological map gives a first impression, but remember that it is not an accurate map.

FIGURE 1 A tourist map of Paris and its monuments uses pictures to identify places.

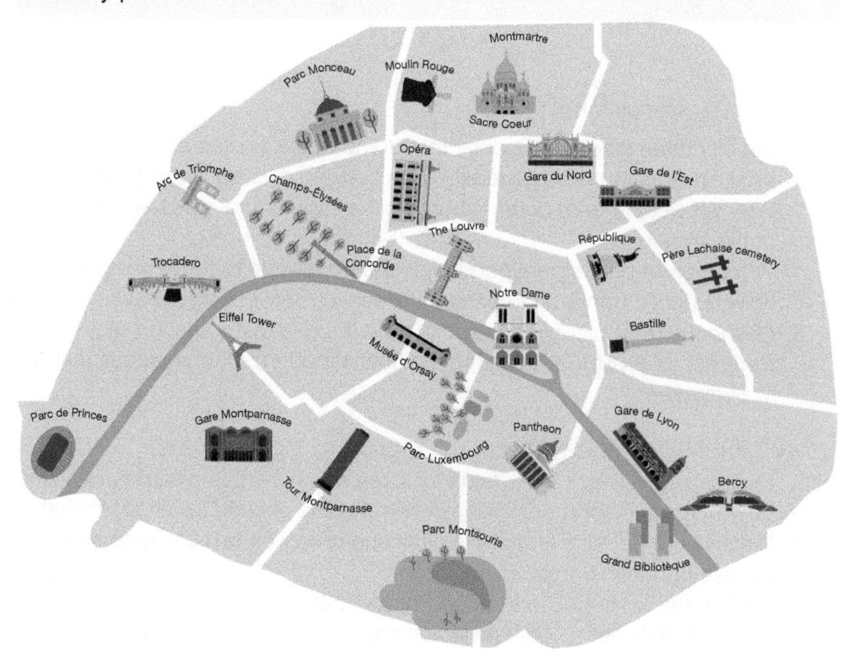

18.5.3 Let me do it

int-3354

Go to learnON to access the following additional resources to help you build this skill:

- a longer explanation of this skill and its application in Geography (Tell me)
- a video showing the step-by-step process of this skill (Show me)
- an activity and interactivity for you to practise this skill (Let me do it)
- self-marking questions to help you understand and use this skill.

Resources

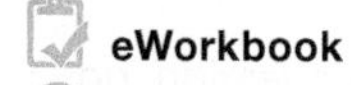

eWorkbook SkillBuilder — Interpreting topological maps (ewbk-9594)

Video eLesson SkillBuilder — Interpreting topological maps (eles-1736)

Interactivity SkillBuilder — Interpreting topological maps (int-3354)

18.6 ICT innovations

LEARNING INTENTION

By the end of this subtopic, you will be able to describe the innovations that led to the digital information age, describe the impact of the internet in connecting people and places, and define the Internet of Things.

18.6.1 Communicating over long distances

In the past, communication between places was often a slow process, especially over long distances. Communication was limited to the physical speed at which a message could be delivered. The Olympic Marathon is based on an ancient Greek legend in which a messenger ran 40 kilometres from Marathon to Athens with news of a victory, before dropping dead from exhaustion! Carrier pigeons flew long distances carrying messages, limited by the amount of extra weight they could carry. Even the 'snail mail' is limited by the physical speed by which people can move letters and parcels by land, sea or air.

This all changed when electricity was invented, as people saw this technology's potential to communicate signals over a distance. The first successful long-distance electrical communication system was the telegraph, developed by Samuel Morse and Alfred Vail. With the use of a hand keypad (see **FIGURE 1**) attached to copper wires, electrical signals were sent using Morse code — a system of dots and dashes (short and longer taps on the keypad) that represented the letters of the alphabet. Soon cables were rolled out over the countryside and eventually underwater, and people were able to connect to far-away places like never before. More scientific developments led to the inventions of the telephone, radio and television, making the world seem smaller and more connected.

FIGURE 1 The telegraph keypad used to communicate dots and dashes over long distances.

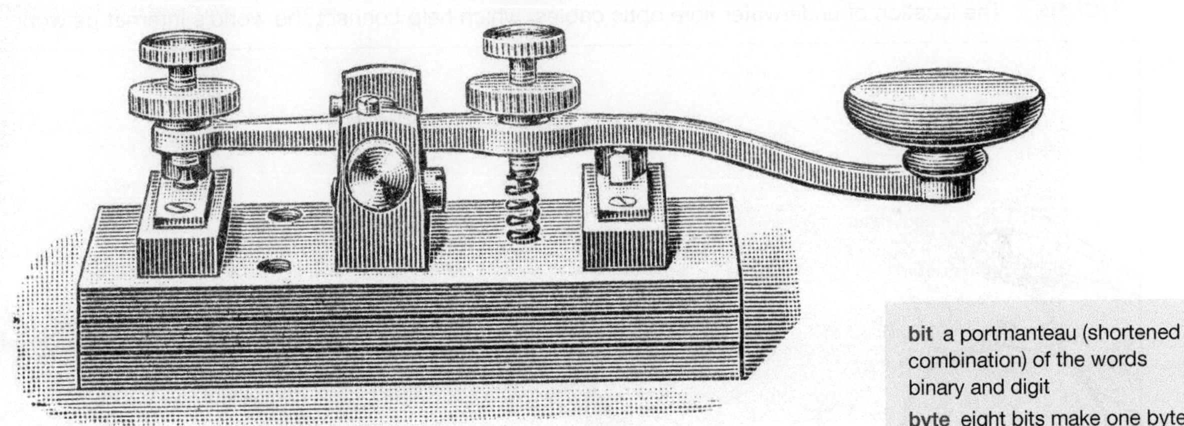

18.6.2 The digital information age

In 1948, a scientist named Claude Shannon developed the idea of sending and receiving information as two digits, 0 and 1, which he called a **bit**. Soon after, the transistor and computer chip were invented. The computer chip, made of billions of transistors, could be made small and cheaply and could process packets of 'bits' called **bytes** at a rate of billions of times a second. Because chips were simple to produce and could make electronics smaller, the cost of these technologies came down, leading to more widely available and accessible computers and other mobile technologies.

Within 50 years of these ground-breaking innovations, computers and communication technologies could send bits and bytes via electrical pulses along copper wires, through light in fibre optical cables (see subtopic 18.12), and as radio waves in wi-fi and digital mobile telephone networks such as **2G**, **3G**, **4G** and **5G**.

bit a portmanteau (shortened combination) of the words binary and digit

byte eight bits make one byte. This is the measurement used to store data such as kilobyte, megabyte and gigabyte.

2G second generation mobile network that moved from analogue to digital with a speed of 0.1 megabits a second

3G third generation digital network with a typical download speed of 8 megabits a second

4G fourth generation digital network with a typical download speed of 15 megabits a second

5G fifth generation digital mobile network with a typical download speed of 200 megabits a second

18.6.3 The internet

Students in Australia today do not know of a world before the internet. Some form of internet existed from the 1960s, connecting computers between places such as universities and military bases, but the internet as we know it now did not come into being until the early 1990s.

The development of the internet evolved from telephone and fibre optic networks under the ocean floor (see **FIGURE 3**) connecting at hubs called internet service providers (ISPs) who follow agreed-upon protocols to communicate easily across the world network.

FIGURE 2 Computer chip technology containing billions of transistors made communication technologies widely accessible.

As the internet has progressed, we have seen more web technologies develop to connect people and places. We now have services such as cloud computing, where people can access and share their digital files from anywhere in the world, that allow people to connect with others with shared interests, experiences and knowledge. For example, a gamer in the United Kingdom can live-stream themselves playing the latest game on a console through a streaming service such as Twitch or Discord, allowing a gamer in Lismore, New South Wales, to watch them live. Or a craftsperson in India can upload a how-to video to YouTube that a fellow crafter in the United States can watch and learn from.

FIGURE 3 The location of underwater fibre optic cables, which help connect the world's internet network

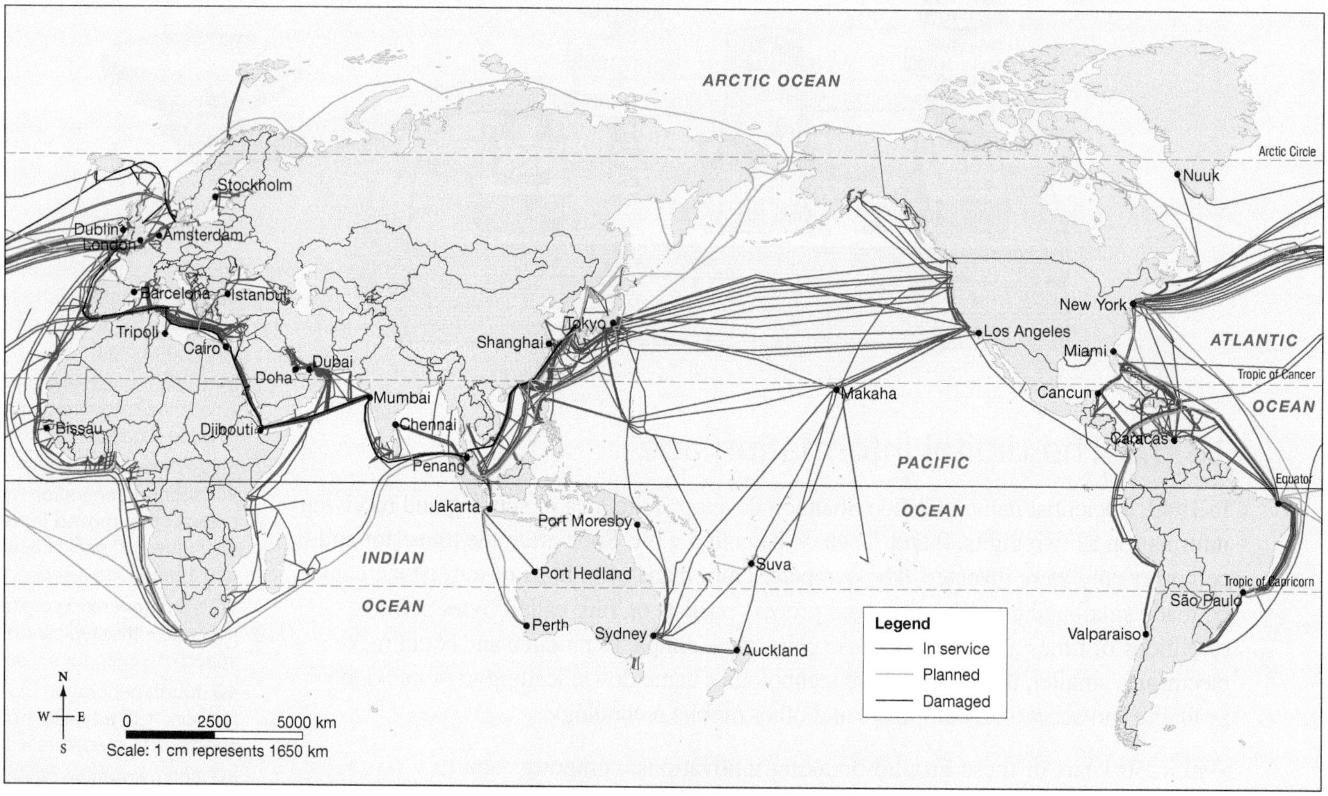

Source: Data by Greg Mahlknecht, www.cablemap.info. Map drawn by Spatial Vision, 2021.

18.6.4 The Internet of Things (IoT)

Another technology that has extended web technologies is called the Internet of Things (IoT). This refers to a number of non-traditional computing devices that send and/or receive data from the internet without the use of input devices such as a keyboard and mouse. Some appliance manufacturers refer to them as smart appliances or internet-connected appliances. If the device can be accessed via a web-enabled device, it is an IoT device. Google's Alexa and Apple's Home Hub are popular IoT devices.

FIGURE 4 Smart appliances are becoming more interconnected through the Internet of Things — they can be controlled remotely as long as they are connected to the internet.

Factories, healthcare, farming and many other industries are benefiting from IoT because sensors and machines can be in one place but be monitored and controlled from another place. For example, health monitoring machines can be placed on an elderly person in their home and the data can be monitored remotely through the internet. This can be helpful when checking on patients in rural areas. It is estimated that there were more than 50 billion IoT devices in use as of 2020, sending and receiving over 4.4 trillion **zettabytes** of data over the internet.

18.6.5 Information and communication technology

Information and communication technology is now used for a variety of purposes to make connections. We often do not realise how convenient our lives have been made by various forms of communication. Some of these include online and telephone banking, storing and sharing of medical records, online shopping and online games.

Video conferencing is another example of an online communication medium that has connected people over long distances — and even within the same city. This service makes it possible for anyone with an internet connection to communicate by video call with someone else anywhere in the world, letting them share their screen and have text-based chats at the same time. During the social distancing restrictions of the COVID-19 pandemic, video conferencing became an important tool to connect people locally through distance learning at schools and collaboration among work colleagues.

zettabyte one trillion gigabytes

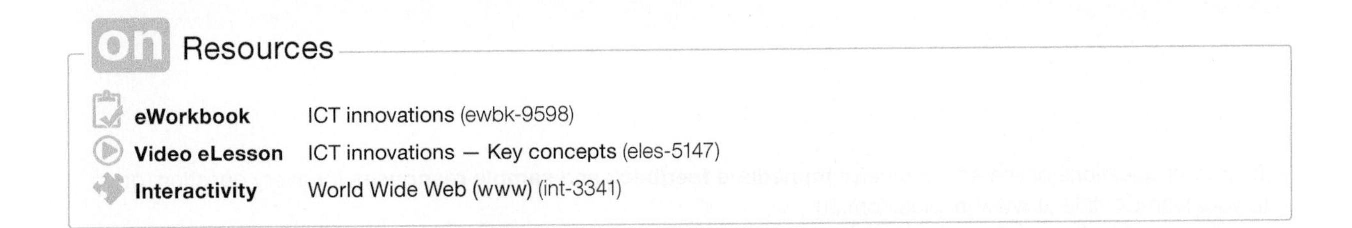
on Resources

eWorkbook ICT innovations (ewbk-9598)

Video eLesson ICT innovations — Key concepts (eles-5147)

Interactivity World Wide Web (www) (int-3341)

18.6 ACTIVITIES

1. Research the Australian Overland Telegraph Line. Outline how it was created and explain its importance in connecting Australia to the rest of the world.
2. Using Google as the search engine, enter the search term **site: mit.edu** (this tells Google to look for only websites with that in their web address).
   a. How many results come up?
   b. How many seconds did it take to conduct this search?
   c. Now click on the search result for the university in the US called MIT. How long did it take for you to connect to this place?
   d. What is MIT's relative location?
   e. Using Google Maps, determine the distance between your school and MIT.
   f. Discuss in small groups or pairs.

18.6 EXERCISE

Learning pathways

■ LEVEL 1	■ LEVEL 2	■ LEVEL 3
1, 2, 4, 10, 15	3, 5, 8, 11, 14	6, 7, 9, 12, 13

Check your understanding

1. What is the name of the nineteenth-century invention that sent electrical signals across long distances?
2. What three innovations started the digital age?
3. What are the three methods in which bits are transmitted as electrical pulses?
4. In what decade did the internet as we know it begin?
5. What is the Internet of Things (IoT)?
6. Refer to **FIGURE 3**. How does the number of submarine cables around Australia compare with the number around Japan?
7. Create a table to show the differences in download speed typical in the generations of mobile networks from 2G to 5G.

Apply your understanding

8. Refer to **FIGURE 3**. Describe the geographical pattern of underwater cables worldwide. Where are the greatest concentrations of cables? Which parts of the seas or oceans seem to lack cables? Why do you think this is?
9. Looking at **FIGURE 3**, what are some risks to New Zealand safely staying connected to the rest of the world?
10. Give two examples of how digital connectivity might have helped businesses to keep operating during the COVID-19 pandemic.
11. What might be some of the benefits and drawbacks of not being connected digitally?

Challenge your understanding

12. Thinking about the Internet of Things (IoT), propose three connected devices that would benefit your school.
13. What safety issues might people need to consider if all their devices are connected to the internet?
14. Predict one way that interconnectivity through technology might improve in the next ten years.
15. Do you think that digital connectivity will make it possible for everyone in Australia to work or go to school completely from home in the future? Support your answer with examples.

To answer questions online and to receive **immediate feedback** and **sample responses** for every question, go to your learnON title at www.jacplus.com.au.

18.7 Connecting through ICT

online only

LEARNING INTENTION

By the end of this subtopic, you will be able to outline how social media connects people, and identify which people are more likely to interconnect with different social media.

The content in this subtopic is a summary of what you will find in the online resource.

Today, we are able to get in touch with people almost anywhere in the world, and the ways in which we connect are constantly evolving. To learn more about how ICT connects people near and far, go to your learnON resources at www.jacPLUS.com.au.

FIGURE 1 Sharing experiences

Contents

- 18.7.1 Death of distance
- 18.7.2 Social media
- 18.7.3 Generations

Resources

eWorkbook	Connecting through ICT (ewbk-9602)
Video eLesson	Connecting through ICT — Key concepts (eles-5148)
Interactivity	Connecting through ICT (int-8567)

18.8 Access to services through ICT

online only

LEARNING INTENTION

By the end of this subtopic, you will be able to explain how people are interconnected through the internet and mobile phones, describe the importance of ICT infrastructure, and describe the 'digital economy' and its benefits.

The content in this subtopic is a summary of what you will find in the online resource.

The growth of information and communication technology (ICT) has had a significant impact on the way people are able to access online services. However, access is not be available to everyone equally. To learn more about ICT infrastructure and the digital economy, go to your learnON resources at www.jacPLUS.com.au.

FIGURE 4 Online classes

Contents

- 18.8.1 Connecting to the internet
- 18.8.2 ICT expenditure
- 18.8.3 Mobile phones
- 18.8.4 The digital economy
- 18.8.5 Opportunities for new services

Resources

eWorkbook	Access to services through ICT (ewbk-9606)
Video eLesson	Access to services through ICT — Key concepts (eles-5149)
Interactivity	Access to services through ICT (int-8568)

18.9 SkillBuilder — Constructing a table of data for a GIS

online only

LEARNING INTENTION

By the end of this subtopic, you will be able to construct a table of data suitable for input into a GIS.

The content in this subtopic is a summary of what you will find in the online resource.

18.9.1 Tell me

Why are there tables within GIS?

GIS, or geographical information systems, uses tables to organise and store information about points, lines and polygons (vector data). These tables have rows and columns, called fields. The GIS software links the rows in the table to the points, lines or polygons on a map. GIS software also stores data as pixels in an image, called raster data. The tables can be drawn with a spreadsheet program and linked to a GIS if there is relevant information about location in the table. However, specialist GIS software is required. Tables are very useful ways of storing large amounts of information, because they help to organise it and make it easier to import.

18.9.2 Show me

eles-1743

How to construct a table of data

Imagine your class has conducted a survey to find out how many mobile phones there are in each home, asking three questions: Where do you live? How many people are in your home? How many mobile phones are in your home?

FIGURE 3 Work out which are text fields and which are number (integer) fields.

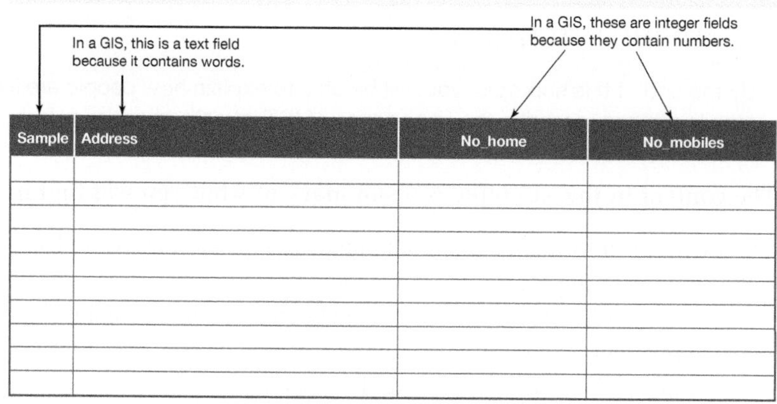

Step 1

Draw a table with rows and columns. For this data there should be 11 rows and four columns.

Step 2

Give each of the columns a heading to represent the data collected. Each heading must be short and use underscores instead of spaces. Create four columns: sample number, address, the number of people in the home and the number of mobile phones.

Step 3

Identify which columns (fields) contain text and which contain numbers (integers).

Step 4

Enter the collected data into the table. A thematic map of the data can then be created using GIS software.

18.9.3 Let me do it

int-3361

Go to learnON to access the following additional resources to help you build this skill:

- a longer explanation of this skill and its application in Geography (Tell me)
- a video showing the step-by-step process of this skill (Show me)
- an activity and interactivity for you to practise this skill (Let me do it)
- self-marking questions to help you understand and use this skill.

on Resources

eWorkbook	SkillBuilder — Constructing a table of data for a GIS (ewbk-9610)
Video eLesson	SkillBuilder — Constructing a table of data for a GIS (eles-1743)
Interactivity	SkillBuilder — Constructing a table of data for a GIS (int-3361)

18.10 ICT accessibility

LEARNING INTENTION

By the end of this subtopic, you will be able to describe the digital divide in access to ICT, and explain the benefits of connectivity and the consequences of being disconnected from ICT.

18.10.1 The digital divide

Even though there is a growth in internet access worldwide, the gap between those able to access information technology and those who not able is widening. This is known as the **digital divide**. The divide is primarily based on internet access, but it includes all forms of information. Factors contributing to the digital divide are shown in **FIGURE 1**. Such divisions can have implications for economic growth and social equity. The importance of being connected via the internet is being considered as a collective human right by the United Nations.

FIGURE 1 Factors contributing to the digital divide

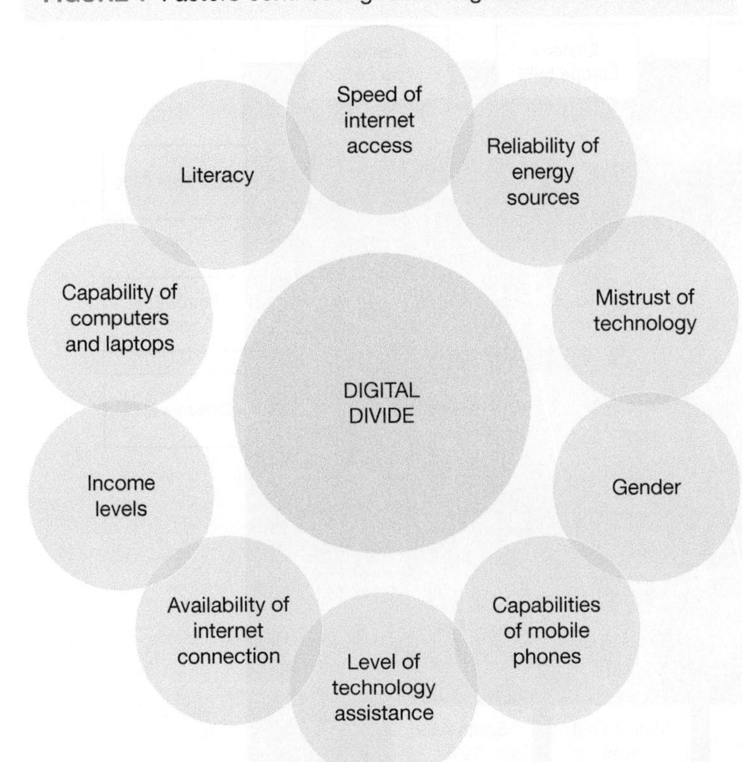

In 2020, 96.5 per cent of the world's population had access to at least second generation (2G) mobile networks, but access to high-speed internet was not equal (see **FIGURE 2**). Even within a country, people in cities usually have access to high-speed internet, whereas rural areas might only have access to dial-up internet or none whatsoever.

Digital disadvantage is a dimension of poverty that leaves people with a sense of powerlessness. If ability to connect to information and communication technology is limited, this affects opportunities for people and places to grow as they miss out on information, knowledge and economic opportunities. Globally, the average level of access to

digital divide a type of inequality between groups in their access to and knowledge of information and communication technology information and communications technologies (ICT) was 42 per cent in 2015. As **FIGURE 3** shows, the region of South Asia (19 per cent) and the continent of Africa (26 per cent) have the lowest levels of connectivity.

FIGURE 2 Quality of fixed internet (broadband) speeds by country, 2020

ARCTIC OCEAN
Arctic Circle
PACIFIC
ATLANTIC
Tropic of Cancer
OCEAN
OCEAN
Equator
INDIAN
OCEAN
Tropic of Capricorn

Average fixed broadband download speed
Megabits per second (mbps)
Less than 20
20 – 49
50 – 99
100 – 149
150 – 199
200 or more
No data

N
W E
S

0 2500 5000 km
Scale: 1 cm represents 2375 km

Source: Average internet performance data from March 2021 based on the Speedtest Global Index™ from Ookla®: https://www.speedtest.net/global-index

FIGURE 3 Regional differences in internet connection

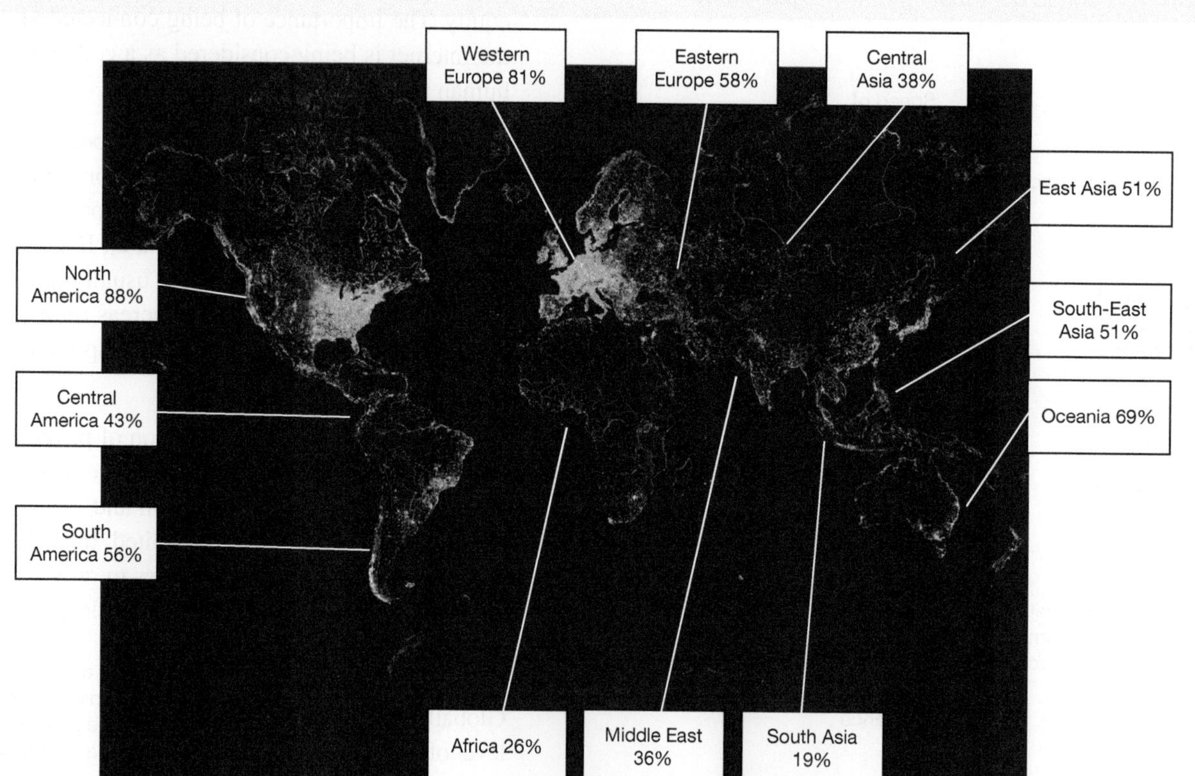

Internet access is only one aspect of the challenges faced by regions — specific areas within a country may face disadvantage in terms of the factors displayed in **FIGURE 1**. For example, regional areas of Australia may have difficulty accessing information technology. Nine out of ten people living in rural Africa don't have access to electricity. This unequal access may restrict the ability of communities to access information, communication and ideas that will enable them to overcome poverty and disadvantage.

18.10.2 Connectivity

Global connectivity has been increasing rapidly over the past three decades. Many countries are working on improving their digital economies because they give access to a variety of resources and markets that can help improve life for their population. We are moving towards a world in which personal mobile phone technology can connect every human being in every village in every country to knowledge, markets, services and communities. This increased connectivity has affected people from villages and cities in all parts of the world in a range of ways, as well as having an impact on the natural environment.

While mobile phones have become the most important form of communication, use across the globe is also uneven. This could be due to a variety of factors, such as access, financial situation and way of life. In countries with high mobile phone use, many people have more than one mobile phone connected. For example, in Hong Kong in 2019, there were 289 mobile phone subscriptions per 100 people. In contrast, Mozambique in 2019 had 49 mobile phone subscriptions per 100 people.

FIGURE 4 Mobile phone use is increasing rapidly in Africa.

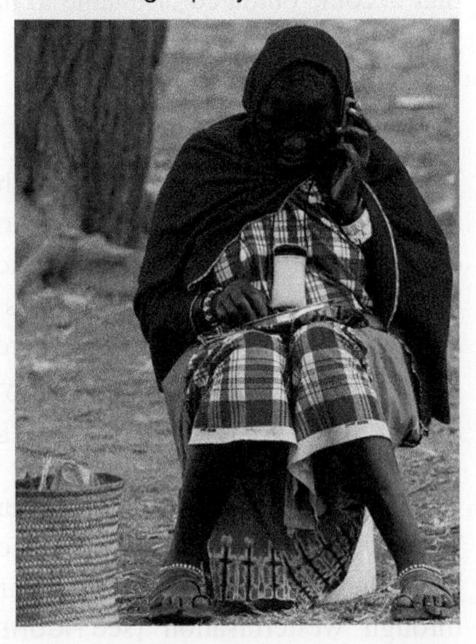

18.10.3 Benefits of connectivity

Economic benefits

The Global Connectivity Index (GCI) can be used to measure and predict a country's economic health. There is a strong link between GCI and a country's gross domestic product (GDP) — the higher a nation's GCI the higher the GDP. A 20 per cent increase in ICT investment will grow GDP by 1 per cent. Singapore has implemented a Smart Nation strategy, aimed at moving from natural resources to developing technological services. Greater connectivity and more integrated markets allow for new jobs and careers all over the world.

Social benefits

The development of many countries can be enhanced with greater connectivity. A country's level of connectivity can be categorised using the terms *Leaders, Followers* or *Beginners*. By investing in technologies such as 4G, 5G and mobile phone infrastructure, countries in Africa and South America can enhance their levels of development much more rapidly, and therefore improve the standard of living for their citizens. These changes will also allow for more connection between people from different cultures and backgrounds and, it is hoped, greater understanding.

Environmental benefits

The climate change conference in Paris in 2015 showed the benefits that can be achieved by an inclusive and connected approach to the environment. These talks involved **delegates** from 196 countries and culminated in the adoption of the United Nations resolution, 'Transforming our world: the 2030 Agenda for Sustainable Development'.

delegate person sent or authorised to represent others; in particular an elected representative sent to a conference

18.10.4 Consequences of being disconnected

Economic impacts

Poorer nations, or those unable to adapt to technological change as fast as other countries, may find themselves at a disadvantage. While implementing new technology and infrastructure is important, if countries invest in the wrong areas, they may have systems that become outdated or unable to handle the growing demands of their citizens.

Social impacts

With rapid technological changes, there is a risk of creating 'haves' and 'have nots'. Those countries that are unable to keep pace with the 'leaders' may fall behind. Countries such as Portugal, Italy, Greece and Spain are regarded as developed nations. However, they are now *followers* when it comes to ICT. Beginner nations, such as those in Africa, are well behind the leaders and will likely remain there. These changes and increased connectivity also bring the possibility of loss of cultural identity through 'westernisation' (see **FIGURE 5**).

Environmental impacts

The greater push for ICT resources and infrastructure puts pressure on the natural environment. The need for certain resources, such as copper, gold and nickel, will require continued mining and disruption to the natural landscape. The need for more electricity to run devices, and the impact of building and maintaining batteries, also put pressure on non-renewable resources. The disposal and treatment of existing and outdated equipment is also an issue that needs addressing. E-waste is a growing global concern (see **FIGURE 6**). Many individuals, groups and governments have put in place plans to deal with this; an example is Planet Ark, which runs e-waste collection points at schools and nominated businesses.

FIGURE 5 Some impacts of western culture

FIGURE 6 E-waste sorting

18.10.5 Sustainable Development Goals

The United Nations Sustainable Development Goals were adopted in 2015 and support efforts to promote greater global connectivity. Improved global internet access is expected to provide the tools and knowledge needed to increase education and health opportunity, to create jobs, fight injustice and lift communities out of poverty.

Why bridging the gap is important

Economic equality

Some social welfare services are administered electronically, so access to a telephone and the internet is important. The telephone also provides security, and can be used in emergency situations. In addition, access to the internet can be important for career development and accessing civic information.

Social mobility

Computers play an important role in learning and education. The digital divide is unfair for children in lower socio-economic groups. The higher the educational qualification held by an individual, the more likely they are to have internet access at home.

Democracy

Some people think that access to the internet creates a healthier democracy. It is thought that it increases public participation in elections and decision-making processes. Online information allows voters to research candidates and their policies to learn about election results, government decisions and issues that affect them directly. It also provides a way for people to lobby decision makers and to protest decisions they disagree with.

Economic growth

Information and communication infrastructure and active use could stimulate economic growth, particularly in less-developed nations. Economic changes and improvements tend to be associated with information technology. Certain industries can give a country's economy a competitive advantage.

fdw-0024

FOCUS ON FIELDWORK

Technology — Evaluating sources

At different times during our day, we are more connected to places, information and people than at other times. Being connected, however, does not mean we are receiving reliable information. Some social media networks and sites flag content that is unreliable or not based in fact, but not all do.

Learn how to assess the reliability of information using the **Technology — Evaluating sources** fieldwork activity in your Resources tab.

 Resources

eWorkbook	ICT accessibility (ewbk-9614)
Video eLesson	ICT accessibility — Key concepts (eles-5150)
Interactivity	The digital divide (int-3342)
Fieldwork	Technology — Evaluating sources (fdw-0024)

18.10 ACTIVITIES

1. Refer to **FIGURE 3**. Research which countries are in South Asia. How do you account for low connectivity levels in these countries?
2. Propose an initiative to improve the level of connectivity currently found in one community. What challenges need to be overcome? What change will it introduce to the community?

18.10 EXERCISE

Learning pathways

■ LEVEL 1	■ LEVEL 2	■ LEVEL 3
1, 2, 7, 8, 14	3, 4, 9, 10, 13	5, 6, 11, 12, 15

Check your understanding

1. Define the term *digital divide* in your own words.
2. List four factors that contribute to the digital divide.
3. Who generally does not have access to high-speed internet?
4. Which two regions of the world have the lowest internet connectivity?
5. How are most people in the world connected to the internet?
6. Based on **FIGURE 3**, describe the general pattern of digital connectivity by region across the world.

Apply your understanding

7. Refer to **FIGURE 5**.
   a. What culture is the cartoon claiming is dominant?
   b. Explain two negative impacts that are being portrayed in this cartoon.
8. **FIGURE 6** shows a negative environmental impact of ICT, which is e-waste. How is it showing the process of recycling parts?
9. Explain how internet connectivity can improve social mobility.
10. Discuss one example of how connectivity can help to break down barriers to economic equality.
11. What might the cartoonist of **FIGURE 5** be implying about western culture by drawing the invading forces as well-known cartoon characters?
12. Discuss the following statement: 'The first step in bridging the digital divide is access, but that is only a small first step to equality.'

Challenge your understanding

13. What will be the consequences for developing countries whose connectivity to the internet is not improved in the next ten years?
14. The United Nations developed 17 sustainable development goals. Investigate which of the goals is linked to ICT access and explain how it is linked.
15. Predict how a complete, week-long shutdown of Australia's NBN might affect our political climate and economy.

To answer questions online and to receive **immediate feedback** and **sample responses** for every question, go to your learnON title at www.jacplus.com.au.

18.11 Impacts of ICT in developing countries

LEARNING INTENTION

By the end of this subtopic, you will be able to explain how access to a mobile phone helps people in developing countries, give examples of the impact M-Pesa had for Kenyan citizens.

18.11.1 Connections in Kenya

While there is a difference between the developed and developing worlds, it is important to note that access to the internet and mobile phone networks is improving. This is largely thanks to non-government agencies and to service providers. The people of Kenya, for example, no longer need to take long journeys into town or to wait in long queues just to transfer money. Instead, they can type a couple of numbers, hit a button, and pay for anything they want within seconds via mobile phone.

FIGURE 1 A customer at an M-Pesa agent in Kenya

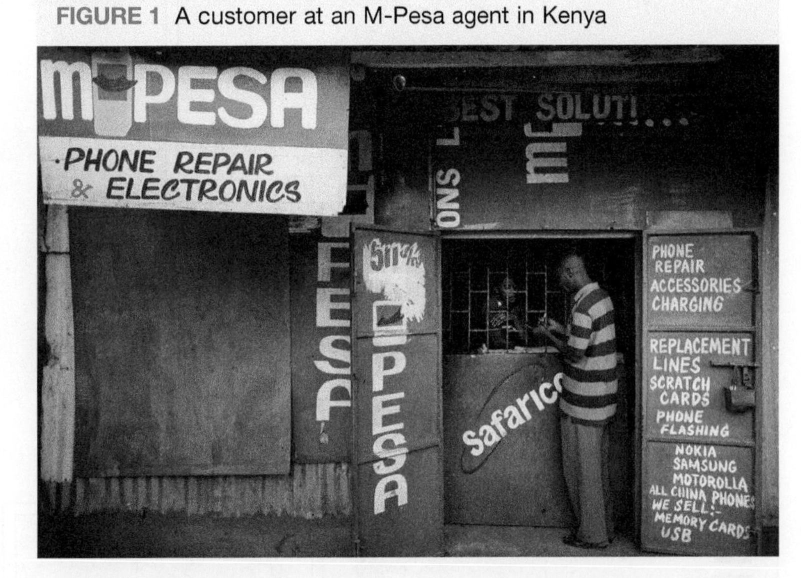

Most people in Kenya live in rural and remote places in the countryside. Landline access is very limited, and it is also considered costly to install, owing to the vast distances. This can leave families disconnected from one another, as the primary earner in the family often has to work in a distant township in order to provide money for the family. Those left in the rural villages are often self-employed farmers or tradespeople. Without communication, families are often disconnected.

Improving access

Mobile phone coverage and access has been significantly improved in Kenya, which has dramatically improved people's lives. The UK-based **non-government organisation** (NGO) called Financial Deepening Challenge Fund (FDCF) has worked with Vodafone to set up M-Pesa, meaning 'mobile money'. A customer can go to an M-Pesa agent, such as a supermarket, and:

- deposit and withdraw money
- transfer money to other users and non-users
- pay bills
- purchase phone credit
- transfer money between the service and a bank account (see **FIGURE 1**).

People do not need a bank account or even a permanent address to use M-Pesa. They receive a text message to confirm a transaction, and the money can then be stored on the phone or it can be forwarded to someone else. At any time, the money can be transferred back to cash through an M-Pesa agent or an ATM. **FIGURE 2** shows mobile phone coverage and population in Kenya, indicating the coverage is significant in highly populated areas.

non-government organisation (NGO) non-profit group run by people (often volunteers) who have a common interest and perform a variety of humanitarian tasks at a local, national or international level

FIGURE 2 (a) Population density and (b) mobile phone network coverage in Kenya

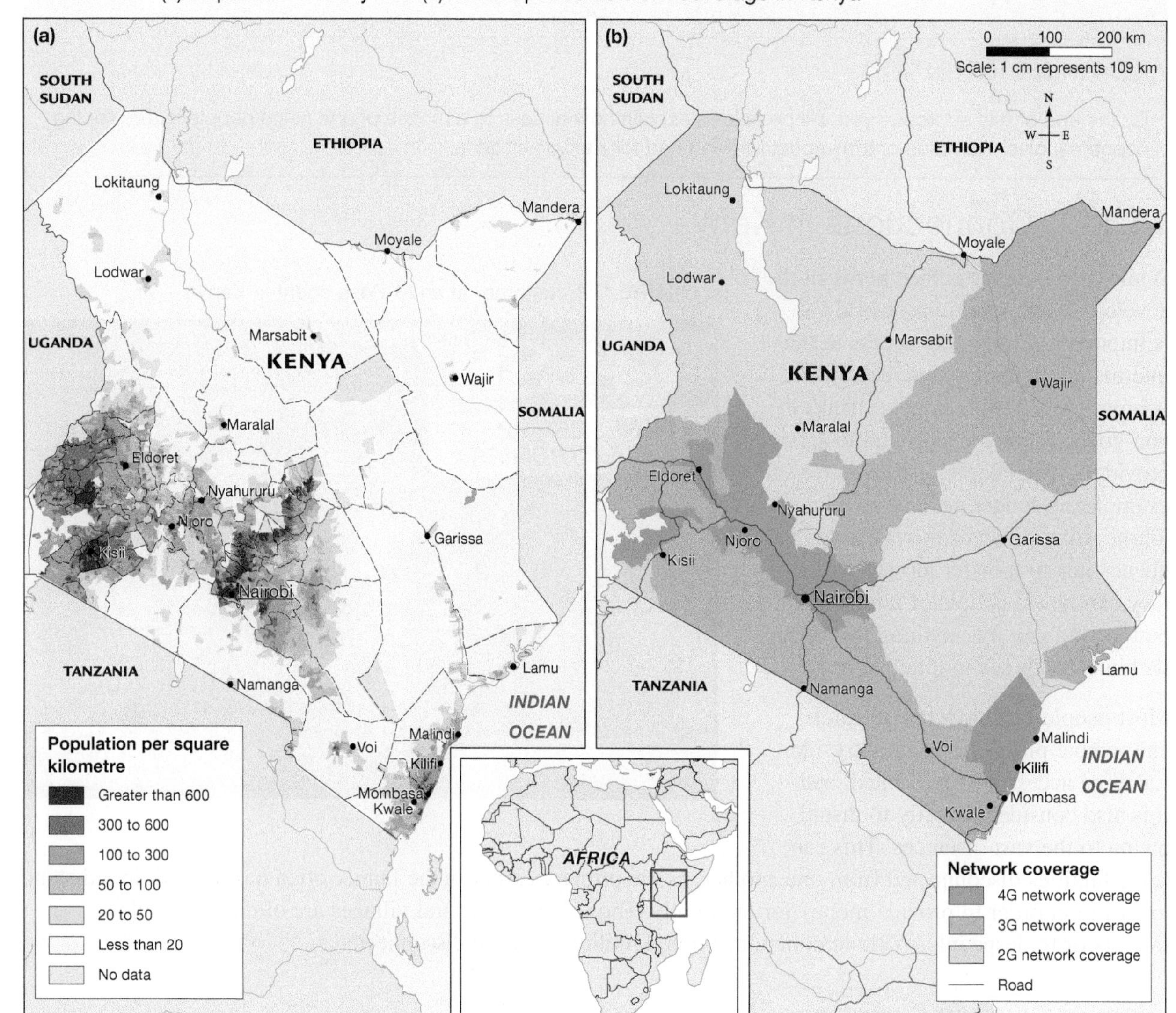

Source: JUMIA (2018); Spatial Vision

How successful has it been?

Today there are around 25 million subscribers and more than 173 000 M-Pesa agents in Kenya. These agents include petrol stations, supermarkets and other retail outlets. In 2020, over 72 per cent of Kenya's population was using M-Pesa. It has been reported that M-Pesa has increased spending levels in Kenya, helping the economy grow, and has lifted 194 000 households (2 per cent of Kenyan households) out of poverty.

Taxi drivers have benefited too. They no longer need to drive around with a lot of cash in their cars, because their passengers are able to pay using M-Pesa. It has also been seen as an improvement in personal safety, because it is a secure and easy way to transfer money.

There are now more mobile phone providers, covering 90 per cent of Kenya, mobile phone packages are more affordable, and smartphones are now cheaper. Access to mobile phones has meant that small business operators are now able to advertise to a larger audience, and are no longer dependent on word-of-mouth advertising to connect them to more people and places. Clients can now contact business operators with ease. For those working away from home, it is a safe and easy way to send money back to families in the countryside. It has enabled **Maasai** herdsmen to go to the markets and make purchases using their mobile phones. M-Pesa has eliminated the need to carry large sums of cash to markets, and has reduced the number of thefts. By 2018, several local digital start-ups and international ICT companies were calling Kenya home. Kenya is bridging the digital divide.

M-Pesa demonstrates how dreaming big but thinking locally can have a significant effect on the economic and social structure of a place, just through the use of a mobile phone.

FIGURE 3 A local market in Kenya

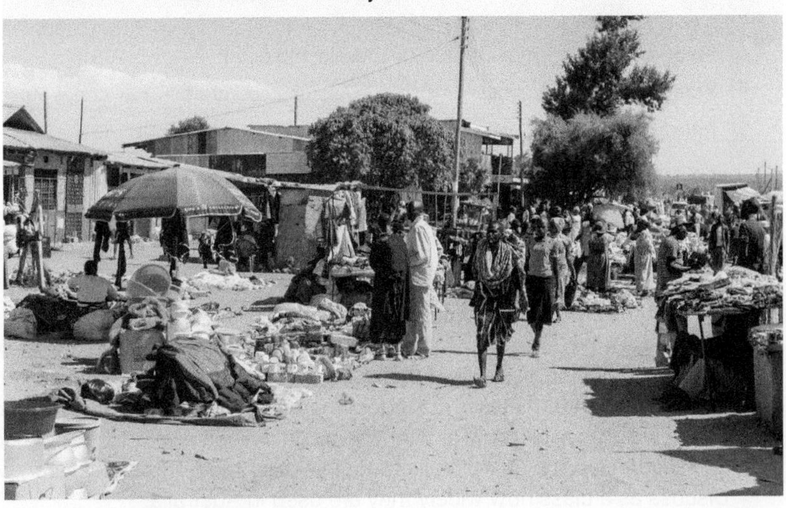

FIGURE 4 M-Pesa has had a significant effect on the economic and social structure of Kenya.

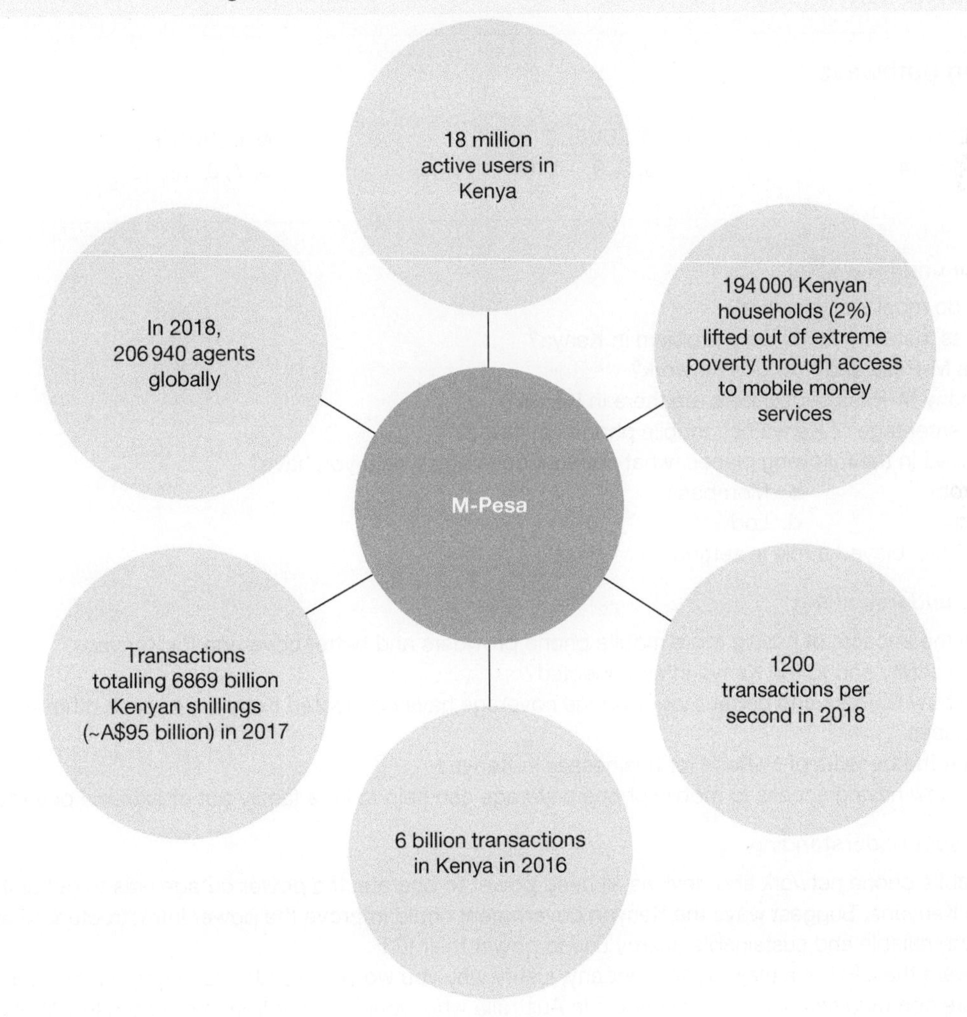

Maasai ethnic group of semi-nomadic people in Kenya and Tanzania, whose way of life depends on their cattle

on Resources

eWorkbook Impacts of ICT in developing countries (ewbk-9618)

Video eLesson Impacts of ICT in developing countries — Key concepts (eles-5151)

Interactivity Impacts of ICT in developing countries (int-8570)

18.11 ACTIVITIES

1. Research how another developing country has used Kenya's example to improve its ICT. Create a presentation on what they did, and how it has improved the interconnections for its people.
2. In small groups, suggest a list of possible criteria that you could use to judge how effective technology has been in improving people's lives in Kenya. Share and discuss your group's criteria with the class and select the three criteria that the class considers the most effective for judging. Did you have to make many changes to create the class list? Why or why not? How difficult was it to reach a consensus?
3. Investigate services in Australia that are similar to M-Pesa. Research what they are and how they operate. Discuss as a class how widely they are used in Australia.

18.11 EXERCISE

Learning pathways

■ LEVEL 1	■ LEVEL 2	■ LEVEL 3
1, 2, 4, 11, 14	3, 5, 9, 10, 13	6, 7, 8, 12, 15

Check your understanding

1. Where do most Kenyans live?
2. Why was installing landlines a problem in Kenya?
3. What is M-Pesa and how does it work?
4. How many M-Pesa subscribers are there in Kenya?
5. What percentage of Kenya has mobile phone coverage?
6. If you lived in the following places, what network coverage would you have?
   a. Nairobi
   b. Mombasa
   c. Lamu
   d. Lodwar
7. Which NGO played a role in setting up M-Pesa?

Apply your understanding

8. Explain the impacts of having more mobile phone providers and better coverage for Kenyans.
9. How are safety and ICT in Kenya interconnected?
10. Explain how M-Pesa and better mobile phone coverage have connected Kenyans to each other and to other places.
11. What are the benefits of M-Pesa for businesses in Kenya?
12. Explain how having access to mobile phone coverage can help to lift a family out of extreme poverty.

Challenge your understanding

13. The mobile phone network and devices all need power to operate. If a power outage was to occur it would cripple Kenyans. Suggest ways the Kenyan government could improve the power infrastructure to ensure constant, reliable and sustainable energy use to power their ICT.
14. If you were the CEO of a major tech company, justify why you would consider setting up an office in Kenya.
15. Are there communities or groups of people in Australia who would benefit from a program like M-Pesa? Justify your response.

To answer questions online and to receive **immediate feedback** and **sample responses** for every question, go to your learnON title at www.jacplus.com.au.

18.12 Connectivity in Australia

LEARNING INTENTION

By the end of this subtopic, you will be able to define and use the term *digital economy*, and explain the effectiveness of Australia's NBN network compared to other places.

18.12.1 The digital economy

When you use your debit card to buy a movie ticket, send a text to a friend on your mobile or buy something on eBay or Amazon, you are participating in the **digital economy**. According to the Australian Bureau of Statistics (ABS), digital activity contributed $96.9 billion to Australia's economy in 2018.

In 2020, it was estimated that approximately $35.5 billion was spent on **e-commerce** alone. Just in the month of November 2020, 5.5 million households shopped online. **TABLE 1** shows the most popular online purchases made. The digital economy connects business, services and people regardless of where they live, provided they can access the internet. People can order groceries online from their local supermarket and purchase fashion items located in another state or country. As a result, postal and delivery services are experiencing high growth.

TABLE 1 Most popular online purchases, November 2020

Rank	Product
1	Games
2	Clothing
3	Fashion accessories
4	Books
5	Sporting
6	Alcohol
7	Pet products
8	Athleisure

Source: © Australia Post. Inside Australian Online Shopping update, December 2020.

18.12.2 Digital infrastructure — NBN

In order for Australia's digital economy to thrive, good infrastructure needs to be in place. Australia had an ageing copper wire system that serviced telephones for many decades. That network was proving to be too slow, old and expensive to fix. This was particularly the case in rural areas. The old network's inefficiencies were leaving people who lived outside major urban areas with far less efficient and reliable internet access. In a bid to make the internet more accessible and faster, the federal government in the late 2000s started the development of the National Broadband Network (NBN). This major infrastructure project aimed to connect more people, businesses and places so they could participate in and enhance Australia's digital economy.

The NBN uses three main technologies to deliver fast, reliable internet (see **FIGURE 1**): fibre optic cable (**FIGURE 2**), fixed wireless, and satellite. The plan is to ensure that all Australians have some sort of access to this new high-speed broadband network.

digital economy the network of economic and social activities that are supported by information and communication technology, such as the internet and mobile and sensor networks

e-commerce business conducted online

FIGURE 1 NBN technologies

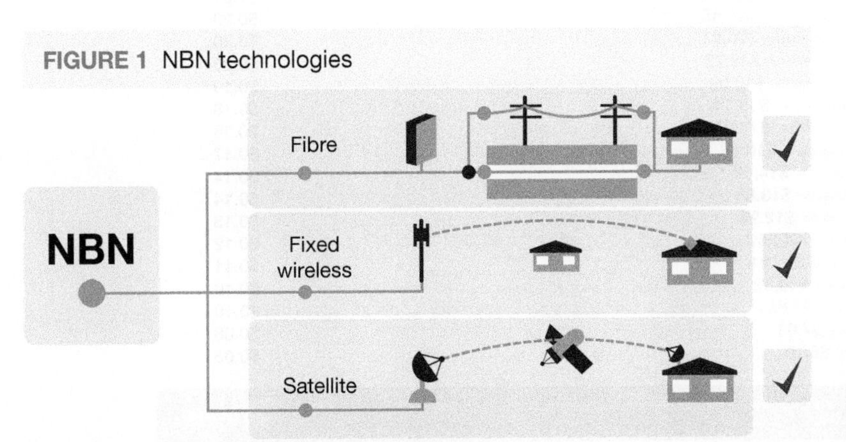

FIGURE 2 Fibre optic cable

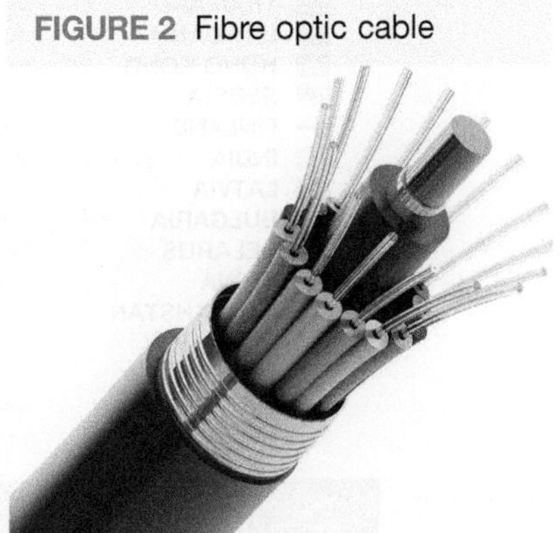

18.12.3 Assessing Australia's connectivity

According to the ABS, 86.1 per cent of Australian households were connected to the internet in 2018. **NBN Co** reported that at the end of 2020, 11.9 million business and homes could now access the NBN, with 7.9 million already connected to the network. However, while those figures are positive, Australians pay more for internet speeds that are slower compared to other countries, ranking as the fourth most expensive fibre network in the world (see **FIGURE 3**). This has implications for access by low-income households to quality internet. Furthermore, the global average internet speed is 70Mbs compared to the NBN's 45.92Mbs. These high costs and lower speeds put Australia's competitiveness in the digital economy at a disadvantage.

NBN Co publicly owned corporation of the Australian government, tasked to design, build and operate Australia's National Broadband Network as the nation's wholesale broadband provider

FIGURE 3 Price of the internet around the world

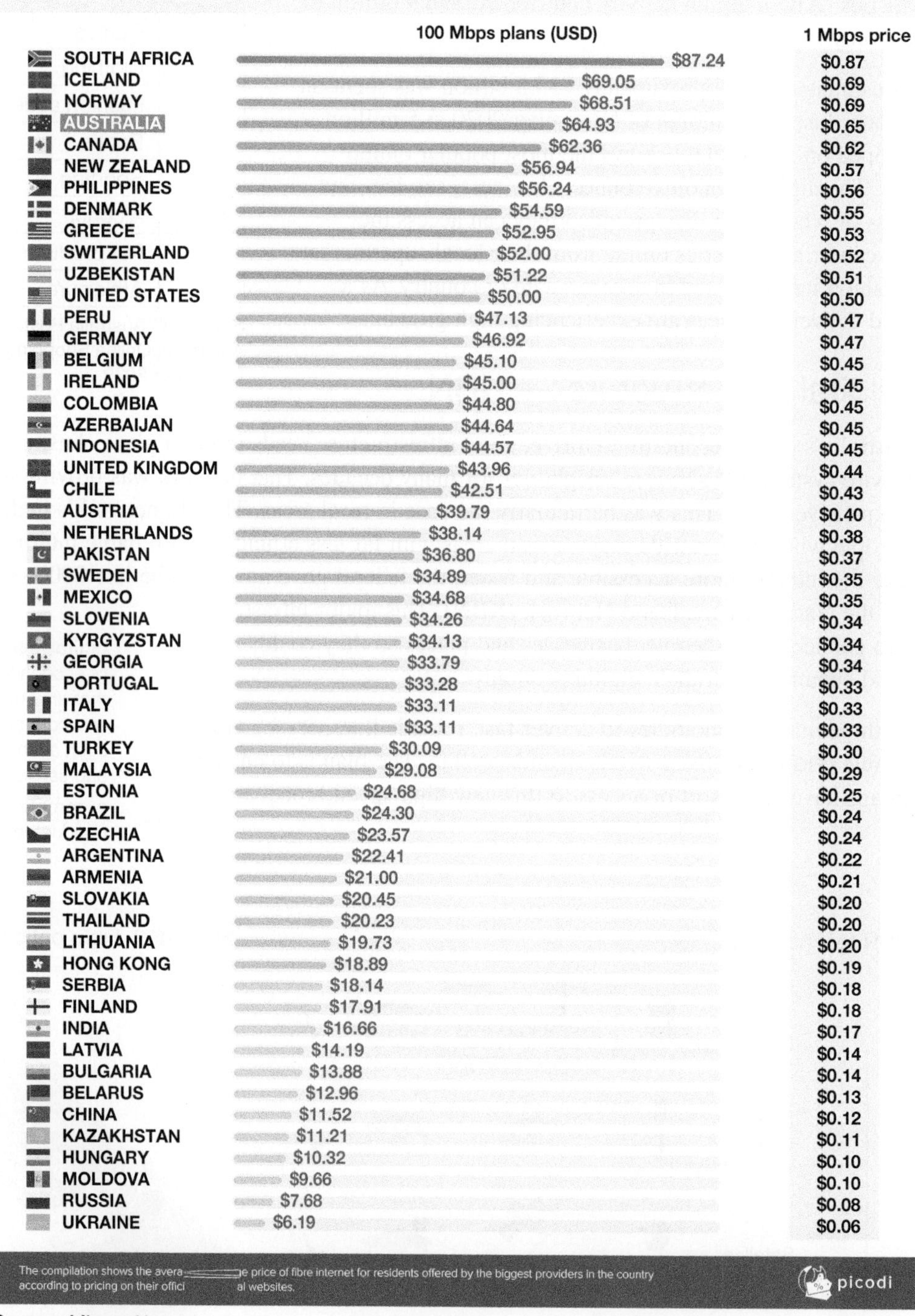

Source: Mirage.News

Rural and remote communities

The digital divide between rural and urban areas is slowly shrinking, but improvements are still needed to ensure equitable access. **TABLE 2** shows internet access based on remoteness from 2018, highlighting that the difference between rural and urban access has shrunk. However, nearly a quarter of remote and very remote households still do not have access to the internet due to high levels of remoteness in this vast country. The Sky Muster satellite network (**FIGURE 5**) aims to give access to households and business in these areas, but a limiting factor will be the high costs Australians generally pay to receive slower internet speeds than other countries, thus maintaining a digital divide.

TABLE 2 Household access to the internet based on remoteness, 2018

	With access (%)	Without access (%)
Remoteness category		
Major cities	87.9	12.0
Inner regional	82.7	17.3
Outer regional	80.7	19.2
Remote or very remote	77.1	22.9
Section of community		
Urban	86.3	13.7
Rural	84.1	15.9

Source: Australian Bureau of Statistics

FIGURE 4 Australia's remoteness areas

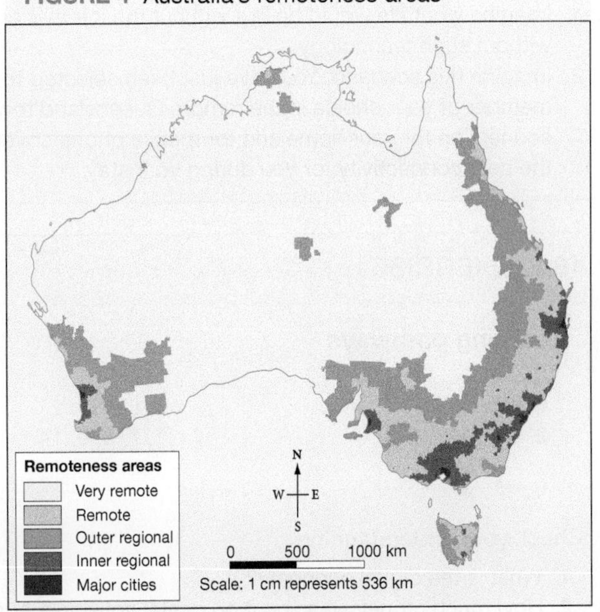

Source: Australian Bureau of Statistics (2016)

FIGURE 5 The Sky Muster satellite was launched in 2015 to improve connectivity for remote Australia.

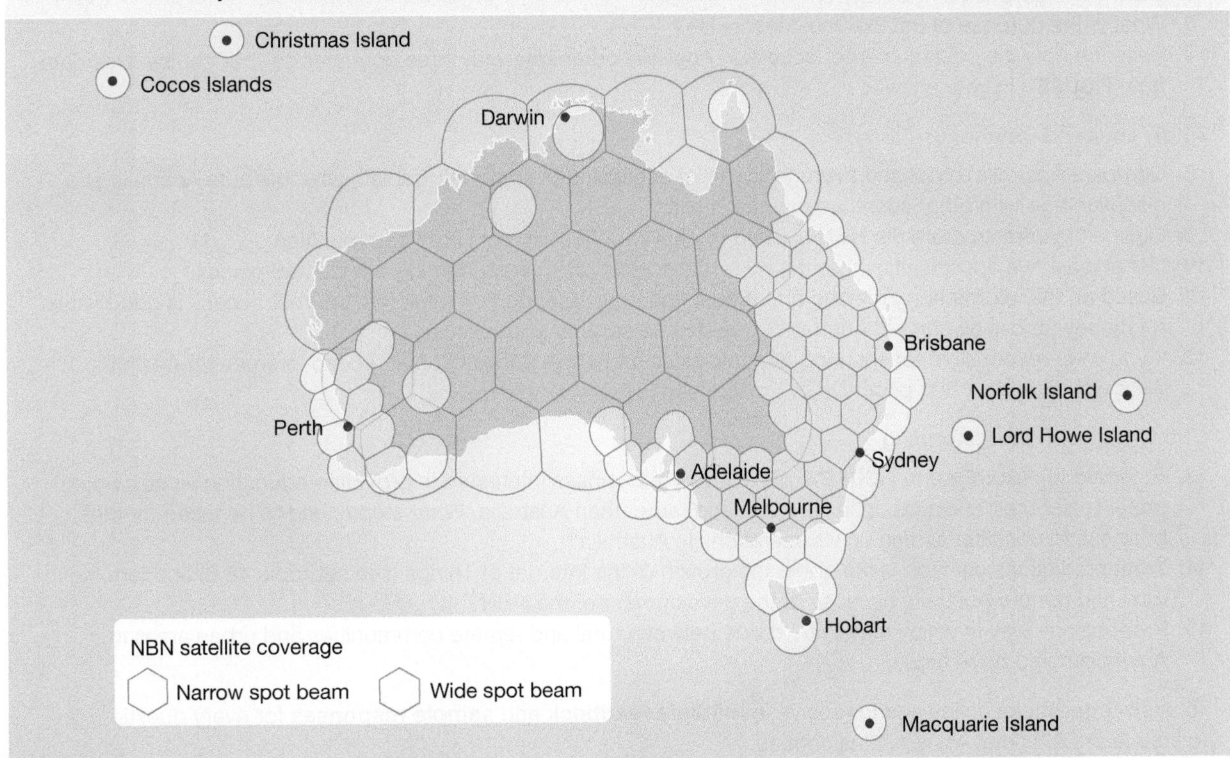

Source: NBN Co

on Resources

eWorkbook Connectivity in Australia (ewbk-9622)
Video eLesson Connectivity in Australia — Key concepts (eles-5152)
Interactivity Connectivity in Australia (int-8571)

18.12 ACTIVITIES

1. Imagine what life would be like without the internet and mobile phones. Write a letter to a friend about your life without such technology.
2. Imagine this scenario. You have just been selected to 'live remotely' for two years with a friend or family member of your choice in the remote Queensland town of Eromanga. Research your choice of internet connection for your home and for mobile phone coverage. Justify your choices in terms of what will provide the best connectivity for you during your stay.

18.12 EXERCISE

Learning pathways

■ LEVEL 1	■ LEVEL 2	■ LEVEL 3
2, 3, 4, 10, 15	1, 7, 8, 9, 14	5, 6, 11, 12, 13

Check your understanding

1. What is the digital economy?
2. How much money was spent on digital activity in Australia in 2018?
3. List the most popular online purchases made by Australians in late 2020.
4. What are the three forms of NBN technologies for households and businesses to connect to?
5. Outline the purpose of the NBN.
6. What is the purpose of the Sky Muster satellite?
7. Describe how the cost of internet access in Australia compares with three countries of your choice. Use data from **FIGURE 3** in your answer.

Apply your understanding

8. Compare Australia's cost and average speed of accessing the internet. Explain why this puts Australia at a disadvantage with other countries that are similar.
9. Outline the infrastructure the NBN is using to deal with Australian remote communities.
10. List at least five opportunities that the NBN has created by interconnecting people and places.
11. Based on the information provided about cost and coverage, do you think that internet access is good value for money across all parts of Australia? Use data to support your answer.
12. If you were responsible for ensuring Australians all had equal access to fast, cheap internet connections, what would be your first priority? Give reasons to justify your choice.

Challenge your understanding

13. Referring to **FIGURE 3** and **FIGURE 4**, explain why a business interested in providing education videos would prefer to set their business up in New Zealand rather than Australia. What steps need to be taken for that business to consider setting up its operations in Australia?
14. What predictions can you make about the growth of the Internet of Things (see subtopic 18.6) and farming in rural and remote Australia because of the development of the NBN?
15. Propose one way in which the digital divide between rural and remote communities and urban areas in Australia could be reduced.

To answer questions online and to receive **immediate feedback** and **sample responses** for every question, go to your learnON title at www.jacplus.com.au.

18.13 Investigating topographic maps — Singapore is a well-connected place

LEARNING INTENTION

By the end of this subtopic, you will be able to interpret a topographic map to describe how Singapore is connected to the world.

18.13.1 Singapore's strategic location

Singapore plays an important role in connecting places around the world. It has a very strategic location in the heart of South-East Asia, and is an important trade and transport hub. Singapore is an island that borders Indonesia and Malaysia by sea, and its export-oriented economic policy encourages flows of trade and investment.

Singapore is a major transport node. Annually more than 55 million passengers pass through Changi International Airport and more than 140 000 vessels pass through the Port of Singapore. The airport is the sixth busiest for international traffic, providing connectivity to more than 300 cities worldwide. There are more than 27 000 aircraft movements occurring every month and almost one flight every 90 seconds.

FIGURE 1 Changi airport and port facilities connect Singapore to more than 300 countries.

FIGURE 2 The Strait of Malacca links the Pacific Ocean and the Indian Ocean.

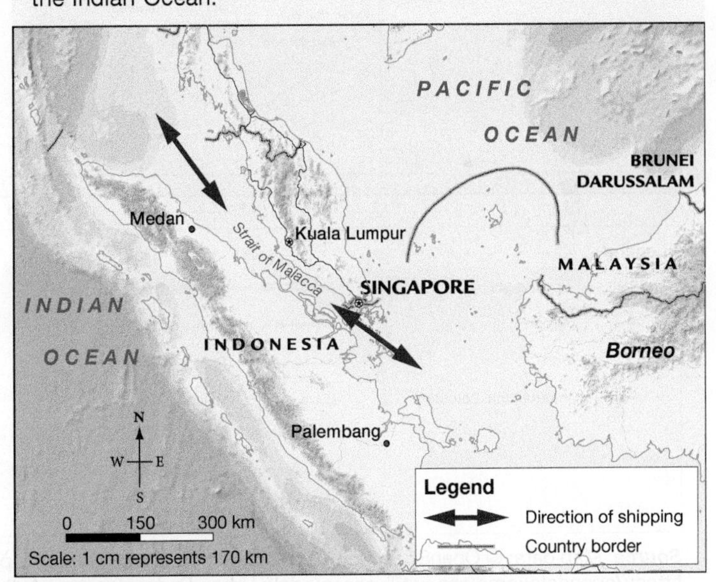

Source: Spatial Vision

FIGURE 2 shows Singapore's strategic location at the southern end of the Strait of Malacca, the shortest shipping route between the Indian Ocean and the Pacific Ocean. This stretch of water is the biggest oil corridor in the world and accounts for 25 per cent of all traded goods. At one point the strait narrows to only 2.8 kilometres, creating shipping congestion and an increased risk of piracy. The Port of Singapore handles a wide range of cargo including containers, motor vehicles and crude oil, which is refined into petrochemical products.

Singapore consists of 63 islands and is linked to neighbouring mainland Malaysia by road and rail connections. The Causeway Link buses shown in **FIGURE 4** enable day workers from Malaysia to commute to Singapore. Singapore is also a popular tourist destination and in 2018, the Singapore Tourism Board reported that more than 18.5 million international tourists visited the nation.

Interconnections exist between Singapore and its multicultural population. Chinese groups comprise over 74 per cent of the population, and Malay and Indian groups form over 20 per cent combined. As a former British colony, a member of the Commonwealth of Nations and a founding nation of Association of South-East Asian Nations (ASEAN), Singapore has many links to cultures and economies around the world.

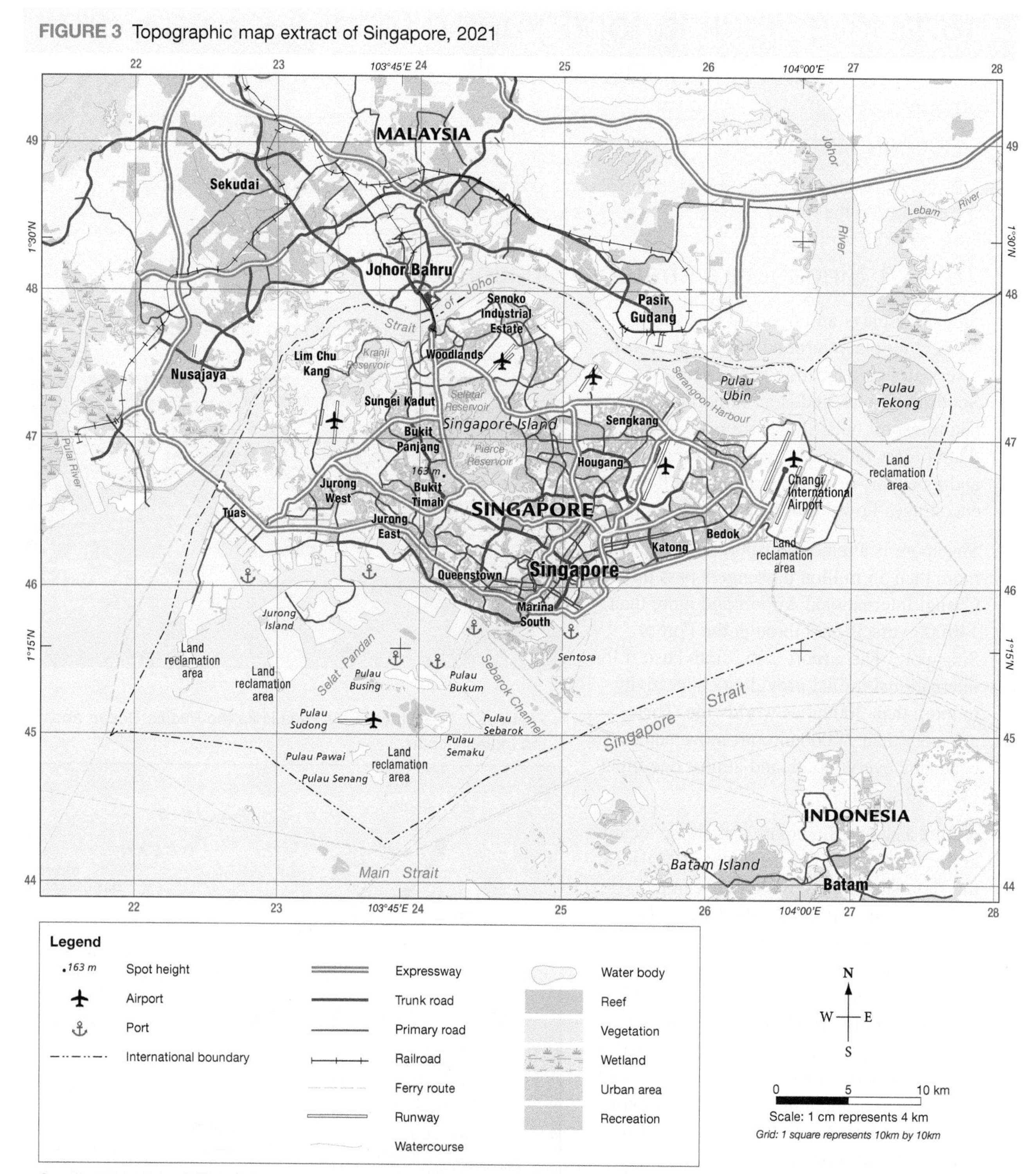

FIGURE 3 Topographic map extract of Singapore, 2021

Source: Map data: © OpenStreetMap contributors, https://openstreetmap.org. Data is available under the Open Database Licence, https://opendatacommons.org/licenses/odbl/; Urban Redevelopment Authority (2017). Data is available under the Singapore Open Data Licence v1, https://data.gov.sg/open-data-licence.

FIGURE 4 Causeway Link buses connect Singapore with Malaysia.

on Resources

eWorkbook	Investigating topographic maps — Singapore is a well-connected place (ewbk-9626)
Digital document	Topographic map of Singapore (doc-36262)
Video eLesson	Investigating topographic maps — Singapore is a well-connected place — Key concepts (eles-5153)
Interactivity	Investigating topographic maps — Singapore is a well-connected place (int-8572)
Google Earth	Singapore (gogl-0095)

18.13 EXERCISE

Learning pathways

■ LEVEL 1	■ LEVEL 2	■ LEVEL 3
1, 2, 3, 8	4, 5, 6, 11	7, 9, 10, 12

Check your understanding

1. Which country is Singapore's closest neighbour?
2. Name the body of water that separates Singapore from its closest neighbour.
3. Identify the country that is found at the following area reference locations:
   a. AR2548
   b. AR2644
   c. AR2545.
4. What is the area reference for Changi International Airport?
5. What is the direction of Batam from Singapore?
6. What is the direction of Johor Bahru from Singapore?
7. Describe the location of ports within Singapore Island as shown on the map.

Apply your understanding

8. Suggest two advantages of road and rail access to Malaysia for residents of Singapore.
9. Use the scale for the map of Singapore to measure the following:
   a. the length of Singapore
   b. the width of Singapore.
10. Estimate the land area (in square kilometres) of Singapore.
11. What is the distance between Singapore and Batam Island along the ferry route shown?

Challenge your understanding

12. If you were responsible for choosing the site for a new international airport for Singapore, where would you locate it? Give reasons for your choice.

To answer questions online and to receive **immediate feedback** and **sample responses** for every question, go to your learnON title at www.jacplus.com.au.

18.14 Thinking Big research project — Electric vehicle report

online only

The content in this subtopic is a summary of what you will find in the online resource.

Scenario

The use of electric vehicles (EVs) around the world is increasing. As technology such as autonomous driving and fast charging continue to improve, it is likely that in Australia the popularity of EVs will increase. In March 2019, the Australian Labor Party (ALP) made a target of 50 per cent of all car sales in Australia being electric by 2030. Your local council has asked you to prepare a report on the viability of electric vehicle use in your suburb.

Task

You are to research and assess EV use and availability and compare it to petrol and diesel alternatives. Your detailed report outlining the viability of EVs in your suburb or town will be presented at the next council meeting.

Go to your Resources tab to access the resources you need to complete this research project.

on Resources

ProjectsPLUS Thinking Big research project — Electric vehicle report (pro-0175)

18.15 Review

18.15.1 Key knowledge summary

18.2 Connecting people with places

- The invention of steam and internal combustion engines led to improvements in transport technologies, making long-distance travel faster.
- Improvement in ship and plane technologies have made travel to far-flung places faster.
- Trains and automobiles helped connect people and places within a country, allowing people to live further from where they worked.
- New transport technologies such as the Hyperloop and driverless cars will make travelling long distances faster and safer.

18.3 Public transport

- Public transport is an important form of transport that connects people to places at a relatively low cost.
- Singapore has the best public transport system in the world because of its availability, affordability and convenience in using contactless ticketing systems.
- Young people depend on public transport than most other generations.
- Australia's two largest cities by population, Sydney and Melbourne, have significantly higher rates of public transport use than the other state capitals.

18.4 Transport accessibility

- A good public transport system makes a place more liveable by connecting many people to other places.
- People who live in the outer suburbs of cities and in regional and rural areas tend to have poor access to public transport, often leading them to rely on the car.
- Vulnerable people and those with a disability often find accessing public transport difficult. This has been improving with technological innovations such as buses that lower to the curb and mobile phone technology that gives real-time tracking and information.

18.6 ICT innovations

- The telegraph was the first communication technology to connect people to places over long distances using electricity.
- The development of the binary digit (bit), transistors and computer chips brought in the digital age, where information could be sent over long distances reliably using fibre, radio waves and even copper wires.
- The internet is responsible for connecting more people and places than ever before, leading to other web technologies that allow people to communicate, share ideas and content regardless of where they are in the world.
- The Internet of Things (IoT) connects non-traditional computers to the internet, making appliances and sensors accessible anywhere in the world.

18.7 Connecting through ICT

- The 'death of distance' is a term that has developed due to distance no longer being a limiting factor for communication and connection between people thanks to ICT and transport technologies.
- Social media uses the internet to connect people with shared interests, families and friends, and business opportunities, regardless of where they are in the world.
- Different generations interact with communication technologies differently; those identified as Generation Z and Alpha have not known the world without the internet.

18.8 Access to services through ICT

- Connecting to the internet relies on available infrastructure. Many developed countries have good connection while developing countries do not. This is due to the amount of money countries spend on ICT.

- Mobile phones are an important technology that connects people worldwide. Many poor countries rely on this communication tool.
- The internet has created a digital economy in which people buy, sell and trade goods online, giving businesses access to more customers and giving consumers more opportunity to find a bargain.

18.10 ICT accessibility

- Not all people have equal access to the internet and other ICT, limiting their opportunities in what is called a 'digital divide'.
- Most of the world has some access to the internet, but the digital divide also includes the quality of access. Some countries still rely on second generation (2G) mobile phone networks as their main source of connectivity.
- There are many economic, social and environmental benefits to being connected, and consequences of being disconnected, such as poor economic growth and social concerns about falling behind in the ICT race.

18.11 Impacts of ICT in developing countries

- Developing countries usually have poor access to ICT.
- Kenya, a country on the African continent, is helping its citizens with a mobile phone payment system called M-Pesa (mobile money).
- As a result, more Kenyans are connected to the internet through better mobile phone coverage networks, their economy is growing and 2 per cent of households have been lifted out of poverty.

18.12 Connectivity in Australia

- The digital economy is where people, businesses and services trade online.
- It is an important part of Australia's economy and requires good infrastructure to support it.
- The National Broadband Network (NBN) has helped Australia's digital infrastructure.
- The NBN was initiated by the government to provide internet access to as many Australians as possible using fibre optic, radio waves and satellite technologies.
- Australia's connectivity is good, but in comparison to similar countries, our internet access is more expensive for slower internet speeds.

18.15.2 Key terms

2G second generation mobile network that moved from analogue to digital with a speed of 0.1 megabits a second

3G third generation digital network with a typical download speed of 8 megabits a second

4G fourth generation digital network with a typical download speed of 15 megabits a second

5G fifth generation digital mobile network with a typical download speed of 200 megabits a second

airliner an aircraft that transports people and goods

bit a portmanteau (shortened combination) of the words binary and digit

byte eight bits make one byte. This is the measurement used to store data such as kilobyte, megabyte and gigabyte.

delegate person sent or authorised to represent others; in particular an elected representative sent to a conference

digital divide a type of inequality between groups in their access to and knowledge of information and communication technology

digital economy the network of economic and social activities that are supported by information and communication technology, such as the internet and mobile and sensor networks

e-commerce business conducted online

expenditure the amount of money spent

gross domestic product (GDP) the value of all the goods and services produced within a country in a given period of time. It is often used as an indicator of a country's wealth.

infrastructure the features that are built by humans to allow other activities to occur; for example, roads, electricity networks and water supply pipes

live streaming technology that lets you watch, create and share videos in real time over the internet

liveable city an informal name given to cities that rank highly on a reputable annual survey of living conditions

Maasai ethnic group of semi-nomadic people in Kenya and Tanzania, whose way of life depends on their cattle

mass production producing a large volume of one product by automating or standardising the process

mobility the ability to move or be moved freely and easily

NBN Co publicly owned corporation of the Australian government, tasked to design, build and operate Australia's National Broadband Network as the nation's wholesale broadband provider

non-government organisation (NGO) non-profit group run by people (often volunteers) who have a common interest and perform a variety of humanitarian tasks at a local, national or international level

patron a customer, especially a regular one

public transport transport of passengers by group travel systems available for use by the general public, typically managed on a schedule, operated on established routes, and that charges a posted fee for each trip. Public transport modes include city buses, trolleybuses, trams (or light rail), passenger trains, rapid transit (e.g. metro/subway/underground) and ferries.

social network allows like-minded individuals to be in touch with each other using websites and web-based applications

zettabyte one trillion gigabytes

18.15.3 Reflection

Complete the following to reflect on your learning.

Revisit the inquiry question posed in the Overview:

What are the different ways that technology connects people and places?

1. Now that you have completed this topic, what is your view on the question? Discuss with a partner. Has your learning in this topic changed your view? If so, how?
2. Write a paragraph in response to the inquiry question, outlining your views.

Subtopic	Success criteria			
18.2	I can identify the major transport technologies that helped connect people and places.			
	I can describe how transport technologies developed.			
	I can outline future innovations in transport technologies.			
18.3	I can identify places that have the best public transport systems.			
	I can outline who uses public transport in Australia.			
	I can name technologies that make public transport more accessible.			
18.4	I can identify groups of people and places that have limited access to public transport.			
	I can describe some ways that technology is helping people living with a disability access public transport.			
18.5	I can read and interpret a topological map.			
18.6	I can describe the innovations that led to the digital information age.			
	I can describe the impact of the internet in connecting people and places.			
	I can define the Internet of Things.			

Subtopic	Success criteria			
18.7	I can outline how social media connects people.			
	I can identify which people are more likely to interconnect with different social media.			
18.8	I can explain how people are interconnected through the internet and mobile phones.			
	I can describe the importance of ICT infrastructure.			
	I can describe the 'digital economy' and its benefits.			
18.9	I can construct a table of data suitable for input into a GIS.			
18.10	I can describe the digital divide in access to ICT.			
	I can explain the benefits of connectivity and the consequences of being disconnected from ICT.			
18.11	I can explain and provide examples of how access to a mobile phone helps people in developing countries.			
18.12	I can define and use the term *digital economy*.			
	I can describe and explain the effectiveness of Australia's NBN network compared to other places.			
18.13	I can interpret a topographic map to describe how Singapore is connected to the world.			

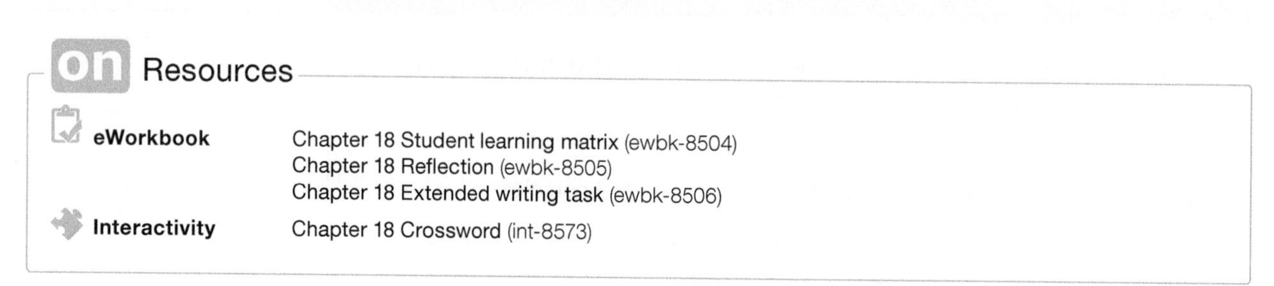

on Resources

eWorkbook Chapter 18 Student learning matrix (ewbk-8504)
Chapter 18 Reflection (ewbk-8505)
Chapter 18 Extended writing task (ewbk-8506)

Interactivity Chapter 18 Crossword (int-8573)

ONLINE RESOURCES

Below is a full list of **rich resources** available online for this topic. These resources are designed to bring ideas to life, to promote deep and lasting learning and to support the different learning needs of each individual.

eWorkbook

- 18.1 Chapter 18 eWorkbook (ewbk-7998) ☐
- 18.2 Connecting people with places (ewbk-9582) ☐
- 18.3 Public transport (ewbk-9586) ☐
- 18.4 Transport accessibility (ewbk-9590) ☐
- 18.5 SkillBuilder — Interpreting topological maps (ewbk-9594) ☐
- 18.6 ICT innovations (ewbk-9598) ☐
- 18.7 Connecting through ICT (ewbk-9602) ☐
- 18.8 Access to services through ICT (ewbk-9606) ☐
- 18.9 SkillBuilder — Constructing a table of data for a GIS (ewbk-9610) ☐
- 18.10 ICT accessibility (ewbk-9614) ☐
- 18.11 Impacts of ICT in developing countries (ewbk-9618) ☐
- 18.12 Connectivity in Australia (ewbk-9622) ☐
- 18.13 Investigating topographic maps — Singapore is a well-connected place (ewbk-9626) ☐
- 18.15 Chapter 18 Student learning matrix (ewbk-8504) ☐
- Chapter 18 Reflection (ewbk-8505) ☐
- Chapter 18 Extended writing task (ewbk-8506) ☐

Sample responses

- 18.1 Chapter 18 Sample responses (sar-0150) ☐

Digital documents

- 18.4 Public transport accessibility, Sydney (doc-36611) ☐
- Public transport accessibility, Brisbane (doc-36612) ☐
- 18.13 Topographic map of Singapore (doc-36262) ☐

Video eLessons

- 18.1 Making connections (eles-1722) ☐
- Transport and technology — Photo essay (eles-5143) ☐
- 18.2 Connecting people with places — Key concepts (eles-5144) ☐
- 18.3 Public transport — Key concepts (eles-5145) ☐
- 18.4 Transport accessibility — Key concepts (eles-5146) ☐
- 18.5 SkillBuilder — Interpreting topological maps (eles-1736) ☐
- 18.6 ICT innovations — Key concepts (eles-5147) ☐
- 18.7 Connecting through ICT — Key concepts (eles-5148) ☐
- 18.8 Access to services through ICT — Key concepts (eles-5149) ☐
- 18.9 SkillBuilder — Constructing a table of data for a GIS (eles-1743) ☐
- 18.10 ICT accessibility — Key concepts (eles-5150) ☐
- 18.11 Impacts of ICT in developing countries — Key concepts (eles-5151) ☐
- 18.12 Connectivity in Australia — Key concepts (int-8570) ☐
- 18.13 Investigating topographic maps — Singapore is a well-connected place — Key concepts (eles-5153) ☐

Interactivities

- 18.2 Connecting people with places (int-8564) ☐
- 18.3 Off the rails (int-3333) ☐
- 18.4 City connections (int-3334) ☐
- 18.5 SkillBuilder — Interpreting topological maps (int-3354) ☐
- 18.6 World Wide Web (www) (int-3341) ☐
- 18.7 Connecting through ICT (int-8567) ☐
- 18.8 Access to services through ICT (int-8568) ☐
- 18.9 SkillBuilder — Constructing a table of data for a GIS (int-3361) ☐
- 18.10 The digital divide (int-3342) ☐
- 18.11 Impacts of ICT in developing countries (int-8570) ☐
- 18.12 Connectivity in Australia (int-8571) ☐
- 18.13 Investigating topographic maps — Singapore is a well-connected place (int-8572) ☐
- 18.15 Chapter 18 Crossword (int-8573) ☐

Fieldwork

- 18.10 Technology — Evaluating sources (fdw-0024) ☐

ProjectsPLUS

- 18.14 Thinking Big research project — Electric vehicle report (pro-0175) ☐

Google Earth

- 18.13 Singapore (gogl-0095) ☐

Teacher resources

There are many resources available exclusively for teachers online.

19 Interconnection through trade

INQUIRY SEQUENCE

To access a pre-test with **immediate** feedback and **sample responses** to every question in this chapter, select your learnON format at www.jacplus.com.au.

19.1 Overview

Numerous **videos** and **interactivities** are embedded just where you need them, at the point of learning, in your learnON title at www.jacplus.com.au. They will help you to learn the content and concepts covered in this topic.

Buy, swap, sell, give. Is the trade that occurs between different countries just a way of getting things?

19.1.1 Introduction

Trade, in the form of buying, swapping, selling, and gifting goods and services (products), is a driving force that interconnects people and places all over the world. Trade has taken place ever since human societies have existed. Trade can bring people together to share the Earth's resources, but there can be problems when those resources are limited, and there can be negative consequences for the environment, people and economies.

STARTER QUESTIONS

1. What goods and services do you need to support your lifestyle? Think about your everyday life at home, at school, and in sport, recreation and hobbies.
2. Where do the goods you purchase come from?
3. What goods and services does Australia trade with the rest of the world?
4. What recent changes have taken place to the way we purchase goods and services?
5. How do you access your goods and services? To what extent do you shop online?
6. Watch the **Interconnection through trade — Photo essay** in your Resources panel. How aware are you of the way trade occurs between countries? What products or information do you buy from other countries?

FIGURE 1 Container terminals, Hong Kong

Resources

eWorkbook Chapter 19 eWorkbook (ewbk-7999)

Video eLessons Trading places (eles-1724)
Interconnection through trade — Photo essay (eles-5155)

19.2 Features of trade

LEARNING INTENTION

By the end of this subtopic, you will be able to explain some of the reasons why, and give examples of how, trade occurs across a range of scales.

19.2.1 Trade in goods and services

The Earth's resources are not distributed evenly. For instance, some places may have an abundance of iron ore deposits and other places may have none. To solve this problem, nations have developed trade, allowing producers and consumers to exchange goods and services (products). The system of trade has existed for a long time. Its earliest form involved bartering at local markets. Merchants also used land and sea routes to access markets in foreign lands and exchange goods for payment. This included Aboriginal Peoples and Torres Strait Islander Peoples, who for centuries traded with communities on the islands that now make up Indonesia and Papua New Guinea. Today we have a highly sophisticated, large-scale, global system of trade (see section 19.2.3 for one example of this global interconnection).

Goods and services, of which there are many, are generated by either processing Earth's resources or people performing tasks for each other. A good can be an item as simple as a loaf of bread or it can be as complex as a car. A service is not something you can hold in your hand; examples of services are education in a school or the advice a doctor gives a patient. What types of goods and services do you use to support your lifestyle?

As seen in **FIGURE 1**, the processing of a resource into more complex goods can involve a series of transitions, in which **value adding** takes place (that is, the value of the good increases) at each level of industry. An important consideration in the production of goods and services is the impact on the environment.

value adding processing a material or product and thereby increasing its market value

FIGURE 1 Value adding across the four levels of industry

Primary industry
Takes natural resources from the Earth or grows them. Example: a farmer grows corn that is then transported to a canning factory.

Secondary industry
Makes products from natural resources. Example: a factory makes tins of corn and sells them to supermarkets.

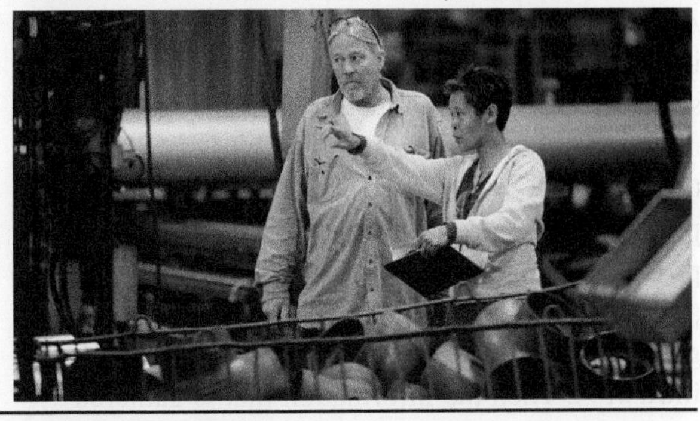

Tertiary industry
Sells products or services. Example: a supermarket sells tins of corn and other products to consumers.

Quaternary industry
Sells knowledge and information. Example: a marketing team works out the best strategies for selling products.

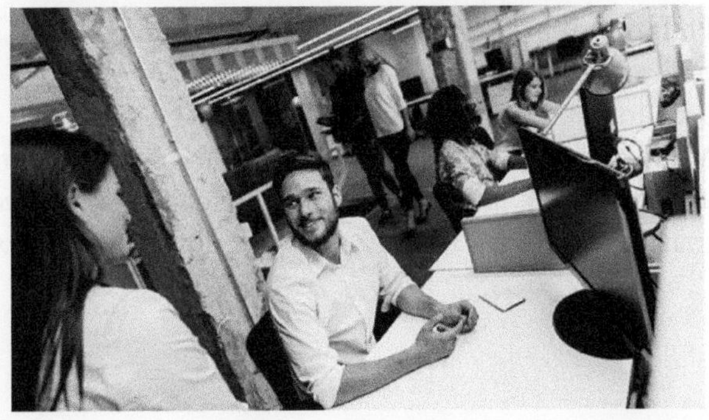

19.2.2 Trade connections within Australia

Throughout Australia and your local area, a system of trade and interconnection occurs. A visit to a local market will reveal a range of goods and services on offer from global, national, regional and local areas, as shown in **FIGURE 2**. Trade has connected people living in Australia since long before British colonisation.

FIGURE 2 The scale of production found at a local market: global products (Fairtrade goods from Nepal); national products (beef and livestock from all states of Australia); regional products (fresh fruit and vegetables from the Hunter Valley); local products (honey from local bees)

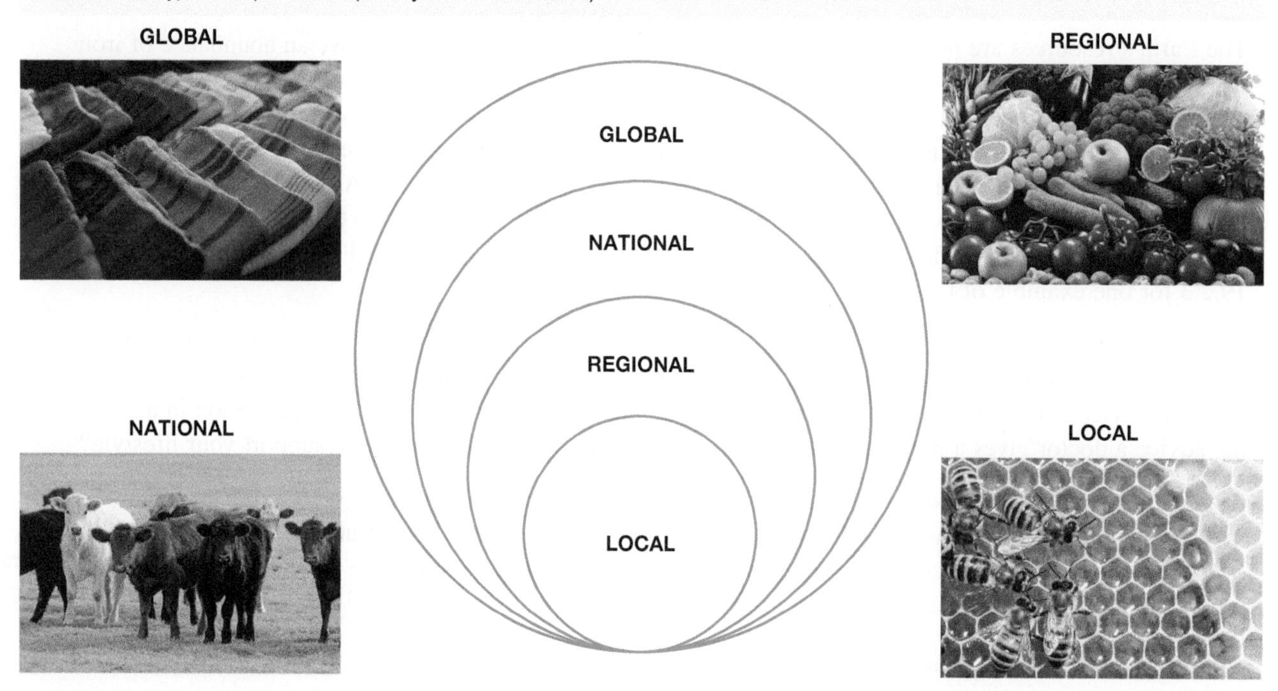

Interstate business

A heavy concentration of Australia's settlement in the eastern areas of the country has led to a strong trade connection between major urban, regional and interstate places. Freight companies transport raw materials such as fresh agricultural produce (**primary industry**) to locations that process the raw materials into a more useful product (**secondary industry**). Manufactured goods are distributed to shops for sale (**tertiary industry**). Then the information that is gathered by companies is used for marketing purposes (**quaternary industry**).

FIGURE 3 Coal is a bulky product to transport.

Freight is transported around Australia using rail, road, air and shipping. Major ports are used to link land-based (rail and road) and sea-based transport systems. Many global trade links are accessed through ports and international airports. When international shipping containers arrive in ports, they are transferred to road and rail transport for distribution within Australia.

primary industry the gathering of natural resources, such as iron ore and timber, or activities such as farming and fishing

secondary industry making raw materials into a more useful product; for example, making furniture from timber

tertiary industry selling goods and services; for example, a furniture business

quaternary industry gathering information about an industry; for example, marketing furniture

FIGURE 4 Cranes lift and stack shipping containers.

FIGURE 5 Road train transporting fuel

FIGURE 6 Inter-regional freight flows for NSW during 2011

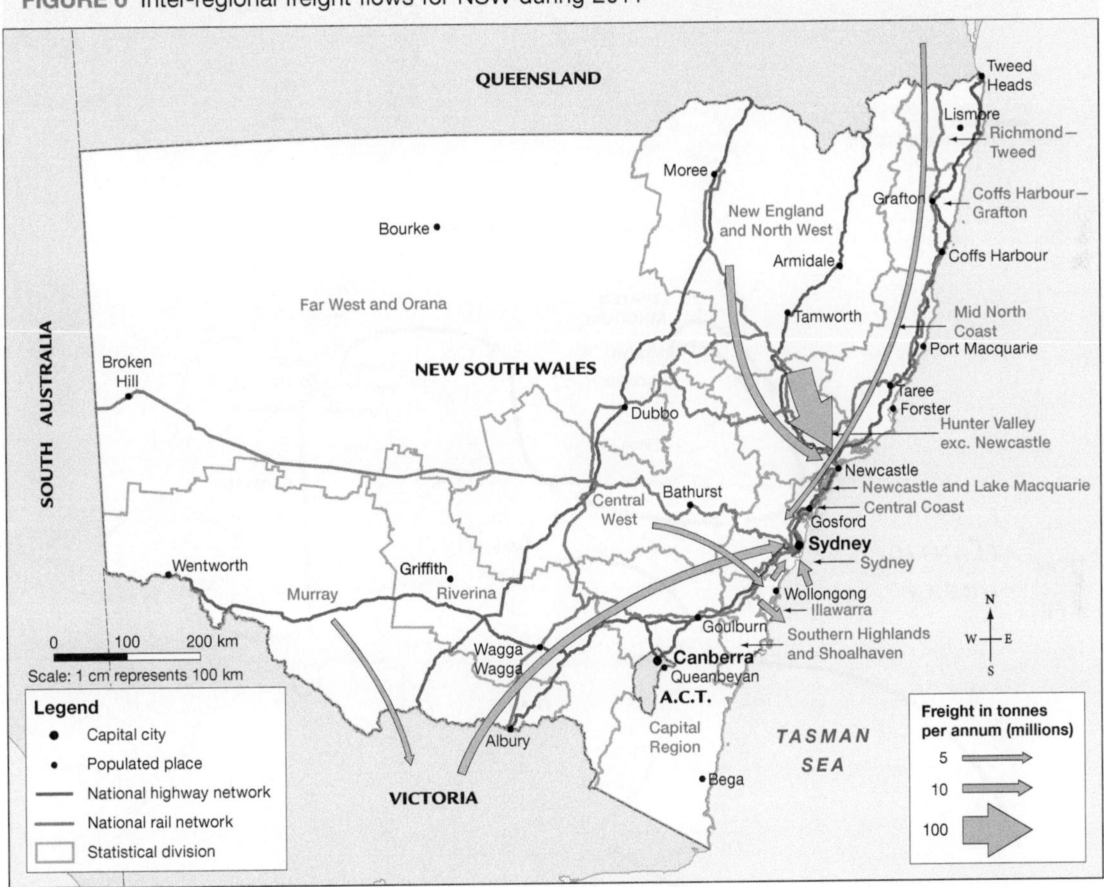

Source: Transport for New South Wales and Australian Bureau of Statistics.

19.2.3 Global trade in action — The production of Airbus 380

A modern example of the interconnection of trade is the production of the Airbus A380 aircraft (see **FIGURE 7**). To construct this plane, component parts were purchased from different countries and transported over land and sea to reach their final assembly place in Toulouse, France (see **FIGURE 9**). Factories located across Europe specialised in manufacturing, testing and assembling specific components of the aircraft to the specifications of Airbus.

FIGURE 7 The Airbus A380

FIGURE 8 Production of the A380 ceased in 2021.

FIGURE 9 The routes taken by the component parts of the Airbus A380

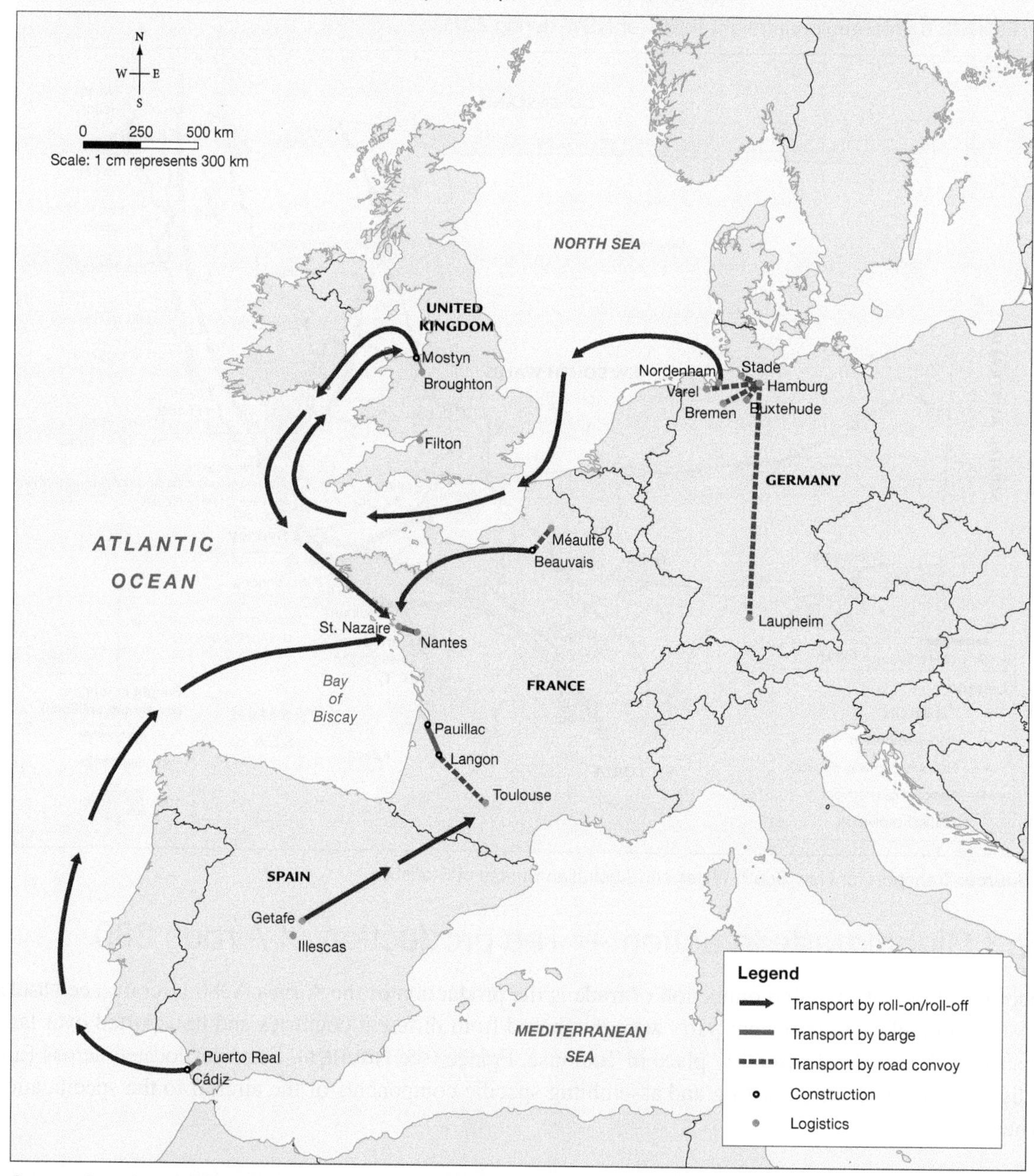

Source: Data from Wikimedia Commons, User: Steff.

on Resources

eWorkbook	Features of trade (ewbk-9634)
Video eLesson	Features of trade — Key concepts (eles-5156)
Interactivity	Up, up and away (int-3339)

19.2 ACTIVITY

Investigate the process of turning raw materials into finished products.

a. Examine **FIGURE 5** and construct a flow chart explaining the transformation of a raw material produced in one primary industry to the sale of a finished product to customers.

b. Prepare a series of images to demonstrate each step in the transformation of this product from a raw material to a finished product. Assign a number to each image and prepare a sentence to explain how value is being added to this product at each stage.

19.2 EXERCISE

Learning pathways

■ LEVEL 1	■ LEVEL 2	■ LEVEL 3
1, 5, 8, 12, 13	3, 4, 7, 10, 14	2, 6, 9, 11, 15

Check your understanding

1. What land-based transport systems operate in Australia?
2. Outline two reasons why goods and services (also referred to as products) are traded.
3. Name the four levels of industry and give an example of a product for each.
4. Consider **FIGURE 1**.
   a. Outline what is meant by the term *value adding* as a product moves through the four levels of industry.
   b. Identify three value-adding processes that take place for wheat grains.
5. List three examples of goods, and three examples of services.
6. At a local market, would you expect goods to come from local, regional, national or global sources? Give reasons for your answer.

Apply your understanding

7. Study **FIGURE 6**. Give two reasons to explain why the largest flow of freight into Sydney travels from the Hunter Valley.
8. Explain why communities have always engaged in trade.
9. It has been claimed that countries such as China and India, each with a growing middle class eager for goods and services, will put a strain on world resources. Explain why this might be the case.
10. Using **FIGURE 1**, suggest which level of industry your teachers work in. Provide a justification for your answer.
11. Why might road transport be important for trade in regional Australia?
12. Explain the difference between a good and a service.

Challenge your understanding

13. How might a change such as growth in Australia's population from 25 million to 40 million people affect Australia's trade?
14. Suggest reasons why component parts of the Airbus A380 must come from different countries.
15. State borders in Australia were closed to most traffic during the COVID-19 pandemic, but trucks carrying freight were given special permits to cross borders. Suggest why this might have been the case, and predict what would happen if state borders were closed to all vehicles, including those carrying freight.

To answer questions online and to receive **immediate feedback** and **sample responses** for every question, go to your learnON title at www.jacplus.com.au.

19.3 SkillBuilder — Constructing and describing a flow map

online only

LEARNING INTENTION

By the end of this subtopic, you will be able to create and describe a flow map.

The content in this subtopic is a summary of what you will find in the online resource.

19.3.1 Tell me

What is a flow map?

A flow map shows the movement of people or objects from one place to another. Arrows are drawn from the point of origin to the destination. Sometimes these lines are scaled to indicate how much of the feature is moving, with thicker lines showing a larger amount.

19.3.2 Show me

How to construct a flow map

FIGURE 1 Net interstate migration

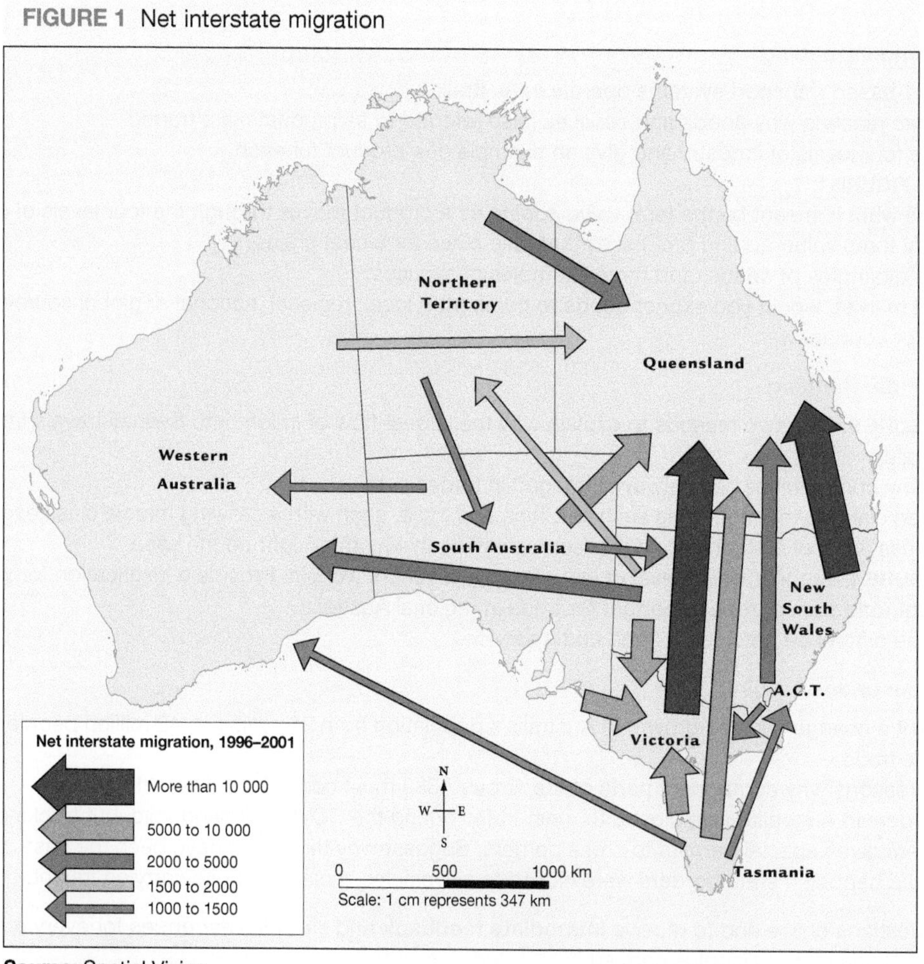

Source: Spatial Vision
Note: Migrations of less than 1000 not shown

Step 1

If you are planning to simply show the flow between places, then you need only identify each place and draw an arrow from the origin to the destination. Writing numbers on the flow lines is another method for creating a basic flow map. (To use this method, go to step 5.) On the other hand, if you want to create a map that provides an instant snapshot of the quantities of a feature being moved, then a scaled flow map is a better option. Your first step is to determine the scale you will use. Look over the data set that you have, and establish no more than five categories that will allow you to represent the data. Notice how these appear in the key of **FIGURE 1**.

Step 2

Draw the legend. Note that you will have to work in millimetres; otherwise, your arrows will dominate the map. In a key, 1 millimetre could be used to represent 1000 people. Label your key appropriately.

Step 3

Rule an arrow from a place of origin to a place of destination. Before you begin, think where you will place each arrow, as it is best to avoid overlapping them.

Step 4

As you draw arrows between the places of origin and destination, use your ruler to keep the arrow widths consistent. Neatness is important. Colour the arrows as you go.

Step 5

Ensure that the completed map includes geographical conventions (BOLTSS).

Step 6

Identify any patterns that are evident in the map. Is there an interconnection between the widest arrows? Is there an interconnection between the narrowest arrows? Write a few sentences to explain any patterns you can identify.

Step 7

Look for any anomalies in the pattern — arrows that stand out as being different. Write a sentence to describe any anomalies.

19.3.3 Let me do it

learnON

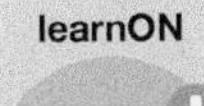

Go to learnON to access the following additional resources to help you build this skill:

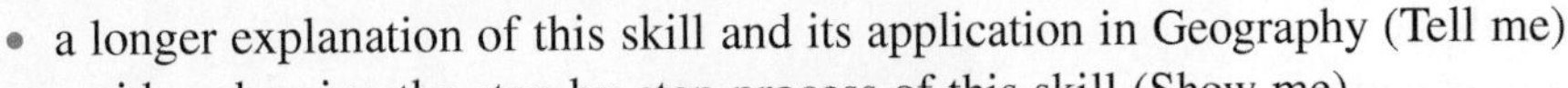

- a longer explanation of this skill and its application in Geography (Tell me)
- a video showing the step-by-step process of this skill (Show me)
- an activity and interactivity for you to practise this skill (Let me do it)
- self-marking questions to help you understand and use this skill.

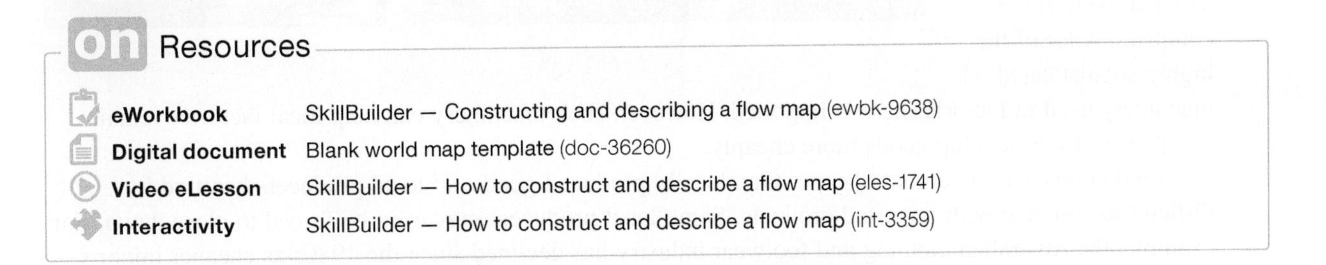

on Resources

eWorkbook	SkillBuilder — Constructing and describing a flow map (ewbk-9638)
Digital document	Blank world map template (doc-36260)
Video eLesson	SkillBuilder — How to construct and describe a flow map (eles-1741)
Interactivity	SkillBuilder — How to construct and describe a flow map (int-3359)

19.4 Australian trade connections

LEARNING INTENTION

By the end of this subtopic, you will be able to identify the interconnections between Australia and other places through the exchange of goods and services.

19.4.1 Australia's global trade

Australia has a long history of trading with other nations. For tens of thousands of years before the European colonisation of Australia, Aboriginal Peoples and Torres Strait Islander Peoples traded with each other, and with the people who lived to the north of our continent — among them, people from places that are now in Indonesia and Papua New Guinea. European settlers were involved in trade with other nations from the beginning of colonisation in Australia. During the nineteenth century, the Australian colonies were a major source of agricultural products such as wool, and minerals such as gold for Britain and its empire.

In recent years Australia has developed strong trading links with our Asian neighbours. Four of our five most important trading partners in 2019–20 were in the Asia–Pacific region. These countries include China, the United States, Japan and South Korea.

Australian businesses sell exports to both consumers and producers in other countries. Australian consumers buy imports that have been manufactured in other parts of the world and brought into this country.

FIGURE 1 People involved with trade

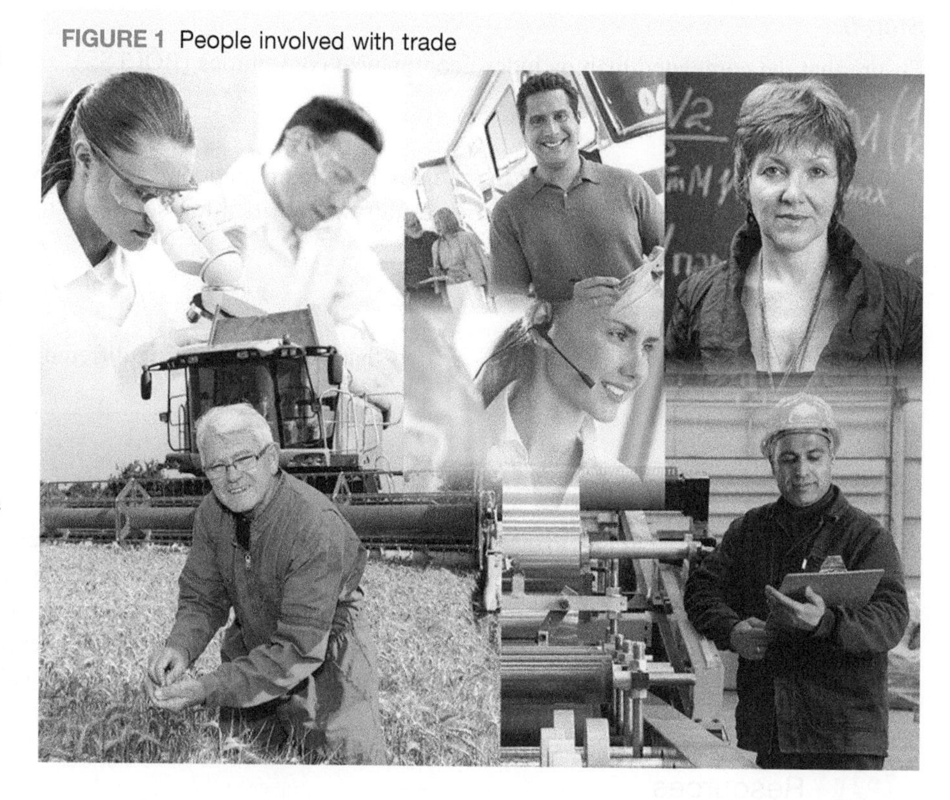

International trade can affect our economic system in several ways:

- Australia has a relatively small population, so if overseas consumers are willing to buy the goods and services we produce, this can help our local businesses to grow and employ more people. A significant proportion of jobs in Australia are directly or indirectly connected with the production of exports.
- Some products cannot be made here as efficiently as they can be in other countries. A lot of the highly sophisticated machinery used in factories here is imported. However, such machinery can help local factories remain competitive by producing goods more cheaply.
- Imported goods are sometimes cheaper than locally produced goods, so local producers can find it difficult to compete with imported products. Some local producers may even be forced to close down. For example, the Australian clothing and footwear industry has declined since the 1980s as cheaper imports from Asian countries have increased dramatically.

19.4.2 Australia's trading partners

China, Japan and the United States were Australia's top two-way **trading partners** during 2019–20. **TABLE 1** shows positive and negative changes to Australia's trading relationship with its partners over the period 2018–19 to 2019–20. Further changes to trade relationships and patterns are expected following COVID-19 disruptions, political leadership changes and tensions between trading partners.

trading partner a participant, organisation or government body in a continuing trade relationship

TABLE 1 Australia's trade in goods and services with its top ten two-way trading partners

Rank	Australia's two-way trading partners	2018–19 A$ million	2019–20 A$ million	% share of total 2019–20	% growth 2018–19 to 2019–20
1	China	235 060	251 231	28.8	6.9
2	United States	76 652	80 787	9.3	5.4
3	Japan	88 312	79 131	9.1	−10.4
4	Republic of Korea	41 407	38 868	4.5	−6.1
5	United Kingdom	30 458	36 676	4.2	20.4
6	Singapore	32 148	31 143	3.6	−3.1
7	New Zealand	30 568	28 689	3.3	−6.1
8	India	30 366	26 245	3.0	−13.6
9	Germany	23 234	21 763	2.5	−6.3
10	Malaysia	25 117	21 649	2.5	−13.6

Source: Department of Foreign Affairs and Trade website – www.dfat.gov.au

FIGURES 2 and **3** show details of Australia's import and export trade during the period 2019–20.

FIGURE 2 Australia's top ten import markets, 2019–20 (A$ billion)

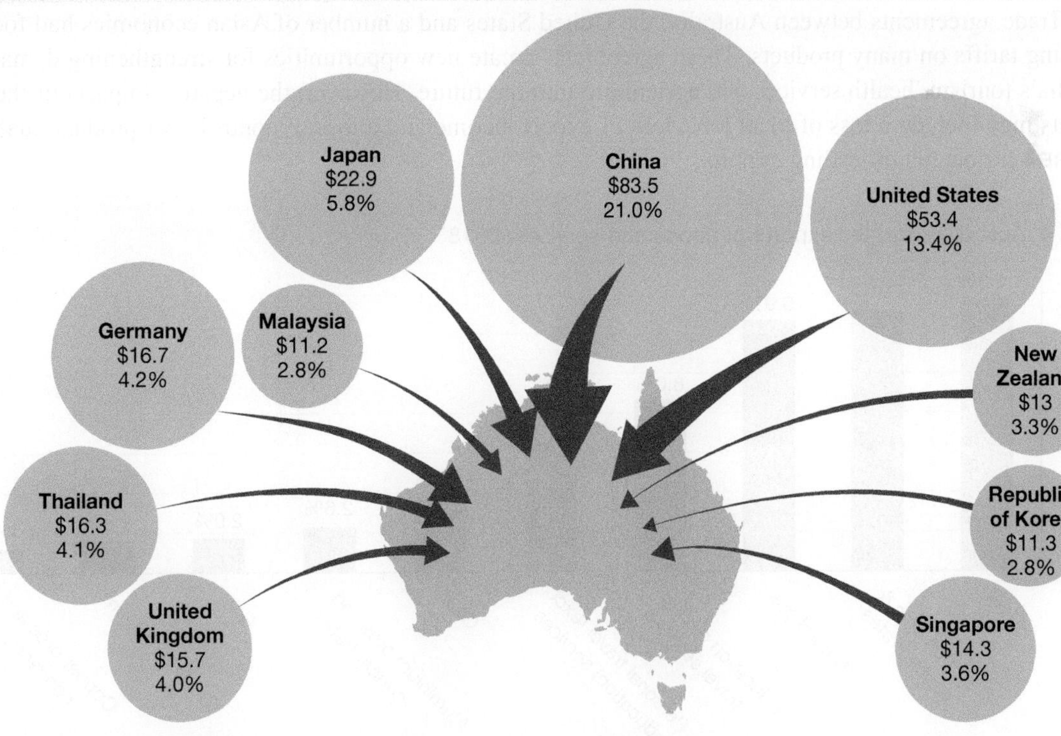

Source: Department of Foreign Affairs and Trade website – www.dfat.gov.au

FIGURE 3 Australia's top ten export markets, 2019–20 (A$ billion)

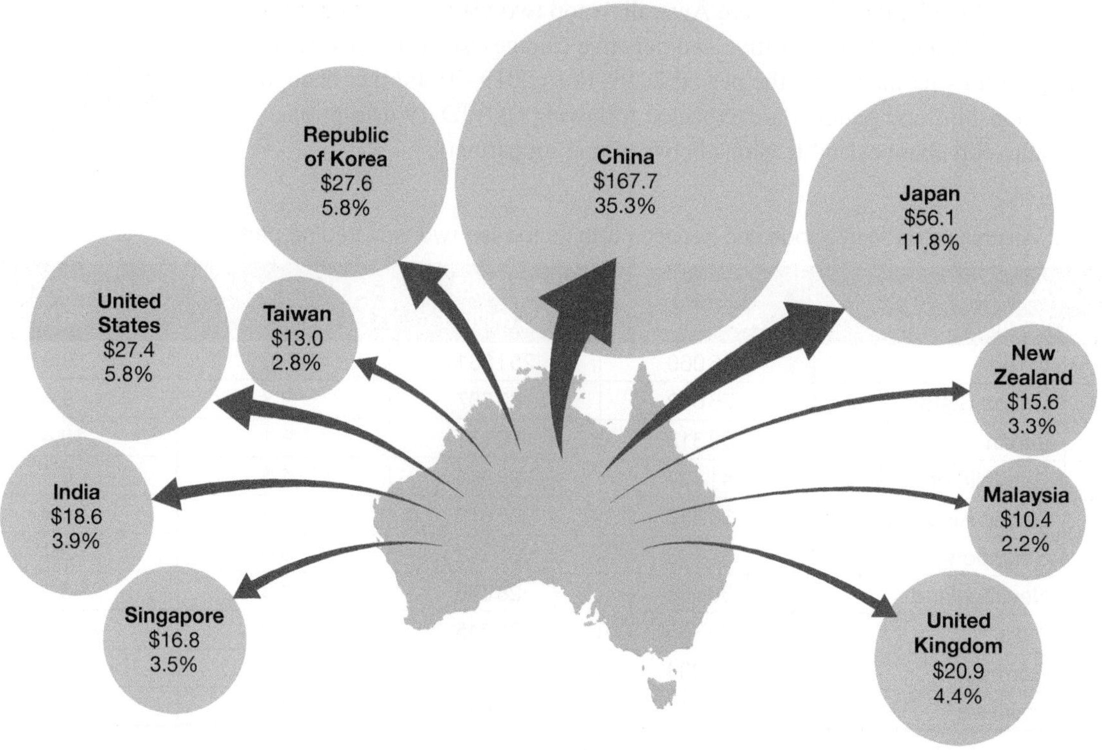

Source: Department of Foreign Affairs and Trade website – www.dfat.gov.au

19.4.3 Australia's types of trade

Exports

Australia's export trade in 2018 was valued at $438 billion and dominated by the mineral products of coal and iron ore. Trade agreements between Australia, the United States and a number of Asian economies had focused on removing tariffs on many products. These agreements create new opportunities for strengthening demand in Australia's tourism, health services and agriculture into the future. However, the negative impacts of these agreements may include a loss of local jobs, loss of export income and a loss of control over product quality. See **FIGURE 4** for details of leading exports.

FIGURE 4 Australia's top ten exports of goods and services, 2018

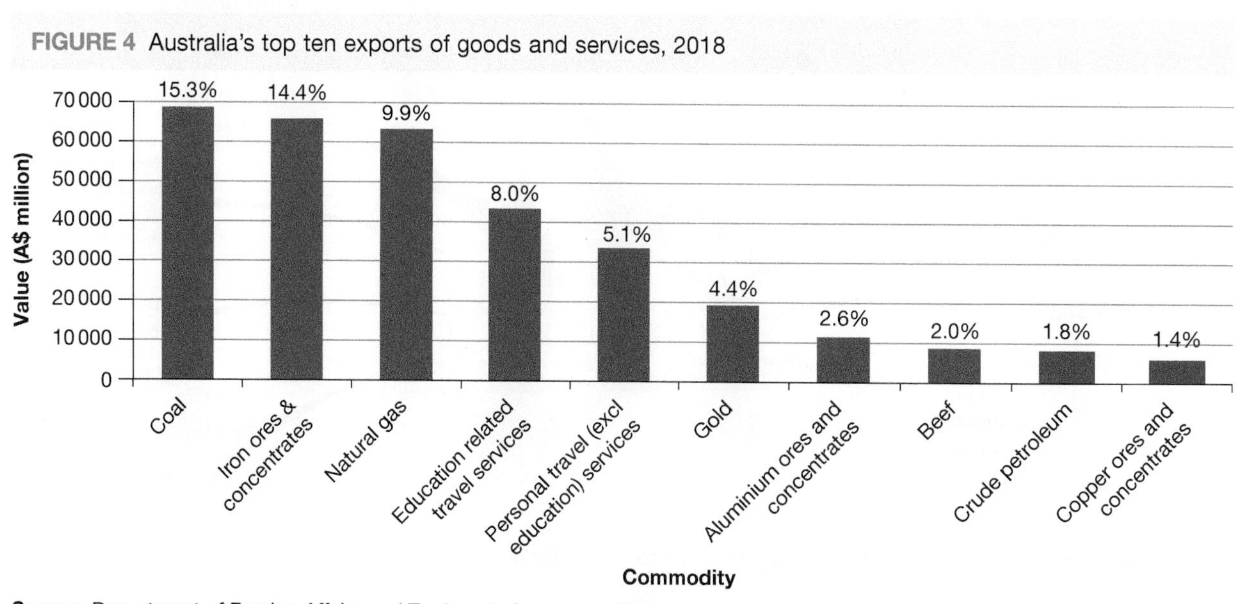

Source: Department of Foreign Affairs and Trade website – www.dfat.gov.au

International students

During 2018, the third highest export earner for Australia was the category of 'education-related travel services'. This sector was valued at over $35 billion, with 693 750 international students from more than 200 countries studying in Australia. In this service industry, students pay for knowledge and skills that they will take back to their home country. During 2020, COVID-19-related international border restrictions heavily disrupted the operation of this industry as students were not permitted to enter or leave their homelands for a period. Many continued their studies remotely from places outside of Australia.

FIGURE 5 COVID-19 led to a drop in international student numbers for Australia during 2020.

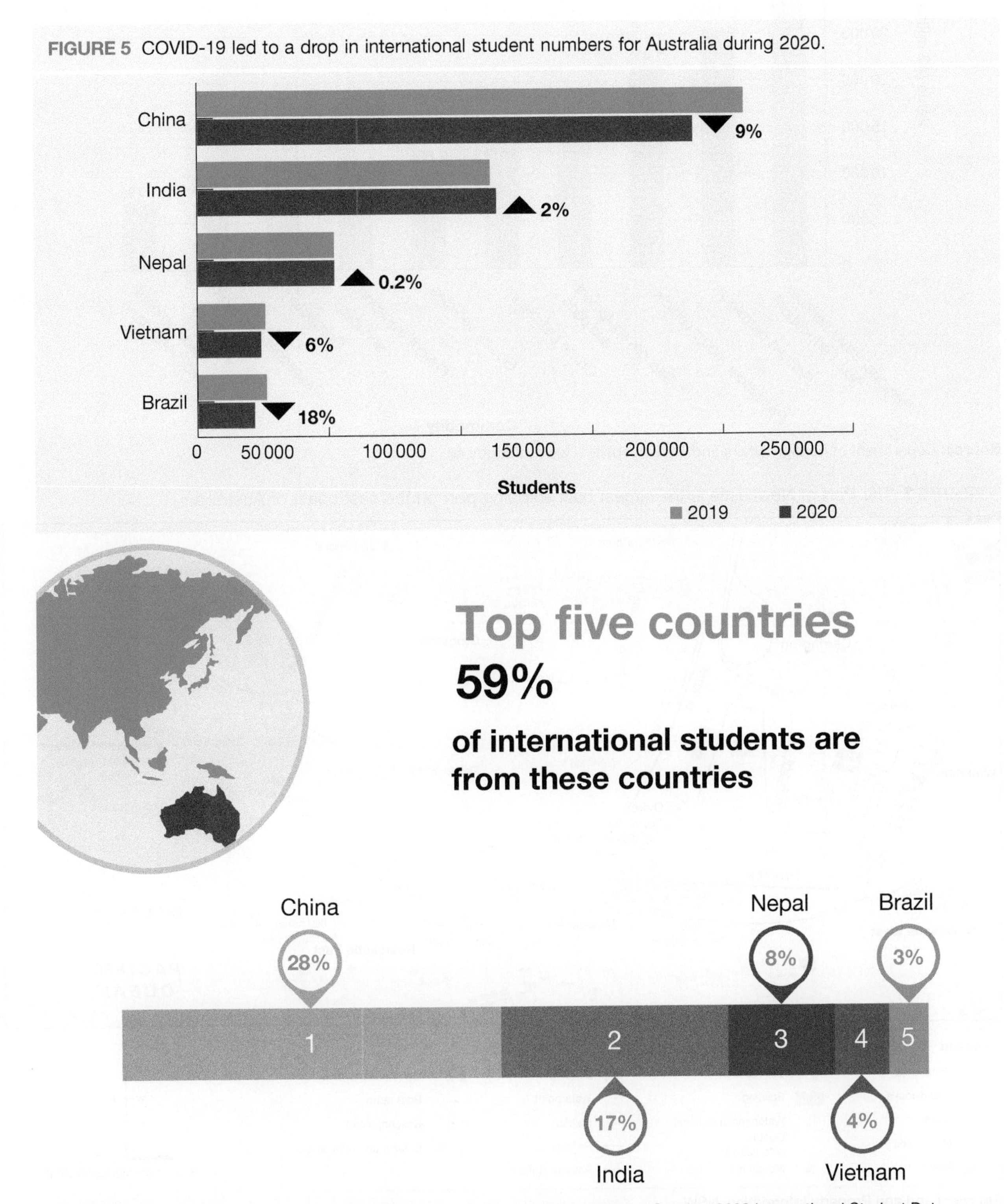

Source: Australian Government Department of Education, Skills and Employment – October 2020 International Student Data

FIGURE 6 Australia's top ten imports of goods and services, 2018 (A$ million)

Commodity	Share
Personal travel services (exc.l education)	10.8%
Refined petroleum	6.1%
Passenger motor vehicles	5.4%
Telecom equipment and parts	3.5%
Crude petroleum	3.3%
Goods vehicles	2.6%
Freight transport services	2.4%
Computers	2.3%
Passenger transport services	1.8%
Medicaments (inc. veterinary)	1.7%

Value (A$ million): 0, 5000, 10 000, 15 000, 20 000, 25 000, 30 000, 35 000, 40 000, 45 000

Commodity

Source: Department of Foreign Affairs and Trade website – www.dfat.gov.au

FIGURE 7 The Port of Newcastle is the largest bulk shipping port on the east coast of Australia.

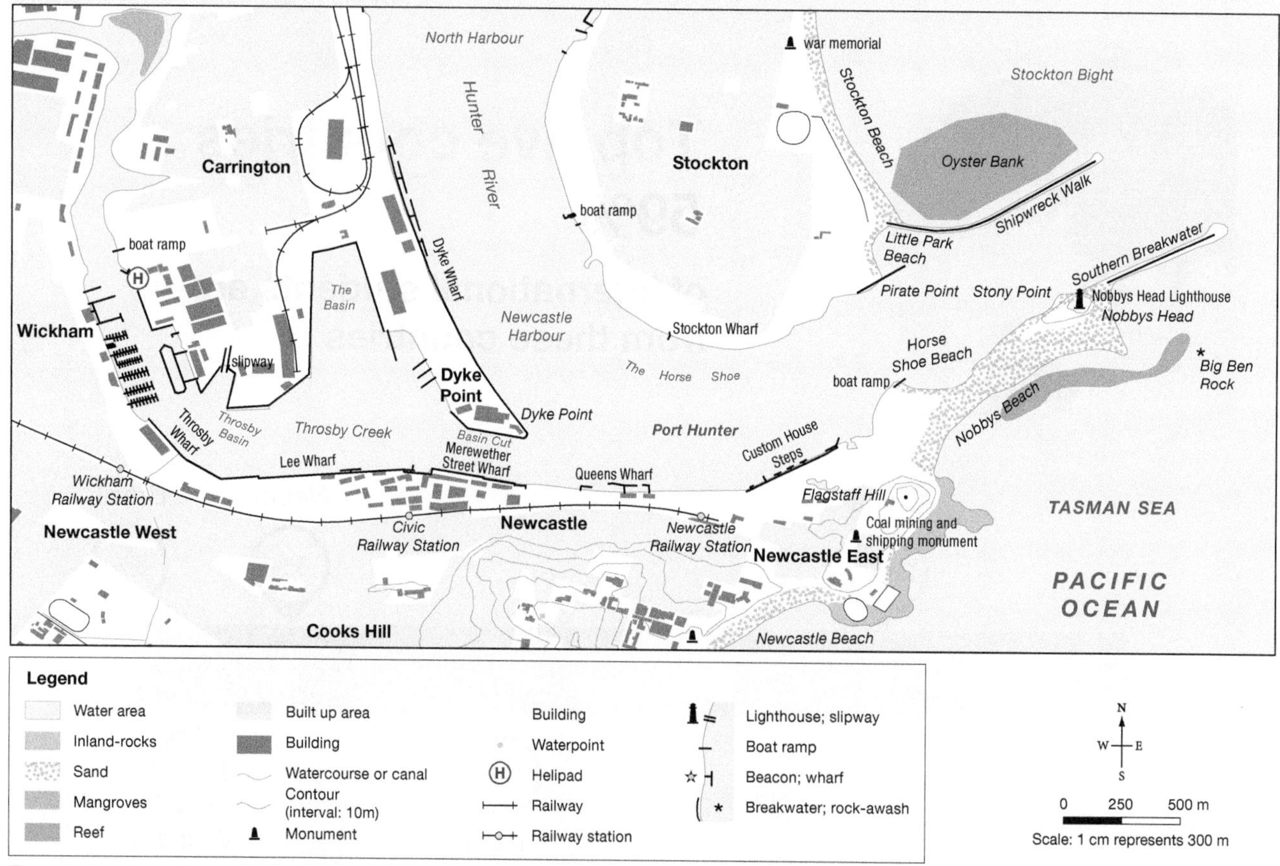

Source: Land and Property Information, NSW.

Imports

Like many countries, Australia is not self-sufficient in all goods and services. In 2018, Australia imported goods and services to a value of $414 991 million. **FIGURE 6** shows the top ten commodities of this trade. **FIGURE 7** shows the infrastructure of the Port of Newcastle, the largest bulk shipping port on the east coast of Australia. In 2020, the port handled an average of more than 400 000 tonnes of import goods and more than 13 million tonnes of export goods each month.

Trade barriers

Australia is one of the 164 members of the World Trade Organization (WTO), which covers 95 per cent of global trade. The organisation promotes free and fair trade between countries and, since 2001, its Doha Development Agenda has aimed to help the world's poor by slashing **trade barriers** such as tariffs, quotas and farm subsidies.

The Department of Foreign Affairs and Trade (DFAT) coordinates trade agreements on behalf of the Australian government, and the Australian Trade Commission (Austrade) promotes the export of goods and services. About 70 per cent of Australia's trade takes place with the member countries of the Asia–Pacific Economic Cooperation (APEC) forum. **Free trade agreements** have been negotiated by Australia with a number of countries in the Asia–Pacific region (see subtopic 19.5).

trade barrier government-imposed restriction (in the form of tariffs, quotas and subsidies) on the free international exchange of goods or services

free trade agreement a treaty between two or more countries or economies that sets out special trading conditions that reduce or remove trade barriers

on Resources

eWorkbook Australian trade connections (ewbk-9642)

Video eLesson Australian trade connections — Key concepts (eles-5157)

Interactivity Trading partners (int-3338)

myWorldAtlas Deepen your understanding of this topic with related case studies and questions.
> Aid, migration and trade

19.4 ACTIVITIES

1. Brainstorm a list of the products that Australia has in abundance and the products that Australia is unable to produce in large quantities and/or at a very cheap price.
2. As a class, discuss what types of jobs Australians perform for the rest of the world. Categorise them according to their stage of production. Discuss what impression of Australia this might give to people around the world.
3. Using the **Australian trade partners** weblink in your Resources tab, describe in detail the trade connection that exists between Australia and one other trading partner.
4. Use the **Trade investment, Ports of Australia, Trade through time** and **Trade by country** weblinks in your online resources to construct an infographic explaining one aspect of Australia's trade relationships.

19.4 EXERCISE

Learning pathways

■ LEVEL 1	■ LEVEL 2	■ LEVEL 3
1, 5, 10, 12, 15	2, 4, 7, 11, 13	3, 6, 8, 9, 14

Check your understanding

1. What are Australia's three most important exports and imports?
2. Using **FIGURES 2** and **3**, compare Australia's imports from and exports to one specific country. Use data in your response.
3. Refer to **FIGURE 7**. Identify three features of Newcastle Harbour that make it ideal for a shipping port.
4. Define the term *trade barrier* in your own words.
5. Refer to **FIGURE 5**. Which country do most of the international students in Australia come from?

Apply your understanding

6. How might a change in the growth of Australia's population affect the country's agricultural exports?
7. Look at the list of goods that Australia imports shown in **FIGURE 6**. What factors could lead to a change in the types of goods imported by the year 2050?
8. What evidence can you provide to confirm the fact that Australia is regarded as mostly a primary industry exporter?
9. Despite having a relatively small population, Australia has many goods and services to trade. Explain why this might be so.
10. What is the purpose of a free trade agreement?
11. How can Australia benefit from being a member of the WTO?
12. Explain why international students are an important part of Australia's trade connections.

Challenge your understanding

13. Prepare a list of reasons why international students from a country of your choice should come to study in Australia.
14. Suggest why there were two countries from which there was an increase in international students coming to Australia between 2019 and 2020. Explain how this might have been possible considering Australia's borders were closed to international arrivals for much of 2020.
15. Predict which of Australia's current top ten exports will become less profitable in the future. Justify your choices.

To answer questions online and to receive **immediate feedback** and **sample responses** for every question, go to your learnON title at www.jacplus.com.au.

19.5 Global food trade

LEARNING INTENTION

By the end of this subtopic, you will be able to identify how food is traded around the world and give examples to demonstrate the global movement of food produced in Australia.

The content in this subtopic is a summary of what you will find in the online resource.

The world's population is unevenly distributed, as is the quantity of food produced. Some places produce an abundance of food, while others struggle to produce enough to maintain food security.

To learn more about global food trade, go to your learnON resources at www.jacPLUS.com.au.

FIGURE 1 Fresh and processed foods are traded all around the globe

Contents

on Resources

 **eWorkbook** Global food trade (ewbk-9646)

 **Video eLesson** Global food trade — Key concepts (eles-5158)

Interactivity Global food trade — Interactivity (int-7932)

19.6 SkillBuilder — Constructing multiple line and cumulative line graphs

online only

LEARNING INTENTION

By the end of this subtopic, you will be able to transform data into multiple line and cumulative line graphs.

The content in this subtopic is a summary of what you will find in the online resource.

19.6.1 Tell me

What are multiple line graphs and cumulative line graphs?

Multiple line graphs consist of a number of separate lines drawn on a single graph. Cumulative line graphs are more complex to read, because each set of data is added to the previous line graph. Both formats show change over time and show trends effectively.

19.6.2 Show me

How to draw multiple line graphs and cumulative line graphs

Step 1

Place all the data into an Excel spreadsheet. Click on the Insert tab and select a multiple line chart. Add the axis labels by clicking on your graph's outer border, selecting the Layout tab in the Chart Tools section, clicking on Axis Titles in the Labels section, and following the steps from there.

Step 2

A cumulative line graph can be generated by selecting the table data and selecting an Area chart from the Charts section under the Insert tab.

Step 3

Check that you have supplied labels on all axes, units of measurement, and a title or caption.

FIGURE 1 Water use by five states in the Murray–Darling Basin, 1920–2020, as a multiple line graph

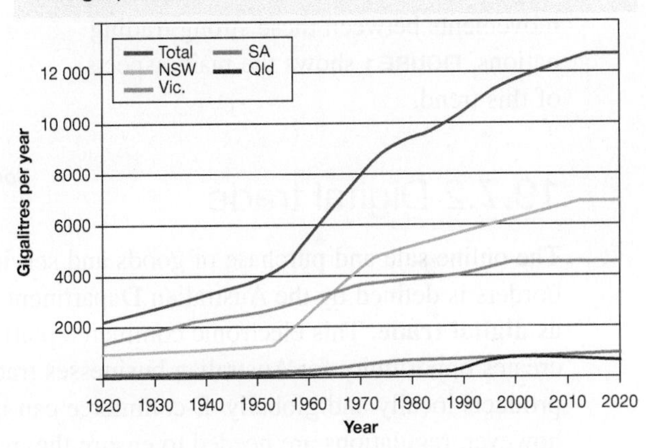

Source: Department of Foreign Affairs and Trade website – www.dfat.gov.au

19.6.3 Let me do it

learnON

Go to learnON to access the following additional resources to help you build this skill:

- a longer explanation of this skill and its application in Geography (Tell me)
- a video showing the step-by-step process of this skill (Show me)
- an activity and interactivity for you to practise this skill (Let me do it)
- self-marking questions to help you understand and use this skill.

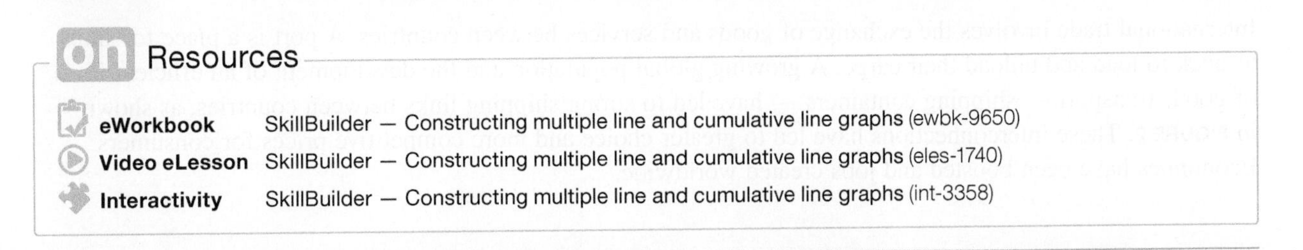

on Resources

eWorkbook SkillBuilder — Constructing multiple line and cumulative line graphs (ewbk-9650)

Video eLesson SkillBuilder — Constructing multiple line and cumulative line graphs (eles-1740)

Interactivity SkillBuilder — Constructing multiple line and cumulative line graphs (int-3358)

19.7 Linking markets to consumers

LEARNING INTENTION

By the end of this subtopic, you will be able to identify how trade links customers to producers and explain how trade moves around the world.

19.7.1 How trade moves around the world

Global trade of resources, manufactured goods and services connects people and places around the world. Information technology enables environments with a competitive advantage in the production of a good or service to connect with destinations that have a need. Transport technology provides an efficient interconnection between source areas, markets and businesses on a global and a local scale.

China is the world's largest exporter of goods, accounting for more than 12 per cent of total exports. The United States is the largest importer, with a share of almost 13 per cent. Global trade is dominated by movements between these strong trading nations. **FIGURE 1** shows the main aspects of this trend.

FIGURE 1 Value of exports and imports (US$ billions) for selected countries, 2018

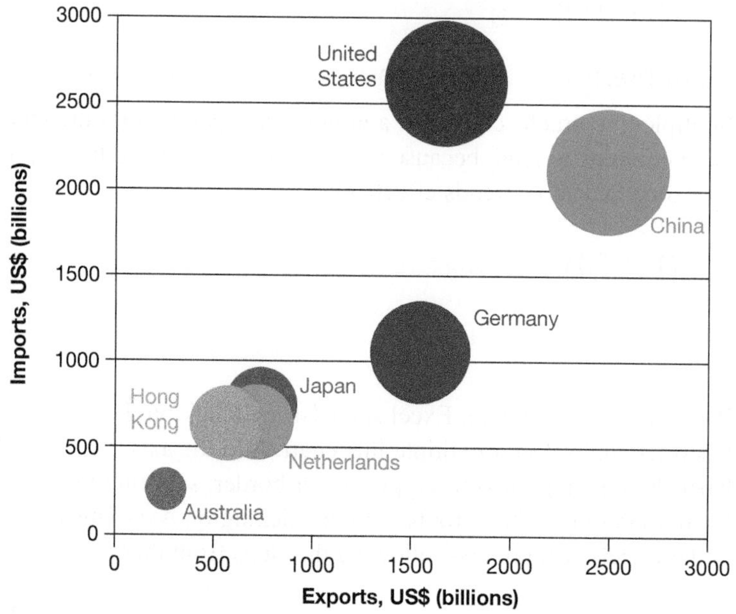

Source: WTO and UNCTAD

19.7.2 Digital trade

The online sale and purchase of goods and services and the transmission of information and data across borders is defined by the Australian Department of Foreign Affairs and Trade (DFAT) as **digital trade**. This electronic commerce platform (also known as e-commerce), creates opportunity for Australian businesses trading components and finished products locally and globally. E-commerce can improve the efficiency of trade; however, regulations are needed to ensure the market is safe and the rights of producers and consumers are protected. The e-commerce market can be challenging because different countries enforce different e-commerce trading regulations. In 2017, the value of Australia's digital exports, our fourth largest export sector, was $6 billion.

digital trade selling or buying goods or services or transmitting information and data across borders

19.7.3 Shipping

Global shipping routes

International trade involves the exchange of goods and services between countries. A port is a place for ships to dock to load and unload their cargo. A growing global population and the development of an efficient form of goods transport — shipping containers — have led to strong shipping links between countries, as shown in **FIGURE 2**. These interconnections have led to greater choice and more competitive prices for consumers. Economies have been boosted and jobs created worldwide.

FIGURE 2 Global shipping routes (number of journeys per year)

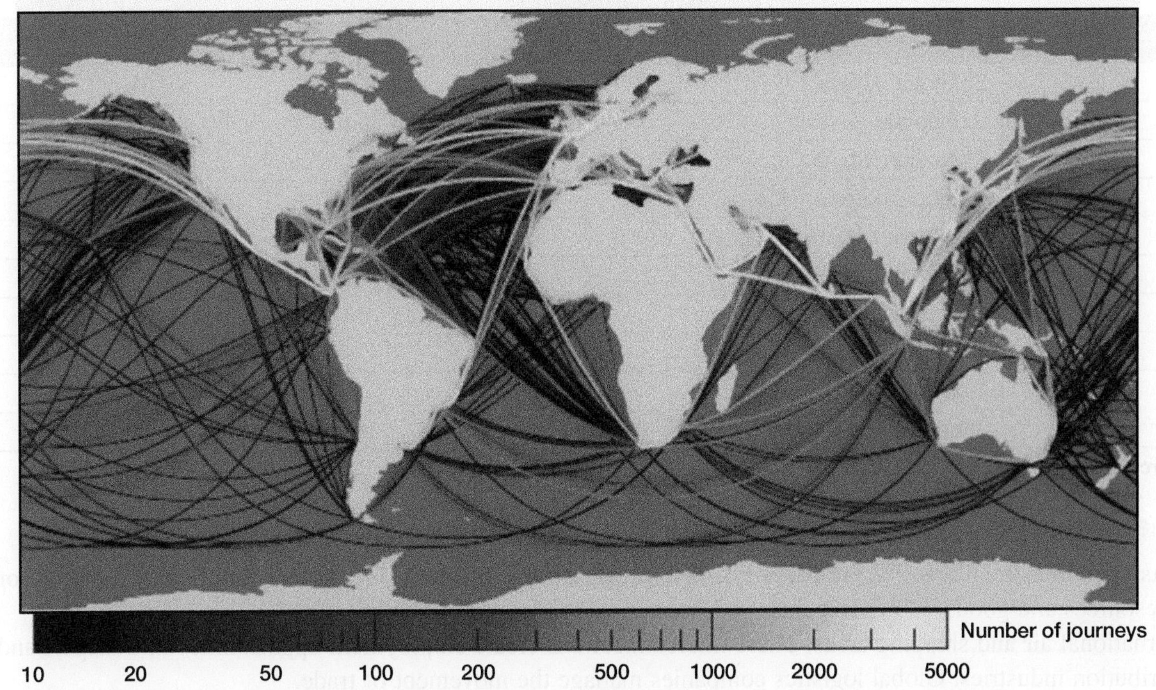

Source: Pablo Kaluza, Andrea Kölzsch, Michael T. Gastner and Bernd Blasius. The complex network of global cargo ship movements. J. R. Soc. Interface (2010) 7, doi:10.1098/rsif.2009.0495. Figure 1a.

The development of the shipping container has created an efficient method of transporting resources and goods from place to place. The introduction of a standardised measure for shipping containers (the TEU, which stands for twenty-foot equivalent) has shaped the design of cargo ships, port handling facilities and trade routes around the world. Ships can carry increasing loads efficiently and more cheaply. For instance, the MSC Gulson (**FIGURE 3**), can carry up to 23 756 containers.

FIGURE 3 World's largest container ship in 2019, MSC Gulson

FIGURE 4 The carrying capacity of container ships has increased 1200 per cent since 1968.

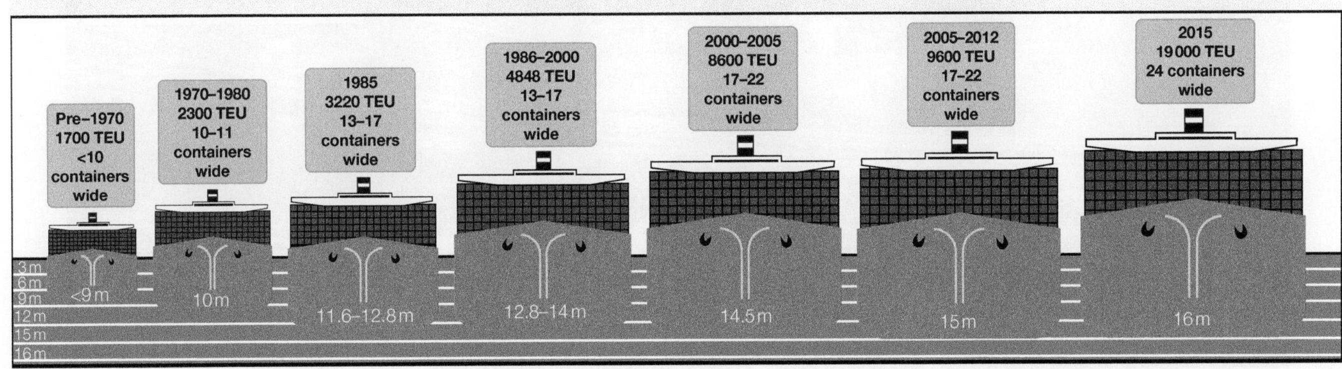

TABLE 1 Largest ten ports in the world by volume of containers

Rank	Port	Volume 2018 (million TEU)
1	Shanghai, China	42.01
2	Singapore	36.60
3	Shenzhen, China	27.74
4	Ningbo-Zhoushan, China	26.35
5	Guangzhou Harbor, China	21.87
6	Busan, South Korea	21.66
7	Hong Kong	19.60
8	Qingdao, China	18.26
9	Tianjin, China	16.00
10	Jebel Ali, Dubai, United Arab Emirates	14.95

Source: © World Shipping Council

What are port areas like?

Ports are commercial areas where water and land transportation links meet. Efficient handling of the transport of resources and goods will ensure that road and rail networks link the source areas within a country to international air and shipping areas. These interconnections create employment opportunities in transport and distribution industries. Global logistics companies manage the movement of trade.

Large ports such as Singapore operate 24 hours a day and 365 days a year to load and distribute goods. Raw materials and manufactured goods are distributed to global markets.

FIGURE 5 Port of Singapore

FIGURE 6 Singapore is strategically located where west-bound and east-bound trade links cross.

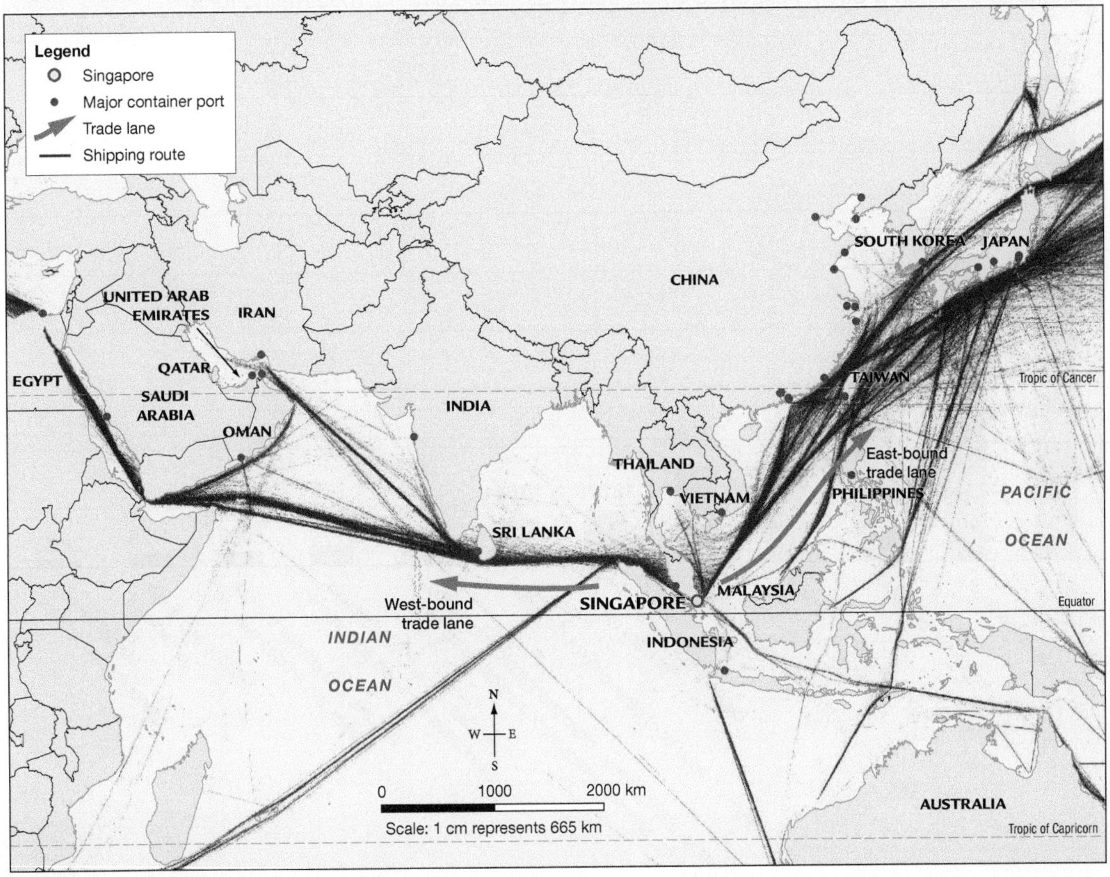

Source: Spatial Vision

19.7.4 Air and land freight

To facilitate the movement of products around Australia and globally, investment in road, rail, airport and port infrastructure is essential to the development of an efficient and interconnected trade network. The Australian government is responsible for the construction and maintenance of transport infrastructure and as **FIGURE 7** shows, inter-modal hubs are being created to further promote efficiency between various modes of transport used to move goods. Investment in well-planned and coordinated facilities ensures that logistics companies can support businesses that engage in trade. **FIGURE 8** shows the range of products Australia exported by air in 2018.

FIGURE 7 Inter-modal points manage the transfer of goods.

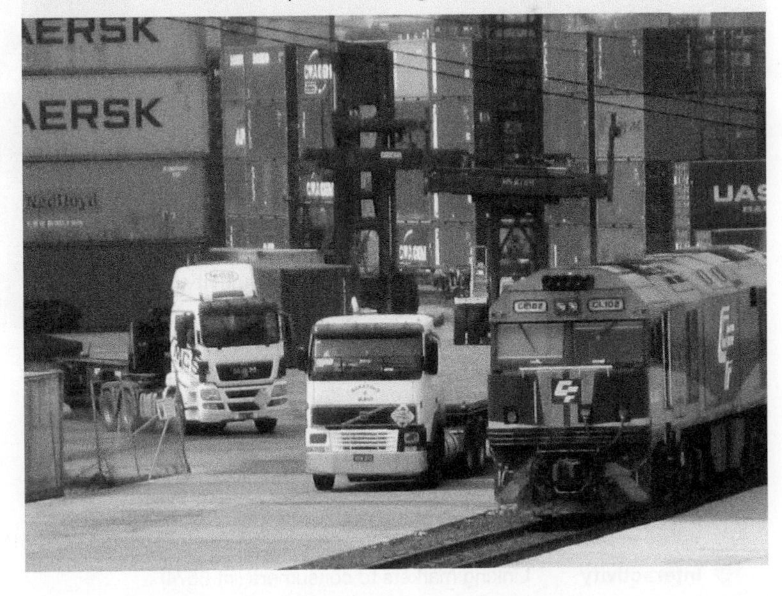

FIGURE 8 Volume of selected products exported by air from Australia, 2018 (in 1000 metric tons)

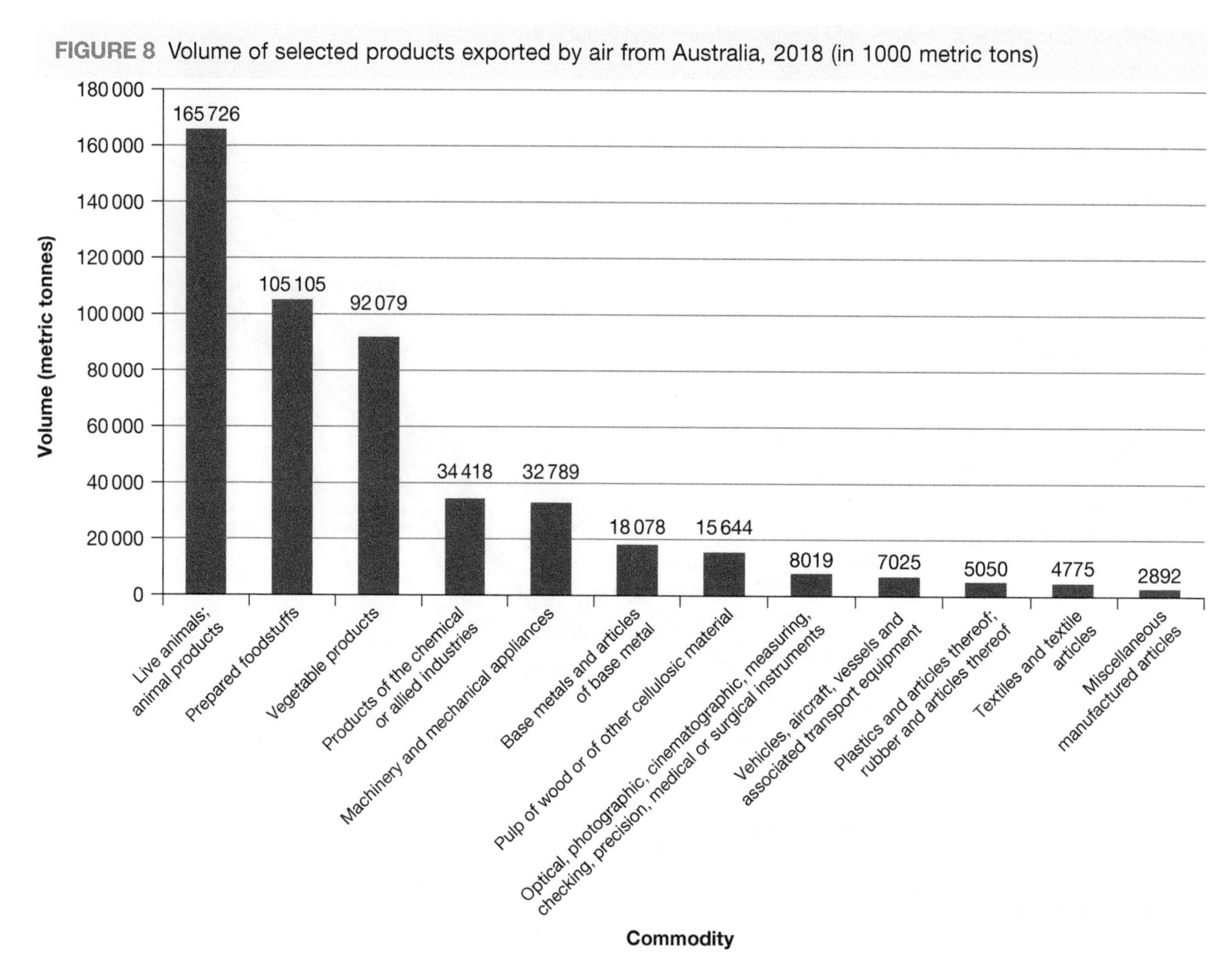

Source: Based on information from IPA/BISOE, ABS Statistics and Statista Ltd

fdw-0025

FOCUS ON FIELDWORK

Trade connection

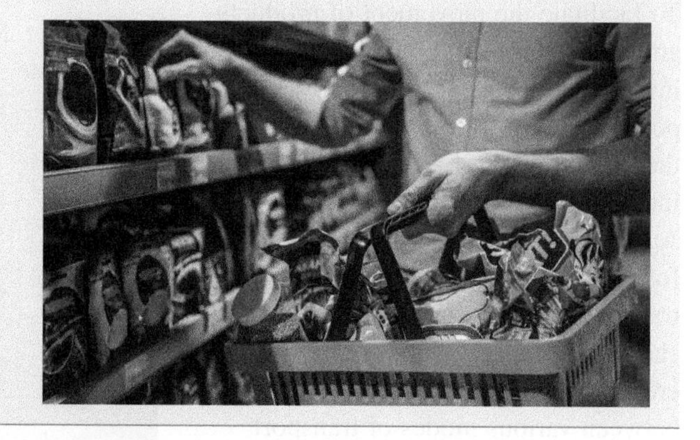

Collecting and mapping data is an important part of fieldwork. You can map data — for example, by using proportional circles on a map — to look for interconnections between the origins of food products that you and your family buy.

Learn how best to record your data about the products you buy using the **Trade connection** fieldwork activity in your Resources tab.

on Resources

eWorkbook	Linking markets to consumers (ewbk-9654)
Video eLesson	Linking markets to consumers — Key concepts (eles-5159)
Interactivity	Linking markets to consumers (int-8574)
Fieldwork	Trade connection (fdw-0025)

19.7 ACTIVITY

Use the **Ports** weblink in your Resources tab and select one of the ports to investigate. Visit the GIS image of your chosen port and capture a screenshot. Use this image to annotate the key areas of the port. Where do the ships enter the port? What types of goods are traded through the port? What volume of traffic passes through this port?

19.7 EXERCISE

Learning pathways

■ LEVEL 1	■ LEVEL 2	■ LEVEL 3
2, 3, 7, 8, 13	1, 4, 5, 10, 14	6, 9, 11, 12, 15

Check your understanding

1. Define the term *inter-modal hub*.
2. Which country exports the greatest amount of goods?
3. Which country imports the greatest amount of goods?
4. Refer to **FIGURE 1** to rank the top three trading economies in US$ billion in 2018.
5. Refer to **FIGURE 2** to describe the shipping routes that have the greatest number of shipping journeys.
6. Outline the four forms of trade described in this unit. Which form of trade accounts for the largest volume of produce?

Apply your understanding

7. What are the advantages to trading companies of using ships that are able to carry large container loads?
8. What natural features would a port need in order to process a large volume of ships each day?
9. Refer to **FIGURE 4**.
   a. How many pre-1970 ships would be needed to carry the same number of containers as a ship built after 2012?
   b. Why have the changes occurred to these ships and what are the impacts (positive and negative)?
10. Refer to **FIGURE 8**.
    a. Explain one challenge faced by one of the listed industries of using air travel to connect producer to market.
    b. Suggest how this challenge might be overcome.
11. What role does a port play in the movement of global trade?
12. The most cost-effective way to transport goods overseas is on a container ship. Discuss.

Challenge your understanding

13. Suggest ways that the import and export of goods from and to Australia could be more effective.
14. Which of the methods of moving freight do you think is the most important to Australia? Is it possible to choose one? Might your view be different depending on where you live? For example, if you live in a remote inland location, road transport might be more vital because it's how you can access food and essential products.
15. How are digital trade and global distribution networks connected? Predict how an increase in online shopping in the future would affect global distribution networks.

To answer questions online and to receive **immediate feedback** and **sample responses** for every question, go to your learnON title at www.jacplus.com.au.

19.8 Impacts of globalisation

LEARNING INTENTION

By the end of this subtopic, you will be able to provide examples of how globalisation has changed manufacturing.

19.8.1 Changing trends

Today, you might purchase a jacket online that was designed in Milan, woven from New Zealand wool and stitched together in China. Technological developments since the 1990s have enabled consumers to purchase goods and services online from markets around the world at affordable prices. The globalised economy brings together raw materials, large-scale manufacturing production, and distribution and promotion services to produce global markets for goods and services. This process is known as **globalisation**.

FIGURE 1 This symbol signifies that a product has been manufactured in Australia by an Australian-owned company.

The Australian clothing industry has produced some very recognisable brand names and distinctive products. Today, the industry faces tough international competition, especially from producers in developing countries who can afford to mass-produce clothing far more cheaply than Australian companies can. As a result, Australian clothing manufacturers tend to focus on high-end, high-quality products rather than attempting to compete with lower-cost producers.

Many multi-national firms have moved production away from the relatively expensive location of Australia, 'offshore' to a relatively cheaper location such as India, China, Malaysia or the Philippines. These places may offer low labour costs and government incentives such as reduced taxation payments to attract investment.

globalisation the process that enables markets and companies to operate internationally, and products and ideas to be freely exchanged

19.8.2 Foreign companies in China

Consider this example of the growth in global business operations: in 1979, there were 100 foreign-owned enterprises in China; in 1998, there were 280 000; and by 2015 there were 835 000. Since 2007, foreign companies have employed 25 million people in China. These companies include Coca-Cola, Pepsi, Nike, Citibank, General Motors, Philips, Ikea, Microsoft and Samsung. China's economy is growing rapidly and the country is destined to remain an engine for global growth for some years to come. This has also affected product development. IKEA products for sale in China have been redesigned with Chinese customers in mind; for example, making bowls deeper for eating rice-based meals.

FIGURE 2 IKEA, Beijing, China

FIGURE 3 Top 20 locations for offshore companies, 2017, by region

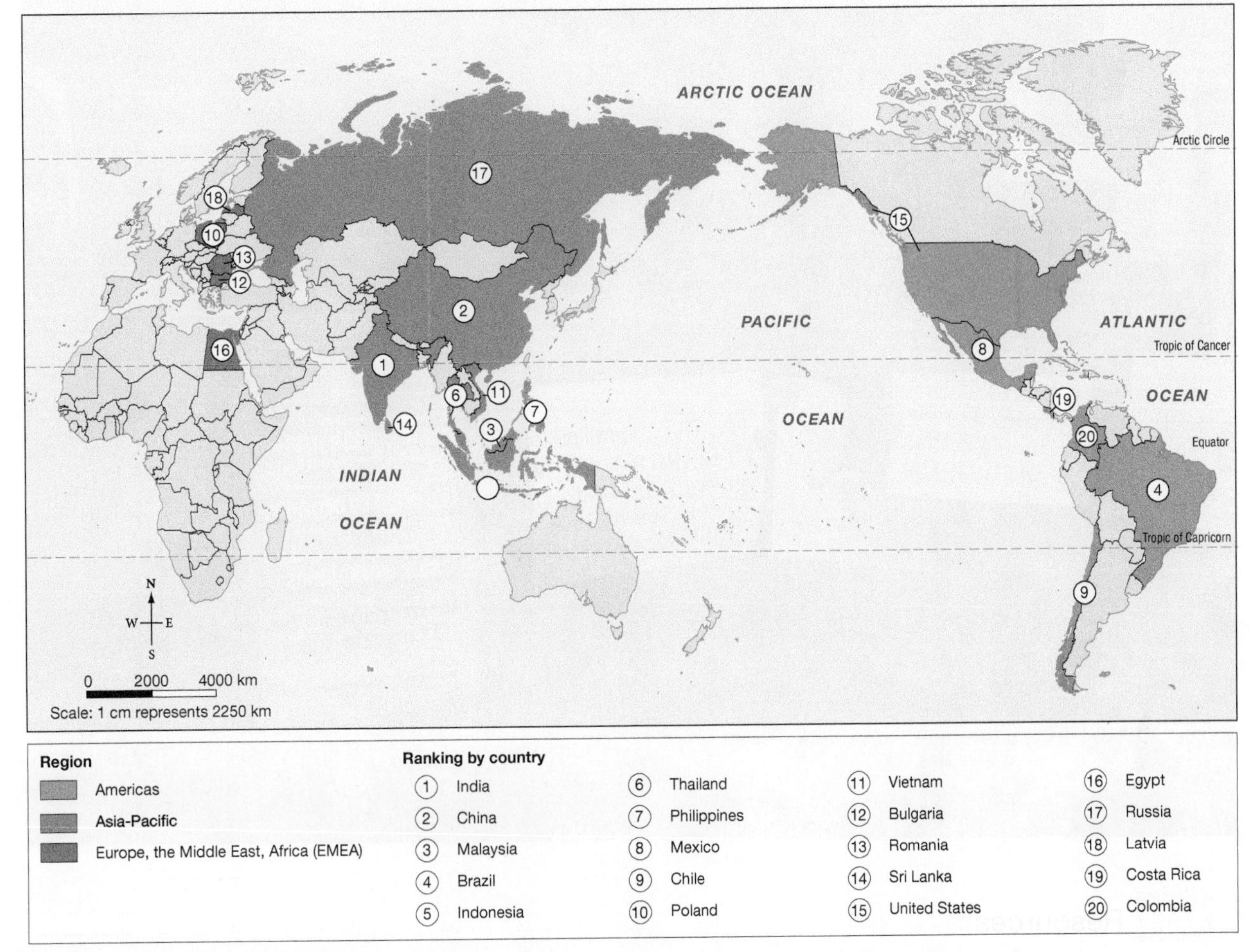

Source: Data from Statista. Map drawn by Spatial Vision.

19.8.3 Sweatshop production

If you buy well-known global brands, then you may be wearing clothing or footwear that was made in a sweatshop.

A sweatshop is any working environment in which the workers experience long hours, low wages and poor working conditions. Typically, they are workshops that manufacture goods such as clothing. Sweatshops are common in developing countries, where labour laws are less strict or are not enforced at all. Workers often use dangerous machinery in cramped conditions and can even be exposed to toxic substances. In the worst cases, child labour may be used.

FIGURE 4 Garment workshop in Vietnam

Sweatshop workers' wages are generally insufficient to sustain reasonable living conditions, and many workers live in poverty. Most are women aged 17 to 24. Such conditions highlight the importance of considering the ethics of our globalised economy and international trade systems.

FIGURE 5 Activists in New York raise awareness of Bangladesh's Rana Plaza garment factory deaths.

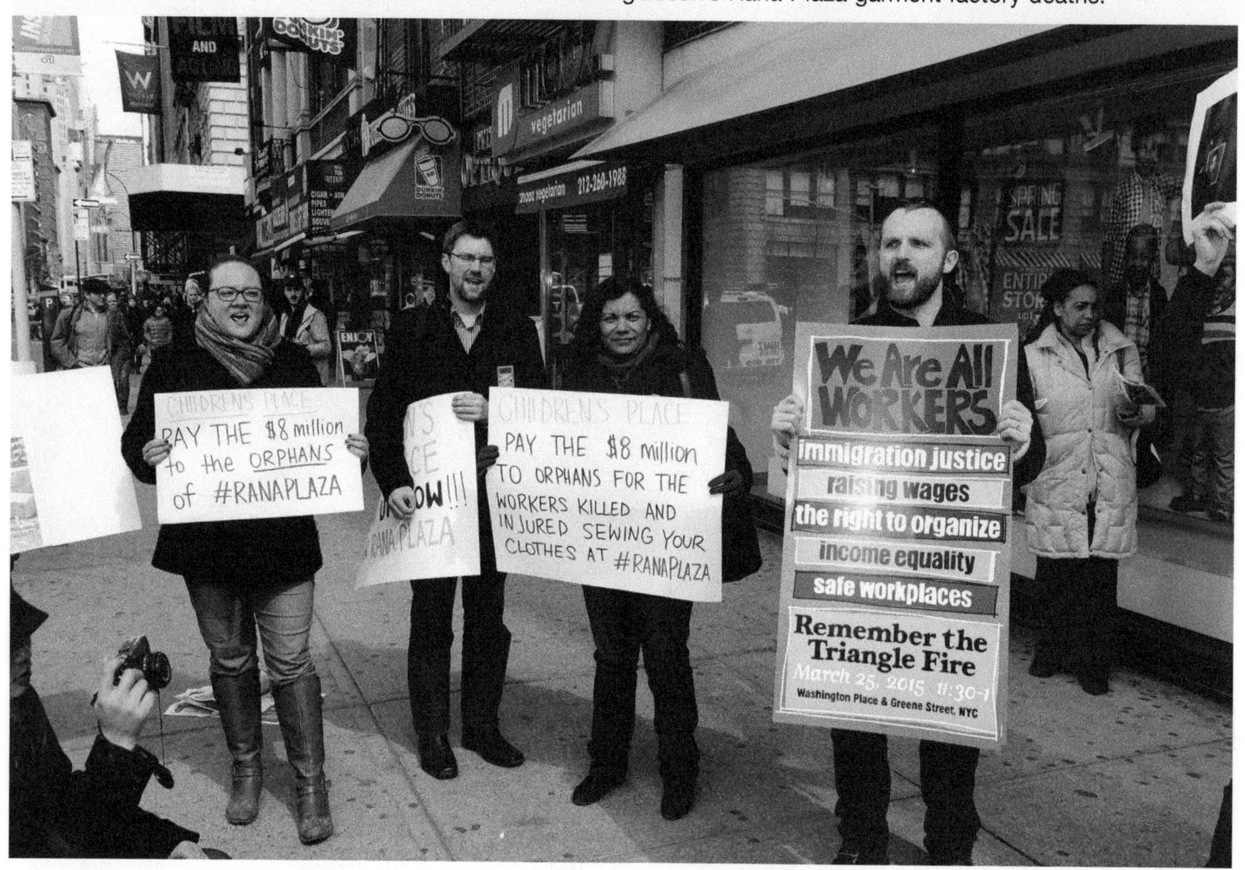

on Resources

eWorkbook	Impacts of globalisation (ewbk-9658)
Video eLesson	Impacts of globalisation — Key concepts (eles-5160)
Interactivity	Impacts of globalisation (int-8575)

19.8 ACTIVITIES

1. If clothing carries the Ethical Clothing Australia (ECA) label, it means the garment was manufactured in Australia and the manufacturer has ensured that all people involved in its production received the legally stated wage rates and conditions — known in Australia as award wages and conditions. Find out which Australian-made garments you can purchase to support fair working conditions. Do an audit of your clothes at home; how many items do you have that are made by ECA brands?
2. It's not only fast-fashion that creates sweatshops and unsustainable trading practices. Technology companies, such as Apple and Samsung, encourage users to buy the latest models of their products, even if the improvements over the previous model are small. This can have a negative effect on the environment with the old phones being discarded and the environmental impact of transporting new models around the world. Discuss the impact of this as a class; should consumers or technology companies look to change their behaviour?

FIGURE 6 (a) Crowds queued outside the Apple store in New York City on the morning of the iPhone 6 release. (b) People queued in the rain and camped in tents for several days outside the Apple store in Sydney in 2015 for the release of the iPhone 6s.

3. Use the **Running free** weblink in your Resources panel to examine 'Who earns what from a sport shoe?'
   a. Calculate the approximate percentages of the sale received by people at each stage of the process.
   b. Find a pair of sports shoes online that you would like to buy. Using the percentages you calculated in part a, determine how much a person at each stage of the production of the shoe earns.

19.8 EXERCISE

Learning pathways

■ LEVEL 1	■ LEVEL 2	■ LEVEL 3
2, 6, 7, 8, 14	1, 4, 9, 10, 13	3, 5, 11, 12, 15

Check your understanding

1. Why have some countries moved their production to offshore places?
2. What are sweatshops?
3. Why might companies move their production facilities from developed to developing countries? Give two reasons.
4. Define the term *globalisation*.
5. List the positive and negative impacts of sweatshop labour for the workers themselves.
6. What is 'child labour'?

Apply your understanding

7. a. What change do you think online shopping has made to the Australian retail industry?
   b. How might internet shopping affect places such as shopping centres?
8. What do you think would happen to the price of clothing if sweatshops were to close down?
9. Look at **FIGURE 3**. Give reasons why most offshore manufacturing companies are located in the Asia–Pacific region.
10. What impact does moving production offshore have on the Australian economy and its people?
11. Are sweatshops ethical or sustainable? Explain your answer.
12. Explain why people will queue for hours (or days) to buy a new model mobile phone.

Challenge your understanding

13. Recommend three ways that people can ensure they are buying ethically produced clothing.
14. Some people know about sweatshop labour, and might even recognise that it is unlikely that clothes they are buying were produced ethically. Suggest why they might buy them anyway.
15. What can Australia and the global community do to ensure that people across the world are paid fairly for their work?

To answer questions online and to receive **immediate feedback** and **sample responses** for every question, go to your learnON title at www.jacplus.com.au.

19.9 Global disruption to the Australian economy

LEARNING INTENTION

By the end of this subtopic, you will be able to explain how events overseas can have an impact on Australia and our economy, and provide examples of global events that have affected the Australian economy.

19.9.1 Our interconnected world

International trade has contributed to economic growth and the generation of wealth in all nations that engage in the import and export of goods and services. Developments in travel and communications have made trade easier and broken down many traditional barriers between countries. The growth of trade between almost all countries has created greater economic interdependence between those countries. As a result of this interconnectedness, both positive and negative economic events can spread quickly between trading partners. The economic growth of China since the 1980s has had a beneficial effect on many other countries that trade with it. On the other hand, the widespread use of electronic communication has made internet fraud and identity theft much easier for criminal groups.

FIGURE 1 International flights have made travel between countries quick and easy, but they come with risks such as the possibility of spreading contagious diseases across the globe.

The ease of travel between countries has also enabled the rapid international spread of infectious diseases. As shown in the case study in this subtopic, the COVID-19 pandemic had a significant impact on business in Australia.

19.9.2 The growth of the Chinese economy

Since the early 1980s, the Chinese government has pursued a number of policies designed to promote rapid economic growth. Economic growth is measured by increases in a country's gross domestic product (GDP). As GDP is the total value of all goods and services produced in a country in any given year, the rate by which GDP increases each year is effectively the rate of economic growth of that country. From 1979 until 2010, China's average annual GDP growth was 9.9 per cent. Since 2010, the rate of growth has slowed somewhat, with a rate of 6.11 per cent recorded for 2019. One way in which China has been able to achieve this level of growth is through a rapid expansion in trade with other countries. As a result of this policy, China has become the world's largest trading nation, with a total trade value of US$4.1 trillion in 2017.

FIGURE 2 China produces high-quality electronic products more cheaply than Australia can.

China has set out to increase its manufacturing capacity to provide all the goods and services required by its own large population and to export to other countries. It has had to import large quantities of raw materials, including the materials to build hundreds of new factories and the fuel to power them. Countries such as Australia have benefited enormously from this growth in the Chinese economy. As we have seen, China is Australia's largest export customer, buying large quantities of Australian iron ore and coal. For a roughly ten-year period in the early 2000s, this generated a mining boom in Australia that contributed significantly to our growth in GDP.

Australia also imports large quantities of consumer goods from China, particularly clothing and other textile products, as well as increasing quantities of electronic goods and other home appliances. Most of these are produced more cheaply than we can produce them ourselves, so Australian consumers benefit from paying lower prices for a wide variety of goods imported from China.

19.9.3 Global financial crisis (GFC)

During the early 2000s, many banks in the United States lent money via mortgage loans to people who were ultimately unable to repay the amounts they had borrowed. In 2006 and 2007 a fall in house prices in the US left many of these people with homes that were valued at less than the amount owing on their loans. When large numbers defaulted on their loans and had to abandon their houses, many of the banks and other financial intermediaries lost a lot of money, severely damaging the reputation of the US financial system. This led to a tightening of credit; banks lent less money and there was a slowing in growth of the US economy. In 2008 the US economy went into **recession**. Around 9 million people lost their jobs in the following two years.

FIGURE 3 Many US homeowners had to abandon their mortgaged homes when house prices fell dramatically during 2006 and 2007.

In response to the problems in the US banking system, other banking systems throughout the world placed restrictions on lending. This led to a recession in much of the rest of the world, leading to decreases in economic growth and high levels of unemployment. Recession was largely avoided in Australia because the government rapidly increased spending to stimulate economic growth.

The global financial crisis (GFC) of 2008 and 2009 occurred because of the close connections between the economic and financial systems of most of the world's countries. International trade and the flow of money between nations means that events that occur in one country can have an influence on economic conditions in other countries.

19.9.4 Natural disasters

A natural disaster can have a serious economic impact on a country. When houses and businesses are destroyed, money and resources are needed to repair and replace them. Therefore, these resources cannot be used for other purposes. In February 2009, the Black Saturday bushfires in Victoria caused damage to the Victorian economy costed at more than $5 billion. In January 2011, floods in Queensland damaged many homes and businesses, and devastated a large area of valuable farming land. The resulting produce shortages forced food prices all over Australia to rise. Rail lines and coal mines were also damaged. Drought throughout much of New South Wales, Queensland and South Australia in 2018

recession a period of decline in economic growth when GDP decreases led to farmers needing government assistance. In February 2019, serious flooding in Queensland destroyed farmland and livestock, threatening future food supplies and placing farmers and graziers under significant financial stress.

A 2012 report by the Centre for Australian Weather and Climate research at the Bureau of Meteorology found that the impact of tropical cyclones in the north of Western Australia costs the Australian economy between $40 and $100 million per year from direct damage alone. In addition, the economy is indirectly affected by the impact of natural disasters on industry; for example, mining and agricultural operations in the north of the state are vulnerable to extreme weather events.

FIGURE 4 The Queensland floods affected food prices all over Australia, as well as some of our export industries.

Natural disasters in other countries can also affect the Australian economy, particularly if they occur in the Asia region to which we are so closely tied.

The 2011 Japanese earthquake and tsunami

In March 2011, the largest earthquake ever to hit Japan occurred under the ocean to the country's east, causing a 40-metre tsunami. As many as 18 000 people are believed to have died. Tens of thousands of buildings were destroyed, and a meltdown at the Fukushima nuclear power station led to serious radioactive pollution. There was an immediate slowdown in the growth of the Japanese economy, but the international economy was also seriously affected. Japan is a major trading nation and the world's third largest economy. Japanese cars, computers and electronics products are assembled in many factories around the world, and they rely on parts imported from Japan. The slowdown in the Japanese economy had an impact on many Japanese-owned businesses globally.

FIGURE 5 The destruction caused by the 2011 tsunami had an impact on Japan's trading partners as well as on its own economy.

Japan is Australia's second largest trading partner, so an event as dramatic as the 2011 tsunami had an impact on Australia's economy, although the effects were largely short term. These effects were positive as well as negative:

- The slowdown in the Japanese economy resulted in a reduction in demand for Australian exports such as coal, iron ore and beef. However, the reconstruction effort in Japan eventually led to a rise in demand for steel, so many of these exports subsequently increased.
- Pollution from the Fukushima nuclear power station led to concerns about the safety of the food supply in that area and so to a rise in imported food. As a significant supplier of food to Japan, Australia exported more food to that country in the period after the tsunami.
- The nuclear meltdown caused Japan to reassess its reliance on nuclear power. As a result, it has been making greater use of coal- and gas-fired power stations. This is likely to result in a higher demand for coal and liquefied natural gas (LNG) from Australia. In 2010, 13 million tonnes of LNG were exported from Australia to Japan; this rose to 24.8 million tonnes in 2016–17.

CASE STUDY

The COVID-19 shock to supply chains

By Associate Professor William Ho

The coronavirus pandemic is only the latest shock to supply chains — but COVID-19 is a wake-up call to businesses in terms of the cost of being under-prepared.

The Australian supermarket chain Woolworths announced that it's closing all of its supermarkets nationwide early for one night, so it can restock stores in an effort to manage panic buying in the face of the COVID-19 pandemic.

But it's not just here in Australia. Community fears over COVID-19 have led to consumers around the world panic-buying goods.

And it highlights the risk this pandemic is posing to supply chains.

In Australia, shoppers have stripped supermarkets of toilet paper, hand sanitiser and dried goods like rice and pasta. Some have even come to blows in the aisles.

FIGURE 6 Panic buying during COVID-19 exposed supply chain risks.

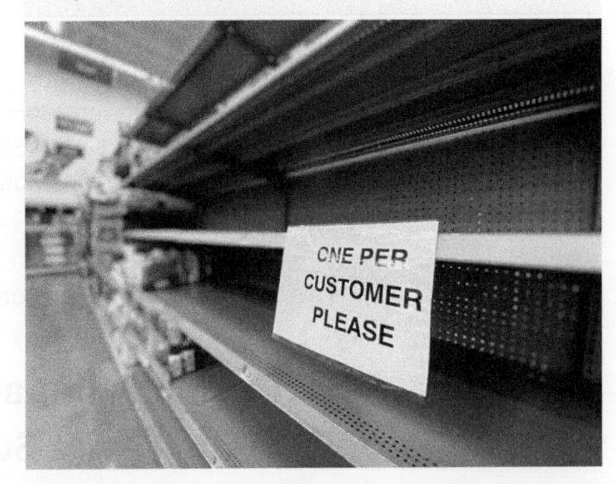

Distributors like the toilet roll subscription service Who Gives A Crap? have been unable to fulfil orders, and the supermarket chains have put in place limits on the number of purchases.

However, with most toilet paper made in Australia, manufacturers assure they can rapidly increase production to manage the surge in demand.

But this panic-buying provides us with a prime example of something called a 'demand risk' to supply chains — a sudden surge in demand that catches manufacturers and retailers by surprise.

This can also affect other sectors on the frontline of the pandemic including the health system, tourism and hospitality, and universities.

The economic fallout of this pandemic is already telling us that businesses were ill-prepared for supply chain risks on this scale. But my research suggests most businesses can do much more to prepare.

Supply chain shock

COVID-19 is only the latest shock to supply chains.

In 2010, for instance, the eruption of Eyjafjallajökull in Iceland cost airlines $A1.7 billion as flights were grounded due to ash.

The following year, the earthquake, tsunami and nuclear disaster in Japan caused car maker Toyota to cut production by 40,000 vehicles, costing the company $US72 million each day.

Nor is COVID-19 the first disease to threaten supply chains. The 2009 H1N1 flu pandemic reduced GDPs around the world by between 0.5 and 1.5 per cent.

But COVID-19's impact on supply chains is making itself felt in a range of industries.

Apple has suffered component shortages for iPhones because of [a] temporary manufacturing plant closure in China.

Car maker Hyundai Motors has closed seven factories in South Korea, which makes up approximately 40 per cent of global output, because of supply shortages from its China-based suppliers.

According to Channel News Asia, Airbus has shut its Tianjin factory for assembling both A320 and A330 aircrafts.

Consumer goods company Procter & Gamble has also faced challenges as a result of its 387 suppliers across China.

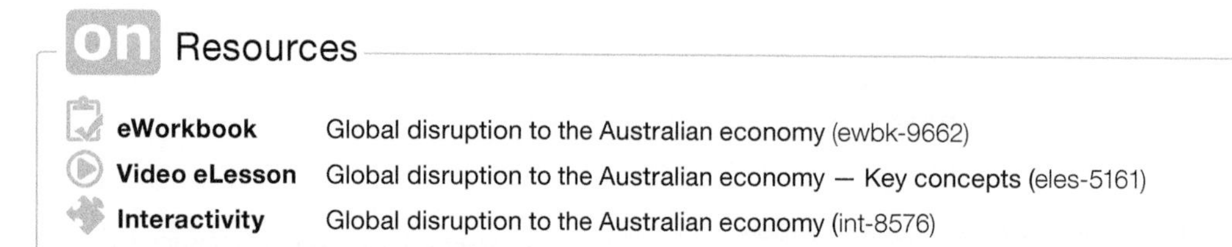

19.9 ACTIVITIES

1. Use the **ABS — COVID-19 impacts** weblink in your Resources tab to research the impacts of the pandemic. Use the reports from the Australia Bureau of Statistics about the impacts of COVID-19 on Australian businesses and the economy to create an infographic about how the pandemic affected one specific part of the Australian economy.

FIGURE 7 COVID-19 impacts

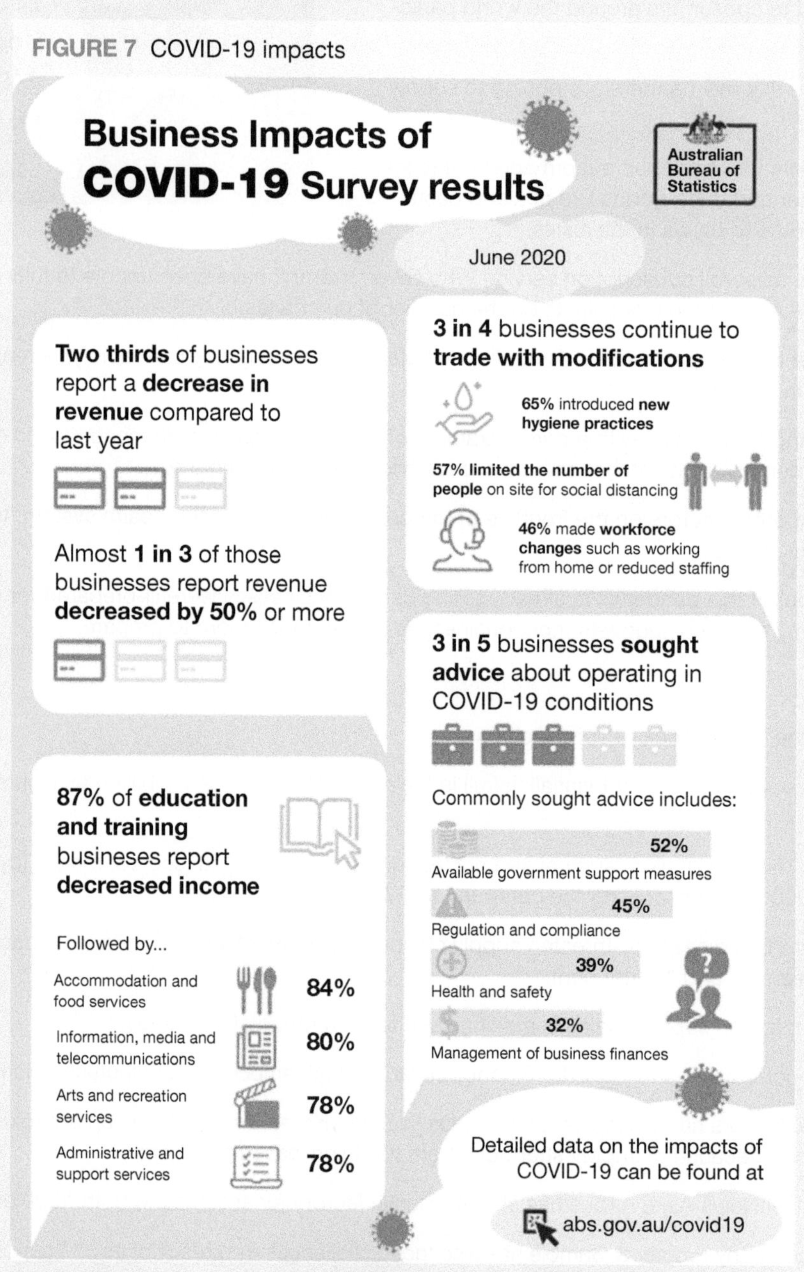

Source: © Commonwealth of Australia. Retrieved from Australian Bureau of Statistics, 5676.0.55.003 — Business Indicators, Business Impacts of COVID-19, June 2020

2. Prepare a collage of images showing the effects the COVID-19 pandemic had on trade in goods and services. For example, the grounding of international airlines and the closing of borders (both international and within Australia) affected tourism businesses. Lockdowns led to significant downturns, and in many cases the closures of businesses in retail, hospitality and other industries.
3. Research the ways COVID-19 disrupted China's trade relations with Australia.
4. The Suez Canal, a major shipping route, was blocked for six days in March 2021. The incident led to global shipping disruption and delays in the supply of imported products. Investigate the cause and impact of this incident, and discuss what other incidents might occur in shipping canals to stop the flow of trade.

19.9 EXERCISE

Learning pathways

■ LEVEL 1	■ LEVEL 2	■ LEVEL 3
2, 3, 6, 10, 15	4, 5, 8, 11, 13	1, 7, 9, 12, 14

Check your understanding

1. What have the policies adopted by Chinese government leaders been designed to do for China's economy?
2. Outline two ways that the economic growth of China has benefited the Australian economy.
3. What two products does China rely heavily on Australia to supply?
4. Describe two ways that a natural disaster such as a major flood can impact production within Australia.
5. Identify one way in which the greater interconnectedness of countries can have a negative effect.
6. What and when was the global financial crisis?
7. Describe two factors that have led to the greater interconnectedness of countries.

Apply your understanding

8. Explain the impact of the 2011 tsunami on Japan's trade with other countries.
9. Evaluate the following statement: 'Greater interconnectedness of countries can't have a positive effect on Australia.'
10. Explain how natural disasters can affect the Australian economy in:
    a. a negative manner
    b. a positive manner.
11. A significant proportion of Australian bananas are grown in the north of Queensland. Explain how a cyclone making landfall in north Queensland might affect the price and availability of bananas in your local supermarket.
12. Explain how the global COVID-19 pandemic affected supply chains in Australia.

Challenge your understanding

13. When COVID-19 restrictions came into place in Australia, some people started panic buying essential products such as toilet paper and food. Does this panic buying suggest that people have a good understanding of the global trade and economic interconnection? Give reasons for your view.
14. Suggest what the impact might be if the Chinese government banned the importation of any Australian goods or resources into China.
15. How would you persuade someone not to panic buy in the event of another pandemic?

To answer questions online and to receive **immediate feedback** and **sample responses** for every question, go to your learnON title at www.jacplus.com.au.

19.10 Investigating topographic maps — Port Botany

LEARNING INTENTION

By the end of this subtopic, you will be able to analyse a topographic map to discuss interconnections between port facilities and trade.

19.10.1 Port Botany

Port Botany, located in Botany Bay, is the major shipping port in Sydney and New South Wales. It is Australia's second largest container port. Some of the bay has been dredged to enable ships requiring deep water to pass through channels leading to the port. Some of the low-lying areas have been reclaimed to build large flat areas for port facilities.

FIGURE 1 A view over the three container terminals, bulk liquids and gas berths, and storage facilities at Port Botany

19.10.2 Trade facilities and links

Today, Port Botany covers an area of 237 hectares and consists of three container terminals, a bulk liquids and gas facility, and a bulk liquids storage and distribution complex connected to underground pipelines. These pipelines supply aviation fuel to Sydney Airport, as well as chemicals and gas to the nearby Botany Industrial Park.

Cargo moves through Port Botany to or from more than 170 countries. The port is adjacent to the Port Botany Line, a dedicated rail freight line for container freight distribution and the collection of export containers from regional areas of New South Wales. Over 80 per cent of import containers passing through Port Botany are delivered within a 40-kilometre radius of Sydney's port.

FIGURE 2 Colour satellite image of Botany Bay, 2017

FIGURE 3 Topographic map extract of Port Botany

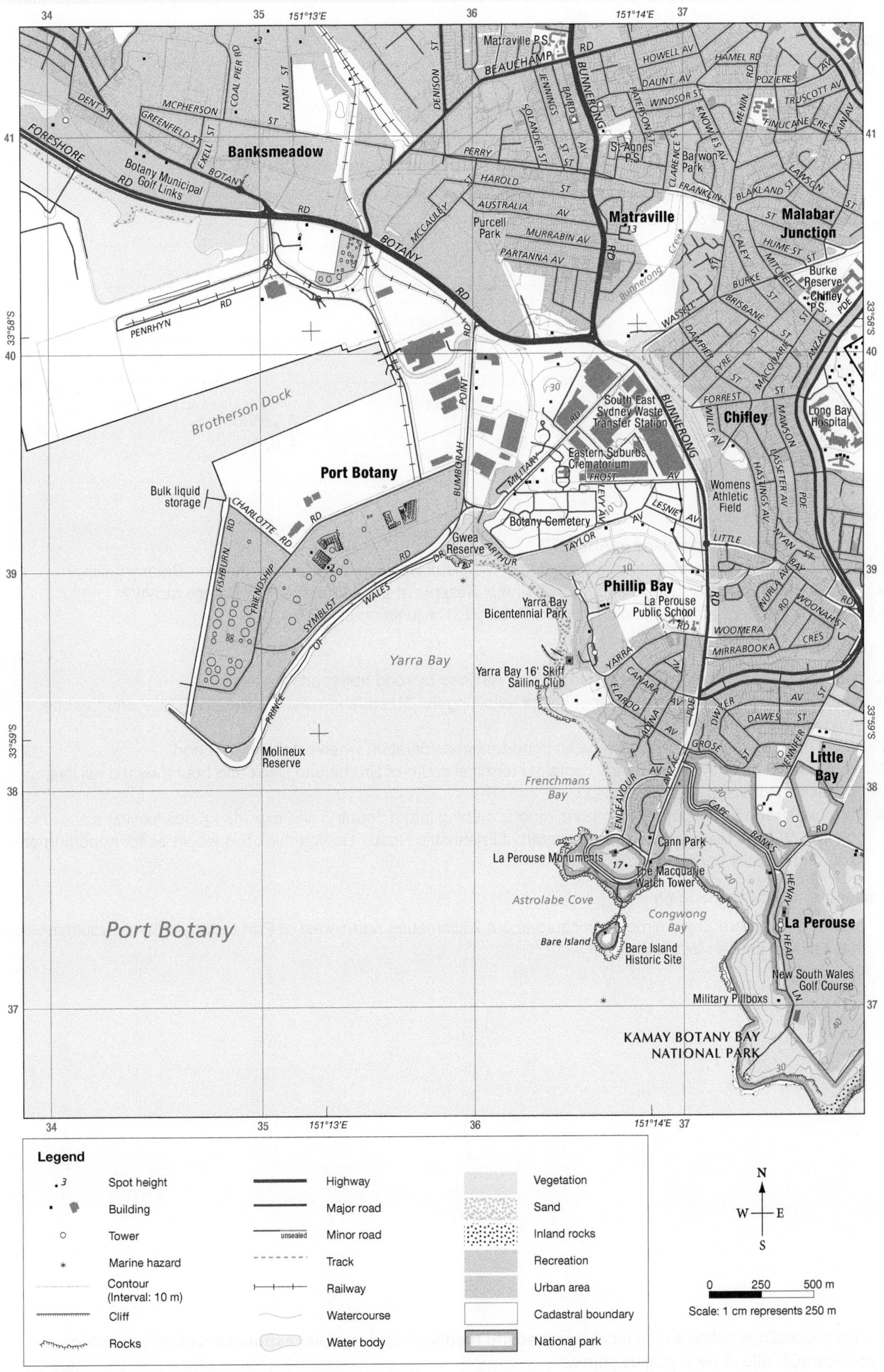

Source: Data based on Spatial Services 2019.

on Resources

eWorkbook	Investigating topographic maps — Port Botany (ewbk-9666)
Digital document	Topographic map of Port Botany (doc-36263)
Video eLesson	Investigating topographic maps — Port Botany — Key concepts (eles-5162)
Interactivity	Investigating topographic maps — Port Botany (int-8577)
Google Earth	Port Botany (gogl-0094)

19.10 EXERCISE

Learning pathways

■ LEVEL 1	■ LEVEL 2	■ LEVEL 3
1, 5, 9	2, 3, 7, 8	4, 6, 10

Check your understanding

1. What is the name of the island in AR3637?
2. What is the grid reference of the gate located in Molineux Reserve at the end of the Port Botany facility?
3. What is the direction of Bare Island from Molineux Reserve at the end of the Port Botany facility?
4. Calculate the length of train track from Port Botany to the junction at GR354413.

Apply your understanding

5. What evidence is there that Port Botany is well serviced by road transport facilities?
6. What is the gradient between the spot height adjoining the Macquarie Watch Tower in AR3637 and the bulk liquid storage facility in AR3439?
7. Why would the gradient of an area be an important consideration when establishing a port?
8. What is the area of the Port Botany container terminal south of Brotherson Dock and bound by the rail line, Friendship Road and Charlotte Road?
9. Ports are used to transport both imports (goods coming into a country) and exports (goods leaving a country). What is the major item stored south of Friendship Road? Do you think this would be for importing or exporting? Justify your answer.

Challenge your understanding

10. Sydney's Kingsford Smith airport is located about 7 kilometres north-west of Port Botany. Why do you think that site was chosen for the airport?

To answer questions online and to receive **immediate feedback** and **sample responses** for every question, go to your learnON title at www.jacplus.com.au.

19.11 Thinking Big research project — Ethical business — Real change or fast fashion?

online only

The content in this subtopic is a summary of what you will find in the online resource.

Scenario

About 20 years ago the clothing industry was taken over by a trend known as fast fashion. Clothing became less expensive, new clothing ranges were released more often and it became commonplace for clothing to be worn only a few times and then discarded. While the fast-fashion business model has delivered increased revenue to clothing manufacturers, it has also significantly increased the impact of the clothing industry on garment workers, animal welfare and the health of our planet.

Some brands are taking steps to improve their practices and finding that increased ethical practices are not only good for the planet, but also good for business.

Task

For this task, you are required to use research skills and creativity to produce an engaging presentation on a fast-fashion brand. In order to do this, you will need to develop a good understanding of what the terms *fast fashion*, *ethics* and *productivity* mean, analyse your chosen brand's current level of ethics, and develop a plan for your chosen brand to increase both their ethical practices and productivity.

Go to your Resources tab to access the resources you need to complete this research project.

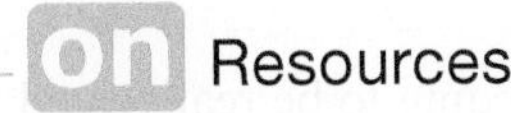

ProjectsPLUS Thinking Big research project — Ethical business — Real change or fast fashion? (pro-0251)

19.12 Review

19.12.1 Key knowledge summary

19.2 Features of trade

- Trade allows producers and consumers to exchange goods and services. This occurs at local, regional, national and international (global) levels.
- The processing of a resource into more complex goods can involve value adding across different levels of industry: primary, secondary, tertiary and quaternary.
- Trade connections between urban, regional and interstate places are facilitated by rail, road, air and shipping freight networks.
- Many global trade links are accessed through ports and international airports.

19.4 Australian trade connections

- Australia is a member nation of the major organisations that control world trade. These include the World Trade Organization (WTO) and Asia–Pacific Economic Cooperation (APEC).
- Australia, with its small population and vast mineral and agricultural resources, is a major exporter of goods.
- Tertiary and secondary education services are now a major export, with students from many nations studying in our schools, colleges and universities.

19.5 Global food trade

- Some countries have an excess of certain goods and services commodities and others have a shortage. There is therefore a need to interconnect in the export and import trade.
- Australia's exports of wheat and other cereal crops, as well as live animals and animal products, are significant aspects of our trade.
- Production levels and export volumes can vary depending on weather conditions experienced during the growing season.

19.7 Linking markets to consumers

- Traded commodities are moved around the world digitally, and through shipping, land and air freight.
- Containerisation has led to efficient movement and distribution of resources and manufactured products.
- Investment in road, rail, airport and port infrastructure is important in creating interconnected trade networks.

19.8 Impacts of globalisation

- Globalisation has led to a change in manufacturing such that goods are now more likely to be produced in developing countries where labour costs are low.
- Offshoring of production and services has occurred in many businesses.
- Sweatshops are a negative aspect of the globalised trade economy.
- The question of ethical trade is increasingly important in our globalised world.

19.9 Global disruption to the Australian economy

- International trade has contributed to economic growth in nations that import and export goods and services. But there are also negative effects of increased global connectedness, such as the rapid international spread of infectious diseases and the rise of internet fraud and identity theft.
- The growth in the Chinese economy has had a significant impact on the global economy, with countries such as Australia benefiting from increased trade with China.
- Natural disasters can have serious impacts on countries' economies, with funds needing to be reallocated to relief and rebuilding. International trade may also be impacted by these events.

19.12.2 Key terms

barter to trade goods in return for other goods or services rather than money
digital trade selling or buying goods or services or transmitting information and data across borders
free trade agreement a treaty between two or more countries or economies that sets out special trading conditions that reduce or remove trade barriers
globalisation the process that enables markets and companies to operate internationally, and products and ideas to be freely exchanged
halal food that is prepared under Islamic dietary guidelines
primary industry the gathering of natural resources, such as iron ore and timber, or activities such as farming and fishing
quaternary industry gathering information about an industry; for example, marketing furniture
recession a period of decline in economic growth when GDP decreases
secondary industry making raw materials into a more useful product; for example, making furniture from timber
tertiary industry selling goods and services; for example, a furniture business
trade barrier government-imposed restriction (in the form of tariffs, quotas and subsidies) on the free international exchange of goods or services
trading partner a participant, organisation or government body in a continuing trade relationship
value adding processing a material or product and thereby increasing its market value

19.12.3 Reflection

Complete the following to reflect on your learning.

Revisit the inquiry question posed in the Overview:

Buy, swap, sell, give. Is the trade that occurs between different countries just a way of getting things?

1. Now that you have completed this topic, what is your view on the question? Discuss with a partner. Has your learning on this topic changed your view? If so, how?
2. Write two paragraphs in response to the inquiry question outlining your views about trade connections: one paragraph about how places are interconnected, and one about how people are interconnected. In your response, make reference to the way that different groups of people might view international trade. For example, how might Aboriginal Peoples feel about the impact of shipping on Port Botany?

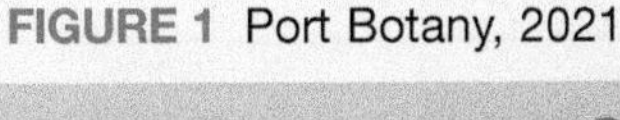

FIGURE 1 Port Botany, 2021

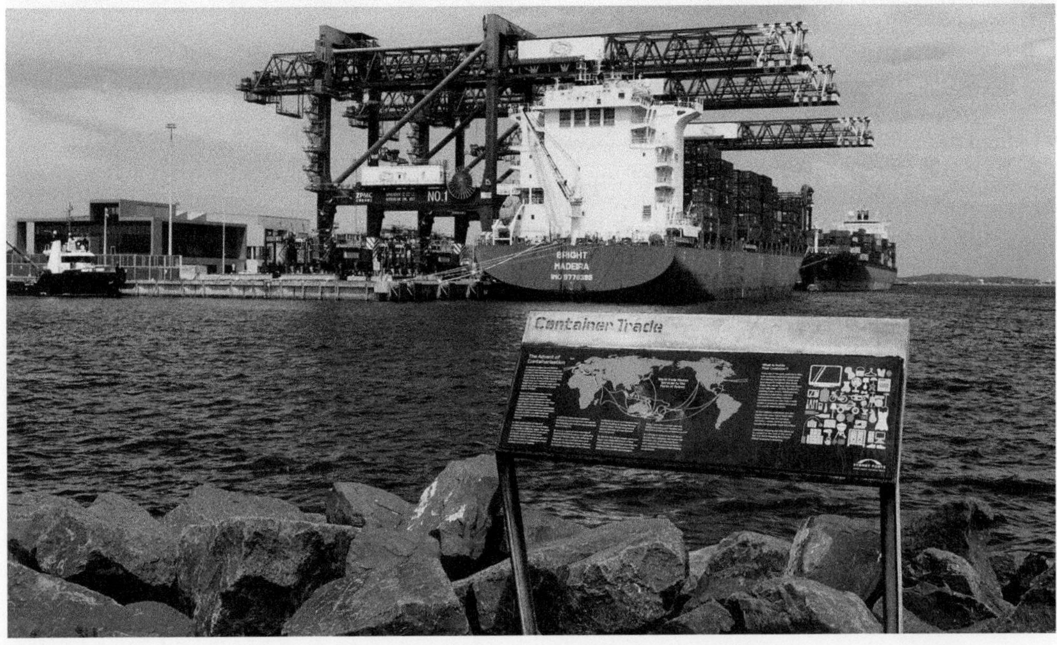

Subtopic	Success criteria			
19.2	I can explain some of the reasons why trade occurs across a range of scales.			
	I can give examples of how trade occurs across a range of scales.			
19.3	I can create and describe a flow map.			
19.4	I can identify the interconnections between Australia and other places through the exchange of goods and services.			
19.5	I can identify how food is traded around the world.			
	I can explain the global movement of food produced in Australia.			
19.6	I can transform data into multiple line and cumulative line graphs.			
19.7	I can identify how trade links customers to producers.			
	I can explain how trade moves around the world.			
19.8	I can provide examples of global companies.			
	I can explain the ways globalisation has changed manufacturing.			
19.9	I can explain how events overseas can have an impact on Australia and our economy.			
	I can provide examples of global events that have affected the Australian economy.			
19.10	I can analyse a topographic map to discuss interconnections between port facilities and trade.			

eWorkbook Chapter 19 Student learning matrix (ewbk-8508)
Chapter 19 Reflection (ewbk-8509)
Chapter 19 Extended writing task (ewbk-8510)

Interactivity Chapter 19 Crossword (int-8578)

ONLINE RESOURCES

on Resources

Below is a full list of **rich resources** available online for this topic. These resources are designed to bring ideas to life, to promote deep and lasting learning and to support the different learning needs of each individual.

eWorkbook

- 19.1 Chapter 19 eWorkbook (ewbk-7999) ☐
- 19.2 Features of trade (ewbk-9634) ☐
- 19.3 SkillBuilder — Constructing and describing a flow map (ewbk-9638) ☐
- 19.4 Australian trade connections (ewbk-9642) ☐
- 19.5 Global food trade (ewbk-9646) ☐
- 19.6 SkillBuilder — Constructing multiple line and cumulative line graphs (ewbk-9650) ☐
- 19.7 Linking markets to consumers (ewbk-9654) ☐
- 19.8 Impacts of globalisation (ewbk-9658) ☐
- 19.9 Global disruption to the Australian economy (ewbk-9662) ☐
- 19.10 Investigating topographic maps — Port Botany (ewbk-9666) ☐
- 19.12 Chapter 19 Student learning matrix (ewbk-8508) ☐
- Chapter 19 Reflection (ewbk-8509) ☐
- Chapter 19 Extended writing task (ewbk-8510) ☐

Sample responses

- 19.1 Chapter 19 — Sample responses (sar-0151) ☐

Digital documents

- 19.3 Blank world map template (doc-36260) ☐
- 19.10 Topographic map of Port Botany (doc-36263) ☐

Video eLessons

- 19.1 Trading places (eles-1724) ☐
- Interconnection through trade — Photo essay (eles-5155) ☐
- 19.2 Features of trade — Key concepts (eles-5156) ☐
- 19.3 SkillBuilder — How to construct and describe a flow map (eles-1741) ☐
- 19.4 Australian trade connections — Key concepts (eles-5157) ☐
- 19.5 Global food trade — Key concepts (eles-5158) ☐
- 19.6 SkillBuilder — Constructing multiple line and cumulative line graphs (eles-1740) ☐
- 19.7 Linking markets to consumers — Key concepts (eles-5159) ☐
- 19.8 Impacts of globalisation — Key concepts (eles-5160) ☐
- 19.9 Global disruption to the Australian economy — Key concepts (eles-5161) ☐
- 19.10 Investigating topographic maps — Port Botany — Key concepts (eles-5162) ☐

Interactivities

- 19.2 Up, up and away (int-3339) ☐
- 19.3 SkillBuilder — How to construct and describe a flow map (int-3359) ☐
- 19.4 Trading partners (int-3338) ☐
- 19.5 Global food trade (int-7932) ☐
- 19.6 SkillBuilder — Constructing multiple line and cumulative line graphs (int-3358) ☐
- 19.7 Linking markets to consumers (int-8574) ☐
- 19.8 Impacts of globalisation (int-8575) ☐
- 19.9 Global disruption to the Australian economy (int-8576) ☐
- 19.10 Investigating topographic maps — Port Botany (int-8577) ☐
- 19.12 Chapter 19 Crossword (int-8578) ☐

Fieldwork

- 19.7 Trade connection (fdw-0025) ☐

ProjectsPLUS

- 19.11 Thinking Big research project — Ethical business — Real change or fast fashion? (pro-0251) ☐

Weblinks

- 19.4 Australian trade partners (web-6309) ☐
- Trade investment (web-6310) ☐
- Ports of Australia (web-6311) ☐
- Trade through time (web-6312) ☐
- Trade by country (web-6313) ☐
- 19.5 Trade (web-4351) ☐
- 19.7 Ports (web-1128) ☐
- 19.8 Running free (web-6314) ☐
- 19.9 ABS — COVID-19 impacts (web-6315) ☐
- 19.11 Good on you (web-4381) ☐
- Good shopping guide (web-5515) ☐

Google Earth

- 19.10 Port Botany (gogl-0094) ☐

myWorld Atlas

- 19.4 Aid, migration and trade (mwa-4543) ☐

Teacher resources

There are many resources available exclusively for teachers online.

20 Connect, produce, consume

INQUIRY SEQUENCE

To access a pre-test with **immediate feedback** and **sample responses** to every question in this chapter, select your learnON format at www.jacplus.com.au.

20.1 Overview

Numerous **videos** and **interactivities** are embedded just where you need them, at the point of learning, in your learnON title at www.jacplus.com.au. They will help you to learn the content and concepts covered in this topic.

A world of products is available for us to buy at the click of a mouse. But what are the costs of our consumer culture for the people and places of the world?

20.1.1 Introduction

For consumers in places such as Australia, there has never been an easier time to buy a wide range of cheap products from around the world to make our lives easier, more interesting and more connected. We upgrade products when a new version or design becomes available — rather than when the old one stops working — and we throw the old products away rather than repairing them.

FIGURE 1 An Atlantic grey seal caught in a piece of fishing net, Horsey Beach, Norfolk, England

The negative impacts of this **consumer culture** are not just on the environment from the growing piles of waste. There is a widening gap between the 'haves' and the 'have-nots', including groups of workers in developing countries who are vulnerable to exploitation from the big global companies that want their products made as quickly and cheaply as possible. Is the convenience of our consumer culture worth the impact it is having on people and places?

consumer culture the influences of and focus on material goods/products in a society

STARTER QUESTIONS

1. Many companies offer customers a chance to donate to a charity when the sale of an item is being finalised. How many examples of this practice can you list?
2. How many businesses can you name that help customers trade in or donate their unwanted items?
3. If you wanted to discard your outdated mobile phone, do you know how to do so responsibly? Do you think about where your e-waste goes?
4. How many United Nations Sustainable Development Goals can you list? What is the purpose of these SDGs?
5. Watch the **Connect, produce, consume — Photo essay** in your Resources tab. Do you think that seeing the impact of rubbish on the environment might encourage people to buy fewer things?

Resources

eWorkbook Chapter 20 eWorkbook (ewbk-8000)

Video eLessons Plugging in (eles-1725)
Connect, produce, consume – Photo essay (eles-5163)

20.2 Environmental impacts of production and consumption

LEARNING INTENTION

By the end of this subtopic, you will be able to describe and list examples of the ways production and consumption influence the way the environment operates.

20.2.1 Climate change

The majority of climate scientists believe that climate change is having an impact on rainfall patterns, based on changes recorded over the past 100 years. Scientific research suggests temperature changes have taken place on land, in the atmosphere and the sea. The concentration of greenhouse gases in the atmosphere has increased, leading to **global warming**.

global warming the observable rising trend in the Earth's atmospheric temperatures, generally attributed to the enhanced greenhouse effect

FIGURE 1 Australia's contribution to global emissions by sector of the economy, March 2020

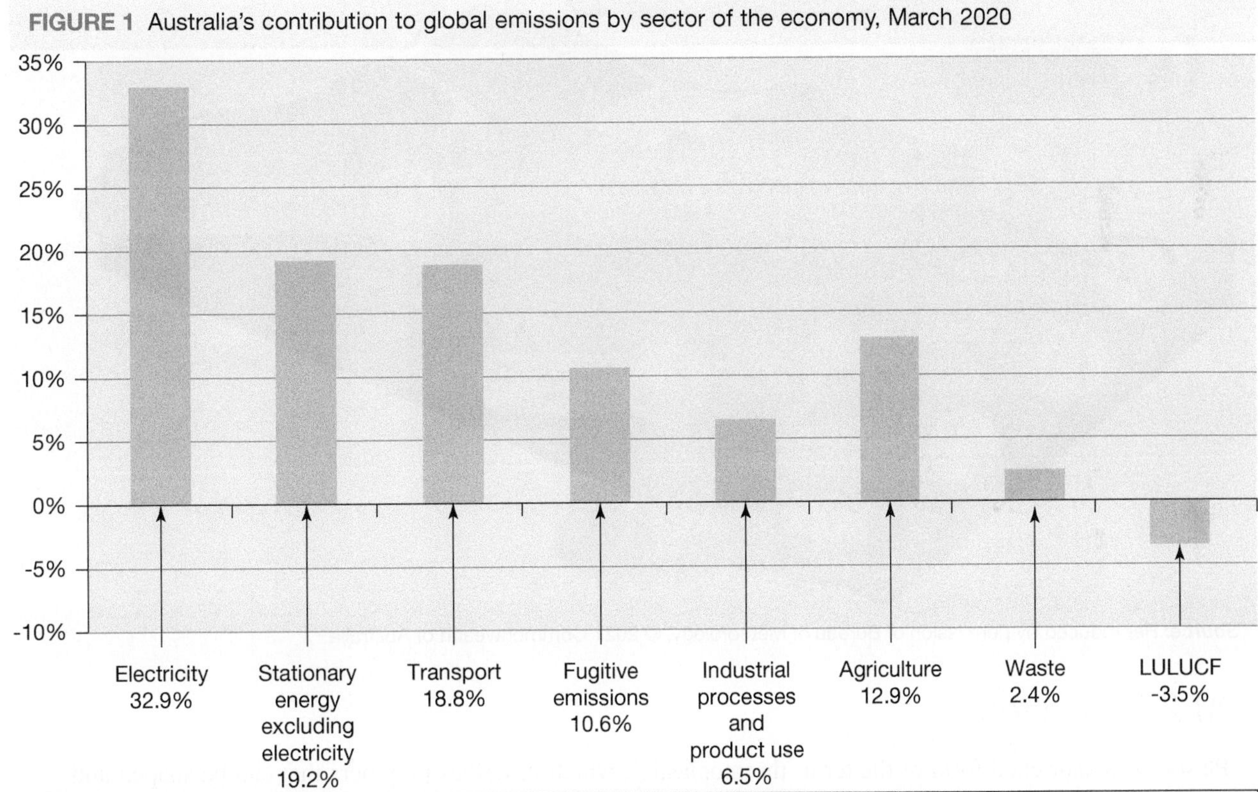

Source: Climate Council. What is climate change and what can we do about it? Published by Climate Council. 16 October 2019. Available at https://www.climatecouncil.org.au/resources/what-is-climate-change-what-can-we-do/

Notes:

- **Electricity** – emissions from burning coal and gas to power our lights, appliances and more
- **Stationary energy** – fuels like gas consumed directly, rather than used for electricity, in industry and in households
- **Transport** – emissions from petrol and diesel used to power cars, trucks and buses, and emissions from aviation fuel used to power planes
- **Fugitive emissions** – gases vented from fossil fuel extraction and transportation
- **Industrial processes** – emissions produced by converting raw materials into metal, mineral and chemical products
- **Agriculture** – greenhouse gases such as methane and nitrous oxide produced by animals, manure management, fertilisers and field burning
- **Waste** — emissions from decaying organic matter
- **Land use, land use change and forestry (LULUCF)** – emissions and removals mainly from forests, but also from croplands, grasslands, wetlands and other lands

It is believed that human activity, including burning fossil fuels such as coal and oil, have led to what is known as the **enhanced greenhouse effect**, which is heating the Earth and its atmosphere. **FIGURE 1** outlines Australia's share of greenhouse gas emissions by key industry sectors.

Electricity is the largest source of greenhouse emissions in Australia, responsible for 32 per cent of emissions. In Australia 84 per cent of our electricity comes from burning fossil fuels. Coal-fired energy represented 56 per cent of total energy generation in 2019.

The production and distribution of goods and services contributes to greenhouse gas emissions globally. The consequences of continued changes to the global climate are outlined in **FIGURE 2**.

FIGURE 2 Consequences of changes in the global climate

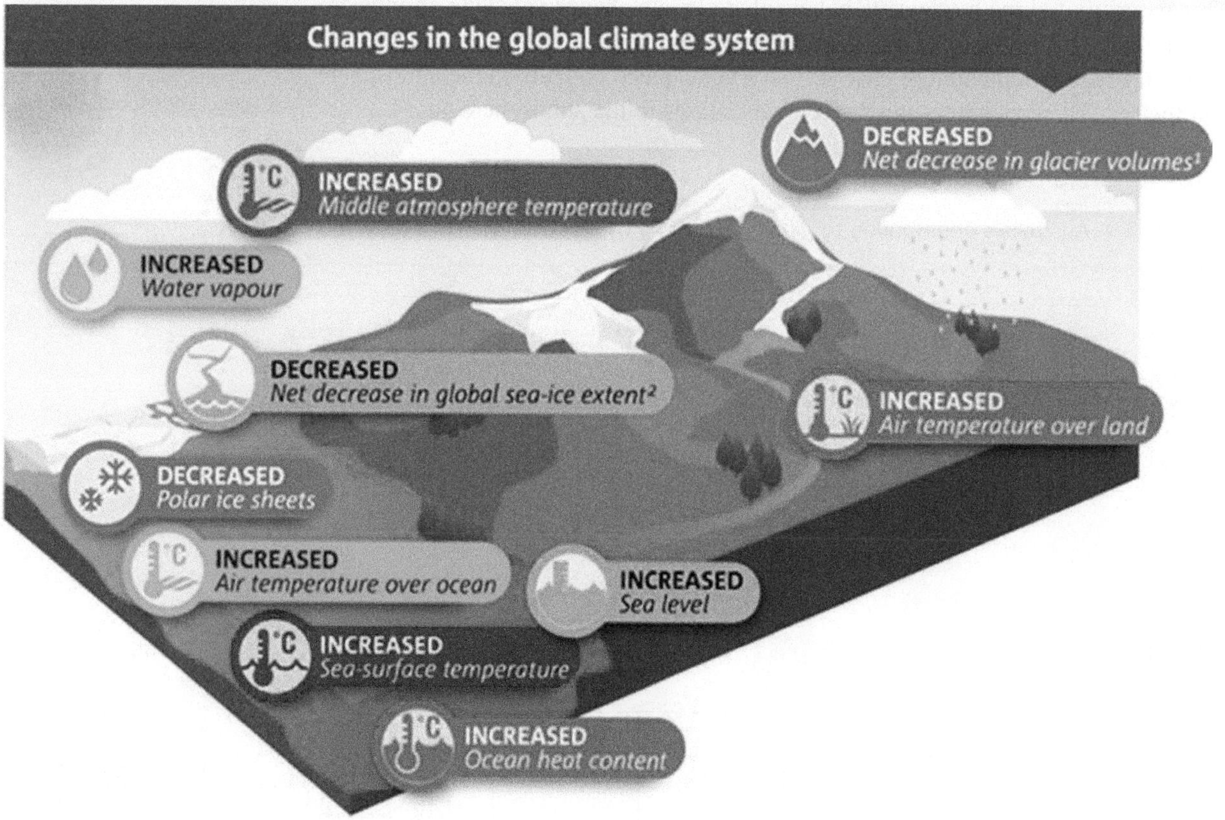

Source: Reproduced by permission of Bureau of Meteorology, © 2021 Commonwealth of Australia.

20.2.2 Plastics

'Plastic' is a shortened form of the term 'thermoplastic', which describes polymers that can be shaped and reshaped using heat. Plastics have different strength and toughness properties, and can be designed to satisfy a specific need. They provide manufacturers with a relatively inexpensive alternative to natural materials such as wood, metal, glass and cotton, and are a versatile product.

The benefits of plastic packaging to producers include a reduction in the overall weight of products and a reduction in food spoilage as perishable products are transported to market. For customers, plastic is a convenient product that can have a single use or more long-term use.

Plastics can be produced from raw materials such as oil and gas, renewable resources such as sugar cane and starch, or minerals such as salt. In 2019, the global production of plastics reached 368 million metric tons. **FIGURE 3** demonstrates the variety of uses of ways plastic is used in production.

enhanced greenhouse effect increasing concentrations of greenhouse gases in the Earth's atmosphere, contributing to global warming and climate change

FIGURE 3 Uses of plastic polymers in 2015 — more than one-third were used for packaging.

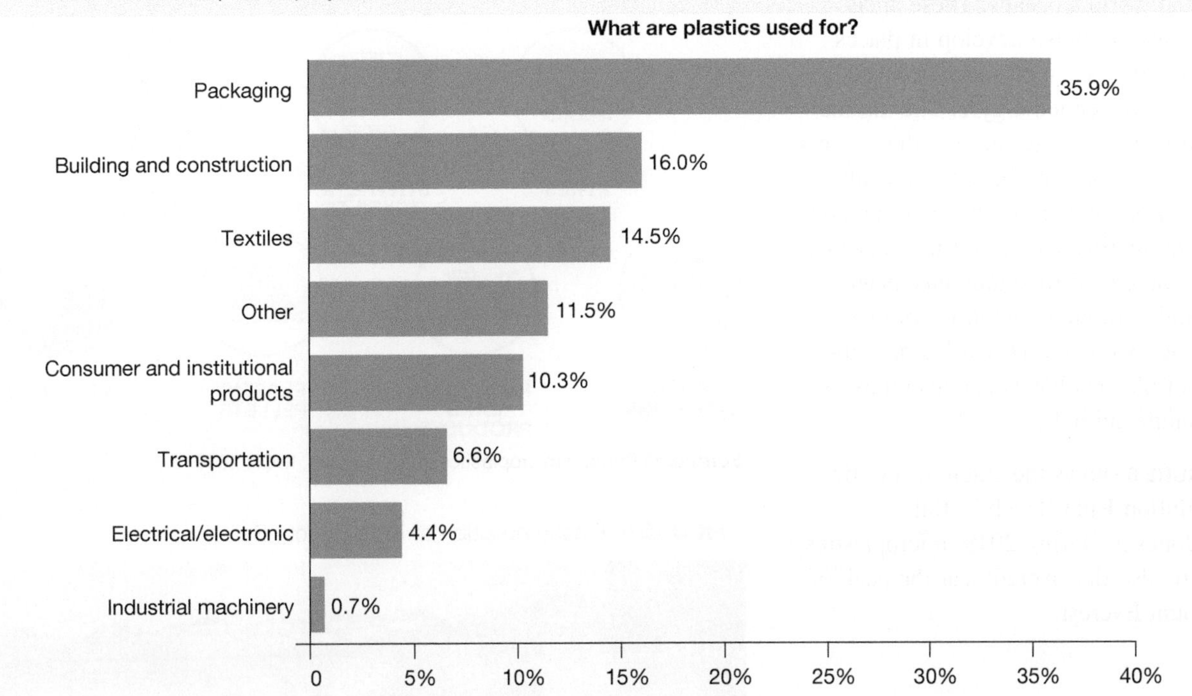

Source: Roland Geyer, Jenna R. Jambeck and Kara Lavender Law, 'Production, use, and fate of all plastics ever made', *Science Advances* 19 Jul 2017: Vol. 3, no. 7, e1700782, DOI: 10.1126/sciadv.1700782

The growing use of plastics in production and consumption is causing widespread environmental challenges. The use of natural resources such as fossil fuels to produce plastics is contributing to the enhanced greenhouse effect. The differing biodegradable properties of types of plastic is contributing to environmental pollution and ingestion of plastic by marine life. Most plastic does not disintegrate; instead it breaks down into smaller pieces referred to as microplastics. **FIGURE 4** shows the differing rates at which plastics break down within the environment and **FIGURE 5** shows the main sources of microplastic pollution.

FIGURE 4 Plastics have different properties to suit their use, which affect how long they last.

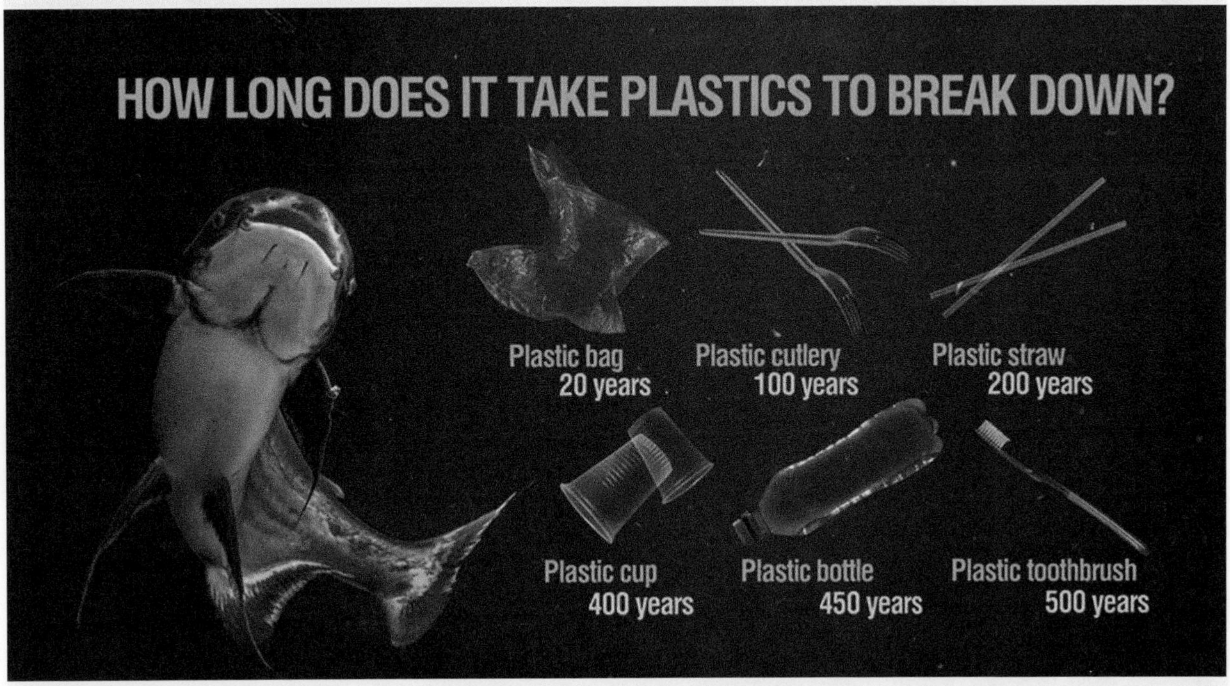

HOW LONG DOES IT TAKE PLASTICS TO BREAK DOWN?

Plastic bag
20 years

Plastic cutlery
100 years

Plastic straw
200 years

Plastic cup
400 years

Plastic bottle
450 years

Plastic toothbrush
500 years

Concentrations of debris have formed in the world's oceans. These areas of plastic rubbish develop in places where the ocean currents circulate in patterns called a gyre. The marine debris concentrates around these gyres or on beaches and becomes a challenge for shipping and marine environments. Many marine species mistake plastics for sources of food and have been found with large quantities of plastic in their stomach. Discarded nets and other plastic objects can also trap and strangle animals.

FIGURE 6 shows the extent of plastic pollution Kuta Beach in Bali, Indonesia. During 2019, microplastics were also discovered near the peak of Mount Everest.

FIGURE 5 Main sources of microplastics

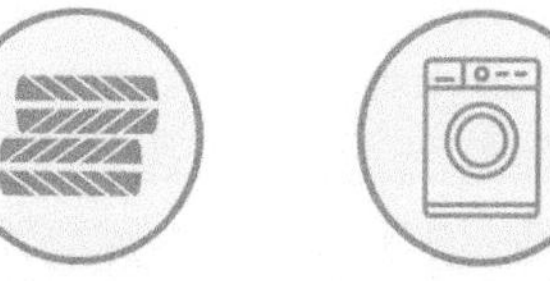

Source: © Primary microplastics in the oceans, IUCN 2017

FIGURE 6 Plastic pollution at Kuta Beach, Bali

FIGURE 7 Marine debris concentrations are commonly referred to as a 'garbage patch'.

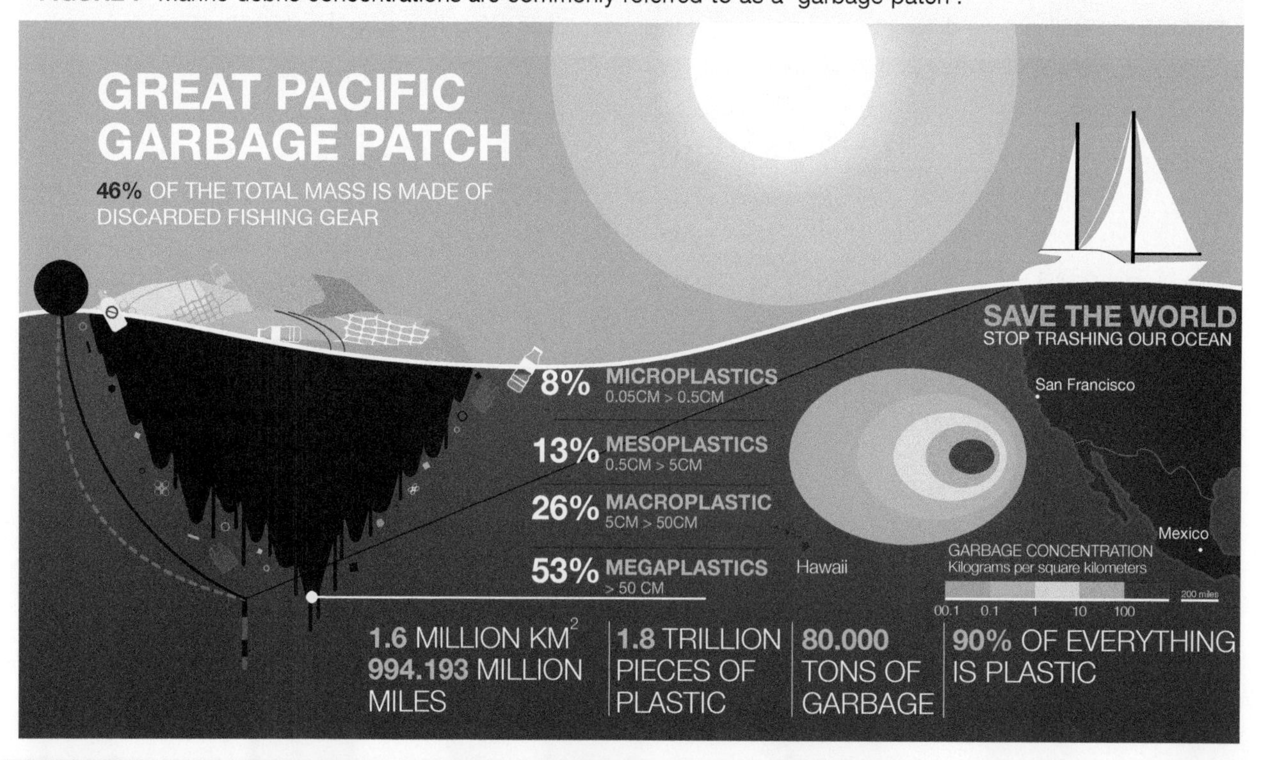

20.2.3 Containerisation

Shipping interconnects places around the world through trade. The use of containers makes the process efficient and more reliable, but what are the environmental costs of this industry?

Containers provide an efficient and standardised method of transport for bulky goods across the world between production sites and consumers. There are many challenges faced by the environment as shipping containers move around the globe, but also many potential benefits (**FIGURE 8**).

FIGURE 8 Some of the sustainable and unsustainable features of shipping containers

MARINE DEBRIS

Wild storms can destabilise cargo ships carrying containers. When the containers fall into the sea, often the contents are scattered along beaches and coastlines. Also, submerged containers are hazardous to other ships moving through shipping channels.

DISRUPTION TO MIGRATORY PATHS

Cargo ships frequently pass through the breeding grounds of whales and lead to vessel strikes. According to the IUCN, the population of the critically endangered North Atlantic right whale was estimated to be fewer than 250 mature whales at the end of 2018. Vessel strikes are reported to be one of many causes of the population decline.

WASTE DISCHARGES

Marine discharges from shipping operations include fuel, oil and human waste which can impact the nutrient levels recorded in coastal locations. Also, ballast water (water taken into the hull of the ship to stabilise it during a voyage) can release contaminated water into the marine ecosystem of a port.

UNSUSTAINABLE

SUSTAINABLE

Containers were used to create a temporary shopping district following the 6.3-magnitude earthquake in Christchurch during 2011.
Around 60 shipping containers were assembled in the town centre, an area that lost 70 per cent of its buildings.

REPURPOSED FOR EMERGENCY USE

An affordable housing project in Havre, France, made using shipping containers offers low-cost student accommodation.

REPURPOSED FOR HOUSING

20.2.4 Food waste

The United Nations reports that 'each year an estimated one third of all food produced ends up rotting in the bins of consumers and retailers, or spoiling due to poor transportation and harvesting practices'. Goal 12 of the United Nations Sustainable Development Goals addresses a need to adopt more sustainable methods of production and consumption.

Suggested improvements to production and consumption methods include:
- coordinated transport and cold storage systems
- better market linkages between producer and consumer
- better processing and packaging of food.

FIGURE 9 In 2016 an average of 13.8 per cent of food was lost before it had reached the consumer.

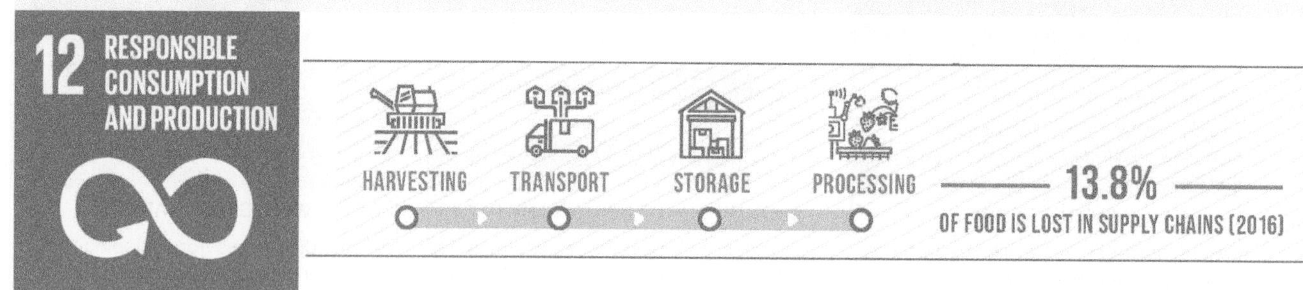

Source: © United Nations

FIGURE 10 Key reasons for food waste at the retail and consumer levels of distribution

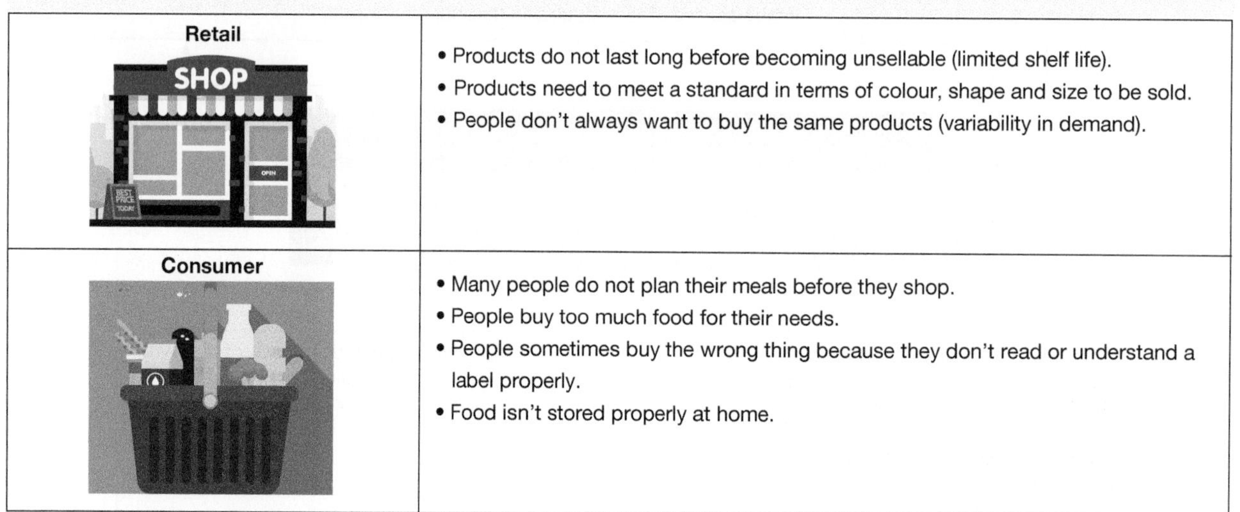

Retail	• Products do not last long before becoming unsellable (limited shelf life). • Products need to meet a standard in terms of colour, shape and size to be sold. • People don't always want to buy the same products (variability in demand).
Consumer	• Many people do not plan their meals before they shop. • People buy too much food for their needs. • People sometimes buy the wrong thing because they don't read or understand a label properly. • Food isn't stored properly at home.

20.2.5 The illegal wildlife trade

Not all trade is legal. The international trade in wildlife has been one of the factors responsible for the massive decline in many species of animals and plants. Millions of live animals and plants are illegally shipped around the world each year to supply the pet trade and to meet the demand for decorative plants. Wild animal and plant products, such as skins, meat, ornaments, animal parts and timber, are traded in large quantities.

In 1973, an international **treaty** known as the Convention on International Trade in Endangered Species (CITES) was drawn up in Washington to prevent international trade that threatened species with extinction. Any trade in products from threatened wildlife now requires a special permit. Unless there are exceptional circumstances, no such permits are issued for species threatened with extinction. These include species of tigers, elephants, rhinoceroses, apes, parrots and all sea turtles.

FIGURE 11 Global exporters and importers of endangered species and products made from endangered species

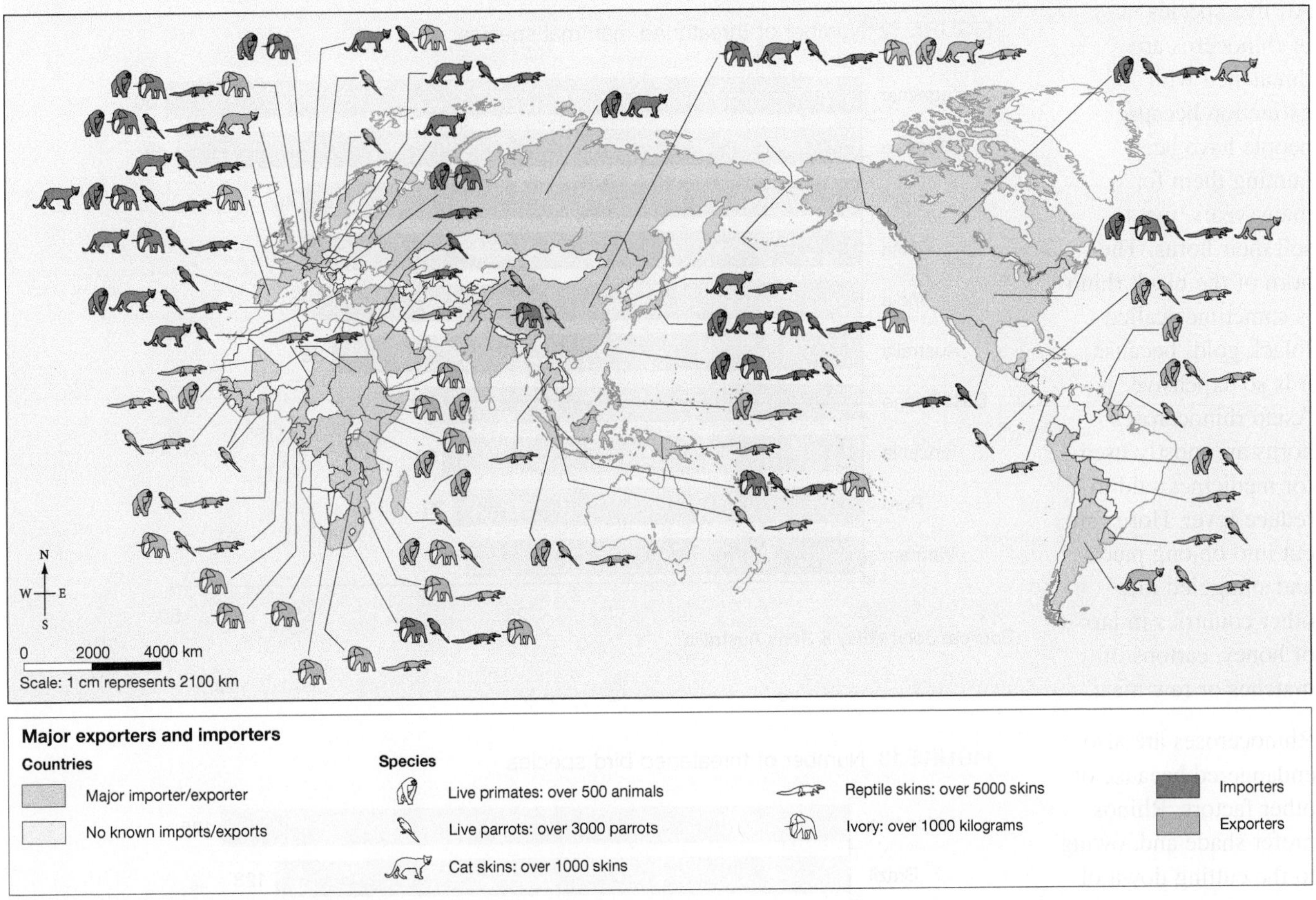

0 2000 4000 km
Scale: 1 cm represents 2100 km

Major exporters and importers

Countries
Major importer/exporter
No known imports/exports

Species
Live primates: over 500 animals
Live parrots: over 3000 parrots
Cat skins: over 1000 skins
Reptile skins: over 5000 skins
Ivory: over 1000 kilograms

Importers
Exporters

Source: © MAPgraphics Pty Ltd, Brisbane

The live pet trade

Trade in threatened wildlife still takes place through **smuggling**. Birds are drugged and stuffed into plastic tubes; snakes are coiled into stockings and posted; lizards are stitched into suitcases. Many of the animals die.

Prices on the **black market** can be very high. Bird traffickers can earn more than $150 000 for taking 30 eggs out of a country in specially designed vests that keep the eggs warm. Overseas collectors will pay up to $50 000 for a breeding pair of endangered red-tailed black cockatoos.

Traditional Chinese medicine

Traditional Chinese medicine (TCM), the most widely practised traditional medicine system in the world, uses more than 1000 plant and animal species. While TCM has been practised for perhaps 5000 years, some of the wild plants and animals used are now threatened or in danger of extinction. Among them are certain orchids, musk deer, rhinoceroses, tigers and some bear species.

treaty a formal agreement between two or more independent states or nations, usually involving a signed document

smuggling importing or exporting goods secretly or illegally

black market any illegal trade in officially controlled or scarce goods

Threatened species

All five species of rhinoceros are threatened with extinction because people have been hunting them for many years, just to sell their horns. The horn of the black rhino is sometimes called 'black gold' because it is so expensive. Asian rhinoceroses' horns are mostly used for medicines said to reduce fever. Horns are cut into oblong pieces and smuggled into other countries in jars of honey, cartons of matches or raw meat.

FIGURE 12 Number of threatened mammal species

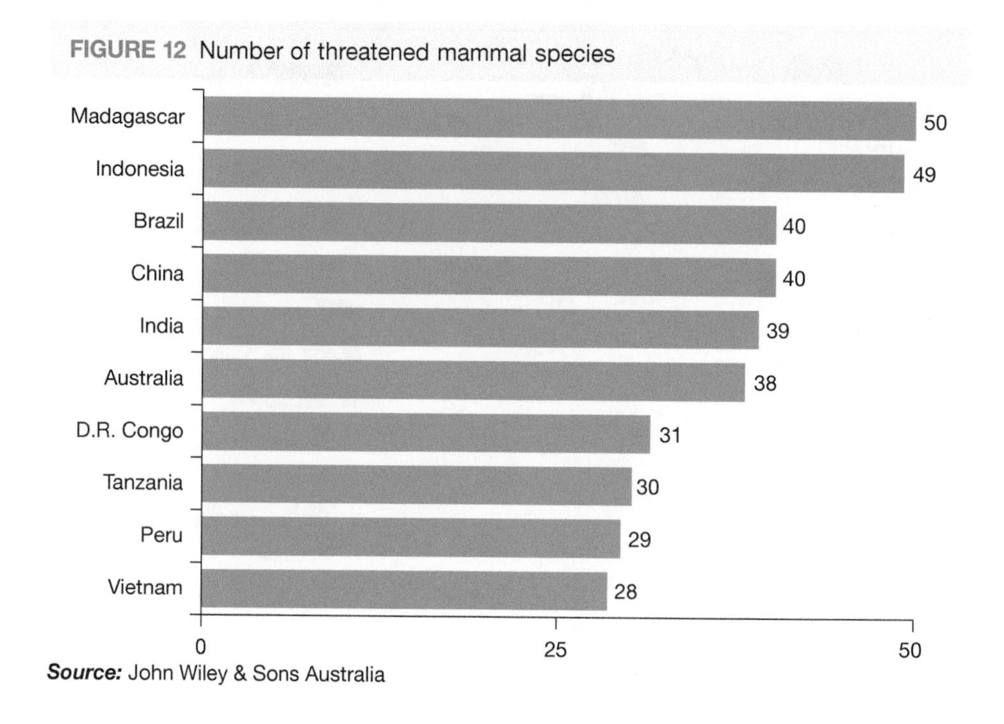

Source: John Wiley & Sons Australia

Rhinoceroses are also endangered because of other factors. Rhinos prefer shade and, owing to the cutting down of trees in grasslands, their source of shelter is being wiped out. Pollution is also endangering these animals; toxic waste and pesticides are left on the grass that the rhinos eat.

FIGURE 13 Number of threatened bird species

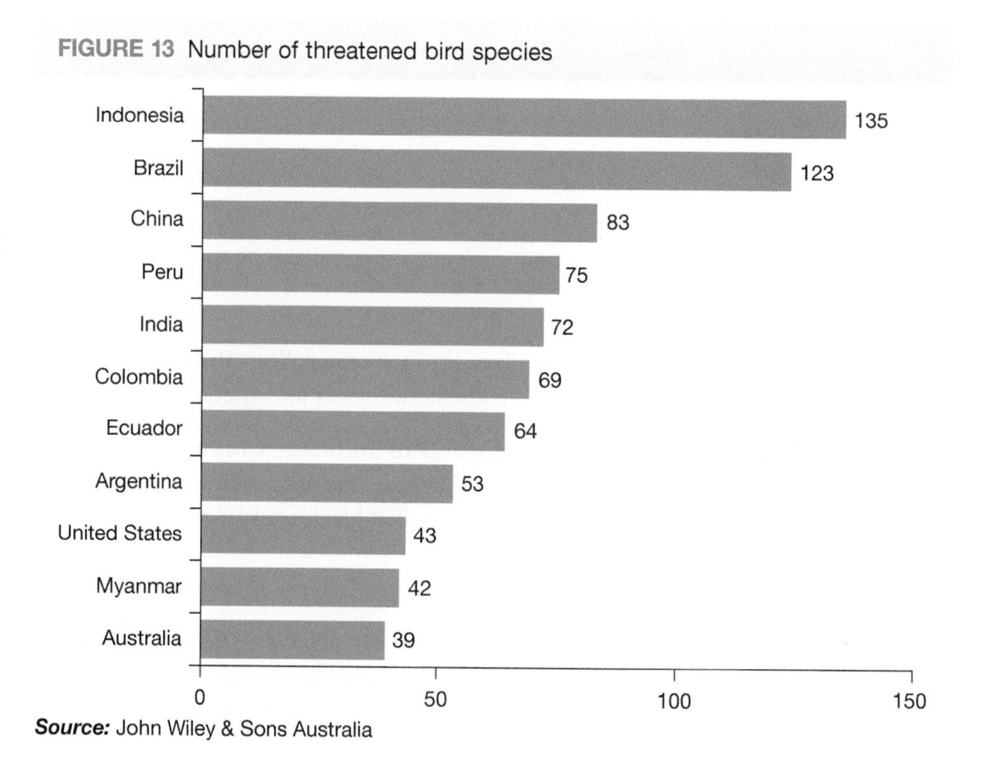

Source: John Wiley & Sons Australia

Although international trade in rhino horn has been banned under CITES since 1977, demand remains high, and this encourages rhino poaching in both Africa and Asia. Criminal syndicates link poachers in places such as South Africa to transit points, smuggling channels and final destinations in Asia. The main market is now Vietnam, where there is a newly emerged belief that rhino horn cures cancer. Rhino horn is used in other traditional Asian medicine to treat a variety of ailments, including fever and blood disorders. It is also used by wealthy Asians as a cure for hangovers.

In October 2011, the World Wide Fund for Nature (WWF) helped to successfully establish a new black rhino population in a safer, more spacious place in South Africa. Nineteen critically endangered black rhinos were transported to this new home. Such relocations reduce pressure on existing wildlife reserves and provide new territory, where rhinos have a better chance of increasing in number. Creating more dispersed and better protected populations also helps keep rhinos safe from poachers.

As of January 2021, it is believed that there are just two northern white rhinos left in the world; both females, living in a reserve in Kenya. The last male of the species died in 2018, which means that the species will likely become extinct unless scientists working with the rhinos can find a successful, artificial solution to reproduction without a surviving male rhino.

FIGURE 14 Conservationists dehorning a black rhino in Zimbabwe to protect it from poachers

Source: © Austral International

Resources

eWorkbook	Environmental impacts of production and consumption (ewbk-9674)
Video eLesson	Environmental impacts of production and consumption — Key concepts (eles-5164)
Interactivity	Environmental impacts of production and consumption (int-8579)

20.2 ACTIVITIES

1. Prepare arguments for a debate on the topic, 'Shipping is a sustainable option for international trade'. Explore the positive and negative points on this topic and suggest strategies to reduce some of the impacts presented in **FIGURE 8**. Use the **Ship tracking** and **Food miles** weblinks in your Resources panel to research the extent and impacts of transporting goods around the world.
2. Look carefully at **FIGURE 11**.
   a. Which categories of wildlife are traded for their skins?
   b. Using an atlas, list the places that are the major exporters of the following:
      - live primates
      - cat skins
      - ivory
      - live parrots
      - reptile skins.
   c. Based on your list in (b), list the continents that are the main sources of wildlife species and wildlife goods.
3. Discuss the ethical considerations involved with waste in the oceans, starting with the following question: 'The most effective ways of reducing marine pollution have to start on land. What are our obligations and duties as global citizens to reduce waste and pollution?' To begin your research use the **Great Pacific garbage patch** and **Microplastics** weblinks in your Resources panel.

FIGURE 15 National parks provide sanctuaries for elephants near Mt Kilimanjaro, Kenya.

4. Use the **IUCN threatened species** weblink in your Resources panel to research the critically endangered North Atlantic right whale. Explain the causes of population decline for this whale species. To what extent do you think shipping vessel strikes have contributed to its declining population numbers?
5. Prepare a digital presentation to highlight the environmental impacts of production and consumption using the examples in the chapter. What other impacts can you include in this presentation? Use the **Climate council** and **UN Sustainable Development Goals** weblinks in your Resources panel to broaden your research.

20.2 EXERCISE

Learning pathways

■ LEVEL 1	■ LEVEL 2	■ LEVEL 3
1, 3, 7, 8, 14	2, 4, 9, 11, 15	5, 6, 10, 12, 13

Check your understanding

1. Identify the largest source of greenhouse emissions in Australia and its percentage contribution to emissions.
2. Define the term *enhanced greenhouse effect*.
3. Identify and describe three impacts that global climate change has on the environment.
4. What is the greatest use made of plastics in production?
5. List two ways that using shipping containers is not sustainable for the environment.
6. In your own words, define the term *supply chain*.
7. What proportion of food was wasted before it reached the consumer in 2016?

Apply your understanding

8. Explain why plastics are dangerous to marine animals.
9. Study **FIGURES** 12 and 13.
   a. Which two countries in the Asia–Pacific region have high numbers of threatened mammals and threatened birds?
   b. What reasons might there be for Indonesia having the greatest number of threatened animal species?
10. Clarify why understanding the difference between long-term climate change and short-term changes in the weather is vital to understanding the impact of global warming.
11. Explain why wild species are traded.
12. International treaties banning trade of endangered animals have not stopped the illegal use of banned animal products in Chinese medicine. Why do you think this is the case?

Challenge your understanding

13. Some countries in Africa allow trophy hunting tours — legal tours for recreational hunters to shoot 'big game' such as elephants and wildebeest. Do you think this is a sustainable and appropriate strategy to minimise the effects of illegal hunting?
14. Suggest one way that the negative impact of shipping containers might be reduced.
15. Propose two ways that international authorities could reduce levels of smuggling and poaching.

FIGURE 16 Wildebeest, Serengeti National Park, Tanzania

To answer questions online and to receive **immediate feedback** and **sample responses** for every question, go to your learnON title at www.jacplus.com.au.

20.3 Social and economic impacts of production and consumption

LEARNING INTENTION

By the end of this subtopic, you will be able to describe some of the social and economic impacts of global production and consumption patterns.

20.3.1 Labour exploitation

Large companies that operate on a global scale often choose manufacturing sites that will deliver the most profitable return on their investment. This can lead to worker exploitation in less economically developed countries, and many negative social impacts for communities.

FIGURE 1 Young boys working in a metal manufacturing workshop, Bangladesh

Workers in many of these manufacturing sites do not have the same rights as Australian workers. They may be forced to work long hours for little pay in poor conditions. They may also be required to work with toxic chemicals and dangerous machinery without the safety equipment and processes that you might expect in Australia. The laws in countries that attract these 'sweatshop' factories are often not enforced. The products are made very cheaply, then sold for high prices in more economically developed countries.

FIGURE 2 Textile workers continued to work in factories during the COVID-19 pandemic.

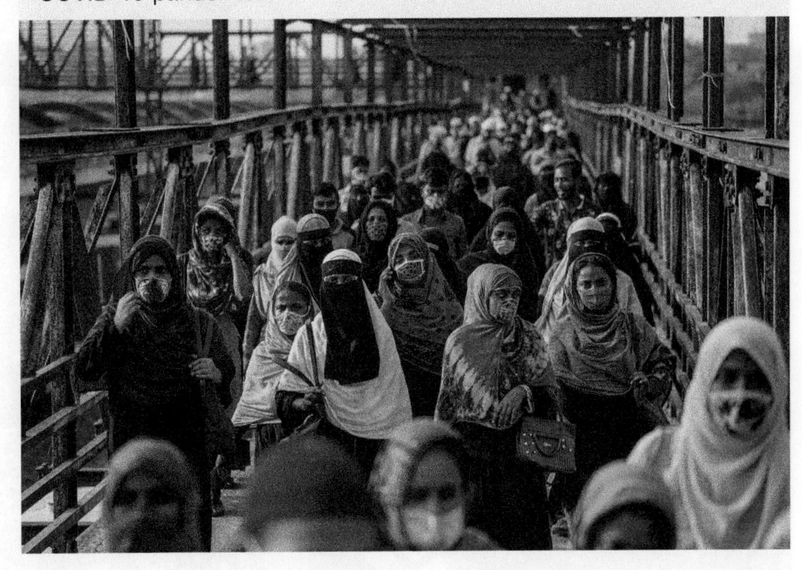

During the COVID-19 pandemic, the demand for some products declined and many global brands cancelled orders that they had placed with manufacturers. This led to further job insecurity for this vulnerable group of workers because fewer orders meant that fewer employees were needed or that there was less profit being made to pay them. **FIGURE 1** shows a group of Bangladeshi textile workers who placed their health at risk by continuing to work in crowded factories during the pandemic.

20.3.2 Throwaway society

In Australia during 2018–19, a total of 3.5 million tonnes of plastic were used and disposed of as waste. Single-use plastic items such as those shown in **FIGURE 3** are convenient packaging options for businesses but generate large amounts of waste.

FIGURE 3 Plastic rubbish

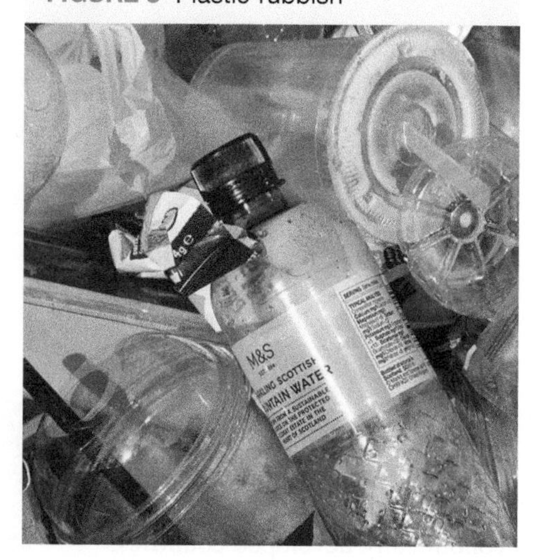

In less economically developed countries, where waste is not separated and recycled, the rubbish of entire cities is often dumped into landfill sites. People known as 'garbage pickers' seek to earn a living by scavenging through the rubbish, looking for plastics, recyclables, rags and metal pieces to sell.

Packaging is the biggest source of waste pollution in our society and we have choices about how items are packaged when we purchase them. Consumers are being encouraged to decline single-use plastic bags and to use re-usable bags and containers instead.

In some tropical countries plastic bags also provide a breeding ground for malaria-carrying mosquitoes and contribute to health issues for the population.

Action is being taken in many places around the world to combat the problem of waste disposal for environmental reasons; however, there are also negative impacts of reducing plastic use. With any change in how we produce or consume goods and services, there is the chance that some people's jobs are no longer required. For example, Kenya has banned plastic bag use and manufacturers face strict penalties if the law is violated. Fines and jail time are imposed for plastic bag use, manufacturing or distribution. This means that people previously employed in making or distributing plastic bags need to find new jobs.

FIGURE 4 Garbage pickers earn a living collecting valued waste.

20.3.3 Disruptions to patterns of production and consumption

When typical production and consumption patterns are disrupted, such as during the COVID-19 pandemic, there are many social and economic impacts. Significant changes to the demand for specific goods and services took place in 2020, and the ability of companies to safely satisfy growing demands for online products was challenging.

Australian businesses that relied on international travel, such as the tourism and education industries, were heavily impacted by COVID-19. Similar impacts were experienced by businesses in the services sector — in Australia, four out of five people work in this sector, so lockdowns significantly impacted many workplaces. As local regulations required businesses to close operations or adapt to 'contactless' transactions to ensure community safety, many people were forced to work remotely and others lost their jobs. Many remote workers reported that they were working longer hours at this time as they were required to respond to emails and work requests in their own time to maintain communication with work colleagues and customers.

FIGURE 5 The use of artificial intelligence is increasing in the services industry.

FIGURE 6 The sharing economy is changing the way producers and consumers interact.

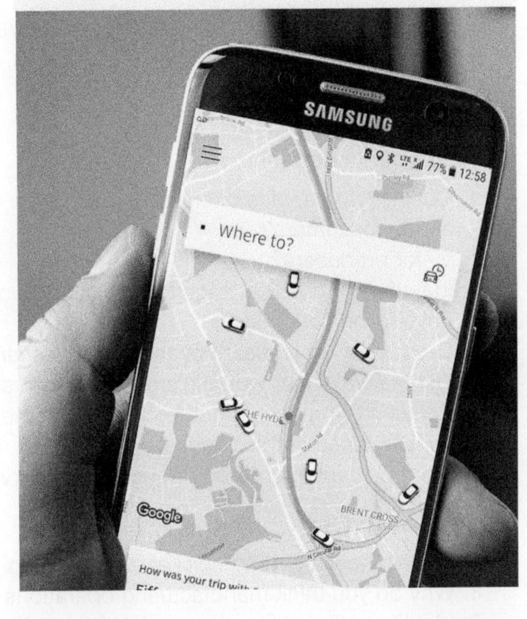

Prior to the pandemic, some workplaces were offering more flexible work options and businesses were using more labour-replacing technology, such as robotics and artificial intelligence. Banking and shopping services increasingly offered automated services and encouraged customers to interact with their 'virtual assistants' on online websites. Some online operations of retail stores operated out of warehouses instead of a physical shop that customers could visit. These changes became far more widespread and in some situations necessary during the pandemic. Some of these changes continued long after businesses were allowed to trade face-to-face again.

Consumers were also beginning to embrace home delivery services and shopping online before the pandemic. The growing **sharing economy** was enabling customers to access a range of goods and services such as cars and holiday accommodation without using the more traditional forms of car hire and hotel booking services. The sharing economy also provided some employment opportunities for people who had lost their jobs because of the pandemic.

In places that experienced extended lockdowns or work-from-home periods, there was also a significant drop in vehicle use and factory operations. This meant less travel and manufacturing — two traditional causes of air pollution and carbon emissions — which led to improvements in air quality and lower greenhouse gas emissions.

sharing economy sharing objects and/or services either for free or for payment, typically arranging the trade online

on Resources

eWorkbook	Social and economic impacts of production and consumption (ewbk-9678)
Video eLesson	Social and economic impacts of production and consumption — Key concepts (eles-5165)
Interactivity	Social and economic impacts of production and consumption (int-8580)

20.3 ACTIVITIES

1. Describe the changes to the working routines of your family during the COVID-19 pandemic. Did you or someone in your home have to work or study remotely? Outline the positive and the negative features of this change to working conditions.
2. Research businesses that performed well economically during the lockdowns. Explain why you think these businesses performed well when so many other businesses were forced to close down and suffered a loss of earnings. Begin your research by investigating places that had extended lockdowns, such as Melbourne, London or Mumbai.
3. Research how Australia's laws and regulations to encourage reduced plastic use compare to other parts of the world. Discuss what this comparison suggests about Australia's attitude to plastic use.
4. Select a large urban centre in a less economically developed country and investigate how waste is disposed of in this place. Prepare a one-page presentation using images to highlight the waste issue that exists in this place.

20.3 EXERCISE

Learning pathways

■ LEVEL 1	■ LEVEL 2	■ LEVEL 3
1, 2, 7, 9, 15	3, 4, 8, 10, 13	5, 6, 11, 12, 14

Check your understanding

1. Outline two benefits of online shopping:
   a. for consumers?
   b. for producers?
2. Identify one possible negative social consequence of global companies selling products that are made in developing countries.
3. Describe what is meant by the term *the sharing economy*. Give examples using your own experience.
4. List the ways that workers are exploited in sweatshop industries.
5. Kenya has imposed strict regulations on the use of plastic. Outline one positive and one negative consequence of this change.
6. What percentage of Australians work in service sector jobs? In your answer, define the term *service sector*.

Apply your understanding

7. Explain how a garbage picker earns a living.
8. Why do you think manufacturing operations that exploit their workers came to be known as 'sweatshops'?
9. Identify and explain an example of Australia being a 'throwaway society'.
10. Explain how people changing their work or shopping habits can have positive effects on the environment.
11. Consider the social and economic impacts of using technology to replace workers. Analyse who benefits and who suffers the most from such a change.
12. If your family's only income came from you working in a crowded factory, what factors would you need to consider when deciding whether or not you would continue working during the COVID-19 pandemic? Discuss how you would evaluate the possible positive and negative consequences for you and your family.

Challenge your understanding

13. Why might labour exploitation be more common in less economically developed countries?
14. Suggest why people might have worked long hours during COVID-19 lockdowns. In your answer, consider whether they may have felt it was necessary for social, environmental or economic reasons, or a combination of factors.
15. In Australia during 2018–19, a total of 3.5 million tonnes of plastic were used and disposed of as waste. Propose one way this could be reduced.

To answer questions online and to receive **immediate feedback** and **sample responses** for every question, go to your learnON title and www.jacplus.com.au.

20.4 SkillBuilder — Describing divergence graphs

online only

LEARNING INTENTION

By the end of this subtopic, you will be able to describe geographical data in a divergence graph.

The content in this subtopic is a summary of what you will find in the online resource.

20.4.1 Tell me

What is a divergence graph?

A divergence graph is a graph that is drawn above and below a zero line (see **FIGURE 1**). The numbers above the line are positive, showing the amount above zero. Negative numbers that are shown indicate that the data has fallen below zero.

20.4.2 Show me

How to describe a divergence graph

eles-1739

FIGURE 1 Australia's tourism trends, 1977–2016

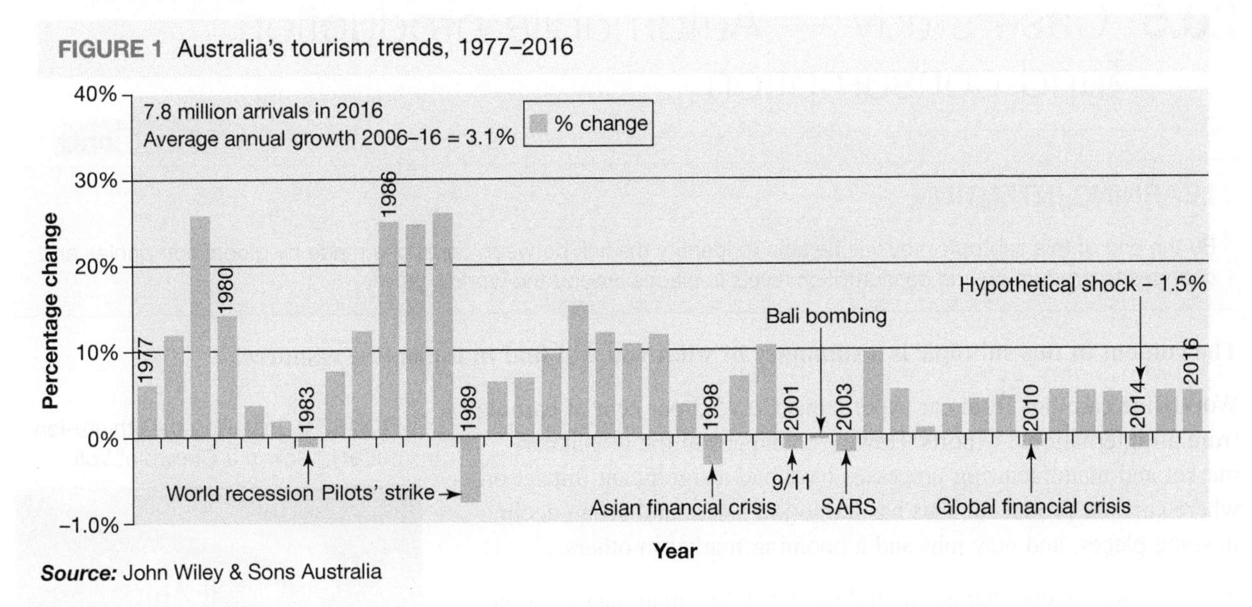

Source: John Wiley & Sons Australia

Step 1

Read the title of the graph carefully to see what data has been graphed, and to check the locations and dates to which the graph refers.

Step 2

Study the labels on both axes and any key or legend provided to add to your knowledge. In **FIGURE 1**, you can see that the time frame of the information is from 1977 to 2016, and that the number of visitors to Australia is represented on the vertical axis as percentage change.

Step 3

Study the shape of the graph; in **FIGURE 1**, you will note that some features are related to global events, such as recessions, a pilots' strike, terrorism and disease outbreaks (SARS). At times the graph flattens out as there is little change, such as in the late 2000s.

Step 4

Write a few sentences to outline the shape of the graph; for example, **FIGURE 1** shows that Australian tourism has experienced fluctuating highs and lows for the past 50 years. Include any events that affected the change in your description. Also look for periods of time where the change was either slow (2010 in **FIGURE 1**) or rapid (1989 in **FIGURE 1**).

20.4.3 Let me do it

int-3357

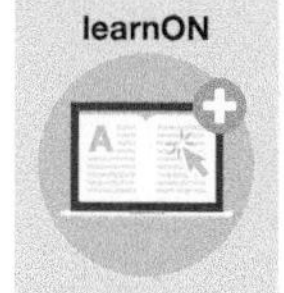

Go to learnON to access the following additional resources to help you build this skill:
- a longer explanation of this skill and its application in Geography (Tell me)
- a video showing the step-by-step process of this skill (Show me)
- an activity and interactivity for you to practise this skill (Let me do it)
- self-marking questions to help you understand and use this skill.

on Resources

eWorkbook SkillBuilder — Describing divergence graphs (ewbk-9562)

Video eLesson SkillBuilder — Describing divergence graphs (eles-1739)

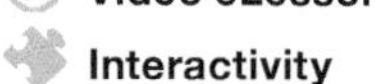

Interactivity SkillBuilder — Describing divergence graphs (int-3357)

20.5 Case study — Automobile production and consumption

online only

LEARNING INTENTION

By the end of this subtopic, you will be able to identify the link between decisions made by global companies and changes to production and consumption levels in places around the world.

The content in this subtopic is a summary of what you will find in the online resource.

Worldwide earnings from car exports make up 5.3 per cent of earnings from all international exports. However, changes in the global car market and manufacturing processes have had a significant impact on where cars are produced. This has led to job losses and urban decline in some places, and new jobs and a booming market in others.

To learn more about changes in global automotive manufacturing, go to your learnON resources at www.jacPLUS.com.au.

FIGURE 6 Installing the lithium-ion battery pack in a Chevrolet Volt

Contents

on Resources

eWorkbook Case study — Automobile production and consumption (ewbk-9686)

Video eLesson Case study — Automobile production and consumption — Key concepts (eles-5166)

Interactivity Case study — Automobile production and consumption — Interactivity (int-8581)

20.6 Case study — E-waste concerns

online only

LEARNING INTENTION

By the end of this subtopic, you will be able to describe and evaluate the impacts that e-waste has on our environment, the economy and society.

The content in this subtopic is a summary of what you will find in the online resource.

Any electrical equipment that is broken, obsolete or no longer wanted is considered to be e-waste. Globally 44.7 million metric tonnes of e-waste were produced in 2016; it is expected this figure will reach 63.7 million metric tonnes by 2025. How e-waste is dealt with has serious economic, environmental and social consequences.

FIGURE 5 E-waste smoke at the Agbogbloshie dump in Ghana.

To learn more about what happens to our devices when they are no longer useful to us, go to your learnON resources at www.jacPLUS.com.au.

Contents

on Resources

eWorkbook Case study — E-waste concerns (ewbk-9690)

Video eLesson Case study — E-waste concerns — Key concepts (eles-5167)

Interactivities E-wasted (int-3343)
Health impacts of e-waste on waste workers and people who live near landfills or incinerators (int-7938)

20.7 Government responses to the impacts of production and consumption

LEARNING INTENTION

By the end of this subtopic, you will be able to explain and give examples of how governments can act to reduce the impacts of production and consumption.

20.7.1 E-waste legislation

Governments pass laws to restrict how waste is disposed of and the safety of those who manage its processing. For example, since 2014, legislation regarding the management of e-waste has been developed and, to varying degrees, adopted across the globe. The coverage by legislation (not including international treaties) has risen from 44 per cent to 66 per cent of the world's population (in 67 countries). India, as a major generator of e-waste, has been leading the way with the adoption of legislation.

The existence of policies or legislation does not necessarily imply successful enforcement or the existence of sufficient e-waste management systems. **TABLE 1** lists some of the more significant attempts at e-waste management around the world.

TABLE 1 Examples of e-waste legislation and treaties around the world

Policy/legislation	Specific actions
Basel Convention 1994	• Keep the production of hazardous waste as low as possible. • Make suitable disposal facilities available. • Reduce and manage international flow of hazardous waste. • Ensure management of waste is controlled in an environmentally friendly way. • Block and punish illegal movement of hazardous waste.
Bamako Convention 1998	• Ban the importation of hazardous and radioactive wastes into Africa (or dumping in the oceans). • Reduce and control transportation of hazardous waste between African countries.
Buy-back policies	Many countries have tried buy-back schemes, with varying degrees of success.
China's e-waste ban, 2002	Although an official ban was placed on e-waste being shipped into China, it continued to be smuggled in or came across the borders by land. In 2017 China strengthened its ban on e-waste.
International Telecommunication Union	Connect 2030 has taken on board the Sustainable Development Goals, especially Goals 3, 7, 11, 12 and 13, where ICT can be applied.
Kenya's e-waste Act	Initiated in 2013 but stalled in parliament, this Act has been replaced by a National E-Waste Management Strategy to cover the period 2019–20 to 2023–24. Its purpose is to prescribe ways to minimise negative impacts of e-waste on the environment and human health.
Global E-waste Statistics Partnership 2017	The International Telecommunication Union, the United Nations University, and the International Solid Waste Association have joined to improve the collection, analysis and publication of worldwide e-waste statistics, with a view to increasing the awareness of the need for further development in the e-waste industry.
India's approach, 2018	Rules were first established in 2011 using the concept of 'extended producer responsibility' whereby the manufacturer is responsible for the safe disposal of electronic goods. In 2018 the emphasis was on regulating the dismantlers and recyclers and providing revised collection targets into the future.

20.7.2 Consumption awareness and responsible e-waste handling

In 2011, the Australian government commenced the National Television and Computer Recycling Scheme (NTCRS). The NTCRS website directs people to places to dispose of e-waste, such as MobileMuster and Planet Ark.

Some states have introduced specific legislation to manage e-waste. On 1 July 2019 Victoria banned the inclusion of e-waste in general garbage collections and curbside collections. E-waste will no longer go to landfill. No matter how restrictive laws and regulation are in any place, the disposal of e-waste is also an individual responsibility. Each individual must be aware of the e-waste being produced by their consumption of modern technological appliances and their method of disposal of no-longer-wanted items.

20.7.3 The role of foreign aid

Overseas aid is the transfer of money, food and services from developed countries such as Australia to less-developed countries in order to help people overcome poverty, resolve humanitarian issues and generally help with their development. This helps to raise living standards and the availability of food and other essential items. If people have access to these essential goods, they are less likely to need to work in dangerous or unsustainable industries.

More than 1 billion people in the world live in poverty and do not have easy access to education and health care. When disasters strike, they lack the resources to get back on their feet. Aid can help families send their children to school, rather than sending them to work to help support the family.

Poverty needs to be addressed by the international community because it can:

- breed instability and **extremism**
- cause people to flee violence and hardship, thus swelling the number of refugees
- increase the likelihood that people will agree to work in dangerous or low paid jobs because it is the only way they can feed their families
- increase the chances of children having to work rather than receiving a basic education.

20.7.4 The Australian Aid program

Australian governments have provided aid to other countries and regions to support **humanitarian principles** and social justice. Apart from showing we care, it is in the interests of our **national security** as it may also help promote stability and prosperity in the region. In addition, it improves our status throughout the world and creates political and economic interconnections with our Asia–Pacific neighbours. Australia's Official Development Assistance (ODA) program is known as Australian Aid.

The Department of Foreign Affairs and Trade (DFAT) manages the Australian government's multi-billion-dollar overseas aid program. To ensure that funds reach those in need, Australian Aid works with Australian businesses, non-government organisations such as CARE Australia, and international agencies such as the United Nations (UN) and the World Bank. In 2018–19, Australia's ODA budget was $4.2 billion, with the majority of this being earmarked for the Indo-Pacific region, of which Australia is a part (see **FIGURE 2**).

extremism extreme political or religious views or extreme actions taken on the basis of those views

humanitarian principles the principles governing our response to those in need, with the main aim being to save lives and alleviate suffering

national security the protection of a nation's citizens, natural resources, economy, money, environment, military, government and energy

FIGURE 1 Australian aid in the Pacific region

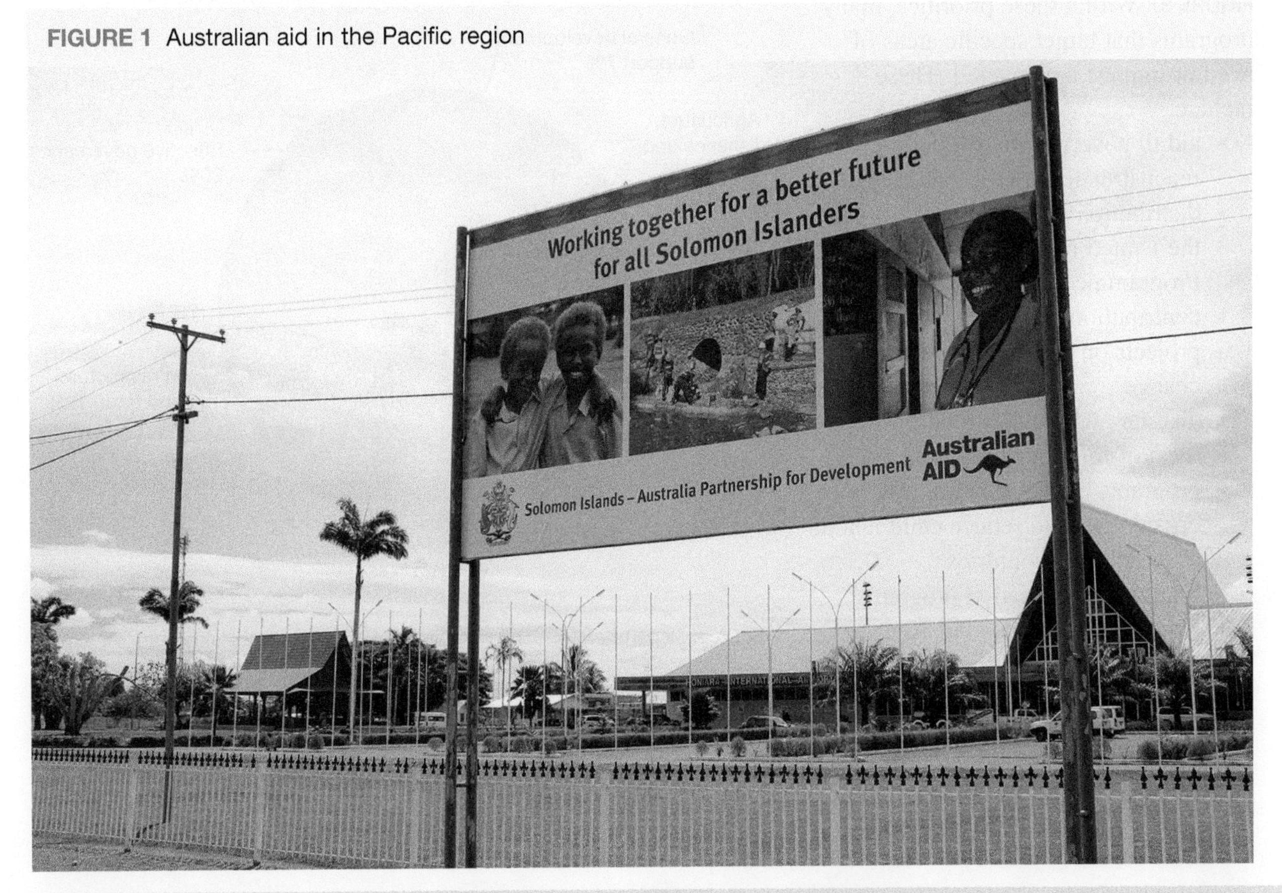

FIGURE 2 Australia's aid 2018–19, by region

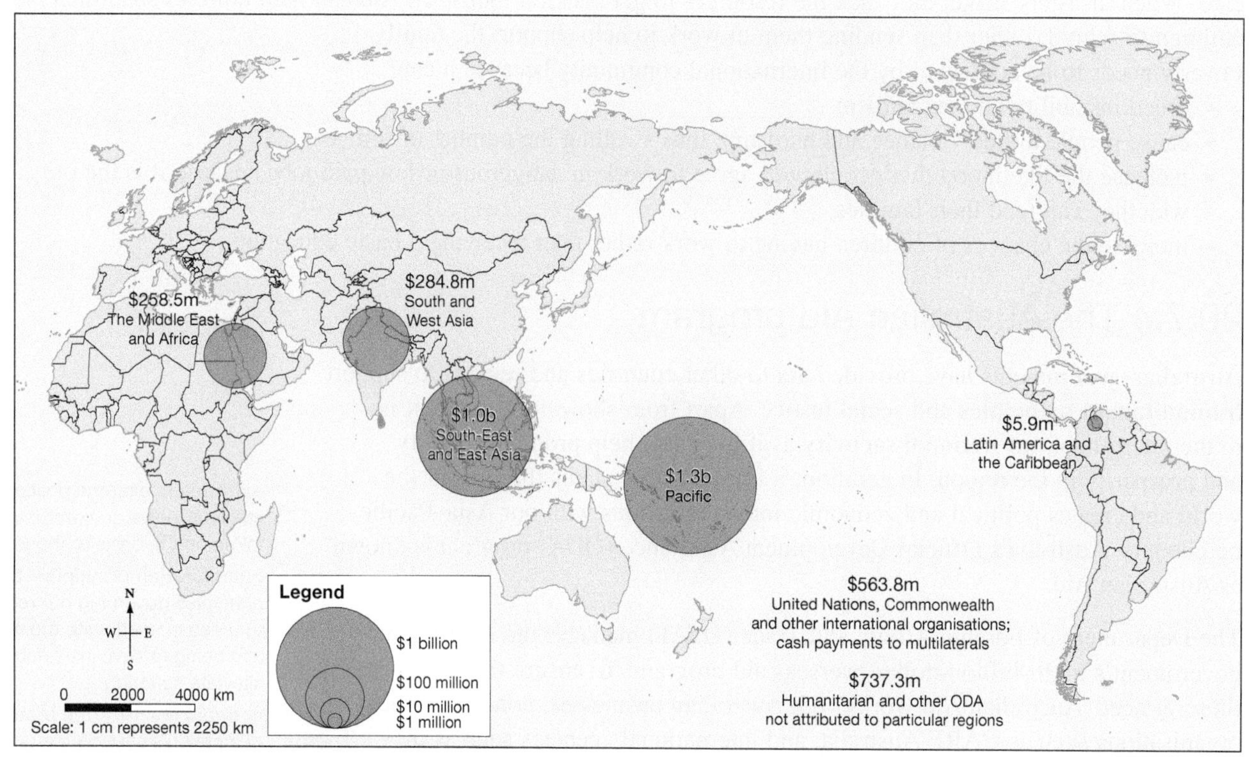

Source: Data from © Commonwealth of Australia, DFAT, Australian Aid Budget Summary 2018–19. Map drawn by Spatial Vision.

There are various investment priorities within Australia's ODA budget (see **FIGURE 3**). Within these priorities, many programs that target specific areas of need or interest are covered. These include:

- aid to governments for post-conflict reconstruction, as in Afghanistan
- distribution of food through the United Nations World Food Programme
- contributions to United Nations projects on refugees and climate change
- disaster and conflict relief in the form of food, medicine and shelter
- programs by non-government organisations to reduce child labour in developing countries
- funding for education programs
- funding for programs to promote gender equality and improve women's economic and social participation
- support for Australian volunteers working overseas.

FIGURE 3 Distribution of Australia's Official Development Assistance (ODA) budget by investment priority, 2018–19

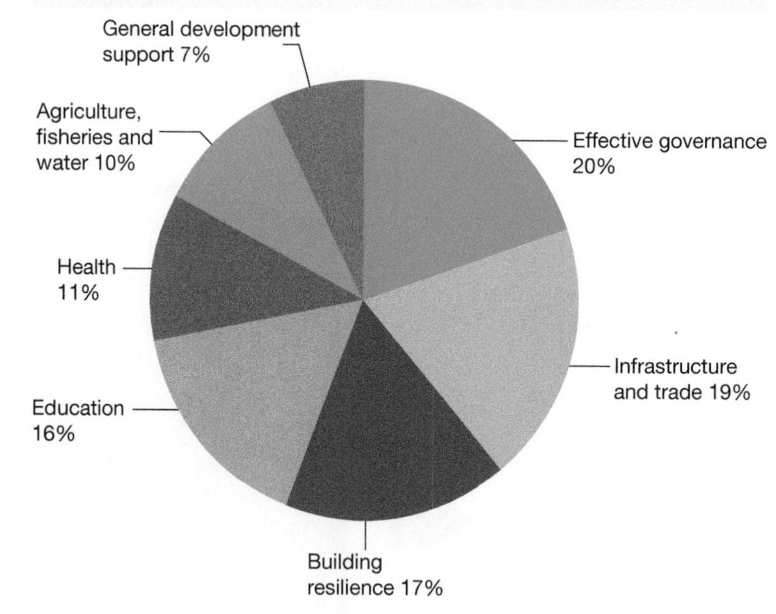

Source: Department of Foreign Affairs and Trade – www.dfat.gov.au

20.7.5 Sustainable Development Goals

Foreign aid programs have become more focused on promoting sustainable economic growth in developing countries and creating opportunity for communities in poverty. This new approach to foreign aid has coincided with the adoption by the United Nations of 17 Sustainable Development Goals (SDGs) as seen in **FIGURE 4**. By 2030, 62 per cent of people around the world are projected to live in extreme poverty, especially in places facing conflict.

FIGURE 4 Sustainable Development Goals

Source: United Nations — UN.org/SustainableDevelopment, Public Domain, https://en.wikipedia.org/wiki/Sustainable_Development_Goals

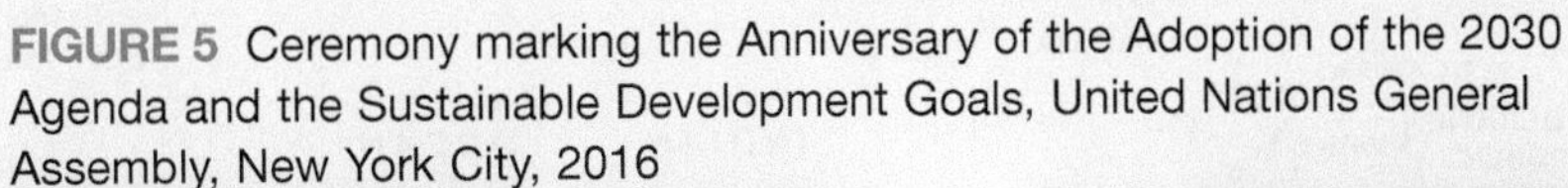

FIGURE 5 Ceremony marking the Anniversary of the Adoption of the 2030 Agenda and the Sustainable Development Goals, United Nations General Assembly, New York City, 2016

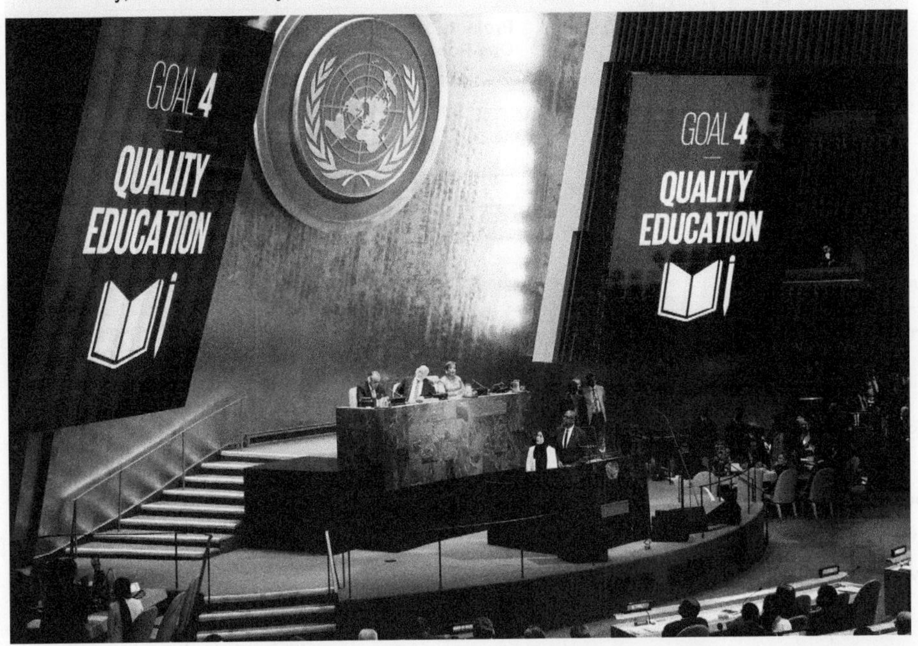

20.7.6 China's One Belt One Road policy

Launched in 2014, China's One Belt One Road (OBOR) initiative aims to interconnect trade, finance, infrastructure and people across Europe and Asia. China now has the second largest economy in the world and OBOR is a way of sustaining this growth. China will become interconnected with 65 countries in Asia, Europe and Africa via these land and sea routes.

FIGURE 6 The extent of China's One Belt One Road initiative

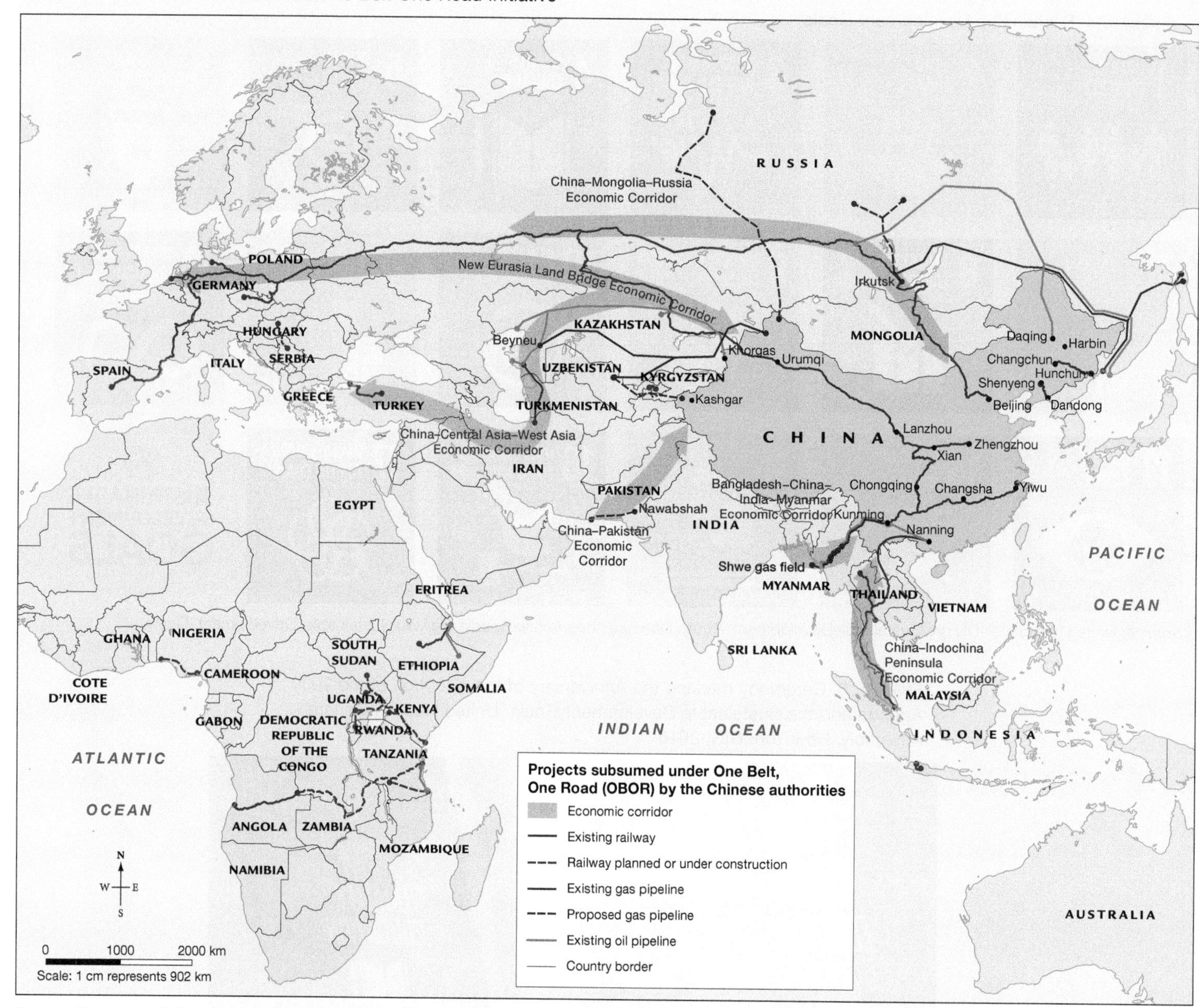

Source: Data from © Commonwealth of Australia, DFAT, Australian Aid Budget Summary 2018–19. Map drawn by Spatial Vision.

on Resources

eWorkbook Government responses to the impacts of production and consumption (ewbk-9694)

Video eLesson Government responses to the impacts of production and consumption — Key concepts (eles-5168)

Interactivity Government responses to the impacts of production and consumption (int-8582)

20.7 ACTIVITIES

1. Brainstorm arguments for and against the following statement: 'Australia should help its less developed neighbours, not just because it benefits Australia but because it is the right thing to do'.
2. Undertake internet research to find out how the Sustainable Development Goals guide the Australian Aid program.
3. Research China's One Belt One Road policy. As a class, discuss the following:
   a. How has Australia reacted to this policy? Have the states and federal governments come to an agreement on how to respond to the policy?
   b. How might this policy (and Australian governments' reactions to it) affect what we can buy and sell in international markets?

20.7 EXERCISE

Learning pathways

■ LEVEL 1	■ LEVEL 2	■ LEVEL 3
1, 3, 7, 10, 13	2, 6, 8, 9, 14	4, 5, 11, 12, 15

Check your understanding

1. What proportion of the world's countries has legislation in place regarding e-waste management?
2. Outline two reasons why foreign aid helps to reduce the negative impacts of production and consumption.
3. Outline the key actions identified in the Basel Convention.
4. List two benefits that the One Belt One Road initiative brings to the countries that China is investing in.
5. What are the underlying principles of Australia's foreign aid program?
6. Which regions of the world receive most of Australia's aid funding and why do you think this is so?

Apply your understanding

7. Which elements of the Australian Aid program do you think will have the greatest impact on the lives of people in the Pacific region? Give reasons for your selection.
8. Explain the benefits of the One Belt One Road policy for China's economy.
9. More than 25 years since the Basel Convention was agreed, how has the world responded to the legislation?
10. Choose two of the Sustainable Development Goals and explain how each goal is connected to the problems created in developing nations by consumer culture.
11. Why are statistics important in addressing the issue of e-waste management?
12. Explain what is meant by the 'need for a global solution to the transboundary issue of waste'.

Challenge your understanding

13. Consider Australia's list of foreign aid priorities. Which two options from this list do you think might provide the most effective help? Explain your answer.
14. Justify Australia spending more on foreign aid in the Pacific region than anywhere else in the world.
15. Of the strategies put in place by governments to regulate the impacts of production and consumption, suggest which will be the most effective in the long term.

To answer questions online and to receive **immediate feedback** and **sample responses** for every question, go to your learnON title at www.jacplus.com.au

20.8 Group responses to production and consumption concerns

LEARNING INTENTION

By the end of this subtopic, you will be able to describe the ways groups work together to address the impacts of production and consumption on the environment, people and the economy.

20.8.1 Problems of trade

The benefits of international trade are not evenly shared around the world, and trade often favours developed countries rather than developing countries. It is the role of governments, organisations and agencies to regulate this trade so that the economic benefits are more evenly distributed.

Australians benefit economically, culturally and politically from international trade, but social justice problems can arise through this trade. For example, if we import 'blood diamonds' (diamonds sourced from mines in war zones that are sold to buy weapons and continue conflicts) from Africa, clothing manufactured in sweatshops in Bangladesh, or carpets from Nepal produced by child labour, we are supporting unethical industries.

FIGURE 1 Children working on a tea plantation, Nanyuki, Kenya

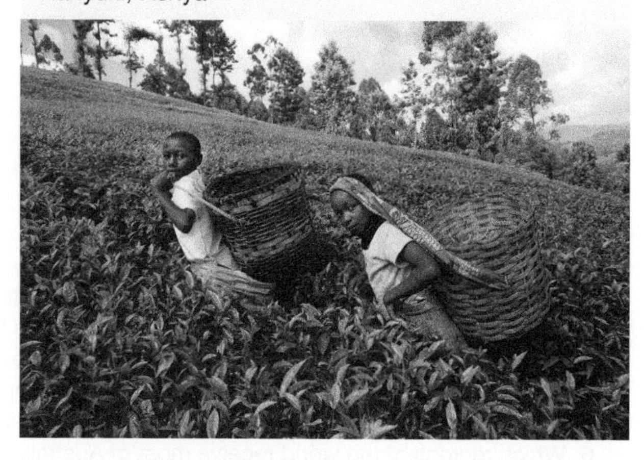

In addition, some countries can make it difficult for other countries to compete fairly, on a 'level playing field'. They do this by:

- imposing *tariffs* — taxes on imports
- imposing *quotas* — limits on the quantity of a good that can be imported
- providing *subsidies* — cash or tax benefits for local farmers or manufacturers.

20.8.2 Fairtrade

The Fairtrade movement aims to improve the lives of small producers in developing nations by paying a fair price to artisans (craftspeople) and farmers who export goods such as handicrafts, coffee, cocoa, sugar, tea, bananas, cotton, wine and fruit. The movement operates through various national and international organisations such as the World Fairtrade Organisation and Fairtrade International.

The Fairtrade labelling system is operated by Fairtrade International, of which Australia is a participating member. This system works to ensure that income from the sale of products goes directly to the farmers, artisans and their communities. Fairtrade International works with 1599 producer groups across 75 countries (see **FIGURE 3**). The number of Fairtrade International farmers and workers is estimated at more than 1.6 million, of which 25 per cent are women.

Fairtrade food items include sugar, chocolate, coffee, tea, wine and rice. Other products include soaps, candles, clothing, jewellery, bags, rugs, carpets, ceramics, wooden handicrafts, toys and beauty products.

In 2017–18, Australia and New Zealand had a combined retail sales total of A$333 million in Fairtrade-certified products, with three in five New Zealanders and two in five Australians purchasing Fairtrade offerings. This included 3 million kilograms of coffee, 10.1 million kilograms of chocolate and 354 000 kilograms of Fairtrade tea. On a global scale, Fairtrade's 1.6 million farmers and their families have benefited from Fairtrade premium-funded infrastructure and community development projects with a value of A$262 million.

FIGURE 2 Workers for the World Fairtrade Organisation

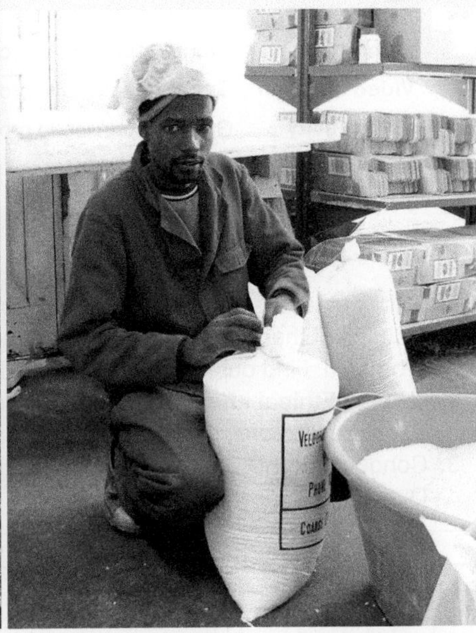

FIGURE 3 Fairtrade in the world, 2017

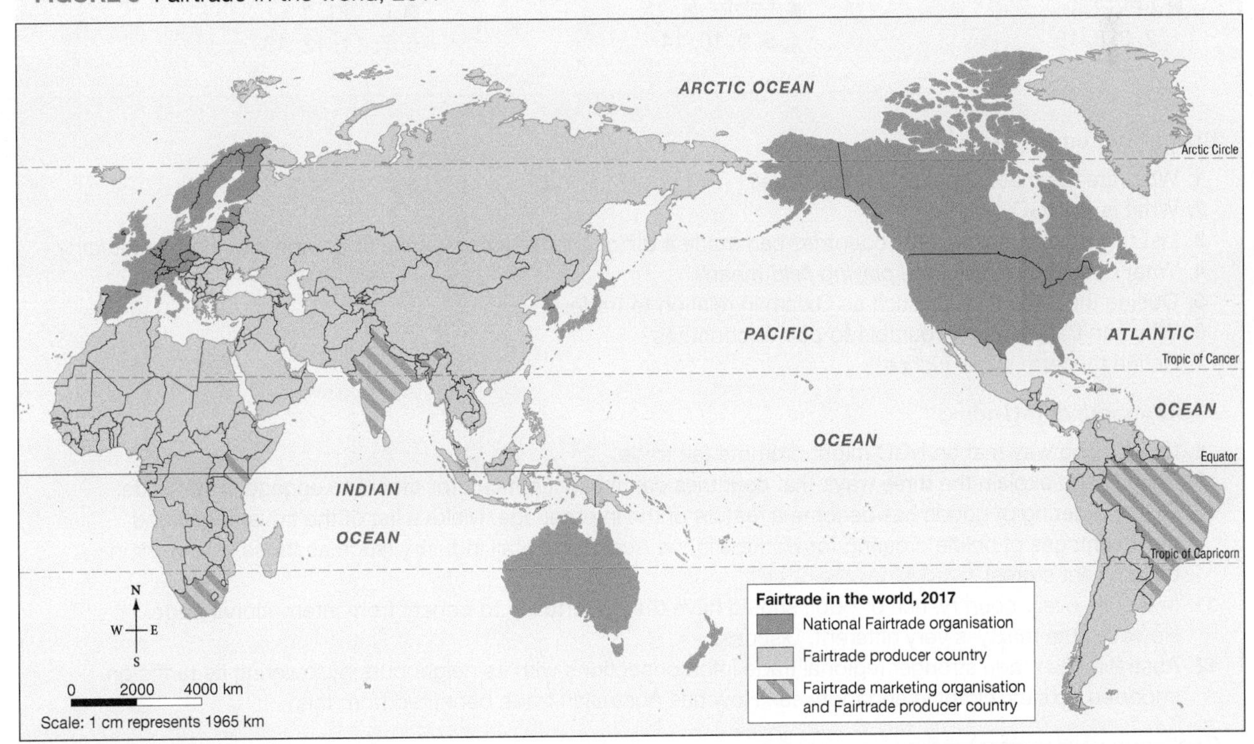

Source: Fairtrade Foundation

20.8.3 Non-government organisations and fair trade

Non-government organisations (NGOs) such as Oxfam and World Vision also support fair trade and oppose socially unjust trade agreements. They oppose attempts by developed countries to:

- block agricultural imports from developing countries
- subsidise their own farmers while demanding that poorer developing countries keep their agricultural markets open.

non-government organisation (NGO) non-profit group run by people (often volunteers) who have a common interest and perform a variety of humanitarian tasks at a local, national or international level

on Resources

eWorkbook	Group responses to production and consumption concerns (ewbk-9698)
Video eLesson	Group responses to production and consumption concerns — Key concepts (eles-5169)
Interactivity	Group responses to production and consumption concerns (int-8583)

20.8 ACTIVITIES

1. Visit your local supermarket and find as many products as you can that carry the Fairtrade symbol. Take pictures of these products and create an annotated world map to show each of the products and where it is produced. Find out more about Fairtrade labelling of products sold in places such as Australia.
2. Use the **Ethical choices** weblink in your Resources tab to learn about ethical shopping for clothing. Prepare a wallet-sized checklist for shoppers (or clothing shop owners) to inform them of ethical clothes shopping tips.
3. Conduct internet research to find out what Oxfam does to promote fair trade. What types of goods does Oxfam sell in Australia?

20.8 EXERCISE

Learning pathways

■ LEVEL 1	■ LEVEL 2	■ LEVEL 3
1, 2, 3, 8, 15	4, 5, 9, 10, 14	6, 7, 11, 12, 13

Check your understanding

1. What are the main principles of fair trade?
2. What is an NGO?
3. List three ways that wealthy countries can made it difficult for other countries to compete in the market fairly.
4. What does the phrase *level playing field* mean?
5. Outline the role of NGOs such as Oxfam in relation to trade.
6. Why can trade be unfavourable to poorer countries?
7. Define the term *social justice*.

Apply your understanding

8. Explain one way that an NGO might promote fair trade.
9. Identify and explain the three ways that countries can make it difficult for others to engage in fair trade.
10. Online ordering of goods has become a feature of the internet age. Make a list of the advantages and disadvantages of online ordering for workers in the Australian retail industry (such as those who work in department stores).
11. In theory, every country, rich or poor, should have the opportunity to benefit from international trade. However, the reality is very different. Discuss.
12. Australia has made stronger regional trade interconnections with its neighbours by lowering its tariffs on imported textiles, clothing and footwear. How has Australian trade benefited from this?

Challenge your understanding

13. Predict whether levels of social inequality across the world will increase or decrease in the next ten years.
14. Based on your understanding of fair trade, will you think differently about what you buy when you next go shopping? Give reasons for your decision.
15. Write a paragraph to persuade someone to buy Fairtrade chocolate rather than other types of chocolate.

To answer questions online and to receive **immediate feedback** and **sample responses** for every question, go to your learnON title at www.jacplus.com.au.

20.9 Reducing your consumption

LEARNING INTENTION

By the end of this subtopic, you will be able to explain the ways an individual can reduce their contribution to the waste and pollution problems that are linked to production and consumption.

20.9.1 Recycling your e-waste

On a global scale, entrepreneurs are tackling the waste problem society is creating. In Australia, consumption of information and communication technology is high. But there are many ways that individuals can act to reduce their technology use, and ensure that older devices are recycled properly (**FIGURE 1**). People are being encouraged to act locally, while thinking globally. This is an important concept when it comes to reducing and managing your own consumption.

An estimated 16 per cent of global e-waste is being recycled and reused. Australians purchase more than 4 million computers each year, and are among the biggest users of technology in the world. However, less than 10 per cent of these computers are recycled.

According to the Australian Bureau of Statistics (ABS), when people were surveyed about their e-waste disposal, most stated that they did not recycle because they did not have enough e-waste to warrant the use of e-waste services or facilities.

FIGURE 1 Mobile phone recycling

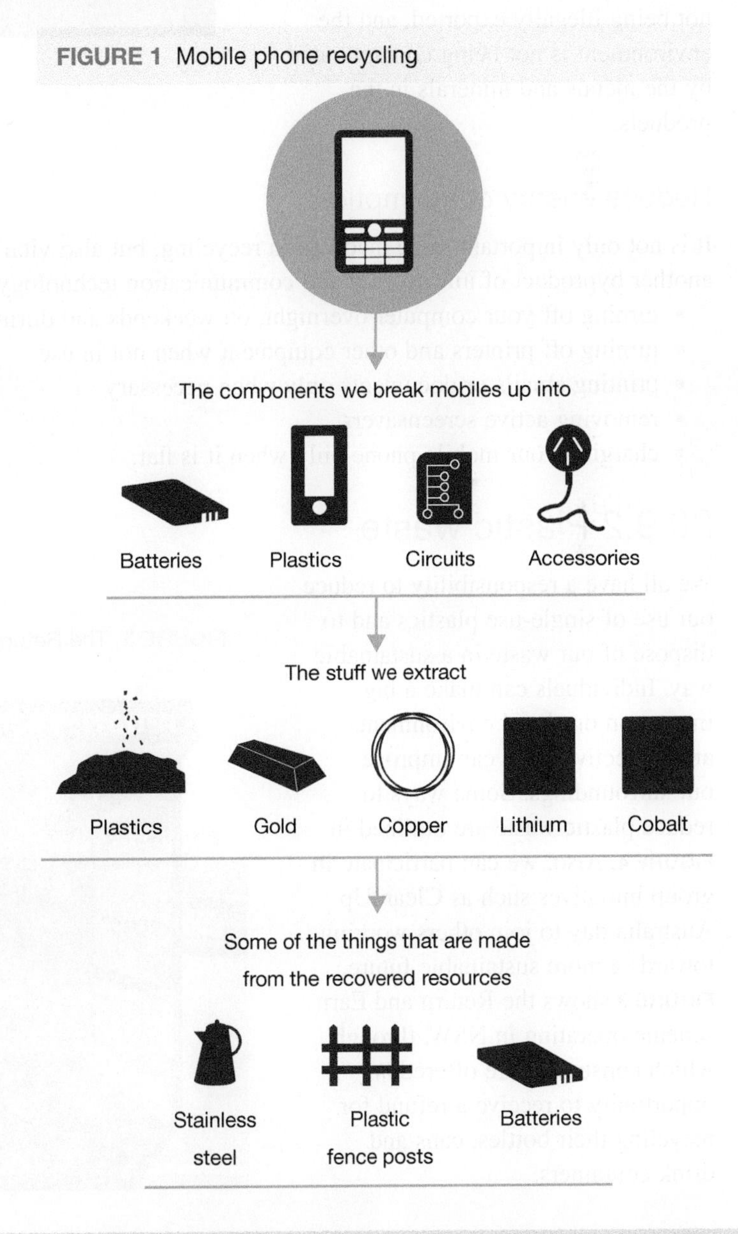

Ziilch

Most of us have an old computer we no longer use, or an old charger for a device we no longer have. We may even have books, furniture or clothing that could be recycled. The website Ziilch has come up with an easy solution to declutter, find new items and save on disposal costs. It is as easy as a click of a button.

This online company connects people in different places, and the platform enables you to exchange goods for free. This promotes the idea of reusing, which reduces consumption and hopefully leads to a more sustainable future. Similar swapping pages are also common on social media. Many Australian towns and suburbs have their own page or group where locals can swap or give away unwanted goods. Sites like Gumtree also give people a place to give away, swap or sell things they no longer need.

MobileMuster

MobileMuster is a simple, easy way to dispose of old mobile phones, batteries and accessories. MobileMuster is voluntarily funded and managed by distributors, network carriers, service providers and manufacturers. It is a free service where you can send your mobile phone or find your nearest drop-off point, of which there are more than 4500 in Australia. This ensures that mobiles are recycled correctly rather than dumped in **landfill**. Over 90 per cent of materials in mobile phones can be recycled (see **FIGURE 1**).

By using a certified recycling program, you can ensure that your e-waste is not being illegally exported, and the environment is not being contaminated by the metals and minerals in the products.

FIGURE 2 A local council advertising e-waste collection

Reduce energy consumption

It is not only important to be involved in recycling, but also vital to reduce your **carbon emissions**, which is another byproduct of information and communication technology. Some strategies include:

- turning off your computer overnight, on weekends and during holidays
- turning off printers and other equipment when not in use
- printing emails or documents only when necessary
- removing active screensavers
- charging your mobile phone only when it is flat.

landfill a method of solid-waste disposal, in which refuse is buried between layers of soil

carbon emissions carbon dioxide that is released into the atmosphere by natural processes or by burning gas, coal or oil; the latter emissions are contributing to climate change

20.9.2 Plastic waste

We all have a responsibility to reduce our use of single-use plastics and to dispose of our waste in a sustainable way. Individuals can make a big impact on our local environments and collectively we can improve our surroundings. Some ways to reduce plastic waste are outlined in **FIGURE 4**. Also, we can participate in group initiatives such as Clean Up Australia day to join others working towards a more sustainable future. **FIGURE 3** shows the Return and Earn scheme operating in NSW, through which consumers are offered an opportunity to receive a refund for recycling their bottles, cans and drink containers.

FIGURE 3 The Return and Earn station in Glen Innes, NSW, 2019

FIGURE 4 Suggestions for ways you can reduce your plastic waste

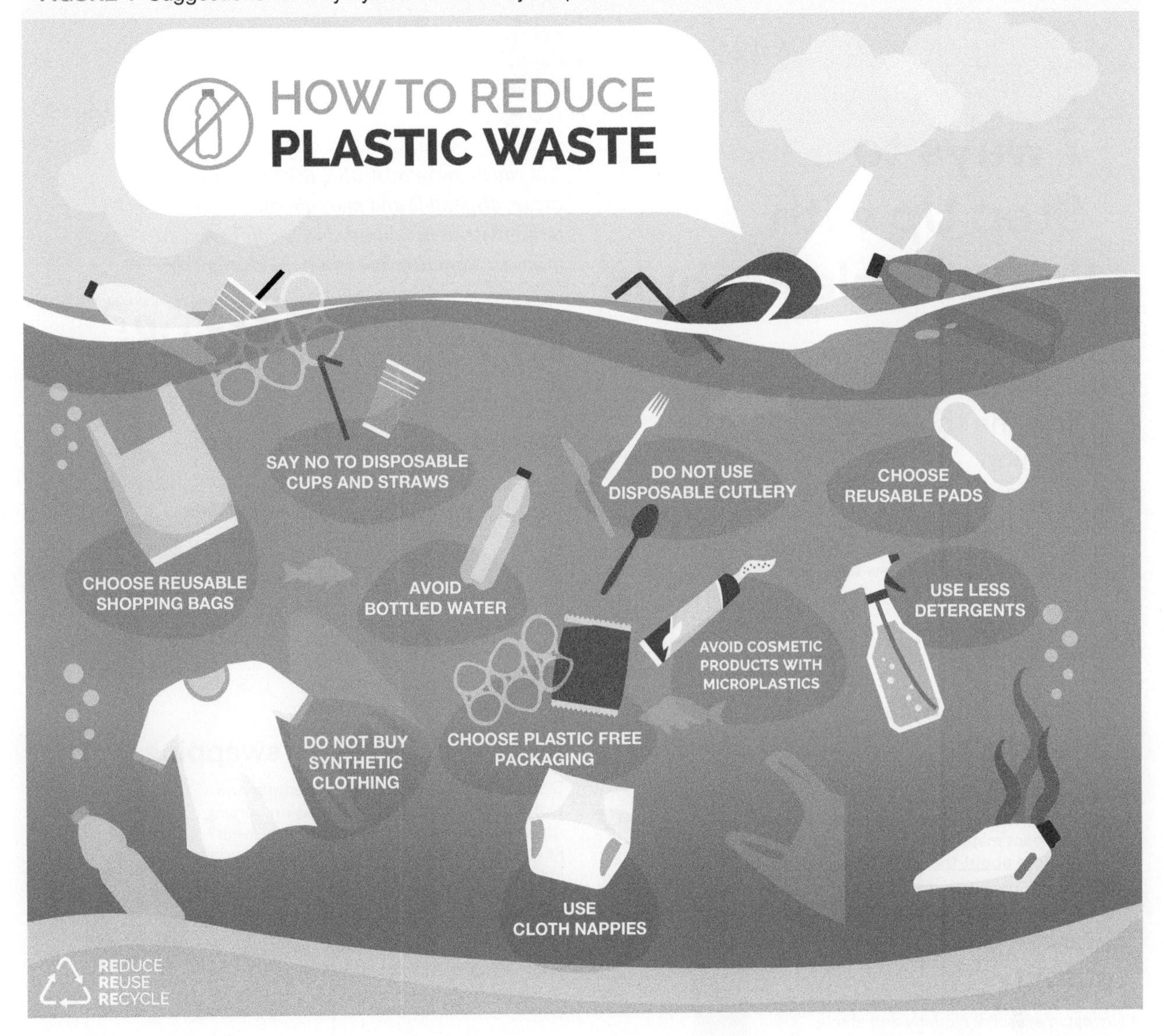

20.9.3 Fashion

The fashion industry is responsible for an estimated 10 per cent of global CO_2 emissions, and chemicals from dyes used in the industry contaminate the environment. More than 500 million kilograms of unwanted clothing ends up in landfill in Australia each year as the average person only wears 40 per cent of their clothing. Clean Up Australia reports on a trend called fast fashion, which describes behaviour by manufacturers to rapidly mass-produce clothing that is designed to be cheap, disposable and reflective of the latest fashion trends.

FIGURE 5 United Nations ActNow for Zero-Waste Fashion initiative

Source: © United Nations, https://www.un.org/actnow

Individuals are encouraged to reduce their spending on fashion items and to choose clothing more responsibly. The United Nations ActNow for Zero-Waste Fashion initiative is one way an individual can participate in the wider movement to change harmful production and consumption practices. Social media links connect people with the efforts being made to change attitudes towards waste and over-consumption.

FIGURE 6 Clean Up Australia promotes ways to reduce the damage caused by unsustainable fashion

7 ways to Step Up with Sustainable Fashion

Did you know in Australia, 6000 kilograms of cheap, disposable and mass-produced 'fast fashion' items are dumped in landfill every 10 minutes. ***Together we can help change this.***

how will you STEP UP?

Follow the three 'R's of fashion

Reduce, reuse and recycle your clothes the right way. From buying quality pieces, renting your next occasion outfit to learning how to sew!

Check out ***our blog*** *for tips on how you can start taking action today.*

Check the labels

Purchase clothes made from eco-friendly threads over synthetic fabrics (i.e. cotton over polyester). Fabrics include recycled or certified GOTS organic cotton, organic linen, hemp and linen, recycled wool, or low impact materials such as tencel.
Learn more about the most sustainable fabrics here.

Choose quality over quantity

Invest in seasonal clothing and staple items that are high quality and will stand the test of time. Take an edit of your current wardrobe and see if you're missing any key staple items that you know you'll wear again and again.

Thrift shop and buy second hand

Shop vintage at your local charity stores including the Salvos, Vinnies or the Sacred Heart Mission. There are also heaps of online platforms that sell second hand including Gumtree, Ebay, Facebook Marketplace. Even your favourite online retailers like ASOS now have Marketplaces allowing consumers to buy second hand.

Don't be impulsive

Start shopping small and fight the urge of buying clothes that you don't need. Try and buy clothes that are going to work all year round. Remember, the most impact you can have in taking action to becoming a more sustainable shopper is by buying less and reusing your existing clothes as many times as you can.

Rentals and clothes swapping

Discover the world of 'Fashion Rentals' and clothes swapping. Have an occasion coming up and need a one time only piece to wear? Rent your next outfit online and reduce your impact. Otherwise, pull out the vino and organise a clothes swap with your friends.

Check out sustainable fashion advocate 'Britts List' for the top 8 online platforms to rent from

Change your mindset

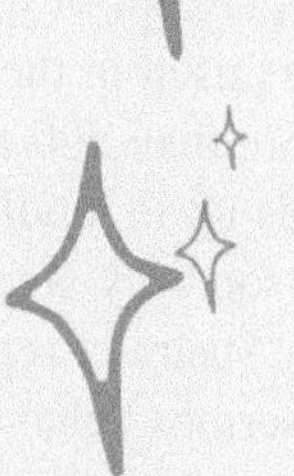

Change your mindset by staying informed and understanding the industry. Follow these ethical instagram accounts, download the Good on You app for ethical brand ratings or take a free online course and learn about sustainable fashion.

cleanup.org.au

Source: Clean Up Australia

Resources

eWorkbook	Reducing your consumption (ewbk-9702)
Video eLesson	Reducing your consumption — Key concepts (eles-5170)
Interactivity	Reducing your consumption (int-8584)

20.9 ACTIVITIES

1. In a group, conduct a class survey in which you interview students and teachers about their recycling habits. Decide on the types of questions you wish to ask and how you will record responses. If you wish to do this online, use the **Survey Monkey** weblink in your Resources tab. After you have conducted your surveys, collate and present your findings in graphic form. Analyse your graphs and write a summary of your findings, ensuring you cover the following questions.
   - Are students more environmentally conscious than teachers?
   - Does age make a difference?

   Prepare a recycling plan for your school.
2. Find out where your nearest Return and Earn centre is. Describe this place in relation to where you live. Is it realistic for you to take your recyclables there?

20.9 EXERCISE

Learning pathways

■ LEVEL 1	■ LEVEL 2	■ LEVEL 3
1, 2, 3, 7, 15	4, 5, 8, 9, 13	6, 10, 11, 12, 14

Check your understanding

1. According to the ABS, what is the main excuse that people give for not recycling?
2. Refer to **FIGURE 2**. Name three byproducts extracted from mobile phones and two products that can be made from recovered resources in mobile phones.
3. List three strategies that could contribute to a reduction in plastic use within your household.
4. What do you understand by the term *fast fashion*?
5. List three things you can do daily to reduce your energy consumption.
6. Define the term *carbon emissions*.

Apply your understanding

7. Create a checklist of factors that a fashion shopper should consider when purchasing new clothes for their wardrobe.
8. How much clothing ends up in landfill in Australia every year? Explain what might be done to reduce this amount.
9. Which of the seven suggestions for how to dress more sustainably shown in **FIGURE 7** do you think would have the biggest impact in reducing the amount of clothing in landfill? Explain your decision.
10. How do you ensure that your old devices are properly recycled?
11. Explain two ways that you could reduce levels of energy consumption in your home.
12. Explain how community sharing and swapping pages on social media help to reduce consumption.

Challenge your understanding

13. Some places provide small refunds for the return of recyclable materials, such as cans or bottles. Should such a system be put in place for plastics? Explore the benefits and problems with such a scheme, and suggest whether it might be successful where you live.
14. To what extent do you think your individual buying habits can have a positive impact on the problems created by consumer culture? Provide a detailed explanation of why you came to this view.
15. Propose one way to reduce the amount of plastic that is thrown away at your school.

To answer questions online and to receive **immediate feedback** and **sample responses** for every question, go to your learnON title at www.jacplus.com.au.

20.10 SkillBuilder — Using advanced survey techniques — Interviews

online only

LEARNING INTENTION

By the end of this subtopic, you will be able to conduct an interview to collect qualitative and quantitative data.

The content in this subtopic is a summary of what you will find in the online resource.

20.10.1 Tell me

What are interviews that survey people's opinions?

Surveys collect primary data, such as data that has been gathered in the field. Conducting a survey means asking questions, recording and collecting responses, and collating the number of responses. Interviews are particularly useful for gathering information on attitudes and values. The information that is gathered can be either quantitative (involving numbers) or qualitative (involving opinions and ideas), or both.

FIGURE 1 An example of an interview sheet

Interview topic: ____________ Date: ____________

Location: ____________ Interviewee name: ____________

1. What are your most common electronic forms of communication?

Leave spaces to write answers →

2. How many computers does your household have? ← A quantitative question

(a) 0 (b) 1–2 (c) 2–3 (d) 3–4 (e) More than 4

3. Who uses computers in your household?

4. How often do you use a computer?

❒ Every day ❒ Every couple of days ❒ Once a week ❒ Never

5. How successful have you been at shopping online? ← A qualitative question

6. How does your use of electronic communication differ from the way other people in your household use electronic communication?

7. How important is your mobile phone for communication with your friends? Mark on the following continuum how important you think your mobile phone is to you.

←——————————————————→

Not important Moderately important Extremely important

20.10.2 Show me

eles-1742

How to construct an interview

Step 1

Determine the purpose of the interview. What information or insights do you want to discover?

Step 2

Begin by developing 10–15 questions that allow the interviewee to express their opinion. No question should be answerable with a simple 'yes' or 'no'.

Step 3

Test your questions on a classmate or family member. Rework any questions to improve clarity of expression and to draw a more extended response from the interviewee. Practise your interview on a friend or family member so that you are confident when talking to members of the public.

Step 4

When conducting an interview, use the following guidelines.

- Have identification to say who you are and what school you are from. Work in pairs or groups.
- Introduce yourself and clearly state where you are from. Speak clearly and in a non-threatening tone of voice. Explain the aim of the survey.
- Do not be offended if a person does not want to participate. Seek permission to record an interview.
- Use a separate interview sheet for each person. Listen carefully to the answers given. Take notes. Be neutral about the responses, even when you do not agree. Never interrupt an answer and always allow plenty of time for the interview.
- Be prepared to ask additional questions if the person has great information.
- Thank the person when you have finished.

Step 5

Quantitative data can be placed in tables. Qualitative data needs to be classified according to the percentage of people with the same or similar viewpoints.

20.10.3 Let me do it

int-3360

learnON

Go to learnON to access the following additional resources to help you build this skill:

- a longer explanation of this skill and its application in Geography (Tell me)
- a video showing the step-by-step process of this skill (Show me)
- an activity and interactivity for you to practise this skill (Let me do it)
- self-marking questions to help you understand and use this skill.

fdw-0026

FOCUS ON FIELDWORK

Survey Monkey

Sometimes, people might not feel comfortable talking about their views or habits face-to-face for a survey. If you are collecting data that people might be more happy to share anonymously, consider using an online survey app or site.

Learn more about using digital survey tools with the **Survey Monkey** fieldwork activity in your Resources tab.

Resources

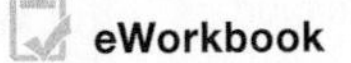

eWorkbook	SkillBuilder — Using advanced survey techniques — Interviews (ewbk-9706)
Video eLesson	SkillBuilder — Using advanced survey techniques — Interviews (eles-1742)
Interactivity	SkillBuilder — Using advanced survey techniques — Interviews (int-3360)
Fieldwork	Survey Monkey (fdw-0026)

20.11 Investigating topographic maps — More sustainable ways of managing waste

LEARNING INTENTION

By the end of this subtopic, you will be able to use the information in a topographic map to discuss ways that waste can be more sustainably managed.

20.11.1 Woodlawn Eco Precinct

Woodlawn Eco Precinct is a former mine site being used as a waste management facility. The facility is located near the settlement of Tarago, New South Wales, 250 kilometres south of Sydney. Here, compost is created from containerised organic and household waste that is delivered to the site by freight train and trucks. The compost is used as landfill to rehabilitate the former open-cut mine site shown in **FIGURE 1**.

More recently, bioreactors are being used to extract organic matter from waste. Sustainable technologies such as wind farms generate enough energy to power more than 4000 homes. Fish farming and horticulture are also being practised at the site. **FIGURE 3** shows the movement of the waste between Sydney and the tip.

FIGURE 1 Former mine site at Tarago, New South Wales, before rehabilitation began

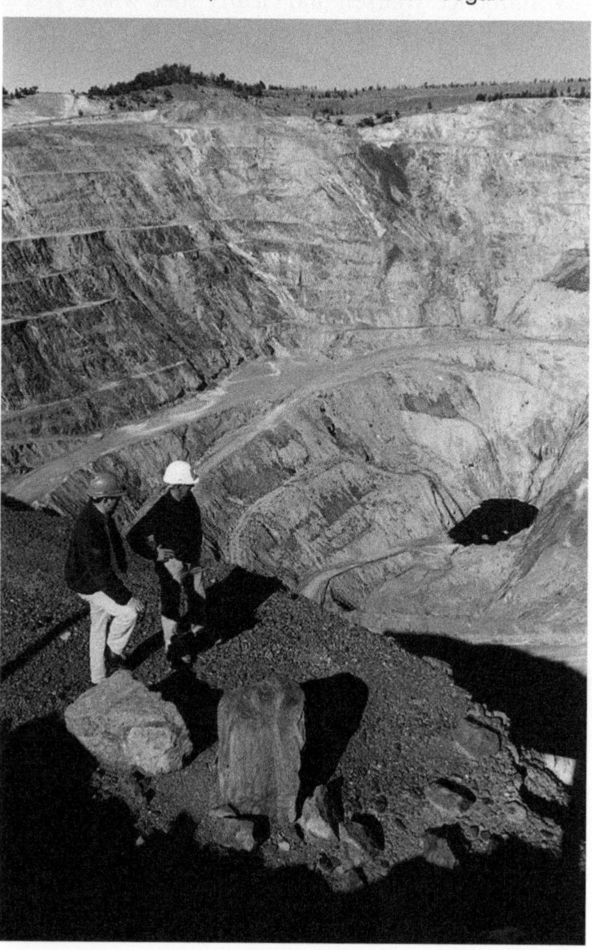

FIGURE 2 Rubbish from Sydney is being transported to Woodlawn.

FIGURE 3 Many forms of transport are used to move 20 per cent of Sydney's waste to Woodlawn.

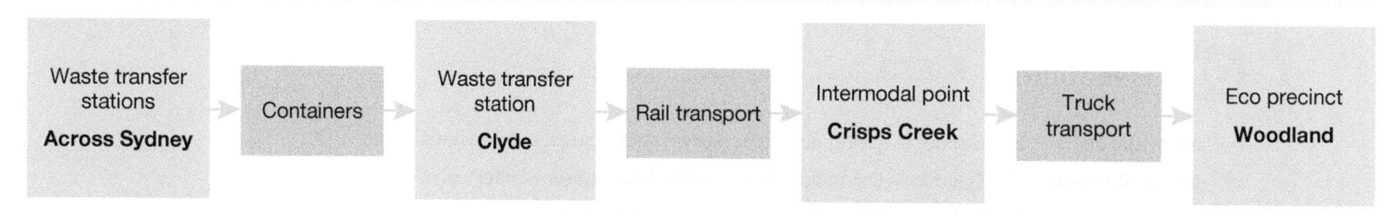

FIGURE 4 Topographic map of Woodlawn

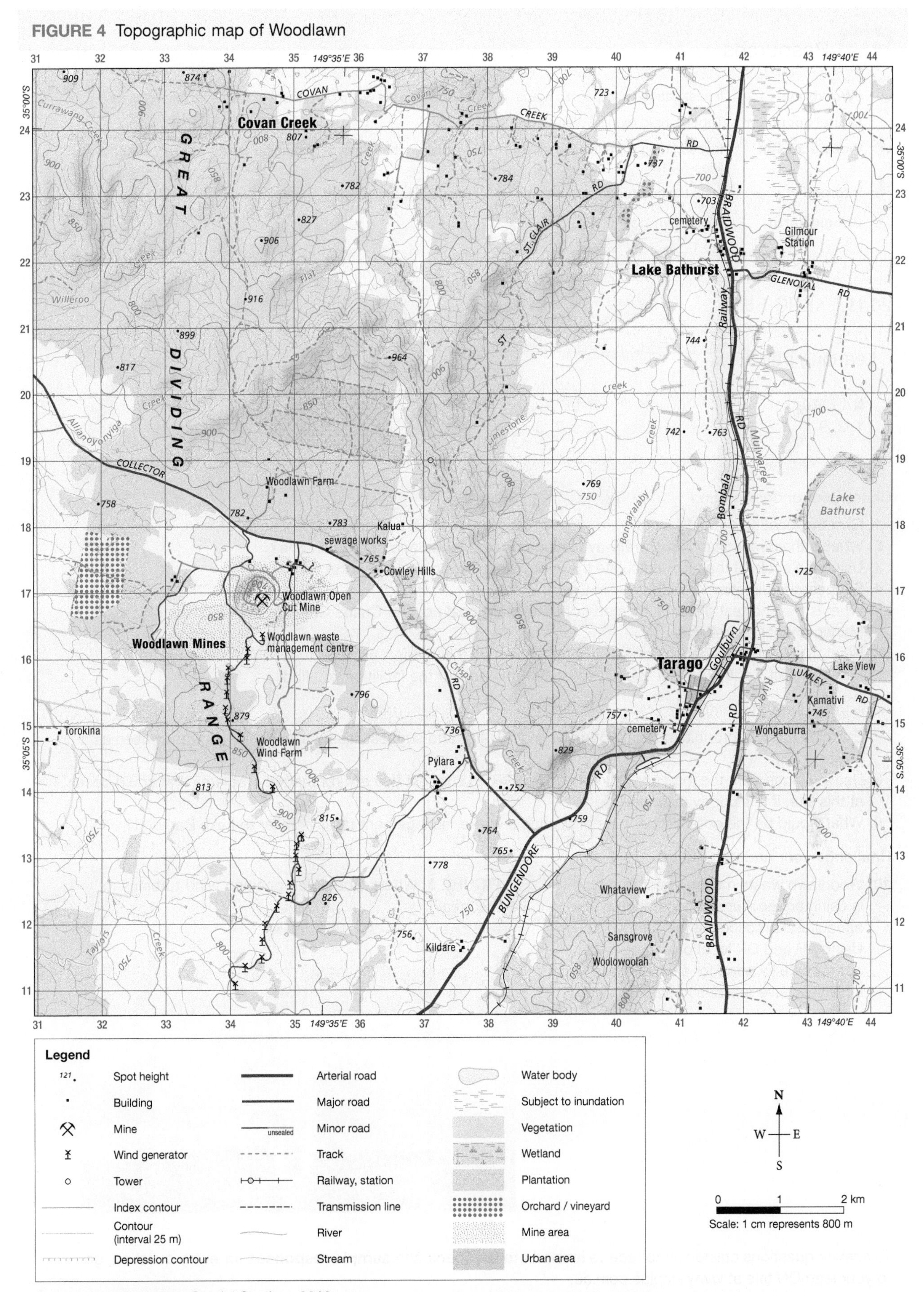

Source: Data based on Spatial Services 2019.

Resources

eWorkbook	Investigating topographic maps — More sustainable ways of managing waste (ewbk-9710)
Digital document	Topographic map of Woodlawn (doc-36264)
Video eLesson	Investigating topographic maps — More sustainable ways of managing waste — Key concepts (eles-5171)
Interactivity	Investigating topographic maps — More sustainable ways of managing waste (int-8585)
Google Earth	Woodlawn (gogl-0144)

20.11 EXERCISE

Learning pathways

■ LEVEL 1	■ LEVEL 2	■ LEVEL 3
1, 2, 9	3, 4, 7, 8	5, 6, 10

Check your understanding

1. Give the location of the Woodlawn mine open pit using at least two area references (AR).
2. What is the direction of Tarago railway station from the Woodland open-cut mine?
3. Using the scale of the map, calculate the width of the Woodlawn open-cut mine.
4. Suggest the direction of water flow in Crisps Creek.
5. What is the shortest distance by road from the intermodal point at Crisps Creek to the Woodlawn Waste Management Centre?
6. What other waste management facilities are shown on this map, in addition to the Woodlawn Waste Management Centre? Give a specific location reference for the facilities you find.

Appy your understanding

7. Examine the topographic map of Woodlawn in **FIGURE 4** and suggest reasons why this site was chosen as a waste dump.
8. Use evidence from the map (for example, AR or GR) to suggest two environmental problems that could occur at this site, if it is poorly managed.
9. What would be the potential dangers of placing a waste management facility closer to Lake Bathurst?

Challenge your understanding

10. Woodlawn waste management facility is using aquaculture, wind farming and agricultural practices at the Tarago site. Using AR and GR locations, suggest an appropriate location for each of these activities. Give reasons for your answer.

FIGURE 5 Where would you position wind turbines in the area?

To answer questions online and to receive **immediate feedback** and **sample responses** for every question, go to your learnON title at www.jacplus.com.au.

20.12 Thinking Big research project — Trash or treasure

on line only

The content in this subtopic is a summary of what you will find in the online resource.

Scenario

Japan first hosted the Summer Olympic Games in Tokyo in 1964. In 2013 Tokyo was declared the venue for the 2020 Summer Olympics. In planning the games, the organisers were keen to make an impression that would mark these games forever. They developed each Olympic medal to contain Japanese electronic waste — mobile phones and other small appliances had component parts melted down to extract their gold, silver and bronze. Recycling of the trash showcased Japanese dedication to sustainability.

Task

In this task you will create a pamphlet to accompany the medals when they are handed to the winning athletes. The pamphlet should explain the background of the medals' production — how the trash of millions was recycled to create the prized Olympic treasures of the athletes of the Tokyo Olympic Games.

Go to your Resources tab to access the resources you need to complete this research project.

Resources

ProjectsPLUS Thinking Big research project – Trash or treasure (pro-0196)

20.13 Review

20.13.1 Key knowledge summary

20.2 Environmental impacts of production and consumption

- Evidence of changes to climatic features such as average temperatures and rainfall patterns is being linked by scientists to growing concentrations of greenhouse gas emissions.
- Human activity such as burning fossil fuels is contributing to an enhanced greenhouse effect.
- Electricity is Australia's largest source of greenhouse gas emissions.
- Plastics are mostly manufactured for packaging uses.
- UN Sustainable Development Goal 12 addresses responsible production and consumption.
- Shipping containers provide an efficient method of transporting goods around the world, but their use impacts the marine environment.
- Approximately 13 per cent of food is lost in the supply chain of goods from farm to market.
- The illegal wildlife trade is linked to a decline in the number of plant and animal species.

20.3 Social and economic impacts of production and consumption

- The COVID-19 pandemic saw an increase in the use of the shared economy and online shopping.
- There is growing use of technology (for example, robotics and artificial intelligence systems) in the services sector.
- Four out of five workers in Australia are employed in the services sector.
- Labour exploitation is more common in less economically developed countries.
- During the COVID-19 pandemic sweatshop workers experienced increased job insecurity.
- Societal attitudes towards single-use plastic are being challenged by the introduction of legislation for manufacturers and consumers.
- Employment opportunities are being created by garbage pickers as they earn cash for recyclables.

20.5 Case study — Automobile production and consumption

- The top ten car manufacturers are found in Asia, Europe and North America.
- There has been a decline in the American and a growth in the Asian car manufacturing industry.
- Overseas competition for cars has led to the demise of the Australian car manufacturing industry with Ford, Toyota, Mitsubishi and Holden moving offshore to lower cost centres in Asia.
- Large employment and economic losses occurred where car manufacturing had been located in Australia. Retraining and counselling programs were introduced to support workers and the community at this time.

20.6 Case study — E-waste concerns

- China is a major producer of electronics followed by the United States.
- Electronic waste contains valuable components that can be recycled.
- Workers that disassemble the electronic waste components are exposed to toxic chemicals and dangerous contaminants. It is very dangerous for their health and the environment.
- In 1994 the Basel Convention was adopted by the European Community, which banned the exports of hazardous waste to countries other than OECD countries.

20.7 Government responses to the impacts of production and consumption

- Australia introduced the National Television and Computer Recycling Scheme to address concerns about responsible disposal of e-waste.
- Foreign aid is a way for more economically developed countries to support and promote economic development in less economically developed countries.
- One Belt One Road is an initiative that the Chinese government is using to sustain its growth.
- UN Sustainable Development Goals (SDGs) drive the efforts of governments and groups towards a more sustainable future. Goal 12 highlights the specific challenges of responsible production and consumption.

20.8 Group responses to production and consumption concerns

- Fairtrade organisations work to promote fair labour practices such as preventing and eliminating child labour and labour exploitation.
- Fairtrade promotes ethical standards in the production and use of ethically sourced raw materials.
- Fairtrade producers are offered a fair price for their products and they are traded using the Fairtrade logo so that consumers can make an ethical choice when they make a purchase.
- Fairtrade International works with 1599 producer groups across 75 countries.
- Non-government organisations such as Oxfam and World Vision support fair trade principles and oppose socially unjust trade agreements.

20.9 Reducing your consumption

- Individuals can use recycling and re-using initiatives to reduce waste disposal problems.
- Electronic waste can be recycled or repurposed by organisations such as Ziilch, Mobile Muster and Planet Ark.
- Plastic waste can be reduced by making smart choices when a product is purchased. Choosing not to use single-use cutlery and straws reduces the amount of waste being generated.
- In Australia more than 500 million kilograms of unwanted clothing ends up in landfill each year.
- Groups such as Clean Up Australia promote more sustainable options to dispose of unwanted clothing.

20.13.2 Key terms

black market any illegal trade in officially controlled or scarce goods

carbon emissions carbon dioxide that is released into the atmosphere by natural processes or by burning gas, coal or oil; the latter emissions are contributing to climate change

consumer culture the influences of and focus on material goods/products in a society

e-waste any electrical equipment that includes computers, toasters, mobile phones and iPads that no longer works or is no longer required

enhanced greenhouse effect increasing concentrations of greenhouse gases in the Earth's atmosphere, contributing to global warming and climate change

extremism extreme political or religious views or extreme actions taken on the basis of those views

global warming the observable rising trend in the Earth's atmospheric temperatures, generally attributed to the enhanced greenhouse effect

humanitarian principles the principles governing our response to those in need, with the main aim being to save lives and alleviate suffering

landfill a method of solid-waste disposal, in which refuse is buried between layers of soil

mercury poisoning a toxic condition caused by the ingestion or inhalation of mercury or a mercury compound. It has various symptoms, including vomiting, nausea, insomnia and fevers.

national security the protection of a nation's citizens, natural resources, economy, money, environment, military, government and energy

non-government organisation (NGO) non-profit group run by people (often volunteers) who have a common interest and perform a variety of humanitarian tasks at a local, national or international level

sharing economy sharing objects and/or services either for free or for payment, typically arranging the trade online

smuggling importing or exporting goods secretly or illegally

treaty a formal agreement between two or more independent states or nations, usually involving a signed document

20.13.3 Reflection

Complete the following to reflect on your learning.

Revisit the inquiry question posed in the Overview:

A world of products is available for us to buy at the click of a mouse. But what are the costs of our consumer culture for the people and places of the world?

1. Now that you have completed this topic, what is your view on the question? Discuss with a partner. Has your learning on this topic changed your view? If so, how?
2. Write a paragraph in response to the inquiry question, outlining your views about the effects of production and consumption.

Subtopic	Success criteria			
20.2	I can describe and list examples of the ways production and consumption influence the way the environment operates.			
20.3	I can describe the social and economic impacts of global production and consumption, and how they are changing.			
20.4	I can represent geographical data in a divergence graph.			
20.5	I can identify the link between decisions made by global companies and changes to production and consumption levels in places around the world.			
20.6	I can describe and evaluate the impacts that e-waste has on our environment, the economy and society.			
20.7	I can explain and give examples of how governments can act to reduce the impacts of production and consumption.			
20.8	I can describe the ways groups work together to address the impacts of production and consumption on the environment, people and the economy.			
20.9	I can explain the ways an individual can reduce their contribution to the waste and pollution problems that are linked to production and consumption.			
20.10	I can conduct an interview to collect qualitative data.			
20.11	I can use the information in a topographic map to discuss ways that waste can be more sustainably managed.			

on Resources

eWorkbook Chapter 20 Student learning matrix (ewbk-8512)
Chapter 20 Reflection (ewbk-8513)
Chapter 20 Extended writing task (ewbk-8514)

Interactivity Chapter 20 Crossword (int-8586)

ONLINE RESOURCES

on Resources

Below is a full list of **rich resources** available online for this topic. These resources are designed to bring ideas to life, to promote deep and lasting learning and to support the different learning needs of each individual.

eWorkbook

- 20.1 Chapter 20 eWorkbook (ewbk-8000) ☐
- 20.2 Environmental impacts of production and consumption (ewbk-9674) ☐
- 20.3 Social and economic impacts of production and consumption (ewbk-9678) ☐
- 20.4 SkillBuilder — Describing divergence graphs (ewbk-9562) ☐
- 20.5 Case study — Automobile production and consumption (ewbk-9686) ☐
- 20.6 Case study — E-waste concerns (ewbk-9690) ☐
- 20.7 Government responses to the impacts of production and consumption (ewbk-9694) ☐
- 20.8 Group responses to production and consumption concerns (ewbk-9698) ☐
- 20.9 Reducing your consumption (ewbk-9702) ☐
- 20.10 SkillBuilder — Using advanced survey techniques — Interviews (ewbk-9706) ☐
- 20.11 Investigating topographic maps — More sustainable ways of managing waste (ewbk-9710) ☐
- 20.13 Chapter 20 Student learning matrix (ewbk-8512) ☐
  - Chapter 20 Reflection (ewbk-8513) ☐
  - Chapter 20 Extended writing task (ewbk-8514) ☐

Sample responses

- 20.1 Chapter 20 Sample responses (sar-0152)

Digital document

- 20.11 Topographic map of Woodlawn (doc-36264) ☐

Video eLessons

- 20.1 Plugging in (eles-1725) ☐
  - Connect, produce, consume — Photo essay (eles-5163) ☐
- 20.2 Environmental impacts of production and consumption — Key concepts (eles-5164) ☐
- 20.3 Social and economic impacts of production and consumption — Key concepts (eles-5165) ☐
- 20.4 SkillBuilder — Describing divergence graphs (eles-1739) ☐
- 20.5 Case study — Automobile production and consumption — Key concepts (eles-5166) ☐
- 20.6 Case study — E-waste concerns — Key concepts (eles-5167) ☐
- 20.7 Government responses to the impacts of production and consumption — Key concepts (eles-5168) ☐
- 20.8 Group responses to production and consumption concerns — Key concepts (eles-5169) ☐
- 20.9 Reducing your consumption — Key concepts (eles-5170) ☐
- 20.10 SkillBuilder — Using advanced survey techniques — Interviews (eles-1742) ☐
- 20.11 Investigating topographic maps — More sustainable ways of managing waste — Key concepts (eles-5171) ☐

Interactivities

- 20.2 Environmental impacts of production and consumption (int-8579) ☐
- 20.3 Social and economic impacts of production and consumption (int-8580) ☐
- 20.4 SkillBuilder — Describing divergence graphs (int-3357) ☐
- 20.5 Case study — Automobile production and consumption (int-8581) ☐
- 20.6 E-wasted (int-3343) ☐
  - Health impacts of e-waste on waste workers and people who live near landfills or incinerators (int-7938) ☐
- 20.7 Government responses to the impacts of production and consumption (int-8582) ☐
- 20.8 Group responses to production and consumption concerns (int-8583) ☐
- 20.9 Reducing your consumption (int-8584) ☐
- 20.10 SkillBuilder — Using advanced survey techniques — Interviews (int-3360) ☐
- 20.11 Investigating topographic maps — More sustainable ways of managing waste (int-8585) ☐
- 20.13 Chapter 20 Crossword (int-8586) ☐

Fieldwork

- 20.10 Survey Monkey (fdw-0026) ☐

ProjectsPLUS

- 20.12 Thinking Big research project — Trash or treasure (pro-0196) ☐

Weblinks

- 20.2 UN Sustainable Development Goals (web-2800) ☐
  - Great Pacific garbage patch (web-6316) ☐
  - Microplastics (web-6317) ☐
  - IUCN threatened species (web-6318) ☐
  - Food miles (web-6319) ☐
  - Ship tracking (web-6320) ☐
  - Climate council (web-6321) ☐
- 20.5 Ford car manufacturers assemble ventilators — COVID-19 (web-6322) ☐
- 20.6 Toxic waste in Africa (web-6323) ☐
- 20.8 Ethical choices (web-1133) ☐
- 20.9 Survey Monkey (web-1146) ☐

Google Earth

- 20.11 Woodlawn (gogl-0144) ☐

Teacher resources

There are many resources available exclusively for teachers online.

21 Fieldwork inquiry — The effects of travel in the local community

21.1 Overview

Numerous **videos** and **interactivities** are embedded just where you need them, at the point of learning, in your learnON title at www.jacplus.com.au. They will help you to learn the content and concepts covered in this topic.

> **LEARNING INTENTION**
>
> By the end of this subtopic, you will be able to plan a fieldwork inquiry on the effects of travel in the local community, collect and record data, and process and communicate the results of your fieldwork.

21.1.1 Scenario and task

Task: Produce a report about the impacts of travel movements around a local school or traffic hotspot, and devise a plan to better manage traffic and pedestrian movement.

People travel for many reasons at the local scale — for example, they may travel to work, to shops, to visit friends and to local sporting venues. Often there are times when traffic congestion occurs, creating danger areas for motorists and pedestrians. Examples of places where such congestion occurs are schools and shopping centres. Undertaking fieldwork allows you to observe and collect original data first-hand.

Go to learnON to access the following additional resources to complete this fieldwork task:

- a sample traffic count data sheet
- a sample survey question sheet
- a report template
- an assessment rubric.

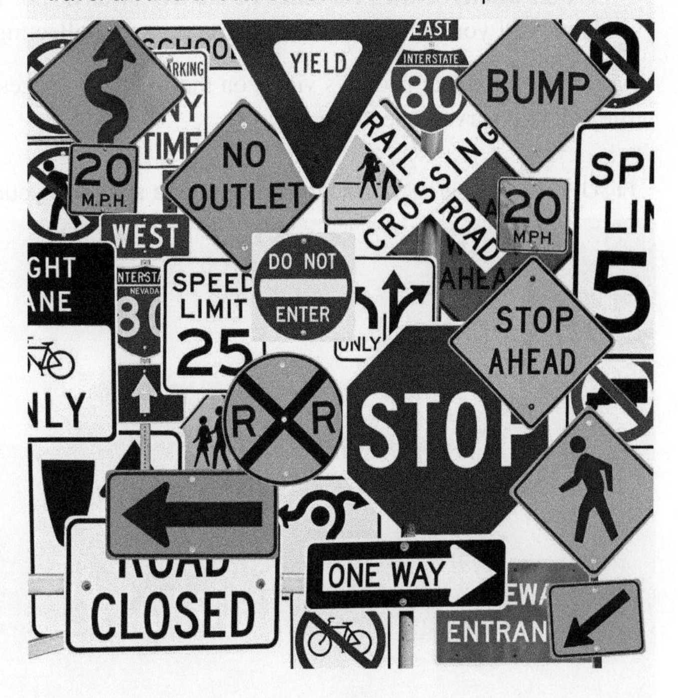

FIGURE 1 Your task is to investigate the impacts of travel around a local school or traffic hotspot.

Your task

Your team has been commissioned by the local council to compile a report evaluating the impacts of travel movements around a local school or traffic hotspot. You will need to collect, process and analyse suitable data and then devise a plan to better manage future traffic and pedestrian movement in the area.

21.2 Inquiry process

21.2.1 Process

Open the ProjectsPLUS application for this chapter located in your Resources tab. Watch the introductory video lesson and then click the 'Start Project' button and set up your project group. You can complete this project individually or invite members of your class to form a group. Save your settings and the project will be launched.

FIGURE 2 What trends, patterns and relationships can you see in the traffic data you collected?

Planning: Navigate to your Research Forum. Research topics have been loaded into the system to provide a framework for your research. You can also add your own new topics.

As part of a class discussion, determine a suitable location for your fieldwork study. This might be your own or a local school, or a nearby shopping centre. Talk about some of the issues related to your fieldwork site and then devise a key inquiry question — for example: *What are the effects of... ? or How can we reduce the impact of... ?* This will be the focus of your fieldwork. You then need to establish the following:

- **What** sort of data and information will you need to study the travel issue at your site?
- **How** will you collect this information?
- **Where** would be the best locations to obtain data?
- **When** would be the best times of the day or day(s) of the week to obtain data?
- **How** will you record the information you are collecting?

If you wish to collect people's views on the issue, or suggestions for improvements, you will need to plan and write suitable survey questions.

FIGURE 3 What would make the roads near a school in your area safer?

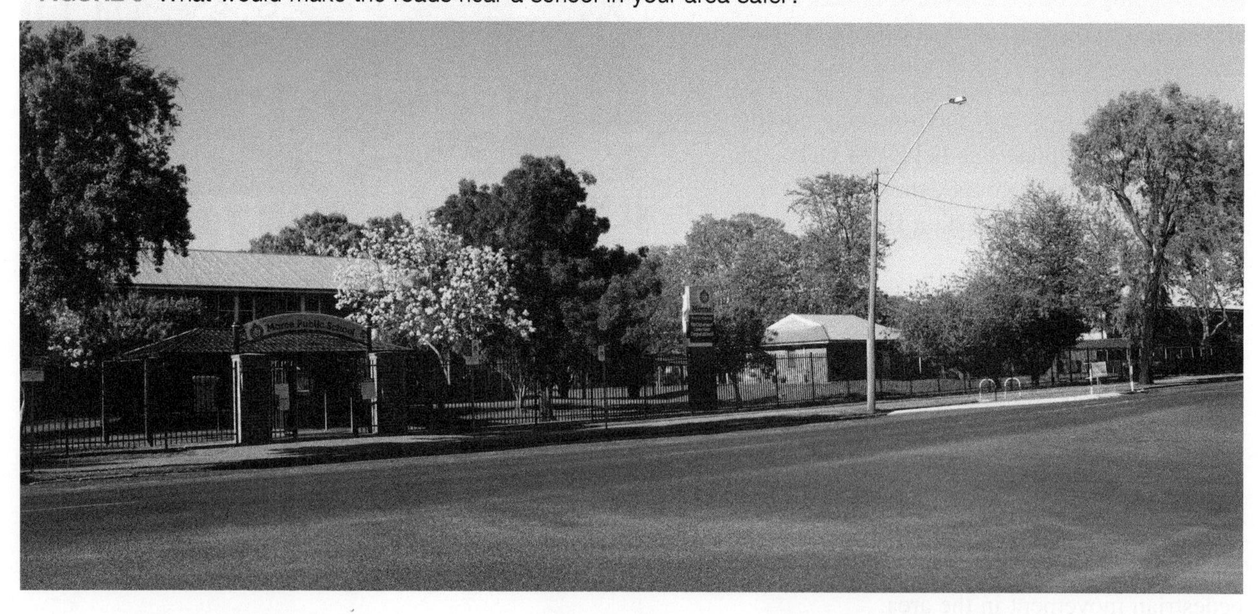

21.2.2 Collecting and recording data

As a class, plan the field trip by identifying and allocating tasks and possible sites to groups or pairs. It is often easier to share data collection. Once everything has been planned, you will need to perform your allocated tasks on the day.

In class, invite your school principal or a member of your local council to be a guest speaker discussing your fieldwork site. They may be able to assist with background information that you may not be able to gain elsewhere. They can also provide a different perception of the effects of travel at your site. Plan a series of questions you would like to ask and be prepared to take notes that you can use in your report.

After the field trip, it may be necessary to collate everyone's data and summarise surveys so that everyone has access to the shared information.

21.2.3 Processing and analysing your information and data

Look at your completed graphs and maps. What trends, patterns and relationships can you see emerging? Within your fieldwork area, are there some places that have a bigger issue with cars and pedestrians than other areas? Is there an interconnection between traffic congestion and time of the day, or day of the week? What have your surveys revealed? What are the major effects of travel at your fieldwork site? How do people perceive the travel issues in this place? Go back to your key inquiry question. To what extent have you been able to answer it? Write your observations up as a fieldwork report using subheadings such as:

- Background and key inquiry question
- Conducting the fieldwork [planning and collecting data]
- Findings [results of data analysis].

Visit your Media Centre and download the report template to help you complete this project. Use the report template to create your report.

21.2.4 Communicating your findings

Now that you have identified a traffic problem and collected and analysed data, it is time to try to solve it. Your completed map and supporting data will form part of your management plan for the future. What have been the main issues that have emerged from your fieldwork research? How can you best manage these issues? Using your base map, create an overlay or annotated map to show possible options for reducing the traffic problem. You will need to support each proposal with data that you have gained from your fieldwork. Possible ideas could include:

- changing parking restrictions
- staggering times of drop-off and pick-up
- introduction of traffic wardens to guide traffic
- creation of a one-way system.

Your teacher may arrange for your completed report to be presented to your school or local council. Considering your audience, what is the best way to present your findings? You might like to produce a PowerPoint presentation or an annotated visual display.

21.3 Review

21.3.1 Reflecting on your work

Think back over how well organised and prepared you were for the fieldwork, the data you collected and how you processed the data for your report. Download and complete the reflection template from your Media Centre.

Print your Research Report from ProjectsPLUS and hand it in with your fieldwork report and reflection notes.

21.3.2 Reflection

Subtopic	Success criteria			
21.2	I can plan a fieldwork inquiry on the effects of travel in my local community.			
21.2	I can collect and record data from a fieldwork inquiry on the effects of travel in my local community.			
21.2	I can process and communicate information from a fieldwork inquiry on the effects of travel in my local community.			

on Resources

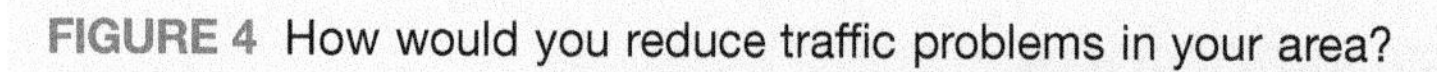

ProjectsPLUS Fieldwork inquiry — The effects of travel in the local community (pro-0149)

FIGURE 4 How would you reduce traffic problems in your area?

GLOSSARY

2G second generation mobile network that moved from analogue to digital with a speed of 0.1 megabits a second
3G third generation digital network with a typical download speed of 8 megabits a second
4G fourth generation digital network with a typical download speed of 15 megabits a second
5G fifth generation digital mobile network with a typical download speed of 200 megabits a second
absolute location the latitude and longitude of a place
absolute water scarcity when annual water supplies drop below 500 cubic metres per person
adobe bricks made from sand, clay, water and straw and dried by the Sun
aerial photograph a photograph taken of the ground from an aeroplane or satellite
aesthetic relating to something beautiful or pleasing to look at
aesthetics (aesthetic qualities) the beautiful or attractive features of something
agriculture the cultivation of land, growing of crops or raising of animals
airliner an aircraft that transports people and goods
alluvium the loose material brought down by a river and deposited on its bed, or on the floodplain or delta
altitude height above sea level
Anthropocene a period of time when humans made significant changes to the Earth's environments
appropriate technology technology designed specifically for the place and the people who will use it. It is affordable and can be repaired locally.
aquifer a body of permeable rock below the Earth's surface that contains water, known as groundwater. Water can move along an aquifer.
archaeological concerning the study of past civilisations and cultures by examining the evidence left behind, such as graves, tools, weapons, buildings and pottery
archipelago a chain or line of islands
arid lacking moisture; especially having insufficient rainfall to support trees or plants
arroyo channel or stream that is normally dry but fills quickly with heavy rain (also called a wadi)
artesian aquifer an aquifer confined between impermeable layers of rock. The water in it is under pressure and will flow upward through a well or bore.
aspect feature or quality, or the direction in which something is facing
atmospheric hazard extreme natural event that begins in the atmosphere (in the air)
Baby Boomer generally refers to people born during the post–World War II baby boom, between 1946 and 1964
backwash the movement of water from a broken wave as it runs down a beach returning to the ocean
barometer an instrument used to measure air pressure
barter to trade goods in return for other goods or services rather than money
basin the land area drained by a river and its tributaries; another name for a river catchment
biodiversity the variety of life in the world or in a particular habitat or ecosystem
biosecurity actions taken to stop specific organisms, animals or plants entering the country
bit a portmanteau (shortened combination) of the words binary and digit
black market any illegal trade in officially controlled or scarce goods
blizzard a strong and very cold wind containing particles of ice and snow that have been whipped up from the ground
blue water the water in fresh-water lakes, rivers, wetlands and aquifers
built environment a place that has been constructed or created by people
byte eight bits make one byte. This is the measurement used to store data such as kilobyte, megabyte and gigabyte.
carbon emissions carbon dioxide that is released into the atmosphere by natural processes or by burning gas, coal or oil; the latter emissions are contributing to climate change
carbon offsetting calculating the amount of carbon released during an activity and investing in strategies to pay off or cancel out the carbon

catchment area the area of land that contributes water to a river and its tributaries
census a regular survey used to determine the number of people living in Australia. It also has a variety of other statistical purposes
change geographical concept concerned with examining, comparing and predicting the impact of processes over time
clearfelling a forestry practice in which most or all trees and forested areas are cut down
climate long-term weather patterns in an area
climate change a change in the world's climate. This can be very long term or short term, and is caused by human activity.
community a group of people who live and work together, and generally share similar values; a group of people living in a particular region
compost a mixture of various types of decaying organic matter such as dung and dead leaves
condensation precipitation that collects into droplets of water from humid air
confluence where two rivers join
constructive wave a gentle backwash that leads to material being deposited on land
consumer culture the influences of and focus on material goods/products in a society
contour line a line joining places of equal height above sea level
convection current a current created when a fluid is heated, making it less dense, and causing it to rise through surrounding fluid and to sink if it is cooled; a steady source of heat can start a continuous current flow
convectional rainfall heavy rainfall as a result of thunderstorms, caused by rapid evaporation of water by the Sun's rays
converging plates a tectonic boundary where two plates are moving towards each other
coral reef (or atoll) reef that partially or completely encircles a lagoon
Country the land and its features, which is bound to the concept of belonging to a place that is fundamental to Aboriginal Peoples' sense of identity and Culture
crust the Earth's outer layer or surface
cultural relating to the ideas, customs and social behaviour of a society
cultural heritage beliefs and ways of living passed down through generations
cumulonimbus clouds huge, thick clouds that produce electrical storms, heavy rain, strong winds and sometimes tornadoes. They often appear to have an anvil-shaped flat top and can stretch from near the ground to 16 kilometres above the ground.
cyclone intense low pressure system producing sustained wind speeds in excess of 65 km/h. It develops over tropical waters where surface water temperature is at least 27 °C.
degradation a decline in quality caused by time or improper use
delegate person sent or authorised to represent others; in particular an elected representative sent to a conference
delta a landform at the mouth of the river where a river splits into smaller streams and sediment is deposited to create an arch of land reaching into the sea
demographic describes statistical characteristics of a population
deposition the laying down of material carried by rivers, wind, ice and ocean currents or waves
desalination the process of removing salt from sea water
desertification the process by which an area becomes a desert
destructive wave a large powerful storm wave that has a strong backwash
developing country nation with a low living standard, undeveloped industrial base, and low human development index relative to other countries
digital divide a type of inequality between groups in their access to and knowledge of information and communication technology
digital economy the network of economic and social activities that are supported by information and communication technology, such as the internet and mobile and sensor networks
digital trade selling or buying goods or services or transmitting information and data across borders
distribution the way things are spread across an area
divergent plates a tectonic boundary where two plates are moving away from each other and new continental crust is forming from magma that rises to the Earth's surface between the two

domesticate to tame a wild animal or plant so it can live with or be looked after by people
downstream nearer the mouth of a river, or going in the same direction as the current
drainage basin the entire area of land that contributes water to a river and its tributaries
Dreaming in Aboriginal cultures, the time when the Earth took on its present form, and cycles of life and nature began; also known as the Dreamtime. Dreaming stories pass on important knowledge laws and beliefs.
e-commerce business conducted online
e-waste any electrical equipment that includes computers, toasters, mobile phones and iPads that no longer works or is no longer required
eastings lines that run up and down a map (north–south)
ecological footprint the total area of land that is used to produce the goods and services consumed by an individual or country
economic a system made up of producing, buying and selling goods and services
ecosystem an interconnected community of plants, animals and other organisms that depend on each other and on the non-living things in their environment
ecotourism tourism that interprets the natural and cultural environment for visitors, and manages the environment in a way that is ecologically sustainable
ecotourist a tourist who travels to threatened ecosystems in order to help preserve them
elevation height of a place above sea level
enhanced greenhouse effect increasing concentrations of greenhouse gases in the Earth's atmosphere, contributing to global warming and climate change
environment geographical concept concerned with why the natural world plays an important part in human life
epicentre the point on the Earth's surface directly above the focus of an earthquake
epiphyte a plant that grows on another plant but does not use it for nutrients
equatorial near the equator, the line of latitude around the Earth that creates the boundary between the northern and southern hemispheres
erosion the wearing down of rocks and soils on the Earth's surface by the action of water, ice, wind, waves, glaciers and other processes
ethnic minority a group that has different national or cultural traditions from the main population
evaporate to change liquid, such as water, into a vapour (gas) through heat
evaporation the process by which water is converted from a liquid to a gas and thereby moves from land and surface water into the atmosphere
evapotranspiration the process by which plants absorb precipitation and release it back into the air as water vapour
expenditure the amount of money spent
extremism extreme political or religious views or extreme actions taken on the basis of those views
facility place or thing that helps people
fault an area on the Earth's surface that has a fracture, along which the rocks have been displaced
field sketch a diagram with geographical features labelled or annotated
fissures cracks, especially in rocks
flood an unusual accumulation of water onto land that is not normally covered by water
floodgate a barrier across a river that is raised when water levels are predicted to be dangerously high
floodplain relatively flat land by a river that experiences flooding
fly in, fly out (FIFO) workers who fly to work in remote places, work 4-, 8- or 12-day shifts and then fly home
focus the point where the sudden movement of an earthquake begins
food security having access to and being able to afford enough nutritious food to stay healthy
formal describes an event or venue that is organised or structured
fossil fuel fuel that comes from the breakdown of living materials, and that is formed in the ground over millions of years. Examples include coal, oil and natural gas.
free trade agreement a treaty between two or more countries or economies that sets out special trading conditions that reduce or remove trade barriers

gale force wind wind with speeds of over 62 km/h
Generation Alpha people born between 2010 and 2025
Generation X people born between 1964 and 1980
Generation Y people born between 1981 and 1994
Generation Z people born between 1995 and 2009
geomorphic describes a process that occurs in the lithosphere
geothermal energy energy derived from the heat in the Earth's interior
geyser a hot spring that erupts, sending hot water and steam into the air
glacier a large body of ice, formed by a build-up of snow, which flows downhill under the pressure of its own weight
global citizen person who is aware of the wider world, who tries to understand the values of others, and tries to make the world a better place
global warming the observable rising trend in the Earth's atmospheric temperatures, generally attributed to the enhanced greenhouse effect
globalisation the process that enables markets and companies to operate internationally, and products and ideas to be freely exchanged
gorge narrow valley with steep rocky walls
green water water that is stored in the soil or that stays on top of the soil or in vegetation
greenbelts plantations of trees for the purpose of conservation
gross domestic product (GDP) the value of all the goods and services produced within a country in a given period of time. It is often used as an indicator of a country's wealth.
groundwater water that seeps into soil and gaps in rocks
groundwater recharge a process in which water moves down from the Earth's surface into the groundwater
habitat the total environment where a particular plant or animal lives, including shelter, access to food and water, and all of the right conditions for breeding
hailstone an irregularly shaped ball of frozen precipitation
hailstorm any thunderstorm that produces hailstones large enough to reach the ground
halal food that is prepared under Islamic dietary guidelines
hard sport tourism tourism in which someone travels to either actively participate in or watch a competitive sport as the main reason for their travel
horticulture the growing of garden crops such as fruit, vegetables, herbs and nuts
host an organism that supports another organism
hotspot an area on the Earth's surface where the crust is quite thin, and volcanic activity can sometimes occur, even though it is not at a plate margin
human features structures built by people
humanitarian principles the principles governing our response to those in need, with the main aim being to save lives and alleviate suffering
humidity the amount of water vapour in the atmosphere
hydroelectric dam a dam that harnesses the energy of falling or flowing water to generate electricity
hydrologic cycle another term for the water cycle
hydrologic hazard extreme natural event that begins in the hydrosphere (in the water)
ice age a geological period during which the Earth is colder, glaciers and ice sheets expand and sea levels fall. The last ice age was approximately 13 000 years ago.
improved drinking water water that is safe for human consumption
industry the production of goods and services
infiltration precipitation absorbed into the ground
informal sector jobs that are not recognised by the government as official occupations and that are not counted in government statistics
infrastructure the features that are built by humans to allow other activities to occur; for example, roads, electricity networks and water supply pipes
inselberg an isolated hill, knob, ridge, outcrop or small mountain that rises abruptly from the surrounding landscape
intensify to become stronger

interconnection geographical concept concerned with how natural places, processes and features can change or affect each other and people, or be changed or affected by people
intermittent creek a creek that flows for only part of the year following rainfall
iridescent bright and colourful
irrigation water provided to crops and orchards by hoses, channels, sprays or drip systems in order to supplement rainfall
islet a very small island
isobar line on a map that joins places with the same air pressure
karst underground landscapes of caves and water channels
katabatic wind very strong winds that blow downhill
kinship system rules and relationships that determine how people interact with each other
La Niña a pattern in the climate that includes warmer ocean waters off Australia's east coast
lagoon a shallow body of water separated by islands or reefs from a larger body of water, such as a sea
landfill a method of solid-waste disposal, in which refuse is buried between layers of soil
landform distinctive, natural feature occurring on the surface of the Earth
landscape visible features occurring across an area of land
landslide a rapid movement of rocks, soil and vegetation down a slope, sometimes caused by an earthquake or by excessive rain
leaching a process that occurs where water runs through the soil, dissolving minerals and carrying them into the subsoil
leeward the side of mountains that faces away from rainbearing winds
levee also known as an embankment, it is a built-up part of the river bank
liquefaction when solids are shaken heavily and act like liquids
literacy rate the proportion of the population aged over 15 who can read and write
lithosphere the Earth's crust, including landforms, rocks and soil
live streaming technology that lets you watch, create and share videos in real time over the internet
liveability an analysis of what it is like to live somewhere, based on a set of criteria
liveable city an informal name given to cities that rank highly on a reputable annual survey of living conditions
livestock animals raised for food or other products
location a point on the surface of the Earth where something is to be found
Maasai ethnic group of semi-nomadic people in Kenya and Tanzania, whose way of life depends on their cattle
mandatory quarantine enforced restriction of movement and separation from the rest of the community, monitored by officials
mantle the layer of the Earth between the crust and the core
mass movement large areas of earth that move
mass production producing a large volume of one product by automating or standardising the process
mature-aged describes individuals aged over 55
meander a curve in a river caused by fast-flowing water eroding the bank of one side of the river and slow-flowing water depositing sediment on the other side of the river
mercury poisoning a toxic condition caused by the ingestion or inhalation of mercury or a mercury compound. It has various symptoms, including vomiting, nausea, insomnia and fevers.
meteorologist a scientist who studies the weather
microclimate specific atmospheric conditions within a small area
mobility the ability to move or be moved freely and easily
monsoon the rainy season in the Indian subcontinent and South-East Asia, followed by a dry season
mosque place of worship for people who follow Islam (Muslims)
mound spring mound formation with water at its centre, which is formed by minerals and sediments brought up by water from artesian basins
national security the protection of a nation's citizens, natural resources, economy, money, environment, military, government and energy
natural disaster an extreme event that is the result of natural processes and causes serious material damage or loss of life

natural hazard an extreme event that is the result of natural processes and has the potential to cause serious material damage and loss of life
natural resources resources (such as landforms, minerals and vegetation) that are provided by nature rather than people
NBN Co publicly owned corporation of the Australian government, tasked to design, build and operate Australia's National Broadband Network as the nation's wholesale broadband provider
neighbourhood a region in which people live together in a community
new Moon the phase of the Moon when it is closest to the Sun and is not normally visible
nomadic a group that moves from place to place depending on the food supply, or on pastures for animals
non-government organisation (NGO) non-profit group run by people (often volunteers) who have a common interest and perform a variety of humanitarian tasks at a local, national or international level
northings lines that run horizontally across a map (east–west)
nutrient an essential element that feeds living things to help them function and grow
Pangaea the name given to all the landmass of the Earth before it split into Laurasia and Gondwana
patron a customer, especially a regular one
perennial permanent or continually available; for example, a stream that flows all year
permafrost permanently frozen ground not far below the surface of the soil
pictorial map a map using illustrations to represent information
place specific area of the Earth's surface that has been given meaning by people
plateau an extensive area of flat land that is higher than the land around it. Plateaus are sometimes referred to as tablelands.
population density the number of people within a given area, usually per square kilometre
precipitation the form of water falling from the sky, such as rain, snow or hail
prevailing wind the main direction from which the wind blows
primary industry the gathering of natural resources, such as iron ore and timber, or activities such as farming and fishing
psychological benefits positive effects on emotional or mental wellbeing
public transport transport of passengers by group travel systems available for use by the general public, typically managed on a schedule, operated on established routes, and that charges a posted fee for each trip. Public transport modes include city buses, trolleybuses, trams (or light rail), passenger trains, rapid transit (e.g. metro/subway/underground) and ferries.
pull factor positive aspect of a place; reason that attracts people to come and live in a place
push factor reason that encourages people to leave a place and live somewhere else
quaternary industry gathering information about an industry; for example, marketing furniture
rain shadow the drier side of a mountain range, cut off from rain-bearing winds
rainfall variability the change from year to year in the amount of rainfall in a given location
recession a period of decline in economic growth when GDP decreases
region any area of varying size that has one or more characteristics in common
relative humidity the amount of moisture in the air
relative location the direction and distance from one place to another
remote a place that is a long way from major towns and cities
rift zone a large area of the Earth in which plates of the Earth's crust are moving away from each other, forming an extensive system of fractures and faults
Ring of Fire the area around the Pacific Ocean where most of the world's earthquakes occur and volcanoes are found
riverine relating to rivers
run-off precipitation not absorbed by soil, which runs over the land and into streams
rural places that are not part of a major town or city
sastrugi parallel wave-like ridges caused by winds on the surface of hard snow, especially in polar regions
scale geographical concept concerned with the different levels or ranges at which issues can be examined
sea change the act of leaving a fast-paced urban life for a more relaxing lifestyle in a small coastal town
seasonal worker person who moves to a place for a short time to do a specific job (e.g. fruit picking)

secondary industry making raw materials into a more useful product; for example, making furniture from timber
seismic waves waves of energy that travel through the Earth as a result of an earthquake, explosion or volcanic eruption
selective logging a forestry practice in which only selected trees are cut down
service program or organisation that helps people
settlement place where humans live together, e.g. houses, villages, towns and cities
shanty town slum settlement housing the poor on the outskirts of major cities in developing nations. Lacks basic services such as clean water, sanitation and electricity.
sharing economy sharing objects and/or services either for free or for payment, typically arranging the trade online
shifting agriculture process of moving gardens or crops every couple of years because the soils are too poor to support repeated sowing
smuggling importing or exporting goods secretly or illegally
soak place where groundwater moves up to the surface
social network allows like-minded individuals to be in touch with each other using websites and web-based applications
soft sport tourism tourism in which someone participates in recreational and leisure activities, such as skiing, fishing and hiking, as an incidental part of their travel
soluble something that can be dissolved
songlines paths across the land that follow the creation journeys of Aboriginal Peoples' ancestors, which are used to teach about Culture and Country; also known as Dreaming Tracks
space geographical concept concerned with location, distribution (spread) and how we change or design places
spatial how things relate to each other in an area or location
species a biological group of individuals having the same common characteristics and being able to breed with each other
spot height a point on a map that shows the exact height above sea level (in metres) at that place
stalactite a feature made of minerals, which forms from the ceiling of limestone caves, like an icicle
stalagmite a feature made of minerals found on the floor of limestone caves
storm surge a sudden increase in sea level as a result of storm activity and strong winds. Low-lying land may be flooded.
subsistence producing only enough crops and raising only enough animals to feed yourself and your family or community
subtropics the areas of Earth just north of the Tropic of Cancer and just south of the Tropic of Capricorn
summit the peak at the top of a mountain
sustainability geographical concept concerned with how to protect the environment and life on Earth
sustainable to use the environment now in such a way that it continues to support our lives and the lives of other creatures into the future
sustainable development economic development that causes a minimum of environmental damage, thereby protecting the interest of future generations
swash the movement of water in a wave as it breaks onto a beach
tectonic relating to force or movement in the Earth's crust
tectonic plate one of the slow-moving plates that make up the Earth's crust. Volcanoes and earthquakes often occur at the edges of plates.
temperate describes the relatively mild climate experienced in the zones between the tropics and the polar circles
tertiary industry selling goods and services; for example, a furniture business
tertiary study the educational level following high school, e.g. university
tornado violent, wildly spinning column of air that forms over land surfaces. It drops from under a cumulonimbus cloud and contacts the ground.
torrential rain heavy rain often associated with storms, which can result in flash flooding
totem a part of the natural world (such as an animal) that has a special meaning or importance for people
tourism moving away from where you normally live/work temporarily (for between 24 hours and 1 year)

trade barrier government-imposed restriction (in the form of tariffs, quotas and subsidies) on the free international exchange of goods or services
trading partner a participant, organisation or government body in a continuing trade relationship
transport the movement of eroded materials to a new location by agents such as wind and water
treaty a formal agreement between two or more independent states or nations, usually involving a signed document
tree change the act of leaving a fast-paced urban life for a more relaxing lifestyle in a small country town, in the bush, or on the land as a farmer
tributary river or stream that flows into a larger river or body of water
troposphere the layer of the atmosphere closest to the Earth. It extends about 16 kilometres above the Earth's surface, but is thicker at the tropics and thinner at the poles, and is where weather occurs.
tsunami large wave that can be triggered by an earthquake under the ocean
tundra landscape without trees, with small shrubs and grass, found in cold or high-altitude places
turbidity the amount of sediment suspended in water
typhoon the name given to cyclones in the Asian region
undernourished not getting enough food for good health and growth
uranium radioactive metal used as a fuel in nuclear reactors
value adding processing a material or product and thereby increasing its market value
virtual water all the water used to produce goods and services. Food production uses more water than any other production.
volcanic loam a volcanic soil composed mostly of basalt, which has developed into a crumbly mixture
water footprint the total volume of fresh water that is used to produce the goods and services consumed by an individual or country
water scarcity when annual water supplies drop below 1000 cubic metres per person. Absolute water scarcity describes the situation when annual water supplies drop below 500 cubic metres per person.
water stress a situation that occurs in a country with less than 1000 cubic metres of renewable fresh water per person
water vapour water in its gaseous form, formed as a result of evaporation
weather short-term changes in the atmosphere at a particular location
weathering the breaking down of rock through the action of wind and water and the effects of climate, mainly by water freezing and cooling as a result of temperature change
weir a barrier across a river, similar to a dam, which causes water to pool behind it. Water is still able to flow over the top of the weir.
wilderness a natural place that has been almost untouched or unchanged by the actions of people
windward the side of mountains that face rain-bearing winds
winter solstice the shortest day of the year, when the Sun reaches its lowest point in relation to the equator
World Heritage List a list of human and natural sites that have been approved by UNESCO as meeting the criteria of being a World Heritage site
zettabyte one trillion gigabytes

INDEX

D

M

N

O

P

Q

R

S

T

U

V

W

Y

Z